EXPLICIT FORMS OF VECTOR OPERATIONS

Cartesian Coordinates

$$\nabla \cdot \mathbf{A} = \frac{\partial A_x}{\partial x} + \frac{\partial A_y}{\partial y} + \frac{\partial A_z}{\partial z}$$

$$\nabla \times \mathbf{A} = \left(\frac{\partial A_z}{\partial y} - \frac{\partial A_y}{\partial z} \right)\hat{\mathbf{x}} + \left(\frac{\partial A_x}{\partial z} - \frac{\partial A_z}{\partial x} \right)\hat{\mathbf{y}} + \left(\frac{\partial A_y}{\partial x} - \frac{\partial A_x}{\partial y} \right)\hat{\mathbf{z}}$$

$$\nabla \Phi = \frac{\partial \Phi}{\partial x}\hat{\mathbf{x}} + \frac{\partial \Phi}{\partial y}\hat{\mathbf{y}} + \frac{\partial \Phi}{\partial z}\hat{\mathbf{z}}$$

$$\nabla^2 \Phi = \frac{\partial^2 \Phi}{\partial x^2} + \frac{\partial^2 \Phi}{\partial y^2} + \frac{\partial^2 \Phi}{\partial z^2}$$

Cylindrical Coordinates

$$\nabla \cdot \mathbf{A} = \frac{1}{\rho}\frac{\partial}{\partial \rho}(\rho A_\rho) + \frac{1}{\rho}\frac{\partial A_\phi}{\partial \phi} + \frac{\partial A_z}{\partial z}$$

$$\nabla \times \mathbf{A} = \left(\frac{1}{\rho}\frac{\partial A_z}{\partial \phi} - \frac{\partial A_\phi}{\partial z} \right)\hat{\boldsymbol{\rho}} + \left(\frac{\partial A_\rho}{\partial z} - \frac{\partial A_z}{\partial \rho} \right)\hat{\boldsymbol{\phi}} + \frac{1}{\rho}\left[\frac{\partial}{\partial \rho}(\rho A_\phi) - \frac{\partial A_\rho}{\partial \phi} \right]\hat{\mathbf{z}}$$

$$\nabla \Phi = \frac{\partial \Phi}{\partial \rho}\hat{\boldsymbol{\rho}} + \frac{1\,\partial \Phi}{\rho\,\partial \phi}\hat{\boldsymbol{\phi}} + \frac{\partial \Phi}{\partial z}\hat{\mathbf{z}}$$

$$\nabla^2 \Phi = \frac{1}{\rho}\frac{\partial}{\partial \rho}\left(\rho \frac{\partial \Phi}{\partial \rho} \right) + \frac{1}{\rho^2}\frac{\partial^2 \Phi}{\partial \phi^2} + \frac{\partial^2 \Phi}{\partial z^2}$$

Spherical Coordinates

$$\nabla \cdot \mathbf{A} = \frac{1}{r^2}\frac{\partial}{\partial r}(r^2 A_r) + \frac{1}{r\sin\theta}\frac{\partial}{\partial \theta}(A_\theta \sin\theta) + \frac{1}{r\sin\theta}\frac{\partial A_\phi}{\partial \phi}$$

$$\nabla \times \mathbf{A} = \frac{1}{r\sin\theta}\left[\frac{\partial}{\partial \theta}(A_\phi \sin\theta) - \frac{\partial A_\theta}{\partial \phi} \right]\hat{\mathbf{r}} + \frac{1}{r}\left[\frac{1}{\sin\theta}\frac{\partial A_r}{\partial \theta} - \frac{\partial}{\partial r}(rA_\phi) \right]\hat{\boldsymbol{\theta}}$$
$$+ \frac{1}{r}\left[\frac{\partial}{\partial r}(rA_\theta) - \frac{\partial A_r}{\partial \theta} \right]\hat{\boldsymbol{\phi}}$$

$$\nabla \Phi = \frac{\partial \Phi}{\partial r}\hat{\mathbf{r}} + \frac{1}{r}\frac{\partial \Phi}{r\,\partial \theta}\hat{\boldsymbol{\theta}} + \frac{1}{r\sin\theta}\frac{\partial \Phi}{\partial \phi}\hat{\boldsymbol{\phi}}$$

$$\nabla^2 \Phi = \frac{1}{r^2}\frac{\partial}{\partial r}\left(r^2 \frac{\partial \Phi}{\partial r} \right) + \frac{1}{r^2\sin\theta}\left(\sin\theta \frac{\partial \Phi}{\partial \theta} \right) + \frac{1}{r^2\sin^2\theta}\frac{\partial^2 \Phi}{\partial \phi^2}$$

ELECTRICITY & MAGNETISM

Munir H. Nayfeh
Morton K. Brussel

*University of Illinois
At Urbana-Champaign*

Dover Publications, Inc.
Mineola, New York

Bibliographical Note

This Dover edition, first published in 2015, is an unabridged republication of the work originally published in 1985 by John Wiley & Sons, New York.

International Standard Book Number

ISBN-13: 978-0-486-78971-2
ISBN-10: 0-486-78971-3

Manufactured in the United States by Courier Corporation
78971301 2015
www.doverpublications.com

To Our Parents

PREFACE

This book is based on lecture notes that we have prepared and taught in our classes on electricity and magnetism and electromagnetic fields for several years. It is designed as a text-book for a two-semester course for students of physics but, with the selective omission of some material, it can well serve a one-semester course. No previous courses on the subject are required beyond the freshman general physics, so that the text can accommodate a wide readership for students in science or engineering.

Influenced by the feedback from the students and other instructors who used the lecture notes, we have chosen to address the lament often heard from students studying electromagnetism: "I really understand the theory; I just can't work the problems." This book presents 300 detailed problem-like examples at various levels of difficulty whose solutions illustrate various techniques and touch on every aspect of the material. From these examples, the student learns how to apply the formalism to concrete situations and practical problems. Self-confidence in analyzing problems is therefore promoted, and in this way competence is made more accessible.

An important feature of this book that sets it apart from the available books at the same level is the chapter on vector algebra. Although it is customary to expect the students to have some knowledge of analytical geometry and vector analysis, many are very uncertain about the mathematics of the coordinate systems, analytical geometry, and vector relationships, analysis and calculus. This chapter is organized to be a self-contained source of such information.

The chapter on the microscopic theory of magnetism gives in a simple, straightforward way, using classical analogies, the Heisenberg explanation of ferromagnetism, which is based on spin-spin interactions. In the existing books only the phenomenological explanation given by Weiss is presented. This chapter, therefore, makes this book the only book at this level with an up-to-date explanation of the phenomenon. This treatment does not require the student to know quantum mechanics, since the classical analogy of the spin is used.

Applications involving discrete quantum mechanical dipoles in external electric or magnetic fields are presented. The results are compared to the cases involving classical dipoles. These examples, also, do not require a knowledge of quantum mechanics.

We explain in Chapters 4, 7, and 9, the use of the method of images for the solution of dielectric, current, and magnetic problems. These topics are not explained in the existing books at this level. Also, magnetic circuits are presented in analogy with electric circuits and are used to design electro- or permanent magnet systems. We employ the very useful methods of coefficients of potential and capacitance more extensively than is customary in solving electrostatic problems. Moreover, the corresponding method of coefficients of resistance is introduced and is used to solve current problems. The presentation of these special techniques may be omitted without a loss of continuity (hence, they are labeled by asterisks for quick identification).

The magnetic scalar potential (magnetic pole) concept is introduced in close analogy with the electrostatic potential (electric charge) concept. Comparisons with the vector potential concept are developed and are further illustrated in a number of examples.

The boundary conditions on the electromagnetic fields and on the scalar electric and magnetic potential, and on the vector magnetic potentials are given special attention. Often the relevant applications are solved with various methods in order to present the various boundary conditions and their interrelationships.

All the material on radiation is placed in one chapter, Chapter 15. This procedure allows us to give comparisons and interrelationships between the various analysis methods employed, resulting in a coherent treatment of the subject.

The exercises presented at the end of each chapter are chosen with the goal of further training the students. The problems fall into two categories. In the first category are extensions of the examples in the chapter. The student thus already has a head start on these problems and has an excellent chance of solving them completely. The other category is more challenging and is intended to further develop the student's independent comprehension of the material.

Our book can also be used for one-semester courses for students at the junior level by selectively choosing a subset of chapters that emphasizes Maxwell's equations and their implications. We have used the book successfully for such courses by omitting Chapters 1, 5, 7, 10, 11, 13 (with the exception of the continuity equation and Faraday's law in Chapters 7 and 11), without any loss of continuity. Moreover, we omitted the following special techniques: magnetic circuits, the application of the method images to dielectrics, current, and magnetic problems, the methods of coefficients of resistance and potential. Also, we omitted Sections 17.1 to 17.4. These are labeled so that they can be conveniently identified.

We are indebted to all of our students, in particular to Walter Mieher and Glen Herrmannsfeldt for proofreading the manuscript, and to all of our colleagues, especially to Professor Robert D. Sard, for their constructive suggestions. Special thanks go to Phyllis Brussel, Hetaf, Hasan, Maha, Ammar, and Osamah Nayfeh.

Munir H. Nayfeh
Morton K. Brussel

CONTENTS

ELECTRICITY
&
MAGNETISM

ONE
VECTOR ANALYSIS

1.1 Properties of Vectors and Coordinate Systems

Ordinary numbers are called *scalars*. They may be real or complex numbers. In contrast to scalars, we have other quantities called *vectors*. These quantities combine with each other differently than scalars. In physics they are used to represent objects that have both *magnitude* and *direction*, the prototype of which is a displacement. Mathematically, vectors are simply quantities that behave and combine according to the following rules.

1. The sum of two vectors $\mathbf{u}$ and $\mathbf{v}$ is another vector: $\mathbf{u} + \mathbf{v} = \mathbf{w}$. The sum is a commutative binary operation; that is, $\mathbf{u} + \mathbf{v} = \mathbf{v} + \mathbf{u}$.
2. Under summation, the associative law holds. For vectors $\mathbf{u}$, $\mathbf{v}$, and $\mathbf{w}$,

$$(\mathbf{u} + \mathbf{v}) + \mathbf{w} = \mathbf{u} + (\mathbf{v} + \mathbf{w}) = \mathbf{u} + \mathbf{v} + \mathbf{w}$$

3. Any vector can be "multiplied" by a scalar to yield another vector.

We shall represent vectors geometrically by directed line segments (i.e., arrows). The magnitude of the vector is proportional to the length of the line segment, and the direction is given by the orientation of the arrow—that is, the direction in which it points. The rules to be followed in performing this (vector) addition geometrically are these (see Fig. 1.1): On a diagram drawn to scale lay out the displacement vector $\mathbf{u}$; then draw $\mathbf{v}$ with its tail at the head of $\mathbf{u}$, and draw a line from the tail of $\mathbf{u}$ to the head of $\mathbf{v}$ to construct the vector sum $\mathbf{w}$. This is a displacement equivalent in length and direction to the successive displacements $\mathbf{u}$ and $\mathbf{v}$. This procedure can be generalized to obtain the sum of any number of successive displacements.

1.1.1 Base Vectors and Coordinate Systems

Choosing a coordinate system in space is essentially equivalent to choosing a set of base vectors. If we choose a cartesian coordinate system (Fig. 1.2), our base vectors are chosen to be along three fixed mutually perpendicular (orthogonal) fixed directions called the x, y, and z directions. If we represent a vector by an arrow, the

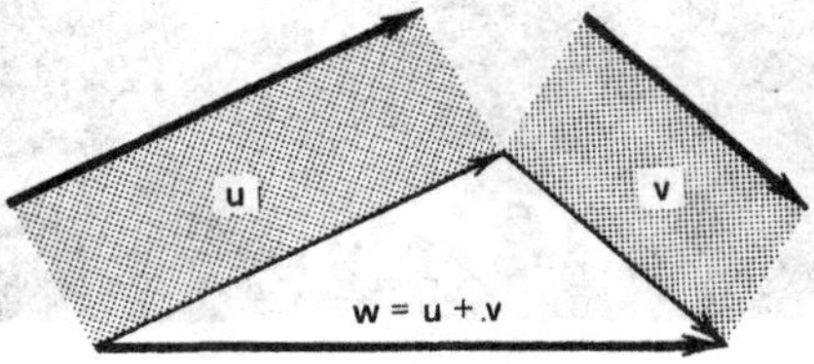

Figure 1.1 Geometrical definition of the sum of two vectors **u** and **v**.

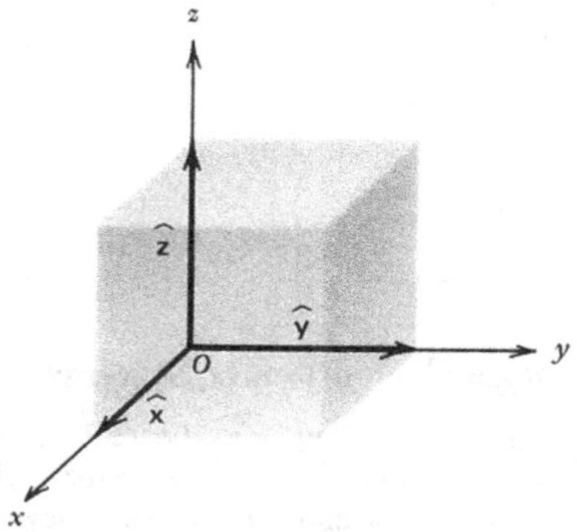

Figure 1.2 Definition of the cartesian coordinate system, showing the associated unit vectors.

perpendicular projections of the arrow upon the three coordinate axes are called the *cartesian components* of the vector in these directions. In terms of these components, the magnitude of a vector **A** is as follows*:

$$\text{Magnitude of } \mathbf{A} \equiv |\mathbf{A}| = (A_x^2 + A_y^2 + A_z^2)^{1/2}$$

A unit vector $\hat{\mathbf{A}}$ is that vector which when multiplied by the magnitude $|\mathbf{A}|$, yields the vector **A**; that is, $\mathbf{A} \equiv |\mathbf{A}|\hat{\mathbf{A}}$. It provides a means for indicating direction. Unit vectors along x, y, and z coordinate axes (cartesian) are denoted by $\hat{\mathbf{x}}$, $\hat{\mathbf{y}}$, $\hat{\mathbf{z}}$, respectively. They provide a convenient and fundamental set of base vectors. In terms of cartesian unit vectors, any vector **A** is represented by

$$\mathbf{A} = A_x\hat{\mathbf{x}} + A_y\hat{\mathbf{y}} + A_z\hat{\mathbf{z}} \tag{1.1}$$

where A_x, A_y, and A_z are the components of **A** along the $\hat{\mathbf{x}}$, $\hat{\mathbf{y}}$, and $\hat{\mathbf{z}}$ directions, respectively.

We shall restrict our attention to cases where the base vectors form an orthogonal set. Moreover the magnitude of each base vector will be taken as unity (orthonormal). The most commonly used base vectors in ordinary three-dimensional space are the unit base vectors ($\hat{\mathbf{x}}$, $\hat{\mathbf{y}}$, $\hat{\mathbf{z}}$). These vectors are considered to be *constant* vectors. Neither their directions nor their magnitudes depend on where they are located with respect to some reference point in space. It is the constancy of this orthonormal set of base vectors that we wish to emphasize by the word *cartesian*.

* Throughout this book scalars are in italics and vectors are in boldface.

The representation of vectors using the unit vectors is very useful in vector manipulations. For example, to add **A** to **B** we simply add the cartesian components:

$$\mathbf{A} + \mathbf{B} = (A_x + B_x)\hat{\mathbf{x}} + (A_y + B_y)\hat{\mathbf{y}} + (A_z + B_z)\hat{\mathbf{z}}$$

It is frequently convenient to use other sets of base vectors whose directions do happen to depend on their locations (curvilinear base vectors). For example, we shall define and often use a spherical coordinate system and a cylindrical coordinate system. For each of these systems we shall find an orthonormal set of associated base vectors that depend on where in space they are located.

Recall that the cartesian unit vector $\hat{\mathbf{x}}$ may be defined as the unit vector that is perpendicular to any plane $x = \text{constant}$. Similarly, for the $\hat{\mathbf{y}}$ and $\hat{\mathbf{z}}$ unit vectors we respectively associate the planes $y = \text{constant}$ and $z = \text{constant}$. Now, there are other surfaces that one can describe that correspond to some geometrical variable being constant. If we can find three surfaces, defined by (three) geometrical variables, that intersect each other perpendicularly at a point, then at this point we can define three associated mutually perpendicular vectors that are normal to these surfaces. In describing the spherical and cylindrical systems, we cite two instances where we find it useful to do so. (There are many others.)

In the cylindrical coordinate system (Fig. 1.3) we define a set of base vectors at a point by considering surfaces, two of which are planes and one of which is a cylinder. The surfaces are denoted by the following equations:

(a) $z = \text{constant}$
(b) $\rho = \text{constant} = \sqrt{x^2 + y^2}$,
(c) $\phi = \text{constant} = \tan^{-1}(y/x)$

In equation (a) the z coordinate specifies a set of parallel planes. It is defined by reference to a reference plane called the $z = 0$ plane. The unit vector $\hat{\mathbf{z}}$ is then a constant vector pointing in the (positive) direction (which may be chosen arbitrarily), perpendicular to the $z = \text{constant}$ planes. The z axis is chosen to be a line pointing in the z direction (for $-\infty < z < +\infty$).

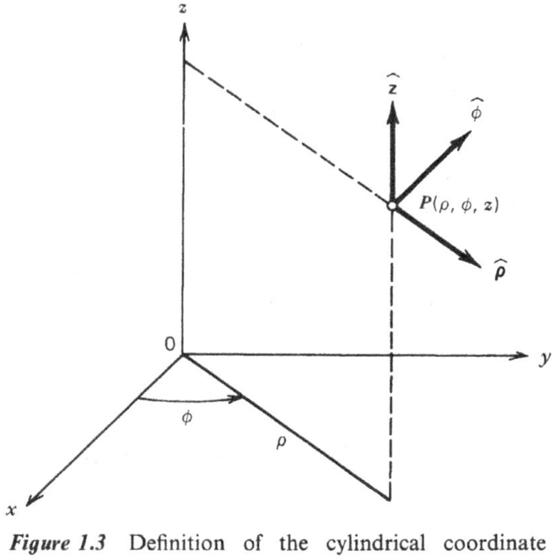

Figure 1.3 Definition of the cylindrical coordinate system, showing the associated unit vectors.

In equation (b) the ρ coordinate is defined with reference to the z axis by a set of cylindrically circular surfaces that intersect the $z =$ constant planes perpendicularly. The distance ρ from a particular surface to the z axis is the radius of the cylindrical surface. The unit vector $\hat{\boldsymbol{\rho}}$ is perpendicular to the cylindrical surface, pointing away from the z axis. Its direction depends upon what geometrical point of the circle resulting from intersection of the planes $z =$ constant and $\rho =$ constant is considered. Thus, in Fig. 1.3, $\hat{\boldsymbol{\rho}}$ is a function of the variable ϕ defined in equation (c) (for $\rho > 0$).

In equation (c) the only surfaces that can perpendicularly intersect the previously defined surfaces at all points of intersection are planes that contain the z axis. One such plane is called the $\phi = 0$ plane and is chosen arbitrarily. The $\hat{\boldsymbol{\phi}}$ unit vectors lie perpendicular to the $\phi =$ constant surfaces, and depend upon the angle ϕ of the plane with reference to the $\phi = 0$ plane.

The intersection of the surfaces described by equations (a), (b), and (c) above locate points in space, just as the intersection of the cartesian coordinate planes do. However, the cylindrical unit vectors are well specified only when a point (not on the z axis) is specified. (The origin is specified by setting $z = 0$ and $\rho = 0$.) Once this has been done by assigning values of (ρ, ϕ, z) or (x, y, z) to the point, any vector may be expressed in terms of the cylindrical unit vectors $(\hat{\mathbf{z}}, \hat{\boldsymbol{\rho}}, \hat{\boldsymbol{\phi}})$ at that point.

One can easily show that these cylindrical unit vectors are related to the cartesian unit vectors by the following relations.

$$\hat{\boldsymbol{\rho}} = \hat{\mathbf{x}} \cos \phi + \hat{\mathbf{y}} \sin \phi \qquad \hat{\boldsymbol{\phi}} = -\hat{\mathbf{x}} \sin \phi + \hat{\mathbf{y}} \cos \phi \qquad \hat{\mathbf{z}} = \hat{\mathbf{z}} \tag{1.2}$$

Remember that $\hat{\boldsymbol{\rho}}$ and $\hat{\boldsymbol{\phi}}$ depend upon the coordinate ϕ. Thus, for any vector $\mathbf{A}$ and for a point at which the unit vectors are $\hat{\mathbf{z}}$, $\hat{\boldsymbol{\rho}}$, $\hat{\boldsymbol{\phi}}$, $\mathbf{A} \equiv A_z\hat{\mathbf{z}} + A_\rho\hat{\boldsymbol{\rho}} + A_\phi\hat{\boldsymbol{\phi}}$ since $\{\hat{\mathbf{z}}, \hat{\boldsymbol{\rho}}, \hat{\boldsymbol{\phi}}\}$ form an orthonormal set. If $\mathbf{A}(\mathbf{r})$ is a vector point field, the natural triad of base vectors used to express $\mathbf{A}$ will be that defined by the location $\mathbf{r}$. Note that the displacement vector $\mathbf{r}$ to a point (z, ρ, ϕ) is given by $\mathbf{r} = \rho\hat{\boldsymbol{\rho}} + z\hat{\mathbf{z}}$.

We shall not describe the spherical coordinate system (Fig. 1.4) in the detail used above for the cylindrical system, except to note that the constant surfaces chosen are as follows:

(a) $r = \text{constant} = \sqrt{x^2 + y^2 + z^2}$, which describes a sphere of radius r with respect to the origin.

(b) $\theta = \text{constant} = \cos^{-1}(z/r)$, which describes a right circular cone with opening angle θ.

(c) $\phi = \text{constant} = \tan^{-1}(y/x)$, which describes a plane containing the axis of the cone in (b).

The unit vectors prescribed by these surfaces are denoted $\hat{\mathbf{r}}$, $\hat{\boldsymbol{\theta}}$, $\hat{\boldsymbol{\phi}}$, respectively, and form an orthonormal set once the point (not at the origin and not at the z axis) located by the intersection of the three orthogonal surfaces is determined. These unit vectors are given in terms of the cartesian unit vectors by the following relations.

$$\begin{aligned}
\hat{\mathbf{r}} &= \hat{\mathbf{x}} \sin \theta \cos \phi + \hat{\mathbf{y}} \sin \theta \sin \phi + \hat{\mathbf{z}} \cos \theta \\
\hat{\boldsymbol{\theta}} &= \hat{\mathbf{x}} \cos \theta \cos \phi + \hat{\mathbf{y}} \cos \theta \sin \phi - \hat{\mathbf{z}} \sin \theta \\
\hat{\boldsymbol{\phi}} &= -\hat{\mathbf{x}} \sin \phi + \hat{\mathbf{y}} \cos \phi
\end{aligned} \tag{1.3}$$

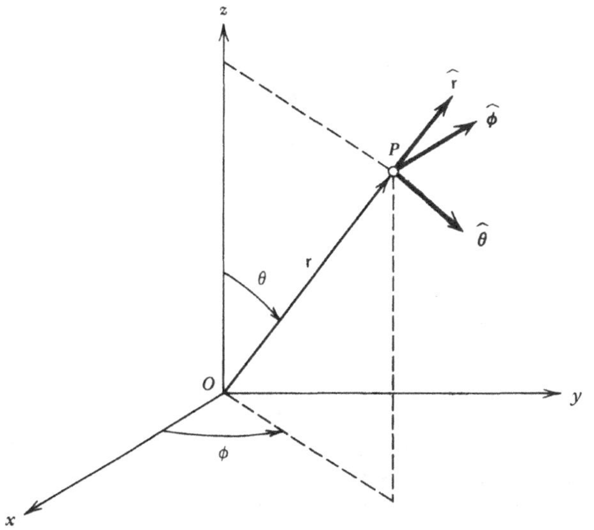

Figure 1.4 Definition of the spherical coordinate system, showing the associated unit vectors.

If **A** is a member of a vector field, **A**(**r**), then at every point given by the displacement **r** one can express **A**(**r**) in terms of the base vectors associated with that point: $\mathbf{A}(\mathbf{r}) = A_r\hat{\mathbf{r}} + A_\theta\hat{\boldsymbol{\theta}} + A_\phi\hat{\boldsymbol{\phi}}$, where A_r is the projection of **A** on $\hat{r}$, and so on. The displacement vector to a point (r, θ, ϕ) is given simply $\mathbf{r} = \hat{\mathbf{r}}r$.

1.1.2 The Scalar Product (Dot Product)

An important concept in vector algebra is that of the scalar product of two vectors. It is denoted by $\mathbf{A} \cdot \mathbf{B}$ and also called *dot product* or *inner product*. It is defined according to the following rule: $\mathbf{A} \cdot \mathbf{B} = |\mathbf{A}||\mathbf{B}| \cos \alpha$, where $|\mathbf{A}|$ and $|\mathbf{B}|$ are the magnitudes of **A** and **B**, and α is the angle between **A** and **B**. It can easily be seen that the scalar product, as defined, has the following properties: Two vectors whose scalar product is zero are said to be orthogonal; that is, if $\mathbf{A} \cdot \mathbf{B} = 0$, **A** is said to be orthogonal to **B**. The unit cartesian vectors $\hat{\mathbf{x}}$, $\hat{\mathbf{y}}$, and $\hat{\mathbf{z}}$ are said to constitute an *orthonormal* set of base vectors because they are orthogonal to each other and their magnitudes are normalized to unity.

1.1.3 The Vector Product (Cross Product)

We have seen that we can assign a scalar to any pair of vectors. The operation that does this is called the scalar product. We now wish to assign a vector quantity to any pair of vectors, **A** and **B**, and so we define what is known as a *vector product* (or *cross product*); it is denoted by $\mathbf{A} \times \mathbf{B}$. The direction of the vector product is taken to be perpendicular to the plane determined by the pair of vectors. Its magnitude is given by the area of the parallelogram whose sides are formed by the vector pair. Therefore, if $\hat{\mathbf{n}}$ is a unit vector perpendicular to the plane formed by the vector pair (**A**, **B**), then the vector product is defined according to the following rule.

$$\mathbf{A} \times \mathbf{B} \equiv |\mathbf{A}||\mathbf{B}| \sin \alpha\, \hat{\mathbf{n}} \tag{1.4}$$

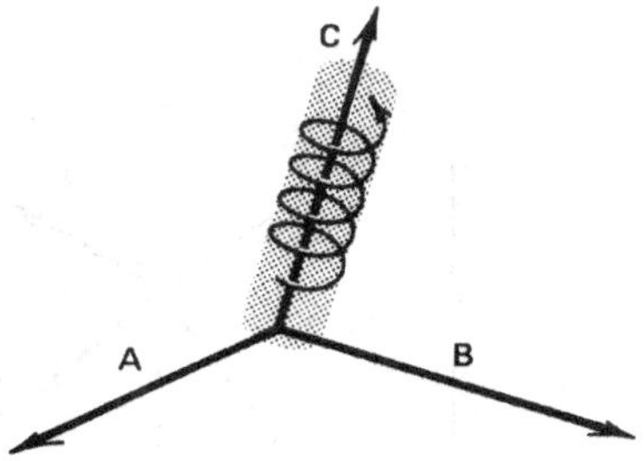

Figure 1.5 Definition of the right-hand screw convention, which gives the sense of the cross product of the vectors **A** and **B**.

So far the definition given remains ambiguous because the normal to the plane of **A** and **B** may point "up" or "down." To specify which way $\hat{\mathbf{n}}$ points, we use the *right-hand screw convention.* We say that if **A** is rotated to the direction of **B**, through the angle α ($\leq 180°$), then the same sence of rotation given to a right-handed screw determines $\hat{\mathbf{n}}$, which points along the direction of advance of the screw as it is rotated (Fig. 1.5).

In terms of cartesian unit vectors the vector product is expressed as

$$\mathbf{A} \times \mathbf{B} = \hat{\mathbf{x}}[A_y B_z - A_z B_y] + \hat{\mathbf{y}}[A_z B_x - A_x B_z] + \hat{\mathbf{z}}[A_x B_y - A_y B_z] \tag{1.5}$$

$$\mathbf{A} \times \mathbf{B} = \begin{vmatrix} \hat{\mathbf{x}} & \hat{\mathbf{y}} & \hat{\mathbf{z}} \\ A_x & A_y & A_z \\ B_x & B_y & B_z \end{vmatrix}, \text{ a determinant} \tag{1.6}$$

Whenever we have a set of three orthonormal vectors, $\hat{\mathbf{e}}_1$, $\hat{\mathbf{e}}_2$, and $\hat{\mathbf{e}}_3$, we say we have a *right-handed system* when $\hat{\mathbf{e}}_i \times \hat{\mathbf{e}}_j = \hat{\mathbf{e}}_k$ where i, j, k are in the order (1, 2, 3), (2, 3, 1), or (3, 1, 2). These are *cyclic* permutations of the integers 1, 2, 3. Note that for a right-handed system given by the triplet $\{\hat{\mathbf{e}}_1, \hat{\mathbf{e}}_2, \hat{\mathbf{e}}_3\}$, one has $\hat{\mathbf{e}}_1 \cdot (\hat{\mathbf{e}}_2 \times \hat{\mathbf{e}}_3) = 1$. The cartesian coordinate system we have used is right-handed if we identify x with 1, y with 2, and z with 3. A *left-handed system* is a mirror image of a right-handed system.

Two useful identities to be remembered are as follows:

1. Triple scalar product

$$\mathbf{A} \cdot (\mathbf{B} \times \mathbf{C}) = (\mathbf{A} \times \mathbf{B}) \cdot \mathbf{C} = (\mathbf{C} \times \mathbf{A}) \cdot \mathbf{B} \tag{1.7}$$

(It is the *volume* of a parallelepiped whose edges are **A**, **B**, and **C**.)

2. Triple vector product

$$\mathbf{A} \times (\mathbf{B} \times \mathbf{C}) = \mathbf{B}(\mathbf{A} \cdot \mathbf{C}) - \mathbf{C}(\mathbf{A} \cdot \mathbf{B}) \tag{1.8}$$

The latter is frequently known as the "back cab" rule. It will be noted in Eq. (1.7) that the dot (·) and the cross (×) may be freely interchanged so long as {**A**, **B**, **C**} remain in cyclic order.

1.2 Elements of Displacement, Area, and Volume; Solid Angle

1.2.1 Element of Displacement

Consider two points in space (x, y, x) and $(x + \Delta x, y + \Delta y, z + \Delta z)$. The first point is displaced relative to the second by the displacement $\Delta\mathbf{r}$; that is,

$$\Delta\mathbf{r} \equiv \hat{\mathbf{x}}\,\Delta x + \hat{\mathbf{y}}\,\Delta y + \hat{\mathbf{z}}\,\Delta z \tag{1.9}$$

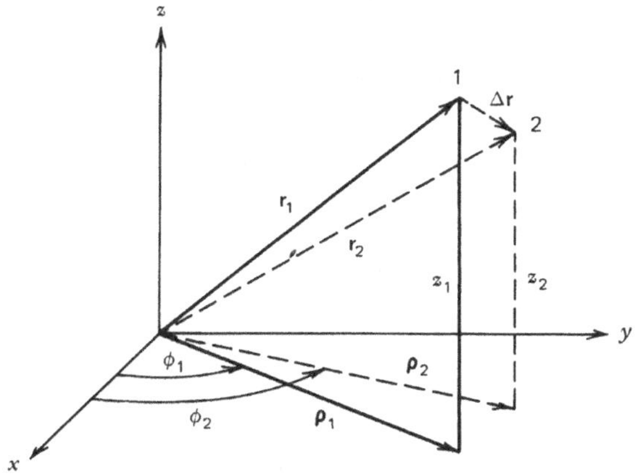

Figure 1.6 The coordinates of two nearby points in cylindrical coordinates that may be used to define the differential displacements in this system.

expressed in a cartesian system. A differential element of displacement is consequently written

$$d\mathbf{r} = dx\,\hat{\mathbf{x}} + dy\,\hat{\mathbf{y}} + dz\,\hat{\mathbf{z}} \tag{1.10}$$

We now wish to express $\Delta\mathbf{r}$ (or $d\mathbf{r}$) in terms of cylindrical and spherical coordinates and their associated unit vectors. We again assume that $\Delta\mathbf{r}$ may be made arbitrarily small, in the limit calling it $d\mathbf{r}$.

Consider Fig. 1.6, where two points are displaced by $\Delta\mathbf{r}$. In the cylindrical coordinate system we have base vectors that are different at the two points 1 and 2. Thus

$$\Delta\mathbf{r} = \rho_2\hat{\boldsymbol{\rho}}_2 + z_2\hat{\mathbf{z}}_2 - (\rho_1\hat{\boldsymbol{\rho}}_1 + z_1\hat{\mathbf{z}}_1) \tag{1.11}$$

where

$$\rho_2 \equiv \rho_1 + \Delta\rho \qquad \hat{\boldsymbol{\rho}}_2 \equiv \hat{\boldsymbol{\rho}}_1 + \Delta\hat{\boldsymbol{\rho}} \qquad z_2 \equiv z_1 + \Delta z \qquad \hat{\mathbf{z}}_2 = \hat{\mathbf{z}}_1 \equiv \hat{\mathbf{z}} \tag{1.12}$$

Substituting Eq. (1.12) in Eq. (1.11) and dropping products of differentials, we obtain

$$\Delta\mathbf{r} = \Delta\rho\,\hat{\boldsymbol{\rho}}_1 + \rho_1\,\Delta\hat{\boldsymbol{\rho}} + \Delta z\,\hat{\mathbf{z}} \tag{1.13}$$

If points (1) and (2) are close enough together then, to good approximation,

$$\Delta\hat{\boldsymbol{\rho}} = \frac{d\hat{\boldsymbol{\rho}}}{d\phi}\,\Delta\phi$$

For $\Delta\rho$ sufficiently small ($\hat{\boldsymbol{\phi}}_1 \approx \hat{\boldsymbol{\phi}}_2 \equiv \hat{\boldsymbol{\phi}}$) we can see that $\Delta\boldsymbol{\rho} = |\hat{\boldsymbol{\rho}}_1|\Delta\phi\,\hat{\boldsymbol{\phi}}$. As a result, Eq. (1.12) becomes

$$\Delta\mathbf{r} = \Delta\rho\,\hat{\boldsymbol{\rho}}_1 + \rho_1\,\Delta\hat{\boldsymbol{\phi}}_1 + \Delta z\,\hat{\mathbf{z}} \tag{1.14}$$

As point 2 approaches point 1, we can write the differential displacement as

$$d\mathbf{r} = dl_\rho\hat{\boldsymbol{\rho}} + dl_{\hat{\phi}}\hat{\boldsymbol{\phi}} + dl_z\hat{\mathbf{z}} \tag{1.15}$$

where

$$dl_\rho = d\rho \qquad dl_\phi = \rho\,d\phi \qquad dl_z = dz \tag{1.16}$$

are the elements of displacement in the ρ, ϕ, and z directions, respectively. Thus

$$d\mathbf{r} = d\rho\,\hat{\boldsymbol{\rho}} + \rho\,d\phi\,\hat{\boldsymbol{\phi}} + dz\,\hat{\mathbf{z}} \tag{1.17}$$

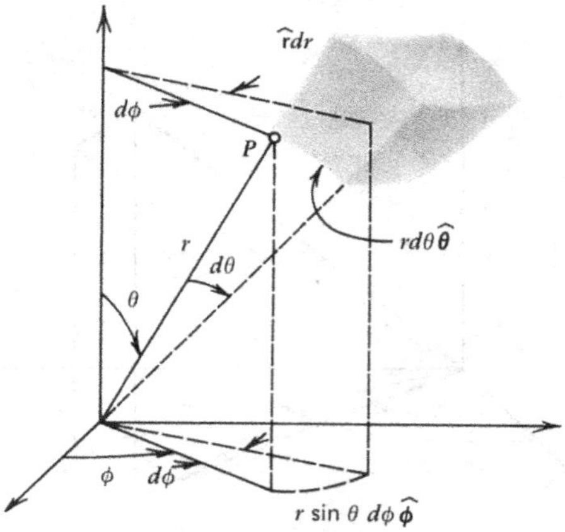

Figure 1.7 Differential displacements in spherical coordinates.

where $\{\rho, \phi, z\}$ are defined at the point where the displacement $d\mathbf{r}$ is made. Geometrically this is a natural result, since close enough to any point of space (z, ρ, ϕ) we can define a "cartesianlike" system in terms of which any element of length may be expressed directly.

Instead of deriving the element of displacement at a point in terms of spherical unit vectors at that point, we shall simply make the remark that at any point the unit vectors $\{\hat{\mathbf{r}}, \hat{\boldsymbol{\theta}}, \hat{\boldsymbol{\phi}}\}$ serve to establish a cartesian system locally (near the point). It will be seen from Fig. 1.7 that the elements of length along the three directions near this point are given by

$$dl_r = dr \qquad dl_\theta = r\,d\theta \qquad dl_\phi = r \sin\theta\,d\phi \tag{1.18}$$

and that

$$d\mathbf{r} = dr\,\hat{\mathbf{r}} + r\,d\theta\,\hat{\boldsymbol{\theta}} + r\sin\theta\,d\phi\,\hat{\boldsymbol{\phi}} = dl_r\,\hat{\mathbf{r}} + dl_\theta\,\hat{\boldsymbol{\theta}} + dl_\phi\,\hat{\boldsymbol{\phi}} \tag{1.19}$$

1.2.2 Element of Surface Area

Having determined expressions for elements of displacement in various coordinate systems, we can now determine elements of surface area. There are three elements of surface area for every coordinate system; these are of the form $dl_1\,dl_2$, $dl_2\,dl_3$, and $dl_3\,dl_1$. For cartesian coordinates, we have

$$dx\,dy \qquad dy\,dz \qquad dz\,dx \tag{1.20}$$

corresponding to the surfaces z = constant, x = constant, and y = constant, respectively. Similarly, for cylindrical coordinates, elements of surface area on the surfaces that define the coordinates are

$$dz\,d\rho \qquad \rho\,d\rho\,d\phi \qquad \rho\,d\phi\,dz \tag{1.21}$$

For spherical coordinates we have

$$r\,dr\,d\theta \qquad r^2\sin\theta\,d\theta\,d\phi \qquad r\sin\theta\,d\phi\,dr \tag{1.22}$$

A direction may be associated with an element of area. This direction is normal to the area. If $dl_i\,dl_j$ is the element of area, the normal direction is given by the cross product $\hat{\mathbf{e}}_i \times \hat{\mathbf{e}}_j$, and we may denote the area as a vector $(\hat{\mathbf{e}}_i\,dl_i) \times (\hat{\mathbf{e}}_j\,dl_j) \equiv \hat{\mathbf{e}}_k\,dl_i\,dl_j$.

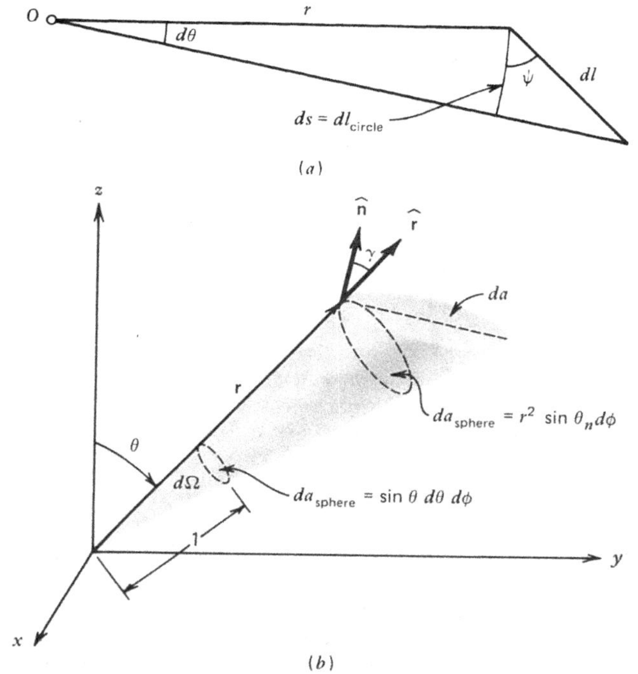

Figure 1.8 Definition of plane and solid angles. (*a*) Plane angle $d\theta$ (*b*) Solid angle $d\Omega$.

Often we shall more simply denote the element of area as $\mathbf{da} \equiv \hat{\mathbf{n}}\, da$, where the sense of $\hat{\mathbf{n}}$ must be specified but is always normal to the surface.

1.2.3 Solid Angle

When an arc element ds of a circle in a plane is referred to its center, we use the concept of an angle $d\theta = ds/r$ where r is the radius of the circle (Fig. 1.8*a*). On the other hand when an element of surface area is referred to an origin, it is often convenient to use the concept of *solid angle* (see Fig. 1.8*b*). The differential element $d\Omega$ of solid angle with respect to the origin is defined as follows:

$$d\Omega \equiv \frac{\mathbf{da} \cdot \hat{\mathbf{r}}}{r^2} = \frac{da\, \hat{\mathbf{n}} \cdot \hat{\mathbf{r}}}{r^2} = \frac{da \cos \gamma}{r^2} \tag{1.23}$$

Here, the surface element da is located at a point displaced from the origin by the vector $\mathbf{r} \equiv \hat{\mathbf{r}} r$, and hence γ is the angle between $\hat{\mathbf{n}}$ and $\hat{\mathbf{r}}$. Since $\mathbf{da} \cdot \hat{\mathbf{r}}$ is just the element of area of a sphere of radius r, then substituting $da = r^2 \sin \theta\, d\theta\, d\phi$, we see that $d\Omega$ is also given by

$$d\Omega = \frac{da_{\text{sphere}}}{r^2} = \frac{r^2 \sin \theta\, d\theta\, d\phi}{r^2} = \sin \theta\, d\theta\, d\phi \tag{1.24}$$

which is an element of area of a unit sphere.

Physically, the solid angle is the "opening angle" of a cone whose sides intercept the area element in question. Thus, just as for ordinary angular elements $d\theta$, where

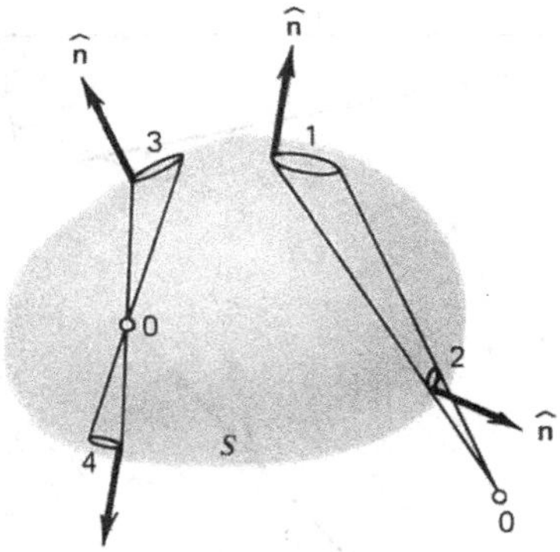

Figure 1.9 Illustrating why the solid angle subtended by a closed surface with respect to an origin inside the surface is 4π and to an origin outside the surface is zero.

we have $r\,d\theta = dl_{\text{circle}} = dl \cos\psi$, for an infinitesimal element of area of a sphere (which approximates a rectangular planar area) we have $da_{\text{sphere}} = r^2 \sin\theta\, d\theta\, d\phi$, $da \equiv r^2\, d\Omega$. The unit of solid angle is known as the *steradian*. It is clearly analogous to the unit of angle, the radian. Any finite solid angle is expressed as $\Omega = \int d\Omega = \int \sin\theta\, d\theta\, d\phi$, where θ and ϕ are the spherical coordinates of the spherical surface element intercepted. If we have a surface that completely encloses the origin, then $\Omega = 4\pi$. If we have a closed surface that does not enclose the origin and if we choose the direction of $d\mathbf{a}$ always to point out from the closed surface (or into the surface), then $\Omega = 0$. This is (see Fig. 1.9) essentially due to the fact that for each positive contribution of solid angle there is an equal contribution of negative solid angle, as seen from the origin.

1.2.4 Element of Volume

Remembering that $\mathbf{A}\cdot(\mathbf{B}\times\mathbf{C})$ is the volume of a parallelepiped, we have that the volume element for a system of base vectors $\{\hat{\mathbf{e}}_1, \hat{\mathbf{e}}_2, \hat{\mathbf{e}}_3\}$ is simply given by the vector triple product:

$$dv = dl_1\, \hat{\mathbf{e}}_1 \cdot (dl_2\, \hat{\mathbf{e}}_2 \times dl_3\, \hat{\mathbf{e}}_3)$$

or

$$dv \equiv |\hat{\mathbf{e}}_1 \cdot (\hat{\mathbf{e}}_2 \times \hat{\mathbf{e}}_3)|\, dl_1\, dl_2\, dl_3$$

where the dl_j are the magnitudes of the elements of displacements along the direction of the respective base vectors $\hat{\mathbf{e}}_j$. Thus, in the cartesian system,

$$dv = |\hat{\mathbf{x}} \cdot (\hat{\mathbf{y}} \times \hat{\mathbf{z}})|\, dx\, dy\, dz = dx\, dy\, dz \tag{1.25}$$

In the cylindrical system we have

$$dv = |\hat{\mathbf{z}} \cdot (\hat{\boldsymbol{\rho}} \times \hat{\boldsymbol{\theta}})|\, dz\, d\rho\, \rho\, d\phi = \rho\, dz\, d\rho\, d\phi \tag{1.26}$$

and in the spherical coordinate system we have

$$dv = |\hat{\mathbf{r}} \cdot (\hat{\boldsymbol{\theta}} \times \hat{\boldsymbol{\phi}})|\, dr\, r\, d\theta\, r \sin\theta\, d\phi = r^2 \sin\theta\, dr\, d\theta\, d\phi \tag{1.27}$$

1.3 Gradient

If we wish to express the change in a scalar function of position $f(\mathbf{r})$ at the location specified by $\mathbf{r}$, then writing f and $\mathbf{r}$ in cartesian components, we find the differential change to be

$$df = dx\frac{\partial f}{\partial x} + dy\frac{\partial f}{\partial y} + dz\frac{\partial f}{\partial z}$$

We now define a linear differential "vector operator" called *del*, and symbolized ∇ as follows:

$$\nabla \equiv \hat{\mathbf{x}}\frac{\partial}{\partial x} + \hat{\mathbf{y}}\frac{\partial}{\partial y} + \hat{\mathbf{z}}\frac{\partial}{\partial z} \tag{1.28}$$

Since $d\mathbf{r} = \hat{\mathbf{x}}\,dx + \hat{\mathbf{y}}\,dy + \hat{\mathbf{z}}\,dz$, then from the definition of the dot product,

$$df = (\nabla f)\cdot d\mathbf{r} \tag{1.29}$$

where ∇f is a vector point function, and is called the *gradient of f*:

$$\text{grad}\, f \equiv \nabla f = \hat{\mathbf{x}}\frac{\partial f}{\partial x} + \hat{\mathbf{y}}\frac{\partial f}{\partial y} + \hat{\mathbf{z}}\frac{\partial f}{\partial z} \tag{1.30}$$

Some applications involve the operation of the above gradient operator on vector fields. If we wish to express the change in a vector field $\mathbf{A}$ at the location specified by $\mathbf{r}$, then writing $\mathbf{A}$ and $\mathbf{r}$ in cartesian components, we find the differential change to be

$$d\mathbf{A} = dx\frac{\partial \mathbf{A}}{\partial x} + dy\frac{\partial \mathbf{A}}{\partial y} + dz\frac{\partial \mathbf{A}}{\partial z} = \left[dx\frac{\partial}{\partial x} + dy\frac{\partial}{\partial y} + dz\frac{\partial}{\partial z}\right]\mathbf{A}$$

Using Eq. (1.30), one can show that the scalar product of $d\mathbf{r}$ and ∇ is

$$d\mathbf{r}\cdot\nabla = \left[dx\frac{\partial}{\partial x} + dy\frac{\partial}{\partial y} + dz\frac{\partial}{\partial z}\right] \tag{1.31}$$

Thus

$$d\mathbf{A} = (d\mathbf{r}\cdot\nabla)\mathbf{A}(\mathbf{r}) \tag{1.32}$$

In words, the scalar operator $(d\mathbf{r}\cdot\nabla)$ acting on a vector point function $\mathbf{A}(\mathbf{r})$ generates the spatial differential of $\mathbf{A}$, $d\mathbf{A}$, at the point in question.

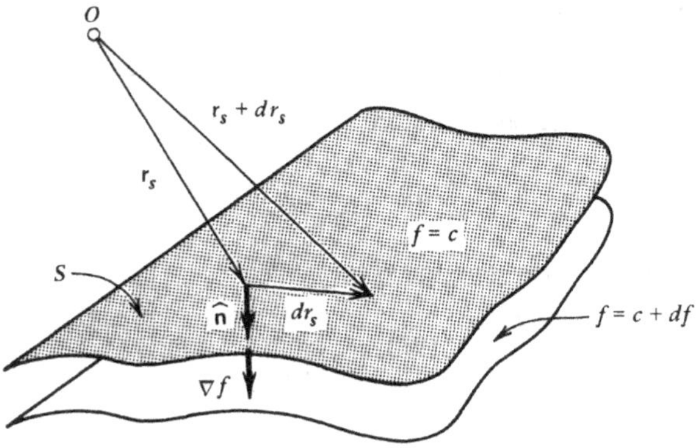

Figure 1.10 The use of the surface $f(\mathbf{r}) = c =$ constant to show that the gradient of the function $f(\mathbf{r})$ is normal to this surface.

The gradient given in Eq. (1.28) has an interesting interpretation (see Fig. 1.10). Suppose we have the scalar function $f(\mathbf{r})$. A surface is generated if we set $f(\mathbf{r})$ equal to a *constant*, c. This being the case, if we find the differential of $f(\mathbf{r})$ when $d\mathbf{r}$ connects two points of the surface $f(\mathbf{r}) = c$, then (calling $d\mathbf{r}_s$ this surface displacement) $f(\mathbf{r} + d\mathbf{r}_s) = f(\mathbf{r}) = c$, so that $df = 0$. This implies that $df = (\nabla f) \cdot d\mathbf{r}_s = 0$, which in turn implies that (∇f) is perpendicular to $d\mathbf{r}_s$. Since $d\mathbf{r}_s$ was defined to lie in the tangent plane to the surface at $\mathbf{r}$, we see that ∇f is *perpendicular* to the surface. Now the value of the change in $f(\mathbf{r})$ when we move to a neighboring point $(\mathbf{r} + d\mathbf{r})$ not in the surface, is given by Eq. (1.29): $df = \nabla f \cdot d\mathbf{r}$. Therefore, if $|d\mathbf{r}| \equiv ds$, $df/ds = (\nabla f) \cdot d\mathbf{r}/ds = \nabla f \cdot \hat{\mathbf{t}}$, where $\hat{\mathbf{t}}$ is a unit vector in the direction of $d\mathbf{r}$. The derivative df/ds is known as the *directional derivative* of f. The maximum value of df/ds will be obtained when $\hat{\mathbf{t}}$ lies along ∇f. In this case $|\nabla f \cdot \hat{\mathbf{t}}| = |\nabla f| = df/ds_{\text{max}}$; that is, the maximum rate of increase in the function $f(\mathbf{r})$ with respect to displacements $d\mathbf{r}$ is given by ∇f, and ∇f points in the direction of maximum increase. These properties of ∇f are summarized by the formula

$$\nabla f = \hat{\mathbf{n}} \frac{df}{dn} \tag{1.33}$$

where $\hat{\mathbf{n}}\, dn$ is the displacement $d\mathbf{r}$ perpendicular to the surface $f =$ constant (in the direction of the maximum increase in f). It is for this reason that it is called the gradient of f. As an example, consider $f \equiv x = c$, which defines a set of planes, one for each value of c, perpendicular to the x axis. Clearly, the fact that $\nabla f = \hat{\mathbf{x}}$ shows that (a) the perpendicular to the planes lies in the x direction, and (b) the rate of increase of f in this direction is unity. As another example, let $f = x^2 + y^2 + z^2 \equiv r^2$. Then one finds $\nabla f = 2r\hat{\mathbf{r}}$, where $\mathbf{r} \equiv x\hat{\mathbf{x}} + y\hat{\mathbf{y}} + z\hat{\mathbf{z}}$. The perpendicular to the surfaces $f = c$ are in the directions of the radii to the points on the spheres given by $x^2 + y^2 + z^2 = r^2$. The rate of change of f in these directions is just $df/dr = 2r$. Finally, note that since ∇f is perpendicular to the surface $f =$ constant, it can be used to generate the unit vectors of coordinate systems. Thus

$$\frac{\nabla f}{|\nabla f|} = \hat{\mathbf{f}} \tag{1.34}$$

is a unit vector perpendicular to the surface $f(x, y, z) =$ constant.

We now find an expression for ∇f where f is any scalar point function in terms of cylindrical coordinates. To do this we turn to our defining equation for the gradient: $d\mathbf{r} \cdot \nabla f = df$. Writing $d\mathbf{r} = \hat{\boldsymbol{\rho}}\, d\rho + \hat{\boldsymbol{\phi}}\rho\, d\phi + \hat{\mathbf{z}}\, dz$ and $\nabla f = \hat{\boldsymbol{\rho}}(\nabla f)_\rho + \hat{\boldsymbol{\phi}}(\nabla f)_\phi + \hat{\mathbf{z}}(\nabla f)_z$, as we can always do at any point $\{\rho, \phi, z\}$, gives

$$df = (\nabla f)_\rho\, d\rho + (\nabla f)_\phi \rho\, d\phi + (\nabla f)_z\, dz$$

Noting that the rules of calculus require

$$df = \frac{\partial f}{\partial \rho} d\rho + \frac{\partial f}{\partial \phi} d\phi + \frac{\partial f}{\partial z} dz$$

and equating the two expressions for df gives

$$(\nabla f)_\rho = \frac{\partial f}{\partial \rho} \qquad (\nabla f)_\phi = \frac{1}{\rho}\frac{\partial f}{\partial \phi} \qquad (\nabla f)_z = \frac{\partial f}{\partial z}$$

or

$$\nabla f = \hat{\boldsymbol{\rho}} \frac{\partial f}{\partial \rho} + \hat{\boldsymbol{\phi}} \frac{1}{\rho} \frac{\partial f}{\partial \phi} + \hat{\mathbf{z}} \frac{\partial f}{\partial z}.$$

Hence, the del (∇) operator in cylindrical coordinates is

$$\nabla = \hat{\boldsymbol{\rho}} \frac{\partial}{\partial \rho} + \hat{\boldsymbol{\phi}} \frac{1}{\rho} \frac{\partial}{\partial \phi} + \hat{\mathbf{z}} \frac{\partial}{\partial z} \tag{1.35}$$

Similarly, one can obtain expressions for ∇f and ∇ in spherical polar coordinates:

$$\nabla f = \hat{\mathbf{r}} \frac{\partial f}{\partial r} + \hat{\boldsymbol{\theta}} \frac{1}{r} \frac{\partial f}{\partial \theta} + \hat{\boldsymbol{\phi}} \frac{1}{r \sin \theta} \frac{\partial f}{\partial \phi}$$

$$\nabla = \hat{\mathbf{r}} \frac{\partial}{\partial r} + \hat{\boldsymbol{\theta}} \frac{1}{r} \frac{\partial}{\partial \theta} + \hat{\boldsymbol{\phi}} \frac{1}{r \sin \theta} \frac{\partial}{\partial \phi} \tag{1.36}$$

1.4 The Divergence of a Vector and Gauss' Theorem

For a vector field, $\mathbf{A}(\mathbf{r})$, we shall define the divergence, written variously as div $\mathbf{A}$ or $\nabla \cdot \mathbf{A}$, by the expression

$$\nabla \cdot \mathbf{A} = \lim_{\Delta V \to 0} \left[\frac{\oint_S \mathbf{A} \cdot d\mathbf{a}}{\Delta V} \right] \tag{1.37}$$

where ΔV is a differential volume and S is its surface. Moreover $\nabla \cdot \mathbf{A}$ may be shown to be equivalently definable as the scalar product of ∇ with the vector $\mathbf{A}$. We will first use the integral definition to derive the divergence of a vector explicitly. Consider a surface S whose surface element is denoted by $d\mathbf{a}$, then the integral

$$F = \int_S \mathbf{A} \cdot d\mathbf{a} \tag{1.38}$$

is called the "*flux* of $\mathbf{A}$ through S." Clearly, as in our discussion of solid angle, S is assumed to have an orientation (with respect to the origin of the coordinate system used), and in general may or may not be closed. The element of flux through $d\mathbf{a}$ is correspondingly given by $dF = \mathbf{A} \cdot d\mathbf{a}$ where $\mathbf{A}$ is evaluated at the center of $d\mathbf{a}$. Now consider the infinitesimal volume element dv, shown in Fig. 1.11. The sides of dv are given by the six spherical-coordinate surfaces: $r = c$, $r = c + dr$; $\theta = c'$, $\theta = c' + d\theta$,

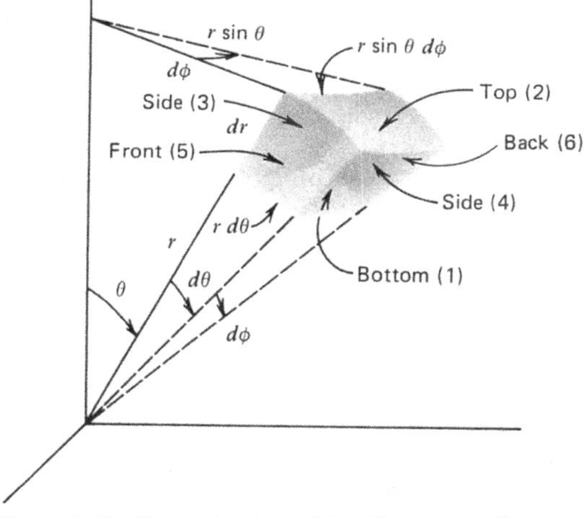

Figure 1.11 Determination of the divergence of a vector in spherical coordinates.

$\phi = c''$, $\phi = c'' + d\phi$, and its volume is $r^2 \sin\theta\, dr\, d\theta\, d\phi$. We compute the fluxes dF_1, $dF_2, \ldots, dF_6$ through the sides 1, 2, ..., 6 of the volume element described above (and illustrated in the figure).

$$dF_1 + dF_2 = -\{r^2 A_r \sin\theta\, d\theta\, d\phi\}_1 + \{r^2 A_r \sin\theta\, d\theta\, d\phi\}_2 \equiv d[r^2 \sin\theta\, d\theta\, d\phi\, A_r]_{2,1}$$

$$dF_3 + dF_4 = -\{r \sin\theta\, d\phi\, dr\, A_\theta\}_3 + \{r \sin\theta\, d\phi\, dr\, A_\theta\}_4 \equiv d\{r \sin\theta\, d\phi\, dr\, A_\theta\}_{4,3}$$

$$dF_5 + dF_6 = -\{r\, d\theta\, dr\, A_\phi\}_5 + \{r\, d\theta\, dr\, A_\phi\}_6 \equiv d\{r\, d\theta\, dr\, A_\phi\}_{6,5}$$

The quantities in the braces can be evaluated at the center of the faces. Thus, for surfaces 1 and 2 only functions of r are different:

$$d\{r^2 \sin\theta\, d\theta\, d\phi\, A_r\}_{2,1} = d\{r^2 A_r\}_{2,1} \sin\theta\, d\theta\, d\phi = \frac{\partial}{\partial r}\{r^2 A_r\} dr \sin\theta\, d\theta\, d\phi$$

Similarly,

$$d\{r \sin\theta\, d\phi\, dr\, A_\theta\}_{4,3} = d\{\sin\theta\, A_\theta\}_{4,3} r\, dr\, d\phi = \frac{\partial}{\partial\theta}\{\sin\theta\, A_\theta\} r\, d\theta\, dr\, d\phi$$

$$d\{r\, dr\, d\theta\, A_\theta\}_{6,5} = \frac{\partial}{\partial\phi}\{A_\phi\} r\, dr\, d\theta\, d\phi$$

Therefore

$$\nabla \cdot \mathbf{A} \equiv \left[\frac{\sum_{j=1}^{6} dF_j}{r^2 \sin\theta\, dr\, d\theta\, d\phi}\right]$$

that is,

$$\nabla \cdot \mathbf{A} = \frac{1}{r^2}\frac{\partial}{\partial r}(r^2 A_r) + \frac{1}{r \sin\theta}\left[\frac{\partial}{\partial\theta}(\sin\theta\, A_\theta) + \frac{\partial A_\phi}{\partial\phi}\right] \tag{1.39}$$

We now use the direct scalar product $\nabla \cdot \mathbf{A}$ to arrive at the divergence in the various coordinate systems. The ∇ operator in spherical coordinates is given by Eq. (1.36), and thus

$$\nabla \cdot \mathbf{A} = \left[\hat{\mathbf{r}}\frac{\partial}{\partial r} + \frac{\hat{\boldsymbol{\theta}}}{r}\frac{\partial}{\partial\theta} + \frac{\hat{\boldsymbol{\phi}}}{r\sin\theta}\frac{\partial}{\partial\phi}\right] \cdot [\hat{\mathbf{r}}A_r + \hat{\boldsymbol{\theta}}A_\theta + \hat{\boldsymbol{\phi}}A_\phi]$$

Expanding the indicated scalar product, we get nine terms:

$$\nabla \cdot \mathbf{A} = \hat{\mathbf{r}} \cdot \frac{\partial}{\partial r}[\hat{\mathbf{r}}A_r + \hat{\boldsymbol{\theta}}A_\theta + \hat{\boldsymbol{\phi}}A_\phi] + \frac{\hat{\boldsymbol{\theta}}}{r} \cdot \frac{\partial}{\partial\theta}[\hat{\mathbf{r}}A_r + \hat{\boldsymbol{\theta}}A_\theta + \hat{\boldsymbol{\phi}}A_\phi]$$
$$+ \frac{\hat{\boldsymbol{\phi}}}{r\sin\theta} \cdot \frac{\partial}{\partial\phi}[\hat{\mathbf{r}}A_r + \hat{\boldsymbol{\theta}}A_0 + \hat{\boldsymbol{\phi}}A_\phi]$$

Now, perform the implied derivatives, noting, for example, that

$$\frac{\partial}{\partial r}(\hat{\mathbf{r}}A_r) = \frac{\partial\hat{\mathbf{r}}}{\partial r}A_r + \hat{\mathbf{r}}\frac{\partial A_r}{\partial r}$$

and so forth. There will be 18 individual terms, but if we remember that $\hat{\mathbf{r}} \cdot \hat{\boldsymbol{\theta}} = \hat{\mathbf{r}} \cdot \hat{\boldsymbol{\phi}} = \hat{\boldsymbol{\theta}} \cdot \hat{\boldsymbol{\phi}} = 0$, and so on, we obtain just 12 nonzero terms:

$$\nabla \cdot \mathbf{A} = \hat{\mathbf{r}} \cdot \frac{\partial\hat{\mathbf{r}}}{\partial r}A_r + \frac{\partial A_r}{\partial r} + \hat{\mathbf{r}} \cdot \frac{\partial\hat{\boldsymbol{\theta}}}{\partial r}A_\theta + \hat{\mathbf{r}} \cdot \frac{\partial\hat{\boldsymbol{\phi}}}{\partial r}A_\phi$$
$$+ \frac{\hat{\boldsymbol{\theta}}}{r} \cdot \frac{\partial\hat{\mathbf{r}}}{\partial\theta}A_r + \frac{\hat{\boldsymbol{\theta}}}{r} \cdot \frac{\partial\hat{\boldsymbol{\theta}}}{\partial\theta}A_\theta + \frac{1}{r}\frac{\partial A_\theta}{\partial\theta} + \frac{\hat{\boldsymbol{\theta}}}{r} \cdot \frac{\partial\hat{\boldsymbol{\phi}}}{\partial\theta}A_\phi$$
$$+ \frac{\hat{\boldsymbol{\phi}}}{r\sin\theta} \cdot \frac{\partial\hat{\mathbf{r}}}{\partial\phi}A_r + \frac{\hat{\boldsymbol{\phi}}}{r\sin\theta} \cdot \frac{\partial\hat{\boldsymbol{\theta}}}{\partial\phi}A_\theta + \frac{\hat{\boldsymbol{\phi}}}{r\sin\theta} \cdot \frac{\partial\hat{\boldsymbol{\phi}}}{\partial\phi}A_\phi + \frac{1}{r\sin\theta}\frac{\partial A_\phi}{\partial\phi}$$

Table 1.1

	$\hat{\mathbf{r}}$	$\hat{\boldsymbol{\theta}}$	$\hat{\boldsymbol{\phi}}$
$\dfrac{\partial}{\partial r}$	0	0	0
$\dfrac{\partial}{\partial \theta}$	$\hat{\boldsymbol{\theta}}$	$-\hat{\mathbf{r}}$	0
$\dfrac{\partial}{\partial \phi}$	$\hat{\boldsymbol{\phi}} \sin \theta$	$\hat{\boldsymbol{\phi}} \cos \theta$	$-[\hat{\mathbf{r}} \sin \theta + \hat{\boldsymbol{\theta}} \cos \theta]$

The values of the partial derivatives of the unit vectors in this expression are given in Table 1.1. This table is easily constructed by using the expressions for $\hat{\mathbf{r}}$, $\hat{\boldsymbol{\theta}}$, and $\hat{\boldsymbol{\phi}}$ given previously in Eq. (1.3). For example, since $\hat{\boldsymbol{\theta}} = \hat{\mathbf{x}} \cos \theta \cos \phi + \hat{\mathbf{y}} \cos \theta \sin \phi - \hat{\mathbf{z}} \sin \theta$ and the cartesian unit vectors are constant, $\partial \hat{\boldsymbol{\theta}}/\partial \theta = \hat{\mathbf{x}}(-\sin \theta)\cos \phi + \hat{\mathbf{y}}(-\sin \theta)\sin \phi - \hat{\mathbf{z}} \cos \theta = -\hat{\mathbf{r}}$. Substitution will yield the same result we got in 1.39.

The operator ∇ in cylindrical coordinates was shown to be [see Eq. (1.35)]

$$\nabla = \hat{\boldsymbol{\rho}} \frac{\partial}{\partial \rho} + \hat{\boldsymbol{\phi}} \frac{1}{\rho} \frac{\partial}{\partial \phi} + \hat{\mathbf{z}} \frac{\partial}{\partial z}.$$

Proceeding by the direct product method to calculate $\nabla \cdot \mathbf{A}$ in cylindrical coordinates and using the fact that $\hat{\mathbf{z}}$ is a constant unit vector and that $\hat{\boldsymbol{\rho}}$ and $\hat{\boldsymbol{\phi}}$ depend on the coordinate ϕ only [see Eq. (1.2)], we find that the only nonzero derivative of the unit vectors are $\partial \hat{\boldsymbol{\rho}}/\partial \phi = \hat{\boldsymbol{\phi}}$ and $\partial \hat{\boldsymbol{\phi}}/\partial \phi = -\hat{\boldsymbol{\rho}}$. Therefore

$$\nabla \cdot \mathbf{A} = \frac{\partial}{\partial \rho} A_\rho + \frac{1}{\rho} A_\rho + \frac{1}{\rho} \frac{\partial}{\partial \phi} A_\phi + \frac{\partial A_z}{\partial z}$$

or

$$\nabla \cdot \mathbf{A} = \frac{1}{\rho} \frac{\partial}{\partial \rho} (\rho A_\rho) + \frac{1}{\rho} \frac{\partial}{\partial \phi} A_\phi + \frac{\partial A_z}{\partial z} \tag{1.40}$$

Finally, in the case of cartesian coordinates the operator ∇ was shown in Eq. (1.28) to be

$$\nabla = \hat{\mathbf{x}} \frac{\partial}{\partial x} + \hat{\mathbf{y}} \frac{\partial}{\partial y} + \hat{\mathbf{z}} \frac{\partial}{\partial z}.$$

Expanding $\nabla \cdot \mathbf{A}$ directly gives

$$\nabla \cdot \mathbf{A} = \frac{\partial A_x}{\partial x} + \frac{\partial A_y}{\partial y} + \frac{\partial A_z}{\partial z} \tag{1.41}$$

The Divergence Theorem (Gauss' Theorem). Finally we prove a relation that is very useful in electrostatics: the divergence theorem, which involves the divergence operation. From our definitions of $\nabla \cdot \mathbf{A}$ we have

$$(\nabla \cdot \mathbf{A})\Delta v \cong \oint_s \mathbf{A} \cdot d\mathbf{a}$$

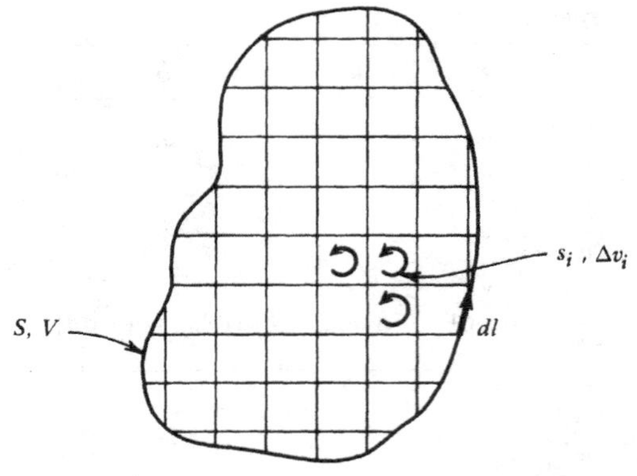

Figure 1.12 Proving the divergence theorem (Gauss' law) by subdividing the volume V of surface S into many infinitesimal volumes Δv_i of surfaces s_i and the application of the definition of the divergence in terms of the infinitesimal volumes.

for small Δv. Consider a volume V with a surface S subdivided into N small volume elements ∇v_i, with surfaces s_i as shown in Fig. 1.12. Then

$$\sum_{i=1}^{N} (\nabla \cdot \mathbf{A})_i \, \Delta v_i = \sum_{i=1}^{N} \oint_{s_i} \mathbf{A} \cdot d\mathbf{a}$$

If we take the limit as $N \to \infty$ and $\Delta v_i \to 0$, on the left we have a volume integral and on the right a surface integral

$$\int_V \nabla \cdot \mathbf{A} \, dv = \oint_S \mathbf{A} \, d\mathbf{a} \tag{1.42}$$

The fact that

$$\lim_{N \to \infty} \sum_i \oint_{s_i} \mathbf{A} \cdot d\mathbf{a}$$

becomes an integral over the surface external to the whole volume can be seen from Fig. 1.12. Note that the net flux through any internal surface is zero, because such a surface is common to two contiguous volume elements whose outward normals point oppositely on the common surface. All that remains are fluxes through surfaces not common to two volume elements—namely, surfaces on the surface of V itself.

Equation (1.42), which is known as the *divergence theorem* (*or Gauss' theorem*), proves very useful when one wishes to relate values of a vector field on the surface of a region to values in the interior. It often will happen that we may want to convert a surface integral to a volume integral, or vice versa, as will be shown in Chapters 3 and 4.

1.5 The Curl and Stokes' Theorem

We now introduce another useful operation involving ∇. For a vector field, $\mathbf{A}(\mathbf{r})$, we shall define the curl written as $\nabla \times \mathbf{A}$ or curl $\mathbf{A}$ by the expression

$$\nabla \times \mathbf{A} = \lim_{\Delta v \to 0} \left[\frac{1}{\Delta v} \oint_S d\mathbf{a} \times \mathbf{A} \right] \tag{1.43}$$

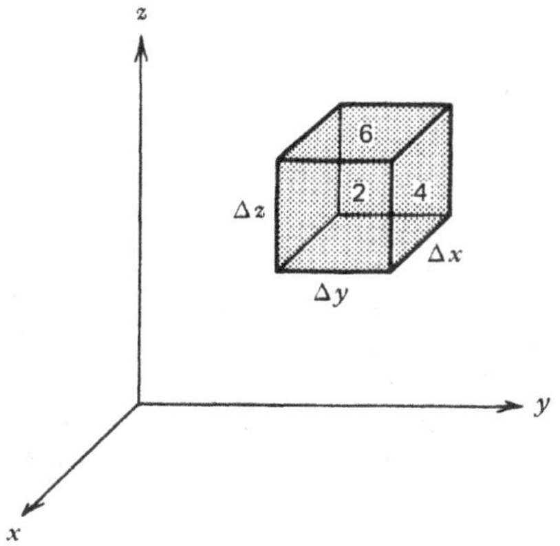

Figure 1.13 Determination of the explicit form of the curl of a vector in cartesian coordinates using a differential rectangular box.

where Δv is a small volume of surface area S. This definition can now be used to determine $\nabla \times \mathbf{A}$ in a cartesian representation. Consider the volume element $dv = dx\,dy\,dz$ shown in Fig. 1.13. The contributions from the opposite sides 1 and 2 to the integral $\int d\mathbf{a} \times \mathbf{A}$ are

$$\int_1 d\mathbf{a} \times \mathbf{A} = \{-\hat{\mathbf{x}}\,dy\,dz \times \mathbf{A}\}_1 = -dy\,dz\{\hat{\mathbf{z}}A_y - \hat{\mathbf{y}}A_z\}_1$$

$$\int_2 d\mathbf{a} \times \mathbf{A} = \{\hat{\mathbf{x}}\,dy\,dz \times \mathbf{A}\}_2 = dy\,dz\{\hat{\mathbf{z}}A_y - \hat{\mathbf{y}}A_z\}_2$$

The sum of these two integrals is the (partial) differential of the expression shown in braces taken between faces 1 and 2:

$$\int_{1+2} d\mathbf{a} \times \mathbf{A} = dy\,dz[\partial\{\hat{\mathbf{z}}A_y - \hat{\mathbf{y}}A_z\}_{2,1}] = dy\,dz\,\frac{\partial}{\partial x}\{\hat{\mathbf{z}}A_y - \hat{\mathbf{y}}A_z\}dx$$
$$= dv\left[\hat{\mathbf{z}}\,\frac{\partial A_y}{\partial x} - \hat{\mathbf{y}}\,\frac{\partial A_z}{\partial x}\right]$$

Similarly, for the opposite sides 3 and 4 and opposite sides 5 and 6, we can readily show that

$$\int_{3+4} d\mathbf{a} \times \mathbf{A} = dv\left[\hat{\mathbf{x}}\,\frac{\partial A_z}{\partial y} - \hat{\mathbf{z}}\,\frac{\partial A_x}{\partial y}\right] \qquad \int_{5+6} d\mathbf{a} \times \mathbf{A} = dv\left[\hat{\mathbf{y}}\,\frac{\partial A_x}{\partial z} - \hat{\mathbf{z}}\,\frac{\partial A_z}{\partial z}\right]$$

Summing all the contributions gives

$$\nabla \times \mathbf{A} = \hat{\mathbf{x}}\left[\frac{\partial A_z}{\partial y} - \frac{\partial A_y}{\partial z}\right] + \hat{\mathbf{y}}\left[\frac{\partial A_x}{\partial z} - \frac{\partial A_z}{\partial x}\right] + \hat{\mathbf{z}}\left[\frac{\partial A_y}{\partial x} - \frac{\partial A_x}{\partial y}\right] \tag{1.44}$$

Similar procedures can be used to determine $\nabla \times \mathbf{A}$ in spherical and cylindrical coordinates. We may also form the cross product of the operator ∇ with the vector point function $\mathbf{A}$ by employing the cross product of two vectors. One can easily show that this operation gives exactly Eq. (1.44).

In determinant form we can write $\nabla \times \mathbf{A}$ in cartesian coordinates as

$$\nabla \times \mathbf{A} = \begin{vmatrix} \hat{\mathbf{x}} & \hat{\mathbf{y}} & \hat{\mathbf{z}} \\ \dfrac{\partial}{\partial x} & \dfrac{\partial}{\partial y} & \dfrac{\partial}{\partial z} \\ A_x & A_y & A_z \end{vmatrix} \tag{1.45}$$

The determinant must be expanded in terms of the top row to have meaning.

Since $\nabla \times \mathbf{A}$ is formed like a cross product of two vectors, ∇ and $\mathbf{A}$, it is expected to have a value independent of the particular coordinate system in which it is represented. We shall see, for example, that $\nabla \times \mathbf{A}$ has a physical meaning and retains that meaning independent of the system of coordinates in which $\mathbf{A}$ is expressed. Thus, we shall assume that $\nabla \times \mathbf{A}$ is a vector, just as we assumed ∇f was a vector, and $\nabla \cdot \mathbf{A}$ a scalar.

One can formally obtain different coordinate representations for $\nabla \times \mathbf{A}$ by writing the operator ∇ and the vector $\mathbf{A}$ in a consistent set of coordinates. We need only be careful to understand that unit vectors in curvilinear coordinates need not have partial derivatives equal to zero (unlike the cartesian $\{\hat{\mathbf{x}}, \hat{\mathbf{y}}, \hat{\mathbf{z}}\}$ unit vectors, whose derivatives are always zero). This method gives the following expression for the curl in cylindrical coordinates:

$$\nabla \times \mathbf{A} = \hat{\boldsymbol{\rho}}\left[\frac{1}{\rho}\frac{\partial A_z}{\partial \phi} - \frac{\partial A_\phi}{\partial z}\right] + \hat{\boldsymbol{\phi}}\left[\frac{\partial A_\rho}{\partial z} - \frac{\partial A_z}{\partial \rho}\right] + \hat{\mathbf{z}}\left[\frac{1}{\rho}\frac{\partial}{\partial \rho}(\rho A_\phi) - \frac{1}{\rho}\frac{\partial A_\rho}{\partial \phi}\right] \tag{1.46}$$

Equation (1.46) indicates that there is no *simple* determinant notation that can be applied to the curl of a vector in cylindrical coordinates. However, one can still write it in a determinant form:

$$\nabla \times \mathbf{A} = \frac{1}{\rho}\begin{vmatrix} \hat{\boldsymbol{\rho}} & \rho\hat{\boldsymbol{\phi}} & \hat{\mathbf{z}} \\ \dfrac{\partial}{\partial \rho} & \dfrac{\partial}{\partial \phi} & \dfrac{\partial}{\partial \mathbf{z}} \\ A_\rho & \rho A_\phi & A_z \end{vmatrix}$$

Similarly, one may derive the following expression for $\nabla \times \mathbf{A}$ in spherical coordinates.

$$\nabla \times \mathbf{A} = \frac{\hat{\mathbf{r}}}{r \sin\theta}\left[\frac{\partial}{\partial \theta}(A_\phi \sin\theta) - \frac{\partial}{\partial \phi}(A_\theta)\right] + \hat{\boldsymbol{\theta}}\left[\frac{1}{r\sin\theta}\frac{\partial A_r}{\partial \phi} - \frac{1}{r}\frac{\partial}{\partial r}(rA_\phi)\right]$$
$$+ \hat{\boldsymbol{\phi}}\left[\frac{1}{r}\frac{\partial}{\partial r}(rA_\theta) - \frac{1}{r}\frac{\partial A_r}{\partial \theta}\right] \tag{1.47}$$

Again, a determinantal notation can be used:

$$\nabla \times \mathbf{A} = \frac{1}{r^2 \sin\theta}\begin{vmatrix} \hat{\mathbf{r}} & r\hat{\boldsymbol{\theta}} & r\sin\theta\hat{\boldsymbol{\phi}} \\ \dfrac{\partial}{\partial r} & \dfrac{\partial}{\partial \theta} & \dfrac{\partial}{\partial \phi} \\ A_r & rA_\theta & r\sin A_\phi \end{vmatrix}$$

Stokes' Theorem. Finally we prove a very useful relation in magnetostatics that involves the curl or a vector: Stokes' theorem. To this end we consider the component of $\nabla \times \mathbf{A}$ in the direction $\hat{\mathbf{e}}$. To compute this we consider a *small* volume

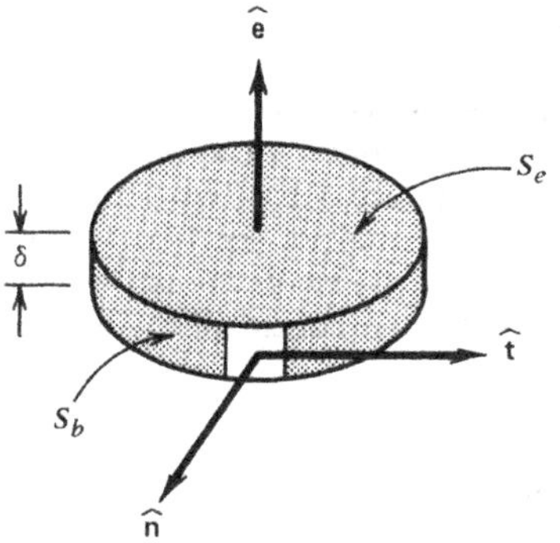

Figure 1.14 Determination of the component of a curl of a vector along a unit vector ê, by applying the definition of the curl to a differential pillbox of negligible height δ, and axis along ê.

element of surface S, shown in Fig. 1.14, in the form of a cylinder whose top and bottom each of area S_e are perpendicular to ê and whose sides of width δ and area S_b is parallel to ê. Using Eq. (1.43) for this volume element, we have

$$\nabla \times \mathbf{A} = \frac{1}{S_e\delta}\oint_S d\mathbf{a} \times \mathbf{A}$$

We now calculate the component of $\nabla \times \mathbf{A}$ along ê. Taking the dot product of ê with $\nabla \times \mathbf{A}$ gives:

$$\hat{\mathbf{e}} \cdot \nabla \times \mathbf{A} = \frac{1}{S_e\delta}\oint_S \hat{\mathbf{e}} \cdot d\mathbf{a} \times \mathbf{A} \tag{1.48}$$

Only the side band will contribute in Eq. (1.48), since $\hat{\mathbf{e}} \times \mathbf{S}_e = 0$. Therefore

$$\hat{\mathbf{e}} \cdot \nabla \times \mathbf{A} = \frac{1}{S_e\delta}\int_{S_b} \hat{\mathbf{e}} \times d\mathbf{a} \cdot \mathbf{A}$$

If δ is sufficiently small, then one writes $d\mathbf{a} = \delta\, dl\, \hat{\mathbf{n}}$, where dl is an element of length along the band, and $\hat{\mathbf{n}}$ is a unit vector normal to the band. Noting that $\hat{\mathbf{n}} \times \hat{\mathbf{e}} = \hat{\mathbf{t}}$, where $\hat{\mathbf{t}}$ is a unit vector tangent to the band, then the triple scalar product $\hat{\mathbf{e}} \times d\mathbf{a} \cdot \mathbf{A}$ can be written as $(\mathbf{A} \cdot \hat{\mathbf{t}})\delta\, dl$. Since $\hat{\mathbf{t}}\, dl$ is $d\mathbf{r}$, then we get

$$\hat{\mathbf{e}} \cdot \nabla \times \mathbf{A} = \frac{1}{S_e}\oint \mathbf{A} \cdot d\mathbf{r}$$

As $\delta \to 0$, S_e is the area bordered by the path of integration and contains the vector $\hat{\mathbf{t}}$. The sense of circulation is related to ê as the turning of a right-hand screw is to its advance. Written formally.

$$\hat{\mathbf{e}} \cdot \nabla \times \mathbf{A} = \lim_{S_e \to 0}\left[\frac{1}{S_e}\oint_C \mathbf{A} \cdot d\mathbf{r}\right] \tag{1.49}$$

where S_e represents an area whose normal is parallel to ê and whose perimeter of length C is the path of integration. Note that the integral $\oint_C \mathbf{A} \cdot d\mathbf{r}$ is called the circulation of **A** around C. With Eq. (1.49) one can readily obtain the components of $\nabla \times \mathbf{A}$ in orthogonal curvilinear coordinates.

We now use Eq. (1.49) to prove Stokes' theorem, which relates the flux of the curl of **A** through a surface to the circulation of **A** around the edge of the surface. Consider an open surface S whose periphery is a closed curve C. We take C to be *simply connected*; that is, it can be continuously shrunk down to a point without the curves leaving the space. The surface S has two sides, one of which we declare the positive side. We now subdivide this surface into vector elements of area $(\hat{\mathbf{e}}_j\, \Delta a_j)$ which are essentially planar if Δa_j are small enough. For each of these area elements we may apply Eq. (1.49) for the component of the curl of **A** in the direction $\hat{\mathbf{e}}_j$; as follows:

$$(\nabla \times \mathbf{A}) \cdot \hat{\mathbf{e}}_j = \frac{1}{\Delta a_j} \oint_{C_j} \mathbf{A} \cdot d\mathbf{r} \qquad \text{or} \qquad (\nabla \times \mathbf{A}) \cdot \Delta a_j = \oint_{C_j} \mathbf{A} \cdot d\mathbf{r}. \tag{1.50}$$

We now form the sum of the expressions in Eq. (1.50) for *all* the surface elements of the surface S. Thus

$$\sum_j (\nabla \times \mathbf{A}) \cdot \Delta \mathbf{a}_j = \sum \oint_{C_j} \mathbf{A} \cdot d\mathbf{r}$$

In going to the limit as $\Delta a_j \to 0$ and the number of elements tends to infinity, the left-hand side becomes an integral over the (open) surface S. The right-hand side becomes the line integral around curve C, since contributions of $\oint \mathbf{A} \cdot d\mathbf{r}$ for all line elements internal to S cancel, and only the contributions of the rim of S (that is, on C) remain. Therefore

$$\int_S (\nabla \times \mathbf{A}) \cdot d\mathbf{a} = \oint_C \mathbf{A} \cdot d\mathbf{r} \tag{1.51}$$

which is known as Stokes' theorem. The positive side of S and the sense in which C is traversed are related via the right-hand convention.

An immediate utilization of Stokes' theorem is the derivation of the criterion for determining whether a field is *conservative* or not. If $\oint \mathbf{A} \cdot d\mathbf{r} = 0$, for all possible closed paths, in a region of space, then it follows that $\nabla \times \mathbf{A} = 0$ everywhere in this region. The converse is also true. Such a vector is called a conservative vector. We may therefore summarize the criteria that determine whether or not a vector field **A** is conservative in a region of space. If in some simply connected region one of the following relations holds,

$$\begin{aligned} &\oint_C \mathbf{A} \cdot d\mathbf{r} = 0 \qquad \text{for arbitrary } C \\ &\nabla \times \mathbf{A} = 0 \\ &\mathbf{A} = \nabla f \qquad \text{for some scalar function } f \text{ (see Example 1.2)} \end{aligned} \tag{1.52}$$

then **A** is a conservative field. One of these criteria being true throughout a simply connected region of space implies that the other two also are true.

1.6 Vector Manipulations of ∇

1.6.1 Single Del Operations

We have discussed the meanings of the gradient, divergence, and curl operations. In so doing we have come to regard the del operator, ∇, as a quantity that acquires meaning only by operating on what is "to the right" of it, but otherwise behaves in many respects like an ordinary vector. We wish here to summarize some of these operations. If f and g are scalar functions, and $\mathbf{A}$ and $\mathbf{B}$ are vector functions of position in space, then

$$\nabla(f + g) = \nabla f + \nabla g \tag{1.53}$$

$$\nabla \cdot (\mathbf{A} + \mathbf{B}) = \nabla \cdot \mathbf{A} + \nabla \cdot \mathbf{B} \tag{1.54}$$

$$\nabla \times (\mathbf{A} + \mathbf{B}) = \nabla \times \mathbf{A} + \nabla \times \mathbf{B} \tag{1.55}$$

$$\nabla(fg) = (\nabla f)g + f(\nabla g) \tag{1.56}$$

$$\nabla \cdot (f\mathbf{A}) = (\nabla f) \cdot \mathbf{A} + f(\nabla \cdot \mathbf{A}) \tag{1.57}$$

$$\nabla \times (f\mathbf{A}) = (\nabla f) \times \mathbf{A} + f(\nabla \times \mathbf{A}) \tag{1.58}$$

$$\nabla(\mathbf{A} \cdot \mathbf{B}) = (\mathbf{B} \cdot \nabla)\mathbf{A} + \mathbf{B} \times (\nabla \times \mathbf{A}) + (\mathbf{A} \cdot \nabla)\mathbf{B} + \mathbf{A} \times (\nabla \times \mathbf{B}) \tag{1.59}$$

$$\nabla \cdot (\mathbf{A} \times \mathbf{B}) = (\nabla \times \mathbf{A}) \cdot \mathbf{B} - (\nabla \times \mathbf{B}) \cdot \mathbf{A} \tag{1.60}$$

$$\nabla \times (\mathbf{A} \times \mathbf{B}) = (\mathbf{B} \cdot \nabla)\mathbf{A} + \mathbf{A}(\nabla \cdot \mathbf{B}) - (\mathbf{A} \cdot \nabla)\mathbf{B} - \mathbf{B}(\nabla \cdot \mathbf{A}) \tag{1.61}$$

These formulas may all be proved by expressing ∇ in cartesian components and comparing both sides of the above equations.

1.6.2 Double Del Operations

The del operator may be applied several times in succession. Considering a scalar point function f we have, for example, $\nabla \cdot \nabla f$ and $\nabla \times \nabla f$. In cartesian coordinates,

$$\nabla \cdot \nabla f = \left(\hat{\mathbf{x}}\frac{\partial}{\partial x} + \hat{\mathbf{y}}\frac{\partial}{\partial y} + \hat{\mathbf{z}}\frac{\partial}{\partial z}\right) \cdot \left(\hat{\mathbf{x}}\frac{\partial f}{\partial x} + \hat{\mathbf{y}}\frac{\partial f}{\partial y} + \hat{\mathbf{z}}\frac{\partial f}{\partial z}\right)$$

Since $\hat{\mathbf{x}} \cdot \hat{\mathbf{x}} = \hat{\mathbf{y}} \cdot \hat{\mathbf{y}} = \hat{\mathbf{z}} \cdot \hat{\mathbf{z}} = 1$ and $\hat{\mathbf{x}} \cdot \hat{\mathbf{y}} = \hat{\mathbf{x}} \cdot \hat{\mathbf{z}} = \hat{\mathbf{y}} \cdot \hat{\mathbf{z}} = 0$, we have

$$\nabla \cdot \nabla f = \left(\frac{\partial^2}{\partial x^2} + \frac{\partial^2}{\partial y^2} + \frac{\partial^2}{\partial z^2}\right)f = \nabla^2 f \tag{1.62}$$

This is called the *Laplacian* operator. The Laplacian in other coordinates can be determined using similar procedures. In cylindrical and spherical coordinates it is given by

$$\nabla^2 f = \frac{1}{\rho}\frac{\partial}{\partial \rho}\left(\rho\frac{\partial f}{\partial \rho}\right) + \frac{1}{\rho^2}\frac{\partial^2 f}{\partial \phi^2} + \frac{\partial^2 f}{\partial z^2} \tag{1.63}$$

$$\nabla^2 f = \frac{1}{r^2}\frac{\partial}{\partial r}\left(r^2\frac{\partial f}{\partial r}\right) + \frac{1}{r^2}\frac{1}{\sin\theta}\frac{\partial}{\partial \theta}\left(\sin\theta\frac{\partial f}{\partial \theta}\right) + \frac{1}{r^2\sin^2\theta}\frac{\partial^2 f}{\partial \phi^2} \tag{1.64}$$

The curl of the gradient of a scalar $\nabla \times \nabla f$ can also be determined using similar procedures. In cartesian coordinates one writes

$$\nabla \times (\nabla f) = \begin{vmatrix} \hat{\mathbf{x}} & \hat{\mathbf{y}} & \hat{\mathbf{z}} \\ \dfrac{\partial}{\partial x} & \dfrac{\partial}{\partial y} & \dfrac{\partial}{\partial z} \\ \dfrac{\partial f}{\partial x} & \dfrac{\partial f}{\partial y} & \dfrac{\partial f}{\partial z} \end{vmatrix}$$

Expanding gives

$$\nabla \times (\nabla f) = \hat{\mathbf{x}}\left(\frac{\partial^2 f}{\partial y\,\partial z} - \frac{\partial^2 f}{\partial z\,\partial y}\right) + \hat{\mathbf{y}}\left(\frac{\partial^2 f}{\partial z\,\partial x} - \frac{\partial^2 f}{\partial x\,\partial z}\right) + \hat{\mathbf{z}}\left(\frac{\partial^2 f}{\partial x\,\partial y} - \frac{\partial^2 f}{\partial y\,\partial x}\right)$$

but for well-behaved, continuous functions $\partial^2 f/\partial y\,\partial z = \partial^2 f/\partial z\,\partial y$, etc., the curl of the gradient of a scalar function vanishes [see Eq. (1.52) defining conservative vectors]; that is,

$$\nabla \times \nabla f = 0 \tag{1.65}$$

With a vector field **f**, one can form the various double del expressions. One can show, however, that $(\nabla \times \nabla) \cdot \mathbf{f}$ and $(\nabla \times \nabla) \times \mathbf{f}$ vanish. The quantity $\nabla \cdot (\nabla \times \mathbf{f})$, the divergence of the curl of a vector can also be shown to be zero by direct calculation in cartesian coordinates; that is,

$$\nabla \cdot \nabla \times \mathbf{f} = 0 \tag{1.66}$$

Equation (1.66) is important in magnetostatics since the divergence of the magnetic field **B** is known to be zero ($\nabla \cdot \mathbf{B} = 0$), then it allows casting of the magnetic field **B** in terms of a vector potential **A**:

$$\mathbf{B} = \nabla \times \mathbf{A}$$

Finally we discuss the curl of the curl of a vector $\nabla \times (\nabla \times \mathbf{f})$. This double del operation has wide application in the propagation of electromagnetic waves, a topic to be discussed in the later chapters of this book. Regarding ∇ as a vector, $\nabla \times (\nabla \times \mathbf{f})$ can be expanded by the usual triple vector product $\mathbf{a} \times (\mathbf{b} \times \mathbf{c}) = \mathbf{b}(\mathbf{a} \cdot \mathbf{c}) - (\mathbf{a} \cdot \mathbf{b})\mathbf{c}$. This gives

$$\nabla \times (\nabla \times \mathbf{f}) = \nabla(\nabla \cdot \mathbf{f}) - \nabla \cdot (\nabla \mathbf{f}) \tag{1.67}$$

where $\nabla \mathbf{f}$ is a second-rank tensor or dyadic (see Example 1.3). In cartesian coordinates we have $\nabla \cdot (\nabla \mathbf{f}) = (\nabla \cdot \nabla)\mathbf{f} = \nabla^2 \mathbf{f}$ where ∇^2 is the laplacian operator.

1.7 Vector Integral Relations

Here we discuss a number of extensions to the divergence theorem and Stokes' theorem. Although we will not need all of these extensions for the development of electricity and magnetism at the level of this book, we include them for the sake of completeness and as a future reference.

Divergence Theorem. The integral relations given below in Eqs. (1.68) to (1.72) are extensions of the divergence theorem.

$$\int_V [\Phi\, \nabla^2 \psi + (\nabla \Phi) \cdot (\nabla \psi)]dV = \int_S (\Phi\, \nabla \psi) \cdot d\mathbf{a} \tag{1.68}$$

This is called Green's first identity or theorem.

$$\int_V (\Phi\, \nabla^2 \psi - \psi\, \nabla^2 \Phi) dV = \int_S (\Phi\, \nabla \psi - \psi\, \nabla \Phi) \cdot d\mathbf{a} \tag{1.69}$$

This is called Green's second identity or symmetrical theorem.

$$\int_V \nabla \times \mathbf{A}\, dV = \oint_S (\hat{\mathbf{n}} \times \mathbf{A}) da = \oint_S d\mathbf{a} \times \mathbf{A} \tag{1.70}$$

$$\int_V \nabla\Phi\, dV = \int_S \Phi\hat{\mathbf{n}}\, da \tag{1.71}$$

$$\int (\nabla \cdot \mathbf{A} + \mathbf{A} \cdot \nabla)\mathbf{B}\, dV = \oint_S \mathbf{B}(\mathbf{A} \cdot \hat{\mathbf{n}})\mathbf{da} \tag{1.72}$$

Equations (1.68) and (1.69) can be proved easily by applying the divergence theorem to the vector $\mathbf{F} = \Phi\, \nabla\psi$ and $\mathbf{F} = \Phi\, \nabla\psi - \psi\, \nabla\Phi$, respectively, where Φ and ψ are scalar functions.

Equations (1.70) and (1.71) can also be proved by applying the divergence theorem to the vector $\mathbf{F} = \mathbf{A} \times \mathbf{C}$ and $\mathbf{F} = \Phi\mathbf{C}$, respectively, where $\mathbf{C}$ is a constant vector.

Stokes' Theorem. The following integral relations are extensions of Stokes' theorem.

$$\oint d\mathbf{r} \times \mathbf{B} = \int_S (\hat{\mathbf{n}} \times \nabla) \times \mathbf{B}\, da \tag{1.73}$$

$$\oint_C \Phi\, d\mathbf{r} = \int_S (\hat{\mathbf{n}} \times \nabla\Phi) da = \int d\mathbf{a} \times \nabla\Phi \tag{1.74}$$

These two relations can be proved by applying Stokes' theorem to the vector $\mathbf{F} = \mathbf{B} \times \mathbf{C}$ and $\mathbf{F} = \Phi\mathbf{C}$, respectively, where $\mathbf{C}$ is a constant vector.

Example 1.1 Velocity Field in a Water Drain

Consider a vector field given by $\mathbf{v}(\mathbf{r}) = \rho\hat{\boldsymbol{\phi}}$ and shown in Fig. 1.15, where ρ is the distance from the z axis ω is a constant, and $\hat{\boldsymbol{\phi}}$ is associated with the angular coordinate ϕ about the z axis. Is $\mathbf{v}$ conservative?

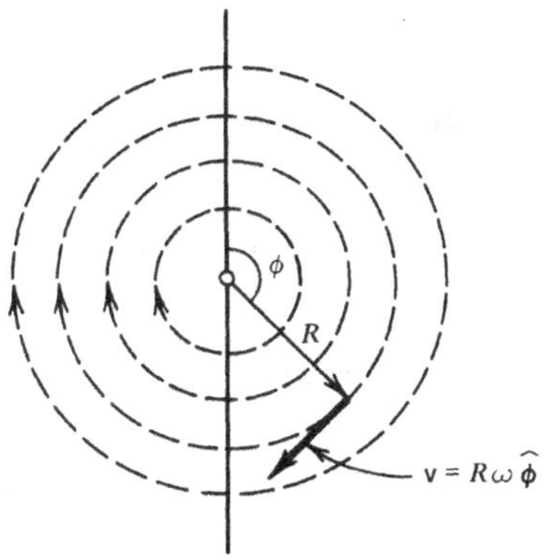

Figure 1.15 Velocity field in a water drain.

That it is clearly not conservative is seen from considering its circulation on a circular path of radius R about the z axis:

$$\oint_{\text{circle}} \mathbf{v} \cdot d\mathbf{r} = \int_{\phi=0}^{2\pi} \omega R\hat{\boldsymbol{\phi}} \cdot R\hat{\boldsymbol{\phi}}\, d\phi = 2\pi\omega R^2$$

The circulation is nonzero, and therefore $\mathbf{v}$ is nonconservative. This example might bring to mind the velocity field of water going down the drain in a sink. The circulation of the water is not in general zero for such a system.

Example 1.2 Conservative Nature of Radial Vectors—Potential Functions

Consider a radial vector field given in spherical coordinates by $\mathbf{A} = f(r)\hat{\mathbf{r}}$ where $f(r)$ is a scalar function that depends on r only. We will show below that this vector is conservative. The criteria that determine whether or not a vector is conservative are given in Eq. (1.52). Substituting $A_r = f(r)$, and $A_\theta = A_\phi = 0$ in $\nabla \times \mathbf{A}$ in spherical coordinates (Eq. 1.47) immediately gives $\nabla \times \mathbf{A} = 0$; thus indicating that $\mathbf{A}$ is conservative.

Radial vector fields are of importance to electrostatics since the electric field produced by a point charge is radial, and hence it is conservative. Because of the importance of this property, we will examine the conservative nature of these vectors from the point of view of the last criterion of Eq. (1.52). If $\mathbf{A}$ is conservative, then it must be written as the gradient of a scalar function Φ; that is, $\mathbf{A} = \nabla\Phi$. The function Φ is called a potential function corresponding to $\mathbf{A}$. To show this we consider a function

$$\Phi = \int_{r_0}^{r} \mathbf{A} \cdot d\mathbf{r} = \int_{r_0}^{r} f(r) dr$$

If the integral exists and is a continuous function of r, then from the expression for the gradient in spherical coordinates we see that

$$\frac{\partial \Phi}{\partial r} = f(r) \qquad \frac{\partial \Phi}{\partial \theta} = \frac{\partial \Phi}{\partial \phi} = 0$$

The result is that there exists a function Φ such that

$$\nabla\Phi = f(r)\hat{\mathbf{r}} \tag{1.75}$$

$\mathbf{A}$ indeed has a potential function.

Example 1.3 Gradient of a Vector—Dyadics

This example deals with the gradient of a vector, which will be useful when we deal with forces on electric dipoles placed in external electric fields. Consider a vector $\mathbf{E} = E_x\hat{\mathbf{x}} + E_y\hat{\mathbf{y}} + E_z\hat{\mathbf{z}}$. Formally, we can define $\nabla\mathbf{E}$ as follows:

$$\nabla\mathbf{E} = \left(\hat{\mathbf{x}}\frac{\partial}{\partial x} + \hat{\mathbf{y}}\frac{\partial}{\partial y} + \hat{\mathbf{z}}\frac{\partial}{\partial z}\right)(E_x\hat{\mathbf{x}} + E_y\hat{\mathbf{y}} + E_z\hat{\mathbf{z}})$$

Expanding, we get

$$\nabla\mathbf{E} = \left(\frac{\partial E_x}{\partial x}\hat{\mathbf{x}}\hat{\mathbf{x}} + \frac{\partial E_y}{\partial x}\hat{\mathbf{x}}\hat{\mathbf{y}} + \frac{\partial E_z}{\partial x}\hat{\mathbf{x}}\hat{\mathbf{z}}\right) + \left(\frac{\partial E_x}{\partial y}\hat{\mathbf{y}}\hat{\mathbf{x}} + \frac{\partial E_y}{\partial y}\hat{\mathbf{y}}\hat{\mathbf{y}} + \frac{\partial E_z}{\partial y}\hat{\mathbf{y}}\hat{\mathbf{z}}\right)$$
$$+ \left(\frac{\partial E_x}{\partial z}\hat{\mathbf{z}}\hat{\mathbf{x}} + \frac{\partial E_y}{\partial z}\hat{\mathbf{z}}\hat{\mathbf{y}} + \frac{\partial E_z}{\partial z}\hat{\mathbf{z}}\hat{\mathbf{z}}\right) \tag{1.76}$$

The quantities $\hat{\mathbf{x}}\hat{\mathbf{x}}$, $\hat{\mathbf{x}}\hat{\mathbf{y}}, \ldots$ are called *unit dyads*. Note that $\hat{\mathbf{x}}\hat{\mathbf{y}}$, for example, is not the same as $\hat{\mathbf{y}}\hat{\mathbf{x}}$; thus we have nine different unit dyads in the gradient. A quantity that can be expanded in the form

$$\Phi = a_{11}\hat{\mathbf{x}}\hat{\mathbf{x}} + a_{12}\hat{\mathbf{x}}\hat{\mathbf{y}} + a_{13}\hat{\mathbf{x}}\hat{\mathbf{z}} + a_{21}\hat{\mathbf{y}}\hat{\mathbf{x}} + a_{22}\hat{\mathbf{y}}\hat{\mathbf{y}} + a_{23}\hat{\mathbf{y}}\hat{\mathbf{z}} + a_{31}\hat{\mathbf{z}}\hat{\mathbf{x}} + a_{32}\hat{\mathbf{z}}\hat{\mathbf{y}} + a_{33}\hat{\mathbf{z}}\hat{\mathbf{z}} \tag{1.77}$$

is called a dyadic, and the nine coefficients a_{ij} are its components.

It is useful to examine the scalar product of a vector $\mathbf{A} = A_x\hat{\mathbf{x}} + A_y\hat{\mathbf{y}} + A_z\hat{\mathbf{z}}$ and a dyadic $\mathbf{\Phi}$ of the form given above. Consider the product $\mathbf{A}\cdot\mathbf{\Phi}$. Formally, we write

$$\mathbf{A}\cdot\mathbf{\Phi} = A_x\hat{\mathbf{x}}\cdot\mathbf{\Phi} + A_y\hat{\mathbf{y}}\cdot\mathbf{\Phi} + A_z\hat{\mathbf{z}}\cdot\mathbf{\Phi}$$

As an example consider the product $\hat{\mathbf{x}}\cdot\mathbf{\Phi}$. The product has nine terms, including, for example, $\hat{\mathbf{x}}\cdot a_{11}\hat{\mathbf{x}}\hat{\mathbf{x}}$, $\hat{\mathbf{x}}\cdot a_{12}\hat{\mathbf{x}}\hat{\mathbf{y}}$, $\hat{\mathbf{x}}\cdot a_{21}\hat{\mathbf{y}}\hat{\mathbf{x}}$, and $\hat{\mathbf{x}}\cdot a_{32}\hat{\mathbf{z}}\hat{\mathbf{y}}$. These individual products can be evaluated using the following rules:

$$\begin{aligned}
\hat{\mathbf{x}}\cdot a_{11}\hat{\mathbf{x}}\hat{\mathbf{x}} &= a_{11}(\hat{\mathbf{x}}\cdot\hat{\mathbf{x}})\hat{\mathbf{x}} = a_{11}\hat{\mathbf{x}} \\
\hat{\mathbf{x}}\cdot a_{12}\hat{\mathbf{x}}\hat{\mathbf{y}} &= a_{12}(\hat{\mathbf{x}}\cdot\hat{\mathbf{x}})\hat{\mathbf{y}} = a_{12}\hat{\mathbf{y}} \\
\hat{\mathbf{x}}\cdot a_{21}\hat{\mathbf{y}}\hat{\mathbf{x}} &= a_{21}(\hat{\mathbf{x}}\cdot\hat{\mathbf{y}})\hat{\mathbf{x}} = 0 \\
\hat{\mathbf{x}}\cdot a_{32}\hat{\mathbf{z}}\hat{\mathbf{y}} &= a_{32}(\hat{\mathbf{x}}\cdot\hat{\mathbf{z}})\hat{\mathbf{y}} = 0
\end{aligned} \tag{1.78}$$

Analogous rules exist for the rest of the products and for the products $\hat{\mathbf{y}}\cdot\mathbf{\Phi}$ and $\hat{\mathbf{z}}\cdot\mathbf{\Phi}$. See Problem 1.20 for a specific example.

Example 1.4 Dirac Delta Function

In this example we introduce a very useful function for dealing with point charges. We will define it here from a mathematical point of view. Its relevence to electromagnetism will be introduced later. The Dirac delta function, given the symbol $\delta(\mathbf{r})$ is defined as follows:

$$\delta(\mathbf{r}) = 0 \qquad \text{for } \mathbf{r} \neq 0 \tag{1.79}$$

$$\int \delta(\mathbf{r}')dv' = 1 \tag{1.80}$$

where the integral is carried out over all space. This definition shows that the delta function is a very high singular mathematical function; it is zero everywhere except at a single point and yet has a nonzero integral ("spike" function). This function is obviously not a continuous one; thus it should not be differentiated as a continuous function. Nevertheless it is a very useful mathematical property if handled cautiously.

Another property of the Dirac delta function comes from its relation to the Laplacian or divergence operator; that is,

$$\nabla\cdot\frac{\mathbf{r}}{r^3} = -\nabla^2\left(\frac{1}{r}\right) = 4\pi\delta(\mathbf{r}) \tag{1.81}$$

It is easy to show by direct differentiation that

$$\nabla\cdot\frac{\mathbf{r}}{r^3} = \nabla\left(\frac{1}{r^3}\right)\cdot\mathbf{r} + \frac{(\nabla\cdot\mathbf{r})}{r^3}$$

Since $\nabla\cdot\mathbf{r} = 3$, $\nabla(1/\mathbf{r}^3) = -3\hat{\mathbf{r}}/r^4$, then $\nabla\cdot(\mathbf{r}/r^3)$ is zero for $r \neq 0$ and becomes indeterminate as $r \to 0$. The nature of the divergence at $r = 0$ can be examined using the divergence theorem. Applying the theorem to a small volume of radius R gives

$$\int_V \nabla\cdot\frac{\mathbf{r}}{r^3}\,dv = \oint_S \frac{\mathbf{r}\cdot\hat{\mathbf{n}}}{r^3}\,da = \frac{1}{R^2}\oint da = 4\pi$$

Since this result is true regardless of how small R, then one can replace the $\nabla\cdot(\mathbf{r}/r^3)$ by $4\pi\delta(\mathbf{r})$.

1.8 Summary

When an element of surface area $d\mathbf{a}$ at $\mathbf{r}$ is referred to the origin, it is often convenient to use the concept of solid angle $d\Omega$:

$$d\Omega = \frac{d\mathbf{a}\cdot\hat{\mathbf{r}}}{r^2} = \sin\theta\,d\theta\,d\phi \tag{1.23),(1.24}$$

If we have a surface S that completely encloses the origin, then $\Omega = \oint_S d\Omega = 4\pi$, whereas if we have a closed surface that does not enclose the origin, then $\Omega = 0$.

The gradient operator ∇ is a linear differential "vector operator" that can operate on scalar as well as vector fields f and $\mathbf{A}$ to give the gradient of a scalar, divergence of a vector, curl of a vector, and gradient of a vector.

$$\nabla f = \hat{\mathbf{x}}\frac{\partial f}{\partial x} + \hat{\mathbf{y}}\frac{\partial f}{\partial y} + \hat{\mathbf{z}}\frac{\partial f}{\partial z} \tag{1.30}$$

$$\nabla \cdot \mathbf{A} = \frac{\partial A_x}{\partial x} + \frac{\partial A_y}{\partial y} + \frac{\partial A_z}{\partial z} \tag{1.41}$$

$$\nabla \times \mathbf{A} = \hat{\mathbf{x}}\left[\frac{\partial A_z}{\partial y} - \frac{\partial A_y}{\partial z}\right] + \hat{\mathbf{y}}\left[\frac{\partial A_x}{\partial z} - \frac{\partial A_z}{\partial x}\right] + \hat{\mathbf{z}}\left[\frac{\partial A_y}{\partial x} - \frac{\partial A_x}{\partial y}\right] \tag{1.44}$$

$$\nabla \mathbf{A} = \left[\frac{\partial A_x}{\partial x}\hat{\mathbf{x}}\hat{\mathbf{x}} + \frac{\partial A_y}{\partial x}\hat{\mathbf{x}}\hat{\mathbf{y}} + \frac{\partial A_z}{\partial x}\hat{\mathbf{x}}\hat{\mathbf{z}}\right] + \cdots \tag{1.76}$$

In the last equation the ellipsis $(\cdots)$ represents three analogous terms for each of the derivatives with respect to y and z. The gradient operator may be applied several times in succession, or may be applied to products of functions. The outcome of such operations can be derived from the above basic differential operations. One important operation is the Laplacian operator acting on a scalar or vector function $\nabla \cdot \nabla f = \nabla^2 f$ or $\nabla^2 \mathbf{A}$. All operations can also be derived in terms of other coordinate systems (e.g., cylindrical and spherical systems).

Some integral identities of the del (gradient) operations can be derived by integrating the differential ones over an arbitrary volume V bounded by a closed surface S, or over an open surface S bounded by a closed curve C. These include the divergence theorem and Stokes' theorem.

$$\int_V \nabla \cdot \mathbf{A}\, dv = \oint_S \mathbf{A} \cdot d\mathbf{a} \tag{1.42}$$

$$\int_S \nabla \times \mathbf{A} \cdot d\mathbf{a} = \oint_C \mathbf{A} \cdot d\mathbf{r} \tag{1.51}$$

A vector $\mathbf{A}$ is said to be conservative if

$$\nabla \times \mathbf{A} = 0 \quad \text{or} \quad \oint_C \mathbf{A} \cdot d\mathbf{r} = 0 \quad \text{(conservative } \mathbf{A}) \tag{1.52}$$

If so, $\mathbf{A}$ can also be written as a gradient of a scalar

$$\mathbf{A} = -\nabla \Phi \quad \text{(conservative } \mathbf{A})$$

Problems

1.1 Determine the unit vector perpendicular to the plane that contains the vectors $\mathbf{A} = 2\hat{\mathbf{x}} - 6\hat{\mathbf{y}} - 3\hat{\mathbf{z}}$ and $\mathbf{B} = 4\hat{\mathbf{x}} + 3\hat{\mathbf{y}} - \hat{\mathbf{z}}$.

1.2 Determine an equation for the plane passing through the points $P_1(2, -1, 1)$, $P_2(3, 2, -1)$, and $P_3(-1, 3, 2)$.

1.3 The position vectors of points P_1 and P_2 are $\mathbf{A} = 3\hat{\mathbf{x}} + \hat{\mathbf{y}} + 2\hat{\mathbf{z}}$ and $\mathbf{B} = \hat{\mathbf{x}} - 2\hat{\mathbf{y}} - 4\hat{\mathbf{z}}$. Determine an equation for the plane passing through P_2 and perpendicular to the line joining the points.

1.4 (a) Show that $\nabla r^n = nr^{n-2}\mathbf{r}$. (b) Find $\nabla \ln|\mathbf{r}|$ and $\nabla(1/r)$.

1.5 Consider the surface defined by the equation $2xz^2 - 3xy - 4x - 7 = 0$. Find a unit vector normal to this surface at the point $(1, -1, 2)$.

1.6 Consider the function $\Phi = x^2yz^3$. In what direction from the point $P(2, 1, -1)$ is the directional derivative of Φ a maximum? What is the magnitude of this maximum?

1.7 Show that $\nabla\cdot(\mathbf{r}/r^3) = 0$ if $r \neq 0$.

1.8 If vectors **A** and **B** are conservative, prove that $\mathbf{A} \times \mathbf{B}$ is solenoidal—that is, that it has a zero divergence.

1.9 (a) Determine the constants, a, b, and c if the vector $\mathbf{A} = (x + 2y + az)\hat{\mathbf{x}} + (bx - 3y - z)\hat{\mathbf{y}} + (4x + cy + 2z)\hat{\mathbf{z}}$ is irrotational (conservative). (b) Determine the function Φ, where $\mathbf{A} = \nabla\Phi$.

1.10 What should the constant a be if the vector $\mathbf{A} = (x + 3y)\hat{\mathbf{x}} + (y - 2z)\hat{\mathbf{y}} + (x + az)\hat{\mathbf{z}}$ is solenoidal (has zero divergence)?

1.11 Show that the vector $\mathbf{E} = \mathbf{r}/r^2$ is conservative. Determine Φ such that $\mathbf{E} = -\nabla\Phi$ and $\Phi(a) = 0$, where $a > 0$.

1.12 Show that the vector $\mathbf{A} = (6xy + z^3)\hat{\mathbf{x}} + (3x^2 - z)\hat{\mathbf{y}} + (3xz^2 - y)\hat{\mathbf{z}}$ is conservative. Find the corresponding potential function Φ such that $\mathbf{A} = \nabla\Phi$.

1.13 Determine $\nabla\cdot\mathbf{r}$ and the surface integral $\oint_S \mathbf{r}\cdot\hat{\mathbf{n}}\, da$, where S is a closed surface.

1.14 Consider the vector $\mathbf{A} = 4x\hat{\mathbf{x}} - 2y^2\hat{\mathbf{y}} + z^2\hat{\mathbf{z}}$ and the region bounded by $x^2 + y^2 = 4$, $z = 0$, and $z = 3$. (a) Determine $\nabla\cdot\mathbf{A}$. (b) Determine the unit vectors normal to the surfaces S_1 $(z = 0)$, S_2 $(z = 3)$ and the curved surface S_3 $(x^2 + y^2 = 4)$. (c) Verify the divergence theorem for **A** taken over the bounded region above.

1.15 Consider the vector $\mathbf{A} = z\hat{\mathbf{x}} + x\hat{\mathbf{y}} - 3y^2z\hat{\mathbf{z}}$ and the surface of the cylinder S defined by $x^2 + y^2 = 16$. (a) Determine the unit vector normal to the surface of the cylinder $\hat{\mathbf{n}}$ as a function of x and y. (b) Evaluate the surface integral $\int_S \mathbf{A}\cdot\hat{\mathbf{n}}\, da$ over the first octant surface of the cylinder between $z = 0$ and $z = 5$.

1.16 Consider the vector $\mathbf{A} = (2x - y)\hat{\mathbf{x}} - yz^2\hat{\mathbf{y}} - y^2z\hat{\mathbf{z}}$, and S the upper half surface of the sphere $x^2 + y^2 + z^2 = 1$. Verify Stokes' theorem, $\int_S \nabla \times \mathbf{A}\cdot\hat{\mathbf{n}}\, da = \oint_C \mathbf{A}\cdot d\mathbf{r}$, where C is the boundary of the surface S.

1.17 Prove that (a) $\oint_S \hat{\mathbf{n}}\, da = 0$ and (b) $\oint \mathbf{r} \times \hat{\mathbf{n}}\, da = 0$ for any closed surface.

1.18 Prove that $\int \nabla \times \mathbf{B}\, dv = \int_S \hat{\mathbf{n}} \times \mathbf{B}\, da$ and (b) $\int \nabla\Phi\, dv = \int \Phi\hat{\mathbf{n}}\, da$.

1.19 Determine $\nabla^2 \ln r, \nabla^2 r^n$ and $\nabla^2(1/r)$ $(r \neq 0.)$

1.20 Consider the dyadic $\boldsymbol{\Phi} = \hat{\mathbf{x}}\hat{\mathbf{x}} + \hat{\mathbf{y}}\hat{\mathbf{y}} + \hat{\mathbf{z}}\hat{\mathbf{z}}$ (see Example 1.3). Determine $\mathbf{r}\cdot(\boldsymbol{\Phi}\cdot\mathbf{r})$ and $(\mathbf{r}\cdot\boldsymbol{\Phi})\cdot\mathbf{r}$. Is there any ambiguity in writing $\mathbf{r}\cdot\boldsymbol{\Phi}\cdot\mathbf{r}$?

1.21 Determine the gradient of **r**.

TWO

ELECTROSTATICS

2.1 Electric Charge

The primordial stuff of electricity is electric charge. It is the essence of electrical phenomena. It is so basic that it is difficult to describe except in the context of the effects that are ascribed to its existence. These effects are only manifest as forces of interaction. We experience something and consequently seek to "explain" our experience in terms of something more elemental. It may be, however, that charge is simply a property of certain of nature's elementary particles and does not exist "outside" of these particles (as in the case of the electron, the μ meson, etc.). However, we will talk of *charge* as if it possesses an independent existence.

So far as we know, electric charge has the following characteristics:

1. There are two kinds of electric charge, denoted arbitrarily as positive and negative charge. The magnitude of the charge is given by a positive real number, and its type is denoted by a plus (+) or minus (−) sign. All charge is equivalent, however, in the sense that charges may be added to each other algebraically just like real (positive or negative) numbers to give other charges. It is found that two charges of the same sign physically repel each other, but two charges of opposite sign attract each other.

2. In nature, the total amount of positive charge just balances the total amount of negative charge; electrical neutrality of objects is the most common occurrence. Moreover, it is not possible to create (or annihilate) positive (or negative) charge without creating or annihilating an equal amount of negative (or positive) charge. This may be regarded as a *principle of conservation of charge.*

3. Physically, we also would like to believe in what may be called "charge symmetry." This means that two worlds that differ only in that all charges in one have opposite signs from the charges of the other would be physically indistinguishable. We see "approximate" manifestations of this effect in the realm of elementary particle physics. There, it has been found that for every *elementary particle* with a positive charge, there exists another "identical" elementary particle that has a negative charge of equal magnitude. Thus, we have electrons and positrons, protons and antiprotons, π^+ mesons and π^- mesons, and so on.

4. Charge is quantized. This means that there seems to be a minimum magnitude (nonzero) to electric charge. This minimum magnitude is associated, for example, with the charge of a positron or electron. Therefore, all charges are integral multiples of this elementary charge. Mathematically, then, charge may be put into correspondence with the integers, if only we agree to call the magnitude of the electron charge unity. So far as we know, all the "elementary particles" of nature have a charge magnitude equal to the unit electron charge or zero charge, though it has recently been suggested that perhaps an elementary particle of subelectronic charge exists. (These particles, named *quarks*, have one-third or two-thirds of the unit electronic charge.) However, if quarks exist, charge will still be quantized.

5. Whatever has charge has mass. One might state this by saying that electricity is a form of energy—or is always associated with energy. Objects with zero (rest) mass have no charge. Just as we talk of gravitational energy, so shall we talk of electric energy.

The lightest charged particle that we know of is the electron. Its (rest) mass is 9.1091×10^{-31} kilogram. The rest mass of a nucleon (proton or neutron) is approximately 1840 times larger.

These characteristics of charge are not its only characteristics. We shall not answer questions such as "What is the geometrical size or shape of the elementary charge?" The characteristics we have listed, however, are those that are important to our understanding of electricity.

2.2 Coulomb's Law

Coulomb's law is associated with Charles Augustin de Coulomb (1736–1806), although Henry Cavendish (1731–1810) earlier (about 1773 to 1785) discovered it.* This law expresses succinctly how two stationary charges affect each other (Fig. 2.1). We write it as

$$\mathbf{F}_{12} = k\frac{q_1 q_2}{r_{12}^2}\hat{\mathbf{r}}_{12} = -\mathbf{F}_{21} \tag{2.1}$$

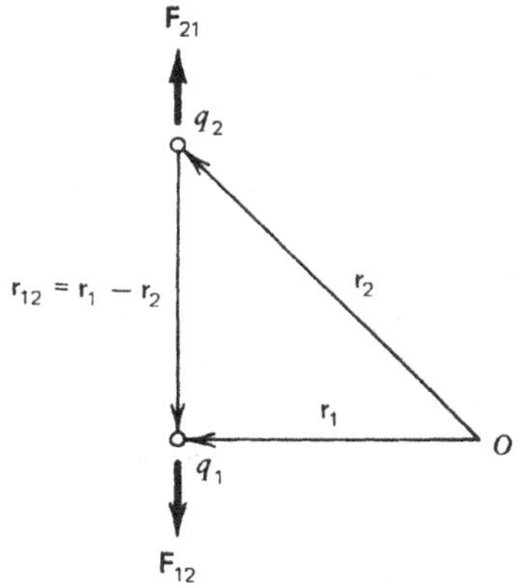

Figure 2.1 Electric force between two charges of the same sign as given by Coulomb's law.

* See R. S. Elliott, *Electromagnetics* (New York: McGraw-Hill, 1966) for a discussion of these historical points.

where q_1 and q_2 represent electric charges of objects (particles) 1 and 2, which may be positive or negative numbers, r_{12} is the distance from particle 1 to particle 2, $\mathbf{F}_{21}$ is the force on particle 2 due to particle 1, k is a positive constant of proportionality, and $\hat{\mathbf{r}}_{12} = -\hat{\mathbf{r}}_{21}$ is a unit vector pointing from q_2 to q_1.

In order for this law to be applied correctly, we must emphasize certain assumptions implicit in its statement.

1. It is often stated that this law is true for "point charges," meaning entities having no finite size. What is practically meant here is that the linear dimensions of the objects 1 and 2 are considered to be "negligible" compared to the distance between them, r_{12}. We shall subsequently discover that if the charges on the objects are distributed with spherical symmetry, the formula is true even if the diameters of the spheres (assumed noninterpenetrating) are not negligible. However, this is a special case.

The idea of point charges runs into difficulty mathematically if $r_{12} \to 0$. We shall adopt the point of view that two particles (*particle* implies a "negligibly" small object) can never have $r_{12} = 0$. Note that for two "nonnegligibly" sized objects the distance r_{12} is in any case ambiguous. The linear dimensions of subatomic elementary particles are of the order of 10^{-13} cm or less.

2. The formula of Coulomb's law is true whether or not charges other than q_1 and q_2 are nearby. It correctly gives the force between the charges q_1 and q_2. Electric forces are then said to be *two-body* forces. These properties are sometimes subsumed under the title of the *superposition principle.*

We now apply the superposition principle for the determination of the force on a charge due to more than one charge. Consider N point charges $1, 2, \ldots, N$ with scalar magnitudes $q_1, q_2, \ldots, q_N$ located at displacements $\mathbf{r}_1, \mathbf{r}_2, \ldots, \mathbf{r}_N$, respectively, from some fixed origin O. The force exerted on the charge q located at displacement $\mathbf{r}$ by all of these other charges is obtained by summing vectorially the forces

$$\mathbf{F}_q = q \sum_{i=1}^{N} \frac{q_i}{4\pi\varepsilon_0} \frac{(\mathbf{r} - \mathbf{r}_i)}{|\mathbf{r} - \mathbf{r}_i|^3} \tag{2.2}$$

Coulomb's law thus states that between charges q_1 and q_2 the force (a) is proportional to q_1, (b) is proportional to q_2, (c) is proportional to $1/r_{12}^2$, and (d) lies along the straight line connecting q_1 and q_2. If $q_1 q_2$ is a negative number, then $\mathbf{F}_{12}$ is in the direction $-\hat{\mathbf{r}}_{12}$ and $\mathbf{F}_{21}$ is in the direction $+\hat{\mathbf{r}}_{12}$. That is, we have attraction; otherwise we have repulsion.

3. We shall use the MKS (meter-kilogram-second) system of units, wherein force is measured in newtons ($1\ \mathrm{N} = 1\ \mathrm{kg \cdot m \cdot s^{-2}}$; that is, in units of mass (M) × length (L) × time^{-2} (T^{-2}). Energy is measured in joules (units of ML^2T^{-2}). In the *Système* International (SI) or MKSA* system of units, charge is measured in coulombs [abbreviation, C] or ampere-seconds (A·s). Charge Q, or charge per unit of time (current), is considered on an equal footing with mass, length, or time. Because the units of M, L, T, and Q, are independently specified, k as it appears in our formula is found from experiment to be

$$k \equiv \frac{1}{4\pi\varepsilon_0} \equiv 10^{-7}c^2 \simeq 9 \times 10^9 \frac{\mathrm{N \cdot m^2}}{\mathrm{C^2}}$$

where c is the speed of light in vacuum which is equal to 2.99792458×10^8 m/s by definition or $\simeq 3 \times 10^8$ m/s. The quantity ε_0 is sometimes called the permittivity of

* Sometimes referred to as "giorgi" units.

free space, and has the value 8.854×10^{-12} $C^2/N \cdot m^2$; that is, with the dimensions $Q^2T^2M^{-1}L^{-3}$. The 4π is inserted to make certain expressions to be encountered simpler (that is, without a factor of 4π), and the system is therefore said to be *rationalized.*

In the CGS (centimeter-gram-second) system of units, $k \equiv 1$, and the unit of charge is set by Eq. (2.1). It has the dimensions dyne$^{1/2}$-centimeter; that is, $M^{1/2}LT^{-1}$. It is called the "statcoulomb" and is equal to $(1/2.998) \times 10^{-9}$ coulomb. For further discussions of the system of units see Appendix I.

4. Coulomb's law will give the correct total force on q_1 due to q_2 if both q_1 and q_2 are stationary. If q_2 is moving with respect to our reference system, then the total force $\mathbf{F}_{12}$ on a stationary q_1 is modified from that predicted by Coulomb's law. However, if the speed of travel are small compared to the speed of light ($c = 3 \times 10^8$ m/s), the modifications are small. An analogous remark pertains to the force $\mathbf{F}_{21}$. At one time it was thought that the electric forces acted instantaneously across the distance r_{21}—that no time interval intervened in the Coulomb interaction of the particles.

We know now that this "action at a distance" concept is not valid, and that the effects on one charge due to another are propagated in time across the intervening space separating them. If the charges are stationary, however, we need not consider these effects. These effects are discussed in Chapters 15 and 17.

5. It is a fascinating fact that Coulomb's law has the same form as Newton's gravitational law. The inverse-square character of this law seems to be verified over a very large range of distances, from the submicroscopic to the macroscopic. For distances which are familiar to us, macroscopic distances, the exponent 2 of the inverse-square dependence has been shown* to be accurate to better than one part in 10^{15}. For submicroscopic distances also, down to the order of at least 10^{-13} cm, we have found that the physical effects predicted by this force law seem true. Because of lack of evidence to the contrary we assume Coulomb's law to be true universally.

2.3 Electric Field

It will be observed that the net force $\mathbf{F}_q$ in Eq. (2.2) depends linearly on the charge magnitude q. This suggests that the presence of the other charges creates a condition in space such that when charge q is placed in that space, it feels a force. The condition is described by saying that electric charge creates (or is associated with) an electric field. If an arbitrary ("test") charge q is placed in this electric field (denoted by $\mathbf{E}$) it will then experience a force $\mathbf{F}_q$, given by

$$\mathbf{F}_q(\mathbf{r}) = q\mathbf{E}(\mathbf{r}) \tag{2.3}$$

From Eq. (2.2) we see immediately that the electric field created by N stationary point charges $q_1, q_2, \ldots, q_N$ is given by

$$\mathbf{E}(\mathbf{r}) = \sum_{i=1}^{N} \frac{q_i}{4\pi\varepsilon_0} \frac{(\mathbf{r} - \mathbf{r}_i)}{|\mathbf{r} - \mathbf{r}_i|^3} \tag{2.4}$$

Even though we have defined the electric field with reference to forces on a charge q, the electric field is an entity independent of q because F is simply proportional to q.

* E. Williams, J. Faller, and H. Hill, *Physical Review Letters*, vol. 26, no. 12, p. 721, 1971.

When there are N stationary point charges, we may then use Eq. (2.4) to calculate the electric field $\mathbf{E}$ at a point in space whose displacement from a reference point 0 is denoted by $\mathbf{r}$. We thus see that once we know where all of a given set of charges are located, we can find $\mathbf{E}$ at any point in space due to the charge distribution. Remember that the charges are assumed to be stationary. To emphasize this, we say that the electric field is an electrostatic field.

Example 2.1 Lines of Force

If there is a curve in space, $\mathbf{r}(s)$, whose tangent vector, $\hat{\mathbf{t}}(\mathbf{r})$ at every point lies parallel to the electric field at that point, we have what is referred to as a *line of force* (or electric field line). If displacements $\mathbf{r}_1$ and $\mathbf{r}$ locate two points on the curve and $\Delta\mathbf{r} \equiv \mathbf{r}_1 - \mathbf{r}$, then the necessary condition for the curve to represent a line of force is that, as $\mathbf{r}_1$ approaches $\mathbf{r}$,

$$\lim_{\Delta\mathbf{r}\to 0} \frac{\Delta\mathbf{r}}{\Delta s} \equiv \frac{d\mathbf{r}}{ds} \equiv \hat{\mathbf{t}}(\mathbf{r}) = \alpha\mathbf{E}(\mathbf{r}) \tag{2.5}$$

where α is a scalar constant and s is taken to measure distance along the curve from some arbitrary point on the curve. If $\mathbf{r}$ is expressed in cartesian coordinates as $\mathbf{r} = x\hat{\mathbf{x}} + y\hat{\mathbf{y}} + z\hat{\mathbf{z}}$,

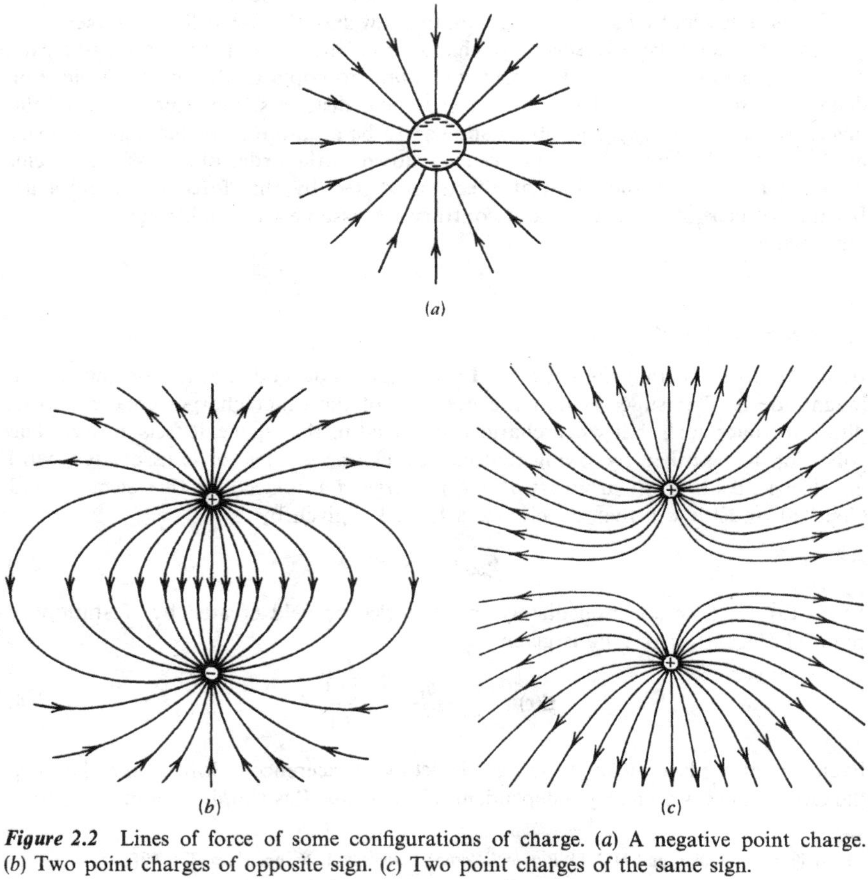

Figure 2.2 Lines of force of some configurations of charge. (*a*) A negative point charge. (*b*) Two point charges of opposite sign. (*c*) Two point charges of the same sign.

then substituting in Eq. (2.5) we find that a necessary condition is that, on the curve,

$$\frac{1}{\alpha}\frac{dx}{ds} = E_x \qquad \frac{1}{\alpha}\frac{dy}{ds} = E_y \qquad \frac{1}{\alpha}\frac{dz}{ds} = E_z$$

or

$$\frac{dx}{E_x} = \frac{dy}{E_y} = \frac{dz}{E_z} \tag{2.6}$$

We now show how this result can be used to find the lines of force of a point charge. From Coulomb's law, the field due to a "point charge" in the x-y plane is given by

$$\mathbf{E} = \frac{q}{4\pi\varepsilon_0}\frac{\hat{\mathbf{r}}}{r^2}$$

where $r^2 = x^2 + y^2$ and $\hat{\mathbf{r}} = [x\hat{\mathbf{x}} + y\hat{\mathbf{y}}]/r$. Substituting for E_x and E_y in Eq. (2.6) gives the following equation for the lines of force in the x-y plane:

$$\frac{dy}{dx} = \frac{E_y}{E_x} = \frac{y}{x} \tag{2.7}$$

The solution to the equation $dy/dx = y/x$ is $y = kx$, where k is any constant. These are just the equations of straight lines passing through the origin (Fig. 2.2*a*). The lines of force for two point charges of opposite and same sign but of same magnitude are shown in Figs. 2.2*b* and 2.2*c*.

2.4 Charge Density

In our discussion to this point we have treated the electric charge as being located on particles. We have considered only point charges—i.e., objects whose linear dimensions were negligible compared to the distance between them. Occasions arise when this procedure is not adequate because it is not convenient or practical to account for these charges individually. In these instances one assumes that there exists a well-behaved function of position in space such that the charge contained in any volume of space V is given by the relation

$$Q = \int_V \rho(\mathbf{r})dv$$

where $\rho(\mathbf{r})$, a function of the displacement $\mathbf{r}$ from some chosen origin is called the *volume charge density*. If electric charge is really smeared out so that it varies continuously in space, then $\rho(\mathbf{r})$ is expressible as the mathematical limit

$$\rho(\mathbf{r}) = \lim_{\Delta v \to 0} \frac{\Delta q}{\Delta v} \equiv \frac{dq}{dv} \tag{2.8}$$

where Δq is the total charge in volume Δv. If our electric measurements are insensitive to linear dimensions smaller than d, then we shall assume that the charge Δq contained in a volume $\Delta v \approx d^3$ defines the *macroscopic charge density*, ρ, at the volume element in question $\rho(\mathbf{r}) = \Delta q/\Delta v = \Sigma q_i/\Delta v$. Here Σq_i denotes the sum of all the constituent charges in Δv. Moreover, we shall assume that this charge density ρ is well behaved for whatever mathematical calculations we might like to perform, and that the charge within Δv may be considered in all such calculations as point charges of magnitude Δq. The density $\rho(\mathbf{r})$ as expressed above is assumed to be defined at a particular time t. It may change in time, however, so that in general we can write $\rho = \rho(\mathbf{r}, t)$.

With the assumptions above, if we are given a charge distribution $\rho(\mathbf{r})$, we can obtain the contribution to the electrostatic field at a point $\mathbf{r}$ from any volume element dv' containing a charge dq and located at position $\mathbf{r}'$ and hence at a distance $\mathbf{r} - \mathbf{r}'$ from the point of observation by using Eq. (2.4):

$$d\mathbf{E}(\mathbf{r}) = \frac{1}{4\pi\varepsilon_0}\frac{dq(\mathbf{r}')}{|\mathbf{r} - \mathbf{r}'|^3}(\mathbf{r} - \mathbf{r}') \tag{2.9}$$

Taking $dq(\mathbf{r}') = \rho(\mathbf{r}')dv'$, then

$$d\mathbf{E}(\mathbf{r}) = \frac{1}{4\pi\varepsilon_0}\frac{\rho(\mathbf{r}')dv'}{|\mathbf{r} - \mathbf{r}'|^3}(\mathbf{r} - \mathbf{r}') \tag{2.10}$$

The total electric field produced by the charge distribution is calculated by integrating Eq. (2.10) over all volume elements where $\rho \neq 0$, or over all space, which proves more convenient. Thus

$$\mathbf{E}(\mathbf{r}) = \frac{1}{4\pi\varepsilon_0}\int_V \frac{\rho(\mathbf{r}')(\mathbf{r} - \mathbf{r}')dv'}{|\mathbf{r} - \mathbf{r}'|^3} \tag{2.11}$$

Occasionally it proves convenient to consider charge as being distributed smoothly over surfaces, or on curves, such that if the surface has an area S, the total charge on it is given by $Q_S = \int_S \sigma(\mathbf{r})da$, and if the charge lies on a curve C, its total charge is given by $Q_l = \int_C \lambda(\mathbf{r})dl$ where

$$\sigma = \lim_{\Delta a \to 0}\frac{\Delta q}{\Delta a} = \frac{dq}{da} \qquad \lambda = \lim_{\Delta l \to 0}\frac{\Delta q}{\Delta l} = \frac{dq}{dl} \tag{2.12}$$

are the *surface* and *line charge densities*, respectively. Thus, in the macroscopic limits, where da is an element of area and dl an element of length small enough to perform arbitrarily accurate calculations, the charge elements dq where

$$dq \equiv \sigma\, da \qquad \text{or} \qquad dq \equiv \lambda\, dl$$

may be used to represent point charge elements for purposes of calculating electric fields. If a charge distribution can be specified completely as a surface charge distribution over S, then

$$\mathbf{E}(\mathbf{r}) = \frac{1}{4\pi\varepsilon_0}\int_S \frac{\sigma(\mathbf{r}')(\mathbf{r} - \mathbf{r}')da'}{|\mathbf{r} - \mathbf{r}'|^3} \tag{2.13}$$

which simply represents a summation of point-charge-like contributions for all the charge elements $dq = \sigma\, da'$ constituting S. Analogously, for a charge lying on curve C,

$$\mathbf{E}(\mathbf{r}) = \frac{1}{4\pi\varepsilon_0}\int_C \frac{\lambda(\mathbf{r}')(\mathbf{r} - \mathbf{r}')dl'}{|\mathbf{r} - \mathbf{r}'|^3} \tag{2.14}$$

If a charge distribution contains densities ρ, σ, and λ distinctly, the electric field is due to the vector sum of their individual contributions. One can express Equations (2.11), (2.13), and (2.14) symbolically by writing

$$\mathbf{E}(\mathbf{r}) = \frac{1}{4\pi\varepsilon_0}\int \frac{dq(\mathbf{r} - \mathbf{r}')}{|\mathbf{r} - \mathbf{r}'|^3} \tag{2.15}$$

where dq may represent $\rho\, dv'$, $\sigma\, ds'$, $\lambda\, dl'$, or appropriate combinations of these, and the integral sign represents a sum over *all* the charge elements of the distribution. Because the combination $\mathbf{r} - \mathbf{r}'$ occurs quite often in this book, we reserve the choice of writing it explicitly or instead using the symbol

$$\boldsymbol{\xi} = \mathbf{r} - \mathbf{r}' \tag{2.16}$$

Since the evaluation of the integral in Eq. (2.15) is troublesome for all but the simplest types of charge distributions, we shall not often employ this method of calculating **E**. One can contemplate, however, evaluating such integrals numerically on modern digital computers for rather arbitrary charge distributions.

We will now calculate the electric field due to a number of symmetric charge distributions that involve line and surface charge distributions. We will not present an example of a symmetric volume charge distribution here, but defer it until we develop a much easier method, Gauss' law, in the next section.

Example 2.2 Line Charge Distribution

Let us find the electrostatic field at a distance ρ from the axis of a straight, thin rod carrying a constant charge density λ per unit length. The physical situation is illustrated in Fig. 2.3.

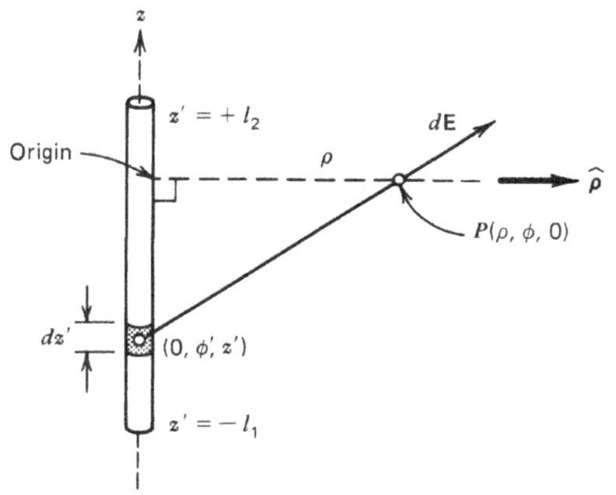

Figure 2.3 The electric field of a finite line of charge.

Since the rod is described as "thin" we assume that $\rho \gg$ thickness of rod. Supposing that the rod has a constant cross section of area a, we apply the Coulomb formula to find the contribution to $\mathbf{E}(\rho, \phi, 0)$ (in cylindrical coordinates) from the length element dz', as follows:

$$d\mathbf{E} = \frac{1}{4\pi\varepsilon_0} \lambda\, dz' \frac{(\rho\hat{\boldsymbol{\rho}} - \hat{\mathbf{z}}z')}{[z'^2 + \rho^2]^{3/2}}$$

Only components along z and ρ exist; that is,

$$dE_z = -\frac{1}{4\pi\varepsilon_0} \frac{\lambda z'\, dz'}{[z'^2 + \rho^2]^{3/2}} \qquad dE_\rho = \frac{1}{4\pi\varepsilon_0} \frac{\lambda \rho\, dz'}{[z'^2 + \rho^2]^{3/2}} \tag{2.17}$$

Integrating over all elements from $z' = -l_1$ to $z' = +l_2$, we get

$$E_z = -\frac{1}{4\pi\varepsilon_0} \lambda \int_{-l_1}^{l_2} \frac{z'\, dz'}{[z'^2 + \rho^2]^{3/2}} \qquad E_\rho = \frac{1}{4\pi\varepsilon_0} \lambda\rho \int_{-l_1}^{l_2} \frac{dz'}{[z'^2 + \rho^2]^{3/2}}$$

These expressions are readily integrated, yielding*

$$\mathbf{E} = \frac{-\lambda}{4\pi\varepsilon_0 \rho} \left[\frac{(\rho\hat{\mathbf{z}} - l_2\hat{\boldsymbol{\rho}})}{\sqrt{l_2^2 + \rho^2}} - \frac{(\rho\hat{\mathbf{z}} + l_1\hat{\boldsymbol{\rho}})}{\sqrt{l_1^2 + \rho^2}} \right] \tag{2.18}$$

* Note that l_1 and l_2 can be positive or negative, depending on the choice of the origin.

If the rod were very long in both directions, which we idealize by stating that $l_1/\rho \to \infty$ and $l_2/\rho \to \infty$, then $E_z \to 0$, and E_ρ becomes $|\mathbf{E}|$. That is,

$$\mathbf{E} \to \frac{\lambda}{2\pi\varepsilon_0 \rho}\hat{\boldsymbol{\rho}} \tag{2.19}$$

It should be noted for this problem that we have not performed an integration in the y or x direction because of our assumption of a thin rod. ρ was assumed to be sufficiently distant from the rod that the rod thickness appeared negligible.*

Example 2.3 The E Field of a Charged Ring

We calculate now the electric field due to a thin ring of charge having a uniform charge density σ for points on the axis of the ring. The geometry is shown in Fig. 2.4. From

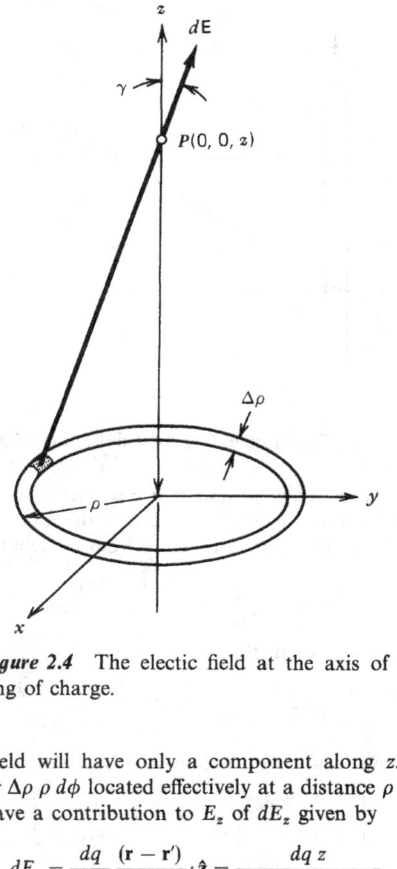

Figure 2.4 The electic field at the axis of a ring of charge.

symmetry, the electric field will have only a component along z. Thus for a differential element of charge, $dq = \sigma\, \Delta\rho\, \rho\, d\phi$ located effectively at a distance ρ from the origin—that is, $(\mathbf{r} - \mathbf{r}') = z\hat{\mathbf{z}} - \rho\hat{\boldsymbol{\rho}}$—we have a contribution to E_z of dE_z given by

$$dE_z = \frac{dq}{4\pi\varepsilon_0} \frac{(\mathbf{r} - \mathbf{r}')}{|\mathbf{r} - \mathbf{r}|^3} \cdot \hat{\mathbf{z}} = \frac{dq\, z}{4\pi\varepsilon_0(z^2 + \rho^2)^{3/2}}$$

* However, we shall find later that the expression for E above, the infinitely long rod, is valid even though ρ is not $\gg$ rod thickness. It suffices that the rod be round, of radius $(a/\pi)^{1/2}$, and that $\rho \geq (a/\pi)^{1/2}$. (See Example 2.7.)

For all charge elements dq, ρ and z remain constant. Therefore, summing over all charges, $\int dq \equiv \Delta Q = 2\pi\rho(\Delta\rho)\sigma$, and we may write

$$E = E_z = \frac{\Delta Q\, z}{4\pi\varepsilon_0(z^2 + \rho^2)^{3/2}} \tag{2.20}$$

It is noteworthy that if point P is far enough away from the ring, then $(z^2 + \rho^2)^{3/2} \cong z^3$, and the field approximates that of a point charge, as it should.

Example 2.4 The E Field of a Charged Sheet

Using the result of Example 2.3, we can also calculate the electrostatic field of a uniformly charged plane sheet of charge density σ for a point sufficiently near the plane (see Fig. 2.5). By "sufficiently near" we mean that we may consider it to be a good approximation to let the plane become infinite in extent.

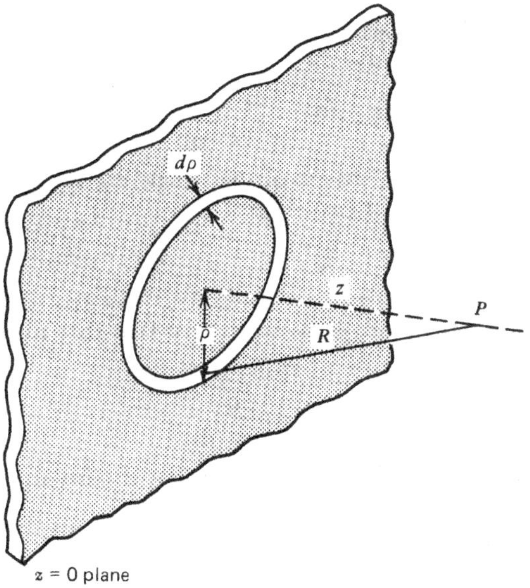

Figure 2.5 The electric field of an infinite surface charge distribution by integrating the field produced by a ring of charge.

Since a ring of charge of thickness $d\rho$ contributes to the field an amount given by Eq. (2.20) of the previous example

$$dE = \frac{1}{4\pi\varepsilon_0}\frac{2\pi\rho\, d\rho\, \sigma}{R^2}\frac{z}{R}$$

and the plane of charge may be considered to consist of such concentric rings, we can find E By integrating the above result from $\rho = 0$ to $\rho = \infty$; that is,

$$E = \frac{2\pi\sigma z}{4\pi\varepsilon_0}\int_{\rho=0}^{\infty}\frac{\rho\, d\rho}{[z^2 + \rho^2]^{3/2}} = \frac{\sigma z}{2\varepsilon_0}\int_{z}^{\infty}\frac{dR}{R^2}$$

Thus

$$E = \frac{\sigma}{2\varepsilon_0} \tag{2.21}$$

2.5 Gauss' Law

We now discuss Gauss' law, which is extremely important in our understanding of vector fields, and of electric fields in particular. In a certain sense this law is even more powerful than Coulomb's law, which, as we saw, was stated with various limitations imposed. We will show below that Gauss' law provides a very powerful method for the solution of electrostatic problems of symmetrical nature.

2.5.1 Integral Form of Gauss' Law

In an electrostatic field arising from some charge distribution with density ρ (see Fig. 2.6) we know that from an element of charge $dq = \rho(\mathbf{r}')dv'$ we have

$$d\mathbf{E}(\mathbf{r}) = \frac{1}{4\pi\varepsilon_0}\frac{dq}{|\mathbf{r}-\mathbf{r}'|^3}(\mathbf{r}-\mathbf{r}') \tag{2.22}$$

Hence at a point given by $\mathbf{r}$, where an element $d\mathbf{a}$ of surface S is located, there is a contribution [see Eq. (1.38)] to the flux dF of

$$dF = \frac{dq}{4\pi\varepsilon_0}\int_S \left[\frac{(\mathbf{r}-\mathbf{r}')\cdot d\mathbf{a}}{|\mathbf{r}-\mathbf{r}'|^3}\right] \tag{2.23}$$

The term in brackets is just an element of solid angle $d\Omega$, which $d\mathbf{a}$ subtends with reference to dq, so the result obtained is

$$dF = \frac{dq}{4\pi\varepsilon_0}\int_S d\Omega \tag{2.24}$$

If the charge element is completely enclosed by a surface, then the integral over the solid angle gives 4π, and thus

$$dF = \frac{dq}{\varepsilon_0} \tag{2.25}$$

The same result is valid for *all* charge elements dq located inside the closed surface

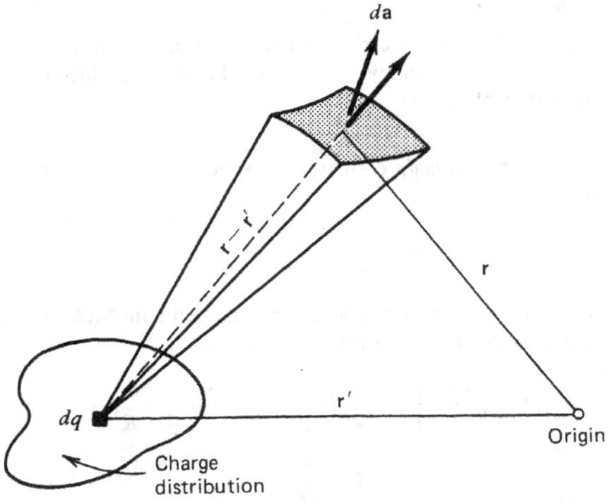

Figure 2.6 Derivation of Gauss's integral law by utilizing the Coulomb field of charge elements.

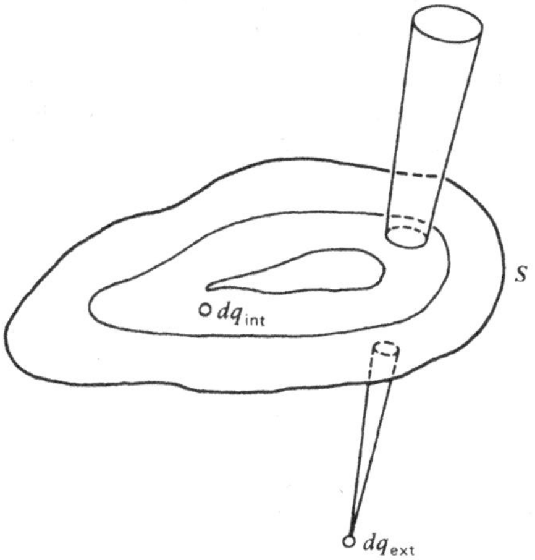

Figure 2.7 Gauss's law emphasizes that the net flux through a closed surface does not depend on the charges that reside outside of it, a concept that follows the properties of solid angles.

S. Therefore the flux due to all these internal elements is

$$F = \int_V \frac{dq}{\varepsilon_0} = \frac{Q_{int}}{\varepsilon_0} \tag{2.26}$$

Now suppose there are charges lying exterior to the closed surface S, as shown in Fig. 2.7. Since the solid angle of a closed surface viewed from outside that surface is zero, the fields due to such charges make no contribution to the net flux through S. (Of course, through any element $d\mathbf{a}$ of S, they would contribute.) Consequently, the total flux of $\mathbf{E}$ through S due to all sources is simply related to the total charge contained in the surface S. This is the statement of Gauss's law, which may now be expressed as follows:

$$\oint \mathbf{E} \cdot d\mathbf{a} = \frac{1}{\varepsilon_0} \int_V dq = \frac{Q_{int}}{\varepsilon_0} \tag{2.27}$$

We now wish to emphasize the following points about Gauss' law:

1. E as written in Eq. (2.27) may be regarded as the *total* electric field acting at a point on the surface of S, even though only contributions to $\mathbf{E}$ for charges internal to S contribute to the net result for F.
2. We have obtained this result (from Coulomb's law) for an electrostatic field $\mathbf{E}$. However, its validity transcends this, and is true for any electric field—even one depending upon time. It is in this sense that we may consider Gauss' law to be even more fundamental than Coulomb's law.
3. The essential feature of Coulomb's law that allowed us to "derive" Gauss' law was its inverse-square central dependence. Any other type of distance dependence would not lead to Gauss' law. Indeed, it is this fact that has been used to verify the truth of the inverse-square dependence. We shall subsequently observe many applications of the use of Gauss' law. A few examples are given below that involve spherical, cylindrical, and plane symmetry.

Example 2.5 Gauss' Law—Spherical Symmetry

We show that for a distribution of charge that has spherical symmetry, Gauss' Law almost immediately yields the electric field. Let $\rho = \rho(r)$ for $r \leq R$ and 0 for $r > R$ (see Fig. 2.8) and the total charge be Q_0. Then the electric field due to this charge distribution cannot depend upon the spherical coordinates (θ, ϕ). Moreover, the electric field itself can have no θ or ϕ components. Therefore $\mathbf{E}(\mathbf{r}) = E(\mathbf{r})\hat{\mathbf{r}}$. We calculate E for the two regions of space, $r \geq R$ and $r \leq R$.

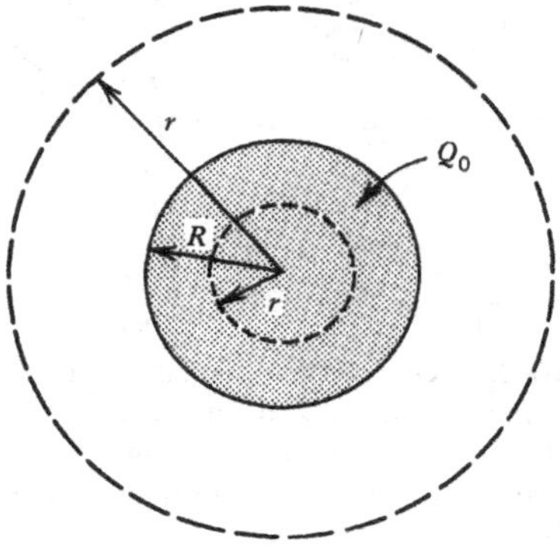

Figure 2.8 Calculation of the electric field of a uniformly charged sphere using Gauss' law, showing two gaussian surfaces.

In the region $r \geq R$, consider a spherical surface of radius $R \geq R$ centered at the origin and apply Gauss' law to this closed surface. Since $\mathbf{E} = E(r)\hat{\mathbf{r}}$, $\mathbf{E}\cdot\hat{\mathbf{n}} = \mathbf{E}\cdot\hat{\mathbf{r}} = E$ is constant on the sphere. Then $\oint \mathbf{E}(\mathbf{r})\cdot d\mathbf{a} = 4\pi r^2 E$. Equating this to Q_0/ε_0, where Q_0 is the total charge inside the Gaussian surface, gives

$$\mathbf{E}(\mathbf{r}) = \frac{1}{4\pi\varepsilon_0}\frac{Q_0}{r^2}\hat{\mathbf{r}} \qquad \text{for } r > R \tag{2.28}$$

This is the form of Coulomb's law for the field of a point charge Q_0 located at $r = 0$. We have found that to an outside observer (at $r \geq R$), the electrical effects of the sphere and point charge of magnitude Q_0 are identical.

To calculate E in the region $r < R$, we form our Gaussian spherical surface with a radius $r \leq R$. As in the case where $r \geq R$, we find that

$$\mathbf{E}(\mathbf{r}) = \frac{1}{4\pi\varepsilon_0}\frac{Q_<}{r^2}\hat{\mathbf{r}} \qquad \text{for } r < R \tag{2.29}$$

where $Q_< = \int_0^r 4\pi r'^2 \rho(r')dr'$ is just the charge enclosed by the sphere of radius $r \leq R$.

We have thus found in a simple fashion that the charge outside the surface of radius R makes no *net* contribution to $\mathbf{E}$. This result can also have been shown directly using Coloumb's law, where we would find that for every element of charge dq in a shell outside r there is another element dq' outside r that cancels the field of the first element.

As an application of the above results, let us consider a uniformly charged sphere, with a constant charge density ρ_0 for $r < R$. Referring to Eqs. (2.28) and (2.29), we note that

$$\mathbf{E}\,(r > R) = \frac{Q_0}{4\pi\varepsilon_0 r^2}\hat{\mathbf{r}} \qquad Q_0 = \frac{4}{3}\pi R^3 \rho_0$$

$$\mathbf{E}(r < R) = \frac{Q_< \hat{\mathbf{r}}}{4\pi\varepsilon_0 r^2} = \frac{\rho_0}{3\varepsilon_0}\mathbf{r} \qquad Q_< = \frac{4}{3}\pi r^3 \rho_0 \tag{2.30}$$

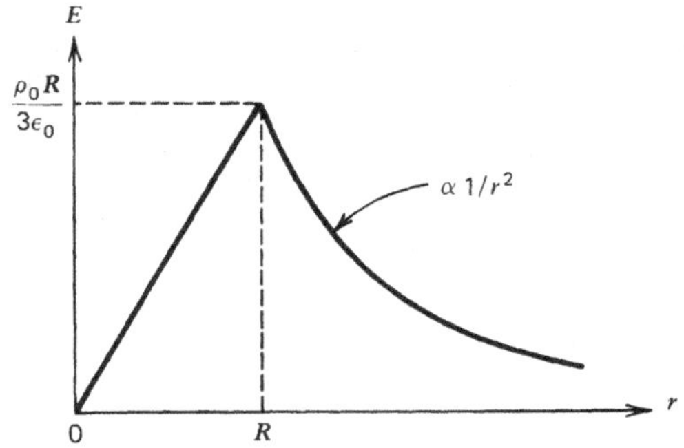

Figure 2.9 A sketch of the field of a uniformly charged sphere showing the continuity of the field at $r = R$.

Figure 2.9, which shows a sketch of the field as a function of r, indicates that the field is continuous at $r = R$.

Example 2.6 Uniformly Charged Infinite Plane—Gauss' Law

Let us find the field of a uniformly charged infinite plane having a surface density σ. Here we set up our coordinates as shown in Fig. 2.10, and remark immediately that E can only have a z component. Also, $\mathbf{E} = E(z)\hat{\mathbf{z}}$. Why? Moreover, again, by symmetry, $\mathbf{E}(z) = -\mathbf{E}(-z)$.

We now construct the gaussian surface shown in Fig. 2.10. Since $\mathbf{E} = \pm E\hat{\mathbf{z}}$, no flux leaves through the sides of the surface, so the only contribution to the flux through the gaussian surface comes from the flat end of the surface. If the area of the flat end is A, then $\mathbf{A} = A\hat{\mathbf{n}}$ for $z > 0$ and $\mathbf{A} = -A\hat{\mathbf{n}}$ for $z < 0$. Therefore

$$EA + EA = 2EA = \frac{Q_{\text{int}}}{\varepsilon_0} = \frac{\sigma A}{\varepsilon_0}$$

implying that

$$\mathbf{E} = \hat{\mathbf{z}}\frac{\sigma}{2\varepsilon_0} \quad \text{for } z > 0, \qquad \mathbf{E} = -\hat{\mathbf{z}}\frac{\sigma}{2\varepsilon_0} \quad \text{for } z < 0$$

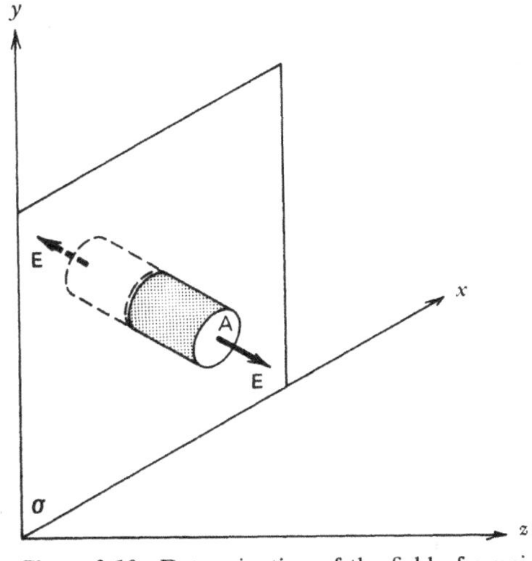

Figure 2.10 Determination of the field of a uniformly charged plane using Gauss' law.

This is exactly the result we got in Eq. (2.21) in Example 2.4. It is instructive to note that the field E does not depend on z. Therefore it is constant on both sides of the $z = 0$ plane; however, the directions are opposite to each other, there being a discontinuity equal to σ/ε_0 at the $z = 0$ plane. This field is due not only to the charge inside the gaussian surface but to all the charge, even though only the charge inside the surface contributes to the net flux of the gaussian surface.

Example 2.7 Gauss' Law—Line Charge

For this example we assume a constant charge density λ per unit length for $\rho \leq \rho_0$, and 0 for $\rho > \rho_0$, where ρ is the cylindrical coordinate measured from the z axis of the cylinder shown in Fig. 2.11. We note that symmetry dictates that **E** does not depend upon z or ϕ and that **E** is of the form $\mathbf{E} = \hat{\boldsymbol{\rho}}E(\rho)$.

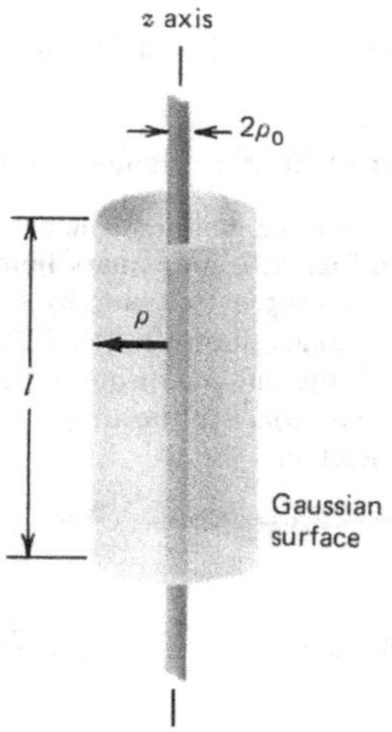

Figure 2.11 Determination of the **E** field of a long line of charge using Gauss' law.

We draw the cylindrical gaussian surface of radius ρ. There is no flux through the sides of this surface in the z = constant planes, since $E_z = 0$. The field on the cylindrical surface is constant, and therefore Gauss' law gives

$$2\pi\rho l E = \frac{Q_{\text{int}}}{\varepsilon_0} = \frac{l\lambda}{\varepsilon_0}$$

or

$$\mathbf{E} = \frac{\lambda}{2\pi\varepsilon_0\rho}\hat{\boldsymbol{\rho}} \qquad \rho > \rho_0 \tag{2.31}$$

If the radius of the gaussian surface is less than ρ_0, we have $Q_{\text{int}} = (\rho^2\lambda/\rho_0^2)l$, and thus

$$\mathbf{E}(\rho) = \frac{\lambda}{2\pi\varepsilon_0\rho_0^2}\rho\hat{\boldsymbol{\rho}} \qquad \rho < \rho_0 \tag{2.32}$$

2.5.2 Derivative Form of Gauss' Law

We have seen that the charge within a closed surface can always be determined if we know the electric field **E** on—and consequently the flux through—the surface. We now want to consider the relation between the charge around a point in space and the electric field near that point. We shall first assume that the charge is characterized at all points in space by a finite and well-defined charge density ρ. At a point in space displaced from our reference point by **r**, we shall seek the relationship between $\rho(\mathbf{r})$ and $\mathbf{E}(\mathbf{r})$. Recalling that

$$\oint_S \mathbf{E} \cdot d\mathbf{a} = \int_V \rho \frac{dv}{\varepsilon_0}$$

and applying the divergence theorem

$$\oint_S \mathbf{E} \cdot d\mathbf{a} = \int_V \nabla \cdot \mathbf{E} \, dv$$

we get

$$\int_V \nabla \cdot \mathbf{E} \, dv = \int_V \rho \frac{dv}{\varepsilon_0}$$

Because this relation is true for all volumes that one can conceive of, the integrands must be equal; that is

$$\int_V \left[\nabla \cdot \mathbf{E} - \frac{\rho}{\varepsilon_0} \right] dv = 0$$

for all V if and only if

$$\nabla \cdot \mathbf{E} = \frac{\rho}{\varepsilon_0} \tag{2.33}$$

This is called the derivative form of Gauss' law.

We wish to emphasize that this differential form of Gauss' law applies only when ρ is a finite, continuous function in space. When ρ is not defined or is not finite—as, for example, when charge occurs on surfaces, curves, or points—$\nabla \cdot \mathbf{E}$ becomes infinite. Physically, this means that a finite electric flux is diverging from an infinitesimal volume element. Thus, if one considers charge distributions that include point charges, as one approaches the location of the point charge,

$$\rho = \varepsilon_0 \nabla \cdot \mathbf{E} = \varepsilon_0 \lim_{\Delta v \to 0} \frac{1}{\Delta v} \oint_S \mathbf{E} \cdot d\mathbf{a} \to \infty$$

Note, however, that the flux emanating from the charge remains finite; that is,

$$\lim_{\Delta v \to 0} \oint_S \mathbf{E} \cdot d\mathbf{a} = \frac{q}{\varepsilon_0}$$

where q is the magnitude of the point charge. Since we presume a point charge to create a spherically symmetric field [point charges have no (angular) spatial structure whatsoever] the total field as the charge is approached is just that given by Coulomb's law that is, ($\mathbf{r}'$ locates q):

$$\mathbf{E}(\mathbf{r})_{\mathbf{r} \to \mathbf{r}'} = \frac{1}{4\pi\varepsilon_0} \frac{q(\mathbf{r} - \mathbf{r}')}{|\mathbf{r} - \mathbf{r}'|^3} \to \infty$$

This is the limiting form of any electric field near enough to an isolated point charge (since the fields due to other charges nearby remain finite when measured at the location of q). Similar behavior is observed in the vicinity of line charges.

It is clear from the above discussion that the charge density corresponding to a point charge can be represented by a Dirac delta function (see Example 1.4). Thus, for a point charge q at $\mathbf{r} = \mathbf{r}'$,

$$\rho(\mathbf{r}) = q\delta(\mathbf{r} - \mathbf{r}') \qquad \text{and} \qquad \nabla \cdot \mathbf{E} = \frac{q}{\varepsilon_0}\,\delta(\mathbf{r} - \mathbf{r}')$$

We shall now consider an example in which we encounter a discontinuity in **E** and use Gauss' law to interpret the result.

Example 2.8 Determining a Charge Distribution for a Given Electric Field—Derivative and Integral Forms of Gauss' Law

Let us suppose that we have an electric field given by $\mathbf{E} = \hat{\mathbf{z}}E_0 z$, for $0 \le z \le l$ and $\mathbf{E} = \hat{\mathbf{z}}k$, for $z < 0$ and $z > l$, where k is a constant as shown in Fig. 2.12*a*. We wish to find the charge that is responsible for producing this field.

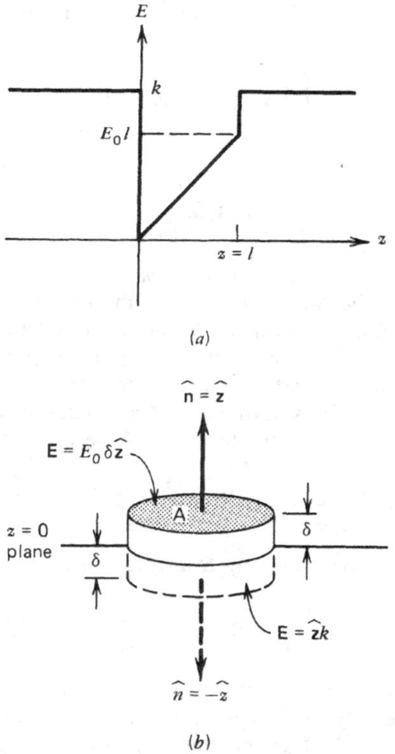

Figure 2.12 Determination of the charge density from the knowledge of **E** field produced by it. (*a*) The **E** field. (*b*) A pillbox at the interface $z = 0$ that may be used to determine the charge density at $z = 0$.

From Gauss' law, we have $\rho = \varepsilon_0(\nabla \cdot \mathbf{E})$. Since $\mathbf{E}$ depends only on z here, then $\rho = 0$ for $z < 0$ and $z > l$. But for $0 < z < l$ the density is

$$\rho = \varepsilon_0 \frac{dE_z}{dz} = \varepsilon_0 E_0 \qquad 0 < z < l$$

We therefore have an infinite slab of uniform charge density. We note, however, that dE_z/dz does not exist at $z = 0$ and $z = l$, because $\mathbf{E}$ is discontinuous there. Consequently, in order to find the charge at these two surfaces, we apply Gauss' integral law at these surfaces. To this end we construct a "pill box" at the $z = 0$ plane flat area A and height 2δ as shown in Fig. 2.12*b*. The contribution to $\oint \mathbf{E} \cdot d\mathbf{a}$ comes only from the surfaces of the pill box parallel to the $z = 0$ plane; hence

$$\oint_S \mathbf{E} \cdot d\mathbf{a} = AE_0\delta - Ak = \frac{Q_{\text{int}}}{\varepsilon_0}$$

The charge inside the pill box, Q_{int}, is the charge that resides on the surface (at $z = 0$) since δ is very small. By the planar symmetry of this problem, the charge on the surface is characterized by a constant charge density σ. The charge inside the pill box as $\delta \to 0$ is therefore $Q_{\text{int}} = \sigma A$, yielding $\sigma(z = 0) = -\varepsilon_0 k$. In a similar fashion, by applying Gauss' integral law to the surface at $z = l$, we would find the surface charge density, $\sigma(z = l) = \varepsilon_0(k - E_0 l)$.

The feature of this problem worth emphasizing is that whenever a discontinuity in $\mathbf{E}$ exists, the volume charge density must diverge, giving rise to a surface (or line, or point) charge density. To find the value of this surface density, one employs Gauss' law in integral form "around" the discontinuity.

2.6 Conductors and Insulators

Most of us know that certain materials "conduct electricity" well and others do not. What we mean by conducting electricity is that in the former materials, elements of charge (electrons, for example) can move freely from one point to another. Actually, most materials will allow charge movement under certain conditions (high temperature, high pressure, etc.), and we do not wish to catalog these materials and their properties now. We wish to emphasize only that by the name *conductor* we imply a medium that contains charge elements and, further, that these elements are free to move under the influence of an applied electric field. (Such charge not being bound to a particular location of the conductor is consequently referred to as "free charge.") We also wish to imply that if such a medium is insulated from other conductors and is subjected to an applied electric field, then, within a very short time, the freely moving charges inside the medium will so rearrange their positions as to annul the effect of the original electric field in the interior of the conducting medium. Inside an isolated "conductor" no *steady* electric field can persist.

It is difficult to define all our terms simultaneously. For example, by *insulate* we imply that there are media that are the antithesis of conductors, called *insulators*, through which charge does not or cannot pass freely. Most nonmetallic solids serve the purpose well enough. Almost by definition, a vacuum (i.e., charge-free space) acts as an insulator. Most gases—as, for example, air under normal conditions (beware of thunderstorms!)—behave as insulators. Materials are classified as conductors because charge moves in them much more readily than in insulators. Various materials at temperatures approaching absolute zero offer no resistance at all to this movement of charge and are called *superconductors*.

By our definition, then, if a long enough time interval elapses (e.g., 10^{-9} second), an isolated conductor will have a macroscopically zero electric field in its interior after a steady electric field is applied. (By "steady" we mean that it does not vary with time, although it turns out that for our present definition we may allow it to vary, but not too rapidly.) A macroscopic field is one averaged over macroscopic dimensions (in space and time). For our present purposes, if we consider media having "normal" densities (i.e., greater than about 10^{-1} kg/m^3), we may consider a *macroscopic space* dimension to be of the order of 10^{-5} meter, and a *macroscopic time* interval to be of the order of one picosecond (10^{-12} second). "Macroscopic" thus is meant to imply the antithesis of "microscopic," i.e., of atomic dimensions. Under the conditions of macroscopic electrostatics, atomic electric fields are "averaged out," yielding a smoother behavior in time and space. The following conclusions may then be drawn:

1. In the interior of the isolated conductor, the macroscopic charge density will be zero: $\rho = (\nabla \cdot \mathbf{E}) = 0$ (anywhere) inside the conductor.

2. Statement 1 implies that a net charge can exist only at the conductor surface. Surface charge densities exist there. Actually, the charge will exist in a region near the surface of the conductor, and the electric field always penetrates slightly into the conductor. But to talk like this is to talk microscopically, and we know that close enough to the "surface" the charges are not completely free to move (e.g., they cannot easily leave the conductor). Macroscopically, it is an excellent approximation to assume that the charge lies on the surface.

3. The external electrostatic field at the conductor surface is perpendicular to the surface. If it were not, forces would be exerted on the charge to move it laterally,* creating a nonstatic condition, violating our preconditions. Furthermore, if the conductor is finite and insulated, the charge will so arrange itself as to annul any lateral component of the electric field and thereby achieve a static condition.

4. The magnitude of the electrostatic field at the conductor surface is σ/ε_0, where σ is the surface charge density. We prove this as follows: If an E field exists outside the conductor and is zero inside, $\nabla \cdot \mathbf{E}$ diverges to infinity at the surface, implying that an infinite charge density exists there. In other words, a surface charge density exists there. In order to find its magnitude we apply Gauss' law to a disklike volume element encompassing an element of area Δa of the surface (Fig. 2.13). As the thickness of the disk shrinks to zero, the electric flux through the sides of the disk perpendicular to the surface become negligibly small since the field at the surface itself approaches perpendicularity at the surface. Since the field within the conductor is zero, the total flux out of the disk is from the surface element $\Delta \mathbf{a}$ outside the conductor. If Δa is small enough, the flux of $\mathbf{E}$ through it is given by $\mathbf{E} \cdot \Delta \mathbf{a}$, where $\mathbf{E}$ is taken as the value at the center of Δa. As a result, we obtain

$$\oint \mathbf{E} \cdot d\mathbf{a} = \mathbf{E} \cdot \Delta \mathbf{a} = \mathbf{E} \cdot \hat{\mathbf{n}}\, \Delta a = \frac{Q_s}{\varepsilon_0} = \sigma \frac{\Delta a}{\varepsilon_0}$$

Thus, on the surface of the conductor,

$$\mathbf{E} \cdot \hat{\mathbf{n}} = \frac{\sigma}{\varepsilon_0} \qquad \text{or} \qquad \sigma = \varepsilon_0 \mathbf{E} \cdot \hat{\mathbf{n}} \tag{2.34}$$

Equations (2.34) is a special case of a more general relation valid at a charged plane interface. Let us suppose that in a small region near the interface the finite fields $\mathbf{E}_1$ and $\mathbf{E}_2$ exist in media 1 and 2, respectively (refer to Fig. 2.14). We construct the gaussian surface whose top and bottom areas have magnitude Δa and whose

* That is, in a direction perpendicular to the normal to the surface at any point; tangential to the surface.

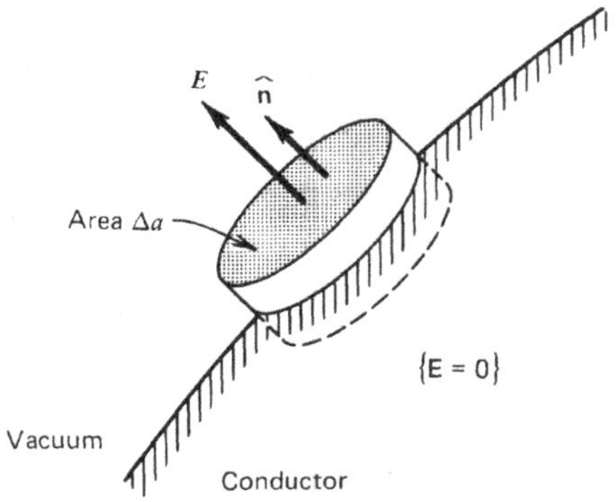

Figure 2.13 Applying Gauss' law to a pillbox at a highly conducting interface to determine the properties of the **E** field just outside the conductor.

side area is negligible compared to the top and bottom (achieved by letting the height 2δ go to zero). Gauss' law then indicates that

$$\oint \mathbf{E} \cdot d\mathbf{a} = \mathbf{E}_1 \cdot (-\hat{\mathbf{n}})(\Delta a) + \mathbf{E}_2 \cdot \hat{\mathbf{n}}(\Delta a)$$

The charge associated with a *finite* volume charge density ρ inside the gaussian pill box approaches zero as the thickness $2\delta \to 0$; only the surface charge remains $\sigma\,\Delta a$. Therefore,

$$\mathbf{E}_2 \cdot \hat{\mathbf{n}} - \mathbf{E}_1 \cdot \hat{\mathbf{n}} = \frac{\sigma}{\varepsilon_0} \tag{2.35}$$

Thus, the difference in the normal components of **E** at the interface equals the surface charge density (divided by ε_0). The result in Eq. (2.34) is a special case of this general result when one of the media (medium 1) is a conductor, in which case $\mathbf{E}_1 \equiv 0$.

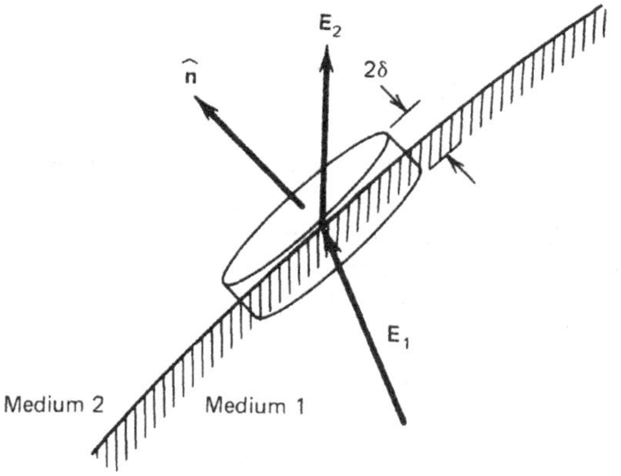

Figure 2.14 Applying Gauss' law to a pillbox at an interface that may have surface charges to determine the relation between the **E** field just across the interface. (Boundary conditions.)

2.7 Electric Potential

It has been shown in Example 1.2 that the curl of a radial vector $\mathbf{A} = f(r)\hat{\mathbf{r}}$ is zero, and that it is expressible as the gradient of a scalar, and hence it is a conservative vector. Here we shall show that any electrostatic electric field can be written as a gradient of a scalar function. The minus sign is chosen for convenience. Since

$$\mathbf{E}(\mathbf{r}) = -\nabla\Phi(\mathbf{r}) \tag{2.36}$$

the vector must be conservative; that is [see Eq. (1.52)],

$$\nabla \times \mathbf{E} = 0 \tag{2.37}$$

To prove this we first show that the electrostatic field due to a point charge q may be so written. It can be shown straightforwardly that:

$$\frac{\mathbf{r} - \mathbf{r}'}{|\mathbf{r} - \mathbf{r}'|^3} = -\nabla \frac{1}{|\mathbf{r} - \mathbf{r}'|} \tag{2.38}$$

where ∇ operates on the variable $\mathbf{r}$. Thus the $\mathbf{E}$ field of a point charge becomes

$$\mathbf{E} = -\nabla\left[\frac{1}{4\pi\varepsilon_0}\frac{q}{|\mathbf{r} - \mathbf{r}'|}\right]$$

Therefore $\mathbf{E}$ is representable as the gradient of a scalar function Φ, as follows:

$$\Phi = \frac{1}{4\pi\varepsilon_0}\frac{q}{|\mathbf{r} - \mathbf{r}'|} \tag{2.39}$$

The function Φ is called the electrostatic potential of the charge q. Since the relation between $\mathbf{E}$ and Φ is a linear one and $\mathbf{E}$ is decomposable, always, into electrostatic fields from individual charge elements q, it will be true that $\mathbf{E} = -\nabla\Phi$ for any electrostatic field. Thus the potential corresponding to the field of a general charge distribution [Eq. (2.15)] is

$$\Phi = \frac{1}{4\pi\varepsilon_0}\int \frac{dq}{|\mathbf{r} - \mathbf{r}'|} \tag{2.40}$$

Having now shown that $\mathbf{E} = -\nabla\Phi$, we may integrate this equation in order to determine Φ in terms of $\mathbf{E}$. In taking $d\Phi = \nabla\Phi \cdot d\mathbf{r}$,

$$\Phi(\mathbf{r}) - \Phi(\mathbf{r}_0) = -\int_{\mathbf{r}_0}^{\mathbf{r}} \mathbf{E} \cdot d\mathbf{r} \tag{2.41}$$

If we examine Eqs. (2.40) and (2.41), we see that $\Phi(\mathbf{r})$ is only given to within an arbitrary constant. This is related to the fact that only potential differences have physical significance. If the charge distribution that creates the electric field is localized in some finite region,* one usually specifies points "at infinity"† as being where the potential is assigned a zero value: $\Phi(\infty) \equiv 0$. Then

$$\Phi(\mathbf{r}) = -\int_{\infty}^{\mathbf{r}} \mathbf{E} \cdot d\mathbf{r} \tag{2.42}$$

The relation $\mathbf{E} = -\nabla\Phi$ immediately implies that *surfaces of constant potential,* given by $\Phi(\mathbf{r}) =$ constant, are at every point perpendicular to $\mathbf{E}$ at that point [see

* If dq is given by $\rho(\mathbf{r}')dv'$, then $\rho(\mathbf{r}')$ should approach zero faster than $(r')^{-2}$ as $r' \to \infty$.

† "At infinity" is that region where the field $\mathbf{E}$ (and hence any electrostatic force) is negligibly small.

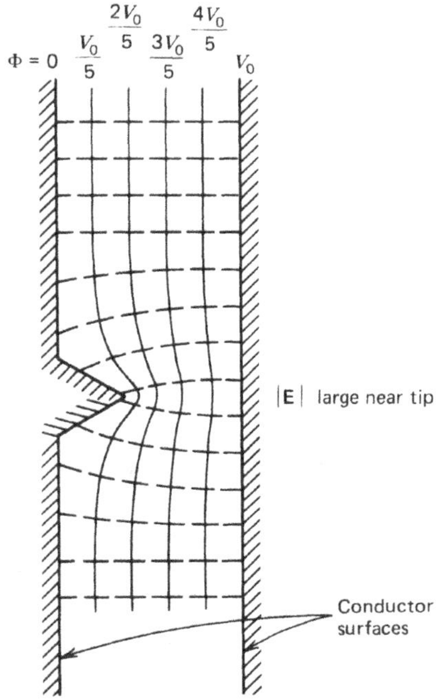

Figure 2.15 Illustration of equipotential surfaces and lines of force in the region between two highly conducting electrodes showing that they are normal to each other.

Eq. (1.33)]. These are called equipotential surfaces, as shown in Fig. 2.15. Similarly, a region that is at a constant potential would be called an equipotential volume. An isolated conductor, under electrostatic conditions, constitutes an equipotential volume. Its surface is an *equipotential surface.* This, of course, is consistent with our notions of potential relative to work: Since $\mathbf{E} = 0$ inside a conductor, no work is done against $\mathbf{E}$ in the displacement of charge inside the conductor.

If one agrees to draw only those equipotential surfaces (to represent a picture of potential) differing successively by constant increments, $\Delta\Phi$, then when the surfaces are close together, $\mathbf{E}$ is large compared to the case where the surfaces are far apart. This is just another way of stating that when $|\nabla\Phi|$ is large, $|\mathbf{E}|$ is large. (Note that $|\nabla\Phi| = \Delta\Phi/\Delta S$, where ΔS is the "distance" between surfaces $\Phi = c$ and $\Phi = c + \Delta\Phi$.) Since $\mathbf{E} = -\nabla\Phi$ and $\nabla\Phi$ is perpendicular to the surfaces $\Phi =$ constant, the streamlines of $\mathbf{E}$ are at every point perpendicular to the equipotential surfaces passing through those points.

Example 2.9 Two "Point Charges" of Opposite Sign—Electric Dipole

In order to find the potential of two point charges we need only add the potentials at $\mathbf{r}$ for each individual charge. Using the notation of Fig. 2.16,

$$\Phi(\mathbf{r}) = \Phi_q + \Phi_{-q} = \frac{q}{4\pi\varepsilon_0}\left(\frac{1}{r} - \frac{1}{r_-}\right) \tag{2.43}$$

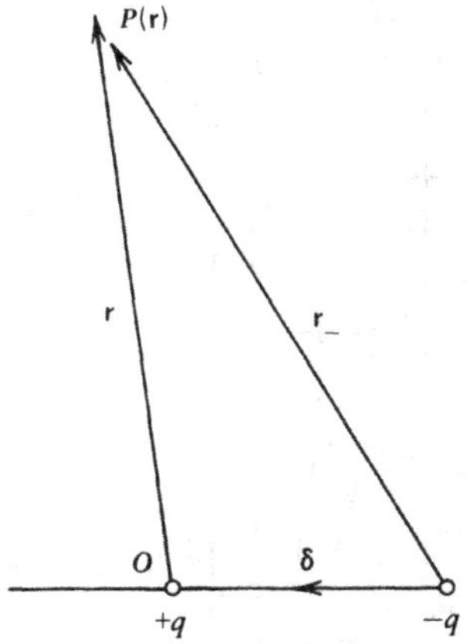

Figure 2.16 Potential and electric field of an electric dipole.

Taking $\mathbf{r}_- = \mathbf{r} + \boldsymbol{\delta}$ gives

$$\Phi(\mathbf{r}) = \frac{q}{4\pi\varepsilon_0}\left(\frac{1}{r} - \frac{1}{|\mathbf{r}+\boldsymbol{\delta}|}\right)$$

Let us now suppose that $\delta \ll r$, and expand $|\mathbf{r}+\boldsymbol{\delta}|^{-1}$ in powers of δ/r. Keeping the lower-order terms gives

$$|\mathbf{r}+\boldsymbol{\delta}|^{-1} \simeq \frac{1}{r}\left(1 - \frac{\mathbf{r}\cdot\boldsymbol{\delta}}{r^2}\right) + \frac{1}{2r}\left[3\left(\frac{\hat{\mathbf{r}}\cdot\boldsymbol{\delta}}{r}\right)^2 - \left(\frac{\delta}{r}\right)^2\right]$$

which, upon substitution gives

$$\Phi(\mathbf{r}) = \frac{q}{4\pi\varepsilon_0 r}\frac{\mathbf{r}\cdot\boldsymbol{\delta}}{r^2} - \frac{q}{8\pi\varepsilon_0 r}\left[3\left(\frac{\hat{\mathbf{r}}\cdot\boldsymbol{\delta}}{r}\right)^2 - \left(\frac{\delta}{r}\right)^2\right]$$

Remember, however, that we assumed $r/\delta \gg 1$, and that the dipole field as $r \to 0$ is defined as the field one would obtain if one allowed δ to approach zero and q to approach infinity such that $q\delta$ remained constant. One then always assumes that as $r \to 0$, $\delta \to 0$ more rapidly, so as to maintain the relation $\delta/r \ll 1$. Thus

$$\Phi = \frac{q}{4\pi\varepsilon_0 r}\frac{\mathbf{r}\cdot\boldsymbol{\delta}}{r^2} = \frac{\mathbf{p}\cdot\mathbf{r}}{4\pi\varepsilon_0 r^3} \tag{2.44}$$

where $\mathbf{p} = q\boldsymbol{\delta}$ is the dipole moment.

It is interesting to note that we may rewrite this dipole potential as the differential of the monopole potential:

$$\Phi(r) = \frac{\mathbf{p}\cdot\mathbf{r}}{4\pi\varepsilon_0 r^3} = -\frac{\mathbf{p}\cdot\nabla(1/r)}{4\pi\varepsilon_0}$$

or

$$\Phi = \frac{(-q)}{4\pi\varepsilon_0}(\boldsymbol{\delta}\cdot\nabla)\left(\frac{1}{r}\right) = (-\boldsymbol{\delta}\cdot\nabla)\left(\frac{+q}{4\pi\varepsilon_0 r}\right) \tag{2.45}$$

Noting that for a scalar function f [see Eq. (1.29)], $df = (d\mathbf{r}\cdot\nabla)f$, it appears that the dipole potential is the differential of the monopole potential of q, $q/4\pi\varepsilon_0 r$, with respect to the displacement $d\mathbf{r} \equiv -\boldsymbol{\delta}$. The dipole potential is thus a difference between two monopole potentials (small compared to either).

Let us now find the electric field **E**. Choosing spherical coordinates and setting $\mathbf{p} = p\hat{\mathbf{z}}$, we have $\Phi = (p \cos \theta)/4\pi\varepsilon_0 r^2$. Thus

$$\mathbf{E} = -\nabla\Phi = -\hat{\mathbf{r}}\frac{\partial\Phi}{\partial r} - \hat{\boldsymbol{\theta}}\frac{1}{r}\frac{\partial\Phi}{\partial\theta} - \hat{\boldsymbol{\phi}}\frac{1}{r\sin\theta}\frac{\partial\Phi}{\partial\phi}$$

Note that

$$\frac{\partial\Phi}{\partial r} = \frac{-2p\cos\theta}{4\pi\varepsilon_0 r^3} \qquad \frac{1}{r}\frac{\partial\Phi}{\partial\theta} = -\frac{p\sin\theta}{4\pi\varepsilon_0 r^3} \qquad \frac{\partial\Phi}{\partial\phi} = 0$$

Consequently the dipole electric field is

$$\mathbf{E} = \frac{p}{4\pi\varepsilon_0 r^3}\left[2\cos\theta\,\hat{\mathbf{r}} + \sin\theta\,\hat{\boldsymbol{\theta}}\right]. \tag{2.46}$$

which can also be written in the following form (show it).

$$\mathbf{E} = \frac{1}{4\pi\varepsilon_0}\frac{1}{r^3}\left[3(\hat{\mathbf{r}}\cdot\mathbf{p})\hat{\mathbf{r}} - \mathbf{p}\right] \tag{2.47}$$

Finally, one can verify that $\nabla\cdot\mathbf{E} = 0$ at all points except at $r = 0$, where **E** diverges. Here the charge density for the dipole is zero everywhere except at $r = 0$. To find the charge there ($\rho \to \infty$) we might try to apply Gauss' law. We thus imagine that "dipole field" to originate from a point in space with a highly singular charge distribution that cannot be well investigated with Gauss' law. Microscopically, electric dipole fields do not exist as $r \to 0$. Macroscopically, they appear often.

Example 2.10 Potential on the Axis of a Uniformly Charged Ring

In this example we meet a case where finding the potential, or electric field, at arbitrary points in space is a difficult task. It is easy, however, to find the potential on the axis of a uniformly charged ring of charge density λ per unit length and radius ρ' (see Fig. 2.17).

We take the zero of potential at infinity since the charge distribution is finite and apply the formula for points on the z axis given by Eq. (2.40). Noting that all charge elements dq are equidistant from points on the z axis, we immediately write

$$\Phi(\mathbf{r} = \hat{\mathbf{z}}z) = \frac{Q}{4\pi\varepsilon_0|\hat{\mathbf{z}}z - \hat{\boldsymbol{\rho}}\rho'|} = \frac{Q}{4\pi\varepsilon_0(z^2 + \rho'^2)^{1/2}} \tag{2.48}$$

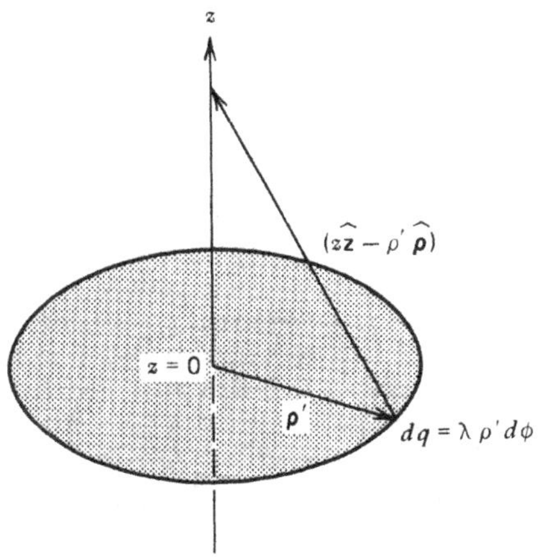

Figure 2.17 The electric potential at the axis of a ring of charge.

where $Q = 2\pi\rho'\lambda$ is the charge on the ring. Knowledge of Φ on the z axis does *not* permit us to determine the derivatives of Φ perpendicular to this axis; that is, we cannot determine E_ρ or E_ϕ (cylindrical components). We can only determine $E_z = -\partial\Phi/\partial z$. However, when the ring is uniformly charged, we know that E_ρ and E_ϕ are zero on the z axis by symmetry considerations. Therefore, we know that $\partial\Phi/\partial\rho$ and $\partial\Phi/\partial\phi$ are both zero on the axis. It is then easy to show (in agreement with Example 2.3) that we can get Eq. (2.20) by taking $\mathbf{E} = -\hat{\mathbf{z}}\,\partial\Phi/\partial z$.

Example 2.11 Potential on the Axis of a Thin Disk

From the preceding example it is easy to obtain the potential on the axis of a uniformly charged disk of surface charge density σ and radius a. We need only sum over (by integration) the contributions of all the rings constituting the disk. The radius of a particular ring is denoted by ρ' and the charge thereon by $Q_{\text{ring}} = \sigma 2\pi\rho'\,d\rho'$ (see Fig. 2.18). Then, from a particular ring we get the potential contribution from Eq. (2.48) as follows:

$$d\Phi = \frac{\sigma 2\pi\rho'\,d\rho'}{4\pi\varepsilon_0(z^2 + \rho'^2)^{1/2}}$$

Summing over all rings from $\rho' = 0$ to $\rho' = a$, the radius of the disk, we obtain

$$\Phi(\mathbf{r} - \hat{\mathbf{z}}z) = \frac{\sigma}{2\varepsilon_0}\int_0^a \frac{\rho'\,d\rho'}{\sqrt{\rho'^2 + z^2}} = \frac{\sigma}{2\varepsilon_0}\int_{|z|}^{\sqrt{a^2+z^2}} \frac{u\,du}{u}$$

Integrating gives

$$\Phi = \frac{\sigma}{2\varepsilon_0}[\sqrt{a^2 + z^2} - |z|] \tag{2.49}$$

As in the preceding example, the electric field on the axis, $E_z = -\partial\Phi/\partial z$, is easily obtained also:

$$E_z = \frac{\sigma}{2\varepsilon_0}\left[1 - \frac{z}{\sqrt{z^2 + a^2}}\right] \qquad \text{for } z > 0$$

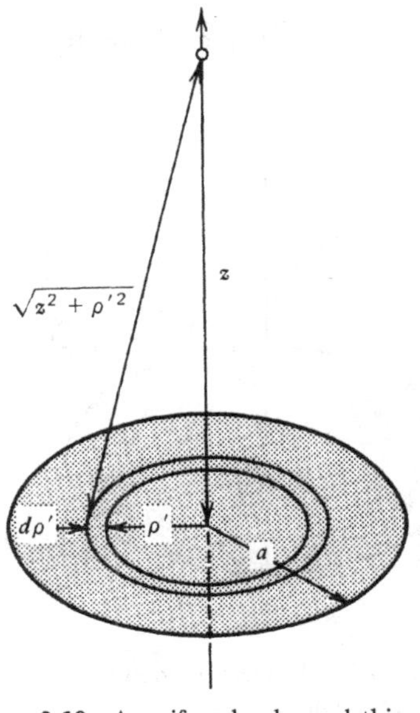

Figure 2.18 A uniformly charged thin disk.

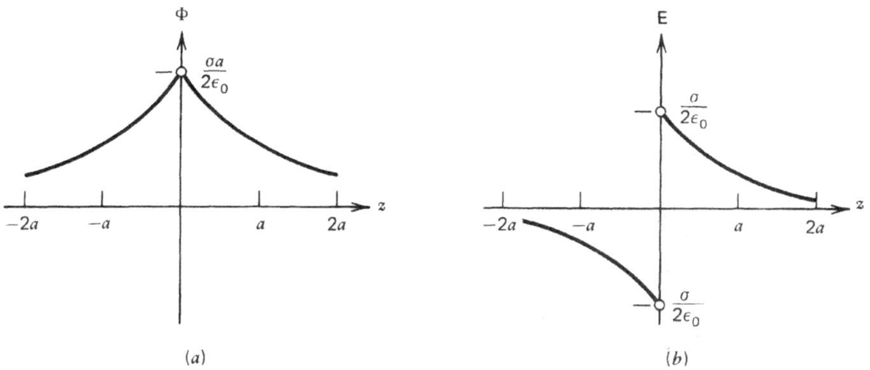

Figure 2.19 Sketches of (*a*) the potential and (*b*) the electric field on the axis of a uniformly charged thin disk.

and

$$E_z = -\frac{\sigma}{2\varepsilon_0}\left[1 + \frac{z}{\sqrt{z^2 + a^2}}\right] \qquad \text{for } z < 0$$

Figures 2.19*a* and 2.19*b* show a sketch of the potential and the electric field as a function of z. The figures show that the potential is continuous as expected while the electric field is not. In fact, one can easily show that on the z axis

$$\lim_{z\to 0}[E(z) - E(-z)] = \frac{\sigma}{\varepsilon_0}$$

which agrees with the boundary condition given by Eq. (2.35).

Example 2.12 Determination of Φ from $\Phi = -\int \mathbf{E}\cdot d\mathbf{r}$

Examples 2.12*a* and 2.12*b* show how to calculate the electrostatic potential by first determining the corresponding electric field and then using Eq. (2.42). This method can be applicable only to problems of high symmetry where the electric field can be easily determined using Gauss' law. Because this method involves integration over the electric field, then the result can be given only to within an arbitrary constant. This effect, as we noted above, is related to the fact that only potential differences have physical significance. The arbitrariness is used to specify the value of the potential at some reference point.

(a) The Spherical Capacitor

Let us assume we have two conducting, concentric, spherical shells of outer radii R_1 and R_2 (refer to Fig. 2.20). Suppose charges Q_1 and Q_2 are placed upon these shells, respectively, and we wish to find the potential function at all points due to the resultant charge distribution.

We define the fields E_1, E_2, and E_3 and potentials Φ_1, Φ_2, and Φ_3 in the $r > R_2$, $R_3 < \mathbf{r} < R_2$, and $R_1 < r < R_3$ regions, respectively. The electric fields in the different regions are radial because of the spherical symmetry. They are easily found by applying Gauss' law to spherical surfaces in the various regions and concentric with the capacitor. Therefore

$$\mathbf{E}_1 = \frac{Q_1 + Q_2}{4\pi\varepsilon_0}\cdot\frac{\hat{\mathbf{r}}}{r^2} \qquad r > R_2$$

$$\mathbf{E}_3 = \frac{Q_1}{4\pi\varepsilon_0}\frac{\hat{\mathbf{r}}}{r^2} \qquad R_1 < r < R_3$$

Inside the conducting shell the electric field vanishes; therefore $\mathbf{E}_2 = 0$.

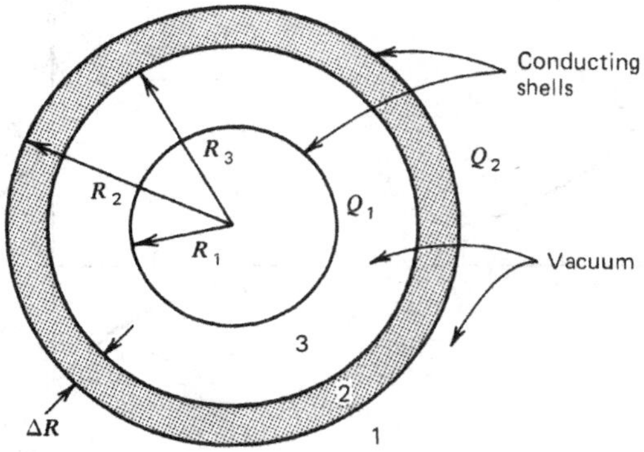

Figure 2.20 Potential of a spherical capacitor utilizing Gauss' law.

The potentials can now be easily determined. Substituting $\mathbf{E}_1$ in Eq. (2.42) we get

$$\Phi_1 = -\int_{\infty}^{r} \frac{Q_1 + Q_2}{4\pi\varepsilon_0} \frac{dr}{r^2} = \frac{Q_1 + Q_2}{4\pi\varepsilon_0 r}$$

In region 2 the electric field is zero, and therefore the corresponding potential is constant:

$$\Phi_2 = \text{constant} = \frac{Q_1 + Q_2}{4\pi\varepsilon_0 R_2}$$

Finally, the potential in region 3 is determined by substituting $\mathbf{E}_3$ in Eq. (2.42), as follows:

$$\Phi_3 = \frac{Q_1 + Q_2}{4\pi\varepsilon_0 R_2} - \int_{R_3}^{r} \frac{Q_1}{4\pi\varepsilon_0} \frac{dr}{r^2} = \frac{Q_1 + Q_2}{4\pi\varepsilon_0 R_2} + \frac{Q_1}{4\pi\varepsilon_0}\left(\frac{1}{r} - \frac{1}{R_3}\right)$$

(b) Charged Spherical Shell

In this example we derive the potential of a uniformly charged spherical shell of charge density ρ_0 (see Fig. 2.21). From Gauss' law, we know that the electric fields $\mathbf{E}_1$, $\mathbf{E}_2$, and $\mathbf{E}_3$ in regions 1, 2, and 3, respectively, are given by

$$\mathbf{E}_1 = \frac{Q_{\text{shell}}}{4\pi\varepsilon_0 r^2}\hat{\mathbf{r}} = \frac{\rho_0}{3\varepsilon_0}[R_2^3 - R_1^3]\frac{\hat{\mathbf{r}}}{r^2} \qquad r > R_2$$

$$\mathbf{E}_2 = \frac{\frac{4}{3}\pi(r^3 - R_1^3)\rho_0}{4\pi\varepsilon_0 r^2}\hat{\mathbf{r}} = \frac{\rho_0}{3\varepsilon_0}\left[1 - \frac{R_1^3}{r^3}\right]\mathbf{r} \qquad \mathbf{R}_1 \le \mathbf{r} \le \mathbf{R}_2$$

and the field $\mathbf{E}_3 = 0$ for $r \le R_1$.

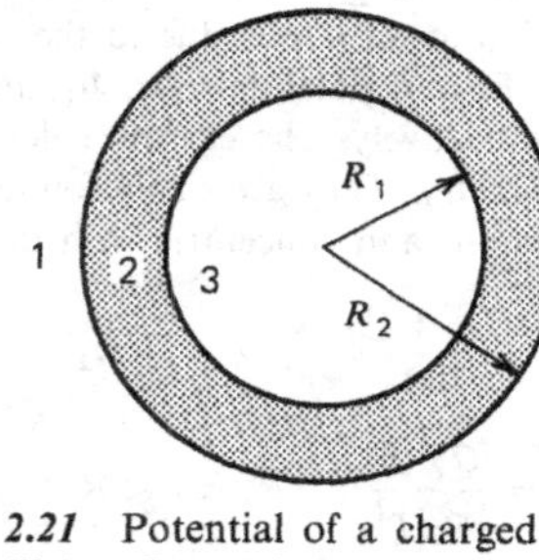

Figure 2.21 Potential of a charged spherical shell utilizing Gauss' law.

Applying the definition $\Phi(r) = -\int_\infty^r \mathbf{E}\cdot d\mathbf{r}$, we obtain the corresponding potentials:

$$\Phi_1 = -\int_\infty^r \frac{\rho_0}{3\varepsilon_0}(R_2^3 - R_1^3)\frac{\mathbf{r}\cdot d\mathbf{r}}{r^3} = \frac{\rho_0}{3\varepsilon_0}[R_2^3 - R_1^3]\frac{1}{r}$$

$$\Phi_2 = \Phi_1(R_2) - \int_{R_2}^r \frac{\rho_0}{3\varepsilon_0}\left[1 - \frac{R_1^3}{r^3}\right]\mathbf{r}\cdot d\mathbf{r} = \frac{\rho_0}{3\varepsilon_0}\left[\frac{3}{2}R_2^2 - \frac{R_1^3}{r} - \frac{r^2}{2}\right] \tag{2.50}$$

$$\Phi_3 = \Phi_2(R_1) = \frac{\rho_0}{3\varepsilon_0}\cdot\frac{3}{2}[R_2^2 - R_1^2]$$

The preceding expressions can be used to find the potential of a uniformly charged sphere merely by setting $R_1 = 0$. In this case $(R_2 \equiv R)$,

$$\Phi_1(r \geq R) = \frac{\rho_0}{3\varepsilon_0}\frac{R^3}{r} = \frac{1}{4\pi\varepsilon_0}\frac{Q_{\text{sphere}}}{r} \tag{2.51}$$

$$\Phi_2(r \leq R) = \frac{\rho_0}{3\varepsilon_0}\cdot\frac{1}{2}(3R^2 - r^2) \tag{2.52}$$

The corresponding electric field of the sphere is

$$\mathbf{E}_1 = \frac{\rho_0}{3\varepsilon_0}R^3\frac{\mathbf{r}}{r^3} \qquad \text{for } r > R$$

$$\mathbf{E}_2 = \frac{\rho_0}{3\varepsilon_0}\mathbf{r} \qquad \text{for } r < R \tag{2.53}$$

Example 2.13 Potential Due to Spherically Symmetric Charge Distribution $\rho(r)$

In this example we consider a general case where the charge density is not uniform. However, we will consider a spherically symmetric distribution where the density is a function of r only.

To calculate the potential at a point of observation P located at distance r, the charge will be divided into spherical shells. Consider two shells as shown in Fig. 2.22, each of thickness dr'. Shell 1 encloses the point of observation, whereas shell 2 does not.

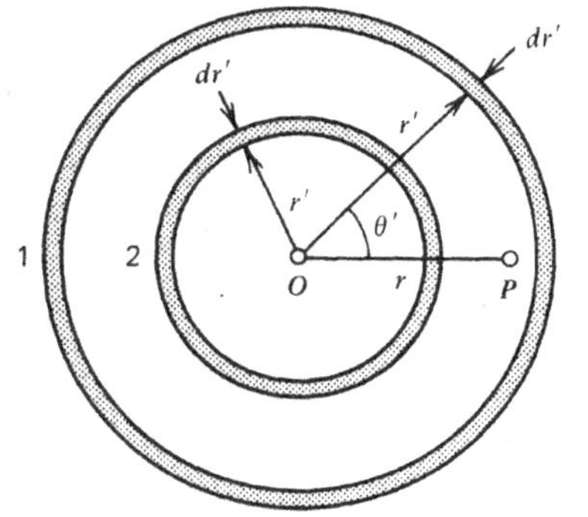

Figure 2.22 Spherically symmetric charge distribution showing two rings of charge. One encloses the observation point P and the other does not.

The potential due to a thin, charged shell was considered in Example 2.12*b*. The potential $d\Phi_2$ due to shell 2 is the same as if all the charge in the shell is concentrated in a point charge at the origin [see Eq. (2.50)]. Therefore

$$d\Phi_2 = \frac{1}{4\pi\varepsilon_0}\frac{\rho(r')4\pi r'^2}{r}\,dr' \tag{2.54}$$

The total potential due to all of the charge at distances smaller than r is therefore determined by integrating the above expression from 0 to r:

$$\Phi_2 = \int_0^r d\Phi_2 = \frac{1}{\varepsilon_0 r}\int_0^r \rho(r')r'^2\,dr' \tag{2.55}$$

To calculate the potential due to shell 1, we consider an area element on the shell at angle θ' with respect to the observation vector $\mathbf{r}$. Then the effect of the shell can be calculated by integrating over the surface of the shell:

$$d\Phi_1 = \frac{1}{4\pi\varepsilon_0}\int \frac{\rho(r')r'^2\,d\Omega'}{(r'^2 - 2rr'\cos\theta' + r^2)^{1/2}}\,dr' \tag{2.56}$$

where $d\Omega' = d\phi'\sin\theta'\,d\theta'$. The integration over ϕ' gives 2π, and then the integration over θ' yields

$$d\Phi_1 = \frac{1}{\varepsilon_0}\rho(r')r'\,dr' \tag{2.57}$$

Note that this result can also be arrived at using the results of Example 2.12*b*. The potential due to all the charge at $r' > r$ is then determined by integrating over r':

$$\Phi_1 = \frac{1}{\varepsilon_0}\int_r^\infty r'\rho(r')dr' \tag{2.58}$$

Adding both contributions, Φ_1 and Φ_2, gives

$$\Phi(r) = \frac{1}{\varepsilon_0 r}\int_0^r \rho(r')r'^2\,dr' + \frac{1}{\varepsilon_0}\int_r^\infty \rho(r')r'\,dr' \tag{2.59}$$

The electric field $\mathbf{E} = -\nabla\Phi$ can now be easily evaluated. The result is

$$\mathbf{E} = \frac{\mathbf{r}}{\varepsilon_0 r^3}\int_0^r \rho(r')r'^2\,dr' \tag{2.60}$$

It will now be shown how the above result can be obtained more easily from Gauss' law. We take a spherical surface of radius r and center at the origin. Because the charge distribution is spherically symmetric, then the electric field at r is radial and independent from angles, and therefore applying Gauss' law gives

$$\mathbf{E} = \frac{\mathbf{r}}{\varepsilon_0 r^3}\int_0^r \rho(r')r'^2\,dr'$$

which is of course what we arrived at in Eq. (2.60). The potential $\Phi(r)$ is now determined from $\mathbf{E}$ by using $\Phi(\mathbf{r}) = -\int_\infty^r \mathbf{E}\cdot d\mathbf{r}$:

$$\Phi(r) = -\int_\infty^r \left[\frac{1}{\varepsilon_0 r^2}\int_0^r \rho(r')r'^2\,dr'\right]dr$$

The integral is of the form $\int u\,dv$, where u is $\int_0^r \rho(r')r'^2\,dr'$ and $dv = -dr/(\varepsilon_0 r^2)$, and therefore can be integrated by parts:

$$\Phi = uv\Big|_\infty^r - \int_\infty^r v\,du.$$

This yields exactly what we got using direct integration [Eq. (2.59)].

We now use this result to determine the potential due to a specific charge distribution: an exponential charge distribution, $\rho(r) = \rho_0 e^{-\alpha r}$, where ρ_0 and α are constants. Substituting for ρ in Eq. (2.59) gives

$$\Phi(r) = \frac{\rho_0}{\varepsilon_0 r}\int_0^r r'^2 e^{-\alpha r'}\,dr' + \frac{\rho_0}{\varepsilon_0}\int_r^\infty r' e^{-\alpha r'}\,dr'$$

The integration can be easily carried out, with the result

$$\Phi(r) = \frac{2\rho_0}{\varepsilon_0 \alpha^3 r}(1 - e^{-\alpha r}) - \frac{\rho_0 e^{-\alpha r}}{\varepsilon_0 \alpha^2}$$

In terms of the total charge $Q = \int \rho\, dv = 8\pi\rho_0/\alpha^3$, the potential is

$$\Phi(r) = \frac{Q}{4\pi\varepsilon_0 r}(1 - e^{-\alpha r}) - \frac{Q\alpha}{8\pi\varepsilon_0} e^{-\alpha r}$$

This result indicates that the potential $\Phi(r)$ has an exponential behavior. In the limit of $\alpha r \gg 1$, which can be achieved at large distances, the potential becomes $\Phi(r) = Q/4\pi\varepsilon_0 r$ which is identical to a potential produce when the total charge Q is concentrated at the origin. Near the origin such that $\alpha r \ll 1$, the potential becomes independent of r; that is, $\Phi(r) = Q\alpha/8\pi\varepsilon_0$.

Example 2.14 Equipotential Surfaces of a Dipole Field

We now explain the surfaces of constant potential of an electric dipole. The dipole is taken to be along the z axis and composed of two point charges of same magnitude and of opposite

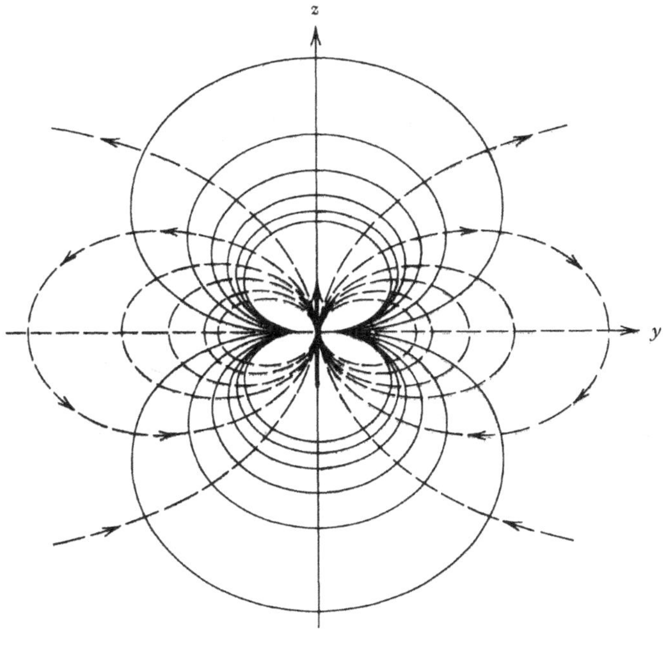

Figure 2.23 Equipotential surfaces and lines of force of an electric dipole. (Equipotential surfaces are indicated by solid lines; lines of force, by dashed lines.)

signs separated by a distance l. The electric dipole was treated in Example 2.9, and the potential was given in Eq. (2.44): $\Phi(\mathbf{r}) = \mathbf{p}\cdot\mathbf{r}/4\pi\varepsilon_0 r^3$, where $\mathbf{p} = ql\hat{\mathbf{z}}$. In the y-z plane, Φ reduces to $\Phi(r, \theta) = p\cos\theta/(4\pi\varepsilon_0 r^2)$ or $r(\theta) = \alpha(\cos\theta)^{1/2}$ where $\alpha = (p/4\pi\varepsilon_0\Phi)^{1/2}$. This relation between r and θ can now be plotted for different values of Φ, and hence different values of α, as shown in Fig. 2.23.

2.8 The Multipole Expansion

We turn now to the problems of characterizing the electrostatic potentials and fields of an arbitrary charge distribution, localized in a rather small region of space. One may think of this charge distribution as the charge distribution of a molecule, whose linear dimensions are of the order of 10^{-10} meter.

Consider Fig. 2.24, where we show a charge distribution that is localized in a volume V and characterized by a density ρ. We choose an origin, O, in or near this charge distribution. The displacement $\mathbf{r}'$ locates an element of charge relative to O. The displacement $\mathbf{r}$ locates a point in space outside the charge distribution, where we wish to determine the potential $\Phi(\mathbf{r})$:

$$\Phi(\mathbf{r}) = \frac{1}{4\pi\varepsilon_0}\int \frac{dq}{|\mathbf{r} - \mathbf{r}'|}$$

In fact, for any $\mathbf{r}'$, we shall assume that $r'/r \ll 1$. Then, we may expand the term in the integrand,

$$\frac{1}{|\mathbf{r} - \mathbf{r}'|} = \frac{1}{r\left[1 - \dfrac{2\mathbf{r}\cdot\mathbf{r}'}{\mathbf{r}^2} + \left(\dfrac{\mathbf{r}'}{\mathbf{r}}\right)^2\right]^{1/2}}$$

Now we apply the binomial expansion

$$(1 + x)^{-1/2} = \left[1 - \frac{1}{2}x - \left(\frac{1}{2}\right)\left(-\frac{3}{2}\right)\frac{x^2}{2!} + \cdots\right]$$

valid for $|x| < 1$. Letting

$$x = \left(-\frac{2\mathbf{r}\cdot\mathbf{r}'}{r^2} + \frac{r'^2}{r^2}\right)$$

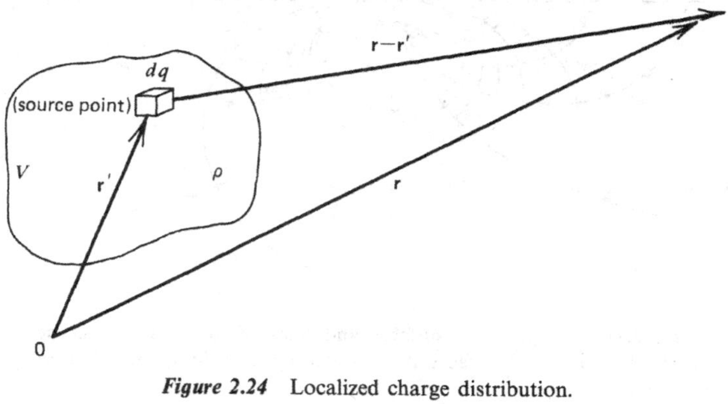

Figure 2.24 Localized charge distribution.

we have

$$\frac{1}{|\mathbf{r}-\mathbf{r}'|} \cong \frac{1}{r}\left[1 - \frac{1}{2}\left(-\frac{2\mathbf{r}\cdot\mathbf{r}'}{r^2} + \frac{r'^2}{r^2}\right) - \frac{1}{2}\left(-\frac{3}{2}\right)\frac{1}{2!}\left(-\frac{2\mathbf{r}\cdot\mathbf{r}'}{r^2} + \frac{r'^2}{r^2}\right)^2 + \cdots\right]$$

Grouping the powers of r'/r in ascending order, we find

$$\frac{1}{|\mathbf{r}-\mathbf{r}'|} \cong \frac{1}{r}\left\{1 + \frac{\hat{\mathbf{r}}\cdot\mathbf{r}'}{r} + \frac{1}{2}\left[3\left(\frac{\hat{\mathbf{r}}\cdot\mathbf{r}'}{r}\right)^2 - \left(\frac{r'}{r}\right)^2\right] + \cdots\right\}$$

Hence the potential becomes

$$\Phi(\mathbf{r}) = \frac{1}{4\pi\varepsilon_0}\int \frac{dq}{r}\left\{1 + \hat{\mathbf{r}}\cdot\frac{\mathbf{r}'}{r} + \frac{1}{2}\left[3\left(\frac{\mathbf{r}'\cdot\hat{\mathbf{r}}}{r}\right)^2 - \left(\frac{r'}{r}\right)^2\right] + \cdots\right\}$$

We now rewrite the potential in a more revealing form by writing it as a sum of three integrals corresponding to three potential contributions $\Phi^{(0)}$, $\Phi^{(1)}$, and $\Phi^{(2)}$, respectively.

$$\begin{aligned}\Phi(\mathbf{r}) &= \Phi^{(0)} + \Phi^{(1)} + \Phi^{(2)} + \cdots \\ &= \frac{1}{4\pi\varepsilon_0 r}\int dq + \frac{1}{4\pi\varepsilon_0 r^2}\hat{\mathbf{r}}\cdot\int \mathbf{r}'\,dq + \frac{1}{4\pi\varepsilon_0 r^3}\int\left[\frac{3(\hat{\mathbf{r}}\cdot\mathbf{r}')^2 - r'^2}{2}\right]dq + \cdots \qquad (2.61)\end{aligned}$$

It will be observed that successive terms of this expansion differ by a factor of the order of R'/r, where R' is a linear dimension characteristic of the charge distribution. Therefore, the dominant term of the distribution, when $r \gg R'$, will be the first nonvanishing term. If R' were of atomic dimensions and r was a macroscopic distance, then $R'/r \leq 10^{-4}$.

In certain cases, the point of observation may be enclosed by the charge distribution. If $r \ll R'$, where R' is the smallest dimension of the distribution, then the potential in the region $r < R'$ is represented by a series expansion in terms of r/r' rather than r'/r as in the previous case. The result is

$$\Phi(\mathbf{r}) = \frac{1}{4\pi\varepsilon_0}\left[\int\frac{dq}{r'} + \mathbf{r}\cdot\int\frac{\mathbf{r}'}{r'^3}\,dq + \frac{1}{2}\int\left(\frac{3(\mathbf{r}\cdot\mathbf{r}')^2}{r'^5} - \frac{r^2}{r'^3}\right)dq\right] \qquad (2.62)$$

We will not discuss the interior problem [given by Eq. (2.62)] any further except in Example 2.17 and Problem 2.2. The exterior problem, however, will now be discussed in detail. Let us now consider the terms $\Phi^{(0)}$, $\Phi^{(1)}$, and $\Phi^{(2)}$ in Eq. (2.61) separately.

Monopole Term: $\Phi^{(0)}$. The potential $\Phi^{(0)}$ is called the monopole potential; it can be written as follows:

$$\Phi^{(0)} = \frac{1}{4\pi\varepsilon_0 r}\int dq = \frac{Q}{4\pi\varepsilon_0 r}$$

where Q is the total net charge of the charge distribution. The presence of this term simply indicates that far enough away the charge distribution in the lowest-order approximation looks like a point charge, Q.

Dipole Term: $\Phi^{(1)}$—Dipoles in External Fields. The contribution $\Phi^{(1)}$ is called the dipole term; it can be written as follows:

$$\Phi^{(1)} = \frac{\hat{\mathbf{r}}\cdot\int\mathbf{r}'\,dq}{4\pi\varepsilon_0 r^2} = \frac{\hat{\mathbf{r}}\cdot\mathbf{p}}{4\pi\varepsilon_0 r^2}$$

where $\mathbf{p}$ is the vector called the *dipole moment* of the charge distribution, defined as

$$\mathbf{p} = \int \mathbf{r}'\, dq = \int \rho(\mathbf{r}')\mathbf{r}'\, dv' \tag{2.63}$$

and ρ is the charge density. The importance of the dipole term is that when $Q = 0$, it is the term that dominates the expansion. Because of the importance of "dipoles" in the discussion of electrical properties of matter, we catalog several properties of the dipole field, and of the dipole moment $\mathbf{p}$.

1. In general, the dipole moment, $\mathbf{p}$, depends upon one's choice of origin. In fact, it can always be set equal to zero for some origin if the total, net, charge of the distribution, Q, is not zero. If, however, $Q = 0$, then $\mathbf{p}$ has a value independent of origin.

2. The prototype dipole consists of two point charges of equal magnitude but opposite sign with a relative displacement $\boldsymbol{\delta}$. We see that the dipole moment can be expressed as

$$\mathbf{p} = \int \mathbf{r}'\, dq = \mathbf{r}_-(-q) + \mathbf{r}_+(+q) = q(\mathbf{r}_+ - \mathbf{r}_-)$$

where $\mathbf{r}_+$ and $\mathbf{r}_-$ are, respectively, the vector positions of q and $-q$ charges from an origin O. One often considers such dipoles as giving rise to dipole fields. However, it must be emphasized that the dipole potential $\Phi^{(1)}$ is the potential of this dipole only in the limit as $\delta/r \to 0$. The potential of two such point charges is not $\Phi^{(1)}$ when $\delta/r \approx 1$. Note that we separate any charge distribution into its positive and negative charge components, such the $dq = dq_+ + dq_-$; that is,

$$\mathbf{p} = \int \mathbf{r}'\, dq = \int \mathbf{r}'\, dq_+ + \int \mathbf{r}'\, dq_- = \langle \mathbf{r}'_+ \rangle Q_+ + \langle \mathbf{r}'_- \rangle Q_-$$

where $\langle \mathbf{r}'_+ \rangle$ and $\langle \mathbf{r}'_- \rangle$ represent the average displacements of the total positive charge Q_+ and the total negative charge Q_- from the origin. If the net charge of the distribution is zero, then $Q_- = -Q_+$, and

$$\mathbf{p} = Q_+[\langle \mathbf{r}'_+ \rangle - \langle \mathbf{r}'_- \rangle] \equiv Q_+ \boldsymbol{\delta}$$

3. The energy required to place a dipole in an external electrostatic field of potential function Φ is

$$U^{(1)} = -q\Phi(\mathbf{r}_-) + q\Phi(\mathbf{r}_+)$$

where $\mathbf{r}_+ = \mathbf{r}_- + \boldsymbol{\delta}$. If $\boldsymbol{\delta}$ is small enough, we can approximate $\Phi(\mathbf{r}_- + \boldsymbol{\delta})$ by the first two terms of the Taylor expansion $\Phi(\mathbf{r}_- + \boldsymbol{\delta}) = \Phi(\mathbf{r}_-) + (\boldsymbol{\delta} \cdot \nabla)\Phi(\mathbf{r}_-)$. Therefore $U^{(1)} = -q\Phi(\mathbf{r}_-) + q\Phi(\mathbf{r}_-) + q\boldsymbol{\delta} \cdot \nabla\Phi(\mathbf{r}_-)$. Taking $\mathbf{E} = -\nabla\Phi(\mathbf{r})$ gives

$$U^{(1)} = -\mathbf{p} \cdot \mathbf{E}(\mathbf{r}) \tag{2.64}$$

In asserting that $\boldsymbol{\delta}$ is small enough, we imply that $\mathbf{E}$ is in fact constant "over the dipole." Then $U^{(1)}$ has the simple interpretation as the work necessary to displace a positive charge q by $\boldsymbol{\delta}$ in the field $\mathbf{E}$. However, it does not include the energy required to form the dipole in the absence of $\mathbf{E}$.

4. The force $\mathbf{F}$ on a dipole immersed in an external field, $\mathbf{E}$, is $\mathbf{F} = \mathbf{F}_- + \mathbf{F}_+$ where $\mathbf{F}_-$ and $\mathbf{F}_+$ are the forces acting on the $-q$ and q charges as shown in Fig. 2.25. Writing $\mathbf{F}_\pm = \pm q\mathbf{E}(\mathbf{r}_\pm)$, $\mathbf{F}$ becomes

$$\mathbf{F} = -q\mathbf{E}(\mathbf{r}_-) + q\mathbf{E}(\mathbf{r}_+) = -q\mathbf{E}(\mathbf{r}_-) + q\mathbf{E}(\mathbf{r}_- + \boldsymbol{\delta})$$

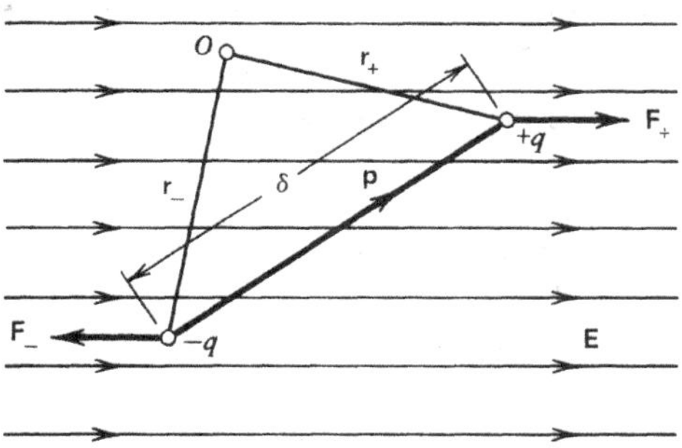

Figure 2.25 Schematic diagram of an electric dipole interacting with an external electric field by representing the dipole by two separated charges.

Keeping the first two terms in the Taylor expansion gives

$$\mathbf{F} \approx -q\mathbf{E}(\mathbf{r}_-) + q\mathbf{E}(\mathbf{r}_-) + q(\boldsymbol{\delta} \cdot \nabla)\mathbf{E}(\mathbf{r}_-)$$

or

$$\mathbf{F} = (\mathbf{p} \cdot \nabla)\mathbf{E} \tag{2.65}$$

It is to be noted from Eq. (2.65) that the force on a dipole is zero if the field in which it is immersed is uniform.

5. The torque $\boldsymbol{\tau}$ on the dipole when placed in a uniform field $\mathbf{E}$ is just

$$\boldsymbol{\tau} = \mathbf{r}_- \times \mathbf{F}_- + \mathbf{r}_+ \times \mathbf{F}_+$$

where $\mathbf{F}_-$ and $\mathbf{F}_+$ are the forces exerted by the field on the $-q$ and q charges, as shown in Fig. 2.25. Substituting for $\mathbf{F}_\pm = \pm q\mathbf{E}$ gives

$$\boldsymbol{\tau} = (\mathbf{r}_+ - \mathbf{r}_-) \times q\mathbf{E} = \boldsymbol{\delta} \times q\mathbf{E} = \mathbf{p} \times \mathbf{E} \tag{2.66}$$

The torque is in such a direction as to align the dipole moment along the field $\mathbf{E}$. If the field $\mathbf{E}$ is not uniform, one can directly show that

$$\boldsymbol{\tau} = \mathbf{p} \times \mathbf{E} + \mathbf{r} \times \mathbf{F} \tag{2.67}$$

where $\mathbf{F}$ is the force acting on the dipole, and $\mathbf{r}$ is its displacement from the origin—about which the torque is computed.

6. Since a dipole may be conceived of two equal and opposite "monopoles" separated by the displacement $\boldsymbol{\delta}$, then referring to Fig. 2.25 we get

$$\Phi^{(1)} = \Phi^{(0)}(\mathbf{r}) - \Phi^{(0)}(\mathbf{r} + \delta)$$

Using the Taylor expansion gives

$$\Phi^{(1)} = \Phi^{(0)}(\mathbf{r}) - \Phi^{(0)}(\mathbf{r}) - (\boldsymbol{\delta} \cdot \nabla)\Phi^{(0)}(\mathbf{r}) = -\boldsymbol{\delta} \cdot \nabla\Phi^{(0)}(\mathbf{r}) \tag{2.68}$$

This expression was originally encountered in Example 2.9. In a similar fashion, it is not hard to show that we may construct a multipole of order n from a multipole of order $(n - 1)$. In fact,

$$\Phi^{(n)} = \Phi^{(n-1)}(\mathbf{r}) - \Phi^{(n-1)}(\mathbf{r} + \boldsymbol{\delta}^{(n)}) = -\boldsymbol{\delta}^{(n)} \cdot \nabla\Phi^{(n-1)} \tag{2.69}$$

where $\boldsymbol{\delta}^{(n)}$ is the displacement of two multipoles of order $(n - 1)$ from each other (see Example 2.16).

Example 2.15 Some Dipole Moments

We determine in this example the dipole moments of a number of charge distributions (shown in Fig. 2.26).

(1) *A Point Charge.* The dipole moment of the point charge q located a displacement $\mathbf{r}'$ from the origin is $\mathbf{p} = q\mathbf{r}'$. If we had chosen the origin at the location of the charge, the dipole moment would have been zero.

(2) *N Point Charges.* The dipole moment of N point charges $q_1, q_2, \ldots, q_N$ is given by

$$\mathbf{p} = \sum_{j=1}^{N} \mathbf{r}'_j q_j$$

where $\mathbf{r}'_j$ is the displacement of charge q_j from the origin. Only if

$$\sum_{j=1}^{N} q_j = 0$$

is $\mathbf{p}$ independent of the choice of origin.

(3) *A Circular Ring of Charge.* The dipole moment of a circular ring uniformly charged with respect to an origin at the center of the ring is given by

$$p = \int \mathbf{r}'\, dq = \int_{\phi=0}^{2\pi} R\hat{\boldsymbol{\rho}}\lambda R\, d\phi = \lambda R^2 \int_0^{2\pi} \hat{\boldsymbol{\rho}}\, d\phi = 0$$

where R is the radius of the ring and λ is the linear charge density.

(4) *Rod of Charge.* The rod shown has the charge density

$$\rho = \alpha\left(z - \frac{l}{2}\right) \qquad \text{for } 0 \le z \le l$$

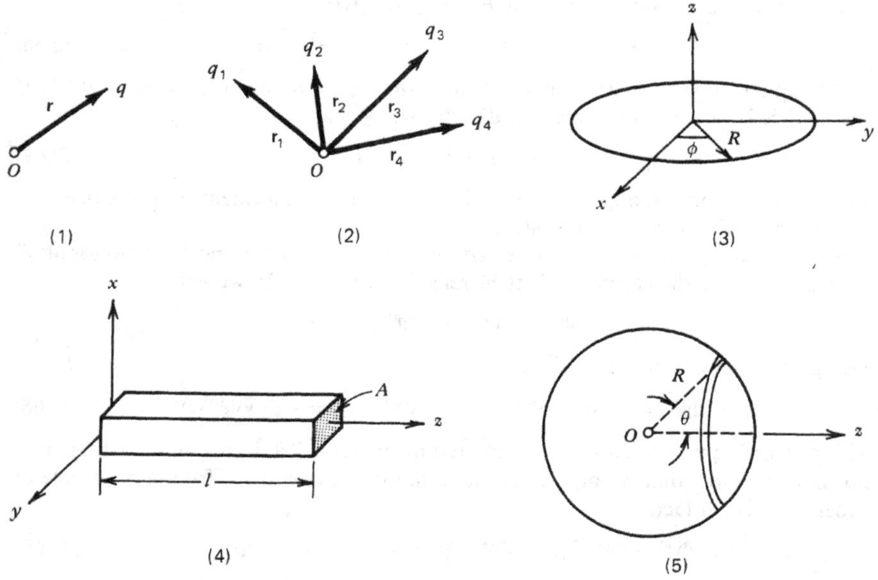

Figure 2.26 Some charge distributions: (1) point charge, (2) collection of point charges, (3) a ring of charge, (4) rectangular rod of charge, and (5) sphere of angular charge distribution.

Its dipole moment is given by

$$\mathbf{p} = \int \mathbf{r}'\,dq = \hat{\mathbf{z}} \int_{z=0}^{l} z(\rho A\,dz) = \hat{\mathbf{z}}\alpha A \int_{z=0}^{l} z\left(z - \frac{l}{2}\right)dz = \frac{1}{12}\hat{\mathbf{z}}l^3\alpha A \tag{2.70}$$

Note that the total charge in the rod $\int dq$ can be easily shown to be zero, and therefore $\mathbf{p}$ is independent of our choice of origin.

(5) *Sphere with an Angular Charge Distribution.* We consider the dipole moment of a sphere with a surface charge density given by $\sigma = \sigma_0 \cos\theta$. A point on the surface is located by $\mathbf{r}' = R\hat{\mathbf{r}}$, which is equal to $z'\hat{\mathbf{z}} + \rho'\hat{\boldsymbol{\rho}}$ in cylindrical coordinates, where $\rho'^2 + z'^2 = R^2$. The charge dq on a ring at an angle θ' is $dq = 2\pi R^2\sigma \sin\theta'\,d\theta'$. Therefore $\mathbf{p} = \int \mathbf{r}'\,dq = \int (\hat{\mathbf{z}}z' + \hat{\boldsymbol{\rho}}\rho')dq$. The integral $\int \hat{\rho}\rho'\,dq$ vanishes; thus

$$\mathbf{p} = \int \hat{\mathbf{z}}z'\,dq = \hat{\mathbf{z}}2\pi R^3\sigma_0 \int_{\theta=0}^{\pi} \cos^2\theta' \sin\theta'\,d\theta' = \hat{\mathbf{z}}\frac{4\pi R^3\sigma_0}{3} = \hat{\mathbf{z}}\sigma_0 V \tag{2.71}$$

where V is the volume of the sphere. This dipole moment is independent of the choice of the origin because the total charge on the sphere is zero.

Quadrupole Term; $\Phi^{(2)}$. The contribution to the potential $\Phi^{(2)}$ is called the quadrupole term;

$$\Phi^{(2)} = \frac{1}{4\pi\varepsilon_0}\cdot\frac{1}{2r^5}\int [3(\mathbf{r}\cdot\mathbf{r}')^2 - r^2r']dq \tag{2.72}$$

We have written this term so that the integrand is completely symmetric with respect to r and r'. Expanding the integrand in cartesian coordinates, and noting that the integration over the charge distribution depends on the primed coordinates (x', y', z'), not on the coordinates of the point of observation, $(x, y, z)]$ give

$$\begin{aligned}\int [3(\mathbf{r}\cdot\mathbf{r}')^2 - (rr')^2]dq &= (3x^2 - r^2)\int x'^2\,dq + 3xy\int x'y'\,dq + 3xz\int x'z'\,dq \\ &+ 3xy\int x'y'\,dq + (3y^2 - r^2)\int y'^2\,dq + 3yz\int y'z'\,dq \\ &+ 3zx\int z'x'\,dq + 3zy\int z'y'\,dq + (3z^2 - r^2)\int z'^2\,dq\end{aligned}$$

The "off-diagonal" elements of the array of terms are equal. The array of integrals is a matrix called the *quadrupole matrix*. Its elements will be denoted by Q_{xx}, Q_{xy}, Q_{xz}, etc. A knowledge of these elements completely specifies $\Phi^{(2)}$, just as a knowledge of the components of the dipole moment (p_x, p_y, p_z) specifies $\Phi^{(1)}$. Equation (2.72) can now be written as

$$\Phi^{(2)} = \frac{1}{4\pi\varepsilon_0}\sum_{i,j=1}^{3}\frac{Q_{ij}}{2r^5}(3x_ix_j - \delta_{ij}r^2) \tag{2.73}$$

where

$$Q_{ij} = \int x_i'x_j'\,dq \tag{2.74}$$

and we use $x_1 = x$, $x_2 = y$, and $x_3 = z$, and δ_{ij} is the Kronecker delta; it is equal to 1 if $i = j$ and zero if $i \neq j$.

Occasionally the terms above are regrouped, so that instead of the diagonal terms

$$(3x^2 - r^2)\int x'^2\,dq + (3y^2 - r^2)\int y'^2\,dq + (3z - r^2)\int z'^2\,dq$$

one writes the equivalent expression (show this)

$$3x^2\int\left(x'^2 - \frac{1}{3}r'^2\right)dq + 3y^2\int\left(y'^2 - \frac{1}{3}r'^2\right)dq + 3z^2\int\left(z'^2 - \frac{1}{3}r'^2\right)dq$$

In terms of these integrals (denoted Q'_{xx}, Q'_{yy}, Q'_{zz}) the quadrupole matrix is said to be "reduced," a word chosen because $Q'_{xx} + Q'_{yy} + Q'_{zz} = 0$, whereas $Q_{xx} + Q_{yy} + Q_{zz} = \int r'^2\,dq$.

Thus the potential in (2.72) can alternatively be written as follows:

$$\Phi^{(2)} = \frac{1}{4\pi\varepsilon_0}\frac{1}{2r^5}\sum_{i,j=1}^{3} 3x_i x_j Q'_{ij} \tag{2.75}$$

where 1, 2, and 3, denote x, y, and z, and

$$Q'_{ij} = \int\left(x'_i x'_j - \frac{1}{3}\delta_{ij}r'^2\right)dq \tag{2.76}$$

It is however possible to simplify the description of this matrix yet further by properly choosing the coordinate axes. If they are chosen judiciously, the (off-diagonal) terms Q_{xy}, Q_{xz}, Q_{yz}, can be made to equal zero. It turns out that this can always be accomplished by choosing the axes to be perpendicular to planes of symmetry. A case of special importance is one where there is *rotational symmetry* about an axis, which we call the z axis. Rotational symmetry here implies that one cannot distinguish one orientation of the distribution about the z axis from any other. In such a case $Q_{xy} = 0$, $Q_{xz} = 0$, and $Q_{yz} = 0$. Moreover, $Q_{xx} = Q_{yy}$, since the rotational symmetry renders the x and y coordinates indistinguishable. In terms of the reduced quadrupole matrix, one sees in this case that $Q'_{xx} + Q'_{yy} = -Q'_{zz}$, implying that

$$Q'_{xx} = Q'_{yy} = -\frac{1}{2}Q'_{zz} \tag{2.77}$$

Thus, all of the nonzero quadrupole matrix elements are expressible in terms of Q'_{zz}, sometimes called the *quadrupole moment* of the distribution, as follows:

$$Q'_{zz} = \int\left(z'^2 - \frac{1}{3}r'^2\right)dq = \frac{1}{3}\int r'^2(3\cos^2\theta' - 1)dq \tag{2.78}$$

where θ' denotes the angle between the z axis and the charge element dq. If $Q'_{zz} > 0$, one has a cigar-shaped ellipsoid, whereas if $Q'_{zz} < 0$, one has a saucer-shaped (oblate) ellipsoid of charge. In this sense, the quadrupole moment is a measure of the deformation of the ellipsoid from spherical symmetry.

Example 2.16 Quadrupole Distributions of Point Charges

In this example we determine the quadrupole matrix of the point charge distribution shown in Figs. 2.27*a* and 2.27*b*. In the case where one is dealing with point charges alone, the quadrupole potential $\Phi^{(2)}$ reduces to the sum:

$$\Phi^{(2)} = \frac{1}{4\pi\varepsilon_0}\cdot\frac{1}{2r^5}\sum_i [3(\mathbf{r}'_i\cdot\mathbf{r})^2 - r_i'^2 r^2]q_i$$

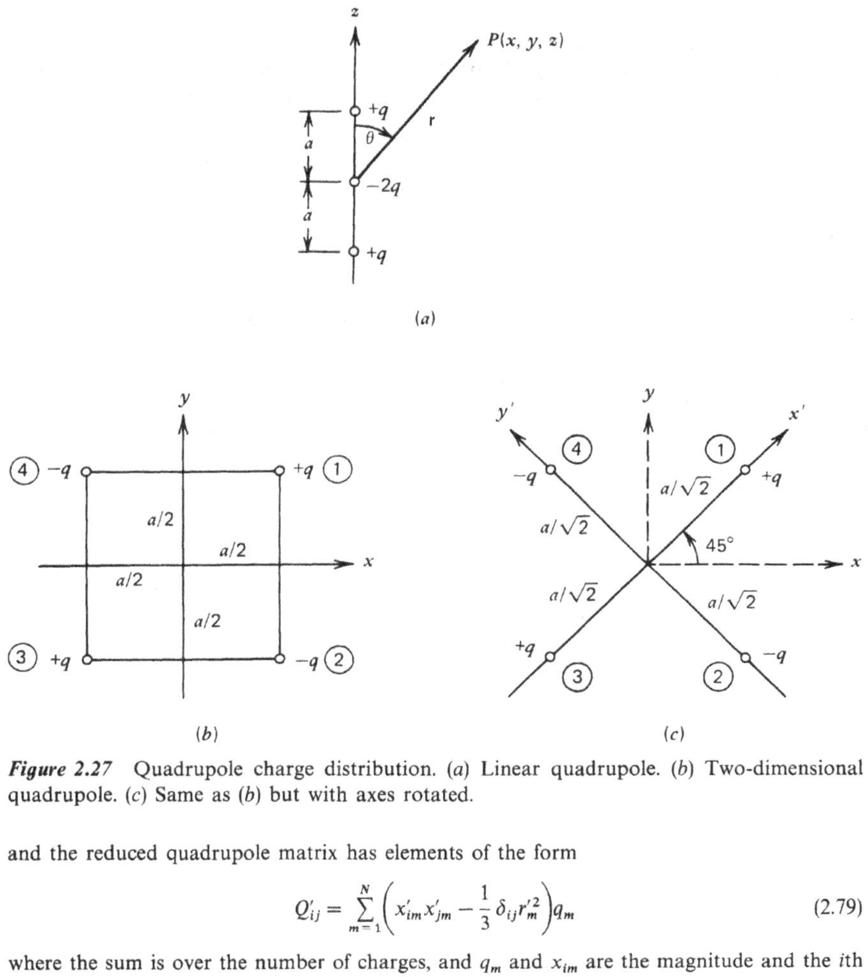

Figure 2.27 Quadrupole charge distribution. (*a*) Linear quadrupole. (*b*) Two-dimensional quadrupole. (*c*) Same as (*b*) but with axes rotated.

and the reduced quadrupole matrix has elements of the form

$$Q'_{ij} = \sum_{m=1}^{N} \left(x'_{im} x'_{jm} - \frac{1}{3} \delta_{ij} r'^2_m \right) q_m \tag{2.79}$$

where the sum is over the number of charges, and q_m and x_{im} are the magnitude and the ith coordinate of the mth charge.

1. The set of charges shown in Fig. 2.27*a* has zero monopole and dipole moments. Its reduced quadrupole matrix elements are easily calculated by using Eq. (2.79):

$$Q'_{xy} = Q'_{xz} = Q'_{yz} = 0$$

$$Q'_{yy} = Q'_{xx} = q\left(-\frac{1}{3}a^2\right) + 0 + q\left(-\frac{1}{3}a^2\right) = -\frac{2}{3}qa^2$$

$$Q'_{zz} = 2q\left[a^2 - \frac{1}{3}a^2\right] = \frac{4}{3}qa^2$$

Note that $Q'_{xx} + Q'_{yy} + Q'_{zz} = 0$. The potential $\Phi^{(2)}$ is therefore (for $r > a$)

$$\Phi^{(2)} = \frac{1}{4\pi\varepsilon_0} \cdot \frac{1}{2r^5} \left[3x^2\left(-\frac{2}{3}qa^2\right) + 3y^2\left(-\frac{2}{3}qa^2\right) + 3z^2\left(\frac{4}{3}qa^2\right) \right]$$

$$= \frac{qa^2}{4\pi\varepsilon_0 r^5}[3z^2 - r^2] = \frac{qa^2}{4\pi\varepsilon_0 r^3}[3\cos^2\theta - 1]$$

This array of charges, called a "linear quadrupole," has rotational symmetry about the z axis, and its "quadrupole moment" is $Q'_{zz} = 4qa^2/3$.

2. Another point quadrupole charge distribution is shown in Fig. 2.27*b*. The monopole and dipole moments are here zero. The reduced quadrupole matrix has matrix elements:

$$Q'_{xx} = \sum q_i\left(x_i^2 - \frac{1}{3}r_i^2\right) = 0$$

$$Q'_{yy} = Q'_{zz} = 0$$

$$Q_{xy} = \sum q_i x_i y_i = qa^2$$

$$Q_{xz} = Q_{yz} = 0$$

that is, only $Q_{xy} = Q_{yx}$ are not zero. The potential $\Phi^{(2)}$ is thus given by

$$\Phi^{(2)} = \frac{1}{4\pi\varepsilon_0}\cdot\frac{1}{2r^5}\cdot[6xyQ_{xy}] = \frac{1}{4\pi\varepsilon_0}\frac{3xy}{r^5}qa^2$$

If this array of charges were referred to the new set of axes shown—that is, x' and y' in Fig. 2.27*c*—we would have had $Q'_{xx} = 2q[\frac{1}{2}a^2] = qa^2$, $Q'_{yy} = -qa^2$, $Q'_{zz} = 0$, and $Q'_{xy} = Q'_{xz} = Q'_{yz} = 0$. Therefore we would have

$$\Phi^{(2)} = \frac{1}{4\pi\varepsilon_0}\cdot\frac{1}{2r^5}[3x'^2\cdot qa^2 - 3y'^2\cdot qa^2] = \frac{1}{4\pi\varepsilon_0}\cdot\frac{3qa^2}{2r^5}[x'^2 - y'^2]$$

The expressions for the potentials have different forms. However, if the transformations between the different coordinate systems are made, or

$$x' \to \frac{1}{\sqrt{2}}(x + y) \qquad \text{and} \qquad y' \to \frac{1}{\sqrt{2}}(-x + y)$$

one finds that the latter expression for $\Phi^{(2)}$ becomes the former.

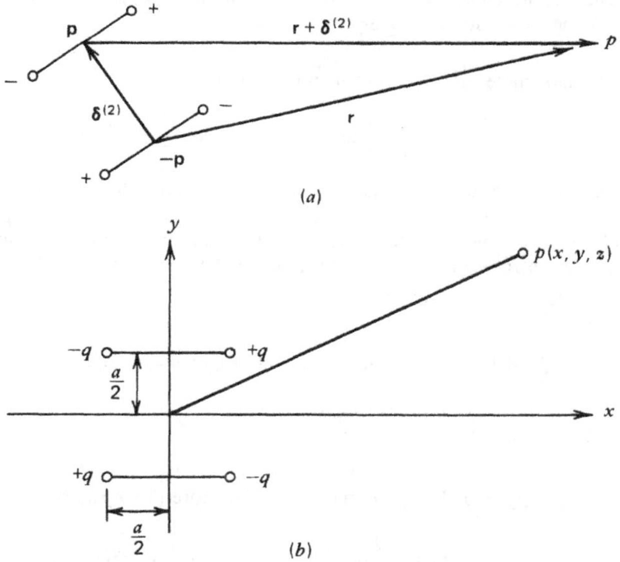

Figure 2.28 Determination of the fields of a quadrupole as a differential of a dipole field. (*a*) General quadrupole distribution. (*b*) Four-charge two-dimensional quadrupole.

3. It is interesting to point out that the "pure" quadrupole distributions shown are simply pairs of dipoles, one displaced from the other with the signs of the charges reversed. If $\boldsymbol{\delta}^{(2)}$ is the displacement of the one dipole from the other (see Fig. 2.28*a*), we have, from Eq. (2.69),

$$\Phi^{(2)} = \Phi^{(1)}(\mathbf{r}) - \Phi^{(1)}(\mathbf{r} + \boldsymbol{\delta}^{(2)}) = (-\boldsymbol{\delta}^{(2)} \cdot \nabla)\Phi^{(1)}(\mathbf{r}) \tag{2.80}$$

This result is analogous to the derivation of the dipole field from the field of a simple monopole [see Eq. 2.68)]. For the quadrupole of Fig. 2.28*b*,

$$\Phi^{(0)}(r) \equiv \frac{q}{4\pi\varepsilon_0 r} \qquad r = \sqrt{x^2 + y^2 + z^2} \qquad \boldsymbol{\delta}^{(1)} = \delta^{(1)}\hat{\mathbf{x}} = a\hat{\mathbf{x}} \qquad \boldsymbol{\delta}^{(2)} = \delta^{(2)}\hat{\mathbf{y}} = a\hat{\mathbf{y}}$$

Substituting these expressions in Eq. (2.80) gives

$$\Phi^{(2)} = -\frac{\partial}{\partial y}\Phi^{(1)}(\mathbf{r}) = -a\frac{\partial}{\partial y}\left[-a\frac{\partial}{\partial x}\left(\frac{q}{4\pi\varepsilon_0 r}\right)\right]$$

or

$$\Phi^{(2)} = a^2\frac{\partial^2}{\partial x\,\partial y}\left\{\frac{q}{4\pi\varepsilon_0\sqrt{x^2 + y^2 + z^2}}\right\} = \frac{a^2 q}{4\pi\varepsilon_0}\frac{3xy}{r^5}$$

just as previously obtained.

Example 2.17 Potential by Multipole Expansion

This example explains the use of the multipole expansion in calculating the potential due to an angular charge distribution. Consider a spherical shell of radius R, carrying a surface charge distribution $\sigma = \sigma_0 \cos\theta$, where θ is the angle with respect to the z axis. From Eq. (2.61), the potential outside the sphere ($r > R$) is

$$\Phi(\mathbf{r}) = \Phi^{(0)}(\mathbf{r}) + \Phi^{(1)}(\mathbf{r}) + \Phi^{(2)}(\mathbf{r}) + \cdots$$

where

$$\Phi^{(0)}(\mathbf{r}) = \frac{1}{4\pi\varepsilon_0 r}\int_{s'} \sigma_0 \cos\theta'\, da'$$

$$\Phi^{(1)}(\mathbf{r}) = \frac{1}{4\pi\varepsilon_0}\frac{\mathbf{r}}{r^3}\cdot\int_{s'} \mathbf{r}'\,\sigma_0\cos\theta'\, da'$$

$$\Phi^{(2)}(\mathbf{r}) = \frac{1}{4\pi\varepsilon_0}\cdot\frac{1}{2}\int_{s'}\left[\frac{3(\mathbf{r}\cdot\mathbf{r}')^2}{r^5} - \frac{r'^2}{r^3}\right]\sigma_0\cos\theta'\, da'$$

where θ' is the angle between $\mathbf{r}'$ and the z axis, and s' is the surface of the shell. The potential $\Phi^{(0)}(\mathbf{r}) = 0$ since the total charge on the sphere is zero.

The potential $\Phi^{(1)}$ can be evaluated as follows: One first writes $\mathbf{r}'$ in terms of θ' and ϕ', as follows:

$$\mathbf{r}' = (R\sin\theta'\cos\phi')\hat{\mathbf{x}} + (R\sin\theta'\sin\phi')\hat{\mathbf{y}} + R\cos\theta'\,\hat{\mathbf{z}}$$

Thus the integral in $\Phi^{(1)}$ becomes

$$\sigma_0\int \mathbf{r}'\cos\theta'\, da' = R^3\sigma_0\int_{\theta'=0}^{\pi}\int_{\phi'=0}^{2\pi}[(\sin^2\theta'\cos\theta'\cos\phi')\hat{\mathbf{x}}$$
$$+ (\sin^2\theta'\cos\theta'\sin\phi')\hat{\mathbf{y}} + \cos^2\theta'\sin\theta'\,\hat{\mathbf{z}}]d\theta'\,d\phi'$$

The first two terms vanish because

$$\int_0^{2\pi}\sin\phi'\,d\phi' = \int_0^{2\pi}\cos\phi'\,d\phi' = 0$$

and the last term gives $\mathbf{p} = (4\pi/3)\sigma_0 R^3 \hat{\mathbf{z}}$. Therefore

$$\Phi^{(1)}(\mathbf{r}) = \frac{1}{4\pi\varepsilon_0}\frac{\mathbf{p}\cdot\mathbf{r}}{r^3} = \frac{1}{4\pi\varepsilon_0}\frac{p\cos\theta}{r^2}$$

The potential $\Phi^{(1)}(\mathbf{r})$ is the first nonvanishing contribution; it is a dipole potential with a dipole moment $\mathbf{p}$ along the z axis and moment equal to σ_0 times the volume bounded by the shell. This potential agrees with the direct calculations of the dipole moment of the shell [see Eq. (2.71)].

Using a similar procedure one can show that $\Phi^{(2)}(\mathbf{r})$ as well as the other higher multipoles vanish, indicating that the field of this charge distribution is a dipole field; that is,

$$\Phi(\mathbf{r}) = \frac{1}{4\pi\varepsilon_0}\frac{\mathbf{p}\cdot\mathbf{r}}{r^3} \qquad r > R \tag{2.81}$$

Inside the sphere—that is, for $r < R$—we use Eq. (2.62). Thus

$$\Phi(\mathbf{r}) = \Phi^{(0)}(\mathbf{r}) + \Phi^{(1)}(\mathbf{r}) + \Phi^{(2)}(\mathbf{r}) + \cdots$$

where

$$\Phi^{(0)}(\mathbf{r}) = \frac{1}{4\pi\varepsilon_0}\int_{s'}\frac{\sigma_0\cos\theta'}{r'}\,da'$$

$$\Phi^{(1)}(\mathbf{r}) = \frac{1}{4\pi\varepsilon_0}\mathbf{r}\cdot\int_{s'}\frac{\mathbf{r}'}{r'^3}\sigma_0\cos\theta'\,da'$$

$$\Phi^{(2)}(\mathbf{r}) = \frac{1}{4\pi\varepsilon_0}\cdot\frac{1}{2}\int_{s'}\left[\frac{3(\mathbf{r}\cdot\mathbf{r}')^2}{r'^5} - \frac{r^2}{r'^3}\right]\sigma_0\cos\theta'\,da'$$

Substituting R for the magnitude of $\mathbf{r}'$, and taking it outside the integral gives

$$\Phi^{(0)} = \frac{1}{4\pi\varepsilon_0 R}\int\sigma_0\cos\theta'\,da' = 0$$

since the integral is just the total charge on the sphere. Doing the same thing in the expression for $\Phi^{(1)}$ gives

$$\Phi^{(1)} = \frac{1}{4\pi\varepsilon_0}\frac{\mathbf{r}}{R^3}\cdot\int\mathbf{r}'\sigma_0\cos\theta'\,da' = \frac{1}{4\pi\varepsilon_0}\frac{\mathbf{p}}{R^3}\cdot\mathbf{r}$$

where $\mathbf{p}$, the dipole moment, is as defined above. Again, as in the region exterior to the sphere, $\Phi^{(2)}$ and the other higher-order terms vanish. Therefore

$$\Phi(\mathbf{r}) = \frac{1}{4\pi\varepsilon_0}\frac{\mathbf{p}}{R^3}\cdot\mathbf{r} = \frac{\sigma_0}{3\varepsilon_0}z \qquad r < R \tag{2.82}$$

It is apparent that the potential inside the shell depends only on z and is independent of the size of the shell, unlike the potential outside the shell. The corresponding electric field is $-(\sigma_0/3\varepsilon_0)\hat{\mathbf{z}}$, which is uniform and directed along the negative z axis.

The previous discussion indicates that the electric multipoles can be used to approximate the electric field or potential of an arbitrary charge distribution. This process, in fact, is equivalent to the approximation of the charge distribution itself by a combination of a point charge, a point dipole, a point quadrupole, and so forth, as shown schematically in Fig. 2.29. The applications of such approximation, thus, go beyond just the electric field or the potential of the distribution.

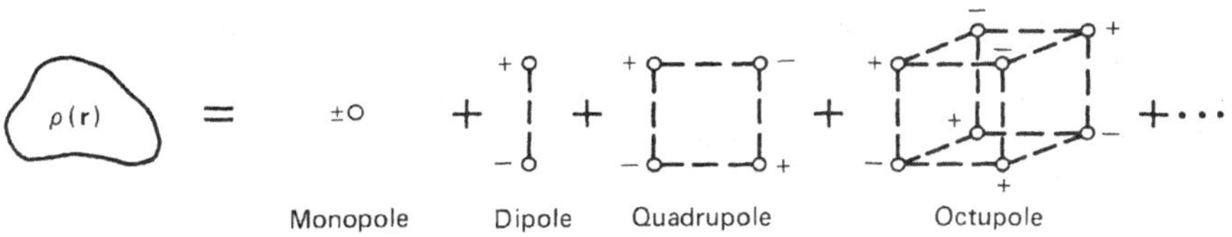

Figure 2.29 Schematic diagram of the representation of a localized charge distribution by its various multipoles.

2.9 Summary

Electrostatics is the subject matter that deals with electric charges that are at rest. Coulomb's law defines the electrostatic force law between a point charge q_0 at the origin and a point charge q located at $\mathbf{r}$; that is,

$$\mathbf{F} = \frac{1}{4\pi\varepsilon_0} \frac{q_0 q \hat{\mathbf{r}}}{r^2}$$

where

$$\frac{1}{4\pi\varepsilon_0} = 9 \times 10^9 \text{ N}\cdot\text{m}^2/\text{C}^2$$

in MKS units. Writing

$$\mathbf{F} = q\mathbf{E}$$

defines the electrostatic field $\mathbf{E}$ associated with the charge q_0

$$\mathbf{E} = \frac{1}{4\pi\varepsilon_0} \frac{q_0 \hat{\mathbf{r}}}{r^2}$$

Coulomb's law and the electric field can be generalized to many point charges or continuous charge distributions that may residue in volumes, on surfaces, or along lines such that the element of charge dq is given by

$$dq = \rho\, dv,\ \sigma\, da, \text{ or } \lambda\, dl$$

where ρ, σ, and λ are the volume, surface, and line charge densities, respectively. For a point charge q_i located at $\mathbf{r}_i$,

$$\rho(\mathbf{r}) = q_i \delta(\mathbf{r} - \mathbf{r}_i)$$

where δ is the Dirac delta function. Since forces add vectorially, then

$$\mathbf{E} = \frac{1}{4\pi\varepsilon_0} \int \frac{dq(\mathbf{r} - \mathbf{r}')}{|\mathbf{r} - \mathbf{r}'|^3} \tag{2.15}$$

This electrostatic field is said to be conservative. In other words,

$$\nabla \times \mathbf{E} = 0 \tag{2.37}$$

Its divergence, on the other hand, depends linearly on the charge density

$$\nabla \cdot \mathbf{E} = \frac{\rho}{\varepsilon_0} \tag{2.33}$$

This is often called the differential form of Gauss's law, and it is one of four fundamental laws of electromagnetism as we understand them today (Maxwell's equations). It is even satisfied by time-dependent fields. The curl property, however, is true only for electrostatics and will be modified later when time-varying sources are considered.

The integral law of Gauss' law follows from the differential law by integrating both sides over an arbitrary volume V bounded by the surface S, and applying the divergence theorem to the left side; that is,

$$\oint_S \mathbf{E}\cdot\hat{\mathbf{n}}\,da = \frac{1}{\varepsilon_0}\int \sigma\,dv = \frac{Q_{\text{int}}}{\varepsilon_0} \tag{2.27}$$

where Q is the total charge enclosed by S. In cases of symmetry where the electric field is of constant magnitude over all elements of the surface, and of constant direction relative to the surface, Gauss' law simplifies the calculation of the electric field at the surface. A powerful implication of Gauss' law, along with the fact that the electric field inside a conductor is zero, is the fact that charge on a conductor must reside on its outer surface, with the fields just outside the conductor being $\mathbf{E} = (\sigma/\varepsilon_0)\hat{\mathbf{n}}$. The law also shows that the fields just below and just above a surface charge distribution, $\mathbf{E}_2$ and $\mathbf{E}_1$, are discontinuous, with the discontinuity being

$$(\mathbf{E}_2 - \mathbf{E}_1)\cdot\hat{\mathbf{n}} = \frac{\sigma}{\varepsilon_0} \tag{2.35}$$

At distances from the charge distribution that are large compared to the largest dimension of the distribution, the charge distribution can be approximated by a combination of a point charge Q, a point dipole $\mathbf{p}$, a point quadrupole Q_{ij}, etc., where

$$Q = \int \rho(\mathbf{r}')dr' \qquad \mathbf{p} = \int \mathbf{r}'\rho(\mathbf{r}')dr' \qquad Q'_{ij} = \int (x'_i x'_j - \frac{1}{3}\delta_{ij}r'^2)dq$$

$$\Phi = \frac{1}{4\pi\varepsilon_0}\left(\frac{Q}{r} + \frac{\mathbf{p}\cdot\hat{\mathbf{r}}}{\mathbf{r}^2} + \frac{1}{2\mathbf{r}^5}\sum_{i,j=1}^{3} 3x_i x_j Q'_{ij}\right) \tag{2.63)–(2.76}$$

where 1, 2, and 3 denote x, y, and z, respectively. The applications of such approximation in fact goes beyond just the electric or the potential of the distribution.

When a point charge is placed in an electrostatic potential Φ of corresponding field $\mathbf{E}$, the charge experiences a force $\mathbf{F} = q\mathbf{E}$, and the potential energy U of the charge is

$$U = q\Phi$$

If a dipole $\mathbf{p}$ is placed in such a field, we have for the force and energy

$$\mathbf{F} = (\mathbf{p}\cdot\nabla)\mathbf{E} \qquad U = -\mathbf{p}\cdot\mathbf{E} \tag{2.64),(2.65}$$

The dipole will also experience a torque $\boldsymbol{\tau}$ as follows:

$$\boldsymbol{\tau} = \mathbf{p}\times\mathbf{E} + \mathbf{r}\times\mathbf{F} \tag{2.67}$$

Problems

2.1 Four point charges, $q = 2\times 10^{-5}$ C each, are on the corners of a square of length 4 m. Find the force on a point charge $q_0 = 10^{-4}$ C located 3 m just above the center of the square.

2.2 Determine the electric field at the center of curvature of a uniformly charged semicircular rod with a total charge q.

2.3 A filamentary charge is distributed along the z axis with a charge density $\lambda = \lambda_0$ for $|z| > 5$ and $\lambda = 0$ for $|z| < 5$. Find $\mathbf{E}$ on the x axis, 2 m from the origin.

2.4 A filamentary charge is distributed along the z axis with a charge density $\lambda = \lambda_0$ for $|z| < d$ and $\lambda = 0$ for $|z| > d$. Determine the $\mathbf{E}$ field in the $x-y$ plane a distance R from the z axis.

2.5 Determine, by direct integration, the **E** field on the axis of a uniformly charged disk of radius a and charge density σ.

2.6 A circular disk of radius a lies in the $z = 0$ plane with its center at the origin. The disk has a surface charge density $\sigma = \sigma_0/\rho$. Determine the electric field at $z = h$ along the z axis. Discuss the nature of the field when $h \gg a$.

2.7 A sheet of charge lies in the $z = 0$ plane and occupies the area $0 \leqq x < 2$ m and $0 \leqq y < 2$ m. The surface charge density is $\sigma = 2x(x^2 + y^2 + 4)^{3/2}$ C/m^2. Determine the electric field 2 m above the sheet on the z axis.

2.8 A circular disk of radius a has a nonuniform charge density $\sigma = \sigma_0 \sin^2 \phi$. Determine **E** on its axis at $z = h$.

2.9 Two infinite, uniform sheets of charge in the $y - z$ plane of charge densities σ and σ' are located at $x = 1$ and $x = -1$, respectively. Determine the **E** in all regions for both cases $\sigma' = \sigma$, and $\sigma' = -\sigma$.

2.10 A sheet of uniform charge density $\sigma = -10^{-7}$ C/m^2 occupies the $y = 2$ m plane. Next to it a line of uniform charge density $\lambda = 0.4$ μC/m lies parallel to the x axis at $y = -1$ m and $z = 2$ m. Determine the region in which **E** will be zero.

2.11 Use Gauss' law to determine the **E** field produced by a very long cyclinder of charge of volume density $\rho = 5re^{-2r}$ C/m^3, where r is the distance from the axis of the cyclinder.

2.12 Use Gauss' law to determine the **E** field produced by a spherical charge distribution of density $\rho = \alpha/r^2$, where α is a constant.

2.13 Consider a closed surface S. Calculate the net flux that crosses it due to the following charge distributions enclosed by it. (a) Three point charges: $q_1 = 3 \times 10^{-8}$ C, $q_2 = 1.5 \times 10^{-7}$ C, and $q_3 = -7 \times 10^{-8}$ C. (b) A circular disk of charge of radius 2 m and $\sigma = (\sin^2 \phi)/\rho$ C/m^2. (c) A circular disk of charge of radius 2 m and $\sigma = \sin \phi$ C/m^2. (d) Two point charges $q_1 = 20 \times 10^{-7}$ C and $q_2 = -20 \times 10^{-7}$ C.

2.14 Determine the electric flux density 5 m from a point charge, $q = 3 \times 10^{-8}$ C.

2.15 Find the charge densities that produce the following fields in V/m.

(a) $\mathbf{E} = 10 \sin\theta\, \hat{\mathbf{r}} + 2\cos\theta\, \hat{\boldsymbol{\theta}}$.

(b) $\mathbf{E} = \dfrac{1}{2}\alpha\left(\rho - \dfrac{a^2}{\rho}\right)\hat{\boldsymbol{\rho}}$ for $a \leq \rho \leq b$ $\qquad$ $\mathbf{E} = \dfrac{1}{2}\dfrac{\alpha}{\rho}(b^2 - a^2)\hat{\boldsymbol{\rho}}$ for $\rho > b$.

where α, a, and b are constants.

2.16 A thin ring has its inner and outer diameters equal to ρ_0 and $\rho_0 + W$, respectively. The ring has a uniform surface charge density σ. Determine the potential at the center of the ring. Does it depend on ρ_0?

2.17 A charge is distributed uniformly along a straight line of finite length $2l$. (a) Determine the potential at distance r_1 just above the midpoint. (b) What is the potential in the limit $l \gg r_1$? (c) Determine for two points, r_1 and r_2, just above the midpoint the potential difference Φ_{12} if $l \gg r_1 > r_2$. How does this potential difference compare with the potential of an infinite line charge?

2.18 (a) Determine the **E** field associated with the potential

$$\Phi = \frac{a\cos\theta}{r^2} + \frac{b}{r}$$

(b) What is the charge distribution responsible for this potential? (c) Find the charge distribution that gives rise to the potential $\Phi(\mathbf{r}) = -qe^{-\alpha r}/r$, where q and α are constants.

2.19 This problem is given to illustrate the power of linear superposition as applied to calculations of electric fields and potentials. A spherical cavity of radius a is hollowed out from the interior of a uniformly charged sphere of radius R and charge density ρ_0. The distance between the centers of the sphere and the cavity is $d\hat{\mathbf{z}}$. (a) Determine the electric field at a distance $r < R$ from the center of the sphere, assuming that there is no cavity. (b) Assuming that the cavity is filled with charge of uniform density ρ'_0 and the sphere is not, determine the electric field at distance $r' < a$ from the center of the cavity. (c) Determine the actual electric field inside the cavity. Sketch the lines of force of this field. (d) Determine the actual potential at a point inside the cavity relative to that at the origin of the cavity.

2.20 Determine the potential in spherical coordinates due to two equal but opposite point charges placed on the y axis at $y = \pm l/2$ with $l \ll r$, the distance to the observation point.

2.21 Show that the potential at large distances from a linear octupole shown in Fig. 2.30 is

$$\frac{6qa^3P_3(\cos\theta)}{4\pi\varepsilon_0 r^4},$$

where $P_3 = (5\cos^3\theta - 3\cos\theta)/2$.

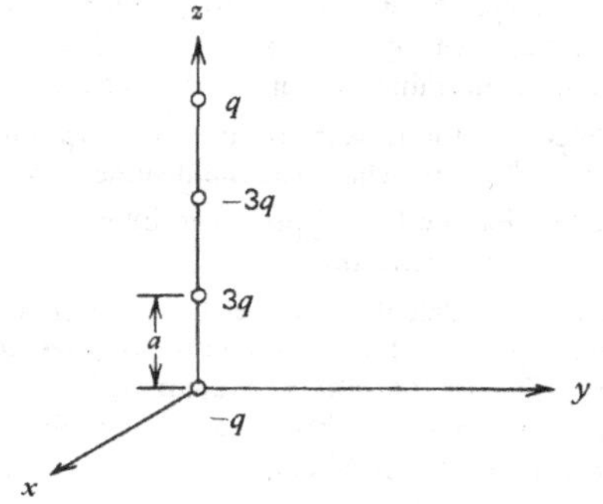

Figure 2.30 Linear octupole.

2.22 Consider two thin, coaxial, coplanar, uniformly charged rings with radii a and b ($a > b$) and charges q and $-q$, respectively. Determine the potential at large distances from the rings. Compare this potential with that of a linear quadrupole (see Fig. 2.27*a*).

2.23 Consider half a sphere that has a uniform surface charge distribution σ. Calculate the dipole moment of the distribution relative to an origin at the center of curvature.

2.24 Consider an electric dipole $\mathbf{p} = p_0\hat{\mathbf{x}}$ located at the origin and placed in external potential $\Phi = \frac{1}{2}\alpha_1 x^2 + \alpha_2 x + \alpha_3$. (a) What was the energy needed to place the dipole in the potential? (b) Determine the force acting on the dipole. (c) Determine the torque acting on the dipole (sec Example 1.3). α_1, α_2 and α_3 are constants.

2.25 Determine the force and torque acting on an electric dipole of moment $\mathbf{p}$ due to a point charge q.

2.26 A dipole of moment $\mathbf{p}_1$ lies at the origin and a dipole of moment $\mathbf{p}_2$ lies at a point whose position vector is $\mathbf{r}$. Determine the force between the two dipoles. For which orientation of the dipoles does the force maximize?

2.27 Two charges q and $-q$ are placed on the x axis at d and $-d$, respectively. A dipole of moment $\mathbf{p}$ is placed on the z axis at l. (a) Determine the force acting on the dipole. (b) Show that this force can be produced by replacing the two charges with a dipole of moment $2qd(1 + d^2/l^2)^{-5/2}$ placed at the origin and directed along the x axis.

2.28 Use the multipole expansion given in Eq. (2.62) to find the potential near the center of a thin sphere of radius R that has a uniform surface charge density σ.

2.29 Calculate the first three multipole moments of the following charge distributions. (a) A uniform filamentary distribution extending from $z = z_0$ to $z = -z_0$ and of total charge q. (b) The uniformly distributed ring of charge (discussed in Example 2.15) with an additional point charge $q = -2\pi R\lambda$ placed at its center.

2.30 Use Eq. (2.69) and the dipole field to calculate the potential of the quadrupole of Fig. 2.27*a*.

THREE

ELECTROSTATIC BOUNDARY VALUE PROBLEMS

3.1 Poisson's and Laplace's Equations

Often, we are confronted with situations where we do not know, a priori, the charge distributions, and consequently we cannot directly determine **E** or Φ. An important example of this is when we have a system of conductors whose relative potentials are known, but the charge densities on the conductor surfaces are not. In this instance, as we will see later, there exists a solution, producing well-behaved electric fields and charge densities. Our explicit formula for **E** and Φ demand that we know the charge distributions on the conductor surfaces, yet we cannot obtain these without knowing the electric fields at the conductor surfaces ($\sigma = \varepsilon_0 E$ there). It turns out that we can resolve this problem. What allows us to resolve it is basically that we employ two criteria simultaneously to solve two problems. For example, although each of two linear algebraic equations may contain two unknowns and neither can be solved alone for both unknowns, the two in combination can be solved. In the case of electrostatics, two relations that can be solved simultaneously are as follows:

$$\nabla \cdot \mathbf{E} = \frac{\rho}{\varepsilon_0} \tag{3.1}$$

$$\mathbf{E} = -\nabla\Phi \tag{3.2}$$

These two equations may be further combined into one equation—namely,

$$\nabla \cdot (-\nabla\Phi) = \rho/\varepsilon_0$$

—which is customarily written as

$$\nabla^2\Phi = -\frac{\rho}{\varepsilon_0} \tag{3.3}$$

This latter equation, which is called *Poisson's equation*, summarizes the equations of electrostatics. It relates the partial derivatives of the potential function at a point to the charge density, ρ, at that point. The symbol ∇^2 ("del squared"), called the Laplacian operator, is a linear, scalar operator [see Eqs. (1.62) to (1.64)]. In terms of cartesian coordinates,

$$\nabla^2\Phi \equiv \frac{\partial^2\Phi}{\partial x^2} + \frac{\partial^2\Phi}{\partial y^2} + \frac{\partial^2\Phi}{\partial z^2} \tag{1.62}$$

$\nabla^2\Phi$ may be expressed in other coordinates also, such as cylindrical or spherical coordinates, using the gradient and divergence expressions in those coordinates. They are given in Eqs. (1.63) and (1.64), respectively.

If $\rho = 0$ in some region of space, then in that region Poisson's equation becomes

$$\nabla^2\Phi = 0 \tag{3.4}$$

This is known as *Laplace's equation*. The study of the properties and solutions of these partial differential equations occupy an important place in mathematical physics. In problems with conductors, ρ is usually zero in the region between the conductors, so we seek a solution of Laplace's equation, $\nabla^2\Phi = 0$, in these regions that will have the correct boundary conditions at the surfaces of the conductors. Having obtained Φ, we can then obtain $\mathbf{E} = -\nabla\Phi$ and subsequently the surface charge densities on the conductors.

The assumption that the charge density ρ is zero in the space not occupied by conductors has a greater application than might be supposed. It will be true if that space is *free space* (obviously) or is occupied by a simple dielectric (to be discussed later), and if there is no agent that continually injects charge into that region. In the case of free space, any charge that may have existed there will have been swept out by the existing (static) electric fields, so ultimately one has a situation wherein charge resides only on bounding conductor surfaces. The fact that charge cannot exist in *stable* equilibrium in free space under the influence of electrostatic fields alone is often given the name *Earnshaw's theorem*. In establishing this theorem it suffices to remark that surrounding any such point in space where a charge could reside in stable equilibrium one could locate a small Gaussian surface S through which the electrostatic flux would be finite. This follows since $\mathbf{E}$ must point toward or away from the point everywhere on the sphere in order that restoring forces exist around the point. This implies that a net charge exists inside the sphere, which contradicts the assumption of free space. Therefore stable equilibrium is not possible.

It is interesting to note here that these arguments about electrostatic equilibrium in free space are similar to those involved in showing that the macroscopic charge density must be zero inside a conductor. The main difference, macroscopically, between free space and the interior of a conducting medium is that the conductor possesses a large available supply of elementary charges that can be used to annul any existing static fields. This occurs by a rearrangement of charge on its surfaces. Such a field is usually not possible in free space.

3.2 Uniqueness of Solutions to Electrostatic Problems

We address ourselves now to problems involving conductors in free space. We shall first indicate that a unique solution to the electrostatic field exists in this space so long as certain boundary conditions are specified. We shall assume that a mathematical solution to Eq. (3.4) indeed exists. This is not obvious, although physically we know that if we have a system of conductors, with an arbitrary charge or potential on each, then there will exist some electrostatic field solution in the space.

Let us assume the following. (1) The region of interest, V, is charge free space where $\nabla \cdot \mathbf{E} = \nabla^2\Phi = 0$. (2) The boundaries of V are completely specified by various surfaces (such as the surfaces of conductors), collectively called S. One boundary may correspond to a surface of a sphere at "infinity," where by definition the potentials will vary as $1/r$, and the fields as $1/r^2$.* We consider three uniqueness conditions:

I. Potential Boundaries. We take all bounding surfaces to be surfaces on which the potential Φ is uniquely specified. Then the potential in the intervening region is uniquely specified.

Proof

(i) Assume that there exist two solutions to the equation $\nabla^2\Phi = 0$, Φ_1 and Φ_2, satisfying the same potential boundary conditions. We shall show that they are identical.

(ii) Consider the function $\Phi_0 = \Phi_2 - \Phi_1$. It is a solution of Laplace's equation in this region since $\nabla^2\Phi_0 = \nabla^2(\Phi_2 - \Phi_1) = \nabla^2\Phi_2 - \nabla^2\Phi_1 = 0$. Moreover, the potential Φ_0 vanishes on all the bounding surfaces since Φ_1 and Φ_2 have identical values there by hypothesis.

(iii) Consider the vector function $[\Phi_0 \mathbf{E}_0]$ where $\mathbf{E}_0 \equiv -\nabla\Phi_0$. Then, by the divergence theorem,

$$\int_V \nabla \cdot (\Phi_0 \mathbf{E}_0) dv = \oint_S (\mathbf{E}_0 \Phi_0) \cdot d\mathbf{a}$$

The surface integral in this equation vanishes since $\Phi_0 \equiv 0$ on S. If a part of the bounding surface of V is "at infinity," $\oint_S (\mathbf{E}_0 \Phi_0) \cdot d\mathbf{a} \to 0$ either because the integral varies as $(1/R)\cdot(1/R^2)\cdot 4\pi R^2 \sim 1/R$, and so the integral vanishes or we assume that as $R \to \infty$, Φ_1 and Φ_2 becomes identical. Thus

$$\int_V \nabla \cdot (\Phi_0 \mathbf{E}_0) dv = 0$$

Using the property $\nabla \cdot (\Phi_0 \mathbf{E}_0) = \nabla\Phi_0 \cdot \mathbf{E}_0 + \Phi_0 \nabla \cdot \mathbf{E}_0$ in the integrand of the volume integral, and noting that $\nabla \cdot \mathbf{E}_0 = 0$ anywhere inside V, and that $\mathbf{E} = -\nabla\Phi_0$, we obtain

$$-\int_V (\nabla\Phi_0)^2 \, dv = 0$$

(iv) For this volume integral to vanish at any point in V, the integrand itself should vanish identically. Thus

$$\mathbf{E} = -\nabla\Phi_0 \equiv 0$$

i.e., $\Phi_0 =$ constant. Since Φ_0 is zero and continuous on the boundaries of V it follows that $\Phi_0 = 0$, or $\Phi_2 \equiv \Phi_1$.

A physical argument for the validity of this uniqueness property is that if Φ_0 is zero on all boundaries, then it either must be zero everywhere in between or it must have an extremum somewhere in between. If it has an extremum, then through some Gaussian surface surrounding the region of this extremum, which is at a lower or

* That is, at a distance far enough from all the charge (the charge must therefore be localized) the charge looks like a single point charge.

higher equipotential, there must be a nonzero net flux. (The difference in potential $d\Phi_0 = -\mathbf{E}_0 \cdot d\mathbf{r} \neq 0$.] The latter implies the existence of charge inside the Gaussian surface, and contradicts the hypothesis of charge-free space. Therefore Φ_0 must be zero everywhere. Note that this proof nowhere is dependent upon the assumption that surface S is that of a conductor.

II. Charged Boundaries. Assume the surface of conductors to be the bounding surfaces of a region V, and that each conductor has its (total surface) charge well specified. If V is an open region, let the potential at infinity approach that of a point charge equal to the total net charge of the system. Then the electric field in V is uniquely specified, and the potential is specified to within a constant value that is determined if the potential at one point is specified. (This point is sometimes called "ground" or "common," and is assigned the potential zero.)

Proof

(i) Again assume two solutions Φ_1 and Φ_2 are possible in V having the same boundary conditions (including the condition at infinity, if required.) Then $\Phi_0 \equiv \Phi_2 - \Phi_1$ is also a solution, but to a problem with all conductors having zero net charge. Clearly, if $\Phi_2 \neq \Phi_1$, the charge densities will differ on the conductors even though the total charges are equal.

(ii) Now, as in condition I above, we again find that

$$\int_V E_0^2\, dv = \oint_S \Phi_0 \mathbf{E}_0 \cdot d\mathbf{a}$$

$\oint_S$ again indicates all bounding surfaces, including possibly one "at infinity." If we consider the conductor bounding surfaces, we note that Φ_0 is constant there, and so, for each conductor surface S_i,

$$\int_{S_i} (\Phi_0 \mathbf{E}_0) \cdot d\mathbf{a} = \Phi_0 \int_{S_i} \mathbf{E}_0 \cdot d\mathbf{a}$$

Since the net charges on each conductor vanishes, then the surface integral vanishes. If the surface at infinity is considered, then the surface integral at infinity vanishes too because Φ_1 and Φ_2 and $\mathbf{E}_1$ and $\mathbf{E}_2$ behave identically there. Thus, as before, $\int_V \mathbf{E}_0^2\, dv = 0$, implying $\mathbf{E}_0 = 0$, implying $\mathbf{E}_1 = \mathbf{E}_2$.

III. Mixed Boundaries. Other uniqueness conditions are possible. For example, one may specify for each conductor either the potential or the total charge (not both at once). Again, it is found that the resulting fields are unique. The proof proceeds similarly to those given above. Similarly, uniqueness will result if, on the boundaries of $\mathbf{V}$, the normal component of $\mathbf{E} = -\nabla\Phi$ is everywhere given.

Though we have proved uniqueness for the case where $\nabla^2\Phi = \nabla \cdot \mathbf{E} = 0$ (Laplace's equation), similar arguments show the uniqueness of solutions even when a charge density exists between the boundaries. We shall thus feel completely free to suppose that a unique solution always exists, although we must be careful not to impose contradictory or insufficient conditions on our problem. By the word "solution" we simply mean that we have found the electrostatic field or potential throughout the region of space of concern.

The uniqueness property of electrostatic problems is of great utility in obtaining solutions, because one can then more plausibly use the power of guesswork and intuition. If a field is obtained by whatever argument, and that field satisfies the electrostatic equations and the physical boundary conditions, then that field is the true solution to the electrostatic problem and no further arguments need to be made. (See Section 3.5, The Method of Images.)

3.3 Boundary Conditions

One often desires the solution to an electrostatic problem in a restricted volume of space, V. For example, in using the differential electrostatic equations, we would like to exclude regions where there are point charges, surface charges, or line charges, so that $\mathbf{E}$ will everywhere be a continuous function in V. We shall see that it is possible to find a solution in V if we also know certain properties that we must obtain for the solution at the boundaries of V. Thus it is crucially important to the subject to specify the boundary conditions.

It will suffice to give these boundary conditions at surfaces separating two regions of space. We shall assume that point charges, and line and surface charge densities, exist. Consider first the behavior of the electric field near an isolated point charge. The magnitude of the field grows arbitrarily large as the point charge is approached. Therefore it will always be true that the electric field on the surface of a sphere centered at the point charge will approach that of the point charge alone as the radius of the sphere, r, shrinks to zero:

$$\mathbf{E} \to \frac{1}{4\pi\varepsilon_0}\frac{q\mathbf{r}}{r^3} \qquad \text{as } r \to 0 \tag{3.5}$$

This is the boundary condition for point charges.

Similarly, if there exists an isolated line charge density of magnitude λ the field at a radial distance ρ from the line satisfies Eq. (2.19); that is,

$$\mathbf{E} \to \frac{1}{2\pi\varepsilon_0}\frac{\lambda\hat{\boldsymbol{\rho}}}{\rho} \qquad \text{as } \rho \to 0 \tag{2.19}$$

Now consider any surface separating two regions labeled 1 and 2. The surface may carry a surface charge density σ. The boundary condition on the normal components of $\mathbf{E}$ at that surface have already been derived [see Eq. 2.35)]. It states that

$$(\mathbf{E}_2 - \mathbf{E}_1)\cdot\hat{\mathbf{n}} = \frac{\sigma}{\varepsilon_0} \tag{2.35}$$

where $\mathbf{E}_2$ and $\mathbf{E}_1$ are the values of $\mathbf{E}$ in regions 2 and 1, respectively, as a point on the surface is approached. This result follows directly from an application of Gauss' integral law. It asserts that a discontinuity in the normal components of $\mathbf{E}$ at a surface exists only if a surface charge density exists at the surface.

A relation between the tangential components of $\mathbf{E}$ across a boundary can be determined using $\nabla \times \mathbf{E} = 0$. Integrating this equation over an open area S and using Stokes' theorem gives $\int_S \nabla \times \mathbf{E}\cdot d\mathbf{a} = \oint_C \mathbf{E}\cdot d\mathbf{r} = 0$, where C is a closed loop bounding S. The equation $\oint_C \mathbf{E}\cdot d\mathbf{r} = 0$ always implies that the tangential components of $\mathbf{E}$ across any interface separating two regions of space are continuous. The proof consists in evaluating this integral (circulation of $\mathbf{E}$) around the closed rectangular loop (Fig. 3.1), which straddles the surface. The long side of the loop, of length l, is chosen small enough so that $\int_l \mathbf{E}\cdot d\mathbf{r} = \mathbf{E}\cdot\mathbf{l}$, where $\mathbf{E}$ equals the value at the center of $\mathbf{l}$. The short sides of the loop may be chosen small enough so that as long as $\mathbf{E}$ is *finite* across the interface surface, their contributions to the line integral may be neglected. Then

$$\oint_C \mathbf{E}\cdot d\mathbf{r} \approx \mathbf{E}_1\cdot(-\mathbf{l}) + \mathbf{E}_2\cdot\mathbf{l} = 0 = \mathbf{E}_1\cdot(-\hat{\mathbf{t}})l + \mathbf{E}_2\cdot(\hat{\mathbf{t}})l$$

where $\hat{\mathbf{t}}$ is a unit vector parallel to the surface. Consequently

$$(\mathbf{E}_1\cdot\hat{\mathbf{t}}) = (\mathbf{E}_2\cdot\hat{\mathbf{t}}) \tag{3.6}$$

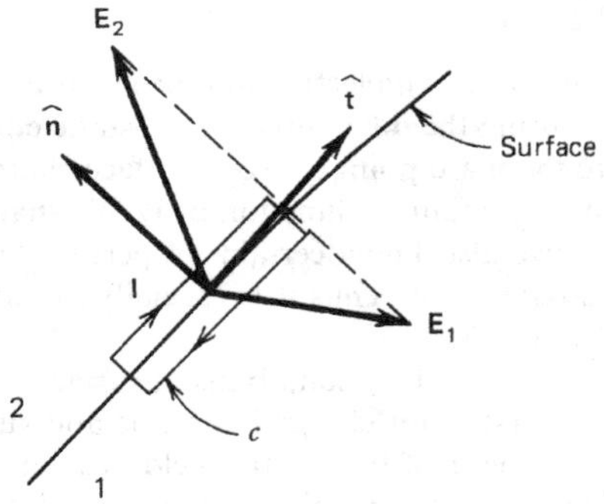

Figure 3.1 Application of Stokes' theorem to the area of the rectangle at the interface of two regions to determine conditions on the component of **E** tangent to the interface.

This equation indicates that the tangential component of **E** is continuous. Moreover, it can also be written in terms of a cross product:

$$\hat{\mathbf{n}} \times (\mathbf{E}_2 - \mathbf{E}_1) = 0$$

where $\hat{\mathbf{n}}$ is a unit vector normal to the interface. Finally Eqs. (3.6) and (2.35) can be combined into one single vector relation, as follows:

$$\mathbf{E}_2 - \mathbf{E}_1 = \frac{\sigma}{\varepsilon_0}\hat{\mathbf{n}} \tag{3.7}$$

The boundary condition on the potential function Φ is simply that Φ is continuous across any boundary where the electric fields remain finite. This is an immediate consequence of the definition of $\Delta\Phi$ as the work required to displace a unit point charge between two points. If **E** is finite, then as the displacement $\Delta\mathbf{r}$ goes to zero, $\mathbf{E}\cdot\Delta\mathbf{r} \to 0$, implying $\Delta\Phi \to 0$, or

$$\Phi_1 = \Phi_2 \tag{3.8}$$

The conditions in Eqs. (2.35) and (3.6)—or combined in Eq. (3.7)—and Eq. (3.8) are the boundary conditions needed to solve electrostatic boundary value problems in vacuum. However, it is to be noted that the three conditions are not independent of each other. In fact, the continuity of the potential is equivalent to the continuity of the tangential components of the electric field. We would like to reemphasize that **E** must be finite for the potential to be continuous. When **E** is not finite, such as when a dipole layer (which will be discussed in Chapter 4) is crossed, the potential will not be continuous (see Example 4.1).

3.4 Problems Involving Laplace's Equation

In this and following sections we consider electrostatic problems where the charge is confined to surfaces of conductors or localized at discrete points, or both. In the space between conductors and away from the point charges the electrostatic potential satisfies Laplace's equation: $\nabla^2\Phi = 0$. In this section we study the solution of this equation for various physical configurations.

3.4.1 Laplace's Equation in One Dimension

We shall study first some simple types of solutions of this equation. These solutions arise in problems of very high geometrical symmetry where the potential Φ is a function of a single variable. In these cases Laplace's equation reduces to an ordinary differential equation with very simple solutions.

Consider a cartesian geometry where Φ is only a function of z; then

$$\frac{d^2\Phi}{dz^2} = 0 \tag{3.9}$$

with a solution

$$\Phi(z) = az + b \tag{3.10}$$

where a and b are constants to be evaluated from the boundary conditions.

In the case of spherical geometry where the potential is a function of r only, the potential satisfies the following equation:

$$\frac{1}{r^2}\frac{d}{dr}\left(r^2\frac{d\Phi}{dr}\right) = 0 \tag{3.11}$$

Multiplying by $r^2 \neq 0$ and integrating gives

$$\frac{d\Phi}{dr} = -\frac{a}{r^2}$$

where a is a constant. Integrating again gives

$$\Phi = \frac{a}{r} + b \tag{3.12}$$

where b is another constant. The constants a and b are to be evaluated from the boundary conditions.

When the potential is a function of ρ and independent of ϕ and z of the cylindrical coordinates, then

$$\frac{1}{\rho}\frac{d}{d\rho}\left[\rho\frac{d\Phi}{d\rho}\right] = 0 \tag{3.13}$$

Multiplying by $\rho \neq 0$ and integrating gives

$$\frac{d\Phi}{d\rho} = \frac{a}{\rho}$$

where a is a constant. Integrating again gives

$$\Phi = a \ln \rho + b \qquad \text{or} \qquad \Phi = a \ln\left(\frac{\rho}{\rho_0}\right) \tag{3.14}$$

where b or ρ_0 is another constant, which along with the constant a can be evaluated from the boundary conditions.

Other problems involving other single independent variables such as θ in spherical coordinates and ϕ in cyclindrical coordinates will be discussed later in Examples 3.3 and 3.4.

Example 3.1 Long, Uniformly Charged, Conducting Rod

In this example (see Fig. 3.2) we approximate the long conductor as one extending in both directions to infinity. The conductor has, on its surface, a surface charge density σ. We are required to find the potential at a distance ρ from the axis of the rod. It is apparent that this potential should depend only upon the coordinate ρ. (The electrostatic field of this charge distribution depends only on ρ, and points in the ρ direction by symmetry arguments.)

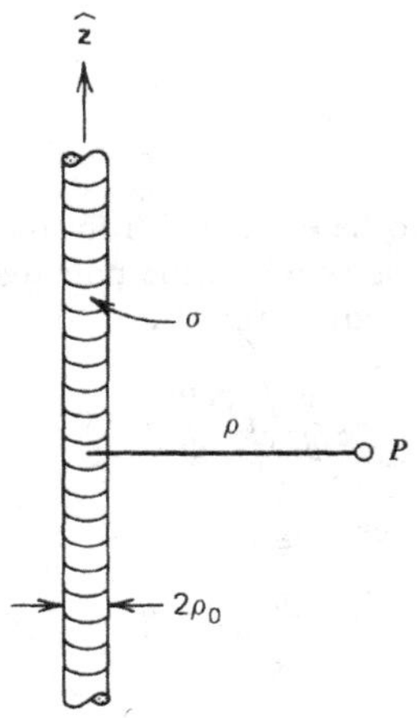

Figure 3.2 Long, uniformly charged rod.

The charge extends to infinity, and therefore we expect Φ not to vanish at ∞; that is, $\Phi(\infty) \neq 0$. Consequently we choose the zero of potential to be on the axis of the rod (the symmetry axis). Since the conductor is an equipotential volume, the whole volume of the rod has zero potential.

In the region outside the conductor, we must have $\nabla^2\Phi = 0$. Since Φ and E depend only on the cylindrical coordinate ρ, then Φ satisfies Eq. (3.13), whose solution is given by $\Phi = a \ln(\rho/\rho_0)$. To determine the constants our solution is made to satisfy the correct physical conditions on the boundary of the conductor. We therefore require that

$$\Phi(\rho = \rho_0) = 0,$$

which has already been satisfied. Moreover, recalling that at the surface of a conductor $E = \sigma/\varepsilon_0$, then

$$-\left.\frac{\partial\Phi}{\partial\rho}\right|_{\rho=\rho_0} = \frac{\sigma}{\varepsilon_0}$$

implying that

$$-\frac{a}{\rho_0} = \frac{\sigma}{\varepsilon_0} \qquad \text{or} \qquad a = -\frac{\rho_0\sigma}{\varepsilon_0}$$

Our final solution is therefore

$$\Phi = -\frac{\rho_0\sigma}{\varepsilon_0}\ln\left(\frac{\rho}{\rho_0}\right) = \frac{-\lambda}{2\pi\varepsilon_0}\ln\left(\frac{\rho}{\rho_0}\right) \qquad \rho > \rho_0$$

where λ is the charge per unit length of the rod.

Example 3.2 Spherical Capacitor

Consider two concentric spherical shells of radii R_1 and R_2 (where $R_2 > R_1$). The inner and outer shells are kept at potentials V_1 and V_2, respectively. Because of the spherical geometry we use spherical polar coordinates with the origin at the center of the shells. Moreover, because the shells are concentric, we expect the potential between them to be independent of the angles θ and ϕ. Therefore, the potential is given by Eq. (3.12); that is, $\Phi = a/r + b$. The boundary condition at $r = R_1$ gives $V_1 = a/R_1 + b$ and that at $r = R_2$ gives $V_2 = a/R_2 + b$. These two equations can now be solved simultaneously for a and b and hence, for the potential and the field;

$$\Phi = -\left(\frac{V_2 - V_1}{R_2 - R_1}\right)\frac{R_2 R_1}{r} + \frac{R_2 V_2 - R_1 V_1}{R_2 - R_1} \qquad \mathbf{E} = -\left(\frac{V_2 - V_1}{R_2 - R_1}\right)\frac{R_2 R_1}{r^2}\hat{\mathbf{r}}$$

Using this expression, one can calculate the charge q_1 on the inner shell, as follows:

$$q_1 = \varepsilon_0 \int \mathbf{E}\cdot\hat{\mathbf{r}}|_{R_1}\, da = -4\pi\varepsilon_0\left(\frac{V_2 - V_1}{R_2 - R_1}\right)R_1 R_2$$

and thus $\mathbf{E}$ is written in terms of the total charge as follows:

$$\mathbf{E} = \frac{1}{4\pi\varepsilon_0}\frac{q_1}{r^2}\hat{\mathbf{r}}$$

The charge on the outer shell can be determined using this field and that in the region $r > R_2$, which can be determined by solving the boundary value problem of the outer region. (Do it.)

Example 3.3 Wedge Capacitor

Consider the capacitor shown in Fig. 3.3. It consists of two large plates in the shape of a wedge forming an angle β. The plates are insulated from each other and kept at 0 and V volts. This capacitor can be best described by cylindrical coordinates. Because the plates are large, we expect the potential to be independent of z and ρ. In this case the Laplacian ∇^2 reduces to $(1/\rho^2)d^2/d\phi^2$ and the potential between the plates Φ satisfies the equation

$$\frac{1}{\rho^2}\frac{d^2\Phi}{d\phi^2} = 0 \tag{3.15}$$

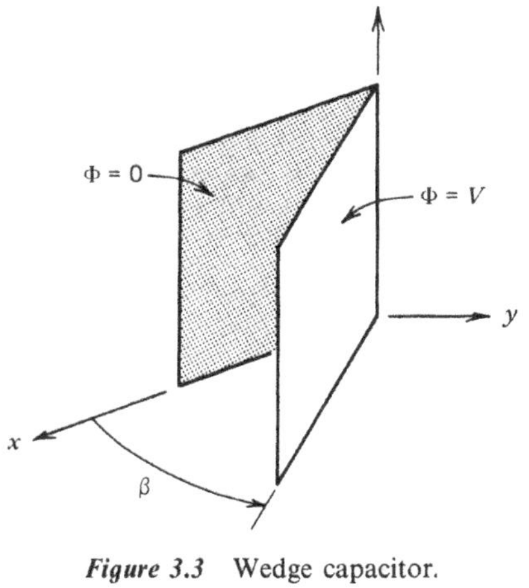

Figure 3.3 Wedge capacitor.

The most general solution for Φ is

$$\Phi = a\phi + b \tag{3.16}$$

where a and b are constants to be evaluated from the boundary conditions. At $\phi = 0$, $\Phi = 0$; and therefore $b = 0$. At the other plane ($\phi = \beta$), $\Phi = V$; and therefore $a = V/\beta$. Substituting for a and b in Eq. (3.16) gives

$$\Phi = \frac{V}{\beta}\phi$$

To calculate the electric field, we take the gradient of Φ. This gives

$$\mathbf{E} = -\frac{V}{\beta\rho}\hat{\boldsymbol{\phi}}$$

where $\hat{\boldsymbol{\phi}}$ is a unit vector in the ϕ direction.

Example 3.4 Coaxial Conic Capacitor

The capacitor of Fig. 3.4 consists of two coaxial cones whose apexes are placed at the origin and whose axes are along the z axis. The apex angles are θ_1 and θ_2 ($\theta_2 > \theta_1$). The cones are insulated from each other and the inner and outer cones are kept at V and zero potentials respectively.

To determine the potential and the electric field between the cones we start with Laplace's equation in spherical polar coordinates. Taking the cones to be sufficiently large that end effects can be neglected results in the situation where the potential is independent of r and ϕ.

$$\frac{1}{r^2 \sin\theta}\frac{d}{d\theta}\left(\sin\theta\frac{d\Phi}{d\theta}\right) = 0 \tag{3.17}$$

which, upon multiplication by $r^2 \sin\theta$, gives

$$\frac{d}{d\theta}\left(\sin\theta\frac{d\Phi}{d\theta}\right) = 0$$

which, upon integration, gives

$$\frac{d\Phi}{d\theta} = \frac{a}{\sin\theta}$$

where a is a constant. Integrating again gives

$$\Phi = a\ln\left(\tan\frac{\theta}{2}\right) + b \tag{3.18}$$

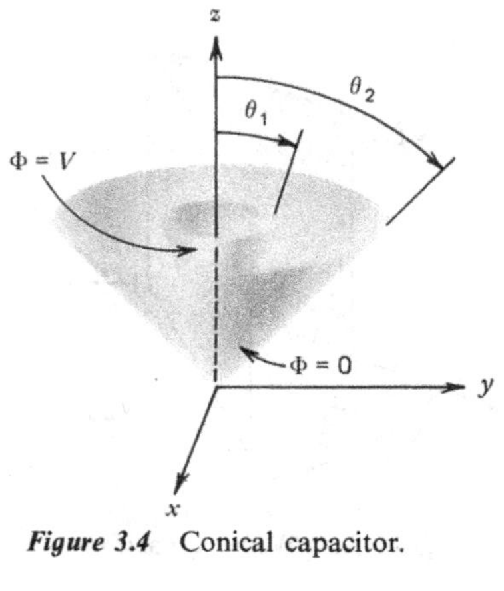

Figure 3.4 Conical capacitor.

where b is another constant. The constants can now be evaluated. At $\theta = \theta_1$, $\Phi = V$, and at $\theta = \theta_2$, $\Phi = 0$; thus

$$a \ln\left(\tan\frac{\theta_1}{2}\right) + b = V \quad \text{and} \quad a \ln\left(\tan\frac{\theta_2}{2}\right) + b = 0$$

The above equations for a and b when solved simultaneously yield:

$$a = \frac{V}{\ln\left[\dfrac{\tan(\theta_1/2)}{\tan(\theta_2/2)}\right]} \quad \text{and} \quad b = -a\ln\left(\tan\frac{\theta_2}{2}\right)$$

Substituting for a and b in Eq. (3.18) gives

$$\Phi = V\frac{\ln\left[\dfrac{\tan(\theta/2)}{\tan(\theta_2/2)}\right]}{\ln\left[\dfrac{\tan(\theta_1/2)}{\tan(\theta_2/2)}\right]}$$

3.4.2 Laplace's Equation in Two Dimensions—Spherical Coordinates

We will now consider some problems that are less symmetric than the ones discussed above in Section 3.4.1. These involve potentials that are functions of two variables. In spherical coordinates we choose geometries of azimuthal symmetry and thus choose the dependence of the potential to be on r and θ. In this case Laplace's equation reduces to

$$\frac{1}{r^2}\frac{\partial}{\partial r}\left(r^2\frac{\partial\Phi}{\partial r}\right) + \frac{1}{r^2\sin\theta}\frac{\partial}{\partial\theta}\left(\sin\theta\frac{\partial\Phi}{\partial\theta}\right) = 0 \tag{3.19}$$

The dependence of Φ on r and θ can be determined by the method of separation of variables. This method assumes $\Phi(r, \theta)$ to be a product of two functions. One depends on r only, and the other depends on θ only and, as a result, transforms the above partial differential equation to two ordinary differential equations. Substituting $\Phi(r, \theta) = Y(r)P(\theta)$ in Eq. (3.19) gives

$$\frac{P(\theta)}{r^2}\frac{d}{dr}\left(r^2\frac{dY}{dr}\right) + \frac{Y}{r^2\sin\theta}\frac{d}{d\theta}\left(\sin\theta\frac{dP}{d\theta}\right) = 0 \tag{3.20}$$

Dividing by $Y(r)P(\theta)$ and multiplying by r^2 we get:

$$\frac{1}{Y}\frac{d}{dr}\left(r^2\frac{dY}{dr}\right) = -\frac{1}{P\sin\theta}\frac{d}{d\theta}\left(\sin\theta\frac{dP}{d\theta}\right) \tag{3.21}$$

Since the left-hand side of this equation depends only on r and the right-hand side depends only on θ, then the only way that both sides can be equal for all values of r and θ is for both sides to be equal to a constant K, which is called the *separation constant*. Thus

$$\frac{1}{\sin\theta}\frac{d}{d\theta}\left(\sin\theta\frac{dP}{d\theta}\right) + KP = 0 \tag{3.22}$$

$$\frac{d}{dr}\left(r^2\frac{dY}{dr}\right) = KY \tag{3.23}$$

Equation (3.22) is called *Legendre's equation*; it has solutions that behave well for all values of θ including 0 and π only if $K = n(n+1)$, where n is a positive integer. The corresponding solutions are labeled by n, written as $P_n(\theta)$ and called Legendre's polynomials or *zonal harmonics.* Table 3.1 gives the explicit dependence of a few of them on θ. An important property of these polynomials is that they are orthogonal with each other; that is,

$$\int_{-1}^{1} P_n(x)P_m(x)dx = \frac{2\delta_{nm}}{2n+1} \tag{3.24}$$

where $x = \cos\theta$. The *Kronecker delta function*, δ_{nm}, is zero for $n \neq m$ and is unity for $n = m$.

Table 3.1

n	$P_n(\cos\theta)$
0	1
1	$\cos\theta$
2	$\frac{1}{2}(3\cos^2\theta - 1)$
3	$\frac{1}{2}(5\cos^3\theta - 3\cos\theta)$

Since K has been determined, the equation for $Y(r)$ can now be solved for different values of n:

$$\frac{d}{dr}\left(r^2\frac{dY}{dr}\right) = n(n+1)Y \tag{3.25}$$

This equation permits two linearly independent solutions for each n; these are labeled by n and are given by the following equation. (You can check them by substitution.)

$$Y_n = r^n \qquad \text{and} \qquad Y_n = r^{-(n+1)} \tag{3.26}$$

The overall equation $\Phi_n(r,\theta) = P_n(\theta)Y_n(r)$ then has two linearly independent solutions, which are as follows:

$$\Phi_n(r,\theta) = r^nP_n(\theta) \qquad \text{and} \qquad r^{-(n+1)}P_n(\theta) \tag{3.27}$$

The most general solution is then written as a linear combination of all the possible solutions—namely, $n = 0, 1, \ldots, \infty$. That is,

$$\Phi(r,\theta) = \sum_{n=0}^{\infty}[A_nr^n + B_nr^{-(n+1)}]P_n(\theta) \tag{3.28}$$

where A_n and B_n are constants to be evaluated from the boundary conditions.

This expansion can be interpreted as a multipole expansion of similar nature to Eqs. (2.61) and (2.62). The terms having the solution $Y_n = r^{-(n+1)}$ correspond to terms in which $r'/r \ll 1$, whereas those having $Y_n = r^n$ correspond to the case $r/r' \ll 1$. Further details of this nature can be seen in some of the examples in the text.

We consider below four examples where the potential is a function of r and θ. These examples arise in quite different physical situations and thus help explain the foregoing technique.

*Example 3.5 Nonconcentric Spherical Capacitor

Let us consider again the spherical capacitor treated above in Example 3.2. In here we take the inner shell to have a charge q and the outer shell to be kept at zero potential (mixed boundary conditions). When the shells are concentric, as discussed in Example 3.2, the potential is only a function of r. A dependence on θ can be introduced by displacing the centers of the shells by an amount δ, as shown in Fig. 3.5. In order to simplify the derivation we consider only the case where the displacement is very small—namely, $\delta \ll R_1$.

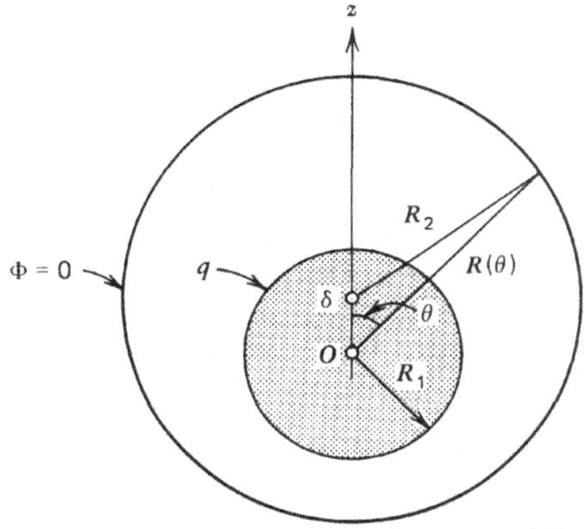

Figure 3.5 Nonconcentric spherical capacitor.

We choose a spherical polar coordinate system whose origin is at the center of the inner sphere. Using the law of cosines one can easily show that the surface of the outer sphere is described approximately in this coordinate system by $R(\theta) \approx R_2 + \delta \cos\theta$.

Because there is only a slight departure from spherical geometry, order δ, then the θ dependence of the potential will be of order δ. Since the potential has azimuthal symmetry, as in the concentric case, then

$$\Phi(r, \theta) = A_0 + \frac{B_0}{r} + \delta\left(A_1 r + \frac{B_1}{r^2}\right)\cos\theta \tag{3.29}$$

We will now use the boundary conditions to evaluate the constants A_0, A_1, B_0, and B_1.

1. The potential of the inner shell is constant [$\Phi(R_1, \theta) = \text{constant}$] but unknown at this stage. Therefore

$$A_0 + \frac{B_0}{R_1} + \delta\left(A_1 R_1 + \frac{B_1}{R_1^2}\right)\cos\theta = \text{constant}$$

Equating the coefficient of $\cos\theta$ to zero gives

$$\frac{B_1}{A_1} = -R_1^3 \tag{3.30}$$

2. The potential of the outer shell is zero [$\Phi(R, \theta) = 0$]. Therefore

$$A_0 + \frac{B_0}{(R_2 + \delta\cos\theta)} + \delta A_1\left[R_2 + \delta\cos\theta - \frac{R_1^3}{(R_2 + \delta\cos\theta)^2}\right]\cos\theta = 0$$

Neglecting terms of order δ^2 and higher, we get

$$A_0 + \frac{B_0}{R_2}\left(1 - \frac{\delta\cos\theta}{R_2}\right) + \delta A_1\left(R_2 - \frac{R_1^3}{R_2^2}\right)\cos\theta = 0$$

Equating the coefficients of cos θ gives

$$\frac{B_0}{A_1} = (R_2^3 - R_1^3) \qquad \text{and} \qquad \frac{A_0}{B_0} = -\frac{1}{R_2} \tag{3.31}$$

3. The last boundary condition states that the total charge on the inner shell is q; that is,

$$\int_{S_1} \mathbf{E}\cdot\hat{\mathbf{n}}\, da = \frac{q}{\varepsilon_0} \tag{3.32}$$

The electric field at the surface of the inner shell is normal and is given in terms of B_0 by

$$\mathbf{E} = -\left.\frac{\partial \Phi}{\partial r}\right|_{r=R_1} = \frac{B_0}{R_1^2}\,\hat{\mathbf{n}} - \delta\left(A_1 - \frac{2B_1}{R_1^3}\right)\cos\theta\,\hat{\mathbf{n}} \tag{3.33}$$

and which, upon substitution in Eq. (3.32), gives

$$B_0 = \frac{q}{4\pi\varepsilon_0} \tag{3.34}$$

Equations (3.30), (3.31), and (3.34) are now solved simultaneously for the constants A_0, B_0, A_1, and B_1, and then substituted in Eq. (3.29). The result is

$$\Phi = \frac{q}{4\pi\varepsilon_0}\left(\frac{1}{r} - \frac{1}{R_2}\right) + \frac{\delta q}{4\pi\varepsilon_0(R_2^3 - R_1^3)}\left(r - \frac{R_1^3}{r^2}\right)\cos\theta \tag{3.35}$$

Example 3.6 Angle-Dependent Charge Distribution

We will give an example in which angular dependence in the potential is caused by the introduction of angular dependence in the charge distribution on a surface boundary. Consider a shell of radius R that has a surface charge distribution $\sigma = \sigma_0 \cos\theta$, where σ_0 is a constant. Inside and outside the shell the potential satisfies Laplace's equation and therefore the potentials in both regions are represented by Eq. (3.28), as follows:

$$\begin{aligned} \Phi_1 &= \sum_{n=0}^{\infty} [A_n r^n P_n(\cos\theta) + B_n r^{-(n+1)}] P_n(\cos\theta) \qquad r < R \\ \Phi_2 &= \sum_{n=0}^{\infty} [A'_n r^n P_n(\cos\theta) + B'_n r^{-(n+1)}] P_n(\cos\theta) \qquad r > R \end{aligned} \tag{3.36}$$

The coefficients of the expansions can now be determined using the following four boundary conditions.

1. The potential Φ_1 should be finite as $r \to 0$, which requires $B_n = 0$, for $n \geq 0$.
2. The potential Φ_2 should be zero as $r \to \infty$; thus $A'_n = 0$ for $n \geq 0$.
3. The electric fields at the surface of the shell are related by

$$(\mathbf{E}_2 - \mathbf{E}_1)\cdot\hat{\mathbf{n}} = \frac{\sigma_0}{\varepsilon_0}\cos\theta \tag{3.37}$$

where $\mathbf{E}_2$ and $\mathbf{E}_1$ are the electric field vectors just outside and just inside the shell surface, and $\hat{\mathbf{n}}$ is a unit vector normal to the surface and pointing into region 2. Calculating the fields from Eqs. (3.36) and substituting them in Eq. (3.37), we obtain

$$nA_n R^{n-1} P_n(\cos\theta) + (n+1)B'_n R^{-(n+2)} P_n(\cos\theta) = \frac{\sigma_0}{\varepsilon_0}\cos\theta \tag{3.38}$$

Since the various $P_n(\cos\theta)$ are linearly independent functions, we equate their coefficients on both sides, which gives $B'_0 = 0$ and

$$A_1 + \frac{2B'_1}{R^3} = \frac{\sigma_0}{\varepsilon_0} \tag{3.39}$$

$$A_n = -\frac{(n+1)}{n} B'_n R^{-(2n+1)} \qquad n \geq 2 \tag{3.40}$$

4. The potential at the shell is continuous ($\Phi_1 = \Phi_2$). This gives $A_0 = 0$ and

$$A_1 - \frac{B_1'}{R^3} = 0, \tag{3.41}$$

$$A_n = B_n' R^{-(2n+1)} \qquad n \geq 2 \tag{3.42}$$

The simultaneous solution of Eqs. (3.39) to (3.42) yields the nonzero coefficients $A_1 = \sigma_0/3\varepsilon_0$ and $B_1' = \sigma_0 R^3/3\varepsilon_0$, and thus

$$\Phi_1(r, \theta) = \frac{\sigma_0}{3\varepsilon_0} r \cos\theta \qquad r < R \tag{3.43}$$

$$\Phi_2(r, \theta) = \frac{\sigma_0 R^3}{3\varepsilon_0} \frac{1}{r^2} \cos\theta \qquad r > R \tag{3.44}$$

It is apparent that the potential inside the shell is due to a uniform field along the negative z axis, $\mathbf{E} = -\sigma_0/3\varepsilon_0 \hat{\mathbf{z}}$. The potential outside the sphere, however, is due to an electric dipole of moment $(4\pi/3)R^3\sigma_0$ which is just the product of the volume of the shell by the maximum charge density σ_0. Finally, we note that these results are precisely what we arrived at when the problem was solved using the method of multipole expansion (see Example 2.17).

Example 3.7 A Conducting Sphere in an Electric Field

A conducting sphere carrying a charge Q is placed in an electric field that is initially *uniform* and along the z axis, $\mathbf{E} = E_0 \hat{\mathbf{z}}$, in the absence of the sphere or far away from it, as shown in Fig. 3.6*a*. The electrostatic potential associated with the electric field at far distances is

$$\Phi = -\int \mathbf{E} \cdot d\mathbf{r} = -E_0 z + C = -E_0 r \cos\theta + C \tag{3.45}$$

where C is a constant. This condition constitutes an angle-dependent boundary condition on the potential of the combined problem (sphere + electric field); it introduces θ dependence in the potential at arbitrary r. However, there is still symmetry about the z axis, and therefore the potential will be a function of r and θ only. Another boundary condition in this problem is the fact that the potential at the surface and inside the sphere is constant since the electric field in a conductor is zero.

Outside the sphere the potential satisfies Laplace's equation and thus is given by

$$\Phi(r, \theta) = \sum_{n=0}^{\infty} [A_n r^n + B_n r^{-(n+1)}] P_n(\cos\theta) \tag{3.46}$$

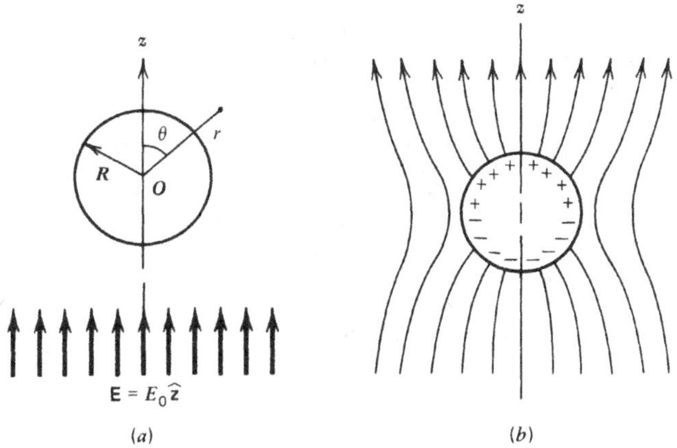

Figure 3.6 Conducting sphere in an initially uniform electric field. (*a*) Schematic diagram of the sphere and the lines of force in the absence of the sphere. (*b*) Lines of force in the presence of the sphere.

Using the boundary condition of Eq. (3.45) derived above for $r \to \infty$ gives:

$$\sum_{n=0}^{\infty} A_n r^n P_n(\cos\theta) = -E_0 r \cos\theta + C \tag{3.47}$$

Equating coefficients of $P_n(\cos\theta)$ in this equation gives

$$A_1 = -E_0,\ A_0 = C, \text{ and } A_n = 0 \qquad \text{for } n \geq 2 \tag{3.48}$$

Substituting these values into Eq. (3.46) and using the boundary condition $\Phi(r, \theta) = \Phi_0 = \text{constant}$ at $r = R$ gives

$$\Phi(r, \theta) = \Phi_0 = \sum_{n=0}^{\infty} B_n R^{-(n+1)} P_n(\cos\theta) + C - E_0 R \cos\theta \tag{3.49}$$

Equating the coefficients of $P_n(\cos\theta)$ on both sides of this equation gives

$$C + \frac{B_0}{R} = \Phi_0,\ B_1 = E_0 R^3, \text{ and } B_n = 0 \qquad \text{for } n \geq 2$$

Thus

$$\Phi(r, \theta) = \Phi_0 + B_0\left(\frac{1}{r} - \frac{1}{R}\right) - E_0 r \cos\theta + \frac{E_0 R^3}{r^2}\cos\theta \tag{3.50}$$

The third boundary condition is the statement about the total charge on the sphere, which will now be used to evaluate B_0. First, one calculates the electric field on the surface by taking the normal derivative of Φ: $E = -\partial\Phi/\partial r$ evaluated at $r = R$. The surface charge density, $\sigma = \varepsilon_0 E$ is then calculated.

$$\sigma = \frac{B_0 \varepsilon_0}{R^2} + 3\varepsilon_0 E_0 \cos\theta \tag{3.51}$$

Thus

$$Q = \int \sigma\, da = \frac{\varepsilon_0 B_0}{R^2}\int da + 3\varepsilon_0 E_0 \int \cos\theta\, da$$

The second integral vanishes and the integral $\int da$ gives $4\pi R^2$, yielding $B_0 = Q/4\pi\varepsilon_0$. Hence

$$\Phi(r, \theta) = \Phi_0 + \frac{Q}{4\pi\varepsilon_0}\left(\frac{1}{r} - \frac{1}{R}\right) - E_0 r \cos\theta + \frac{E_0 R^3}{r^2}\cos\theta \tag{3.52}$$

This result shows that the potential consists of four terms. The first and third terms are associated with the external field $E_0\hat{\mathbf{z}}$. The second and the fourth terms are the result of the introduction of the sphere in the external field. The second is only present when the sphere has a net charge Q, whereas the fourth is a dipole field produced by a dipole moment equal to $4\pi R^3 \varepsilon_0 E_0 \hat{\mathbf{z}}$, indicating that the sphere has been polarized by the external field. Figure 3.6*b* gives the lines of force of the electric field in the presence of the sphere with no net charge.

We would like now to present a very simple idea for arriving at the above results. Consider the case where the sphere does not have a net charge. The net field inside the sphere must be zero. Therefore the external field E_0 must be annulled by the field produced by charges induced on the conducting surface. Since the previous example showed that we could produce a uniform field inside a sphere by placing a charge density $\sigma = \sigma_0 \cos\theta$ on its surface, it is easy to annul the field $\mathbf{E}$ by choosing σ_0 so that

$$E_0\hat{\mathbf{z}} - \frac{\sigma_0 \hat{\mathbf{z}}}{3\varepsilon_0} = 0$$

that is,

$$\sigma_0 = 3\varepsilon_0 E_0$$

Consequently, $\sigma = 3\varepsilon_0 E_0 \cos\theta$. Now, external to the sphere, such a charge density produces a dipole potential that is superimposed upon the potential associated with the field E_0. Also, from the previous example, the potential function inside the sphere is a constant. Outside the sphere,

$$\Phi(r > R) = -\mathbf{E}_0 \hat{\mathbf{z}} + \frac{R^3}{r^2} E_0 \cos\theta = -E_0 r \cos\theta\left(1 - \frac{R^3}{r^3}\right)$$

Clearly, at $r = R$, $\Phi = 0$; that is, the potential of the sphere is zero.

Example 3.8 Angle-Dependent Potential Boundary

Two concentric spheres have radii R_1 and R_2 ($R_2 > R_1$). The potential everywhere on the surface of the smaller sphere is zero. On the surface of the larger sphere the potential is given by

$$V(R_2, \theta) = V_0 \cos\theta \tag{3.53}$$

where V_0 is a constant.

This angle-dependent potential boundary introduces angular dependence in the potential between the spheres, whose explicit form can be found by solving Laplace's equation in this region. Thus, between the spheres, Φ is given by the following expansion [see Eq. (3.28)].

$$\Phi(r, \theta) = \sum_{n=0}^{\infty} [A_n r^n + B_n r^{-(n+1)}] P_n(\cos\theta)$$

Note that the present region of interest does not include $r = 0$ or $r = \infty$, as some of the previous cases did. The boundary condition $\Phi(R_1, \theta) = 0$ gives

$$A_n + B_n R_1{}^{-(2n+1)} = 0 \qquad \text{for all } n \tag{3.54}$$

and the boundary condition $\Phi(R_2, \theta) = V_0 \cos\theta$ gives

$$A_1 + \frac{B_1}{R_2^3} = \frac{V_0}{R_2}, \qquad A_n + B_n R_2^{-(2n+1)} = 0 \qquad \text{for } n \neq 1 \tag{3.55}$$

Equations (3.54) and (3.55) are now solved simultaneously; they yield

$$A_n = B_n = 0 \qquad \text{for } n \neq 1$$

and

$$A_1 = \frac{V_0 R_2^2}{R_2^3 - R_1^3} \qquad B_1 = -\frac{V_0 R_2^2 R_1^3}{R_2^3 - R_1^3} \tag{3.56}$$

Therefore

$$\Phi(r, \theta) = \frac{V_0 R_2^2}{R_2^3 - R_1^3}\left(r - \frac{R_1^3}{r^2}\right)\cos\theta \tag{3.57}$$

The electric field at the surface of the inner sphere can be calculated from the relation $\mathbf{E} = -\nabla\Phi$:

$$\mathbf{E}(R_1, \theta) = \frac{3V_0 R_2^2}{R_2^3 - R_1^3}\hat{\mathbf{r}} \cos\theta \tag{3.58}$$

which shows that it is purely normal indicating that the inner sphere is a conductor as it should be. We leave the determination of the potential and the electric field outside the larger sphere as an exercise.

3.4.3 Laplace's Equation in Two Dimensions—Cylindrical Coordinates

We now turn our attention to boundary value problems where the geometrical configuration is cylindrical in nature, and where the potential is a function of more than one coordinate. Here we will consider potentials that are functions of ρ and ϕ

only. Such potentials arise in cases where there is a symmetry along the z axis. In regions excluding point charges, the potential satisfies the equation

$$\frac{1}{\rho}\frac{\partial}{\partial\rho}\left(\rho\frac{\partial\Phi}{\partial\rho}\right)+\frac{1}{\rho^2}\frac{\partial^2\Phi}{\partial\phi^2}=0 \tag{3.59}$$

The method of separation of variables used above to solve for the potential in spherical coordinates applies here. We write Φ as the product of two functions, $\Phi = R(\rho)Y(\phi)$, and substitute it into Eq. (3.59):

$$\frac{\rho}{R}\frac{d}{d\rho}\left(\rho\frac{dR}{d\rho}\right)=-\frac{1}{Y}\frac{d^2Y}{d\phi^2} \tag{3.60}$$

Both sides of the equation will be taken equal to K^2, which is the separation constant. The equation for Y,

$$\frac{d^2Y}{d\phi^2}+K^2Y=0 \tag{3.61}$$

has the solutions $\cos K\phi$ and $\sin K\phi$. The magnitude of K has to be restricted in order to make these solutions single-valued functions of ϕ. Or, in other words, for the solution to make sense physically it should be the same after a rotation of 2π, or

$$\cos K(\phi+2\pi)=\cos K\phi \qquad \text{and} \qquad \sin K(\phi+2\pi)=\sin K\phi \tag{3.62}$$

which requires that $K = n$, where n is zero or a positive integer. Dropping the negative integers will not result in neglecting any possible solutions, since $\cos(-n\phi)$ is identical to $\cos(n\phi)$ and $\sin(-n\phi) = -\sin(n\phi)$. An important property of these solutions is the fact that they are orthogonal; that is,

$$\begin{gathered}\int_0^{2\pi}\cos(m\phi)\cos(n\phi)d\phi=\int_0^{2\pi}\sin(m\phi)\sin(n\phi)d\phi=\pi\delta_{mn}\\ \int_0^{2\pi}\sin(m\phi)\cos(n\phi)d\phi=0\end{gathered} \tag{3.63}$$

where δ_{mn} is the Kronecker delta introduced in Eq. (3.24).

The radial dependence of the potential can now be obtained. Setting the left-hand side of Eq. (3.60) equal to $K^2 = n^2$, we get:

$$\frac{d}{d\rho}\left(\rho\frac{dR}{d\rho}\right)-\frac{n^2R}{\rho}=0 \tag{3.64}$$

For $n = 0$, the potential satisfies the same equation that we encountered in the case where the potential has no angular dependence [see Eq. (3.13)]; namely,

$$\frac{d}{d\rho}\left(\rho\frac{dR}{d\rho}\right)=0$$

which has the solutions $R(\rho) = \text{constant}$ and $R(\rho) = \ln\rho$. For $n \neq 0$ the equation has the two solutions ρ^n and ρ^{-n}. Therefore the most general solution is

$$\begin{aligned}\Phi=&\sum_{n=1}^{\infty}(A_n\cos(n\phi)+B_n\sin(n\phi))\rho^n\\ &+\sum_{n=1}^{\infty}(A_n'\cos(n\phi)+B_n'\sin(n\phi))\rho^{-n}+A_0+A_0'\ln\rho\end{aligned} \tag{3.65}$$

where A_n, A'_n, B_n, and B'_n, for $n \geq 0$, are constants to be evaluated from the boundary conditions. It is to be noted that this result has more types of expansions than the corresponding expansions in spherical coordinates. Moreover, the solution of the angular equation in spherical coordinates involves Legendre polynomials, while this expansion involves the harmonics of $\sin\phi$ and $\cos\phi$. Also, the radial dependence of this expansion contains a $\ln\rho$ term. Below we consider some applications of the cylindrical expansion in order to explain further the nature of these terms.

*Example 3.9 The Nonaxial Cylindrical Capacitor

Consider a cylindrical capacitor. The cylinders are displaced so that there is a distance δ between the axes of the cylinders, as shown in Fig. 3.7. The inner shell of radius ρ_1 is kept at potential V_1 and the outer shell of radius ρ_2 is kept at potential V_2. Further, we take $\delta \ll \rho_1$ in order to simplify the problem.

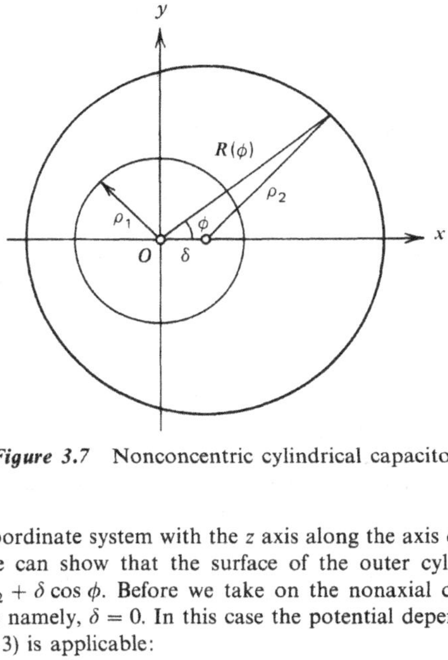

Figure 3.7 Nonconcentric cylindrical capacitor.

Take a cylindrical coordinate system with the z axis along the axis of the inner shell. From the law of cosines one can show that the surface of the outer cylinder is approximately described by $R(\phi) = \rho_2 + \delta\cos\phi$. Before we take on the nonaxial case, it is instructive to consider the axial case: namely, $\delta = 0$. In this case the potential depends only on ρ and thus the solution of Eq. (3.13) is applicable:

$$\Phi(\rho) = A_0 + A'_0 \ln\rho \tag{3.66}$$

With $\Phi(\rho_1) = V_1 = A_0 + A'_0\ln\rho_1$ and $V_2 = A_0 + A'_0\ln\rho_2$, the constants A_0 and A'_0, and hence Φ, take the following values:

$$\begin{aligned} A_0 &= V_1 - \frac{V_2 - V_1}{\ln(\rho_2/\rho_1)}\ln\rho_1 \\ A'_0 &= \frac{V_2 - V_1}{\ln(\rho_2/\rho_1)} \\ \Phi(\rho) &= V_1 + \frac{V_2 - V_1}{\ln(\rho_2/\rho_1)}\ln\left(\frac{\rho}{\rho_1}\right) \end{aligned} \tag{3.67}$$

It can be easily shown that the term proportional to $\ln \rho$ is due to a charge density. If we calculate $\mathbf{E} = -\nabla\Phi = -\hat{\rho}(\partial\Phi/\partial\rho)$ at $\rho = \rho_1$ and then use Gauss' law, we obtain:

$$\sigma_1 = \frac{1}{\rho_1}\varepsilon_0 \frac{(V_1 - V_2)}{\ln(\rho_2/\rho_1)}$$

This solution is modified when δ is not zero. A first-order correction of the order δ can be written using Eq. (3.65); that is,

$$\Phi(\rho, \phi) = A_0 + A_0' \ln \rho + \delta\rho(A_1 \cos\phi + B_1 \sin\phi) + \frac{\delta}{\rho}(A_1' \cos\phi + B_1' \sin\phi) \quad (3.68)$$

We now apply the boundary conditions.

1. At $\rho = \rho_1$, $\Phi(\rho_1, \phi) = V_1$. This gives

$$V_1 = A_0 + A_0' \ln \rho_1 + \delta\rho_1(A_1 \cos\phi + B_1 \sin\phi) + \frac{\delta}{\rho_1}(A_1' \cos\phi + B_1' \sin\phi)$$

Since the sine and cosine functions are linearly independent, then we equate on both sides of the equation the coefficients of $\cos\phi$ and also the coefficients of $\sin\phi$. This yields:

$$A_1 = -\frac{A_1'}{\rho_1^2} \qquad B_1 = -\frac{B_1'}{\rho_1^2} \qquad V_1 = A_0 + A_0' \ln \rho_1 \quad (3.69)$$

2. At $\rho = R(\phi) = \rho_2 + \delta\cos\phi$, $\Phi(R, \phi) = V_2$. This gives

$$A_0 + A_0' \ln \rho_2 + \frac{A_0'\,\delta\cos\phi}{\rho_2} + \delta\rho_2(A_1 \cos\phi + B_1 \sin\phi) + \frac{\delta}{\rho_2}(A_1' \cos\phi + B_1' \sin\phi) = V_2$$

where $1/(\rho_2 + \delta\cos\phi)$ was expanded as $(1 - \delta\cos\phi/\rho_2)/\rho_2$ and $\ln(\rho_2 + \delta\cos\phi)$ as $\ln\rho_2 + \delta\cos\phi/\rho_2$, and terms of order δ^2 and higher were neglected. Equating coefficients of $\cos\phi$ and $\sin\phi$, we get

$$A_1 = -\frac{1}{\rho_2^2}(A_1' + A_0) \qquad B_1 = -\frac{B_1'}{\rho_2^2} \qquad V_2 = A_0 + A_0' \ln \rho_2 \quad (3.70)$$

Examining Eqs. (3.69) and (3.70) shows that the relations between A_0 and A_0' are the same ones derived in the coaxial case, and therefore they take the same values calculated above. The relations between B_1 and B_1' derived from both boundary conditions can be satisfied only when $B_1 = B_1' = 0$. On the other hand, there is a nonzero solution for A_1 and A_1':

$$A_1 = -\frac{A_0}{\rho_2^2 - \rho_1^2} \qquad A_1' = \frac{\rho_1^2}{\rho_2^2 - \rho_1^2} A_0$$

Therefore the potential is

$$\Phi(\rho, \phi) = V_1 + \frac{V_2 - V_1}{\ln(\rho_2/\rho_1)} \ln\left(\frac{\rho}{\rho_1}\right) - \frac{\delta}{\rho_2^2 - \rho_1^2}\left[V_1 - \frac{V_2 - V_1}{\ln(\rho_2/\rho_1)}\right]\left[\rho - \frac{\rho_1^2}{\rho}\right]\cos\phi \quad (3.71)$$

The fact that the potential depends only on the harmonic $\cos\phi$, which is the same harmonic involved in the geometry $R = \rho_2 + \delta\cos\phi$ indicates that all the other harmonics could have been dropped early in the solution. Consider the following special case of the above results. When the two cylinders are kept at the same potential (namely, $V_1 = V_2 = V$), then Eq. (3.71) becomes

$$\Phi(\rho, \phi) = V - \frac{\delta V}{\rho_2^2 - \rho_1^2}\left(\rho - \frac{\rho_1^2}{\rho}\right)\cos\phi \quad (3.72)$$

In the coaxial case, $\delta = 0$; hence Eq. (3.72) yields $\Phi(\rho, \phi) = V$. Equation (3.68) shows that the corresponding charge on the inner cylinder is zero, as expected. In the nonaxial case we find it has the density

$$\sigma_1 = \frac{-2\delta V \varepsilon_0}{(\rho_2^2 - \rho_1^2)} \cos\phi \qquad (3.73)$$

Example 3.10 An Angular Charge Distribution

This example involves an angular charge distribution held at a cylindrical surface. Consider a long cylinder of radius ρ_0 with a surface charge density $\sigma = \sigma_1 \sin 2\phi + \sigma_2 \cos\phi$, where ϕ is the angle measured from the x axis and σ_1 and σ_2 are constants.

In the regions outside and inside the cylinder, the electrostatic potential satisfies Laplace's equation. Moreover, the expansion for the inside region should not include ρ^{-n} and $\ln\rho$ terms since the potential at the center of the cylinder should be finite. On the other hand, the expansion for the outside region should not include ρ^n terms. Therefore, the potentials inside and outside the cylinder, Φ_1 and Φ_2, respectively, are given by the following expansions:

$$\Phi_1(\rho, \phi) = A_0 + \sum_{n=1}^{\infty} (A_n \cos n\phi + B_n \sin n\phi)\rho^n \qquad \rho < \rho_0 \qquad (3.74)$$

$$\Phi_2(\rho, \phi) = A'_0 \ln\rho + \sum_{n=1}^{\infty} (A'_n \cos n\phi + B'_n \sin n\phi)\rho^{-n} \qquad \rho > \rho_0 \qquad (3.75)$$

As was discussed in Example 3.9, the term $A'_0 \ln\rho$ arises when the total charge on the cylinder is nonzero. In this particular example the total charge Q on the cylinder is

$$Q = \int \sigma\, da = \int_0^{2\pi} L\sigma \rho_0\, d\phi$$

where L is the length of the cylinder. The integration gives

$$Q = \left[\frac{1}{2} L\sigma_1\rho_0 \cos 2\phi + L\sigma_2\rho_0 \sin\phi\right]_0^{2\pi} = 0$$

Therefore the constant A'_0 equals zero. Since potentials are arbitrary within a constant, one can pick $A_0 = 0$ to make $\Phi = 0$ at $\rho = 0$.

The rest of the constants can now be evaluated from the boundary conditions at the surface of the cylinder:

1. At $\rho = \rho_0$, the potential is continuous: $\Phi_1(\rho_0, \phi) = \Phi_2(\rho_0, \phi)$ or

$$\sum_{n=1}^{\infty} (A_n \cos n\phi + B_n \sin n\phi)\rho_0^n = \sum_{n=1}^{\infty} (A'_n \cos n\phi + B'_n \sin n\phi)\rho_0^{-n} \qquad (3.76)$$

Equating coefficients of $\cos n\phi$ and coefficients of $\sin n\phi$ on both sides gives

$$A_n = \rho_0^{-2n} A'_n \qquad \text{and} \qquad B_n = \rho_0^{-2n} B'_n \qquad \text{for } n \geq 1 \qquad (3.77)$$

2. At $\rho = \rho_0$, the normal components of the fields satisfy the relation

$$(\mathbf{E}_2 - \mathbf{E}_1)\cdot\hat{\boldsymbol{\rho}} = \frac{\sigma}{\varepsilon_0} \qquad (3.5)$$

where $(\mathbf{E}_i)\cdot\hat{\boldsymbol{\rho}} = -\partial\Phi_i/\partial\rho$. Substituting for Φ_i from Eqs. (3.74) and (3.75) yields

$$\sum_{n=1}^{\infty} (A_n \cos n\phi + B_n \sin n\phi) n\rho_0^{n-1} + \sum_{n=1}^{\infty} (A'_n \cos n\phi + B'_n \sin n\phi) n\rho_0^{-(n+1)}$$

$$= \frac{\sigma_1}{\varepsilon_0} \sin 2\phi + \frac{\sigma_2}{\varepsilon_0} \cos\phi$$

Equating coefficients of cos $n\phi$ and coefficients of sin $n\phi$ on both sides of this equation gives

$$\begin{aligned} A_1 + \frac{A_1'}{\rho_0^2} &= \frac{\sigma_2}{\varepsilon_0} \quad \text{and} \quad A_n = -\rho_0^{-2n}A_n' \quad \text{for } n \geq 2 \\ B_2 + \frac{B_2'}{\rho_0^4} &= \frac{\sigma_1}{2\varepsilon_0\rho_0} \quad \text{and} \quad B_n = -\rho_0^{-2n}B_n' \quad \text{for } n \neq 2 \end{aligned} \tag{3.78}$$

Solving the relations between the coefficients in Eqs. (3.77) and (3.78) yields the following nonzero coefficients:

$$A_1 = \frac{\sigma_2}{2\varepsilon_0} \qquad B_2 = \frac{\sigma_1}{4\varepsilon_0\rho_0} \qquad A_1' = \frac{\sigma_2}{2\varepsilon_0}\rho_0^2 \qquad B_2' = \frac{\sigma_1\rho_0^3}{4\varepsilon_0}$$

Therefore the potentials are given by

$$\Phi_1(\rho, \phi) = \frac{\sigma_2}{2\varepsilon_0}\rho\cos\phi + \frac{\sigma_1}{4\varepsilon_0\rho_0}\rho^2\sin 2\phi \tag{3.79}$$

$$\Phi_2(\rho, \phi) = \frac{\sigma_2}{2\varepsilon_0}\frac{\rho_0^2}{\rho}\cos\phi + \frac{\sigma_1\rho_0^3}{4\varepsilon_0}\frac{\sin 2\phi}{\rho^2} \tag{3.80}$$

The fact that the determined potentials depend only on the harmonics cos ϕ and sin 2ϕ, which are the same harmonics involved in the given charge distribution, indicates that all the other harmonics could have been dropped early in the solution—that is, in Eqs. (3.74) and (3.75). This of course would simplify the algebra.

3.4.4 Laplace's Equation in Three Dimensions—Rectangular Coordinates

Laplace's equation in rectangular coordinates,

$$\frac{\partial^2\Phi}{\partial x^2} + \frac{\partial^2\Phi}{\partial y^2} + \frac{\partial^2\Phi}{\partial z^2} = 0 \tag{3.81}$$

can also be solved by the method of separation of variables. We take Φ to be equal to the product $F_1(x)F_2(y)F_3(z)$. Upon substitution of the product we get

$$\frac{1}{F_3(z)}\frac{d^2F_3(z)}{dz^2} + \frac{1}{F_2(y)}\frac{d^2F_2(y)}{dy^2} = -\frac{1}{F_1(x)}\frac{d^2F(x)}{dx^2} \tag{3.82}$$

Both sides of the equation are taken to be equal to a separation constant α^2, yielding

$$\frac{d^2F_1(x)}{dx^2} + \alpha^2F_1(x) = 0 \tag{3.83}$$

which has, for $\alpha^2 > 0$, cos αx and sin αx solutions, and

$$\frac{1}{F_2(y)}\frac{d^2F_2(y)}{dy^2} = \alpha^2 - \frac{1}{F_3(z)}\frac{d^2F_3(z)}{dz^2} \tag{3.84}$$

Again the x and the y dependence can be separated by equating both sides of Eq. (3.84) to a second separation constant, $-\beta^2$. This gives

$$\frac{d^2F_2}{dy^2} + \beta^2F_2 = 0 \tag{3.85}$$

$$\frac{d^2F_3}{dz^2} - (\alpha^2 + \beta^2)F_3 = 0 \tag{3.86}$$

The equation for F_2 is similar to that for F_1 and therefore, for $\beta^2 > 0$, has $\cos \beta y$ and $\sin \beta y$ solutions. On the other hand, the equation for F_3 has $\sinh \gamma z$ and $\cosh \gamma z$ solutions, where $\gamma = (\alpha^2 + \beta^2)^{1/2}$.

The nature of the separation constants are not known at this point. However, the physical boundary conditions can be satisfied if one takes α and β to be proportional to integers m and n, respectively. So we will take α and β to be proportional to integers and thus the most general solution consists of eight infinite expansions, with the following combinations: $\sinh \gamma z \sin \alpha x \cos \beta y$, $\sinh \gamma z \sin \alpha x \sin \beta y$, $\sinh \gamma z \cos \alpha x \cos \beta y$, $\sinh \gamma z \cos \alpha x \sin \beta y$, and the other four possibilities are the above ones with $\sinh \gamma z$ replaced by $\cosh \gamma z$.

A special case arises when the separation constants are zeros. In this case Eqs. (3.83), (3.85), and (3.86) give $F_1 = a_1 x + b_1$, $F_2 = a_2 y + b_2$ and $F_3 = a_3 z + b_3$, where a_i and b_i are constants, and thus the most general solution is

$$\Phi = A_1 xyz + A_2 xy + A_3 xz + A_4 yz + A_5 x + A_6 y + A_7 z + A_8 \tag{3.87}$$

where A_1 to A_8 are constants.

Below are two examples of the boundary value problem in cartesian coordinates.

Example 3.11 Conducting Box

Consider a box of dimensions a, b, and c, as shown in Fig. 3.8. The top face at $z = c$ is kept at a voltage $V_1(x, y)$, and the other faces are isolated from it and kept at zero potential. Because the potential on the faces passing through the origin is zero, then only one of the eight expansions will satisfy this condition—namely, $\sinh \gamma z \sin \alpha x \sin \beta y$.

The separation constants α and β can now be restricted by invoking the conditions at $x = a$ and $y = b$. For the potential to be zero at these planes it is required that $\sin \alpha a = \sin \beta b = 0$, which yields $\alpha = n\pi/a$ and $\beta = m\pi/b$, where m and n are positive integers. Therefore, for the potential inside the box, we write

$$\Phi(x, y, z) = \sum_{n=1}^{\infty} \sum_{m=1}^{\infty} A_{mn} \sinh \gamma_{mn} z \sin \frac{n\pi x}{a} \sin \frac{m\pi y}{b} \tag{3.88}$$

where $\gamma_{mn} = \pi(n^2/a^2 + m^2/b^2)^{1/2}$ and A_{mn} are constants to be evaluated from the rest of the

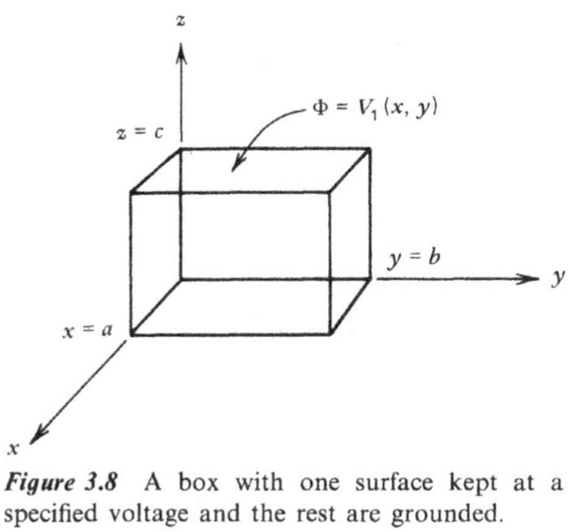

Figure 3.8 A box with one surface kept at a specified voltage and the rest are grounded.

boundary conditions. The last boundary condition is at the top face of the box: $\Phi(x, y, c) = V_1(x, y)$ is a given function; thus,

$$V_1(x, y) = \sum_{n=1}^{\infty} \sum_{m=1}^{\infty} A_{mn} \sinh \gamma_{mn} c \sin \frac{n\pi x}{a} \sin \frac{m\pi y}{b} \tag{3.89}$$

which is a *double Fourier series.* Since the sine functions are linearly independent [see Eq. (3.63)], then multiplying both of its sides by $\sin(n\pi x/a)\sin(m\pi y/b)$ and integrating over x and y from 0 to a and 0 to b, respectively, we get

$$A_{mn} = \frac{4}{ab \sinh \gamma_{mn} c} \int_0^a dx \int_0^b dy\, V_1(x, y) \sin \frac{n\pi x}{a} \sin \frac{m\pi y}{b} \tag{3.90}$$

In the case $V_1(x, y) = V_0 =$ a constant, then $A_{mn} = 16V_0/(mn\pi^2 \sinh \gamma_{mn} c)$, where m and n are odd integers, and therefore

$$\Phi(x, y, z) = \sum_{n=1,3,\ldots} \sum_{m=1,3,\ldots} \frac{16V_0}{mn\pi^2} \frac{\sinh \gamma_{mn} z}{\sinh \gamma_{mn} c} \sin \frac{n\pi x}{a} \sin \frac{m\pi y}{b} \tag{3.91}$$

We consider now the case where two faces are kept at specified nonzero potentials. Let $\Phi(x, b, z) = V_2(x, z)$ in addition to $\Phi(x, y, c) = V_1(x, y)$. The rest of the faces are kept at zero potential. The potential can be determined by superimposing the potential for the case where all the faces are grounded except the $y = b$ face, and the solution derived above, Eq. (3.91). The solution for the ungrounded $y = b$ face can be determined by the same procedure used to arrive at Eq. (3.91) with the roles of z and y are interchanged. We finally note that the method of superposition can be generalized to include more general boundary conditions.

Example 3.12 Three Intersecting Planes

Consider three conducting planes intersecting at right angles, as shown in Fig. 3.9. The planes are kept at potential V_0. Because this volume is not enclosed, the potential is described by the expansion in the case where the separation constants are zeros. Using Eq. (3.87) and the boundary conditions at $x = 0$, $y = 0$, and $z = 0$ gives

$$V_0 = A_4 yz + A_6 y + A_7 z + A_8 \tag{3.92}$$

$$V_0 = A_3 xz + A_5 x + A_7 z + A_8 \tag{3.93}$$

$$V_0 = A_2 xy + A_5 x + A_6 y + A_8 \tag{3.94}$$

These relations give $A_8 = V_0$ and $A_2 = A_3 = A_4 = A_5 = A_6 = A_7 = 0$. Thus

$$\Phi(x, y, z) = A_1 xyz + V_0 \tag{3.95}$$

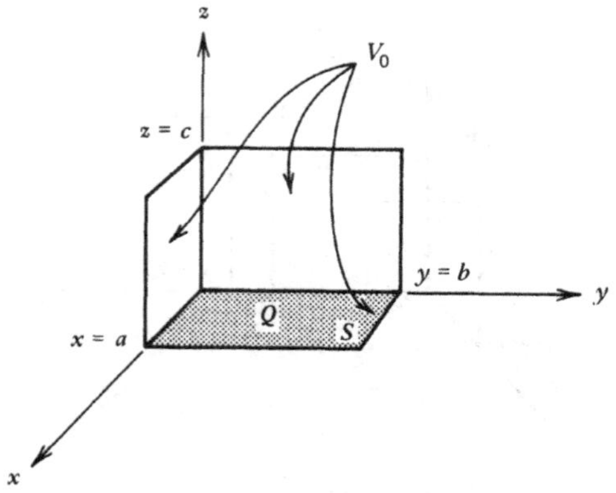

Figure 3.9 Half of a box with the surfaces kept at potential V_0.

The knowledge of the total charge on one of the planes allows the evaluation of A_1. Let us take the total charge on the x-y plane to be Q and its area to be S. Therefore

$$Q = \int_S \mathbf{E} \cdot \hat{\mathbf{n}}\, da = -A_1 \int xy\, dx\, dy = -\frac{1}{4} A_1 S^2 \tag{3.96}$$

or $A_1 = -Q/4S^2$, and hence

$$\Phi(x, y, z) = -\frac{1}{4}\frac{Q}{S} xyz + V_0 \tag{3.97}$$

3.5 The Method of Images

We now relax the condition employed above, which assumes that all charges exist on the surface of conductors, and consider electrostatic problems where the charge density is not zero in the space not occupied by conductors. However, only point charges and line charges will be considered in detail. The generalization to continuous charge distributions can then easily be made. We note that these problems are not best treated by the boundary value problem methods developed above. To solve these problems we introduce the very powerful method of images; in fact, it is an impressive illustration of the power of application of the uniqueness theorems discussed above in Section 3.2, Uniqueness of Solutions to Electrostatic Problems. To find an electrostatic solution for these problems, we must require that the solution (1) satisfies the equipotential conditions only at the conductor surfaces bounding the region and (2) satisfies Laplace's or Poisson's equation everywhere in the region. Once a solution has been obtained, we then know it is the true solution. If charges reside in the space outside the conducting surfaces, the requirement that Poisson's equation be satisfied is equivalent to the statement that a portion of the solution must be that due to the charges in the space. The remainder of the solution will "force" the correct boundary condition at the conductor surfaces. This remainder, or other part of the solution, will in fact arise from the charge distribution on the conductor surfaces. However, it turns out that often simpler, "unreal," charge distributions *outside* of the region where a solution is sought will provide the correct boundary conditions on the surfaces of interest.

3.5.1 Point Charge and Plane

We illustrate these points by considering a *point charge*, q, at distance z' above a large, plane, conducting sheet, which effectively may be considered infinite in extent (see Fig. 3.10). Since this sheet is grounded, its potential is zero.* We wish to find the potential and electric field in the space containing q. In the figure this is the region of space $z \geq 0$. We locate the origin just beneath the charge q, on the surface of the conductor.

Cylindrical coordinates are appropriate for this problem since there is clearly symmetry about the z axis. We note for this problem that realizing a zero potential for the plane $z = 0$ is simply achieved by adding to the potential due to q, a potential due to an *imaginary image* charge, $-q$, located a distance z' beneath the plane $z = 0$. Consequently the potential and electric field for $z \geq 0$ is simply that

* For the problem at hand this is not essential; rather it is a convenient choice. If we had left the potential to be arbitrary, it would simply mean that the potential of the point charge at infinity was chosen to be nonzero also.

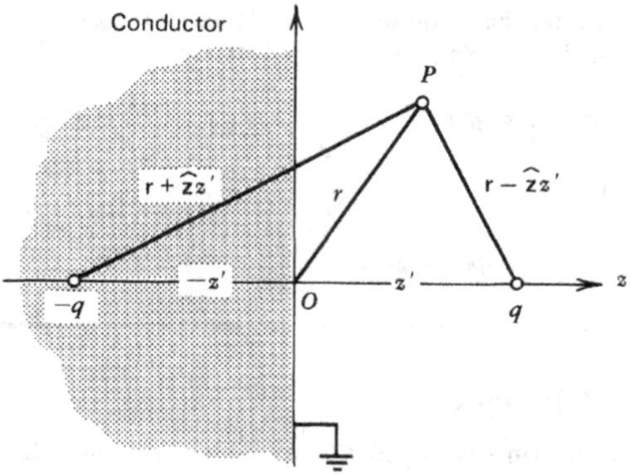

Figure 3.10 A point charge q near a large, grounded conducting plate. Also shown is an image charge $-q$ whose field when added to that of the real one gives the correct field.

due to two point charges q and $-q$ separated by a distance $2z'$. We examine this solution. It may be written

$$\Phi(\mathbf{r}) = \frac{1}{4\pi\varepsilon_0} \cdot \left(\frac{q}{|\mathbf{r} - \hat{\mathbf{z}}z'|} + \frac{-q}{|\mathbf{r} + \hat{\mathbf{z}}z'|} \right) \qquad z \geq 0 \tag{3.98}$$

In cylindrical coordinates, $\mathbf{r} = \rho\hat{\boldsymbol{\rho}} + z\hat{\mathbf{z}}$, and $|\mathbf{r} \pm z'\hat{\mathbf{z}}|^2 = \rho^2 + (z \pm z')^2$. The electric field may be obtained from the expression $\mathbf{E}(\mathbf{r}) = -\nabla\Phi(\mathbf{r})$ or, more directly, from Coulomb's formula for two point charges:

$$\mathbf{E}(r) = \frac{1}{4\pi\varepsilon_0} \left[\frac{q(\mathbf{r} - z'\hat{\mathbf{z}})}{|\mathbf{r} - z'\hat{\mathbf{z}}|^3} - \frac{q(\mathbf{r} + z'\hat{\mathbf{z}})}{|\mathbf{r} + z'\hat{\mathbf{z}}|^3} \right] \qquad z \geq 0 \tag{3.99}$$

For $z = 0$, $\mathbf{r} = \rho\hat{\boldsymbol{\rho}}$; thus we get

$$\mathbf{E}(\rho) = \frac{q}{4\pi\varepsilon_0} \left[\frac{\rho\hat{\boldsymbol{\rho}} - z'\hat{\mathbf{z}}}{(\rho^2 + z'^2)^{3/2}} - \frac{\rho\hat{\boldsymbol{\rho}} + z'\hat{\mathbf{z}}}{(\rho^2 + z'^2)^{3/2}} \right] = \frac{q}{4\pi\varepsilon_0} \left[\frac{-2z'\hat{\mathbf{z}}}{(\rho^2 + z'^2)^{3/2}} \right] \tag{3.100}$$

From this result, we can obtain the actual surface charge density on the $z = 0$ face of the conductor. It is given by

$$\sigma = \varepsilon_0(\mathbf{E} \cdot \hat{\mathbf{z}})_{z=0} = \frac{-q}{2\pi} \frac{z'}{(\rho^2 + z'^2)^{3/2}} \tag{3.101}$$

The induced charge, as expected, is negative. It has its maximum value at $\rho = 0$, and falls off as $1/\rho^3$ as ρ becomes large compared to z' (Fig. 3.11). It is this induced charge plus the original charge q that produces the actual solution, even though we have imagined the solution to be due to q and its "image charge," $-q$. One may note that the unique solution in a given region of space may be given by nonunique charge distributions outside this region. Only when the charge distribution is restricted to the physical boundaries of this region is the charge distribution unique.

It is easy to verify that the total induced surface charge Q is equal to $-q$, implying that all the lines of force terminate on the conductor. Integrating the charge density over the area of the conducting plane gives

$$Q = \int 2\pi\sigma\rho \, d\rho = -qz' \int_{\rho=0}^{\infty} \frac{\rho \, d\rho}{[\rho^2 + (z')^2]^{3/2}} = -q$$

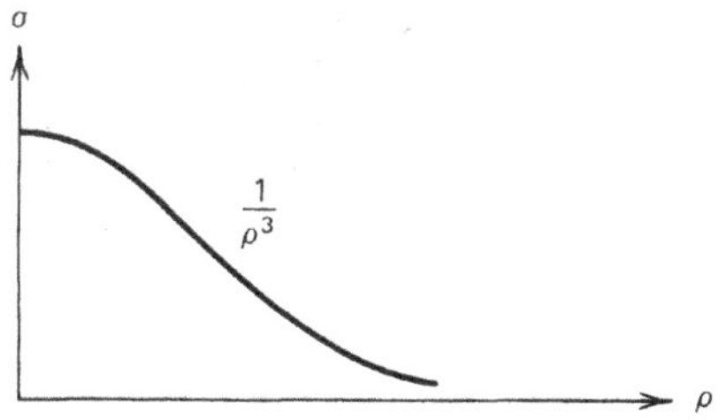

Figure 3.11 The induced surface charge density on a grounded conducting plate due to the presence of a point charge nearby.

The charge q experiences a force that attracts it to the conductor surface. The trajectory of a charged particle near a conducting plane surface is shown in Fig. 3.12. The force has a magnitude determined by the electric field of the induced surface charge. Since this field is identical with the field of the image charge, the force is easily obtained as the force between two point charges:

$$\mathbf{F} = -\hat{\mathbf{z}}\frac{q^2}{4\pi\varepsilon_0(2z')^2} \tag{3.102}$$

This *image force* contributes in large measure to preventing electrons from leaving the surfaces of conductors and is associated with the "work function" of conducting materials.

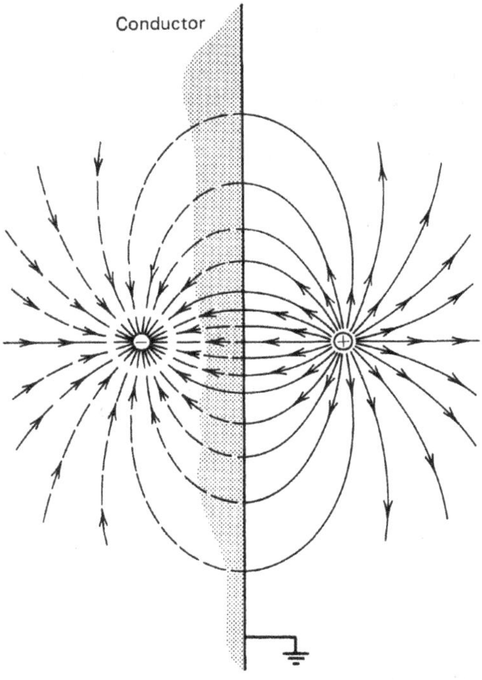

Figure 3.12 The lines of force between a point charge and a nearly grounded conducting plate.

With knowledge of the solution to the "point and plane" problem other problems involving conducting planes may be solved. Utilizing the superposition principle, the solution to any fixed charge distribution (near an "infinite" plane conductor) can be simply constructed by superimposing the solutions for the individual charge elements, each real charge element having its (mirror) image in the plane. The solution can then be formally written, as follows:

$$\Phi(\mathbf{r}) = \frac{1}{4\pi\varepsilon_0}\int_V \frac{\rho}{\xi}\,dv + \frac{1}{4\pi\varepsilon_0}\int_{V_I} \frac{\rho_I\,dv_I}{\xi_I} \tag{3.103}$$

The subscript I refers to the mirror image distribution. (See Example 3.13).

Example 3.13 A Dipole and a Conducting Plane

The method of superposition can be illustrated by considering an electric dipole near a large grounded conducting plane shown in Fig. 3.13. The dipole, of moment **p**, is at a distance z_0 from the plane, and at an angle θ_0 with respect to the normal to the plane.

We use a coordinate system whose origin is just beneath the dipole on the surface of the plane, and the z axis goes through the dipole. To find the system of images needed to produce a zero potential on the $z = 0$ plane, we represent the dipole by two charges q and $-q$ separated by distance $\mathbf{l}$ such that $\mathbf{p} = q\mathbf{l}$ where $\mathbf{l}$ is in the direction of $\mathbf{p}$. Satisfying the potential requirement at $z = 0$ requires introducing two image charges. The first image charge, of magnitude $-q$, is the image of the q charge of the dipole and is placed beneath the plane at an equidistance. The magnitude of the second image charge is, on the other hand, q; it is the image of the $-q$ charge of the dipole and is placed beneath the plane at an equidistance.

It is clear that the system of image charges constitute a dipole placed at distance z_0 just beneath the plane. In terms of the real dipole moment, the image dipole moment, $\mathbf{p}'$, has the components: $p'_z = p\cos\theta_0$ and $p'_x = -p\sin\theta_0$. Therefore, in general, the component of the dipole that is normal to the plane has an image dipole of the same magnitude and direction placed at the same distance beneath the plane. The component of the dipole that is parallel to the plane, on the other hand, has an image dipole of the same magnitude but of opposite direction, and its position is also at the equidistance just beneath the plane.

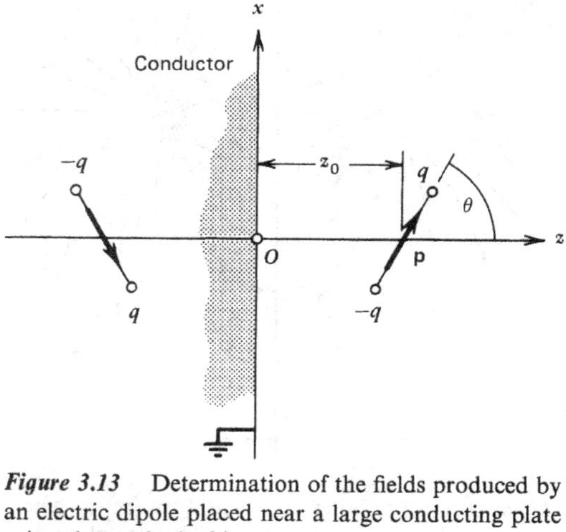

Figure 3.13 Determination of the fields produced by an electric dipole placed near a large conducting plate using the method of images.

We are now in a position to calculate the potential in the $z > 0$ region:

$$\Phi = \frac{1}{4\pi\varepsilon_0}\frac{\mathbf{p}\cdot(\mathbf{r} - \hat{\mathbf{z}}z_0)}{|\mathbf{r} - \hat{\mathbf{z}}z_0|^3} + \frac{1}{4\pi\varepsilon_0}\frac{\mathbf{p}'\cdot(\mathbf{r} + \hat{\mathbf{z}}z_0)}{|\mathbf{r} + \hat{\mathbf{z}}z_0|^3} \tag{3.104}$$

The force exerted by the dipole on the plane is equal to the force exerted by the dipole on its image. In Chapter 2 it was shown that a dipole moment $\mathbf{p}$ in an external field $\mathbf{E}$ experiences a force $\mathbf{p}\cdot\nabla\mathbf{E}$ (see Eq. (2.65)]. In this problem one takes $\mathbf{E}$ to be the electric field produced by the dipole at the location of its image. Let us calculate the force in the case where the dipole is perpendicular to the plane; that is, $\theta_0 = 0$. According to Eq. (2.47), $\mathbf{E}$ is:

$$\mathbf{E} = \frac{1}{4\pi\varepsilon_0}\left[\frac{3(\mathbf{r} - \mathbf{r}')\cdot\mathbf{p}}{|\mathbf{r} - \mathbf{r}'|^5}(\mathbf{r} - \mathbf{r}') - \frac{\mathbf{p}}{|\mathbf{r} - \mathbf{r}'|^3}\right] \tag{2.47}$$

where $\mathbf{r}' = \hat{\mathbf{z}}z_0$. At $\mathbf{r} = -z_0\hat{\mathbf{z}}$, the location of the image dipole, the field components in the x and y directions vanish and the z component is given by

$$E_z = \frac{p}{4\pi\varepsilon_0}\left[\frac{3(z - z_0)^2}{[x^2 + y^2 + (z - z_0)^2]^{5/2}} - \frac{1}{[x^2 + y^2 + (z - z_0)^2]^{3/2}}\right] \tag{3.105}$$

Differentiating E_z with respect to z and evaluating the result at $(0, 0, -z_0)$ gives

$$\frac{\partial E_z}{\partial z} = \frac{-3p}{32\pi\varepsilon_0 z_0^4}$$

Therefore the force between the dipoles

$$\mathbf{F} = \mathbf{p}'\cdot\nabla\mathbf{E} = p'\frac{\partial E_z}{\partial z}\hat{\mathbf{z}} = -\frac{3p^2}{32\pi\varepsilon_0 z_0^4}\hat{\mathbf{z}} \tag{3.106}$$

3.5.2 Point Charge and Sphere

We consider the problem of finding the solution to the problem of a point charge q_0, external to a grounded ($\Phi \equiv 0$) conducting sphere of radius R (Fig. 3.14). The question is whether we can find a convenient set of imaginary point charges, located inside the sphere, that will force the potential of the sphere to be zero when the potential of q_0 (located at z_0) is added. The answer is yes, but it can be only guessed at beforehand.

By the symmetry of the problem, the image charge distribution must be symmetric about the line connecting q_0 to the center of the sphere at 0. We attempt to find a single image charge, of magnitude q' and located at $z = z'$, that will satisfy out

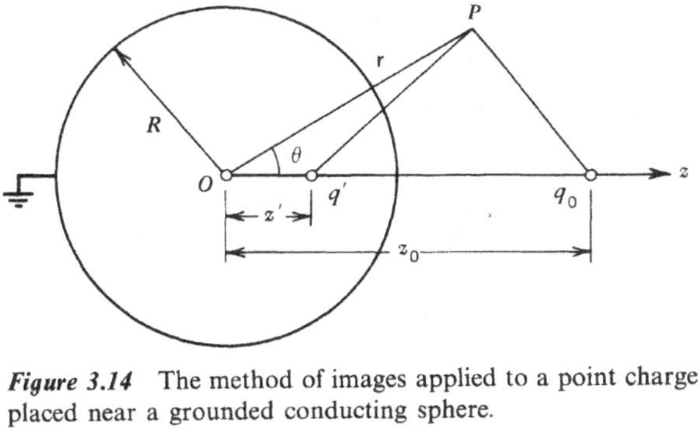

Figure 3.14 The method of images applied to a point charge placed near a grounded conducting sphere.

the zero-potential condition. The sign of the image charge must be opposite to that of q_0 if we are to get "potential cancellation."

The geometry is sketched in Fig. 3.14. We require that

$$\Phi(\mathbf{r} = \mathbf{R}) \equiv \frac{1}{4\pi\varepsilon_0}\left[\frac{q_0}{|\mathbf{R} - \mathbf{z}_0|} + \frac{q'}{|\mathbf{R} - \mathbf{z}'|}\right] = 0 \tag{3.107}$$

where $\mathbf{R} = \hat{\mathbf{r}}R$. We must try to solve this equation for q' and z', given q_0, R, and z_0. A solution is immediately suggested if we write it as

$$\frac{q_0/R}{\left|\hat{\mathbf{r}} - \dfrac{z_0}{R}\hat{\mathbf{z}}\right|} = \frac{-q'/z'}{\left|\hat{\mathbf{r}}\dfrac{R}{z'} - \hat{\mathbf{z}}\right|} \tag{3.108}$$

[This factorization is the only one that makes sense; if we factored out R on both sides, we would have found (see below) no solution to be possible.] To obtain equality, the numerators and denominators of Eq. (3.108) may be set equal, respectively, to obtain

$$\frac{q_0}{R} = \frac{-q'}{z'} \qquad \text{and} \qquad \hat{\mathbf{r}} - \frac{z_0}{R}\hat{\mathbf{z}} = \frac{R}{z'}\hat{\mathbf{r}} - \hat{\mathbf{z}} \tag{3.109}$$

These two simultaneous equations are easily solved. For example, squaring the denominators gives

$$1 - 2\frac{z_0}{R}(\hat{\mathbf{r}}\cdot\hat{\mathbf{z}}) + \left(\frac{z_0}{R}\right)^2 = 1 - 2\frac{R}{z'}(\hat{\mathbf{r}}\cdot\hat{\mathbf{z}}) + \left(\frac{R}{z'}\right)^2$$

and so equality is achieved if we let $z_0/R = R/z'$; that is $z_0 z' = R^2$. The two relations in Eq. (3.109) therefore give

$$q' = -\frac{R}{z_0}q_0 \qquad z' = \frac{R^2}{z_0} \tag{3.110}$$

These values are realizable and show that indeed (only) one image charge whose location and magnitude are now determined is required to solve the problem. We may now write expressions for Φ and $\mathbf{E} = -\nabla\Phi$ for any point $r \geq R$. The potential

$$\Phi(\mathbf{r}) = \frac{1}{4\pi\varepsilon_0}\left[\frac{q_0}{|\mathbf{r} - \mathbf{z}_0|} + \frac{q'}{|\mathbf{r} - \mathbf{z}'|}\right] \tag{3.111}$$

simply that of two point charges q_0 and q'. The electric field is obtained from Eq. (3.111) as follows:

$$\mathbf{E}(\mathbf{r}) = \frac{1}{4\pi\varepsilon_0}\left[\frac{q_0(\mathbf{r} - \mathbf{z}_0)}{|\mathbf{r} - \mathbf{z}_0|^3} + \frac{q'(\mathbf{r} - \mathbf{z}')}{|\mathbf{r} - \mathbf{z}'|^3}\right] \tag{3.112}$$

In terms of the angle θ, one finds that the field, as $\mathbf{r}$ approaches the surface from outside the sphere, is given by

$$\mathbf{E}(\mathbf{r} \to \mathbf{R}) \to -\frac{q_0}{4\pi\varepsilon_0}\frac{\mathbf{R}}{z_0^3}\frac{z_0^2/R^2 - 1}{[z_0^2/R^2 - 2(z_0/R)\cos\theta + 1]^{3/2}}$$

As expected, it is perpendicular to the spherical surface. Consequently, the surface charge density σ there is $\sigma(\mathbf{R}) = \varepsilon_0[\mathbf{E}(\mathbf{r} = \mathbf{R})\cdot\hat{\mathbf{R}}]$, or

$$\sigma(\mathbf{R}) = \frac{-q_0}{4\pi R^2}\cdot\left(\frac{R}{z_0}\right)^3\frac{1}{[z_0^2/R^2 - 2(z_0/R)\cos\theta + 1]^{3/2}} \tag{3.113}$$

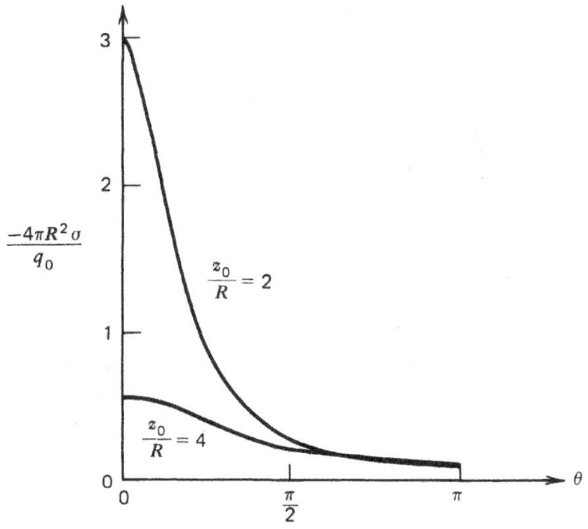

Figure 3.15 The induced surface charge density on a grounded sphere due to the presence of a point charge nearby.

Figure 3.15 gives the induced charge density as a function θ for two values of z_0/R. As expected, the charge density is maximum at $\theta = 0°$ and minimum at $\theta = 180°$.

The total charge on the (outer) surface of the conductor is just q'. This may be verified by direct integration of the above expression for σ over the spherical surface. It is a necessary consequence of the fact that the total flux due to the image charge over a surface just enclosing the sphere must equal the total flux from the surface charge.

The force exerted on the point charge by the induced charge on the sphere is

$$\mathbf{F} = \frac{q_0 q' \hat{\mathbf{z}}}{4\pi\varepsilon_0 (z_0 - z')^2} = \frac{q_0^2 (R/z_0) \hat{\mathbf{z}}}{4\pi\varepsilon_0 z_0^2 (1 - R^2/z_0^2)^2} \tag{3.114}$$

When $(R/z_0) \ll 1$,

$$F \to \frac{q_0^2}{4\pi\varepsilon_0} \frac{R}{z_0^3}$$

like the force due to a dipole of moment $q_0 R$. As z_0 approaches R, that is, as $(z_0/R) \to (1 + \delta)$ where $\delta \ll 1$,

$$F \to \frac{q_0^2}{4\pi\varepsilon_0} \left(\frac{1}{2\delta}\right)^2$$

as in the point charge and plane case. Figure 3.16*a* gives the line of force between the charge and the sphere.

Finally, we have supposed in this problem that our sphere was at *zero potential*, and thus had a charge q'. If we are interested in a problem where the sphere is given a potential $\Phi = V$, then the *method of superposition* can be used. It is clear in this case that the charge on the sphere will differ from q'—say, by an amount q''. In other words, a charge q'' is required to bring the spherical surface up to a potential

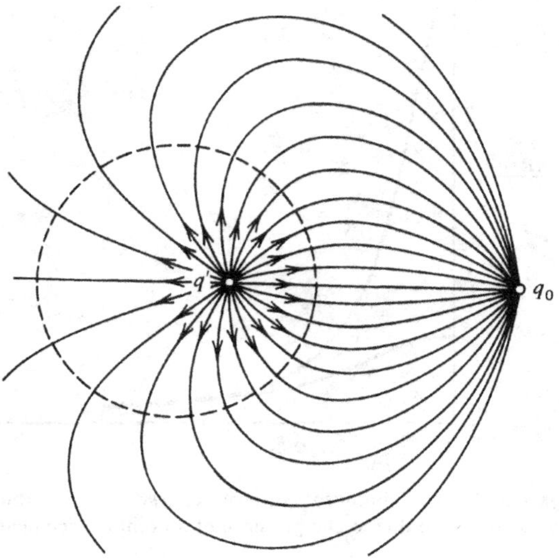

(a)

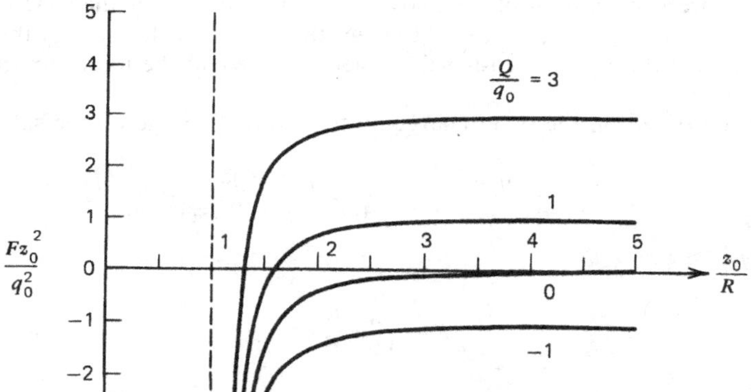

(b)

Figure 3.16 (*a*) The lines of force between a point charge and a grounded conducting sphere. (*b*) The force between the point charge and a charged sphere as a function of distance between them.

V. In terms of image charges, we can immediately discern that q'' should be placed at the origin, and is given via the equation

$$V = \frac{q''}{4\pi\varepsilon_0 R} \tag{3.115}$$

In terms of surface charge, q'' is distributed uniformly over the sphere. The total surface charge density would then be given by $\sigma = \sigma\,(\Phi = 0) + (V\varepsilon_0)/R$, and thus the total charge on the sphere is

$$Q = q' + q'' \tag{3.116}$$

To summarize this case, *two image charges* are required to force the potential of the sphere to assume the value $V \neq 0$. One of these is at the center of the sphere. The potential outside the sphere is given by

$$\Phi = \frac{1}{4\pi\varepsilon_0}\left[\frac{q_0}{|\mathbf{r} - \mathbf{z}_0|} + \frac{q'}{|\mathbf{r} - \mathbf{z}'|} + \frac{q''}{r}\right] \tag{3.117}$$

Figure 3.16*b* gives the force between the charge q and the charged sphere as a function of the distance between them for a number of values of Q/q_0. It shows that even when $q_0 Q > 0$, the two will attract each other at short distances.

If we placed a *neutral* conducting sphere near a point charge q, we would find that the potential of the sphere would be nonzero, and the potential outside the sphere would again require *two* image charges for its calculation. The potential of the sphere would be equal to $(-1/4\pi\varepsilon_0)(q/R)$. (Why?)

If we had placed a point charge at a distance z' from the center of the sphere—i.e., inside the sphere—we would have a problem almost identical to the one discussed above (the "external" problem). In fact all the formulas given there are valid if we understand that in this case the image is charge q_0 and is located at z_0, whereas the interior charge q' is located at z', both specified. R is then the distance from the origin to the *interior* surface of the sphere. The only possible distinction in the problems comes if we let the sphere be at a potential $V \neq 0$. This is simply achieved by having a charge $q' = 4\pi\varepsilon_0 RV$ uniformly distributed over a spherical surface of radius greater than R. In the actual problem this charge would be located on the outer surface of the conducting spherical shell. The $\mathbf{E}$ field inside the sphere would not be affected by such a charge distribution.

An idea central to the idea of the image technique is that one can "replace" any closed equipotential surface determined from some (point or distributed) charge distribution with a conducting shell tailored to the geometry of the equipotential surface and having the correct potential. In the example above of the point and sphere, one may conceive of the problem first as that of the point charges alone. Having located this spherical equipotential surface, one replaces the equipotential surface by a conducting shell at the appropriate potential. The image charges will now appear on the shell in the form of a charge density. One should be cautioned, however, that for a conducting shell of finite thickness, an equipotential surface can be fit by only one side of the shell, and the fields will be identical to the original fields only on this side of the surface.

Example 3.14 A Dipole Near a Conducting Sphere

When an electric dipole is brought near a grounded sphere, then the potential and the field of the dipole are modified. The method of images can be used to determine the changes. We will represent the dipole by two point charges separated by a small distance as shown in Fig. 3.17

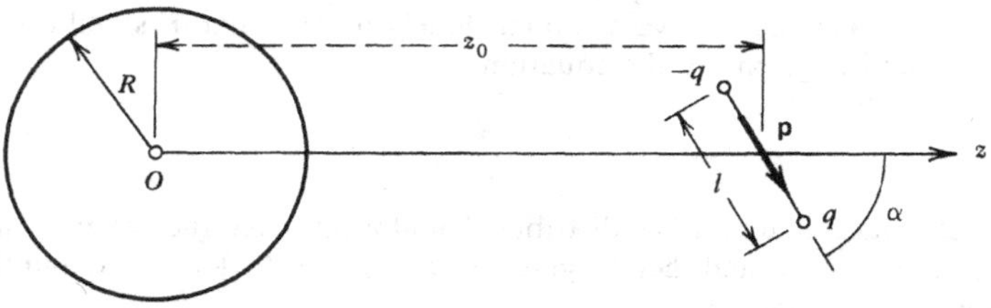

Figure 3.17 A dipole near a grounded conducting sphere.

and use the rules developed above for treating a point charge near a sphere. Having done this, we will derive rules for treating a dipole due to an arbitrary charge distribution.

The dipole moment **p** is taken to be at any angle α with the z axis, at a distance z_0 from the center of a sphere of radius R, and is represented by q and $-q$ charges separated by a distance l, such that $\mathbf{p} = q\mathbf{l}$.

We resolve the dipole moment into two components, one going through the center of the sphere and the other perpendicular to this direction, as shown in Figs. 3.18 and 3.19, respectively. Consider first Fig. 3.18. According to Eq. (3.110) the image charges q_1' and q_2' are

$$q_1' = -\frac{R}{z_0 + l/2} q \cos \alpha \qquad q_2' = \frac{R}{z_0 - l/2} q \cos \alpha$$

and located at

$$b_1 = \frac{R^2}{z_0 + l/2} \qquad b_2 = \frac{R^2}{z_0 - l/2}$$

respectively. In the dipole limit l can be taken to be small, and therefore the magnitudes and positions can be expanded in powers of l/z_0. The first-order terms are

$$q_1' \approx -\frac{qR}{z_0}\left(1 - \frac{l}{2z_0}\right)\cos \alpha, \qquad q_2' \approx \frac{qR}{z_0}\left(1 + \frac{l}{2z_0}\right)\cos \alpha$$

$$b_1 \approx \frac{R^2}{z_0}\left(1 - \frac{l}{2z_0}\right) \qquad b_2 \approx \frac{R^2}{z_0}\left(1 + \frac{l}{2z_0}\right)$$

Because q_2' is different from q_1', then the image of the parallel component of the dipole consists of (1) an image dipole of moment p' located at a distance $b = \frac{1}{2}(b_1 + b_2)$ from the center of the sphere

$$\mathbf{p}' = \left(\frac{R}{z_0}\right)^3 \hat{\mathbf{z}} p \cos \alpha, \qquad b = \frac{R^2}{z_0} \tag{3.118}$$

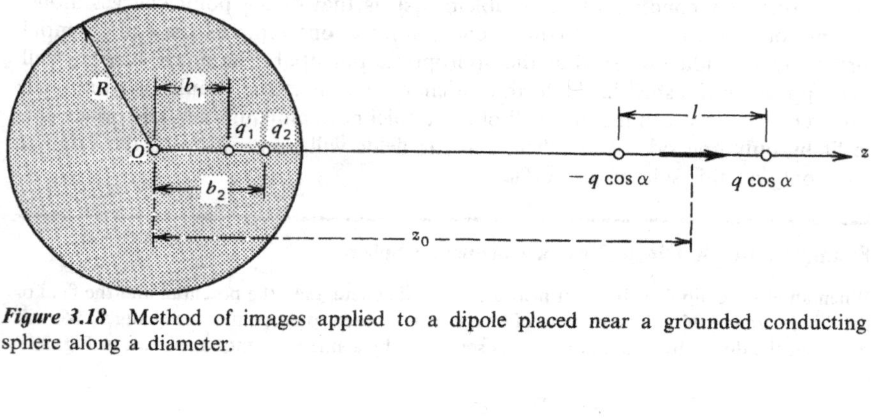

Figure 3.18 Method of images applied to a dipole placed near a grounded conducting sphere along a diameter.

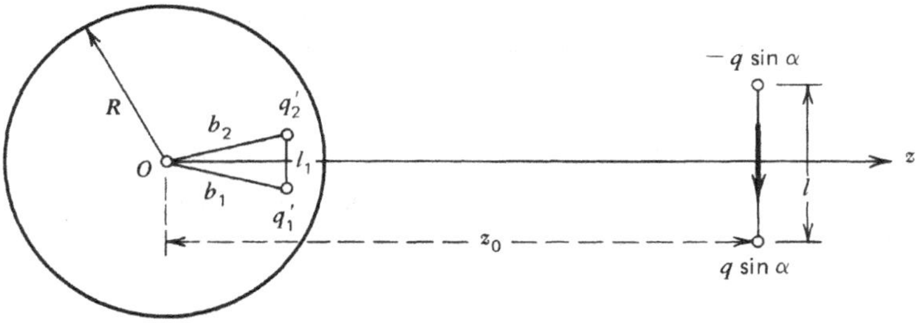

Figure 3.19 Method of images applied to a dipole normal to a diameter of a grounded conducting sphere.

and (2) a charge located at a distance R^2/z_0 from the center of the sphere and of magnitude q_1, given by

$$q_1 = q_1' + q_2' = \frac{Rp\cos\alpha}{z_0^2} \tag{3.119}$$

Figure 3.19 shows the two charges comprising the perpendicular component of the dipole. They are symmetric with respect to the center of the sphere; therefore their image charges will be scaled by the same factor. This results in an image dipole only. To evaluate the image dipole and its location, we again use the rules for a charge near a sphere. The image charges to zero order in l/z_0 for $q \sin\alpha$ and $-q\sin\alpha$ are $q_1' = (-Rq/z_0)\sin\alpha$ and $q_2' = (Rq/z_0)\sin\alpha\Lambda$ respectively. Both are located to zero order in l/z_0 at $b_1 = b_2 = b = R^2/z_0$ from the center of the sphere. The separation between the image charges $l'/l = b/z_0 = R^2/z_0^2$. The dipole moment $\mathbf{p}'$ is then equal to

$$p' = \left(\frac{R}{z_0}\right)^3 p\sin\alpha \tag{3.120}$$

and located at $b = R^2/z_0$ from the center of the sphere.

Example 3.15 A Dipole at the Center of a Conducting Sphere

The method of images can be used to solve for the effect of introducing an electric dipole at the center of a grounded sphere (Fig. 3.20). We represent the dipole by two point charges separated by a small distance l. Because one charge is above and the other is below the center of the sphere, then their image charges are located at opposite sides of the sphere, and consequently such an image system does not constitute an image dipole.

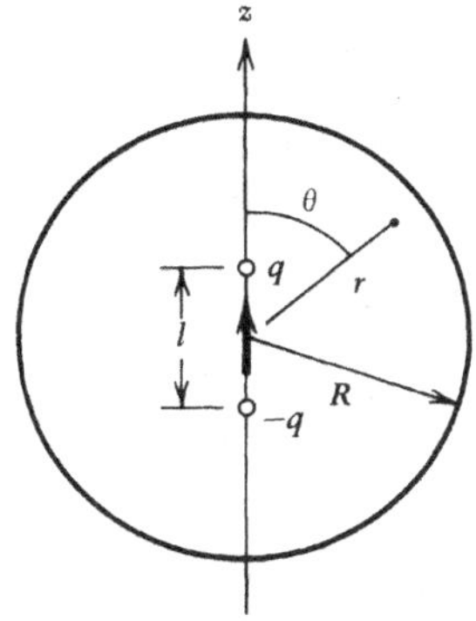

Figure 3.20 A dipole placed at the center of a grounded sphere.

Let us consider the effect of the charge q located at distance $b = l/2$. Its image charge is $-(R/b)q$ and located at $d = 2R^2/l$. The potential due to these two charges in the internal region is

$$\Phi_1(r, \theta) = \frac{1}{4\pi\varepsilon_0}\left[\frac{-Rq/b}{(r^2 + d^2 - 2rd\cos\theta)^{1/2}} + \frac{q}{(r^2 + b^2 - 2rb\cos\theta)^{1/2}}\right] \tag{3.121}$$

Note that d is large while b is small because l is small; therefore the first term is expanded in powers of $1/d$ and the second term is expanded in powers of $1/r$. Keeping terms of order $(1/d)^2$ and $(1/r)^2$ reduces Eq. (3.121) to

$$\Phi_1(r, \theta) = \frac{q}{4\pi\varepsilon_0}\left(\frac{1}{r} - \frac{1}{R}\right) + \frac{1}{8\pi\varepsilon_0}\frac{ql}{r^2}\cos\theta - \frac{1}{8\pi\varepsilon_0}\frac{qlr}{R^3} \tag{3.122}$$

Similarly, the potential due to the charge $-q$ placed at $-l/2$ is calculated from the above result by changing q to $-q$ and l to $-l$. The result is

$$\Phi_2(r, \theta) = \frac{-q}{4\pi\varepsilon_0}\left(\frac{1}{r} - \frac{1}{R}\right) + \frac{1}{8\pi\varepsilon_0}\frac{ql}{r^2}\cos\theta - \frac{1}{8\pi\varepsilon_0}\frac{qlr}{R^3} \tag{3.123}$$

Adding Eqs. (3.122) and (3.123) we get

$$\Phi(r, \theta) = \frac{1}{4\pi\varepsilon_0}\frac{p\cos\theta}{r^2} - \frac{1}{4\pi\varepsilon_0}\frac{pr}{R^3}\cos\theta \tag{3.124}$$

This result indicates that in addition to the potential produced by the dipole itself, first term, there is a potential that is due to a constant electric field: $(1/4\pi\varepsilon_0)\mathbf{p}/R^3$. This extra field is caused by an induced charge distribution on the surface of the grounded sphere. Taking the normal gradient of $\Phi(r, \theta)$ and using $\sigma = -\varepsilon_0\, \partial\Phi/\partial r$ at $r = R$ gives

$$\sigma = \frac{3p}{4\pi R^3}\cos\theta \tag{3.125}$$

3.5.3 Parallel Cylinders

We shall find the potential due to two parallel cylinders (both parallel to the z axis) by working backward—that is, by recognizing that the equipotentials due to two line charges of strength λ (charge/unit length) and $-\lambda$ are right cylindrical surfaces.

Consider, then, two line charges as shown in Fig. 3.21 a distance $2a$ apart. Locate the origin on the z axis midway between the line charges. The potential of a single line charge has been derived in Example 3.1 and has the form

$$\Phi(\rho) = -\frac{\lambda}{2\pi\varepsilon_0}\ln\left(\frac{\rho}{\rho_0}\right)$$

where ρ is the distance from the line, and ρ_0 is where the potential is assigned a potential equal to zero. In the present problem, both lines must have the same zero potential, which then must lie on the plane $x = 0$. The potential due to a superposition of the two line charges is then

$$\Phi = \Phi_+ + \Phi_- = \frac{-\lambda\ln\rho_+}{2\pi\varepsilon_0} + \frac{\lambda\ln\rho_-}{2\pi\varepsilon_0}$$

Thus

$$\Phi(\mathbf{r}) = \frac{\lambda}{2\pi\varepsilon_0}\ln\left(\frac{\rho_-}{\rho_+}\right) \tag{3.126}$$

which is seen to be zero when $\rho_- = \rho_+$—that is, on the plane $x = 0$.

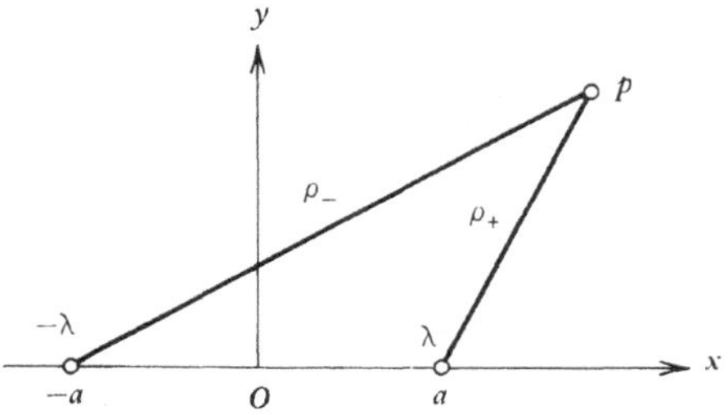

Figure 3.21 Two infinitely long line charges of opposite sign along the z axis are used to explain the method of images of cylindrical problems.

Equipotential surfaces are generated for this problem when the logarithm of Eq. (3.126) is constant—that is, when

$$\frac{\rho_-}{\rho_+} = m \tag{3.127}$$

Values of m between zero and one yield negative values of potential. If $m > 1$, the potential will be positive. The shapes of the equipotential surfaces are easily seen to be circular cylinders whose centers lie on the x axis. The equation for such cylinders is

$$(x - x_0)^2 + y^2 = R^2 \qquad \text{or} \qquad x^2 - 2xx_0 + y^2 = R^2 - x_0^2 \tag{3.128}$$

with x_0 locating the center of the circle and R its radius. To show that Eq. (3.127) defines a circular cylinder, we square it and, as seen from Fig. 3.21, use

$$\rho_-^2 = (x + a)^2 + y^2 \text{ and } \rho_+^2 = (x - a)^2 + y^2$$

Therefore Eq. (3.127) gives $(x + a)^2 + y^2 = m^2\,[(x - a)^2 + y^2]$. Rearranging terms,

$$x^2 - 2x\left[\left(\frac{m^2 + 1}{m^2 - 1}\right)a\right] + y^2 = -a^2$$

which is just the equation of a circle whose center and radius are given by Eq. (3.128) above; that is,

$$x_0 = \left(\frac{m^2 + 1}{m^2 - 1}\right)a \tag{3.129}$$

$$R^2 = x_0^2 - a^2 = \left(\frac{2ma}{m^2 - 1}\right)^2 \tag{3.130}$$

The potential of the cylinder characterized by m is given by

$$\Phi_m = \frac{\lambda}{2\pi\varepsilon_0}\ln m \tag{3.131}$$

Figure 3.22 gives a plot of the cylinders for selected values of m. With these results, we can now consider inverse problems. Suppose, for example, that we wish to find the potential due to a line charge λ and a conducting circular cylinder of radius R whose center is a distance δ from the charge. Consult Fig. 3.23, which shows the line inside the cylinder. To find the potential we seek the location of the

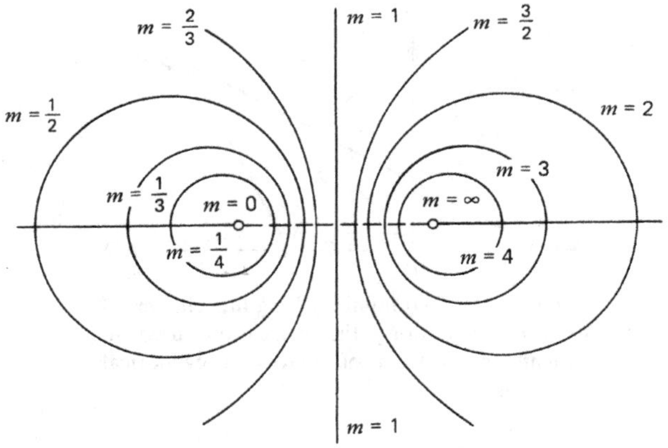

Figure 3.22 Surfaces of constant potentials produced by two parallel line charges of opposite sign.

(image) charge $-\lambda$ that will yield an equipotential on the cylinder. Thus we wish to find x_0, a, and m, given R and δ. In fact, since

$$x_0 - a = \delta \tag{3.132}$$

we have only two unknowns to determine. Substituting Eq. (3.132) into Eq. (3.130) gives $R^2 = x_0^2 - a^2 = \delta(x_0 + a)$, which results in

$$x_0 + a = \frac{R^2}{\delta} \tag{3.133}$$

Equations (3.132) and (3.133) are now solved for a and x_0:

$$a = \frac{\delta}{2}\left(\frac{R^2}{\delta^2} - 1\right) \quad \text{and} \quad x_0 = \frac{\delta}{2}\left(\frac{R^2}{\delta^2} + 1\right).$$

The value of m is then fixed by either Eq. (3.129) or (3.130). Substituting for a and x_0 in Eq. (3.129) gives

$$\frac{m^2 + 1}{m^2 - 1} = \frac{x_0}{a} = \frac{(R/\delta)^2 + 1}{(R/\delta)^2 - 1}$$

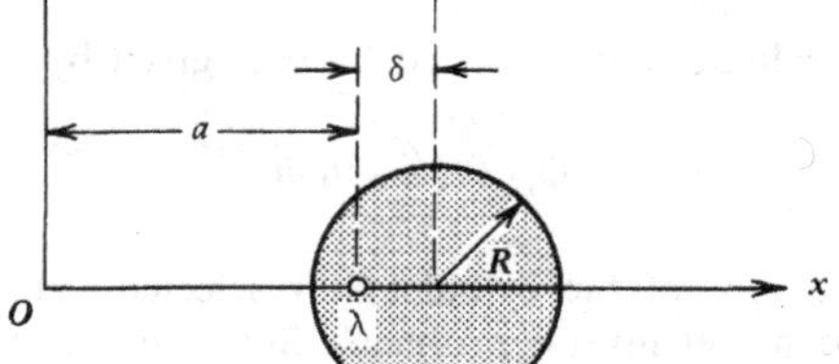

Figure 3.23 A line charge placed inside a conducting cylinder, parallel to its axis.

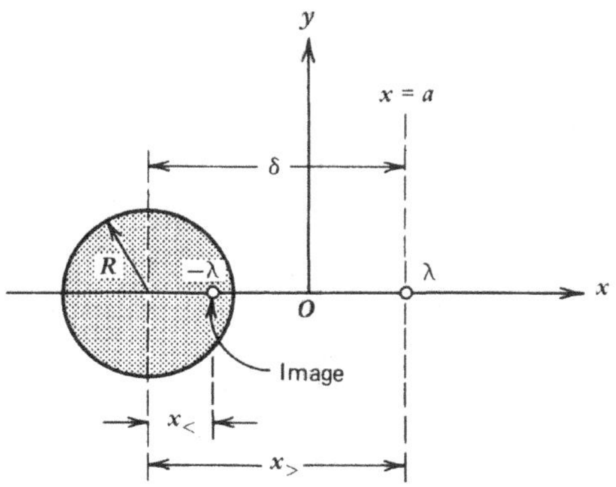

Figure 3.24 A line of charge placed outside a cylinder, parallel to its axis.

and therefore

$$m = \frac{R}{\delta}$$

The problem is now solved. The potential (inside the cylinder) is expressed in terms of the line charges, whose locations we have determined. For example, if $R/\delta = 2$, then $a = 3\delta/2$, $x_0 = 5\delta/2$, and $m = 2$. The potential of the cylinder is given by $\Phi = (\lambda/2\pi\varepsilon_0) \ln 2$, but we are free to add a constant potential to the interior region to set it at any value we please. The potential outside the cylinder is independent of the interior problem.

Had we desired to solve the problem where $|R/\delta| < 1$ (a line charge outside the cylinder, as in Fig. 3.24) we would apply the same equations as before. Here, however, $\delta < 0$ and $m = -R/\delta$, a positive number. Everything proceeds as before. The potential of the cylinder will be negative.

If we had taken as our origin the location of the center of the cylinder then defining $x_< = x_0 - a$, and $x_> = x_0 + a$, we see that $x_< x_> = x_0^2 - a^2 = R^2$, which is reminiscent of the image problem for the point and the sphere. The interior problem specifies that $m = R/x_<$. The exterior problem specifies that $m = R/x_>$.

Finally, in Fig. 3.25 we consider the problem of two cylinders of radii R_1 and R_2 whose axes are separated by a distance Δ. Each cylinder is an equipotential surface, so we now have, from Eq. (3.130),

$$R_1^2 = x_{01}^2 - a^2 \qquad \text{and} \qquad R_2^2 = x_{02}^2 - a^2$$

or

$$\Delta = x_{02} - x_{01} = (R_2^2 + a^2)^{1/2} - (R_1^2 + a^2)^{1/2}$$

From this equation, we find a to be given by

$$a = \left[\left(\frac{R_2^2 + R_1^2 - \Delta^2}{2\Delta}\right)^2 - R_2^2\right]^{1/2}.$$

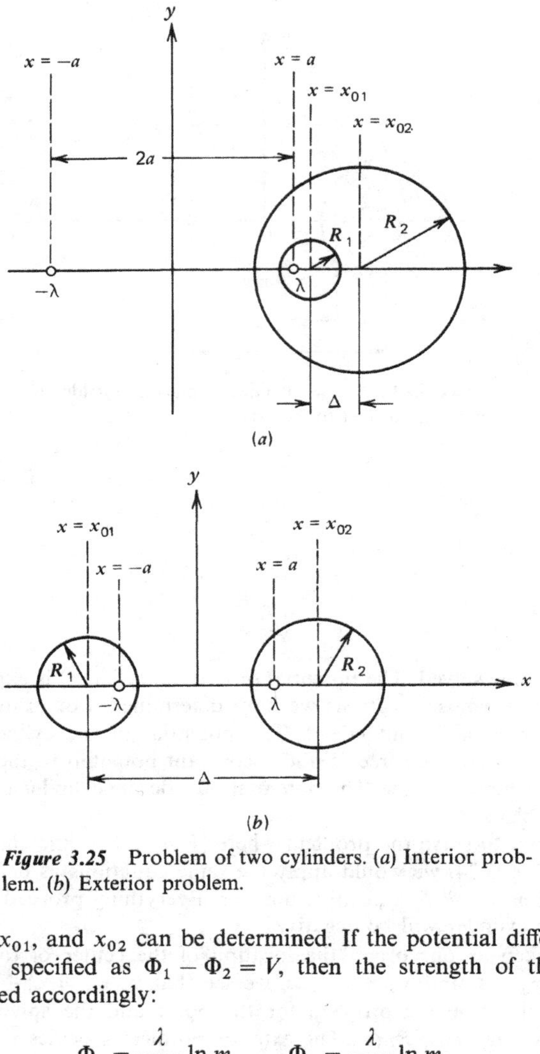

Figure 3.25 Problem of two cylinders. (*a*) Interior problem. (*b*) Exterior problem.

Hence, m_1, m_2, x_{01}, and x_{02} can be determined. If the potential difference between the cylinders is specified as $\Phi_1 - \Phi_2 = V$, then the strength of the (image) line charge is adjusted accordingly:

$$\Phi_1 = \frac{\lambda}{2\pi\varepsilon_0} \ln m_1 \qquad \Phi_2 = \frac{\lambda}{2\pi\varepsilon_0} \ln m_2$$

or

$$\Phi_1 - \Phi_2 = \frac{\lambda}{2\pi\varepsilon_0} \ln \frac{m_1}{m_2}.$$

The line charge must have the magnitude

$$\lambda = V\left[\frac{1}{2\pi\varepsilon_0} \ln\left(\frac{m_1}{m_2}\right)\right]^{-1}$$

The two kinds of problems treatable are sketched in Fig. 3.25. The potentials in the regions of interest are then given by

$$\Phi = \frac{\lambda}{2\pi\varepsilon_0} \ln\left(\frac{\rho_-}{\rho_+}\right)$$

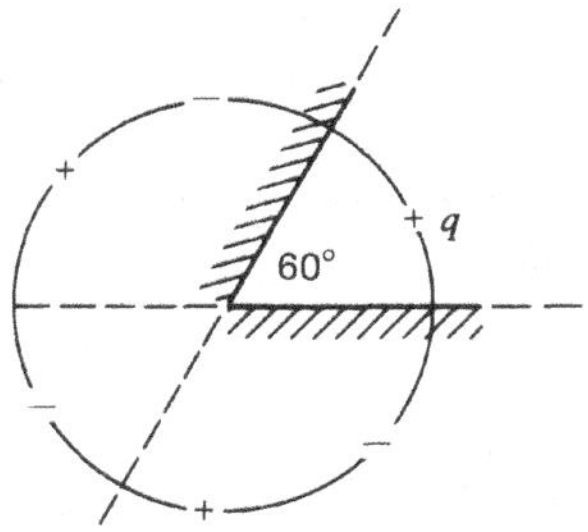

Figure 3.26 A point charge between two grounded conducting plates making an angle between them.

3.5.4 Point Charge and Two Conducting Surfaces

One may also solve problems by images for point charges near two conducting surfaces, e.g., two intersecting, conducting planes. In general, one requires more than one image charge to produce the correct equipotential boundary conditions here. If, in fact, the angle of intersection between the planes is $180°/n$, where $n = 1, 2, 3, \ldots$, then the number of image charges required is $2n - 1$. An illustration is shown in Fig. 3.26. If the angle differs from $180°/n$, an infinite number of images may be required, or it may not be possible to use the method at all. Other methods of solution may then be preferable.

Another example of this sort is a large plane with a small-radius semispherical or cylindrical boss (see Problem 3.12).

3.6 Poisson's Equation

We now go back to the case where the charge is not localized on conductors or distributed over discrete point charges. In this case the potential satisfies Poisson's equation:

$$\nabla^2\Phi = -\frac{\rho}{\varepsilon_0} \tag{3.3}$$

Given a general charge distribution in addition to some boundary conditions, the potential can be found by first solving the homogeneous part of Eq. (3.3), Laplace's equation $\nabla^2\Phi = 0$. This solution is to be added to the particular solution of the Poisson's equation (the Coulomb law)

$$\Phi(\mathbf{r}) = \frac{1}{4\pi\varepsilon_0}\int \frac{\rho\, dv'}{|\mathbf{r} - \mathbf{r}'|} + \text{solution of Laplace's equation} \tag{3.134}$$

where the integration is carried over the given charge distribution. The overall solution is then made to satisfy the boundary condition by choosing the appropriate constants in the solution of Laplace's equation.

In Chapter 2 a number of applications of Eq. (2.40) were given. They involved various charge distributions. In this section, however, we will consider a special class of distributions where the charge density and the potential depend only on one variable. Moreover, we will assume that the charge distribution is bounded, meaning that either the charge occupies a limited region of space or falls off sufficiently rapidly at large distances. Examples of the latter case include exponentially decaying distributions, or distributions dropping off faster than $1/r^2$. Such restrictions allow us to arrive at the solution in a more direct way, as we will show in some examples below.

Example 3.16 An Infinite Slab of Uniform Charge Distribution

We examine here the potential function, Φ, associated with a charge distribution shown in Fig. 3.27 and given by

$$\rho = \rho_0 \qquad \text{for } -\frac{l}{2} \leq z \leq \frac{l}{2}$$

and

$$\rho = 0 \qquad \text{for } |z| > \frac{l}{2}$$

From the symmetry of the distribution, the potential function will depend only on the distance from the $z = 0$ plane. Because of the infinite extent of the distribution, we cannot take $\Phi(\infty) = 0$, but it is convenient to take the potential to be zero at $z = 0$.

There are three regions for which we must solve Poisson's (or Laplace's) equation. In each of these regions, $\Phi(\mathbf{r}) = \Phi(z)$, and so the "Laplacian of Φ," $\nabla^2\Phi$, simply becomes $d^2\Phi/dz^2$. We give below a table of the equations and their corresponding solutions for the different regions of space:

Region	Equation	Solution
$z \geq \frac{l}{2}$	$\frac{d^2\Phi_2}{dz^2} = 0$	$\Phi_2 = C_1 z + C_2$
$\lvert z\rvert \leq \frac{l}{2}$	$\frac{d^2\Phi_0}{dz^2} = -\frac{\rho_0}{\varepsilon_0}$	$\Phi_0 = -\frac{\rho_0}{\varepsilon_0}\frac{z^2}{2} + K_1 z + K_2$
$z \leq -\frac{l}{2}$	$\frac{d^2\Phi_1}{dz^2} = 0$	$\Phi_1 = C_1' z + C_2'$

The five constants C_1, C_2, C_1', C_2', and K_1 are as yet undetermined [$K_2 = 0$ since $\Phi(z = 0) = 0$]. These constants can be determined if we realize that certain relations must exist between the various solutions at the boundaries, i.e., at $z = \pm l/2$.

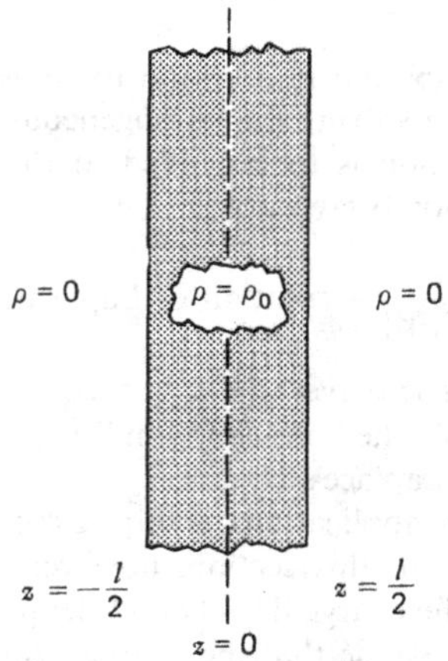

Figure 3.27 An infinite, uniformly charged slab.

1. The first boundary condition is the continuity of the potential across the boundary region. At $z = -l/2$, $\Phi_1 = \Phi_0$, which gives

$$-\frac{l}{2}C_1' + C_2' = -\frac{l^2}{8}\frac{\rho_0}{\varepsilon_0} - \frac{l}{2}K_1$$

and at $z = l/2$, $\Phi_0 = \Phi_2$, which gives

$$\frac{l}{2}C_1 + C_2 = -\frac{l^2}{8}\frac{\rho_0}{\varepsilon_0} + \frac{l}{2}K_1$$

2. Remembering that there is no surface charge density (in this problem), the normal components of **E**, $-\partial\Phi/\partial z$, are everywhere continuous; hence, from the boundary at $z = -l/2$ and at $z = l/2$, we find that

$$C_1' = \frac{l}{2}\frac{\rho_0}{\varepsilon_0} + K_1 \qquad \text{and} \qquad C_1 = -\frac{l}{2}\frac{\rho_0}{\varepsilon_0} + K_1$$

3. The symmetry of the configuration requires that the electric field be zero at $z = 0$ [that is, $\mathbf{E}(z) = -\mathbf{E}(-z)$]; therefore $K_1 = 0$. Solving the above equations simultaneously results in the following expressions for the potentials:

$$\Phi_0 = -\frac{\rho_0 z^2}{2\varepsilon_0} \qquad \Phi_1 = \frac{\rho_0 l}{2\varepsilon_0}z + \frac{\rho_0 l^2}{8\varepsilon_0} \qquad \Phi_2 = -\frac{\rho_0 l}{2\varepsilon_0}z + \frac{\rho_0 l^2}{8\varepsilon_0}$$

Example 3.17 Uniformly Charged Sphere

We now analyze a case where the charge is distributed in a spherically symmetric way. Let a charge q be distributed over a sphere of radius R with a constant volume charge density ρ, and thus for $r > R$ the charge density is zero. In the region $r \leq R$ the potential satisfies Poisson's equation

$$\frac{1}{r^2}\frac{d}{dr}\left(r^2\frac{d\Phi}{dr}\right) = -\frac{\rho}{\varepsilon_0} \tag{3.135}$$

and in the region $r > R$, the potential satisfies Laplace's equation

$$\frac{1}{r^2}\frac{d}{dr}\left(r^2\frac{d\Phi}{dr}\right) = 0 \tag{3.136}$$

On can easily show that Eq. (3.135) has the solution

$$\Phi(r) = \frac{-\rho r^2}{6\varepsilon_0} + \frac{A_1}{r} + B_1 \qquad r \leq R \tag{3.137}$$

and Eq. (3.136) has the solution

$$\Phi(r) = \frac{A_2}{r} + B_2 \qquad r \geq R \tag{3.138}$$

The potential has to satisfy the following boundary conditions: (1) $\Phi\ (r \to \infty) = 0$; (2) $\Phi\ (r \to 0)$ is finite since there are no point charges at the center of the sphere; (3) the two potentials should match at $r = R$; and (4) the total charge of this distribution is $(4\pi/3)R^3\rho$. The first condition gives $B_2 = 0$. The second condition requires that $A_1 = 0$. A relation between B_1 and A_2 is now found by imposing condition 3:

$$-\frac{\rho R^2}{6\varepsilon_0} + B_1 = \frac{A_2}{R} \tag{3.139}$$

Finally, the last condition can be used to evaluate A_2. Taking a gaussian surface, which is a shell of radius $r > R$ with its center at the center of the charge distribution, gives:

$$\oint \mathbf{E}\cdot\hat{\mathbf{n}}\, da = \frac{4\pi R^3}{3\varepsilon_0}\rho \tag{3.140}$$

The electric field outside the sphere can be found by taking the gradient of Eq. (3.138), yielding $\mathbf{E} = A_2\hat{\mathbf{r}}/r^2$ and thus $\oint \mathbf{E}\cdot\hat{\mathbf{n}}\,da = 4\pi A_2$. Replacing the left-hand side of Eq. (3.140) with $4\pi A_2$ gives $A_2 = (R^3/3\varepsilon_0)\rho$. Substituting this value for A_2 in Eq. (3.139) gives $B_1 = \rho R^2/2\varepsilon_0$. Therefore the potential is

$$\Phi(r) = \frac{\rho R^2}{2\varepsilon_0}\left(1 - \frac{r^2}{3R^2}\right) \qquad r < R \tag{3.141}$$

or

$$\Phi(r) = \frac{\rho R^3}{3\varepsilon_0}\frac{1}{r} \qquad r > R \tag{3.142}$$

Equation (3.141) indicates that the potential inside the sphere is a quadratic function of r with the potential at the origin larger than that at the edge of the sphere. It is to be noted that the electric field is continuous at $r = R$. For $r \le R$, $\mathbf{E} = (\rho r/3\varepsilon_0)\hat{\mathbf{r}}$ and for $r \ge R$, $\mathbf{E} = \rho R^3/(3\varepsilon_0 r^2)\hat{\mathbf{r}}$, giving $\mathbf{E} = \rho R/(3\varepsilon_0)\hat{\mathbf{r}}$ at $r = R$. This continuity is a direct result of the absence of surface charge at $r = R$.

Example 3.18 Exponential Charge Distribution

Consider a spherically symmetric charge distribution (of total charge q), which has the radial dependence $\rho(r) = \rho_0 e^{-\alpha r}$. We mention here that this density describes the electronic charge distribution in the ground state of hydrogen. The potential at an arbitrary r satisfies Poisson's equation:

$$\frac{1}{r^2}\frac{d}{dr}\left(r^2\frac{d\Phi}{dr}\right) = \frac{-\rho(r)}{\varepsilon_0} \tag{3.135}$$

or

$$\frac{d\Phi}{dr} = -\frac{1}{r^2}\int \frac{\rho(r)}{\varepsilon_0} r^2\,dr$$

The integral can be easily evaluated:

$$\frac{d\Phi}{dr} = \frac{\rho_0}{\varepsilon_0\alpha^3}e^{-\alpha r}\left(\frac{2}{r^2} + \frac{2\alpha}{r} + \alpha^2\right) + \frac{C}{r^2} \tag{3.143}$$

where C is a constant. The potential can now be found be integrating again. The result is

$$\Phi(r) = \frac{q}{4\pi\varepsilon_0 r}e^{-\alpha r} - \frac{\alpha q}{8\pi\varepsilon_0}e^{-\alpha r} - \frac{C}{r} + D \tag{3.144}$$

where D is another constant, and $\alpha^3 q/8\pi$ was substituted for ρ_0.

We now impose the boundary conditions in order to evaluate C and D. Since the charge density goes to zero as $r \to \infty$, we set the potential zero as $r \to \infty$. Therefore we take $D = 0$. To evaluate C, we apply Gauss' law on a closed shell of radius r and center at the origin. The electric field $\mathbf{E} = -d\Phi/dr\,\hat{\mathbf{r}}$ is given by the negative of the right-hand side of Eq. (3.143). Therefore

$$\oint \mathbf{E}\cdot\hat{\mathbf{n}}\,da = \int \rho\,dv = 4\pi\rho_0\int_0^r e^{-\alpha r}r^2\,dr$$

which gives $C = q/4\pi\varepsilon_0$. Thus the potential is

$$\Phi(r) = \frac{q}{4\pi\varepsilon_0 r}(1 - e^{-\alpha r}) - \frac{q\alpha}{8\pi\varepsilon_0}e^{-\alpha r} \tag{3.145}$$

and the electric field is

$$\mathbf{E} = \frac{q\hat{\mathbf{r}}}{4\pi\varepsilon_0 r^2}[1 - (\alpha r + 1)e^{-\alpha r}] - \frac{q\alpha^2}{8\pi\varepsilon_0} e^{-\alpha r}\hat{\mathbf{r}} \tag{3.146}$$

Example 3.19 Localized, Nonuniform, Spherically Symmetric Distribution

Let us now consider a localized, nonuniform, spherically symmetric charge distribution. Consider a concentric shell of charge of radii R_1 and R_2 $(R_2 > R_1)$. The charge density is given by $\rho = \beta/r^2$ where β is a constant and r is the distance from the center of the shell.

This example involves three regions of space where the potential will have distinct functional dependence. The potential satisfies Laplace's equation $\nabla^2\Phi_1 = 0$ for $R < R_1$, Poisson's equation $\nabla^2\Phi_2 = -\rho/\varepsilon_0$ for $R_1 < r < R_2$, and Laplace's equation $\nabla^2\Phi_3 = 0$ for $r > R_2$. The solutions for Φ_1 and Φ_3 are easily derived:

$$\Phi_1(r) = \frac{A_1}{r} + B_1 \qquad r < R_1 \tag{3.147}$$

$$\Phi_3(r) = \frac{A_3}{r} + B_3 \qquad r > R_2 \tag{3.148}$$

The potential inside the shell satisfies the following radial equation.

$$\frac{1}{r^2}\frac{d}{dr}\left(r^2\frac{d\Phi_2}{dr}\right) = -\frac{\beta}{\varepsilon_0 r^2} \qquad R_1 < r < R_2 \tag{3.149}$$

This equation can be easily integrated twice:

$$\Phi_2(r) = \frac{-\beta}{\varepsilon_0}\ln r - \frac{A_2}{r} + B_2 \qquad R_1 < r < R_2 \tag{3.150}$$

The six constants A_i and B_i are now determined from the following four boundary conditions.

1. Because the charge distribution is bounded, then the potential vanishes as $r \to \infty$, which requires that $B_3 = 0$.

2. Because of the requirement that the potential be finite as $r \to 0$, it is necessary that $A_1 = 0$.

3. The continuity of the potential at $r = R_1$ and $r = R_2$ gives the following relations:

$$B_1 = -\frac{\beta}{\varepsilon_0}\ln R_1 - \frac{A_2}{R_1} + B_2 \tag{3.151}$$

and

$$\frac{A_3}{R_2} = -\frac{\beta}{\varepsilon_0}\ln R_2 - \frac{A_2}{R_2} + B_2 \tag{3.152}$$

4. The last boundary condition is that the charge distribution and the total charge be given. This fact can be utilized by applying Gauss' law. First, we apply it at a spherical surface whose center at the origin and radius $r > R_2$:

$$\oint \mathbf{E}\cdot\hat{\mathbf{n}}\,da = \frac{1}{\varepsilon_0}\int\frac{\beta}{r^2}\,dv = \frac{4\pi\beta}{\varepsilon_0}(R_2 - R_1)$$

Substituting $\mathbf{E} = -\nabla\Phi_3 = (A_3/r^2)\hat{\mathbf{r}}$ gives $A_3 = \beta(R_2 - R_1)\varepsilon_0$. Similarly, we apply Gauss' law at a spherical surface of radius r, where $R_1 < r < R_2$. The electric field is

$$\mathbf{E} = -\nabla\Phi_2 = \left(\frac{\beta}{\varepsilon_0 r} + \frac{A_2}{r^2}\right)\hat{\mathbf{r}}$$

and the charge inside the surface is $4\pi\beta(r - R_1)$, thus Gauss' law gives $A_2 = \beta R_1/\varepsilon_0$. Substituting for A_2 and A_3 in Eqs. (3.151) and (3.152), we get $B_2 = \beta/\varepsilon_0(1 + \ln R_2)$ and

$B_1 = \beta \ln(R_2/R_1)/\varepsilon_0$. Substituting the determined values of A_i and B_i in Eqs. (3.147), (3.148), and (3.150) gives the following expressions for the potential:

$$\Phi_1(r) = \frac{\beta}{\varepsilon_0} \ln \frac{R_2}{R_1}$$

$$\Phi_2(r) = \frac{\beta}{\varepsilon_0}\left(1 - \ln \frac{r}{R_2} - \frac{R_1}{r}\right)$$

$$\Phi_3(r) = \frac{\beta}{\varepsilon_0} \frac{R_2 - R_1}{r}$$

Example 3.20 Charge Distribution Due to a Given Potential

Since Poisson equation relates the potential to the charge density, then it can be used to determine the charge density of a given potential. Consider the potential $\Phi(r) = (q/4\pi\varepsilon_0 r)e^{-\alpha r}$. The charge due to this potential is $\rho = -\varepsilon_0 \nabla^2 \Phi(r)$:

$$\rho = -\frac{1}{4\pi} \nabla^2 \left[\frac{q}{r} + \frac{q(e^{-\alpha r} - 1)}{r}\right]$$

where we added and subtracted $q/4\pi\varepsilon_0 r$, the potential of a point charge at the origin, to $\Phi(r)$. Evaluating, we get

$$\rho = q\delta(\mathbf{r}) - \frac{q}{4\pi} \alpha^2 \frac{e^{-\alpha r}}{r}$$

where $\delta(\mathbf{r})$ is the Dirac delta function. This potential corresponds to the point charge q at the origin and a spherically symmetric volume distribution. The total charge of the volume distribution is $\int \rho \, dv = -q$, indicating that the total charge is zero.

3.7 Electrostatic Shielding

We now discuss the concept of electrostatic shielding, which has considerable practical importance. The concept makes possible the creation of a region in space in which the electric field is vanishingly small. Such "field-free" regions are often required in experiments or for reliable operation of electronic devices.

By way of illustration, let us consider a charge-free cavity V inside a conductor, as shown in Fig. 3.28*a*. The cavity has well-defined boundaries, and therefore the potential is completely prescribed inside the cavity if the potential of the conductor, Φ_c, is specified. Then the potential inside the cavity, Φ, is simply the constant Φ_c. The condition of continuity of Φ at the boundary, and Laplace's equation $\nabla^2\Phi = 0$, are satisfied inside the cavity. Thus, $\mathbf{E} = 0$ there, and there is no charge on S. Whatever happens outside of the conductor is irrelevant to the inside.

This phenomenon illustrates *electrostatic shielding*. For example, if we placed a charged object inside the cavity, the electrostatic forces it would experience would depend not at all on conditions outside the conductor. These forces would depend only upon the charge, shape, and location of the object in the cavity.

In fact, if we apply Gauss' law to a closed surface inside the conductor as shown in Fig. 3.28*b*, we know that since $\mathbf{E} = 0$ everywhere in the conductor, then $\oint_S \mathbf{E} \cdot d\mathbf{a} = 0$, and consequently $Q_{\text{int}} = 0$. Therefore we conclude that an opposite charge is induced on the inner surface of the conductor, which acts to annul the field of the charged object. The total charge is equal to the object charge. If the conductor is electrically neutral, an equal charge must appear on the outer surface of the conductor. The manner in which this charge is distributed will depend *only* on

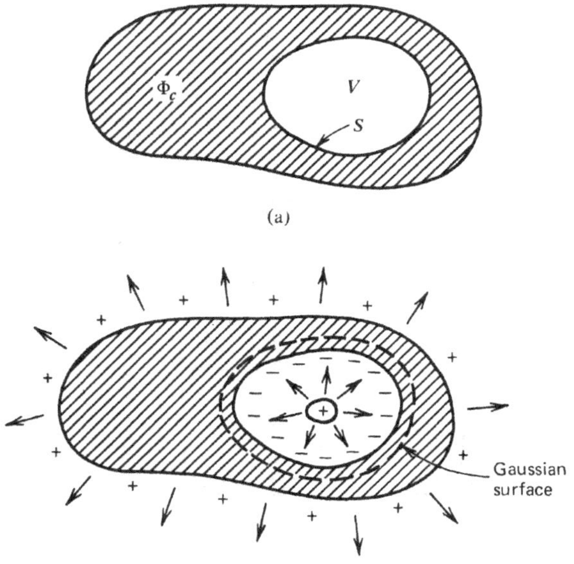

Figure 3.28 Illustrating electrostatic shielding. (*a*) A conductor with a cavity. (*b*) A positive charge inside the cavity.

external conditions, not on the distribution of the internal charge (by the uniqueness condition). So, in a sense, the region external to the conductor also is "shielded" from the internal region. Only if the *total* charge in the cavity of the outer conductor surface is changed will the external field be affected.* The *position* of the cavity charge is irrelevant to the "external world." Physically, this is not unreasonable, because the induced charge on the interior surface of the conductor does not permit the cavity field to penetrate the conductor. Therefore, the surface charge on the outer surface is "unaware" of the existence of the internal cavity. (If there are no lines of force between two objects, they do not interact.) In fact, if the object inside the cavity were another conductor, then upon touching this conductor to the interior surface, all the charge on the object would be transferred to the containing conductor and would appear on its surface. (Why?)

The latter property can be used to remove the net charge that may exist on a conductor, or to transfer charge to a conductor. For example, if a conductor is constructed with a hollow interior that is linked to the outside via a hole, as shown in Fig. 3.29, the interior $|\mathbf{E}|$ field can be made arbitrarily small by making the hole small compared to appropriate other dimensions. If, now, another charged conductor is inserted through the hole into the cavity where $\mathbf{E}$ initially was zero and is made to touch the interior surface of the cavity, the charge will flow from the initially charged conductor to the exterior surface of the larger one. This will occur irrespective of the conditions exterior to the large conductor. A device like this is sometimes called a *Faraday cup.* The same kind of phenomena are utilized when one wants to "charge up" the high-voltage terminal of a Van de Graaff electrostatic

* Even this effect can be eliminated if the (outer surface of the) conductor is grounded—i.e., kept at some constant reference potential.

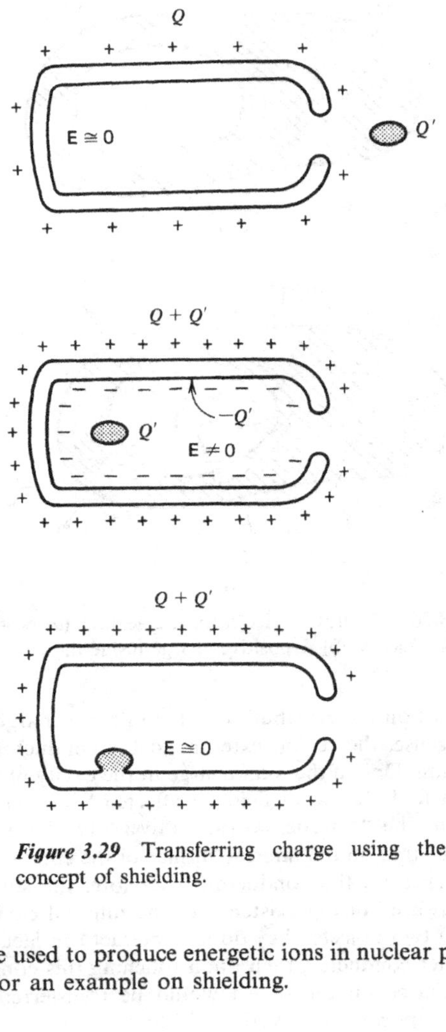

Figure 3.29 Transferring charge using the concept of shielding.

accelerator, a device used to produce energetic ions in nuclear physics (see Fig. 3.30). See Problem 3.26 for an example on shielding.

3.8 Summary

In the case of electrostatics, two relations that can be solved simultaneously are $\nabla \cdot \mathbf{E} = \rho/\varepsilon_0$ and $\mathbf{E} = -\nabla\Phi$. They may be combined into one equation—namely, $\nabla \cdot (-\nabla\Phi) = \rho/\varepsilon_0$—customarily written as

$$\nabla^2\Phi = -\frac{\rho}{\varepsilon_0} \tag{3.3}$$

and called Poisson's equation. If $\rho = 0$ in some region of space, then in that region Poisson's equation becomes what is known as Laplace's equation,

$$\nabla^2\Phi = 0 \tag{3.4}$$

A unique solution to the electrostatic field exists in the space free of charge, that is Laplace's equation has a unique solution, so long as certain boundary conditions are satisfied: (1) The potential on each bounding surface is specified; (2) the net charge on each

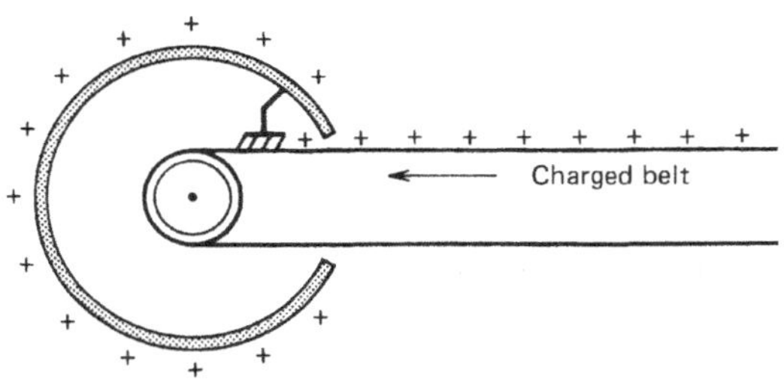

Figure 3.30 Terminal of Van de Graaff machine. The belt charge removed by brushes flows to outside surface of terminal.

bounding surface is specified; or (3) a mixture of (1) and (2), but for a given boundary one may specify either the potential or the total charge (not both at once).

The method of separation of variables can be used to solve Laplace's equation in the three-coordinate systems, which gives solutions composed of products of functions each of which depend on only one coordinate. In case of some degree of symmetry, where one can argue that the potential is a function of one coordinate—i.e., r in spherical, ρ in cylindrical, and z in cartesian coordinates—Laplace's equation reduces to a one-dimensional differential equation with straightforward solutions. In case of lower symmetry—i.e., the potential is a function of r and θ in spherical systems and ρ and ϕ in cylindrical systems—the solution is generally a linear combination of an infinite set of zonal harmonics and cylindrical harmonics, respectively. The uniqueness theorem will indicate if enough of these harmonics have been used to describe a given configuration.

The method of images is a powerful technique for solving electrostatic problems in a volume V where the charge density is not zero in the space not bounded on surfaces. It is applied to problems involving point charges (can be generalized to general charge densities) near large conducting plates or conducting spheres. Also it is applied to long line charges parallel to long conducting cylinders. In the technique one introduces a configuration of fictitious image charges outside the volume V such that the potential of these image charges plus that of the real ones inside V satisfies the prescribed boundary condition for the potential on the boundary of the space V considered. If this is satisfied, then the uniqueness theorem ensures that the chosen configuration does indeed give the correct potential and, hence, the correct field.

The solution of Poisson's equation in general can be determined by adding the solution of Laplace's equation to the particular solution (Coulomb's law); that is,

$$\Phi = \frac{1}{4\pi\varepsilon_0}\int_V \frac{\rho\, dv'}{|\mathbf{r} - \mathbf{r}'|} + \text{solution of Laplace's equation} \tag{3.134}$$

where the integration in Coulomb's law is carried over the given charge distribution. The overall solution is then made to satisfy the boundary conditions by choosing constants in the solution of Laplace's equation.

When Laplace's or Poisson's equation is solved in more than one region, the differential properties of the electrostatic field require certain relationships between the fields and also between the potentials at the common boundaries of the regions. The equation $\nabla \times \mathbf{E} = 0$ implies that

$$\mathbf{E}_{1t} = \mathbf{E}_{2t} \qquad \text{or} \qquad \Phi_1 = \Phi_2 \tag{3.6),(3.8}$$

when one is crossing the boundary, where t means the component of $\mathbf{E}$ tangent to the boundary. On the other hand, the equation $\nabla \cdot \mathbf{E} = \rho/\varepsilon_0$ implies that

$$E_{2n} - E_{1n} = \frac{\sigma}{\varepsilon_0} \tag{2.35}$$

where n means the component of $\mathbf{E}$ normal to the boundary, and σ is the surface charge density at the boundary. Near an isolated point charge q, the electric field should approach

$$\mathbf{E} \to \frac{1}{4\pi\varepsilon_0}\frac{q\mathbf{r}}{r^3} \qquad \text{as } \mathbf{r} \to 0 \tag{3.5}$$

where $\mathbf{r}$ is the distance from the point charge. Similarly, near an isolated long line charge of λ per unit length, the field should approach

$$\mathbf{E} \to \frac{1}{2\pi\varepsilon_0}\frac{\lambda\boldsymbol{\rho}}{\rho^2} \qquad \text{as } \boldsymbol{\rho} \to 0 \tag{2.19}$$

where $\boldsymbol{\rho}$ is the distance from the line.

Problems

3.1 An infinite conducting plate is grounded. A conducting cone of angle θ_1 and large height is placed normal to the plate with its vertex at the plate. The cone is insulated from the plane and kept at potential V. (a) Determine the potential and the electric field in the region between the plate and the cone. (b) Determine the charge density on the plate.

3.2 Consider two large metallic plates forming a wedge capacitor, as shown in Fig. 3.31. The plate at $\phi = 0$ is kept at zero volts while the plate at $\phi = \beta$ is kept at V volts. Neglect fringing effects. (a) Write down the differential equation satisfied by the potential inside the capacitor and determine the potential. (b) Determine the charge density and the total charge residing on the plates.

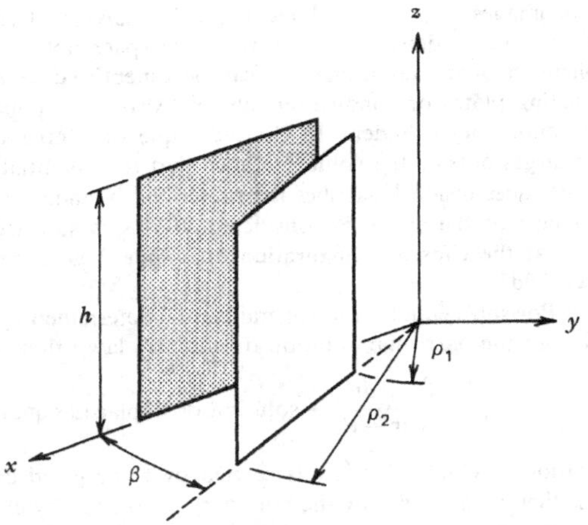

Figure 3.31

3.3 Consider the conic capacitor discussed in Example 3.4. Determine the electric field inside the capacitor and the charge distribution on the cones.

3.4 Consider a spherical surface of radius R that is kept at a potential $\Phi(R, \theta) = V_0 \cos\theta$, where V_0 is a constant and θ is measured with respect to a z axis passing through its center. (a) Write down expressions for the electric potential in the regions $r < R$ and $r > R$. (b) Write down the boundary conditions and determine the potential in both regions. (c) Determine the electric field on the surface of the sphere. Can the spherical surface be a conductor? (d) Determine the charge density at $r = R$. (e) What is the electric dipole moment of the sphere?

3.5 Consider the nonconcentric spherical capacitor discussed in Example 3.5. Determine the field between the spheres and the charge distribution on the spheres. What is the total charge residing on the outer sphere?

3.6 An insulating thin spherical shell of radius R is kept at a potential $V = V_0 \cos 2\theta$, where V_0 is a constant and θ is the angle relative to a diameter of the sphere. Determine (a) the potential everywhere and (b) the charge distribution on the shell.

3.7 An insulating thin spherical shell of radius R has a surface charge distribution given by $\sigma = \sigma_0(\cos\theta - 1)^2$. Determine the potential produced by the sphere everywhere.

3.8 A long, insulating cylinder of radius ρ_0 has a charge density $\sigma = \sigma_0 \cos 3\phi$, where σ_0 is a constant and ϕ is the angle from the x axis. Determine (a) the potential inside and outside the cylinder and (b) the electric field inside and outside the cylinder.

3.9 A plane rectangular trough shown in Fig. 3.32 is of lengths y_0 and z_0. The sides oa, oc, and ab are connected together and grounded. The side cb is isolated and kept at potential V_0. Calculate the potential inside the trough in the y-z plane (see Example 3.11).

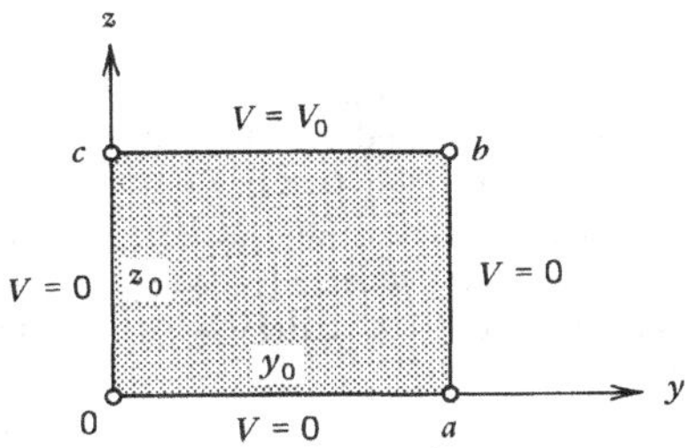

Figure 3.32

3.10 Two grounded conducting planes intersect at 45° and a point charge q lies between them. Determine the positions of the image charges that will give the electric field between the planes.

3.11 A charge q is placed at a distance $l/2$ from the center of a grounded conducting sphere of radius R such that $l \ll R$. Determine the charge distribution on the sphere and the force exerted on the charge.

3.12 A conducting plate has a semispherical boss of radius R with a center on the plate as shown in Fig. 3.33. The plate is grounded and a point charge q is brought next to it at a distance $D > R$. The charge is on the normal to the plate passing through the center of the boss. (a) Determine the image charge needed to replace the plate. (b) Determine the potential on the side of the charge. (c) Determine the charge induced on the boss. (d) Determine the force between the charge and the plate.

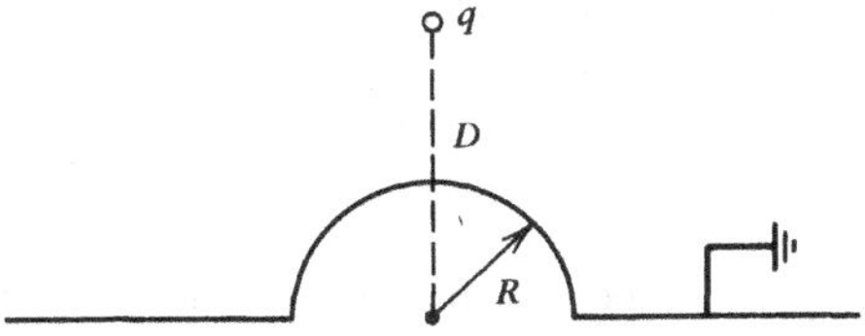

Figure 3.33

3.13 An electric dipole of moment **p** is placed near a conducting sphere of radius R_0 at a right angle with the line between the dipole and the center of the sphere. Determine the force between the dipole and the sphere (a) when the sphere is grounded, and (b) when the sphere is isolated and carries a charge q.

3.14 Calculate the torque acting on the dipole of Problem 3.13.

3.15 An electric dipole of moment **p** is placed near a conducting grounded sphere of radius R_0 along the line between the center of the sphere and the dipole. Determine the force and the torque between the sphere and the dipole.

3.16 A long, conducting cylinder of radius R is parallel to a large, grounded conducting plate and at a distance d from it. The cylinder carries a charge λ per unit length. Determine the charge distribution of the plate.

3.17 Consider two parallel, conducting cylinders each of radius R. The axes are placed at a distance Δ from each other. The cylinders carry a charge of $\pm\lambda$ per unit length. Determine (a) the charge distribution on the surfaces of the cylinders and (b) the force per length between them.

3.18 Consider a very long cylinder of radius ρ_0, charged uniformly with a volume charge density α. For $\rho > \rho_0$ the charge density is zero. Determine the potential inside and outside the cylinder.

3.19 A concentric shell of charge of radii R_1 and R_2 ($R_2 > R_1$) has a charge density of $\rho = \beta/r$, where β is a constant and r is the distance from the center of the shell. Determine the potentials everywhere (see Example 3.19).

3.20 Take the electric charge of an atomic nucleus, ze, to be uniformly distributed over the volume of a sphere of radius R_0, where z is the atomic number and e is the magnitude of the charge of an electron. Determine the electric potential at a distance $r \leq R_0$ from the center of the sphere.

3.21 The charge density in the region $-z_0 < z < z_0$ depends only on z; that is,

$$\rho = \rho_0 \cos \frac{\pi z}{z_0}$$

where ρ_0 and z_0 are constants. Determine the potential in all regions of space.

3.22 Consider a periodic volume charge distribution $\rho(x, y, z) = \rho_0 \sin a_1 x \sin a_2 y \sin a_3 z$, where ρ_0, a_1, a_2, and a_3 are constants. Determine the electric potential as a function of x, y, and z.

3.23 Consider a periodic surface charge distribution

$$\sigma = \sigma_0 \cos a_1 x \cos a_2 y$$

where σ_0, a_1, and a_2 are constants. Determine the potential (a) in the x-y plane and (b) at any point in space.

3.24 The electrostatic potential due to a volume charge density is given by

$$\Phi(\mathbf{r}) = \frac{1}{4\pi\varepsilon_0} e^{-\alpha r^3}$$

where α is a constant. Calculate the volume charge density.

3.25 Determine the charge distribution that produces the potential

$$\Phi(\mathbf{r}) = \frac{q}{4\pi\varepsilon_0}\left(\frac{1}{r} + \frac{1}{a}\right)\exp\left[\frac{-2r}{a}\right]$$

where a is a constant.

3.26 Suppose we have a uniform electric field over a region of space and we insert into this space a thin, but very large, conducting plate, as shown in Fig. 3.34. (a) Determine the potential in the regions $z < 0$ and $z > a$. Take the potential of the conductor to be zero

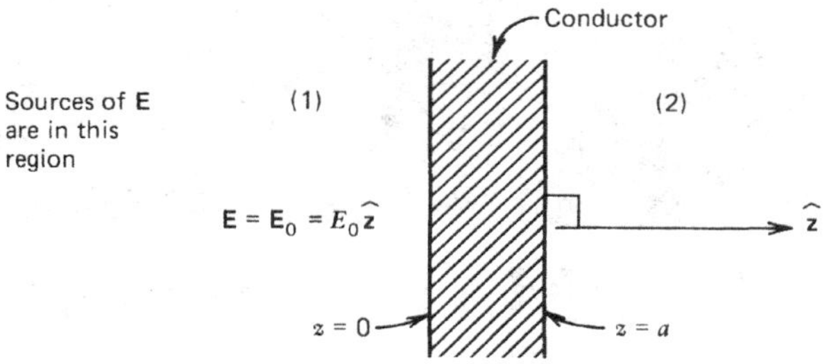

Figure 3.34

(not grounded). (b) Determine the charge density at $z = 0$ and $z = a$. (c) Sketch the lines of force of the electric field. Now the plate is grounded. (d) Determine the charge densities at $z = 0$ and $z = a$, and the electric field in $z > a$ region. (e) Sketch the lines of force of the electric field. Is there any shielding effect? (f) Determine the force per unit area on the plate before and after grounding.

FOUR

FORMAL THEORY OF DIELECTRIC ELECTROSTATICS

In this chapter we shall take on electrostatic problems that involve dielectric media. The treatment thus far concerned itself with charge distribution "in vacuo" where charges resided on conductors or were given in specified volume, surface, or line distributions. When considering charge distributions in the presence of matter, we have to deal with the fact that these distributions induce charge distributions in the atomic and molecular constituents (polarization) that are otherwise neutral. These induced charge distributions produce fields, which in turn may affect the external distributions. It is the purpose of this chapter to describe some techniques for treating these effects on electrostatic problems.

4.1 Polarization and Dipole Moment Density

In this chapter we shall not get into detailed microscopic questions regarding the dielectric properties of various kinds of matter. Rather, we shall assume that matter has certain microscopic properties to be regarded as given, and we shall attempt to describe the electrostatic situation in terms of these properties. Nonetheless, it is useful to keep in mind the relevant underlying microscopic phenomena. We start, for example, by postulating that the macroscopic fields produced by the atomic aggregates of matter, hereafter simply called molecules, are characterizable completely by their monopole and dipole moments, higher multipole effects being negligible. Since we shall be concerned only with dimensions much greater than molecular dimensions and molecules that are electrically neutral, the dominating electrical effects for a collection of neutral molecules will be *dipole effects.*

Our immediate aim is to describe matter in a macroscopic way so that it can be directly related to macroscopic fields, by which we mean fields averaged over macroscopic spacial dimensions and time intervals. Because of this point of view, we tend to focus upon the dipole moments of macroscopically small volume elements, which will in general contain the order of 10^{10} molecules or more. This implies that the electrical condition of these volume elements will tend to vary smoothly within the material. By this we mean that in the material there exist dipole moment densities $\mathbf{P}(\mathbf{r})$ that are smooth functions of position. Thus the vector sum $d\mathbf{p}$ of the dipole moments in a volume dv is

$$d\mathbf{p}(\mathbf{r}) \equiv \mathbf{P}(\mathbf{r})dv \tag{4.1}$$

where $\mathbf{P}$, the *dipole moment density*, is usually called the *polarization* of the medium at dv. It is the (vector) field that is used to characterize the dipole properties of the medium. Those media that are well described in terms of a polarization are called *dielectrics*. It is clear that for a volume V of dielectric, its dipole moment will be given by

$$\mathbf{p} = \int_V \mathbf{P}\,dv \tag{4.2}$$

4.2 Fields Due to a Dielectric Medium

We consider now a medium in which the polarization is finite everywhere and is continuous except possibly at the bounding surface. Consider Fig. 4.1. For every element dv', located by $\mathbf{r}'$, of this medium, we have a dipole potential contribution to the potential at a point $\mathbf{r} \equiv (x, y, z)$. Using Eq. (2.44), we get

$$d\Phi_P(\mathbf{r}) = \frac{1}{4\pi\varepsilon_0}\frac{(\mathbf{r}-\mathbf{r}')\cdot d\mathbf{p}}{|\mathbf{r}-\mathbf{r}'|^3} = \frac{1}{4\pi\varepsilon_0}\frac{d\mathbf{p}\cdot\boldsymbol{\xi}}{\xi^3} \tag{4.3}$$

Using Eq. (4.1), we can write this potential in terms of the polarization

$$d\Phi_P(\mathbf{r}) = \frac{1}{4\pi\varepsilon_0}\frac{\mathbf{P}(\mathbf{r}')\cdot\boldsymbol{\xi}\,dv'}{\xi^3}$$

The term $\boldsymbol{\xi}/\xi^3$ is now written as $-\nabla(1/\xi)$; thus

$$d\Phi_P(\mathbf{r}) = -\frac{\mathbf{P}(\mathbf{r}')}{4\pi\varepsilon_0}\cdot\nabla\left(\frac{1}{\xi}\right)dv' \tag{4.4}$$

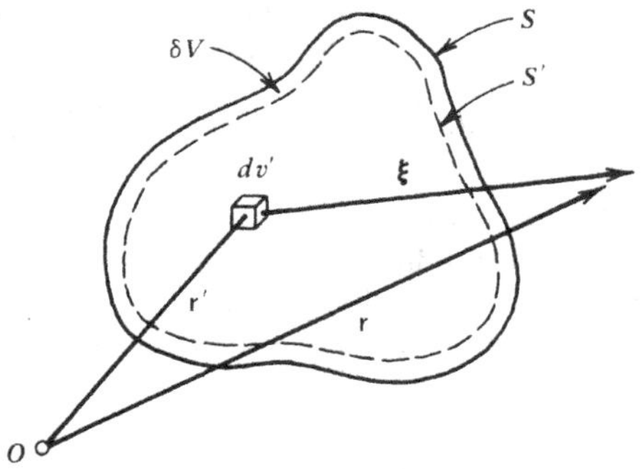

Figure 4.1 A polarized piece of material. Inside S' the polarization is continuous and may be discontinuous at S.

where the operator ∇ involves derivatives with respect to the coordinates of the point $\mathbf{r}(x, y, z)$. The total contribution from the medium, which we assume to occupy a volume V, is therefore the finite result

$$\Phi_P(\mathbf{r}) = -\int_V \frac{\mathbf{P}(\mathbf{r}')}{4\pi\varepsilon_0} \cdot \nabla\left(\frac{1}{\xi}\right) dv' \tag{4.5}$$

This potential can be rewritten in terms of ∇' by noting that

$$\nabla[\xi^{-1}] = -\nabla'[\xi^{-1}] \tag{4.6}$$

whereas ∇' involves derivatives with respect to the coordinates of the point $\mathbf{r}'(x', y', z')$. Thus

$$\Phi_P(\mathbf{r}) = \int_V \frac{\mathbf{P}(\mathbf{r}')}{4\pi\varepsilon_0} \cdot \nabla'\left(\frac{1}{\xi}\right) dv' \tag{4.7}$$

Consider now a volume V' (shown in Fig. 4.1) contained in V so defined that everywhere in V' the polarization is continuous and equal to its value in V; hence $V \approx V' + \delta V$, where V has the surface of the medium as a boundary (where $\mathbf{P}$ is discontinuous). The volume V' is defined so that $\delta V \to 0$; thus V' is bounded by a surface S' that lies just inside the surface S that encloses V. Since $\mathbf{P}$ is finite everywhere in V, and ξ is continuous everywhere in V, we may write

$$\Phi_P(\mathbf{r}) = \int_V \frac{\mathbf{P}(\mathbf{r}')}{4\pi\varepsilon_0} \cdot \nabla'\left(\frac{1}{\xi}\right) dv' = \int_{V'} (\text{the same integrand}) \tag{4.8}$$

Using the identity given in Eq. (1.57) we find

$$\nabla' \cdot \left[\frac{\mathbf{P}(\mathbf{r}')}{\xi}\right] = \frac{\nabla' \cdot \mathbf{P}(\mathbf{r}')}{\xi} + \mathbf{P}(\mathbf{r}') \cdot \nabla'\left(\frac{1}{\xi}\right)$$

which, when substituted into Eq. (4.8), gives

$$\Phi_P(\mathbf{r}) = \int_{V'} \nabla' \cdot \left[\frac{\mathbf{P}(\mathbf{r}')}{4\pi\varepsilon_0 \xi}\right] dv' + \int_{V'} -\frac{\nabla' \cdot \mathbf{P}(\mathbf{r}')}{4\pi\varepsilon_0 \xi} dv' \tag{4.9}$$

Using the divergence theorem transforms Φ_P to

$$\Phi_P(\mathbf{r}) = \oint_{S'} \frac{P(\mathbf{r}') \cdot d\mathbf{a}'}{4\pi\varepsilon_0 \xi} + \int_{V'} -\frac{\nabla' \cdot \mathbf{P}(\mathbf{r}')}{4\pi\varepsilon_0 \xi} dv' \tag{4.10}$$

As $V' \to V$, S' becomes S, and the second integral involves only the continuous interior values of $\mathbf{P}$, whose derivatives are finite. Thus

$$\Phi_P(r) = \frac{1}{4\pi\varepsilon_0}\left[\oint_S \frac{\mathbf{P}(\mathbf{r}') \cdot \hat{\mathbf{n}}\, da'}{\xi} + \int_V -\frac{\nabla' \cdot \mathbf{P}(\mathbf{r}')}{\xi} dv'\right] \tag{4.11}$$

where $\hat{\mathbf{n}}$ is the outward normal from V on S.

Now electrostatic fields are produced by charges, and in any physical problem there is a unique charge distribution* that produces a given electrostatic field. If we were given the charge densities associated with this charge distribution in V, we would write the potential as

$$\Phi_P(\mathbf{r}) = \frac{1}{4\pi\varepsilon_0} \int \frac{dq}{\xi}$$

* We are assuming that $\{E, \Phi\}$ is defined throughout all of space.

or

$$\Phi_P(\mathbf{r}) = \frac{1}{4\pi\varepsilon_0}\int_V \frac{\rho_P\,dv'}{\xi} + \frac{1}{4\pi\varepsilon_0}\oint_S \frac{\sigma_P\,da'}{\xi} \tag{4.12}$$

We are thus led to the supposition that the polarization **P** associated with the volume of material V gives rise to the physical charge distribution with volume and surface charge densities

$$\rho_P = -\nabla'\cdot\mathbf{P}(\mathbf{r}') \qquad \text{and} \qquad \sigma_P = \mathbf{P}(\mathbf{r}'_S)\cdot\hat{\mathbf{n}} \tag{4.13}$$

where $d\mathbf{a}' \equiv \hat{\mathbf{n}}\,da'$, $\hat{\mathbf{n}}$ being the outward normal from V at the location of the surface area da', $\mathbf{r}'$ and $\mathbf{r}'_S$ denoting points in the volume and on the surface where **P** is being evaluated.

The electric field produced by the polarized material can also be written in terms of the charge densities of Eq. (4.13); that is,

$$\mathbf{E}_P(\mathbf{r}) = \frac{1}{4\pi\varepsilon_0}\int_V \frac{\rho_P\boldsymbol{\xi}}{\xi^3}\,dv' + \frac{1}{4\pi\varepsilon_0}\oint_S \frac{\sigma_P\boldsymbol{\xi}}{\xi^3}\,da' \tag{4.14}$$

In conclusion, we assert that a material region in which the polarization **P** is known can always be represented, insofar as exterior determinations of macroscopic electrostatic potentials and fields are concerned, by charge densities ρ_P and σ_P given in Eq. (4.13). We call these charge densities the *polarization* or *bound charge*, in order to distinguish it from other "*free*" charge that is not conveniently associated with a polarization process. This free charge will, of course, make its own additional contribution to the electrostatic field. By superposition, $\Phi(\mathbf{r}) = \Phi_P(\mathbf{r}) + \Phi'(\mathbf{r})$, $\mathbf{E}(\mathbf{r}) = \mathbf{E}_P(\mathbf{r}) + \mathbf{E}'(\mathbf{r}) = -\nabla\Phi_P(\mathbf{r}) - \nabla\Phi'(\mathbf{r})$, where as we have shown above $\{\Phi', \mathbf{E}'\}$ are due to all other charges. We emphasize that $\{\Phi, \mathbf{E}\}$ are static macroscopic fields, and represent a space and time average over macroscopic volume elements.

These charge densities can be used to determine the dipole moment or any other moment of the polarized material. For example, using Eq. (2.63), $\mathbf{p} = \int \mathbf{r}'\,dq$, one finds (See also definition 4.2)

$$\mathbf{p} = \oint \sigma_p\mathbf{r}'\,da' + \int \rho_p\mathbf{r}'\,dv'$$

or, explicitly,

$$\mathbf{p} = \oint_S \mathbf{r}'(\mathbf{P}\cdot\hat{\mathbf{n}})da' - \int_V \mathbf{r}'(\nabla\cdot\mathbf{P})dv' \tag{4.15}$$

Polarization effects are likely to contribute significant macroscopic fields only when the matter density is reasonably high—i.e., in liquids or solids. The effects from gases are smaller by a factor roughly proportional to matter density. Also, it is useful to characterize matter by a polarization only if the (macroscopic groups of) molecular constituents acquire dipole moments. In most metals, for example, the atoms acquire no discernible dipole moments (at least macroscopic groups of them acquire no net dipole moment), and so we assign no dielectric properties to most metals. Most insulators, however, will have macroscopic dipole (dielectric) properties.

It is helpful to envision the process whereby **P** gives rise to ρ_P and σ_P (see Fig. 4.2). The creation of **P** means the creation of macroscopic dipole elements, which in turn means the creation of a charge separation—positive charges being displaced from negative charges. One may regard the dielectric medium electrically as being

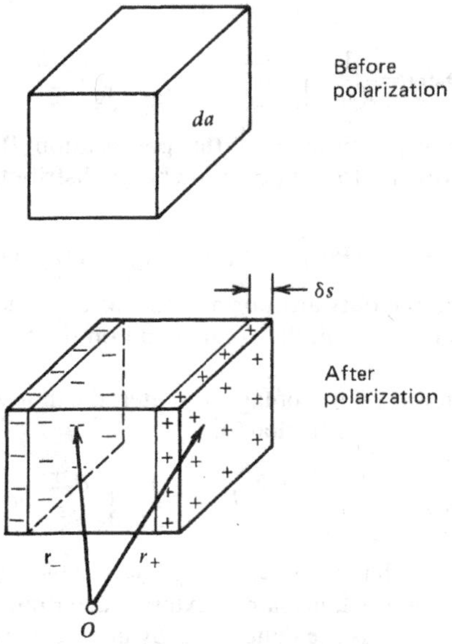

Figure 4.2 The process whereby **P** gives rise to ρ_P and σ_P. (*a*) A block of dielectric before polarization. (*b*) The block after polarization.

composed of two superimposed macroscopic charge densities: ρ_+ and ρ_- (see Example 4.2 for a specific case). The ρ_+ arises from the average charge density of the positively charged constituents of the medium (the atomic nuclei), and the ρ_- from the negatively charged constituents (the electrons). If the medium is unpolarized, $\rho_+ = -\rho_-$ everywhere in the medium, and therefore in every little macroscopic volume element, dv, of the medium, there is charge neutrality. On the other hand, when the medium is polarized, the ρ_+ charge becomes displaced with respect to the ρ_- charge by a tiny displacement $\delta\mathbf{s}$ of the order of 10^{-13} m, and is therefore infinitesimally small compared to the dimensions of macroscopic volume elements dv ($\approx 10^{-3}$ cm)^3. Therefore, across any surface element *imagined* in the medium, $d\mathbf{a}$, a charge equal to $[\delta\mathbf{s}\cdot d\mathbf{a}]\rho_+$ will move. But, as shown below, $\rho_+\delta\mathbf{s}$ is just the polarization $\mathbf{P}(\mathbf{r})$ of the medium. Therefore $\mathbf{P}\cdot d\mathbf{a}$ is the charge that moves across element da in the polarization process.

Let us see why in our macroscopic model the polarization $\mathbf{P}$ is equal to $\rho_+\delta\mathbf{s}$. Consider a cubical volume element, dv, of the medium (shown in Fig. 4.2). Before the medium is polarized, the charge in dv contributes no dipole moment. After the medium becomes polarized, the ρ_+ charge in dv will have moved by an average displacement $\delta\mathbf{s}$ relative to the ρ_- charge. Orienting dv so that $\delta\mathbf{s}$ lies along an edge, we can see from the figure that the charge originally in dv redistributes itself so that there are uncompensated regions of positive and negative charge around the two end surfaces of dv. The dipole moment thus acquired by the charge originally in dv is equal to $[\rho_+ \, da \, \delta s\mathbf{r}_+ + \rho_- \, da \, \delta s\mathbf{r}_-]$, where da is the area of a face of dv and $\mathbf{r}_+$ and $\mathbf{r}_-$ are vectors that locate the centers of the uncompensated charge regions. Consequently

$$d\mathbf{p} = \rho_+ \delta s \, da[\mathbf{r}_+ - \mathbf{r}_-] = \rho_+ \, dv[\delta\mathbf{s}]$$

Since, by definition, $d\mathbf{p} = \mathbf{P}\,dv$, we obtain the result

$$\mathbf{P} = \rho_+ \delta\mathbf{s} \tag{4.16}$$

For any volume element dv of surface area S, the charge that moves out of dv in the polarization process is then equal to $\oint_S \mathbf{P}\cdot d\mathbf{a}$. Since the material was electrically neutral originally, the charge remaining in dv is $-\oint \mathbf{P}\cdot d\mathbf{a}$. Therefore the charge density in dv is

$$\rho_P = -\frac{1}{dv}\oint_S \mathbf{P}\cdot d\mathbf{a} \tag{4.17}$$

This is just the defining expression for the divergence of $\mathbf{P}$ [see Eq. (1.37)]. Thus, there is a polarization charge density given by $\rho_P = -\nabla\cdot\mathbf{P}$, as was given in Eq. (4.13). If $\rho_P \neq 0$, a net charge equal to $\rho_P\,dv$ must have passed through $S(dv)$ in the process of polarization, then on actual surfaces of a dielectric, because the charge that moves across the surface is uncompensated, one will observe a surface charge density σ_P equal to $\sigma_P = (\mathbf{P}\cdot\hat{\mathbf{n}})$, as was given in Eq. (4.13), where $\hat{\mathbf{n}}$ is the outwardly directed unit vector perpendicular to the surface. In the model we have used, the charge on the surface actually resides in a layer of thickness $(\delta\mathbf{s}\cdot\hat{\mathbf{n}})$, which is negligible.

In conclusion, we see that a knowledge of the polarization $\mathbf{P}$ allows one to deduce the macroscopic charge densities that exist by virtue of that polarization. These charge densities create the associated macroscopic fields. The truth of the latter statement is established by considering the equations of electrostatics in their microscopic derivative forms, Eqs. (2.33) and (2.37): $\nabla\cdot\mathbf{E} = \rho/\varepsilon_0$ and $\nabla\times\mathbf{E} = 0$. Taking the space and time average of these equations over a volume element dv, we obtain

$$\langle\nabla\cdot\mathbf{E}\rangle = \frac{\langle\rho\rangle}{\varepsilon_0} \qquad \langle\nabla\times\mathbf{E}\rangle = 0 \tag{4.18}$$

Now we can take $\langle\nabla\cdot\mathbf{E}\rangle = \nabla\cdot\langle\mathbf{E}\rangle$, and $\langle\nabla\times\mathbf{E}\rangle = \nabla\times\langle\mathbf{E}\rangle$ (see Problem 4.21). This means that the equations governing the macroscopic electrostatic fields are given by

$$\nabla\cdot\langle\mathbf{E}\rangle = \frac{\langle\rho\rangle}{\varepsilon_0} \qquad \nabla\times\langle\mathbf{E}\rangle = 0 \tag{4.19}$$

These are the same forms as the fundamental microscopic equations. Thus the charge densities obtained from knowledge of the polarization are valid in calculating the average electrostatic fields, whether they be inside or outside the polarized medium, although in point of fact it is rare that one wants to know the field $\langle\mathbf{E}\rangle$ inside matter. In ordinary applications one usually has one's instruments outside of the medium itself, perhaps in little cavities carved in it. An exception to this statement concerns the passage of charged particles through matter. The net deflection of these particles does depend on $\langle\mathbf{E}\rangle$, given above.

Example 4.1 Polarization Charges—Cylindrical Electret—Dipole Layer

As a first example we consider a solid right circular cylinder that has a constant polarization $\mathbf{P} = P\hat{\mathbf{z}}$ along its axis (see Fig. 4.3). There do indeed exist materials that can maintain such a polarization in the absence of an applied electric field. Such materials are called *electrets*, and the associated phenomenon of a "residual" polarization is usually called *ferroelectricity* (see

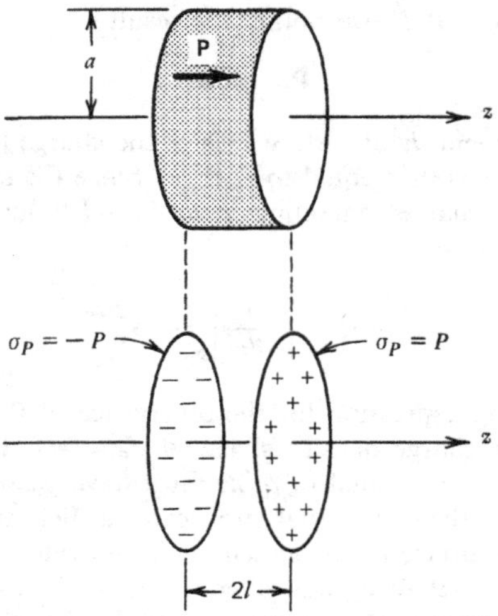

Figure 4.3 A uniformly polarized cylinder along its axis is represented by surface polarizations at its ends.

Chapter 5). In these materials the internal macroscopic electric fields maintain the polarization; i.e., they keep the dipoles aligned. Clearly, such substances are highly *nonlinear*, not "simple."

The macroscopic electric field generated by our electret may be found from the charge densities generated by the polarization $\sigma_P = \mathbf{P}\cdot\hat{\mathbf{z}}$ at end faces of rod, $\sigma_P = 0$ on cylindrical sides of rod, and the volume density $\rho_P = -\nabla\cdot\mathbf{P} = 0$. Consequently the rod is represented electrically by two circular disks of charge, with charge densities $\pm P$, separated by a distance $2l$. The electric field on the axis of a charged single disk with a constant surface charge density was determined in Example 2.11. With a charge density σ_P, Eq. (2.49) gives

$$E = |E_z| = \frac{\sigma_P}{2\varepsilon_0}\left(1 - \frac{|z|}{\sqrt{z^2 + a^2}}\right)$$

With E pointing away from the disk if σ_P is positive, and $|z|$ is the distance from the disk. If we have two disks, with charge densities $-P$ and $+P$, separated by a distance $2l$, and we take the origin at a point on the axis midway between the disks, then the field "outside" the disks on the axis is given by

$$\mathbf{E}_> = \frac{\hat{\mathbf{z}}P}{2\varepsilon_0}\left[1 - \frac{z-l}{\sqrt{(z-l)^2 + a^2}}\right] - \frac{\hat{\mathbf{z}}P}{2\varepsilon_0}\left[1 - \frac{z+l}{\sqrt{(z+l)^2 + a^2}}\right]$$

or

$$\mathbf{E}_> = \frac{\mathbf{P}}{2\varepsilon_0}\left[\frac{z+l}{\sqrt{(z+l)^2 + a^2}} - \frac{z-l}{\sqrt{(z-l)^2 + a^2}}\right] \tag{4.20}$$

Equation (4.20) can alternatively be written in terms of the angles θ_1 and θ_2 shown in Fig. 4.4*a*; that is,

$$\mathbf{E}_> = \frac{\mathbf{P}}{2\varepsilon_0}[\cos\theta_2 - \cos\theta_1] \tag{4.21}$$

Similarly, the internal field between the disks is given on the axis by

$$\mathbf{E}_< = -\frac{\hat{\mathbf{z}}P}{2\varepsilon_0}\left[1 - \frac{l-z}{\sqrt{(l-z)^2 + a^2}}\right] - \frac{\hat{\mathbf{z}}P}{2\varepsilon_0}\left[1 - \frac{l+z}{\sqrt{(l+z)^2 + a^2}}\right] \tag{4.22}$$

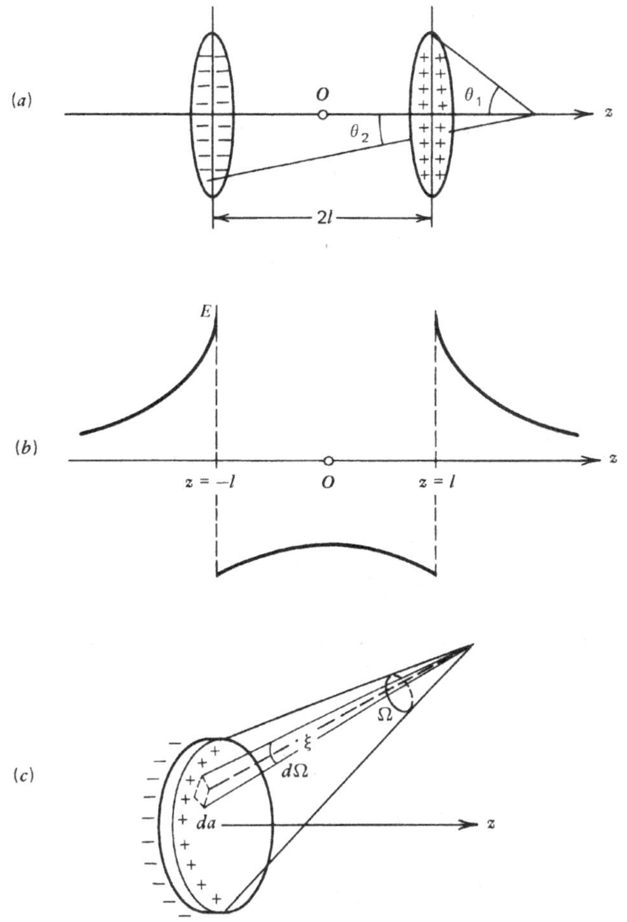

Figure 4.4 Fields and potentials of the uniformly polarized cylinder of Fig. 4.3. (*a*) Schematic diagram of the polarization charges. (*b*) The **E** field at the axis as a function of distance along *z*. (*c*) The limit of a short cylinder.

or

$$\mathbf{E}_< = -\frac{\mathbf{P}}{2\varepsilon_0}(2 + \cos\theta_1 - \cos\theta_2) \tag{4.23}$$

Figures 4.4*b* shows a plot of E as a function of z on the axis of the cylinder.

As the field point approaches the right-hand face of the disk (that is, $z = l$ or $\cos\theta_1 = 0$), we obtain from Eq. (4.21)

$$\mathbf{E}_> (z \to l) = \frac{\mathbf{P}}{2\varepsilon_0}\cos\theta_2 \tag{4.24}$$

and, from Eq. (4.23),

$$\mathbf{E}_< (z \to l) = -\frac{\mathbf{P}}{2\varepsilon_0}[2 - \cos\theta_2] \tag{4.25}$$

Subtracting Eq. (4.25) from Eq. (4.24) gives

$$\mathbf{E}_> - \mathbf{E}_< = \frac{\mathbf{P}}{2\varepsilon_0}(\cos\theta_2 + 2 - \cos\theta_2) = \frac{\mathbf{P}}{\varepsilon_0} \tag{4.26}$$

which shows that the normal component of **E** is discontinuous. The discontinuity is due to the existence of σ_P on the disk.

Finally, we note that as $\theta_2 \to \theta_1 \to \pi/2$, which corresponds to a disk of infinite diameter, then Eq. (4.21) reduces to

$$\mathbf{E}_> = \frac{\mathbf{P}}{2\varepsilon_0}\left(\cos\frac{\pi}{2} - \cos\frac{\pi}{2}\right) = 0, \tag{4.27}$$

and Eq. (4.23) reduces to

$$\mathbf{E}_< = -\frac{\mathbf{P}}{\varepsilon_0} \tag{4.28}$$

These fields are similar to the fields of an infinite-area parallel-plate capacitor to be discussed in more detail in Chapter 6. It is interesting to remark that as $l \to 0$ $(2l = \delta)$ we may regard the resulting disk of area S as a dipole layer, Fig. 4.4c, such that the dipole moment associated with an element of area da of the disk is $\mathbf{dp} = P\delta\, da\, \hat{\mathbf{z}} \equiv P_S\, da\, \hat{\mathbf{z}}$, where P_S might be called the dipole moment per unit area of the surface. The potential due to this dipole layer can be determined using Eq. (2.44)

$$\Phi = \frac{1}{4\pi\varepsilon_0}\int \frac{\mathbf{dp}\cdot\boldsymbol{\xi}}{\xi^3} = \int_S \frac{P_S\, \mathbf{da}\cdot\boldsymbol{\xi}}{4\pi\varepsilon_0 \xi^3} \equiv \int_S \frac{P_S\, d\Omega}{4\pi\varepsilon_0} \tag{4.29}$$

where $d\Omega$ is the element of solid angle subtended by area da with respect to the point of observation and is to be reckoned as positive when the polarization vector associated with da is such that $\mathbf{P}_S \cdot \boldsymbol{\xi} \geq 0$; otherwise it is negative. If P_S is constant over the surface area, then Eq. (4.29) can be easily integrated; hence

$$\Phi = \frac{P_S \Omega}{4\pi\varepsilon_0} \tag{4.30}$$

where Ω is the total solid angle subtended by the disk with respect to the point of observation.

It is interesting to examine the potentials just above and just below the dipole layer, Φ_+ and Φ_- respectively. Using (Eq. 4.30), we get

$$\Phi_+(z \to 0^+) = \frac{P_S \cdot 2\pi}{4\pi\varepsilon_0} = \frac{P_S}{2\varepsilon_0} \qquad \text{and} \qquad \Phi_-(z \to 0^-) = -\frac{P_S}{2\varepsilon_0}$$

These results indicate that $\Phi_+ - \Phi_- = P_S/\varepsilon_0$. The potential is therefore discontinuous in passing through a dipole layer. The electric field is as a consequence highly singular. [At a dipole layer, it goes to infinity; see the discussion following Eq. (3.8).]

Example 4.2 The Polarized Sphere—Two Superimposed, Oppositely Charged Spheres

Consider a polarized dielectric sphere of radius R. We assume the polarization throughout its volume to be constant: $\mathbf{P} = P\hat{\mathbf{z}}$. We wish to find the electric fields generated by this polarization. These fields will be entirely equivalent to the field produced by the charge distribution associated with the polarization:

$$\rho_P = -\nabla\cdot\mathbf{P} = 0 \qquad \text{and} \qquad \sigma_P = \mathbf{P}\cdot\hat{\mathbf{r}} = P(\hat{\mathbf{z}}\cdot\hat{\mathbf{r}}) = P\cos\theta \tag{4.31}$$

One could straightforwardly calculate the potential or field due to such a charge distribution using Coulomb's formulas, though the integrations involved are not trivial. Instead, however, we observe that a shortcut exists [see the discussion preceding and following Eq. (4.15)]. To see it, consider the following problem.

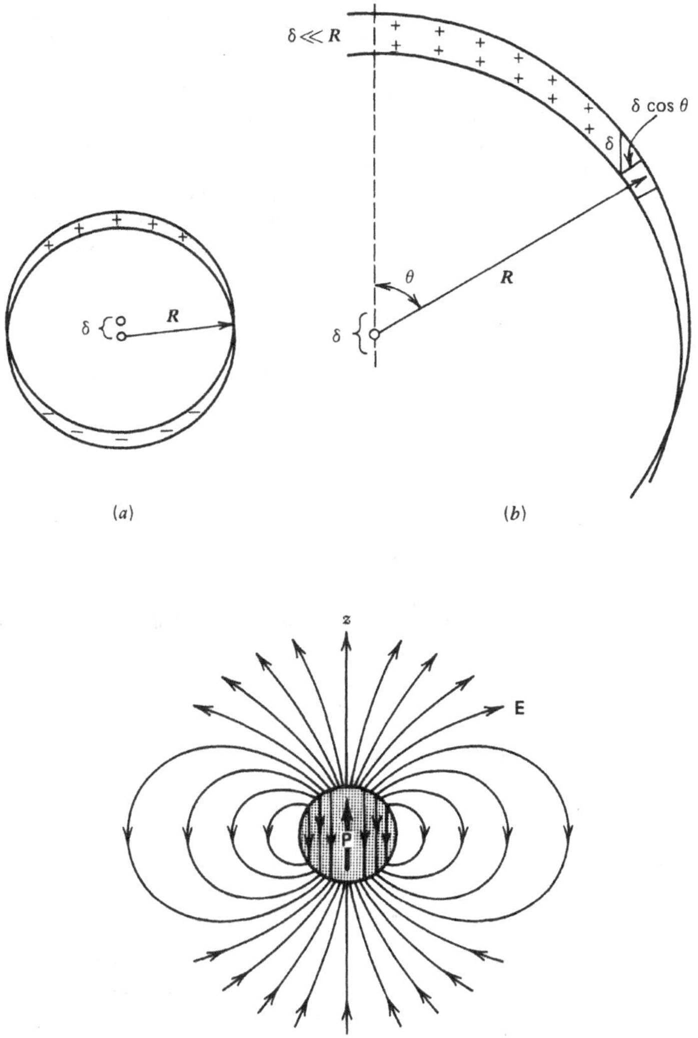

Figure 4.5 The fields and potentials of a polarized sphere. (a) Representing it as two superimposed oppositely charged nonconcentric spheres. (b) Enlargement of a section of the sphere. (c) Lines of force of the **E** field.

Imagine two uniformly charged spheres of radius R displaced from each other by a small distance, $\delta \ll R$ (see Fig. 4.5a). Let the charge densities of the spheres be $\rho_+ = \rho$ and $\rho_- = -\rho$, respectively. In the region where the spheres overlap, the net charge density is zero. Only on the periphery of the superimposed spheres will there be an unbalanced charge. Referring to Fig. 4.5b, we see that the charge in a layer on the surface is given by $dq = \rho\, dv = \rho\, da\, dr = \rho\, da(\delta \cos\theta)$, so $dq/da \equiv \sigma = \rho\delta\cos\theta$. We now observe that if we set P for our original problem equal to $\rho\delta$ [see Eq. (4.16)], we have identical charge distributions, and therefore identical electric fields, for these two problems. However, it is a simple matter to calculate the electric fields or potentials for the problem of two superimposed uniformly charged spheres. Thus, by solving this problem, we will have solved the polarization problem also.

Note that an unpolarized dielectric indeed may be thought of as two superimposed, uniformly charged spheres of opposite charge densities: ρ_+, $\rho_- = -\rho_+$. In the process of polarization, there will occur a displacement of the positive charge from the negative charge by a very small amount, δ, to produce the surface charge density $\rho_+ \delta \cos\theta$. Thus, the problem of the uniformly polarized sphere is in all ways identical to the problem of the two displaced, uniformly charged spheres. We simply recognize that $\rho\delta = P$.

Outside the spheres (at $r > R$), each uniformly charged sphere appears as a point charge [see Eq. (2.51)] of magnitude

$$Q_\pm = \frac{4}{3}\pi R^3 \rho_\pm = \pm\frac{4}{3}\pi R^3 \rho$$

These point charges are separated by the displacement $\boldsymbol{\delta}$, where $\delta/R \ll 1$. The field is a dipole field from a dipole of moment $Q_+\delta$ (see Example 2.9):

$$\Phi(r > R) = \frac{Q\boldsymbol{\delta}\cdot\hat{\mathbf{r}}}{4\pi\varepsilon_0 r^2} = \frac{\frac{4}{3}\pi R^3 \rho\boldsymbol{\delta}\cdot\hat{\mathbf{r}}}{4\pi\varepsilon_0 r^2} = \frac{V\mathbf{P}\cdot\hat{\mathbf{r}}}{4\pi\varepsilon_0 r^2}$$

or

$$\Phi(\mathbf{r}) = \frac{\mathbf{p}\cdot\hat{\mathbf{r}}}{4\pi\varepsilon_0 r^2} \tag{4.32}$$

where $\mathbf{p} \equiv \mathbf{P}V$ is the dipole moment of the sphere, and V is the volume of the sphere. Inside the spheres ($r \leq R$) we use the results of Example 2.12*b*. Using Eq. (2.53) we write:

$$\mathbf{E}_+ = \frac{\rho}{3\varepsilon_0}\mathbf{r}_+ \qquad \mathbf{E}_- = -\frac{\rho}{3\varepsilon_0}\mathbf{r}_-.$$

The total field is the vector sum of these two fields:

$$\mathbf{E} = \mathbf{E}_+ + \mathbf{E}_- = \frac{\rho}{3\varepsilon_0}(\mathbf{r}_+ - \mathbf{r}_-) = -\frac{\rho\boldsymbol{\delta}}{3\varepsilon_0}$$

Using $\mathbf{P} = \rho\boldsymbol{\delta}$, we find that

$$\mathbf{E} = -\frac{\mathbf{P}}{3\varepsilon_0} \tag{4.33}$$

The striking feature found is that the electric field is constant inside the sphere and is a simple dipole field outside. Figure 4.5*c* shows the lines of force of the uniformly polarized sphere.

If now the materials in this example are interchanged, then one gets a cavity in an infinite dielectric that is homogeneous and has uniform polarization $\mathbf{P}$. The electric field in the cavity is the same as in the polarized sphere except in the opposite direction (see Problem 4.16):

$$\mathbf{E} = \frac{\mathbf{P}}{3\varepsilon_0} \tag{4.34}$$

See Problem 4.18 for another method of solving the same problem (boundary value techniques, to be introduced in Section 4.6, the Solution of Electrostatic Boundary Value Problems with Dielectrics).

4.3 Gauss' Law for Dielectrics

We should explicitly indicate that macroscopic fields differ from their microscopic counterparts, even though they satisfy the same differential equations. Certainly, inside a dielectric they are very different; the microscopic fields change greatly from point to point due to the nearness of atomic charges, whereas the macroscopic fields change very smoothly. We shall not so distinguish these fields since it is typographically cumbersome to do so. We must simply be mindful that in material media whose charge densities are governed by the macroscopic field **P**, macroscopic fields $\{\mathbf{E}, \mathbf{P}, \Phi\}$ are created. Thus, we shall continue to write the equations, $\nabla \cdot \mathbf{E} = \rho/\varepsilon_0$, and $\nabla \times \mathbf{E} = 0$ as before, even in material media. There, however, a part of ρ is due to polarization charge, $\rho_P = -\nabla \cdot \mathbf{P}$. Denoting the remaining part of the charge density as ρ_f (f for "free"), we have $\rho = \rho_f + \rho_P$; hence

$$\nabla \cdot \mathbf{E} = \frac{\rho_f}{\varepsilon_0} - \frac{\nabla \cdot \mathbf{P}}{\varepsilon_0} \quad \text{or} \quad \nabla \cdot [\varepsilon_0 \mathbf{E} + \mathbf{P}] = \rho_f$$

Defining a new vector field,

$$\mathbf{D} \equiv \varepsilon_0 \mathbf{E} + \mathbf{P} \tag{4.35}$$

we have

$$\nabla \cdot \mathbf{D} = \rho_f \tag{4.36}$$

The new vector field **D** is called the *electric displacement*. Its divergence is related to the *free* charge rather than to the *total* charge density. Its curl need not be zero, so that it is in general a nonconservative field. Equation (4.36) is the differential form of Gauss' law in the presence of material.

We now drive the integral form of Gauss' law in the presence of material. Consider Fig. 4.6, which shows a dielectric material bounded by a number of surfaces: S_1, S_2, S_3, and S_4. Shown also in the figure is a gaussian surface S enclosing a volume V. In Gauss' law, $\oint \mathbf{E} \cdot d\mathbf{a} = Q/\varepsilon_0$, Q represents the total charge in volume V enclosed by S. With dielectrics, there is a polarization charge to account for. We therefore write $Q = Q_f + Q_P$, where Q_P is the polarization charge enclosed by S and Q_f is the remaining charge inside S. The charge Q_P may be due to surface and volume charge density distributions inside S, or

$$Q_P = \int_{S'} \sigma_P \, da + \int_{V'} -\nabla \cdot \mathbf{P} \, dv \tag{4.37}$$

where S' signifies all surfaces inside S where P is discontinuous, and V' is a "subvolume" of V exclusive of points where $\nabla \cdot \mathbf{P}$ is infinite. We would like to make the following two statements about the polarization charge Q_P. The first statement asserts that the total polarization charge in all of the dielectric material is zero. This can be easily seen by taking S' to include the very outer surface S_1 of the dielectric. Applying the divergence theorem to the second term of the right-hand side of Eq.

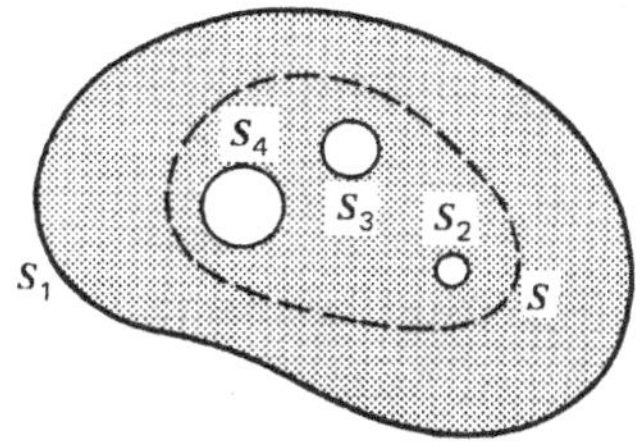

Figure 4.6 A dielectric material bounded by a number of surfaces: S_1, S_2, S_3, and S_4.

(4.37) converts it to a surface integral that exactly cancels out the first term in the same side of the equation. This result is a direct consequence of the fact that the dielectric proper is, by definition, neutral.

The second statement, on the other hand, asserts that the total polarization charge enclosed by a surface that does not include S_1 is not zero. It is

$$Q_P = -\int_S \mathbf{P}\cdot d\mathbf{a} \tag{4.38}$$

The reason is that the dielectric proper is, by definition, neutral, and only the charge that in the process of polarization passes through S will "add" to the net charge in V. But this is simply the integral Q_P defined above in Eq. (4.38). Inserting Q_P into Gauss' law, Eq. (2.27), we have for any closed surface S

$$\oint_S \mathbf{E}\cdot d\mathbf{a} = \frac{Q_f}{\varepsilon_0} - \frac{1}{\varepsilon_0}\oint_S \mathbf{P}\cdot d\mathbf{a}$$

That is

$$\oint\left(\mathbf{E} + \frac{\mathbf{P}}{\varepsilon_0}\right)\cdot d\mathbf{a} = \frac{Q_f}{\varepsilon_0}$$

Multiplying through by ε_0 gives the equivalent of Gauss' law for dielectrics:

$$\oint(\varepsilon_0\mathbf{E} + \mathbf{P})\cdot d\mathbf{a} = Q_f \tag{4.39}$$

It is a law governing the *macroscopic* fields (**E** and **P**). The vector function $\mathbf{D} \equiv \varepsilon_0\mathbf{E} + \mathbf{P}$ is the electric displacement encountered previously in Eq. (4.35). We have obtained the general result that, for an arbitrary closed surface S,

$$\oint_S \mathbf{D}\cdot d\mathbf{a} = Q_f \tag{4.40}$$

where Q_f is the total free charge contained by S. If Q_f can be everywhere characterized by a volume charge density, ρ_f, inside S, then

$$\int(\nabla\cdot\mathbf{D})dv = Q_f \equiv \int \rho_f\, dv \tag{4.41}$$

implying $\nabla\cdot\mathbf{D} = \rho_f$, as before. The integral expression, Eq. (4.41), clearly has a greater generality than the differential one insofar as the charge Q_f is not restricted to characterization as ρ_f. It makes obvious the assertion that the flux of **D** (representable as lines of force of **D**) is continuous in regions of zero (macroscopic) free charge. We assert that **D** ought not to be regarded as a fundamental field of the status of **E**. It is rather a purely mathematical construct related to the way in which one seeks a macroscopic solution for **E** from the basic equations. Its physical conceptualization is difficult; it has no direct connection with the forces exerted on charges.

4.4 The Equations of Electrostatics Inside Dielectrics

The equations for macroscopic electrostatics may now be summarized either by

$$\nabla\cdot\mathbf{E} = \frac{\rho}{\varepsilon_0} \quad \text{and} \quad \nabla\times\mathbf{E} = 0 \quad \text{or} \quad \nabla\cdot\mathbf{D} = \rho_f \quad \text{and} \quad \nabla\times\mathbf{E} = 0 \tag{4.42}$$

with

$$\mathbf{E} = -\nabla\Phi \qquad \text{and} \qquad \mathbf{D} = \varepsilon_0\mathbf{E} + \mathbf{P} \tag{4.43}$$

In free space, $\rho_f \equiv \rho$, $\mathbf{D} = \varepsilon_0\mathbf{E}$, and the equations are identical. In dielectrics, $\rho = \rho_f + \rho_P = \rho_f - \nabla\cdot\mathbf{P}$, and the equations are again completely equivalent. In either case, the **E** fields are conservative, and we may make use of the potential function Φ.

If ρ is known everywhere, one can obtain (in principal at least) a solution to these equations, for then the curl and divergence of **E** would be everywhere specified, and this is sufficient to obtain a solution for **E** (or Φ). But in the presence of dielectrics, ρ is given only if **P** and ρ_f are known. Once **P** and ρ_f are known, the problem of obtaining **E** is no different from that treated heretofore. One calculates σ_P and ρ_P from **P**, and hence **E** (or Φ) by an integration.

Often, it is assumed that ρ_f is under our control and is considered as known; **P**, however, is not explicitly given. The problem then assumes a different form: Although one knows that $\nabla\cdot\mathbf{D} = \rho_f$ and $\nabla\times\mathbf{E} = 0$, one cannot find either **D** or **E** (even if we know ρ_f everywhere) without knowing the relation of **D** to **E**. In other words, to determine **D** or **E** requires knowledge of both the curl and divergence of either **D** or **E** everywhere.*

4.5 The Electric Constitutive Relations

The relations between **P** (or **D**) and **E** are called *constitutive relations*. For example, there exists a class of dielectric materials for which this relation is expressed as

$$\mathbf{P} = \varepsilon_0\chi\mathbf{E} \tag{4.44}$$

or

$$\mathbf{D} = \varepsilon_0(1 + \chi)\mathbf{E} \tag{4.45}$$

which can alternatively be written as

$$\mathbf{D} \equiv \varepsilon_0 K\mathbf{E} \equiv \varepsilon\mathbf{E} \tag{4.46}$$

where χ is a dimensionless parameter called the *electric susceptibility*; $K = 1 + \chi$ is called the *relative permittivity*, or *dielectric constant*, and is "relative" to regions of zero polarization, such as free space, whose dielectric constant is unity; and $\varepsilon = \varepsilon_0 K$ is the permittivity of the material.

It must be emphasized that the constitutive relations do not express laws of physics but merely adequate representations of the behavior of some materials. If χ (or K) does not depend upon location in a piece of material, the material is called *homogeneous*. If χ does not depend on E, then the material is called *linear*, and if χ does not depend upon the direction of **E** in the material, the material is called *isotropic*. Most liquids and gases are homogeneous, isotropic, linear, materials at least at low electric fields. We shall refer to them as "simple" dielectrics. Most noncrystalline solids also are simple in this sense. However, many crystalline substances do not satisfy these conditions. For example, the relationship between **E** and **P** for these may be better represented by the set of equations:

$$\frac{P_x}{\varepsilon_0} = \chi_{xx}E_x + \chi_{xy}E_y + \chi_{xz}E_z \tag{4.47}$$

* If ρ_f is known only in a *restricted* region V, then to find **E** or **D** in this region we must also know the appropriate boundary conditions (see below) for this region.

and analogous relations hold for the y and z components of **P**. The set of nine coefficients determine the relations of **E** to **P**. These coefficients are called *susceptibility coefficients* and they may be functions of **E** in the case of nonlinear crystals. The set of coefficients is known as the *susceptibility tensor*. Such substances may be linear and homogeneous, but they are clearly not isotropic, for even when $\mathbf{E} = \hat{\mathbf{x}}E_x$ one has different polarization components in the y and z directions. Only if $\chi_{xy} = \chi_{xz} = \chi_{yz} = \chi_{yx} = \chi_{zx} = \chi_{zy} = 0$ and $\chi_{xx} = \chi_{yy} = \chi_{zz} = \chi$ does one have a *simple* susceptibility. This nonlinearity observed in some crystals has recently been employed in important practical applications. It has, for example, provided means of increasing the number of wavelengths available from lasers (harmonic generation).

Even if simple, the electric susceptibility will depend on such parameters as temperature and pressure. In Table 4.1 are given representative values of χ at standard temperatures and pressures (STP) for several simple dielectric materials.

It is the task of materials research to understand why a substance has a particular dielectric constant, and how it is affected by environmental conditions such as temperature or pressure. Suffice it to say here that there commonly occur two types of dielectrics: *polar* and *nonpolar*. The molecules of polar dielectrics have permanent dipole moments, which under the influence of (applied) electric fields tend to align themselves with these fields. If the fields are zero, the molecules tend to be randomly oriented, resulting in zero polarization. Such dielectrics generally have considerably larger susceptibilities than the nonpolar kinds, whose dipole moments are "induced" via a distortion and displacement of the molecular electron clouds relative to the atomic nuclei. A substance such as water is polar, whereas atomic substances (nonmolecular) such as monoatomic gases tend to be nonpolar (see Chapter 5). As seen in Table 4.1, the common gases at STP have essentially unity dielectric constants for most applications, primarily because of their atomic densities are so low.

Table 4.1 **Dielectric Constants**[a]

Dielectric Material	K
Porcelains	5–10
Glasses	5–10
Nylon	3.6
Polyethylene	2.25
Teflon	2.1
Lucite	3.0
Neoprene	6.7
Water	78.5
Ethanol	24.3
Methanol	32.6
Benzene	2.28
Mica	7.0
Paraffin	2.2
Mineral Oil	2.15
Air	1.00059
CO_2	1.000985
H_2	1.000065
O_2	1.000531

[a] Compiled from *AIP Handbook*, 3rd ed., 1972 (New York: McGraw-Hill, 1972). Values given are for STP.

There are instances where the existence of a polarization may be due to forces other than those due to imposed electric fields. Thus inertial or gravitational forces may affect a charge separation in atoms or molecules creating an effective polarization. Mechanical stresses may also produce a polarization. This occurs, for example, in quartz, and the associated phenomenon is called the *piezoelectric effect*. It has many practical applications, as in the fabrication of electromechanical transducers where a mechanical signal is to be converted into an electric signal or vice versa.

Finally, we should mention that as the ambient electric field is increased, there will eventually occur a departure from linearity between **P** and **E**. If **E** achieves a certain critical value, the dielectric may "break down," implying that electrons have their bonds to their associated molecules or atoms broken. In this case the dielectric will no longer act as an insulator: The electrons are torn away from their normal positions in the material by the high fields. The critical value of electric field at which this occurs is called the *dielectric strength* of the material. For air at STP, it is about 3×10^6 volts per meter.

In the following we shall mainly consider cases where simple relations hold between **P** and **E**. Note that the relation $\mathbf{P} = \chi\varepsilon_0\mathbf{E}$ relates the total macroscopic field **E** at a point to **P**, so that if a piece of dielectric is placed in an applied electrostatic field $\mathbf{E}_0$, the polarization induced in the dielectric is not $\mathbf{P} = \chi\varepsilon_0\mathbf{E}_0$ but $\mathbf{P} = \chi\varepsilon_0\mathbf{E}$, where **E** includes both $\mathbf{E}_0$ and the field produced by the polarized dielectric itself.

Example 4.3 Conducting Sphere Enclosed by a Dielectric Shell—Gauss' Law

In this example, Fig. 4.7, we shall assume we have a conducting sphere of radius R_1, on which is placed a charge $Q\ (\equiv Q_f)$. In contact and concentric with this sphere is dielectric material having a dielectric constant K that extends out to a radius R_2. We wish to find the fields and charge densities generated everywhere.

Because of the spherical symmetry, it is expected that the electric field at a distance r from the center of the sphere will be radial and independent of θ and ϕ. As a result, Gauss' law can be used to determine the fields easily.

Applying Gauss' law to a spherical surface concentric with the sphere and of radius r, such that $R_1 < r \leq R_2$, and taking $\mathbf{D} = D\hat{\mathbf{r}}$, we obtain $\oint \mathbf{D}\cdot d\mathbf{a} = 4\pi r^2 D = Q_f$, which yields

$$\mathbf{D} = \frac{Q_f}{4\pi r^2}\hat{\mathbf{r}} \quad \text{and} \quad \mathbf{E}_d = \frac{Q_f}{4\pi\varepsilon r^2}\hat{\mathbf{r}} \qquad R_1 < r \leq R_2 \tag{4.48}$$

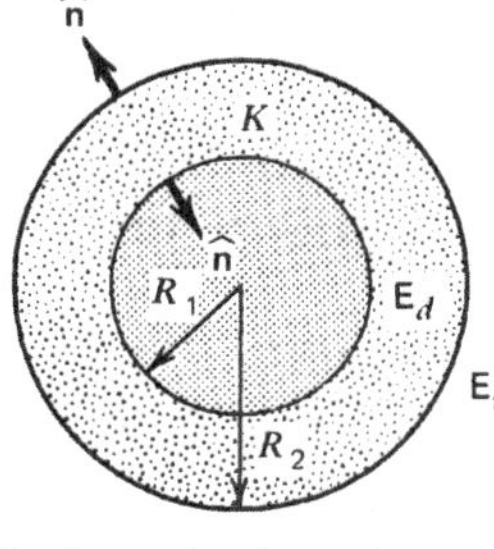

Figure 4.7 A conducting sphere enclosed by a dielectric shell.

where $\varepsilon = \varepsilon_0 K$ is the permittivity of the medium. Similarly, the fields in the vacuum regions can be determined using Gauss' law; that is,

$$\mathbf{D} = \frac{Q_f}{4\pi r^2}\hat{\mathbf{r}} \quad \text{and} \quad \mathbf{E}_v = \frac{Q_f\hat{\mathbf{r}}}{4\pi\varepsilon_0 r^2} \qquad r > R_2 \tag{4.49}$$

The fields inside the sphere, of course, vanish; therefore

$$\mathbf{D} = 0 \quad \text{and} \quad \mathbf{E} = 0 \qquad r < R_1 \tag{4.50}$$

We can also find the polarization surface charge densities, at $r = R_1$. Using Eqs. (4.44), (4.48), and (4.50), we find that the polarization $\mathbf{P}$ at $r = R_1$ is

$$\mathbf{P} = \varepsilon_0 \chi \mathbf{E}_d = (K-1)\frac{Q_f}{4\pi K R_1^2}\hat{\mathbf{r}} \qquad \text{at } r = R_1 \tag{4.51}$$

Hence $\sigma_p = \mathbf{P}\cdot\hat{\mathbf{n}} = \mathbf{P}\cdot(-\hat{\mathbf{r}})$ gives

$$\sigma_p = -(K-1)\frac{Q_f}{4\pi K R_1^2} = -\frac{(K-1)\sigma_f}{K} \tag{4.52}$$

where $\sigma_f = Q_f/4\pi R_1^2$. It is this negative charge density at R_1 that decreases the E field in the dielectric from what it would have been without the dielectric.

The polarization charge at $r = R_2$ can be determined in a similar way.

$$\mathbf{P} = (K-1)\frac{Q_f}{4\pi K R_2^2}\hat{\mathbf{r}} \qquad \text{at } r = R_2$$

and hence

$$\sigma_p = \mathbf{P}\cdot\hat{\mathbf{n}} = \mathbf{P}\cdot(\hat{\mathbf{r}}) = (K-1)\frac{Q_f}{4\pi K R_2^2} \tag{4.53}$$

One can easily show from Eqs. (4.52) and (4.53) that the total polarization charge at R_1 is equal in magnitude and opposite in sign to the total polarization charge at R_2. This must be true because the dielectric is assumed to have no net charge, and there is no volume polarization charge; hence $\rho_P = -\nabla\cdot\mathbf{P} = -(K-1)\varepsilon_0(\nabla\cdot\mathbf{E}_d) = 0$. This is also seen to be true because, by assumption, no free charge exists in the dielectric, and the total charge density in simple dielectrics is proportional to the free charge density.

Finally, the potential of the conductor, Φ_c, is determined by using Eq. (2.42). For $\Phi(\infty) \equiv 0$,

$$\Phi_c \equiv -\int_{\infty}^{R_1} \mathbf{E}\cdot d\mathbf{r} = -\int_{\infty}^{R_2} \mathbf{E}_v\cdot d\mathbf{r} - \int_{R_2}^{R_1} \mathbf{E}_d\cdot d\mathbf{r}$$

Substituting for the fields gives

$$\Phi_c = \frac{Q_f}{4\pi\varepsilon_0 R_2} + \left[\frac{Q_f}{4\pi\varepsilon}\left(\frac{1}{R_1} - \frac{1}{R_2}\right)\right] \tag{4.54}$$

If the voltage, Φ_c, of the sphere had been initially specified rather than Q_f, the relation between Q_f and Φ_c above would have been used in the equation for E_d and E_v [see Eq. (4.55) in the following example].

Example 4.4 The Parallel-Plate Capacitor—Gauss' Law

Consider two parallel conducting plates whose dimensions are very large compared to their separation, d (see Fig. 4.8). The surface area of the plates is A. A dielectric slab of thickness t was inserted between the plates. The dielectric has a permittivity $\varepsilon = K\varepsilon_0$, and the potential difference between the plates is $\Delta\Phi$.

Because the plate dimensions are much larger than the distance between them, we expect the electric field to be perpendicular to the plates and to be constant in the dielectric and in vacuum with values $E_d\hat{\mathbf{z}}$ and $E_v\hat{\mathbf{z}}$, respectively.

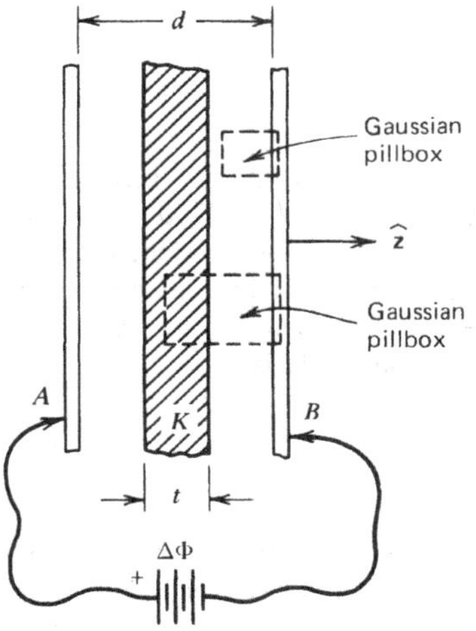

Figure 4.8 Parallel-plate capacitor with a dielectric slab inserted in it.

In order to find the values E_v and E_d we utilize Eq. (2.41)

$$\Delta\Phi = \int_A^B \mathbf{E}\cdot d\mathbf{r} = E_d t + E_v(d - t) \tag{4.55}$$

and Gauss' law, $\oint \mathbf{D}\cdot d\mathbf{a} = Q_f$. Applying Gauss' law to a Gaussian pillbox whose "top" side is inside the conductor, and whose "bottom" side is taken in vacuum (between the dielectric and the conductor) yields $D_v = \sigma_f$. Similarly, if the bottom side of the Gaussian pillbox lies inside the dielectric, $D_d = D_v = \sigma_f$, because the charge associated with the polarized dielectric is, by definition, "nonfree." From the equality of D_d and D_v, we then have $E_v = KE_d$. Inserting this into Eq. (4.55) gives the result: $\Delta\Phi = E_d[t + K(d - t)]$, or

$$E_d = \frac{\Delta\Phi}{t + K(d - t)} \quad \text{and} \quad E_v = \frac{K\,\Delta\Phi}{t + K(d - t)} \tag{4.56}$$

This constitutes a solution to the problem. It is easily checked that as $K \to 1$, $E_d = E_v \to \Delta\Phi/d$, and that as $t \to d$, $E_d \to \Delta\Phi/d$. Other quantities of interest—the charge density on the conducting plates, σ_f, and the charge density on the surface of the dielectric, σ_p—are as follows:

$$\sigma_f = \varepsilon_0 E_v = \frac{K\varepsilon_0\,\Delta\Phi}{t + K(d - t)} \quad \text{and} \quad \sigma_p = \mathbf{P}\cdot\hat{\mathbf{n}} = -\left(\frac{K - 1}{K}\right)\sigma_f. \tag{4.57}$$

Note that, in terms of charge densities,

$$E_v = \frac{\sigma_f}{\varepsilon_0} \quad \text{and} \quad E_d = \frac{\sigma_f - \sigma_p}{\varepsilon_0}$$

4.6 The Solution of Electrostatic Boundary Value Problems with Dielectrics

It follows from the relation $\nabla\cdot\mathbf{D} = \rho_f$, that $\nabla\cdot(\varepsilon_0 K\mathbf{E}) = \rho_f$ in a medium characterizable by a dielectric constant K. If K is simple (constant), then

$$\nabla\cdot\mathbf{E} = \frac{\rho_f}{\varepsilon_0 K} \tag{4.58}$$

Taking $\mathbf{E} = -\nabla\Phi$ gives

$$\nabla^2\Phi = -\frac{\rho_f}{\varepsilon} \tag{4.59}$$

This is Poisson's equation as applied to points in simple dielectric media. Its form is identical to the more general condition $\nabla^2\Phi = -\rho/\varepsilon_0$. Thus, such a dielectric medium acts very much like free space. Evidently, the effects of the polarization charge density are accounted for simply by replacing the ε_0 of free space by $\varepsilon_0 K$ (in free space, $K = 1$). If $\rho_f = 0$ in the media, Laplace's equation, $\nabla^2\Phi = 0$, is satisfied just as in space free of any charge. It is then clear that, for simple media, $\rho_f = 0$ implies that $\rho = 0$, since $\rho = -\varepsilon_0\nabla^2\Phi$. Thus, $\rho_P = 0$ also there.

4.6.1 Uniqueness

Just as Poisson's (or Laplace's) equation has unique solutions for regions of free space when appropriate boundary conditions are specified, so do the analogous equations applied to dielectric materials have unique solutions. If we are considering a finite region of space, a unique solution will exist (for $\nabla\Phi$) in that region if the electrostatic potential is specified on all the surfaces bounding the region. These conditions are no different than they were for the electrostatics of free space (see Section 4.6.2 below). Thus again, as before, if a part of the bounding surface of the region is a conductor, it will be sufficient to specify the total (free) charge on the conductor instead of its potential. In any of these cases, we must know something of the dielectric characteristics of the media within the region of interest to find the solution.

4.6.2 Boundary Conditions for Dielectric Media

Since the boundary conditions are essential to the specification of the solution, we now enunciate these conditions for dielectric media with the help of Fig. 4.9.

From Gauss' law we have seen that

$$\mathbf{E}_2 \cdot \hat{\mathbf{n}} - \mathbf{E}_1 \cdot \hat{\mathbf{n}} = \frac{\sigma}{\varepsilon_0} \tag{2.35}$$

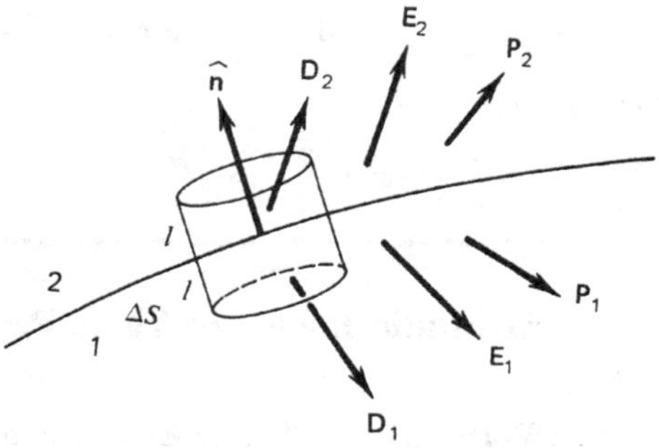

Figure 4.9 Application of Gauss' law to a pillbox at the interface of two dielectric media to determine the boundary conditions on the displacement vector **D**.

where $\hat{\mathbf{n}}$ is a unit vector pointing from medium 1 to medium 2, normal to the separating surface, and σ is the total, net, surface charge density on the surface. We may distinguish the polarization surface charge density from the free charge density contribution to σ by $\sigma = \sigma_p + \sigma_f$. The surface charge contribution from polarization of medium 1 alone is given by $\mathbf{P}_1 \cdot \hat{\mathbf{n}}$, where P_1 is evaluated in medium 1 at a point arbitrarily close to the surface. Similarly, we have a surface charge contribution due to the polarization in medium 2 alone equal to $\mathbf{P}_2 \cdot (-\hat{n})$. The total polarization surface charge is therefore given by

$$\sigma_p = \mathbf{P}_1 \cdot \hat{\mathbf{n}} - \mathbf{P}_2 \cdot \hat{\mathbf{n}} \tag{4.60}$$

Therefore

$$\sigma \equiv (\mathbf{P}_1 - \mathbf{P}_2) \cdot \hat{\mathbf{n}} + \sigma_f$$

which, upon substitution in Eq. (2.35), gives

$$\varepsilon_0(\mathbf{E}_2 - \mathbf{E}_1) \cdot \hat{\mathbf{n}} = \sigma_f - (\mathbf{P}_2 - \mathbf{P}_1) \cdot \hat{\mathbf{n}}$$

or

$$(\varepsilon_0 \mathbf{E}_2 + \mathbf{P}_2) \cdot \hat{\mathbf{n}} - (\varepsilon_0 \mathbf{E}_1 + \mathbf{P}_1) \cdot \hat{\mathbf{n}} = \sigma_f$$

In terms of $\mathbf{D} \equiv \varepsilon_0 \mathbf{E} + \mathbf{P}$, we obtain

$$\mathbf{D}_2 \cdot \hat{\mathbf{n}} - \mathbf{D}_1 \cdot \hat{\mathbf{n}} = \sigma_f \tag{4.61}$$

This is the equation pertinent to the interface between two dielectric media, and relates the normal components of $\mathbf{D}$ "across" the interface.

The boundary condition given by Eq. (4.61) can also be derived directly from Gauss' law for dielectrics [Eq. (4.40)]. Consider the gaussian pillbox shown in Fig. 4.9. Its flat surface area ΔS is taken to be small, and its height $2l$ is taken to be much smaller than the diameter of the flat surface. Using Eq. (4.40) we get

$$\mathbf{D}_2 \cdot \hat{\mathbf{n}}\, \Delta S - \mathbf{D}_1 \cdot \hat{\mathbf{n}}\, \Delta S = \sigma_f\, \Delta S$$

where only the surface charge contributed since l is taken to be very small; the lateral surface area did not contribute for the same reason. Thus

$$(\mathbf{D}_2 - \mathbf{D}_1) \cdot \hat{\mathbf{n}} = \sigma_f$$

which is just Eq. (4.61).

We now consider a number of special cases of Eq. (4.61).

1. When $\sigma_f = 0$, then the normal component of $\mathbf{D}$ is continuous across the interface; that is,

$$\mathbf{D}_2 \cdot \hat{\mathbf{n}} = \mathbf{D}_1 \cdot \hat{\mathbf{n}} \qquad \text{if } \sigma_f = 0 \tag{4.62}$$

2. If the two media are characterizable by dielectric constants K_1 and K_2 or their permitivities ε_1 and ε_2, we have $\varepsilon_2 \mathbf{E}_2 \cdot \hat{\mathbf{n}} - \varepsilon_1 \mathbf{E}_1 \cdot \hat{\mathbf{n}} = \sigma_f$ or, expressed in terms of the corresponding potentials Φ_2 and Φ_1,

$$-\varepsilon_2 \nabla\Phi_2 \cdot \hat{\mathbf{n}} + \varepsilon_1 \nabla\Phi_1 \cdot \hat{\mathbf{n}} = \sigma_f \tag{4.63}$$

3. If medium 1 is a conducting medium, then $\mathbf{E}_1$, $\mathbf{D}_1$, and $\mathbf{P}_1$ are equal to zero and the relevant boundary condition is

$$\mathbf{D} \cdot \hat{\mathbf{n}} = \sigma_f \tag{4.64}$$

where we have identified $\mathbf{D}_2$ with $\mathbf{D}$ and have assumed that the polarization inside the conductor is zero (see page 129).

The equations relating the tangential components of the electrostatic field arise from the equation, still valid in dielectrics, $\nabla \times \mathbf{E} = 0$, or $\oint_C \mathbf{E}\cdot d\mathbf{r} = 0$. It states that the tangential component of $\mathbf{E}$ is continuous across any interface; that is,

$$\mathbf{E}_1 \cdot \hat{\mathbf{t}} = \mathbf{E}_2 \cdot \hat{\mathbf{t}} \tag{4.65}$$

As in the case of free space, the conservative nature of the $\mathbf{E}$ field implies continuity of the potential; that is, $\Phi_1 = \Phi_2$, where Φ_1 and Φ_2 are the potentials in regions 1 and 2 as this interface is approached. Therefore, the boundary conditions on the potential function most often employed are summarized as follows:

$$\begin{gathered} \Phi_1 = \Phi_2 \\ \varepsilon_1(\nabla\Phi_1)\cdot\hat{\mathbf{n}} - \varepsilon_2 \nabla\Phi_2\cdot\hat{\mathbf{n}} = \sigma_f \end{gathered} \tag{4.66}$$

We might now summarize how we attempt to find the electrostatic fields for a problem involving dielectrics where the polarization is not explicitly given. We simply employ to advantage whatever we know. We use Gauss' law for dielectrics where symmetry allows. We use the fact that a potential function Φ exists and finally we use the constitutive relations between $\mathbf{D}$ (and $\mathbf{P}$) and $\mathbf{E}$ wherever possible.

Example 4.5 Point Charge on a Plane Interface—Laplace's Equation In One Dimension

This example deals with a situation where a point charge q is placed on the plane interface of two homogeneous infinite dielectrics 1 and 2 with permittivities ε_1 and ε_2, respectively, as shown in Fig. 4.10. At points away from the point charge, one can see from Eq. (4.59) that the potential satisfies Laplace's equation; thus the potentials Φ_1 and Φ_2 in regions 1 and 2, respectively, can be written as follows:

$$\Phi_1 = \frac{C_1 q}{r} + B_1 \qquad \Phi_2 = \frac{C_2 q}{r} + B_2$$

where C_1, C_2, B_1, and B_2 are constants. Since the potential is due to a localized point charge, then it should go to zero as $r \to \infty$, thus yielding $B_1 = B_2 = 0$.

The fact that the potential is continuous at the boundary gives: $C_1 = C_2 = C$ and $\Phi_1 = \Phi_2 = \Phi$. We can evaluate C by using Gauss' law. The electric displacement $\mathbf{D} = -\varepsilon\,\nabla\Phi$ takes the value $(\varepsilon_1 Cq/r^2)\hat{\mathbf{r}}$ and $(\varepsilon_2 Cq/r^2)\hat{\mathbf{r}}$ in regions 1 and 2, respectively. We take a spherical

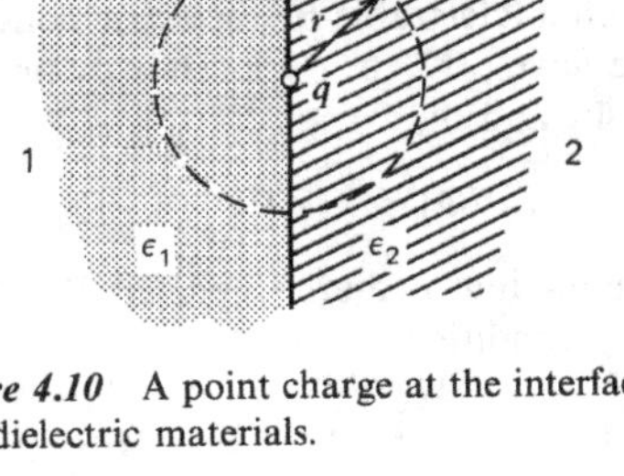

Figure 4.10 A point charge at the interface of two dielectric materials.

surface S with its center at the point charge. Applying Gauss' law on S (Eq. 4.40) yields $C = 1/2\pi(\varepsilon_1 + \varepsilon_2)$, and thus

$$\Phi(\mathbf{r}) = \frac{1}{2\pi(\varepsilon_1 + \varepsilon_2)} \frac{q}{r} \tag{4.67}$$

The electric field and displacement vectors can now be easily determined.

$$\mathbf{E}(\mathbf{r}) = \frac{1}{2\pi(\varepsilon_1 + \varepsilon_2)} \frac{q}{r^2} \hat{\mathbf{r}} \qquad \text{for all } r \tag{4.68}$$

$$\mathbf{D}_i(\mathbf{r}) = \frac{\varepsilon_i}{2\pi(\varepsilon_1 + \varepsilon_2)} \frac{q}{r^2} \hat{\mathbf{r}} \qquad i = 1, 2 \tag{4.69}$$

where $\mathbf{D}_i$ is the displacement vector in the ith region.

The polarization $\mathbf{P} = \mathbf{D} - \varepsilon_0 \mathbf{E}$ may now be evaluated:

$$\mathbf{P}_i(\mathbf{r}) = \frac{\varepsilon_i - \varepsilon_0}{2\pi(\varepsilon_1 + \varepsilon_2)} \frac{q}{r^2} \hat{\mathbf{r}} \tag{4.70}$$

In the absence of the materials, namely when the charge is in vacuum, the electric field is $\mathbf{E} = (1/4\pi\varepsilon_0)(q\mathbf{r}/r^3)$. Thus the presence of the materials weakens the electric field. This is due to the fact that the induced charge in the media screens the charge q. The induced charge component $\rho_p = -\nabla \cdot \mathbf{P}$ is zero for $r \neq 0$; the surface density, $\sigma_p = \mathbf{P} \cdot \mathbf{n}$, however, is not zero close to the surface of the charge.

The total charge induced, $q_p = \int \sigma_p \, da$, is

$$q_p = -\lim_{a \to 0} \left[\frac{\varepsilon_1 - \varepsilon_0}{2\pi(\varepsilon_1 + \varepsilon_2)} \frac{q}{a^2} + \frac{\varepsilon_2 - \varepsilon_0}{2\pi(\varepsilon_1 + \varepsilon_2)} \frac{q}{a^2} \right] 2\pi a^2 = -q \left[1 - \frac{2\varepsilon_0}{\varepsilon_1 + \varepsilon_2} \right]$$

where a is taken to be the radius of q. The assignment of a radius to the point charge is just an intermediate step to facilitate evaluating the induced charge since the final result was derived in the limit of a becoming very small.

The total charge is equal to the free charge plus the polarization charge, or

$$Q = q_p + q = \frac{2\varepsilon_0 q}{\varepsilon_1 + \varepsilon_2}$$

This result gives the screening effect where the total charge, when viewed from inside the dielectric material, appears to be less than q.

A special case of the above results is a situation in which a point charge q is embedded in a single dielectric material. The potential, electric field, displacement vector, polarization, and the charge distribution produced in this case are all given by the corresponding results of the above example in the limit $\varepsilon_1 = \varepsilon_2 = \varepsilon$, where ε is the permittivity of the single medium.

Example 4.6 A Conducting Charged Sphere Between Two Dielectrics—One-Dimensional Problem

Let us take the point charge in the previous example to be distributed over a conducting sphere of radius R and center at the plane interface. As in the point charge case, the potential in both regions is $\Phi(r) = Cq/r$, where $C = 1/2\pi(\varepsilon_1 + \varepsilon_2)$. We can now determine the charge density on the sphere. Since the electric field in a conductor is zero, the charge density on the surface of the sphere is given by Eq. (4.64); thus $\sigma_{1f} = D_{1n}$ and $\sigma_{2f} = D_{2n}$ are the charge densities on the two halves of the sphere. The normal component D_{in} is easily evaluated on the surface of the sphere:

$$D_{in} = -\varepsilon_i \frac{\partial \Phi}{\partial r} \qquad \text{evaluated at } r = R$$

yielding

$$\sigma_{1f} = \frac{q\varepsilon_1}{2\pi R^2(\varepsilon_1 + \varepsilon_2)} \qquad \text{and} \qquad \sigma_{2f} = \frac{q\varepsilon_2}{2\pi R^2(\varepsilon_1 + \varepsilon_2)} \tag{4.71}$$

The total free charge on the sphere can be easily shown to be equal to q, as expected, by integrating the charge densities in Eq. (4.71) over the surface of the sphere:

$$Q_f = 2\pi R^2(\sigma_{1f} + \sigma_{2f}) = q$$

We now calculate the polarization charge densities on the surface of the sphere. First we calculate the polarizations P_1 and P_2 in media 1 and 2, respectively: $\mathbf{P}_i = \mathbf{D}_i - \varepsilon_0 \mathbf{E}$, with $i = 1, 2$. Thus

$$\mathbf{P}_1 = \frac{\varepsilon_1 - \varepsilon_0}{2\pi(\varepsilon_1 + \varepsilon_2)} \frac{q}{r^2} \hat{\mathbf{r}} \tag{4.72}$$

$$\mathbf{P}_2 = \frac{\varepsilon_2 - \varepsilon_0}{2\pi(\varepsilon_1 + \varepsilon_2)} \frac{q}{r^2} \hat{\mathbf{r}} \tag{4.73}$$

The polarization volume charge density $\rho_i = -\nabla \cdot \mathbf{P}_i$ in the materials is zero. The surface densities, however, are not zero and are equal to

$$\sigma_{1p} = \mathbf{P}_1 \cdot \hat{\mathbf{n}}|_{r=R} = -\frac{q(\varepsilon_1 - \varepsilon_0)}{2\pi R^2(\varepsilon_1 + \varepsilon_2)}$$

$$\sigma_{2p} = \mathbf{P}_2 \cdot \hat{\mathbf{n}}|_{r=R} = -\frac{q(\varepsilon_2 - \varepsilon_0)}{2\pi R^2(\varepsilon_1 + \varepsilon_2)}$$

The total polarization charge on the sphere, $q_p = 2\pi R^2(\sigma_{1p} + \sigma_{2p})$, is

$$q_p = -q\left(1 - \frac{2\varepsilon_0}{\varepsilon_1 + \varepsilon_2}\right)$$

It is to be noted that q_p is independent of the radius of the sphere. In fact it is identical to the charge induced on the surface of a point charge (see the previous example). As a result the same screening effect arises here as was encountered in the point charge case, namely, the total charge on the sphere is reduced by the same factor.

Example 4.7 A Long, Dielectric Cylinder in an Electric Field—Two-Dimensional Problem

We consider here a boundary value problem where the potential is a function of two variables. Consider a long, dielectric cylinder of permittivity ε placed in a uniform electric field that is normal to its axis. We choose a cylindrical coordinate system with the origin taken at the axis of the cylinder and the x axis along the electric field.

Since there is no free charge on the cylinder, the potential in the x-y plane satisfies Laplace's equation. The potentials Φ_1 and Φ_2 inside and outside the cylinder, respectively, depend on ρ and ϕ. We expect these potentials not to depend on z because the cylinder is long. Therefore, the potentials are given by the cylindrical harmonics that were derived in Eq. 3.65). We will now give some arguments, based on physical grounds, why only a subset of these terms will contribute:

1. The potential $\Phi_1(\rho, \phi)$ should not blow up as $\rho \to 0$. This implies that it should not have terms of radial dependence $1/\rho^n$ and $\ln \rho$.

2. Far away from the cylinder, the potential should reduce to a uniform electric field in the x direction. Therefore $\Phi_2(\rho, \phi) = -E_0 \rho \cos \phi + V_0$, where V_0 is a constant. Thus $\Phi_2(\rho, \phi)$ should not have terms of radial dependence ρ^n where $n > 1$. Also, because of this boundary condition, Φ_2 and Φ_1 should not include terms of $\cos n\phi$ where $n > 2$ and terms of $\sin n\phi$ where $n \geq 1$. This result is directly related to the fact that $\sin n\phi$ and $\cos n\phi$ are linearly independent functions.

3. Because the cylinder has no free charges, then Φ_2 should not include a $\ln \rho$ term, since such a term is proportional to the total charge on the cylinder.

Using these restrictions we find that the potentials take the following expressions:

$$\Phi_1(\rho, \phi) = A_0 + A_1\rho \cos \phi \qquad \rho < \rho_0 \tag{4.74}$$

$$\Phi_2(\rho, \phi) = B_0 + \frac{B_1}{\rho} \cos \phi + C_1\rho \cos \phi \qquad \rho > \rho_0 \tag{4.75}$$

The constants A_0, A_1, B_0, B_1, and C_1 can now be evaluated using some more conditions. As $\rho \to$ large, $\Phi_2(\rho, \phi) = -E_0\rho \cos \phi + V_0$; therefore

$$B_0 + C_1\rho \cos \phi = -E_0\rho \cos \phi + V_0$$

which gives $B_0 = V_0$ and $C_1 = -E_0$. The continuity of the potential at $\rho = \rho_0$ gives

$$A_0 + A_1\rho_0 \cos \phi = V_0 + \frac{B_1}{\rho_0} \cos \phi - E_0\rho_0 \cos \phi$$

which gives $A_0 = V_0$, and $A_1 = (B_1/\rho_0^2) - E_0$. Another equation between A_1 and B_1 can now be found from the continuity condition of the normal D components:

$$-\varepsilon \frac{\partial \Phi_1}{\partial \rho}\bigg|_{\rho=\rho_0} = -\varepsilon_0 \frac{\partial \Phi_2}{\partial \rho}\bigg|_{\rho=\rho_0}$$

which gives: $-KA_1 = (B_1/\rho_0^2) + E_0$, where $K = \varepsilon/\varepsilon_0$. The relations between A_1 and B_1 give $A_1 = -2E_0/(K+1)$ and $B_1 = \rho_0^2 E_0(K-1)/(K+1)$. Substituting the values of the constants in Eqs. (4.74) and (4.75) gives

$$\Phi_1(\rho, \phi) = V_0 - \frac{2E}{K+1}\rho \cos \phi \tag{4.76}$$

$$\Phi_2(\rho, \phi) = V_0 + \frac{\rho_0^2 E_0}{\rho} \frac{K-1}{K+1} \cos \phi - E_0\rho \cos \phi \tag{4.77}$$

We now discuss an interesting limit of these potentials. If the material of which the cylinder is made has a very high dielectric constant—that is, $K \gg 1$—then Φ_1 and Φ_2 reduce to

$$\Phi_1(\rho, \phi) = V_0 \tag{4.78}$$

$$\Phi_2(\rho, \phi) = V_0 + \frac{\rho_0^2 E_0}{\rho} \cos \phi - E_0\rho \cos \phi \tag{4.79}$$

which are exactly the same potentials we would encounter in the case of a conducting cylinder placed in an electric field (see Problem 4.14). This leads us to infer that the dielectric cylinder in this limit is exactly equivalent to a conducting cylinder.

Example 4.8 Dielectric Sphere in an Electric Field—Two-Dimensional Problem

This example, which is sketched in Fig. 4.11*a*, deals with a dielectric sphere in a uniform electric field. We use spherical polar coordinates with the origin at the center of the sphere. The electric field is taken along the z axis: $\mathbf{E} = E_0\hat{\mathbf{z}}$. The permittivity of the sphere and the external medium are ε_1 and ε_2, respectively.

The potentials Φ_1 and Φ_2 inside and outside the sphere, respectively, satisfy Laplace's equation; moreover, they depend on r and θ and therefore can be written as a series of zonal harmonics [those were derived in Eq. (3.28)].

The following arguments are used to eliminate many of the terms in these expansions.

1. The potential inside the sphere, $\Phi_1(r, \theta)$, should not contain terms of $1/r^n$, where $n \geq 1$, because it should not blow up at the origin.

2. The term proportional to $1/r$ should be dropped from the Φ_2 expansion since the sphere has no net charge.

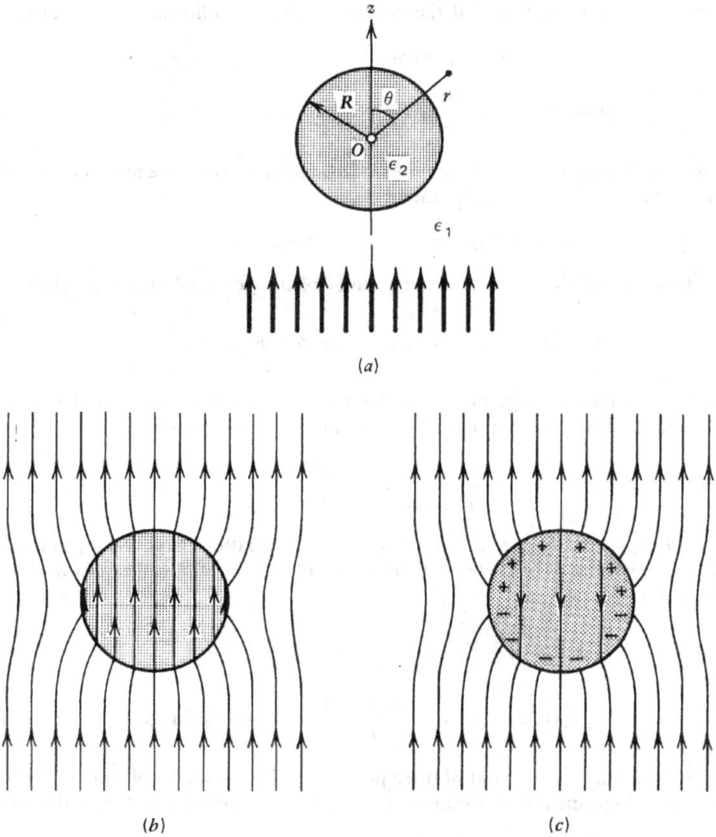

Figure 4.11 Dielectric sphere in a previously uniform electric field. (*a*) Lines of force in the absence of the sphere. (*b*) Lines of electric displacement. (*c*) Lines of electric field.

3. The fact that the spherical symmetry is broken by the presence of the electric field that gives rise to $-E_0 r\cos\theta + V_0$ potential at $r \to \infty$ implies that only the lowest zonal harmonic contributes.

4. The potential outside the sphere should not contain terms of r^n, where $n \geq 2$. Therefore one writes

$$\Phi_1(r, \theta) = A_0 + A_1 r\cos\theta \tag{4.80}$$

$$\Phi_2(r, \theta) = B_0 + B_1 r\cos\theta + \frac{B_2}{r^2}\cos\theta \tag{4.81}$$

where A_0, A_1, B_0, B_1, and B_2 are constants to be determined from the boundary conditions.

The boundary conditions used are as follows:

1. As $r \to \infty$,

$$\Phi_2 = B_0 + B_1 r\cos\theta = -E_0 r\cos\theta + V_0$$

Thus $B_0 = V_0$ and $B_1 = -E_0$.

2. On the surface of the sphere ($r = R$), we have $\Phi_1 = \Phi_2$. Thus

$$A_0 + A_1 R\cos\theta = V_0 - E_0 R\cos\theta + \frac{B_2}{R^2}\cos\theta$$

Equating coefficients of powers of $\cos\theta$ gives $A_0 = V_0$ and

$$A_1 = \frac{B_2}{R^3} - E_0 \tag{4.82}$$

3. On the surface of the sphere, the normal component of **D** is continuous. The continuity is due to the given fact that there is no surface free charge density. Thus $-\varepsilon_1(\partial\Phi_1/\partial r) = -\varepsilon_2(\partial\Phi_2/\partial r)$ at $r = R$, or

$$-\varepsilon_1 A_1 = \varepsilon_2 E_0 + \frac{2\varepsilon_2}{R^3} B_2 \tag{4.83}$$

Equations (4.82) and (4.83) yield

$$A_1 = -\frac{3\varepsilon_2 E_0}{\varepsilon_1 + 2\varepsilon_2} \quad \text{and} \quad B_2 = \frac{\varepsilon_1 - \varepsilon_2}{\varepsilon_1 + 2\varepsilon_2} R^3 E_0$$

Substituting the magnitude of the constants A_0, A_1, B_0, B_1, and B_2 in Eqs. (4.80) and (4.81) gives

$$\Phi_1(r, \theta) = V_0 - \frac{3\varepsilon_2 E_0}{\varepsilon_1 + 2\varepsilon_2} r\cos\theta \qquad r < R \tag{4.84}$$

$$\Phi_2(r, \theta) = V_0 - E_0 r\cos\theta + \frac{\varepsilon_1 - \varepsilon_2}{\varepsilon_1 + 2\varepsilon_2} E_0 R^3 \frac{\cos\theta}{r^2} \qquad r > R \tag{4.85}$$

The electric field inside the sphere $\mathbf{E}_1 = -\nabla\Phi$ is uniform and is given by

$$\mathbf{E}_1 = \frac{3\varepsilon_2}{\varepsilon_1 + 2\varepsilon_2} E_0\hat{\mathbf{z}} = E_0\hat{\mathbf{z}} - \frac{\varepsilon_1 - \varepsilon_2}{\varepsilon_1 + 2\varepsilon_2} E_0\hat{\mathbf{z}} \tag{4.86}$$

When the sphere has a higher dielectric permitivity than the surrounding medium, $\varepsilon_1 > \varepsilon_2$, then $E_1 < E_0$. The reduction of E_0 inside the sphere is attributed to the induced charge on its surface. This induced charge, which will be shown below to have a $\cos\theta$ dependence, produces a uniform *depolarization field* in opposite direction with the external field (see Examples 2.17, 3.6, and 3.7 for calculation of fields due to a $\cos\theta$ charge distribution). The name "depolarization field" is given because it tends to oppose the polarization and disorient the dipoles. On the other hand, E_1 is greater than E_0 in the case $\varepsilon_1 < \varepsilon_2$. In this case the field inside the cavity is strengthened.

The electric field outside the sphere is:

$$\mathbf{E}_2 = E_0\hat{\mathbf{z}} + \frac{\varepsilon_1 - \varepsilon_2}{\varepsilon_1 + 2\varepsilon_2} \frac{E_0 R^3}{r^3} [2\hat{\mathbf{r}}\cos\theta + \hat{\boldsymbol{\theta}}\sin\theta] \tag{4.87}$$

where $\hat{\boldsymbol{\theta}}$ is a unit vector along the θ direction. It consists of the external uniform field $E_0\hat{\mathbf{z}}$ and a field that is due to an electric dipole with a moment

$$\mathbf{p} = 4\pi\varepsilon_2 \frac{\varepsilon_1 - \varepsilon_2}{\varepsilon_1 + 2\varepsilon_2} R^3 E_0\hat{\mathbf{z}} \tag{4.88}$$

Figure 4.11*b* show the effect of the sphere on the lines of force of the initially uniform electric displacement field. Figure 4.11*c* gives the lines of the **E** field. The sphere has a uniform polarization **P** along the z axis (along the external field); it is equal to $\mathbf{p}/V = \mathbf{p}(4\pi R^3/3)^{-1}$, or

$$\mathbf{P} = 3\varepsilon_2 \frac{\varepsilon_1 - \varepsilon_2}{\varepsilon_1 + 2\varepsilon_2} E_0\hat{\mathbf{z}} \tag{4.89}$$

The polarization charge densities in the sphere can now be calculated. The volume density $\rho = -\nabla \cdot \mathbf{P} = 0$, since $\mathbf{P}$ is uniform. The surface charge density is $\sigma_p = \mathbf{P} \cdot \hat{\mathbf{n}} = [3\varepsilon_2(\varepsilon_1 - \varepsilon_2)/(\varepsilon_1 + 2\varepsilon_2)]E_0 \cos\theta$. The dipole field induced outside the sphere is in fact due to this surface charge density. See again examples 2.17, 3.6 and 3.7 which show that a $\cos\theta$ charge distribution produces a dipole field.

Let us consider a special case of the above results. We examine the case where the permittivity of the sphere ε_1 becomes very large, $\varepsilon_1 \gg \varepsilon_2$. In this case the electric field inside the sphere [Eq. (4.86)] vanishes and the field outside the sphere, Eq. (4.87), becomes

$$\mathbf{E}_2(\varepsilon_1 \to \infty) = E_0\hat{\mathbf{z}} + \frac{E_0 R^3}{r^3}[2\hat{\mathbf{r}}\cos\theta + \hat{\boldsymbol{\theta}}\sin\theta]$$

These electric fields are exactly the same fields produced when a conducting sphere is placed in a uniform electric field [see Eq. (3.52)]. This result leads us to conclude that a conductor can be regarded in a sense as a medium of infinitely large polarizability (large dielectric constant)—meaning that the charge displacement is unbounded.

Now we present a different way of solving this problem. We will rely on the fact that when a dielectric sphere is placed in an electric field, it gets polarized along the external field. The resultant field, $\mathbf{E}$, will be a superposition of $\mathbf{E}_0$ and the field produced by a polarization charge induced on the sphere. We assume that the sources of $\mathbf{E}_0$ are undisturbed by the presence of the sphere. The latter field is called the depolarization field, here labeled $\mathbf{E}'$, and $\mathbf{E} = \mathbf{E}_0 + \mathbf{E}'$. Now, we shall assume that the dielectric sphere is polarized by $\mathbf{E}_0$ such that the polarization, $\mathbf{P}$ is $\mathbf{P} = \beta'\mathbf{E}_0$, proportional to $\mathbf{E}_0$ and constant. If this assumption is valid, we shall be led to a self-consistent solution that will be predicted in terms of the dielectric constants. If it is invalid, we will not be able to find a self-consistent solution.

If $\mathbf{P} = \beta'\mathbf{E}_0$, we expect to find that on the surface of the sphere there will be a polarization charge density $\sigma = \beta E_0 \cos\theta = \sigma_0 \cos\theta$, where β is another constant that must be due to polarizations in both dielectric media, but its exact value remains to be determined. In any case $\mathbf{E}'$ will emanate from the surface charge σ_p. Now the field inside the sphere (see again Examples 2.17, 3.6, and 3.7) will be given by

$$\mathbf{E} = E_0\hat{\mathbf{z}} - \frac{\beta E_0}{3\varepsilon_2}\hat{\mathbf{z}} = E_0\hat{\mathbf{z}}\left(1 - \frac{\beta}{3\varepsilon_2}\right)$$

The field outside will be given by

$$\mathbf{E} = E_0\hat{\mathbf{z}} + \frac{\beta E_0 R^3}{3\varepsilon_2 r^3}[2\hat{\mathbf{r}}\cos\theta + \hat{\boldsymbol{\theta}}\sin\theta]$$

where we have recognized that the surface charge distribution produced a dipole field with a dipole moment $p \equiv (\frac{4}{3})\pi R^3\sigma_0 \equiv (\frac{4}{3})\pi R^3\beta E_0$. If these fields are consistent, they will satisfy the required boundary conditions at $r = R$—that is, that the tangential components of the electric field are continuous and the normal components of the displacement vector are also continuous. Using the above assumed field, we find that the tangential condition seems automatically satisfied. The normal condition determines β for us. The result is

$$\beta = 3\varepsilon_2 \frac{\varepsilon_1 - \varepsilon_2}{\varepsilon_1 + 2\varepsilon_2}$$

With β determined, the problem is completely solved, and it is easy to show that this solution is identical to the results of the first method.

Example 4.9 A Dipole at the Center of a Dielectric Sphere

This example deals with a boundary value problem where the potential is a function of an angle as well as a distance. The angular dependence is produced by inserting a dipole of moment $\mathbf{p}$ at the center of a dielectric sphere of radius R and permittivity ε_1. The permittivity of the material external to the sphere is ε_2.

We describe the system by spherical polar coordinates with the origin at the center of the sphere and the z axis along the dipole. Away from the dipole the potential satisfies Laplace's equation, and therefore it can be represented by an expansion of zonal harmonics. Since near the origin the potential is that of the dipole: $(1/4\pi\varepsilon_1)(p \cos\theta/r^2)$, and in the absence of the dipole, the problem is spherically symmetric, then we expect only the lowest-order zones to contribute. Thus the potentials inside and outside the sphere Φ_1 and Φ_2, respectively, are represented by the following expansions:

$$\Phi_1(r, \theta) = A_1 r \cos\theta + \frac{A_2 \cos\theta}{r^2} \qquad r < R \tag{4.90}$$

$$\Phi_2(r, \theta) = B_1 r \cos\theta + \frac{B_2 \cos\theta}{r^2} \qquad r > R \tag{4.91}$$

where A_1, A_2, B_1, and B_2 are to be determined from the boundary conditions:

1. $\Phi_1(r, \theta)$ goes to $(p/4\pi\varepsilon_1)(\cos\theta)/r^2$ as $r \to 0$. Thus

$$\frac{A_2 \cos\theta}{r^2} = \frac{p}{4\pi\varepsilon_1}\frac{\cos\theta}{r^2}$$

yielding $A_2 = p/4\pi\varepsilon_1$.

2. $\Phi_2(r, \theta)$ goes to zero as $r \to \infty$. Thus $B_1 = 0$.

3. The potential at the boundary is continuous: $\Phi_1(R, \theta) = \Phi_2(R, \theta)$. This gives a relation between A_1 and B_2:

$$A_1 R + \frac{p}{4\pi\varepsilon_1}\frac{1}{R^2} = \frac{B_2}{R^2} \tag{4.92}$$

4. The normal component of the displacement vector is continuous at the boundary: $\varepsilon_1\, \partial\Phi_1/\partial r = \varepsilon_2\, \partial\Phi_2/\partial r$ at $r = R$. This condition gives another relation between A_1 and B_2:

$$\varepsilon_1\left(A_1 - \frac{p}{2\pi\varepsilon_1 R^3}\right) = -2\varepsilon_2 \frac{B_2}{R^3} \tag{4.93}$$

Solving Eqs. (4.92) and (4.93) simultaneously for A_1 and B_2 gives

$$A_1 = \frac{2p}{4\pi\varepsilon_1 R^3}\frac{\varepsilon_1 - \varepsilon_2}{\varepsilon_1 + 2\varepsilon_2} \qquad B_2 = \frac{3p}{4\pi\varepsilon_1}\frac{\varepsilon_1}{\varepsilon_1 + 2\varepsilon_2}$$

Thus

$$\Phi_1(r, \theta) = \frac{p}{4\pi\varepsilon_1}\left[\frac{\cos\theta}{r^2} + \frac{2(\varepsilon_1 - \varepsilon_2)}{R^3(\varepsilon_1 + 2\varepsilon_2)} r \cos\theta\right] \qquad r \le R \tag{4.94}$$

$$\Phi_2(r, \theta) = \frac{3p}{4\pi(\varepsilon_1 + 2\varepsilon_2)}\frac{\cos\theta}{r^2} \qquad r \ge R \tag{4.95}$$

Note that as $\varepsilon_1 \to \varepsilon_2$, the potentials reduce to the potential of a dipole in an infinite dielectric material as expected.

*4.7 Method of Images for Dielectric Interfaces

In Example 4.5 we considered a point charge placed at the plane interface of two semi-infinite dielectrics. Because of the symmetry of that situation, Gauss' law and the solution of Laplace's equation in a single variable were applicable, thus making the determination of the fields straightforward.

Now we consider the case where the point charge is placed not at the plane interface but at a certain distance from it. The relocation of the charge breaks the symmetry, and therefore simple methods as mentioned above are not very useful; the fields become dependent on distances and angles.

Previously, in Chapter 3, we saw that the solution for the fields of a point charge q placed near a highly conducting plane surface were conveniently determined by the method of images. It was found that the system is equivalent to the original charge plus an image charge $-q$ placed just under q at distance d on the opposite side of the plane. To solve the dielectric case, we resort to the method of images; however, because of the drastically different boundary conditions needed to be satisfied at the boundary, the system will not be equivalent to the point charge and its image.

We will take the approach of first choosing a reasonable number of image charges with reasonable locations. These choices can then be tested by requiring that the produced fields satisfy the boundary conditions. Consider Fig. 4.12, where a point charge q is placed at a point P, which is at a distance d from the boundary of two semi-infinite homogeneous dielectrics of permittivities ε_1 and ε_2. We take the potential in region 1 to be represented by the charge q and an image charge of magnitude q' located in region 2 at distance d on the x axis. The potential in region 2, however, will be only that of an image charge q'' located at P (that is, at the location of the original charge q). Thus

$$\Phi_1(\mathbf{r}) = \frac{1}{4\pi\varepsilon_1}\left(\frac{q}{r} + \frac{q'}{r'}\right) \qquad \text{for } x < 0$$

$$\Phi_2(\mathbf{r}) = \frac{1}{4\pi\varepsilon_2}\frac{q''}{r} \qquad \text{for } x > 0 \tag{4.96}$$

where $r = [(x + d)^2 + y^2 + z^2]^{1/2}$ and $r' = [(x - d)^2 + y^2 + z^2]^{1/2}$, and q' and q'' magnitudes of the image charges, which are not yet known.

The above solution can now be checked by finding out whether it can satisfy the boundary conditions by determining physical values for these unknown charges:

1. The potential should be continuous at $x = 0$; that is,

$$\Phi_1(x = 0, y, z) = \Phi_2(x = 0, y, z).$$

Thus

$$\frac{1}{\varepsilon_1}(q + q') = \frac{q''}{\varepsilon_2} \tag{4.97}$$

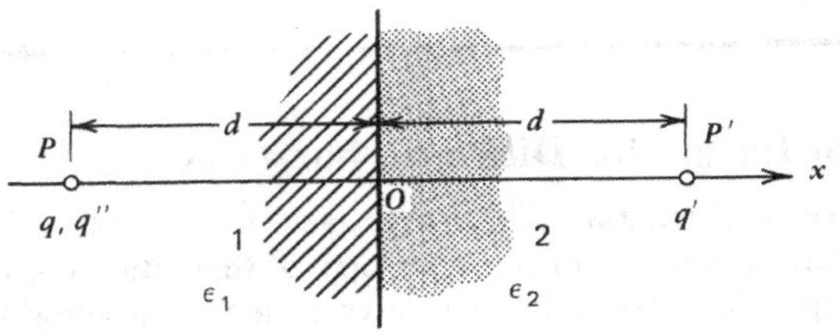

Figure 4.12 Application of the method of images to a point charge placed near a plane interface of two dielectric materials.

2. The normal component of the displacement vector is continuous on the boundary since it has no free charges; that is, $D_{1n}(x = 0, y, z) = D_{2n}(x = 0, y, z)$, or $\varepsilon_1\, \partial\Phi_1/\partial x = \varepsilon_2\, \partial\Phi_2/\partial x$ at $x = 0$. Thus

$$q'' = q - q' \tag{4.98}$$

Solving Eqs. (4.97) and (4.98) simultaneously gives:

$$q' = -\frac{\varepsilon_2 - \varepsilon_1}{\varepsilon_1 + \varepsilon_2} q \qquad \text{and} \qquad q'' = \frac{2\varepsilon_2}{\varepsilon_1 + \varepsilon_2} q \tag{4.99}$$

Thus the potentials are given by

$$\Phi_1 = \frac{q}{4\pi\varepsilon_1}\left[\frac{1}{r} - \left(\frac{\varepsilon_2 - \varepsilon_1}{\varepsilon_1 + \varepsilon_2}\right)\frac{1}{r'}\right] \qquad \text{and} \qquad \Phi_2 = \frac{2q}{4\pi(\varepsilon_1 + \varepsilon_2)}\frac{1}{r} \tag{4.100}$$

Because Φ_1 and Φ_2 satisfy Laplace's equation and the boundary conditions, then the solution is unique, and therefore the above two image charges are sufficient to determine the fields uniquely.

We now give a qualitative description of the lines of force produced by q. Figure 4.13*a* shows the case $\varepsilon_2 > \varepsilon_1$. In this case q' has the opposite sign of q. Therefore in region 1 the lines of force look like the lines of force between two charges of opposite sign (dipole lines). In region 2 the lines of force are described by one charge of same sign as q and emanating from the position of q.

Figure 4.13*b* shows the case $\varepsilon_2 < \varepsilon_1$. In this case q' has the same sign as q. Therefore in region 1 the lines of force are similar to those of two charges of same sign. In region 2 the lines of force are described by a point charge emanating from the position of q.

The presence of charge q in medium 1 produces polarizations P_1 and P_2 in mediums 1 and 2, respectively; $\mathbf{P}_i = -(\varepsilon_i - \varepsilon_0)\nabla\Phi_i$, where $i = 1$ and 2. The polarization charge density has two components: a volume charge density $-\nabla \cdot \mathbf{P}_i$ and a surface charge density $\mathbf{P}_i \cdot \hat{\mathbf{n}}$. Since $\mathbf{P}_i$ is proportional to $\mathbf{E}_i$, which satisfies $\nabla \cdot \mathbf{E}_i = 0$, then the volume charge density vanishes except at the point charge q itself. Using a

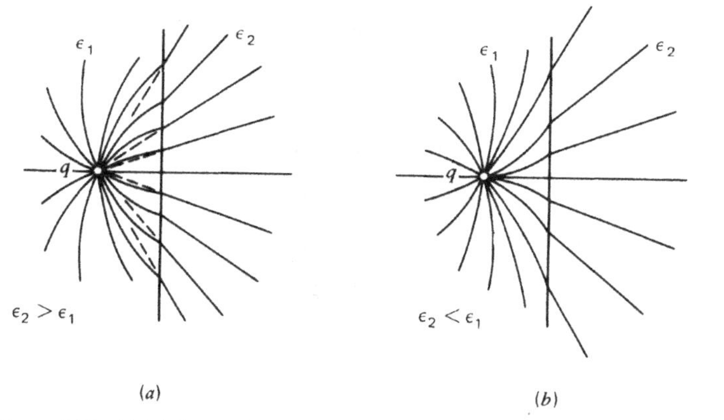

Figure 4.13 Field lines of a point charge implanted in ε_1 material near the plane interface with a material of ε_2. (*a*) $\varepsilon_2 > \varepsilon_1$. (*b*) $\varepsilon_2 < \varepsilon_1$.

unit vector $\hat{\mathbf{x}}$ normal to the interface and pointing from material 1 to 2, the surface charge density is: $\sigma_p = -(\mathbf{P}_2 - \mathbf{P}_1)\cdot\hat{\mathbf{x}}$, which can be easily shown to be

$$\sigma_p = -q\frac{(\varepsilon_2 - \varepsilon_1)}{2\pi\varepsilon_1(\varepsilon_1 + \varepsilon_2)}\frac{d}{(y^2 + z^2 + d^2)^{3/2}} \tag{4.101}$$

***Example 4.10 Forces Between Charges Embedded in Dielectric Materials**

This example shows how the method of images can be used to find the forces between charges embedded in dielectric media. Consider Fig. 4.14, showing two semi-infinite homogeneous dielectrics 1 and 2 with permittivities ε_1 and ε_2, respectively. Two charges q_1 and q_2 are placed in media 1 and 2, respectively, each at a distance d from the interface and with the line between the charges normal to the interface.

The charge q_2 experiences two types of forces; one type is due to the presence of charge q_1 and the other is due to its proximity to the interface. We first calculate the first force. The force is equal to $q_2\mathbf{E}$, where $\mathbf{E}$ is the electric field caused by q_1 at the site of q_2. From the previous example, the field caused by charge q_1 in region 2 is produced by a charge $q = 2\varepsilon_2 q_1/(\varepsilon_1 + \varepsilon_2)$ located at the position of q_1 [see Eq. (4.99)]. Therefore the force exerted by q_1 on q_2 can be easily calculated using Coulomb's law, as follows:

$$\mathbf{F}_{21} = \frac{q_1 q_2}{8\pi(\varepsilon_1 + \varepsilon_2)d^2}\hat{\mathbf{n}} \tag{4.102}$$

where $\hat{\mathbf{n}}$ is a unit vector along the line joining the charges.

Another force acting on q_2 is produced by the induced charge at the interface. This force can alternatively be calculated from the image charges. The field due to charge q_2 in region 2 is produced by q_2 and the image charge $-(\varepsilon_1 - \varepsilon_2)/(\varepsilon_1 + \varepsilon_2)q_2$ placed at the q_1 position. Therefore the force acting on q_2 is

$$\mathbf{F}_{2i} = \frac{1}{16\pi\varepsilon_2}\frac{\varepsilon_2 - \varepsilon_1}{\varepsilon_1 + \varepsilon_2}\frac{q_2^2}{d^2}\hat{\mathbf{n}} \tag{4.103}$$

Adding Eqs. (4.102) and (4.103) gives the total force $\mathbf{F}_2 = \mathbf{F}_{21} + \mathbf{F}_{2i}$, acting on q_2:

$$\mathbf{F}_2 = \left[\frac{1}{16\pi\varepsilon_2}\frac{\varepsilon_2 - \varepsilon_1}{\varepsilon_1 + \varepsilon_2}\frac{q_2^2}{d^2} + \frac{1}{8\pi(\varepsilon_1 + \varepsilon_2)}\frac{q_1 q_2}{d^2}\right]\hat{\mathbf{n}} \tag{4.104}$$

Similarly one can show that the force acting on q_1 is:

$$\mathbf{F}_1 = -\left[\frac{1}{16\pi\varepsilon_1}\frac{\varepsilon_1 - \varepsilon_2}{\varepsilon_1 + \varepsilon_2}\frac{q_1^2}{d^2} + \frac{1}{8\pi(\varepsilon_1 + \varepsilon_2)}\frac{q_1 q_2}{d^2}\right]\hat{\mathbf{n}} \tag{4.105}$$

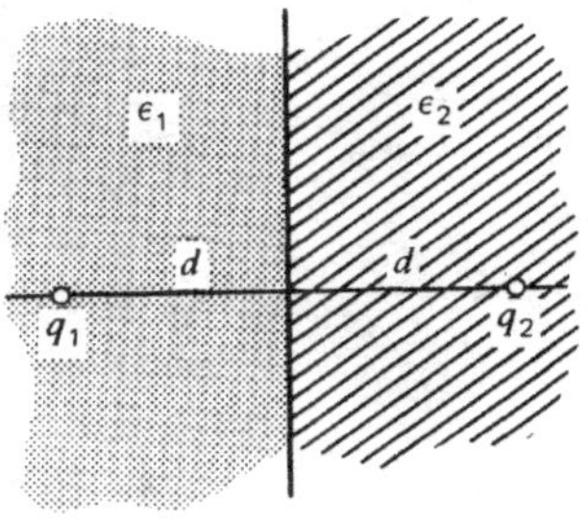

Figure 4.14 Force between two point charges placed on opposite sides of an interface of two dielectric materials.

It appears that the magnitude of the forces acting on q_1 and q_2 are not equal. This is simply explained by the fact that the forces are due not only to q_1 and q_2 interaction but also to the induced charge at the interface.

Thus far the method of images has proved successful in solving problems involving point charges near plane dielectric interfaces. The next situation that will arise involves point charges near spherical dielectric interfaces. Such cases arise when a point charge is brought near a dielectric sphere or a spherical cavity in an infinite dielectric medium. Again one would like to find out if there is a finite set of image charges that can be used to satisfy the boundary conditions, and thus generate a unique solution for the interaction. Unfortunately and contrary to the case where the sphere is a conductor, there is no finite number of charges that can satisfy the boundary conditions in the dielectric case and hence the method is not useful in this respect.

4.8 Forces on Charge Distributions

The electric force on an element of charge dq, placed in an external electric field $\mathbf{E}^{(\varepsilon)}$ is, by definition, $d\mathbf{F} = dq\,\mathbf{E}^{(\varepsilon)}$. The charge element itself has an electric field, and since the charge can exert no net force on itself, we may in general write that $d\mathbf{F} = dq\,\mathbf{E}$ where $\mathbf{E}$ represents the *total* field at the location of the charge element. The total force $\mathbf{F}$ acting on a distribution of charge that is characterized by a charge density ρ is given by

$$\mathbf{F} = \int \mathbf{E}\,dq = \int_V \rho\mathbf{E}\,dv \tag{4.106}$$

where V denotes the volume where the charge density is nonzero. The force on a surface charge, having a surface charge density σ, is similarly expressed as

$$\mathbf{F} = \int_S \sigma\mathbf{E}\,da \tag{4.107}$$

The trouble with Eq. (4.107) however, is that $\mathbf{E}$ on a surface is not well defined. When the surface has a charge density σ, a discontinuity in the $\mathbf{E}$ field exists. In this case $\mathbf{E}$ is ambiguous, but it is not unreasonable to expect that the field that should be used is the average of the field on the two sides of the surface. To verify this conjecture, consider Fig. 4.15, which shows a surface which has a charge density σ, and which separates space into the two regions labeled 1 and 2, with $\mathbf{E}$ fields given just on opposite sides of the boundary as $\mathbf{E}_1$ and $\mathbf{E}_2$. The total field anywhere may

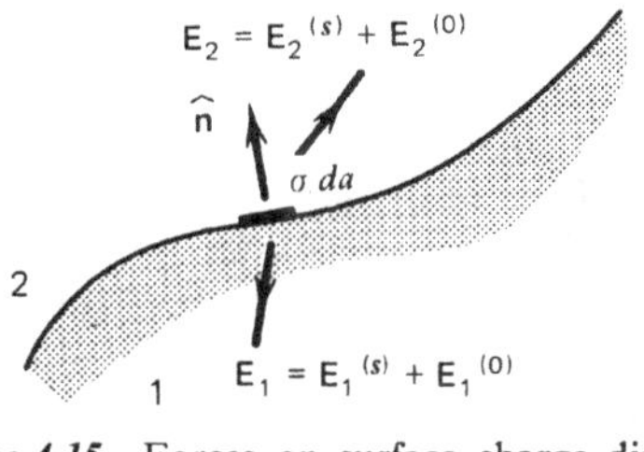

Figure 4.15 Forces on surface charge distributions in terms of the total field of the distribution.

be considered to be due to the $\mathbf{E}^{(s)}$ of the charged surface element da plus fields due to all other charges $\mathbf{E}^{(0)}$ (external to the element itself)

$$\mathbf{E} = \mathbf{E}^{(s)} + \mathbf{E}^{(0)} \tag{4.108}$$

Therefore, we write $\mathbf{E}_i = \mathbf{E}_i^{(s)} + \mathbf{E}_i^{(0)}$, $i = 1$ or 2. Consider two points located just on opposite sides of the boundary at da. To calculate the fields at these points we assume that as the surface is approached, the surface element da appears like a uniformly charged plane with a charge density σ. Using Eq. (2.21) of Example 2, which deals with a uniformly charged plane, we find that

$$\mathbf{E}_1^{(s)} = \frac{\sigma}{2\varepsilon_0}(-\hat{\mathbf{n}}) \qquad \text{and} \qquad \mathbf{E}_2^{(s)} = \frac{\sigma}{2\varepsilon_0}(+\hat{\mathbf{n}}) \tag{4.109}$$

However, the field $\mathbf{E}^{(0)}$ must be continuous across da, and therefore the total field at these points is as follows:

$$\mathbf{E}_1 = -\frac{\sigma}{2\varepsilon_0}\hat{\mathbf{n}} + \mathbf{E}^{(0)} \qquad \text{and} \qquad \mathbf{E}_2 = \frac{\sigma}{2\varepsilon_0}\hat{\mathbf{n}} + \mathbf{E}^{(0)} \tag{4.110}$$

We now provide more information about the nature of Eq. (4.110). Subtracting the first relation from the second, we get

$$\mathbf{E}_2 - \mathbf{E}_1 = \frac{\sigma}{\varepsilon_0}\hat{\mathbf{n}} \tag{4.111}$$

The tangential and normal components of Eq. (4.111) give

$$\mathbf{E}_{2t} = \mathbf{E}_{1t} \qquad \text{and} \qquad E_{2n} - E_{1n} = \frac{\sigma}{\varepsilon_0} \tag{4.112}$$

respectively. The second relation of Eq. (4.112) was previously encountered [see Eq. (2.35)]; it was derived using Gauss' law. The first relation of Eq. (4.112) expresses the continuity of the tangential component of the electric field [see (Eq. 3.6)].

We now calculate the force on the element da. We note that since the net electrostatic force acting on da is due to $\mathbf{E}^{(0)}$ alone, then

$$d\mathbf{F} = \mathbf{E}^{(0)}\sigma\, da \tag{4.113}$$

Adding the equations for $\mathbf{E}_1$ and $\mathbf{E}_2$ given in Eq. (4.110) we obtain

$$\mathbf{E}^{(0)} = \frac{1}{2}(\mathbf{E}_1 + \mathbf{E}_2) \tag{4.114}$$

Substituting this result in Eq. (4.113), we find for the force per unit area

$$\frac{d\mathbf{F}}{da} = \frac{\sigma(\mathbf{E}_1 + \mathbf{E}_2)}{2} \tag{4.115}$$

This verifies our initial guess that the field acting on the surface charge should be taken as the average field of the fields on the two sides.

A common case of a surface charge density is that of a surface density on a charged conductor (see Section 2.6), Conductors and Insulators. In this case, $\mathbf{E}_1 \equiv 0$ and $\mathbf{E}_2 \equiv \mathbf{E} = \hat{\mathbf{n}}E$, which is the field just outside the conductor. We therefore have as the force on an element of surface of the conductor

$$d\mathbf{F} = \sigma\frac{\mathbf{E}}{2}\, da = \frac{\sigma^2}{2\varepsilon_0}\, da\, \hat{\mathbf{n}} \tag{4.116}$$

The pressure $P = dF/da$ on the conductor surface is therefore

$$P = \frac{dF}{da} = \frac{\sigma^2}{2\varepsilon_0} = \frac{1}{2}\varepsilon_0 E^2 \tag{4.117}$$

It always points outward.

Example 4.11 Force Between Two Halves of a Charged Sphere

A conducting spherical shell of radius R is charged to a potential V. The shell is sliced into two separate hemispheres and kept in place (Fig. 4.16). We use here the ideas developed in this section to find the force exerted by the hemispheres on each other. The charge density σ on the surface of the shell is equal to $Q/4\pi R^2$, where Q, the total charge on the shell, is given

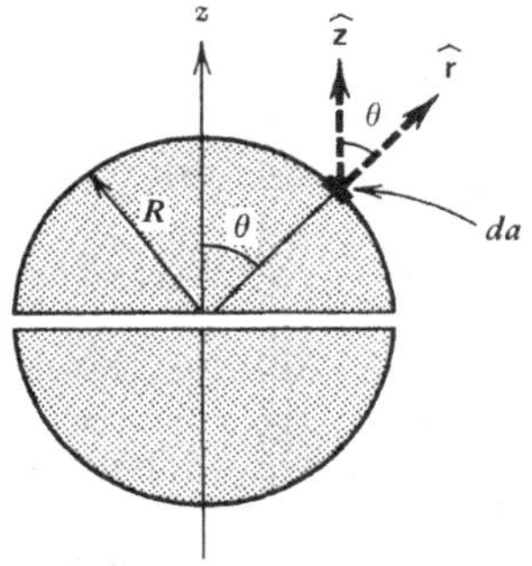

Figure 4.16 Uniformly charged sphere sliced in two halves in place.

by $Q = 4\pi\varepsilon_0 RV$. Thus $\sigma = \varepsilon_0 V/R$. The force on an area element da located at angles θ and ϕ, is given by Eq. (4.116); that is, $d\mathbf{F} = (\sigma^2/2\varepsilon_0)da\,\hat{\mathbf{n}}$, where $\hat{\mathbf{n}} = \hat{\mathbf{r}}$ is a unit vector normal to the differential area. We note that the component of $d\mathbf{F}$ normal to the z axis does not contribute to the total force because of the symmetry about the z axis. Thus the contributing component is

$$dF_z = \frac{\sigma^2}{2\varepsilon_0}\cos\theta\, da$$

The total force is then determined by integrating over the area. Using $da = R^2 \sin\theta\, d\theta\, d\phi$ gives

$$\mathbf{F} = \hat{\mathbf{z}} \int_{\theta=0}^{\pi/2}\int_{\phi=0}^{2\pi} \frac{1}{2\varepsilon_0}\sigma^2 R^2 \cos\theta \sin\theta\, d\theta\, d\phi$$

The integration over ϕ gives 2π and the integration over θ gives $1/2$; therefore

$$\mathbf{F} = \frac{\pi^2\sigma^2 R^2}{\varepsilon_0}\hat{\mathbf{z}} = \pi\varepsilon_0 V^2\hat{\mathbf{z}}$$

Example 4.12 Force on a Conducting Sphere Placed in an Electric Field

Consider a conducting shell of radius R and total charge q placed in an external electric field $E_0\hat{\mathbf{z}}$. The space surrounding the sphere is filled with a dielectric material of permittivity ε. To find the force exerted on the sphere we have to find the charge distribution and the electric field at the surface of the shell. These in turn can be determined by first determining the electric potential outside the sphere.

The potential outside the sphere satisfies Laplace's equation and therefore can be expanded in zonal harmonics. Keeping only terms of up to $P_1(\theta) = \cos\theta$ angular dependence, we write:

$$\Phi(r, \theta) = A_0 + \frac{B_0}{r} + A_1 r\cos\theta + \frac{B_1}{r^2}\cos\theta \tag{4.118}$$

where A_0, B_0, A_1, and B_1 are constants. The term B_0/r is just the Coulomb potential produced by the charge on the sphere; therefore, $B_0 = q/4\pi\varepsilon$. For the evaluation of the rest of the constants see Example 3.7, which deals with a conducting sphere placed in vacuum in the presence of an external electric field. Thus the potential is

$$\Phi(r, \theta) = V_0 + \frac{q}{4\pi\varepsilon r} - E_0 r\cos\theta + \frac{E_0 R^3}{r^2}\cos\theta \tag{4.119}$$

and the electric field, $\mathbf{E} = -\nabla\Phi$, is

$$\mathbf{E}(r, \theta) = \left(\frac{q}{4\pi\varepsilon r^2} + E_0\cos\theta + \frac{2R^3E_0}{r^3}\cos\theta\right)\hat{\mathbf{r}} - E_0\left(1 - \frac{R^3}{r^3}\right)\sin\theta\,\hat{\boldsymbol{\theta}} \tag{4.120}$$

where $V_0 = A_0$. The surface charge density, $\sigma = \varepsilon E(R, \theta)$, is

$$\sigma = \frac{q}{4\pi R^2} + 3\varepsilon E_0\cos\theta$$

The force $d\mathbf{F}$ on a unit area da on the surface of the sphere can be written using Eq. (4.116): $d\mathbf{F} = (1/2\varepsilon)\sigma^2\,da\,\hat{\mathbf{r}}$. Because of the symmetry along the z axis, only the component of $d\mathbf{F}$ along the z axis contributes. Thus $dF_z = (1/2\varepsilon)\sigma^2\,da\cos\theta$. Integrating dF_z over the surface of the sphere gives $\mathbf{F} = E_0 q\hat{\mathbf{z}}$. This result indicates that the force is independent of the radius of the sphere, and it is just equal to the product of q and the field in the dielectric E_0. This relation in fact reaffirms the definition of the electric field in terms of the force per unit charge exerted on a test charge.

4.9 Summary

A dielectric material is said to be polarized if the macroscopic sum of the dipole moments of its atomic constituents is not zero. The polarization $\mathbf{P}$ of the medium is just defined as the volume density of such dipoles

$$\mathbf{P} = \frac{d\mathbf{p}}{dv} \tag{4.1}$$

The electrical properties of a macroscopic piece of polarization $\mathbf{P}$ can be calculated by replacing all of the atomic dipoles by an effective volume and surface charge densities (polarization charges, or bound charges) ρ_p and σ_p:

$$\rho_p = -\nabla\cdot\mathbf{P} \quad \text{and} \quad \sigma_p = \mathbf{P}\cdot\hat{\mathbf{n}} \tag{4.13}$$

The electrostatic potential Φ_p, and hence the electric field take the expressions

$$\Phi_p = \frac{1}{4\pi\varepsilon_0}\int\frac{\rho_p}{|\mathbf{r} - \mathbf{r}'|}dv' + \frac{1}{4\pi\varepsilon_0}\int\frac{\sigma_p}{|\mathbf{r} - \mathbf{r}'|}da' \tag{4.12}$$

and

$$\mathbf{E} = -\nabla\Phi_p$$

In the presence of external charges (free charges), Gauss' differential law becomes

$$\nabla\cdot\mathbf{E} = \frac{\rho_f + \rho_p}{\varepsilon_0} = (\rho_f - \nabla\cdot\mathbf{P})/\varepsilon_0$$

It is convenient to define what is called the displacement vector **D** such that

$$\mathbf{D} = \varepsilon_0 \mathbf{E} + \mathbf{P} \tag{4.35}$$

Thus we get the following for Gauss' law in the presence matter

$$\nabla \cdot \mathbf{D} = \rho_f \tag{4.36}$$

The integral form of Gauss' law in matter immediately follows from the differential one

$$\oint_S \mathbf{D} \cdot \hat{\mathbf{n}}\, da = \int_V \rho_f\, dv = Q_f \tag{4.40}$$

The response of a material to an external field **E** depends on the microscopic structure of the material. In here we classify materials according to their macroscopic response

$$\mathbf{P} = \varepsilon_0 \chi(\mathbf{E})\mathbf{E} \tag{4.44}$$

where χ is called the electric susceptibility. If χ is independent of **E** (magnitude and direction), independent of space, the material is said to be linear (simple). For linear materials $\mathbf{P} = \varepsilon_0 \chi \mathbf{E}$, and hence

$$\mathbf{D} = \varepsilon_0(1 + \chi)\mathbf{E} = \varepsilon \mathbf{E} = \varepsilon_0 K \mathbf{E} \tag{4.46}$$

where ε is the permittivity of the material and K is the relative permittivity of dielectric constant.

The basic equations of electrostatics in the presence of dielectric materials are

$$\nabla \cdot \mathbf{D} = \rho_f \qquad \text{and} \qquad \nabla \times \mathbf{E} = 0$$

In a linear material of permittivity ε, this divergence equation becomes $\nabla \cdot \mathbf{E} = \rho_f/\varepsilon$. The curl equation, on the other hand, implies that $\mathbf{E} = -\nabla\Phi$; hence

$$\nabla \cdot [-\nabla\Phi] = \frac{\rho_f}{\varepsilon} \qquad \text{or} \qquad \nabla^2\Phi = -\frac{\rho_f}{\varepsilon} \tag{4.59}$$

In regions where $\rho_f = 0$, then

$$\nabla^2\Phi = 0 \qquad \text{for } \rho_f = 0$$

When a given space is made up of regions of different dielectric properties, then the fields can be determined using boundary value techniques. The equation for Φ is solved in the different regions independently, followed by matching these solutions at the interfaces of the regions according to the following rules.

$$\mathbf{E}_{1t} = \mathbf{E}_{2t} \qquad \text{or} \qquad \Phi_1 = \Phi_2 \qquad (4.65),(4.66)$$

and

$$D_{2n} - D_{1n} = \sigma_f \tag{4.61}$$

where t and n stand for tangent and normal to the interface, respectively.

In certain geometries the method of images can be used to solve boundary value problems in the presence of dielectrics. The usefulness of this technique, however, is limited.

The electrostatic force on a given charge element dq of a larger surface charge distribution is caused by the electric field $\mathbf{E}^{(0)}$ at the site of the element due to the rest of the distribution

$$d\mathbf{F} = dq\, \mathbf{E}^{(0)} = \sigma\, da\, \mathbf{E}^{(0)} \tag{4.113}$$

where σ and $d\mathbf{a}$ are the density and area of the charge element. In many cases however, it is easier to determine the total electric field at the element that includes the field due to the element itself. In this case the force becomes

$$d\mathbf{F} = \frac{1}{2}\sigma\, da(\mathbf{E}_1 + \mathbf{E}_2) \tag{4.115}$$

where $\mathbf{E}_1$ and $\mathbf{E}_2$ are the total fields on both sides of the element. If the element is a surface of a conductor, where $\mathbf{E}_2 = 0$ and $\mathbf{E}_1 = \mathbf{E}$, then

$$d\mathbf{F} = \frac{1}{2}\sigma\, da\, \mathbf{E} \tag{4.116}$$

From Gauss' law at the surface of a conductor we have $\mathbf{E} = (\sigma/\varepsilon)\hat{\mathbf{n}}$; then

$$\frac{d\mathbf{F}}{da} = \frac{1}{2\varepsilon_0}\sigma^2\hat{\mathbf{n}} \quad \text{or} \quad \frac{d\mathbf{F}}{da} = \frac{\varepsilon_0}{2}E^2\hat{\mathbf{n}} \tag{4.117}$$

Problems

4.1 A hemisphere of dielectric has its flat surface in the x-y plane. It is polarized in the z-direction: $\mathbf{P} = P\hat{\mathbf{z}}$, with P a constant. (a) Find the bound volume charge density. Find the bound surface charge density on the flat and on the hemispherical surface. (b) Find by integration the net bound charge on the hemisphere. (c) Explain why your answer to (b) is expected for physical reasons. (d) Find by integration the dipole moment of the bound charge relative to an origin at the center of the flat surface. Calculate the dipole moment also by using the definition of the polarization $\mathbf{P}$. Do the two answers agree?

4.2 A hemicylinder of radius R and length L has a uniform polarization $\mathbf{P}$ in the direction normal to its rectangular surface. Find its dipole moment and the polarization charge densities.

4.3 The interior of a circular cylinder $x^2 + y^2 = R^2$ is occupied by a polarized material, with the polarization being $\mathbf{P} = (ax^2 + b + cy + a)x\hat{\mathbf{x}} + px\hat{\mathbf{y}}$. Find the volume and surface polarization charge densities.

4.4 A uniform cylindrical volume charge distribution of density α occupies the space between $\rho = \rho_1$ and ρ_2. The charge distribution is surrounded by a cylindrical shell of dielectric material of outer radius ρ_3, and dielectric constant K. (a) Determine the electric field and the displacement vector in all regions of space. (b) Determine the electric polarization and the polarization charges in all regions of space. (c) Calculate the potential difference between $\rho = 0$ and $\rho = \rho_3$. (d) What charge distribution at the axis of the cylinder replacing the original charge and the dielectric will result in the same electric field for $\rho_2 < \rho < \rho_3$?

4.5 Two concentric conducting spheres of inner and outer radii a and b carry q and $-q$ charges, respectively. The space between the spheres is half filled in the form of a hemispherical shell with a dielectric of permittivity ε. (a) Determine the $\mathbf{E}$ field between the spheres. (b) Determine the charge distribution on the inner sphere. (c) Determine the induced surface charge density on the inner hemispherical surface of the dielectric.

4.6 A parallel-plane capacitor of area A and separation d has a slab of dielectric of permittivity ε and thickness $t < d$ inserted between the plates at a distance h from one of the plates. The slab is implanted with external charges by some means prior to its insertion with a volume charge density ρ. Given that the surface charge density at the inner surface of the plate at a distance h from the slab is σ_1. (a) Find $\mathbf{D}$ between the plates. (b) Find the induced volume and surface charge densities. (c) Find the surface charge density at the inner surface of the second plate.

4.7 Consider a concentric spherical capacitor with inner and outer spheres of radii a and b, respectively. The region between the spheres is filled with an inhomogeneous dielectric with $\varepsilon = \varepsilon_0/(c - \alpha r)$ where c and α are constants. A charge q is placed on the inner sphere, and the outer sphere is grounded. (a) Determine the displacement vector $\mathbf{D}$ between the spheres. (b) Determine the volume polarization charge density between the spheres.

4.8 Two parallel capacitor plates enclose a dielectric material that has a spatially varying dielectric constant, $K = e^{\alpha x}$, where α is a constant (see Fig. 4.17). Find the electric field $\mathbf{E}(x)$ inside the plates when they are charged to a potential difference V.

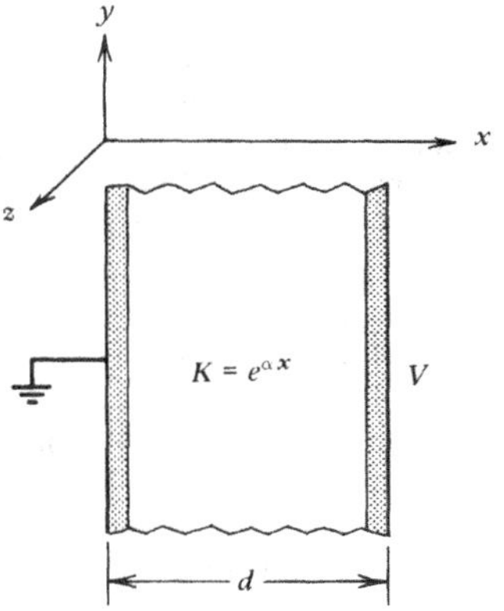

Figure 4.17

4.9 A concentric cylindrical capacitor consists of two conducting cylinders of radii a and b (where $b > a$) and length L, which is large compared to the radii. The charges on the inner and outer cylinders are $+q$ and $-q$. (a) Determine the **E** field between the cylinders if the space there is empty. (b) Show that, by introducing a material whose ε is a function of position, the **E** field between the cylinder can be made constant in magnitude.

4.10 The displacement vector in region $x < 0$ is $\mathbf{D}_1 = 1.5\hat{\mathbf{x}} - 2\hat{\mathbf{y}} + 3\hat{\mathbf{z}}\ \mathrm{C/m^2}$. If ε_0 and $2.5\varepsilon_0$ are the permittivities of regions $x < 0$ and $x > 0$ and there is no free charge at $x = 0$, determine (a) the electric field $\mathbf{E}_2$ in region $x > 0$ and (b) the angles θ_1 and θ_2 which $\mathbf{D}_1$ and $\mathbf{D}_2$, respectively, make with the $x = 0$ plane.

4.11 Apply the boundary conditions on **D** and **E** to an interface between two ideal dielectrics of dielectric constant K_1 and K_2, respectively, to find a "law of refraction" for lines of **E** at the interface. That is, find a relationship between the directions of $\mathbf{E}_1$ and $\mathbf{E}_2$ in terms of K_1 and K_2.

4.12 Two infinite dielectric slabs, 1 and 2, each of unit thickness and with dielectric constants K_1 and K_2, have a face in common and separate two infinite, thin plates of conducting material. The plates are kept at zero and V potentials (see Fig. 4.18). (a) Determine the potential as a function of x in regions 1 and 2. (b) Find the surface density of free and bound charges at $x = 0$. (c) Find the surface density of bound charge at $x = 1$.

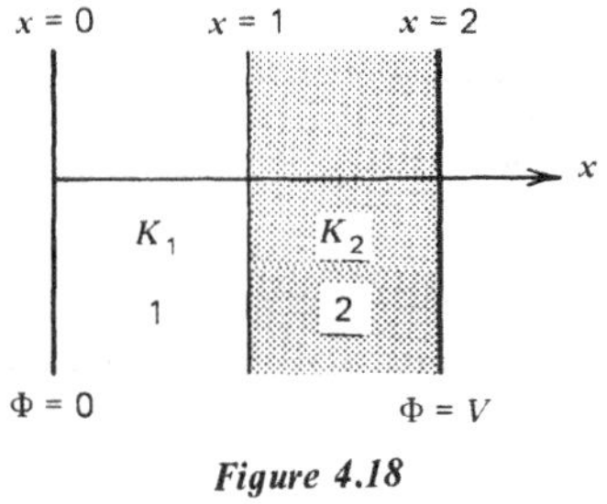

Figure 4.18

4.13 A dielectric sphere of radius a and permittivity ε_1 is surrounded by a dielectric of permittivity ε_2 and inner and outer radii a and b. The potentials inside the sphere and the shell are $\Phi_1 = Ar\theta$ and $\Phi_2 = Aa^2\theta/r$, respectively. Determine the polarization charge densities in both materials and the real (free) charge density at the surface of the sphere.

4.14 A cylindrical shell of dielectric of permittivity ε and inner and outer radii a and b surrounds a very long, conducting cylinder of radius a. The system is placed in an electric field $\mathbf{E}_0$, perpendicular to the axis of the cylinder. (a) Find the potential in the regions $r < a$, $a < r < b$, and $r > b$. (b) Find the charge per unit area at $r = a$ as a function of the angle measured from the direction of $\mathbf{E}_0$.

4.15 A dielectric cylinder of permittivity ε_1 is placed in a homogeneous liquid dielectric of permittivity ε_2. The system is placed in a uniform external electric field. Determine the orientation of the cylinder. What would it be for a thin disk of permittivity ε_1?

4.16 Consider an infinite dielectric that is homogeneous and has a uniform polarization $\mathbf{P}$. A spherical cavity is now introduced in it. Determine the electric field $\mathbf{E}$ in the cavity when the introduction of the cavity (a) does not change the polarization in the surrounding dielectric (as happens in electrets—see Chapter 5) and (b) changes the polarization as a result of the changes in the electric field $\mathbf{P} = (\varepsilon - \varepsilon_0)\mathbf{E}$ (as happens in normal dielectrics).

4.17 A point charge q is brought to a position a distance d away from an infinite plane conductor held at zero potential. (a) Find the force between the plane and the charge by using Coulomb's law for the force between the charge and its image. (b) Find the total force acting on the plane by integrating $\sigma^2/2\varepsilon_0$ over the whole plane. Compare with the results from (a).

4.18 Use the boundary value problem techniques of Section 4.6 to solve for the potential and the fields of a dielectric sphere of radius R and uniform permanent polarization $\mathbf{P}$ [see Example (4.2)].

4.19 Consider a cylinder of wax of length l and radius $a \ll l$. It is uniformly polarized along a direction normal to its axis: $\mathbf{P} = P_0\hat{\mathbf{x}}$. Determine the potentials and the fields everywhere.

***4.20** An electric dipole is placed at a distance h from the plane interface of two semiinfinite dielectrics of permittivities ε_1 and ε_2. The dipole is in the ε_1 material and makes an angle θ with the normal to the interface. Use the method of images for charges to derive the image dipoles needed to solve for the potentials and fields in the dielectrics.

4.21 Consider $\mathbf{E} = \hat{\mathbf{x}}x^n$ in V/m and $\Delta v = \Delta x\,\Delta y\,\Delta z$. (a) Calculate

$$\langle \mathbf{E} \rangle = \frac{1}{\Delta v}\int \mathbf{E}\,dv.$$

(b) Determine $(\partial/\partial x)\langle \mathbf{E} \rangle$. (c) Calculate $\langle \partial \mathbf{E}/\partial x \rangle$ using the same procedure as in (a) and show that it is equal to $(\partial/\partial x)\langle \mathbf{E} \rangle$. Since this result can be generalized to more general functions, then this proves that the operations of averaging and differentiation are indeed interchangeable.

4.22 In a stationary material medium of permittivity ε, a unit positive charge (1 coulomb) is put at various points to find the electric field $\mathbf{E}$. Suppose that the force on the test charge is found to be given (in newtons) as

$$\mathbf{F} = x^2\hat{\mathbf{x}}$$

What is the charge density in the medium? Use the following two methods. (a) Take a small cubic volume element $\Delta x\,\Delta y\,\Delta z$ at the point (x, y, z), compute the net flux out of the surface of the volume element, and find the total charge inside the cubic element and then the charge density at the point (x, y, z). (b) By using the equation $\operatorname{div}\mathbf{D} = \rho_f$.

4.23 A dielectric slab of permittivity ε and uniform charge density ρ_f fills half of the volume of a parallel plate capacitor as shown in Fig. 4.19. (a) Determine the potential everywhere between the plates and sketch it for the case $\varepsilon = 2\varepsilon_0$. (b) Calculate the force per unit area on the conducting surfaces at $x = 0$ and $x = 2d$. (For the former you may assume that there is a *very* small gap between the plate and the dielectric.)

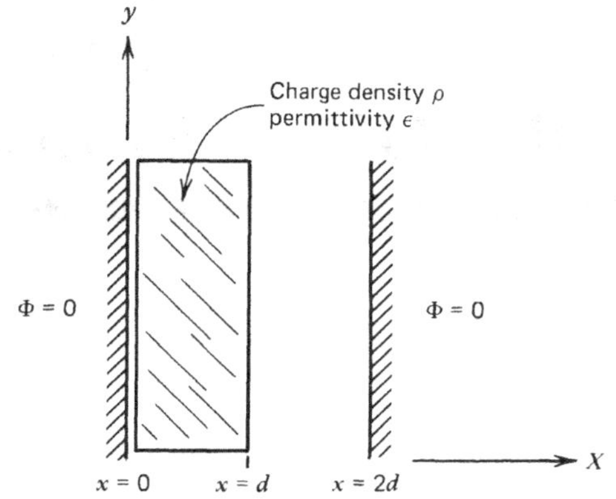

Figure 4.19

FIVE

THE MICROSCOPIC THEORY OF DIELECTRICS

In the previous chapter we took the point of view of dealing with the dielectric properties of matter from a macroscopic limit; namely, we used fields averaged over macroscopic spatial dimensions of the order of 10^{-3} cm which in general contained some 10^{10} molecules or more. In this chapter we examine spatial dimensions of the order of 10^{-7} cm, which in general contain on about 10–100 molecules. The purpose of this examination is to study the fields near individual molecules, or from a microscopic point of view. Because of the averaging involved, we expect the macroscopic field to be different from the field near individual molecules. One of the aims of this chapter is to find how the fields in these two limits are related. This chapter also examines the response of individual atoms and molecules to external fields using a simple model of the atom.

5.1 The Molecular Field

The *molecular or local* electric field at the site of individual molecules in a dielectric material is due to all fields generated by sources external to the dielectric and to all fields produced by all the molecules in the material excluding the *self-field* of the molecule itself. Because it is impossible to treat discretely all fields due to all molecules, we follow a reasonable approach where only the molecules in the vicinity of the molecule in question are treated from the microscopic point of view.

The procedure to be used is described in Fig. 5.1, where a dielectric material is placed in a uniform external electric field $\mathbf{E}_e$. The molecular site in question is labeled *m*, the individual molecules in its vincinity are labeled by dots, and the molecules outside the cavity *c* are shown as a continuum. The fields produced by the bulk of the material can be calculated by using the polarization charges induced in the material. The volume polarization charge density $\rho_p = -\nabla \cdot \mathbf{P} = 0$ for uniform external fields. There are, however, surface charge distributions on the surface of the cavity and on the faces of the dielectric material normal to the external field. There the electric field due to the surface charges on the outer surfaces is $-\mathbf{P}/\varepsilon_0$.

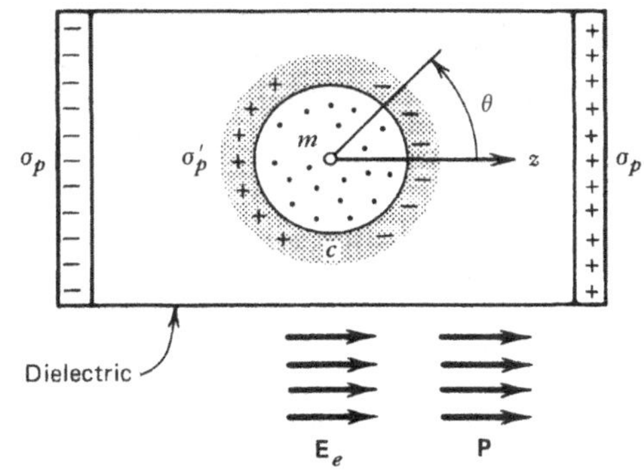

Figure 5.1 A schematic diagram of a model for the calculation of electrical effects at the site of an individual molecule of a dielectric material polarized by an external electric field. In the model the molecule is considered to be in a cavity, along with N other identical molecules, with the rest of the material treated as a continuum.

Thus the molecular field is written as the sum of the following fields

$$\mathbf{E}_m = \mathbf{E} + \mathbf{E}_c + \mathbf{E}_i \tag{5.1}$$

where $\mathbf{E} = \mathbf{E}_e - (\mathbf{P}/\varepsilon_0)$ is the macroscopic field in the materials, $\mathbf{E}_c$ is the electric field due to the charges on the cavity surface, and $\mathbf{E}_i$ is due to all the molecules inside the cavity.

In the presence of an electric field, a neutral molecule gets polarized and thus acts as a dipole. We will just use this fact for now; however, later in this section a simple model will be used to derive this result. Thus according to Eq. (2.46):

$$\mathbf{E}_i = \frac{1}{4\pi\varepsilon_0} \sum_{k=1}^{n} \frac{p}{r_k^2} (2\hat{\mathbf{r}}_k \cos\theta' + \hat{\boldsymbol{\theta}} \sin\theta')$$

where $\mathbf{p}$ is the induced molecular dipole moment, $\mathbf{r}_k$ is the distance of the kth molecule from the site m, θ' is the angle between $\mathbf{p}$ (which is along the external field) and $\mathbf{r}_k$, and n is the number of molecules inside the cavity. In a number of cases the above sum vanishes; these include cubic-crystal lattices and liquids or gases where the positions of the molecules in the cavity are random. In general the sum does not vanish in anisotropic materials; however, it does vanish in many isotropic materials. In this discussion we will only consider materials where it vanishes.

From Example 4.2, one can easily show that the field inside a spherical hole in a dielectric with polarization $\mathbf{P}$ is $(\mathbf{P}/3\varepsilon_0)$. Therefore $\mathbf{E}_m$ becomes

$$\mathbf{E}_m = \mathbf{E} + \frac{\mathbf{P}}{3\varepsilon_0} \tag{5.2}$$

If a molecule of the dielectric is considered to lie in such a spherical hole, its polarization can be determined by $\mathbf{E}_m$. Moreover, its resultant dipole moment $\mathbf{p}$ would be expected to be proportional to $\mathbf{E}_m$ if the material were linear. We write

$$\mathbf{p} = \alpha \mathbf{E}_m \tag{5.3}$$

where α is called the *molecular polarizability*. If we consider that these molecular dipoles do, in fact, constitute the dielectric and there are N such molecules per unit volume, we will have for the polarization

$$\mathbf{P} = N\mathbf{p} = N\alpha\mathbf{E}_m = N\alpha\left(\mathbf{E} + \frac{\mathbf{P}}{3\varepsilon_0}\right) \tag{5.4}$$

If the dielectric were simple, it would also be true that

$$\mathbf{P} = \chi\varepsilon_0\mathbf{E} \tag{4.44}$$

Substituting this in Eq. (5.4) gives

$$\frac{N\alpha}{3\varepsilon_0} = \frac{\chi}{\chi + 3} = \frac{K - 1}{K + 2} \tag{5.5}$$

or

$$\alpha = \frac{3\varepsilon_0}{N}\frac{K - 1}{K + 2}$$

The same relation can be solved for K or χ in terms of α:

$$\chi = K - 1 = \frac{N\alpha/\varepsilon_0}{1 - (N\alpha/3\varepsilon_0)} \tag{5.6}$$

This equation, which is known as the Clausius–Mossotti equation, relates the macroscopic susceptibility χ or the dielectric constant to the microscopic polarizability α and the molecular density N. Its (approximate) validity is appropriate to gases and liquids. For solids, the model on which it was constructed is overly naive.

Example 5.1 The Molecular Polarizability of O_2 and N_2

Equation (5.5) can be used to calculate the molecular polarizability from the macroscopic susceptibility. Consider air, which is composed mainly of N_2 and O_2. From Table 4.1 the dielectric constant K of air is 1.00059. At standard temperature and pressure (STP), the number of molecules per cubic meter in air is equal to $2.7 \times 10^{25}/\text{m}^3$. Therefore

$$\alpha = \frac{3\varepsilon_0}{N}\frac{K - 1}{K + 2} = 1.94 \times 10^{-40}\ \text{C}\cdot\text{m}^2\cdot\text{V}^{-1}$$

Note that in this calculation we have not distinguished between O_2 and N_2, and therefore the above value for α is to be understood as a weighted average of their polarizabilities.

5.2 Interaction of Atoms and Molecules with Electric Fields

Now we turn to the interaction of individual atoms or molecules with electric fields. As we discussed on page 140 we classify molecules into two categories: *polar* and *nonpolar* molecules. In a polar molecule the center of the electronic charge is permanently displaced from the center of the nuclear charge, and thus the molecule, though neutral as a whole, exhibits a permanent electric dipole moment. An example of polar molecules is the water molecule (see Fig. 5.2). On the other hand, when the centers of positive and negative charges are not displaced relative to each other, then the molecule does not exhibit a permanent electric dipole (nonpolar molecules). Examples of nonpolar molecules include O_2, N_2, and H_2. Table 5.1 gives the dipole moment of some polar molecules.

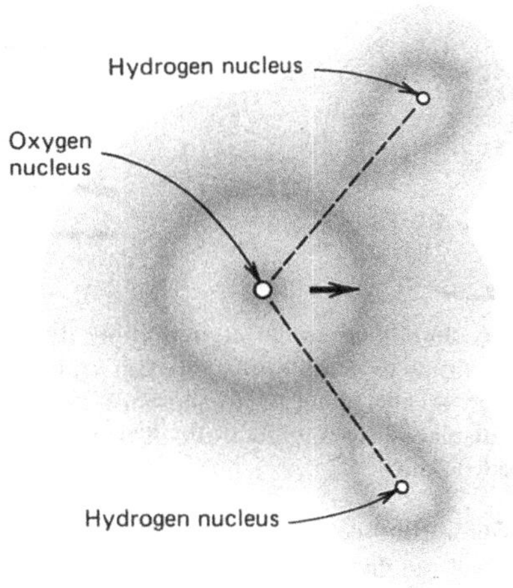

Figure 5.2 Schematic diagram of the atomic and nuclear charge of a water molecule showing the asymmetry in the distribution that gives the molecule its permanent electric dipole moment.

When molecules are placed in an electric field, the field induces a separation of the centers of electronic and nuclear charges along the field in the case of nonpolar molecules and causes additional separation of the centers in the case of polar molecules. This results in what is called *induced polarization* or *induced dipole moment.* The electric field also exerts torques on the permanent dipoles in the case of polar molecules causing a degree of alignment.

5.2.1 Induced Dipoles

We begin by considering nonpolar molecules in an electric field. The derivation of expressions for the induced dipole moment will be approached, however, using a simple atomic model. The applicability of such model to molecules is not universal; however, it may be used in the case of symmetrical diatomic molecules. The total polarizability of the molecule is then determined by simply adding the polarizabilities of its atomic constituents.

In the model, the electronic charge, $-Ze$, is taken to be uniformly distributed over a sphere of radius R and center at the positive nuclear charge Ze, where Z is the atomic number and e is the magnitude of the charge of an electron

Table 5.1 **Permanent Dipole Moments**[a]

Molecule	p	Molecule	p
HCl	3.43	CO	4.0
HBr	2.63	CH_3Cl	6.3
H_2O	6.03	NO	0.33
H_2S	3.06	NO_2	1.33

[a] In units of 10^{-30} coulomb-meters.

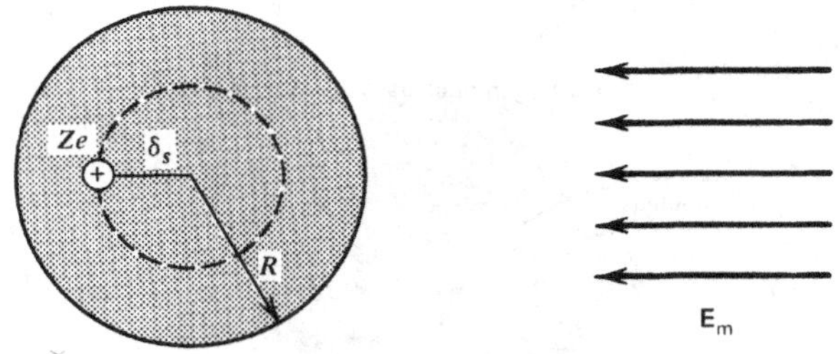

Figure 5.3 A simplified model representing the electronic charge of an atom as a uniformly charged sphere with the nuclear charge at its center. The presence of an external electric field displaces the centers of the distributions resulting in an induced dipole.

(1.6×10^{-19} C). Under the influence of an external field $\mathbf{E}_m$, the positive charge moves *relative* to the center of electron cloud until the attractive Coulomb force between the cloud and the nucleus balances the force ZeE_m. For an amount of separation between the cloud center and the nucleus, $\boldsymbol{\delta}s$ (Fig. 5.3), the balance condition is

$$ZeE_m = \frac{qZe}{4\pi\varepsilon_0(\delta s)^2} \tag{5.7}$$

where q is the charge inside the sphere of radius δs. Taking $q = (4\pi/3)(\delta s)^3\rho = Ze(\delta s)^3/R^3$ in Eq. (5.7), we find that the induced dipole moment $\mathbf{p}_m = Ze\boldsymbol{\delta}\mathbf{s}$ is

$$\mathbf{p}_m = Ze\boldsymbol{\delta}\mathbf{s} = 4\pi\varepsilon_0 R^3\mathbf{E}_m = \alpha\mathbf{E}_m \tag{5.8}$$

where α, which can be called the *deformation polarizability*, is as follows:

$$\alpha = 4\pi\varepsilon_0 R^3 \tag{5.9}$$

For a diatomic molecule this model predicts an induced dipole equal to $2\alpha E_m$.

In Section 5.1, The Molecular Field, the molecular polarization was assumed to be proportional to E_m, and thus this result supports that assumption. Moreover, Eq. (5.6) implies that K, the dielectric constant, is a constant quantity when α is independent of the field. It is interesting to note that even at an extremely strong field of $E_m = 10^{10}$ V/m, one finds from Eq. (5.8) that $\delta s \approx 10^{-12}$ m, which is much smaller than the macroscopic dimension or the atomic size (10^{-10} m).

Example 5.2 Polarizability of Hydrogen

A more realistic model of the electronic charge—e.g., in the hydrogen atom—is a spherically symmetric, exponential charge distribution of the form $\rho = -(e/\pi a_0^3)e^{-2r/a_0}$, where a_0 is a constant (Bohr radius $\approx 10^{-10}$ m) and e is the magnitude of the charge of an electron. This distribution extends over all space compared to the above model, where the charge is uniformly distributed over a sphere.

To calculate the polarization of the hydrogen atom in the $\mathbf{E}_m$ field, we follow the same procedure we used above with regard to the simple model. We calculate the field produced by the charge distribution at a distance δs from the center of the distribution, by applying Gauss' law, to a Gaussian spherical surface with center at the origin and radius δs. The field is expected to be radial. Therefore,

$$4\pi(\delta s)^2 E = \int \rho\, dv = \frac{-e}{\pi a_0^3 \varepsilon_0} 4\pi \int_0^{\delta s} r^2 e^{-2r/a_0}\, dr$$

Thus

$$E = \frac{-e}{\pi a_0^3 \varepsilon_0 (\delta s)^2} G \tag{5.10}$$

where

$$G = \int_0^{\delta s} r^2 e^{-2r/a_0} \, dr \tag{5.11}$$

For $\delta s \ll a_0$, G can be expanded in a Taylor series; keeping the lowest order terms in δs gives $G = \delta s^3/3$ and hence

$$E = \frac{-e}{3\pi\varepsilon_0 a_0^3} \delta s \tag{5.12}$$

The balance condition requires $E = -E_m$; therefore

$$\mathbf{p}_m = e\boldsymbol{\delta} s = 3\pi a_0^3 \varepsilon_0 \mathbf{E}_m = \alpha' \mathbf{E}_m \tag{5.13}$$

If the total charge was distributed over a sphere of radius a_0, then Eq. (5.8) gives $\mathbf{p}_m = 4\pi\varepsilon_0 a_0^3 \mathbf{E}_m$. Therefore the polarizability constant $\alpha' = \frac{3}{4}\alpha$. This example shows that the model is very approximate and yields an order of magnitude only. In fact, accurate quantum mechanical derivations yield $18\varepsilon_0 a_0^3$ for the polarizability of hydrogen.

Example 5.3 Effect of Atomic Interaction on Molecular Polarizability

In applying the atomic model to a diatomic molecule, we noted above that the induced molecular polarization is twice the induced atomic polarization. This example shows that when the actual interaction between the induced atomic dipoles is taken into consideration, the molecular polarizability will not be just twice the atomic polarizability.

Consider a molecule that is made up of two identical spherically symmetric atoms each of polarizability α and at a distance R from each other, as shown in Fig. 5.4. The molecule is placed in an electric field, $\mathbf{E}_m$, parallel to its axis. The polarization of each atom is

$$\mathbf{p} = \alpha(\mathbf{E}_m + \mathbf{E}') \tag{5.14}$$

where $\mathbf{E}'$ is an additional electric field to $\mathbf{E}_m$ produced at each atom by the induced dipole of the other atom. In turn the electric field $\mathbf{E}'$ produced by each atom at the site of the other should be produced by the polarization of the atom given in Eq. (5.14). Using Eq. (2.46) for the field produced by a dipole and taking $\theta = 0$ gives

$$\mathbf{E}' = \frac{2\mathbf{p}}{4\pi\varepsilon_0 R^3} = \frac{2\alpha}{4\pi\varepsilon_0 R^3}(\mathbf{E}_m + \mathbf{E}') \tag{5.15}$$

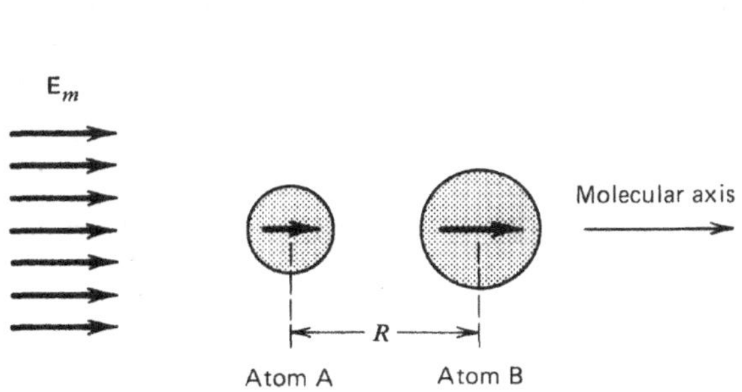

Figure 5.4 Calculation of the polarizability of a diatomic molecule placed in an external field along its axis, taking into account the dipole–dipole interaction between the individual atoms.

which yields

$$\mathbf{E}' = \frac{2\alpha\mathbf{E}_m}{4\pi\varepsilon_0 R^3[1 - (2\alpha/4\pi\varepsilon_0 R^3)]} \tag{5.16}$$

Substituting this result in Eq. (5.14) gives

$$\mathbf{p} = \frac{\alpha\mathbf{E}_m}{1 - (2\alpha/4\pi\varepsilon_0 R^3)} \tag{5.17}$$

The total molecular polarization is twice the above result, and therefore

$$\alpha' = \frac{2\alpha}{1 - (2\alpha/4\pi\varepsilon_0 R^3)} \tag{5.18}$$

This result reduces in the limit, $\alpha/4\pi\varepsilon_0 R^3 \ll 1$, to $\alpha' = 2\alpha$, which is simply the sum of the individual polarizabilities of the atoms neglecting the interaction between them.

When the external electric field is perpendicular to the molecular axis, then the induced atomic dipoles will be normal to the molecular axis, and therefore the additional electric field at each atom produced by the other is different from the above case, and hence the molecular polarizability will also be different. We will leave the determination of such polarizability as an exercise (see problem 5.6).

5.2.2 Permanent Dipoles

When a polar molecule with a permanent electric dipole is placed in an electric field, then the electric field produces two effects: It induces a change in the molecular polarization, and it exerts a torque on the permanent dipole of the molecule that results in a certain degree of alignment along the direction of the field (Fig. 5.5). Since we treated the first effect earlier, let us now turn to the second effect.

The thermal energy of the molecules in a macroscopic piece of a polar dielectric tends to randomize the molecular dipole orientations. In fact, in the absence of an external field the vector sum of the dipole moments of all the molecules vanishes.

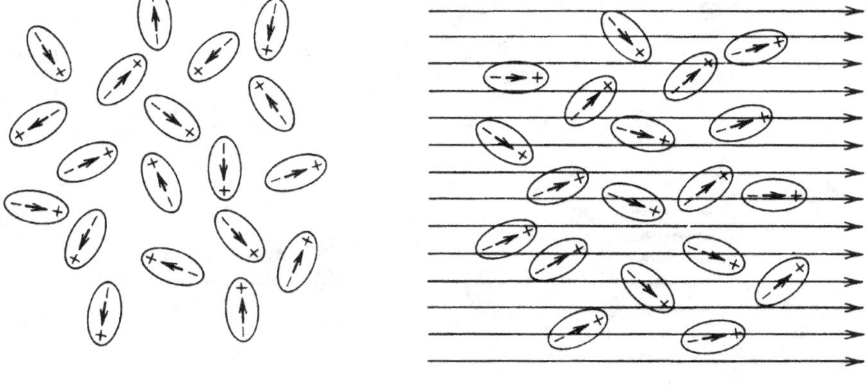

(a) (b)

Figure 5.5 Schematic diagram of a paramagnetic material. (*a*) A random distribution of the permanent dipoles in the absence of external fields. (*b*) Some degree of alignment is produced in the presence of an external field.

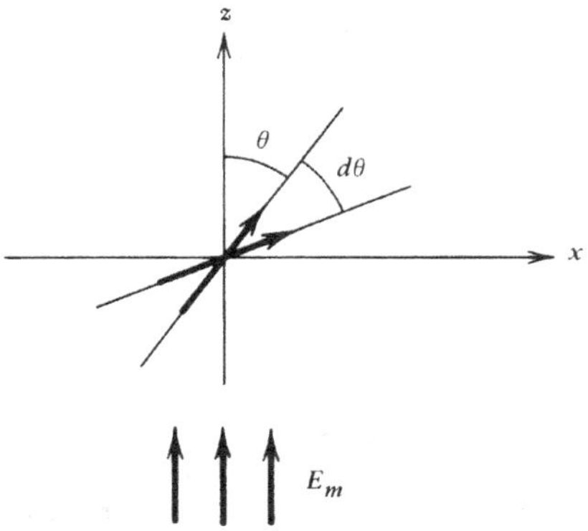

Figure 5.6 Calculation of the paramagnetic properties of materials in external electric fields.

The degree of alignment of the molecular dipoles can be derived quantitatively using statistical methods. Consider an assembly of N_0 polar molecules per unit volume at a temperature T. Classically, each dipole moment can make an arbitrary angle θ with respect to a given direction such as the z axis (see Fig. 5.6). In the absence of an external electric field, the probability that the dipole will be between angles θ and $\theta + d\theta$ is proportional to $2\pi \sin\theta\, d\theta$, which is the solid angle $d\Omega$ subtended by this range of angle. This probability leads to a zero average of the dipoles. When an electric field $\mathbf{E}_m$ is present and is taken along the z axis, this probability becomes also proportional to the Boltzmann distribution $e^{-U/kT}$, where $U \equiv -\mathbf{p}\cdot\mathbf{E}_m = -pE_m \cos\theta$ is the electric energy of the dipole when it is making an angle θ with the electric field [see Eq. (2.64)], k is Boltzmann's constant, and T is the absolute temperature. The Boltzmann factor introduces the dependence of the probability on the electric field and on the temperature in a quantitative, well-defined way.

Before calculating the degree of alignment we would like to consider some limiting cases of the Boltzmann factor. When $U/kT \ll 1$, which arises in cases of high temperature or/and weak electric fields, then $e^{-U/kT} \approx 1$, and therefore the probability becomes proportional to the solid angle only. A first-order correction to this high-temperature limit can be arrived at by writing $e^{-U/kT} \approx 1 - U/kT$. On the other hand, when $|U/kT| \geq 1$, then this factor becomes more important, and directions where U/kT is negative and large in magnitude are weighted more heavily, hence resulting in a large degree of alignment.

Taking into account both probabilities, the solid angle and the Boltzmann factor give for the average dipole moment

$$\langle \mathbf{p} \rangle = \frac{\int \mathbf{p} e^{-U/kT}\, d\Omega}{\int e^{-U/kT}\, d\Omega} \tag{5.19}$$

where the denominator of Eq. (5.19) is introduced to normalize the average. Taking $d\Omega = 2\pi \sin\theta\, d\theta$ and writing $\mathbf{p} = p\cos\theta\,\hat{\mathbf{z}} + p\sin\theta\,\hat{\mathbf{x}}$, where $\hat{\mathbf{z}}$ and $\hat{\mathbf{x}}$ are unit vectors along the z axis and the x axis, respectively, we obtain

$$\langle \mathbf{p} \rangle = \frac{\int_{+1}^{-1} (p\cos\theta\,\hat{\mathbf{z}} + p\sin\theta\,\hat{\mathbf{x}}) e^{(pE_m \cos\theta)/kT}\, d\cos\theta}{\int_{+1}^{-1} e^{(pE_m\cos\theta)/kT}\, d\cos\theta} \tag{5.20}$$

The component along the x axis averages to zero, whereas that in the z direction gives

$$\langle \mathbf{p} \rangle = p\left(\coth \eta - \frac{1}{\eta}\right)\hat{\mathbf{z}} \tag{5.21}$$

where

$$\eta = \frac{pE_m}{kT} \tag{5.22}$$

Equation (5.21) is known as the Langevin formula. At large fields, namely $\eta = (pE_m/kT) \gg 1$, the Langevin formula predicts that $\langle \mathbf{p} \rangle = p\hat{\mathbf{z}}$, which means that complete alignment (saturation) takes place. However, in most cases of dielectric materials the magnitude of p is such that $\eta \ll 1$ at ordinary temperatures even if E_m is taken as high as the dielectric strength of the material. In this limit the Langevin formula gives

$$\langle \mathbf{p} \rangle = \frac{1}{3}\frac{p^2 E_m}{kT}\hat{\mathbf{z}} \tag{5.23}$$

which indicates a linear relationship between the average dipole moment and the field. When η is not small, then $\langle \mathbf{p} \rangle$ becomes a nonlinear function of the field. In Fig. 5.7 the relationship between $\langle \mathbf{p} \rangle$ and η given by Eq. (5.21) is plotted; it shows the three regions discussed above.

In an ensemble of N molecules per unit volume, the total polarization is $\mathbf{P} = N\langle \mathbf{p} \rangle$, and therefore the effective dipole moment per molecule is $\langle \mathbf{p} \rangle$. Using the relation $\langle \mathbf{p} \rangle = \alpha \mathbf{E}_m$ gives what is called the *orientational polarizability*; that is,

$$\alpha = \frac{p^2}{3kT} \tag{5.24}$$

The dielectric constant of the medium can be determined using (Eq. 5.6) as follows:

$$K - 1 = \frac{Np^2}{3\varepsilon_0 kT}\left(1 - \frac{Np^2}{9\varepsilon_0 kT}\right)^{-1} \tag{5.25}$$

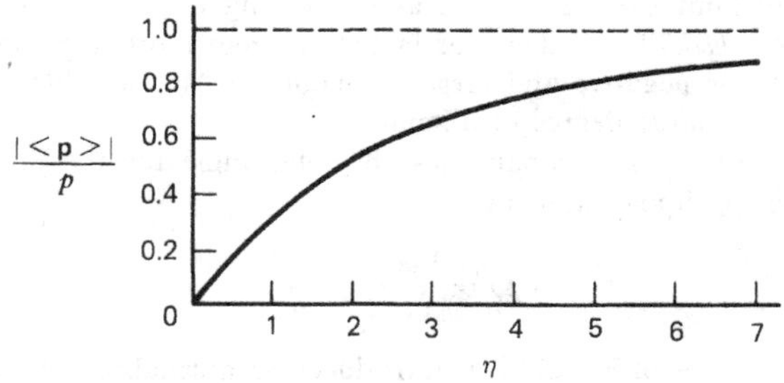

Figure 5.7 The behavior of the average dipole moment of a paramagnetic material per molecule placed in an external electric field E_m as a function of $\eta = pE_m/kT$. It is given by the Langevin function, which tends to 1 for large η (saturates).

When the induced polarizability discussed above α' is considered in addition to the alignment effect, the total polarizability becomes

$$\alpha = \alpha' + \frac{p^2}{3kT} \tag{5.26}$$

which is known as Langevin–Debye equation; it indicates that at higher temperatures α varies as $1/T$.

Example 5.4 Discrete Dipole Orientations*

This example deals with a situation of actual physical interest where the direction of a permanent electric dipole in the presence of an external electric field is restricted to a small number of directions. Consider a solid of lattice separation a kept at absolute temperature T. Some of the atoms in the solid are replaced by some negatively charged impurity ions as shown in Fig. 5.8. Since each negatively charged ion has in its vicinity a positively charged ion, then the solid as a whole is neutral.

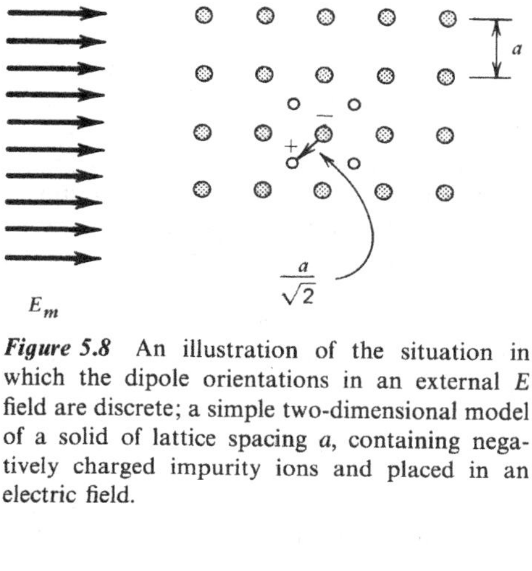

Figure 5.8 An illustration of the situation in which the dipole orientations in an external E field are discrete; a simple two-dimensional model of a solid of lattice spacing a, containing negatively charged impurity ions and placed in an electric field.

The positive ion is smaller than the atom and the negative ion and therefore it moves between lattice sites. There are four possible positions for the positive ion; these are at equidistances from the negative ion as shown in the figure. At each position the ions form a dipole whose moment is $ea/\sqrt{2}$ where e is the magnitude of the charge of the electron.

Since these four positions are equally probable, the average moment vanishes. In the presence of an electric field $\mathbf{E}_m$ in the z direction, however, the average dipole moment will not vanish because of the effect of alignment discussed above. To calculate the electric polarization we use the Boltzmann factor $e^{-U_i/kT}$, where $U_i = (-ea/\sqrt{2})E_m \cos\theta_i$, θ_i is the angle between the dipole and the z axis, and i runs from 1 to 4, the possible positions of the positive ion. Therefore,

$$\langle \mathbf{p} \rangle = \frac{\sum_i \mathbf{p}_i e^{-U_i/kT}}{\sum_i e^{-U_i/kT}} \tag{5.27}$$

* See F. Reif, *Fundamentals of Statistical and Thermal Physics* (New York: McGraw-Hill, 1965).

Substituting for U_i in terms of E_m and θ_i and noting that only the z component of $\mathbf{p}_i$ contributes, we get:

$$\langle \mathbf{p} \rangle = \frac{\sum_i \frac{ea}{\sqrt{2}} \hat{\mathbf{z}} \cos\theta_i e^{g\cos\theta_i}}{\sum_i e^{g\cos\theta_i}} \tag{5.28}$$

where

$$g = \frac{ea}{\sqrt{2}} \frac{E_m}{kT}$$

and $\hat{\mathbf{z}}$ is a unit vector in the z direction; or

$$\langle \mathbf{p} \rangle = \frac{ea}{2} \hat{\mathbf{z}} \frac{e^{g'} - e^{-g'}}{e^{g'} + e^{-g'}} \tag{5.29}$$

where $g' = \sqrt{2}g$. Thus the average dipole moment is

$$\langle \mathbf{p} \rangle = \frac{ea}{2} \hat{\mathbf{z}} \tanh\left(\frac{eaE_m}{kT}\right) \tag{5.30}$$

5.2.3 Ferroelectricity

We now consider some substances that have permanent electric moments even in the absence of external electric fields. These substances are called *ferroelectric* in analogy with the well known ferromagnetic effect exhibited by a number of substances such as iron. Barium titanate, $BaTiO_3$, is an example of a ferroelectric material; it exhibits permanent (spontaneous polarization) in the absence of external electric fields below a somewhat elevated temperature $T_c = 118°C$. Above this temperature it is an ordinary dielectric material, however, with a very large dielectric constant.

For this effect to take place, the molecular constituents must have a certain polarizability. To determine such condition on the polarizability we consider Eq. (5.2). Taking $\mathbf{E} = 0$ gives

$$\mathbf{E}_m = \frac{\mathbf{P}}{3\varepsilon_0} \tag{5.31}$$

This equation indicates that the permanent polarization P produces a molecular field E_m. For this to be self-consistent, the polarization $\mathbf{P}$ should be produced by the same molecular field. Therefore

$$\mathbf{P} = N\alpha \mathbf{E}_m \hat{\mathbf{z}} \tag{5.32}$$

Equations (5.31) and (5.32) can now be solved for α. Eliminating $\mathbf{E}_m$ gives

$$\mathbf{P} = \frac{N\alpha}{3\varepsilon_0} \mathbf{P} \tag{5.33}$$

which has two solutions: $\mathbf{P} = 0$ (a trivial solution) and

$$\frac{N\alpha}{3\varepsilon_0} = 1 \tag{5.34}$$

which is the condition for permanent (spontaneous) polarization. In most materials $N\alpha/3\varepsilon_0$ is less than one; these materials are ordinary dielectrics. This condition, however, is met in the case of some crystals such as $BaTiO_3$, mentioned above.

We now explain the temperature dependence of ferroelectric materials—namely, the temperature threshold. Equation (5.34) gives the condition at the critical temperature or the Curie temperature T_c. At temperatures higher than T_c and as T

approaches T_c, a small correction to this condition is introduced:

$$\frac{N\alpha}{3\varepsilon_0} = 1 - (T - T_c)\delta \tag{5.35}$$

where $\delta \ll 1$ (of the order of 10^{-6} per degree Celsius). Substituting Eq. (5.35) into Eq. (5.6) gives

$$K - 1 = \frac{3 - 3\delta(T - T_c)}{\delta(T - T_c)} \approx \frac{3}{\delta(T - T_c)} \tag{5.36}$$

This is the Curie–Weiss law. It indicates that K is huge for temperatures just above the critical temperature. This result agrees with the experimental observation of ferroelectric substances. Having large dielectric constants such as those of ferroelectric materials near T_c especially when T_c is near room temperature would be very attractive in the construction of capacitors and in other applications. For this reason, there is continuing research to change the behavior of these materials in a controlled fashion (change T_c). For example adding a small amount of $BaLiF_3$ to $BaTiO_3$ lowers T_c to near room temperature.*

Comparing Eqs. (5.25) and (5.36) shows that the temperature dependence of an ensemble of a dipole gas differs from that of a ferroelectric material. Whereas in the case of a dipole gas $K - 1$ varies as $1/T$, it varies as $1/(T - T_c)$ in the case of ferroelectric materials.

When one discusses the stability of ferroelectric materials, one has to worry about the effect of the depolarization field. A depolarization field is produced by the effective surface polarization charge resulting from discontinuities in the polarization at the surfaces of a polarized dielectric material (see Examples 4.8 and 4.2). The effect of such a field is to depolarize the material, since it opposes the polarization creating it.

The depolarization fields in ferroelectric materials can be experimentally eliminated and hence render these materials very useful for some practical applications. To show this, consider Fig. 5.9, which shows a slab of ferroelectric material snugly fitted in the space between two conducting plates. A large potential difference

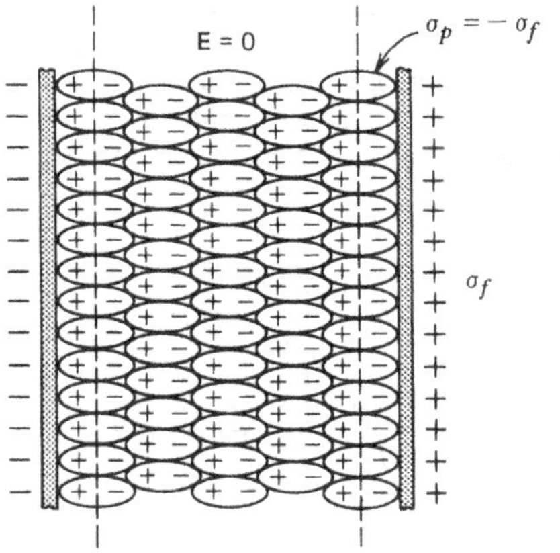

Figure 5.9 A piece of ferroelectric material placed snugly between the plates of a parallel-plate capacitor.

* See for example, I. N. Belyaev et al., Ivz. Akad. Nauk., Ser. Fiz., **33**, (1969) 1144 and A. Benziada-Taibi et al., Journal of the Fluorine Chemistry, 26, (1984) p. 395–404.

is applied to the plate. If σ_f is the free surface charge density on the right plate, then the polarization charge density on the same plate σ_p is of opposite sign. Moreover, if the material is ferroelectric, then $N\alpha/3\varepsilon_0 = 1$, or $K - 1$ becomes very large such that $K - 1 \approx K$. In this case σ_p approaches $-\sigma_f$ [see Eq. (4.57)]. Hence the total charge at the surface of the material or the surface of the plate is $\sigma_p + \sigma_f \approx 0$, indicating complete neutralization, and hence complete elimination of the depolarization field.

If the plates are now brought to the same potential by short-circuiting them with a wire of very small resistance, the following is likely to happen: (1) The state of ferroelectricity remains (energetically favorable), and (2) the free charge density σ_f stays in place, and neutralization continues. The external electric field inside the ferroelectric material in this state vanishes since there is no potential difference between the plates, and there is no depolarization field since there are no surface polarization charges.

If now the same potential difference is established between the plates, there will be no charge flow through the external source and no change in the status of the material. If, on the other hand, a potential difference of the same magnitude but of opposite sign is established between the plates, then the polarization of the ferroelectric material will reverse direction, and all free and polarization charges will change sign. The reversal of the charge densities is induced by a flow of free charge through the external sources. The existence of these two distinct storage states allows a ferroelectric material placed between two-parallel plates to serve as the basic element of a memory device.

If one goes through cycles of reversing of the potential difference between the plates (reversing the external electric field $E_0 \rightarrow -E_0 \rightarrow E_0$), the polarization state of the ferroelectric material attains the same value at the end of each reversal that is $P_0 \rightarrow -P_0 \rightarrow P_0$. However, at intermediate values of E the value of the polarization depends on which half of the cycle the process is in. Figure 5.10 shows this effect, which is called the *hysteresis* effect. Hysteresis effects (meaning that the polarization effect lags the external electric field) also occurs in ferromagnetic materials where the magnetization lags the external magnetic field. See Chapter 9 and 10 for the details of ferromagnetism.

Finally, permanent polarization can be produced to some extent in some kinds of wax. If the wax is melted in the presence of a strong electric field, the permanent dipoles of the wax become oriented along the field. These dipoles stay aligned when the wax is next frozen while keeping the field on. These permanently polarized substances are the analog of permanent magnets, and are simply called *electrets*.

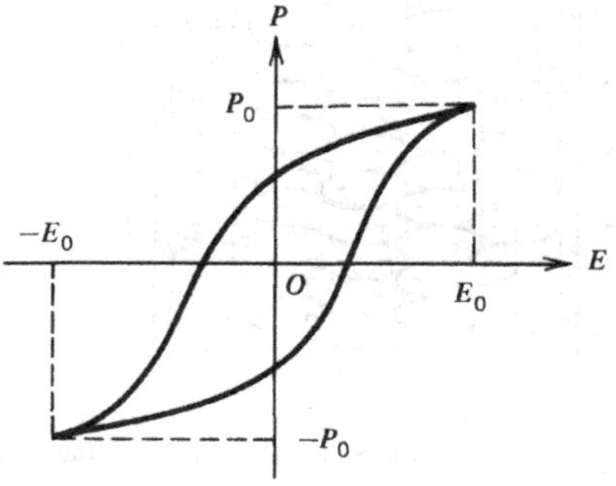

Figure 5.10 Hysteresis loop for a ferroelectric material under the influence of an external **E** field.

Because of the permanent nature of the orientation, the introduction of a small cavity in such a substance is expected not to change the state of polarization of the medium (see Problem 4.16).

5.3 Summary

Molecules are classified into two categories. Polar molecules have permanent dipole moments resulting from the permanent separation of the centers of the electronic charge from that of the nuclear charge. Nonpolar molecules have no permanent dipole moments since these centers are not displaced. When an atom is placed in an electric field $\mathbf{E}_m$, the electronic charge becomes deformed; that is, its center gets displaced relative to the center of the nuclear charge, thus making the atom exhibit an induced dipole moment $\mathbf{p}$, given by

$$\mathbf{p} = \alpha \mathbf{E}_m \tag{5.3}$$

where $\alpha = 4\pi\varepsilon_0 R^3$ is the induced atomic polarizability and $4\pi R^3/3$ is the volume of the electronic charge. The total molecular polarizability can be simply calculated by adding the polarizabilities of the individual atoms in the molecule.

If an isotropic dielectric material of molecular polarizability α is placed in an external electric field $\mathbf{E}$, then each molecule exhibits an induced dipole. The density of these dipoles is just the macroscopic polarization of the medium. If this polarization is $\mathbf{P}$, then the electric field at the site of each molecule is the sum of $\mathbf{E}$ and $\mathbf{P}/3\varepsilon_0$; that is,

$$\mathbf{E}_m = \mathbf{E} + \frac{\mathbf{P}}{3\varepsilon_0} \tag{5.2}$$

The field $\mathbf{E}_m$ is actually the field that induces the dipole moment in the molecule. That is,

$$\mathbf{p}_m = \alpha \mathbf{E}_m \qquad \text{and} \qquad \mathbf{P} = \alpha N \mathbf{E}_m$$

where N is the number density of molecules in the material. A consistent solution of these relations give a constant susceptibility

$$\mathbf{P} = \varepsilon_0 \chi \mathbf{E} \tag{4.44}$$

where

$$\chi = \frac{N\alpha/\varepsilon_0}{1 - \dfrac{N\alpha}{3\varepsilon_0}} \tag{5.6}$$

Another consistent solution arises in the case where $N\alpha/3\varepsilon_0 \geq 1$, even if the external part of the field $\mathbf{E}$ is zero. That is, $\mathbf{E}_m = \mathbf{P}/3\varepsilon_0$ and $\mathbf{P} = \alpha N \mathbf{E}_m$ imply that $N\alpha/3\varepsilon_0 = 1$. In this case the molecules are said to be very polarizable, and the material can exhibit spontaneous polarization in a zero external field, as in ferroelectric materials. Barium titanate is one material that exhibits ferroelectricity.

When polar molecules of permanent dipole moment p are placed in an external field $\mathbf{E}_m$, they exhibit additional polarizability resulting from the tendency of the electric field to align permanent dipoles along its direction. This tendency is opposed by collisions with other molecules, which tend to randomize this direction.

This orientational polarizability, at a given absolute temperature T, is given by the Langevin function

$$\langle p_z \rangle = p\left[\coth \eta - \frac{1}{\eta}\right] \tag{5.21}$$

where $\eta = pE_m/kT$, and $\langle p_z \rangle$ is the average component of $\mathbf{p}$ along the field. At high temperatures or low p we get the Curie law:

$$\langle p_z \rangle \simeq \frac{p^2 E_m}{3kT} \qquad \text{and} \qquad \alpha = \frac{p^2}{3kT} \tag{5.23),(5.24}$$

Problems

5.1 A material (acetamide) of density 1.0 g/cm^3 and molecular weight 59, has a dielectric constant $K = 4$. Determine the polarizability of its individual molecules.

5.2 A sample of polar dielectric of volume $V = 10$ cm^3 kept at temperature $T = 300$ K, has $N = 3.3 \times 10^{22}$/cm^3 molecules, each of which has a dipole moment $p = 1.5 \times 10^{-29}$ C·m. The material is placed in a uniform external electric field $E_0 = 10$ V/cm. (a) Determine the average of the angle between E_0 and p. (b) Determine the induced dipole moment and polarization of the block. (c) Determine the temperature at which the dipole moment becomes twice that in (a). (d) What is the dipole moment of the block if saturation is achieved?

5.3 Consider a nonpolar molecule of polarizability α at a distance R from a molecule having a permanent dipole moment p. (a) Determine the induced dipole moment of the nonpolar molecule for a given relative orientation. (b) What is the interaction energy for this orientation? (c) Show that the interaction energy averaged over all possible orientations is $U(R) = -\alpha p^2/8\pi^2\varepsilon_0^2 R^6$.

5.4 An ensemble of gas at temperature T contains two types of molecules of moments p_1 and p_2. Consider a pair one of each type at a distance R from each other. As a result of collisions with other molecules, their orientation will change such that statistical equilibrium applies. (a) Give an expression for the interaction energy U between the two dipoles for a given relative orientation. (b) Give an expression for the probability that they will assume a given mutual orientation. (c) Taking $U/kT \ll 1$, show that the average of U is

$$U(R) = \frac{p_1^2 p_2^2}{24\pi^2\varepsilon_0^2 kTR^6}$$

5.5 A molecule is made up of two spherically symmetric atoms. An external electric field E_0 was placed on it along the line joining the atoms. If the dipole moment of the molecule is measured to be p_0 when the distance between the atoms is R_0Å,* find (a) the polarizability of each isolated atom and (b) the molecular polarizability as a function of R.

5.6 Determine the molecular polarizability of a diatomic molecule of identical spherically symmetric atoms each of polarizability α and at a distance R from each other if the external field is normal to the molecular axis (see Example 5.3).

5.7 An atom of radius R_0, and a permanent dipole moment p_0 is placed in an external electric field E_0. Plot the polarizability as a function of $1/T$. What is the average dipole moment at very high temperatures?

5.8 The inverse dielectric constant of a ferroelectric material near the Curie temperature was measured to be 0.0035 at $T = -140$°C and 0.0105 at $T = -120$°C. Determine the Curie temperature of the material. Plot the dielectric constant as a function of $1/(T - T_c)$.

5.9 The dielectric constant of carbon disulfide gas (CS_2) at 0°C is 1.0029. Can this gas be a ferroelectric material? The density of liquid CS_2 at 20°C is 380 times higher than the density of gas at 0°C. Assuming that the basic atomic polarizability of CS_2 does not change when it condenses, calculate the dielectric constant of the liquid. Can it be a ferroelectric material?

5.10 Show that $\sigma_p \approx -\sigma_f$ if the dielectric slab in Example 4.4 is ferroelectric.

* 1 angstrom (Å) = 10^{-10} m.

SIX
ELECTROSTATIC ENERGY

In this chapter we take on the problem of calculating the *electrostatic energy* of various charge distributions. These include an assembly of point charges, continuous charge distributions, and charged conductors. We introduce two very useful concepts: the coefficients of potential and the coefficients of capacitance or induction. Also, it will be shown that forces between the charge elements of the distribution can be conveniently determined from the knowledge of the electrostatic energy.

6.1 Electrostatic Energy of an Assembly of Point Charges

We now wish to calculate the energy associated with an electrostatic charge distribution. By this we mean the work necessary to assemble this charge distribution from a condition where all charges are not interacting; i.e., they are all infinitely remote from one another. We assume that in the assembly process no kinetic energy is imparted to the charges; the charges are assembled at rest from positions at rest. Thus the work done is interpreted as creating an increase in the electrostatic potential energy of the system. In assembling the system we give the system an electrostatic energy U.

Consider first the case where we have a number of point charges. We first place charge q_1 in position given by $\mathbf{r}_1$. It requires no work to do this because q_1 is the only charge present in the space we are considering. Next, we place charge q_2 at location $\mathbf{r}_2$ a distance $|\mathbf{r}_1 - \mathbf{r}_2| \equiv r_{12}$ from q_1. This requires an amount of work $[\Phi_1(2)q_2] = q_1 q_2/4\pi\varepsilon_0 r_{12}$, where $\Phi_1(2) = q_1/4\pi\varepsilon_0 r_{12}$ is the potential at the location of charge q_2 due to charge q_1. Next, we place charge q_3 at location $\mathbf{r}_3$. We now do work against the fields of both q_1 and q_2, and so we must perform the work

$$[\Phi_1(3) + \Phi_2(3)]q_3 = \frac{q_3}{4\pi\varepsilon_0}\left(\frac{q_1}{r_{13}} + \frac{q_2}{r_{23}}\right)$$

Continuing this process, we find that when we bring in charge j, there are already $j-1$ charges in place, so the work done at this stage will be

$$\frac{1}{4\pi\varepsilon_0}\left[\frac{q_1}{r_{1j}}+\frac{q_2}{r_{2j}}+\cdots+\frac{q_{j-1}}{r_{(j-1)j}}\right]q_j$$

In summation notation we may therefore represent the indicated series for U as

$$U=\sum_{j=1}^{N}\left(\frac{1}{4\pi\varepsilon_0}\sum_{i=1}^{j-1}\frac{q_i}{r_{ij}}\right)q_j=\sum_{j=1}^{N}\sum_{i<j}^{N}\frac{q_iq_j}{4\pi\varepsilon_0 r_{ij}} \tag{6.1}$$

where N is the total number of point charges. There will be $N(N-1)/2$ terms in the expansion. The sum above is over all possible pairs of N charges. Noting that $r_{ij}=r_{ji}$, we may also write this expansion in terms of $N(N-1)$ terms as

$$U=\frac{1}{2}\sum_{j=1}^{N}\sum_{i\neq j=1}^{N}\frac{q_iq_j}{4\pi\varepsilon_0 r_{ij}} \tag{6.2}$$

where we have simply split each term of the original double sum into two terms and rearranged the terms. Note that the terms with $i=j$ are excluded since they represent self-terms.

The electrostatic energy U can alternately be written in terms of the electrostatic potential. The potential at the location of charge j, $\Phi(j)$, due to all the other charges, is given by

$$\Phi(j)=\frac{1}{4\pi\varepsilon_0}\sum_{i\neq j=1}^{N}\frac{q_i}{r_{ij}}=\sum_{i\neq j=1}^{N}\Phi_i(j)$$

so that

$$U=\frac{1}{2}\sum_{j=1}^{N}q_j\Phi(j) \tag{6.3}$$

It is important to understand three things here.

1. $\Phi(j)$ is the potential at the location of q_j due to the other charges.
2. The factor $\frac{1}{2}$ comes into play because we are using a $\Phi(j)$ that is a potential due to all the other charges in the system, not just the charges that happen to be present when an additional charge is brought up from infinity.
3. When dealing with point charges, we always assume that $r_{ij}\neq 0$. (We cannot do an infinite amount of work.)

6.2 Electrostatic Energy of a Continuous Charge Distribution

The preceding expression for U can be generalized to a continuous charge distribution. We merely employ the identifications:

$$q_j\rightarrow dq(\mathbf{r})\ \text{(located at } \mathbf{r}\text{)},\quad \Phi(j)\rightarrow\Phi(r),\quad \text{and}\ \sum\rightarrow\int$$

Then, the general symbolic expression for U given in Eq. (6.3) becomes

$$U=\frac{1}{2}\int\Phi(\mathbf{r})dq(\mathbf{r}) \tag{6.4}$$

The integral sign merely represents the sum to be performed over all charge elements. For example, if $dq = \rho(r)dv$, we write

$$U = \frac{1}{2}\int_V \rho(\mathbf{r})\Phi(\mathbf{r})dv \tag{6.5}$$

where V is the volume over which $\rho \neq 0$, or in fact any volume containing *all* the charge that has been assembled. If one has a continuously distributed charge, described by volume, surface, and line charge densities (ρ, σ, and λ), and point charges, the total work necessary to assemble this charge distribution is given by

$$U = \frac{1}{2}\int_V \rho(\mathbf{r})\Phi(\mathbf{r})dv + \frac{1}{2}\int_S \sigma(\mathbf{r})\Phi(\mathbf{r})da + \frac{1}{2}\int_C \lambda(\mathbf{r})\Phi(\mathbf{r})dl + \frac{1}{2}\sum_j q_j\Phi(\mathbf{r}_j) \tag{6.6}$$

We do not answer the question regarding the energy required to assemble an individual point charge. We assume either that such charges are inviolate, and so never need to be assembled, or that they are really not points.

Example 6.1 Uniformly Charged Sphere

Consider a uniformly charged sphere of charge Q. Referring to Example 2.12*b* we see that the potential inside a uniformly charged sphere of radius R is given by

$$\Phi(r < R) = \frac{\rho_0}{3\varepsilon_0}\cdot\frac{1}{2}(3R^2 - r^2) \tag{2.52}$$

where ρ_0 is the uniform charge density. Therefore integrating Eq. (6.5) over the volume of the sphere gives

$$U = \frac{1}{2}\int \Phi\rho\, dv = \rho_0\frac{4\pi}{2}\int_{r=0}^{R} r^2\, dr\left[\frac{\rho_0}{6\varepsilon_0}(3R^2 - r^2)\right] = \frac{3}{5}\frac{Q^2}{4\pi\varepsilon_0 R} \tag{6.7}$$

where $Q \equiv (4/3)\pi R^3\rho_0$ is the total charge in the sphere.

We may also calculate this energy directly [analogous to the process yielding Eq. (6.1)] by assuming that the sphere is built up by placing uniformly charged infinitesimal layers on its outer spherical surface. When the sphere has a radius r, its charge is $q = (4/3)\pi r^3\rho_0$, and the potential at its surface will be $q/4\pi\varepsilon_0 r$ [see Eq. (2.51) in Example 2.12*b*]. To add a spherical shell of charge $dq = 4\pi r^2\, dr\, \rho_0$ to its surface will take an amount of work equal to the potential at shell $\times$ charge in shell, or

$$dU = \frac{q}{4\pi\varepsilon_0 r} \times (4\pi r^2\rho_0\, dr) = \frac{4\pi\rho_0^2 r^4}{3\varepsilon_0}dr$$

To build up the sphere to radius R, will therefore require the work

$$U = \int_{r=0}^{R} dU = \int_{r=0}^{R} \frac{\rho_0^2}{4\pi\varepsilon_0}\frac{(4\pi)^2}{3}r^4\, dr = \frac{3}{5}\frac{Q^2}{4\pi\varepsilon_0 R}$$

just as calculated above in Eq. (6.7).

Example 6.2 Classical Radius of an Electron

As an application of the result of Example 6.1, we can find the "*classical*" *radius of an electron.* We ask the question: How much energy does it take to make an electron? If the electron is *not* a point charge, but a uniformly charged glob of electricity, it takes a certain amount of energy to assemble this glob of charge into the small volume of the electron. Let us assume that this energy is given by the proper mass of the electron, which, in units of energy ($E = mc^2$), is approximately 0.5×10^6 eV (electron volts) or 0.8×10^{-13} joules. Then if the electron is a uniformly charged sphere of radius R, $U = mc^2 = (3/5)e^2/4\pi\varepsilon_0 R$, and we can solve for R; that is, $R = (3/5)\, e^2/4\pi\varepsilon_0 mc^2 = 0.6 \times 10^{-15}\ \text{m} \equiv 0.6\ \text{fm}$, where fm stands for a femtometer (or fermi)—that is, 10^{-15} m. This size, 0.6 fm, is of the order of nuclear dimensions, which seems reasonable. The quantity R is called the classical radius of the electron. Unfortunately, we now know that the actual electron radius is much smaller, so that our assumptions are invalid.

6.3 Electrostatic Energy of Conductors; Coefficients of Potential and Capacitance

Now that we have a general expression for U, the work necessary to assemble a charge distribution, we shall consider the special case where all the charge resides on conductors. If $\{Q_j, \Phi_j\}$ represent the {total charge, potential} of the jth conductor of a system of N conductors, we have for the energy of the system, from Eq. (6.4) or Eq. (6.6),

$$U = \frac{1}{2}\int_S \sigma\Phi\, da = \frac{1}{2}\sum_{j=1}^{N} \Phi_j \oint_{S_j} \sigma\, da = \frac{1}{2}\sum_{j=1}^{N} \Phi_j Q_j \tag{6.8}$$

where σ is the surface charge density, and we have split the integration of areas into the individual areas of the different conductor surfaces.

We would like now to introduce two very useful concepts: the coefficients of potential and the coefficients of capacitance or induction. These coefficients relate the potentials of a system of conductors to their various charges. For a system of N conductors, one writes

$$\Phi_j = \sum_{k=1}^{N} P_{jk} Q_k \tag{6.9}$$

and

$$Q_j = \sum_{k=1}^{N} C_{jk} \Phi_k \tag{6.10}$$

where P_{jk} are the *coefficients of potential*, C_{ii} are the *coefficients of capacitance*, C_{jk} ($j \neq k$) are the *coefficients of induction*, and Φ_i and Q_i are the potential and the charge of the ith conductor. We note that these introduced coefficients are independent of the potentials and the charges of the conductors. In other words, the potentials are linearly related to the charges, and the geometrical coefficients P_{ij} and C_{ij} express this relationship. The truth of these statements can be established by using the principle of superposition (the linearity of the equations of electrostatics) and invoking the principle of uniqueness.

Consider a single, isolated conductor having a total charge Q on its surface. The potential of the conductor is given by Φ relative to some point or equipotential region in space, usually taken at infinity. If we increase the charge density everywhere on the surface of the conductor by a factor α, then the linearity of the electrostatic equations tells us that the potential and electrostatic field also increase by the same factor. But in increasing the charge density by a factor α, we have increased the total charge Q by a factor α. Because a given electrostatic field uniquely prescribes the charge on the conductor(s), and conversely, a potential ($\alpha\Phi$) necessitates a total charge on the conductor (αQ). This proves that we can write the potential anywhere in terms of the charge Q on the conductor as

$$\Phi(\mathbf{r}) = P(\mathbf{r})Q \tag{6.11}$$

where $P(\mathbf{r})$ can depend only on the configuration of the conductor relative to the point $\mathbf{r} \equiv (x, y, z)$; that is, it is a geometrical constant independent of the values of Q or Φ. In particular, the potential Φ at the conductor is given by $\Phi = PQ$ where P is evaluated at the conductor.

We will now prove the linearity of Eq. (6.9) in the case of two conductors. On the surface of the conductors, where the potentials are labeled Φ_1 and Φ_2, we have, from Eq. (6.9),

$$\Phi_1 = P_{11}Q_1 + P_{12}Q_2 \tag{6.12}$$

$$\Phi_2 = P_{21}Q_1 + P_{22}Q_2 \tag{6.13}$$

A proof proceeds in three steps:

1. Consider the case where conductor 1 alone has a charge: call it Q_1. Then, by uniqueness [see Eq. (6.11)],

$$\Phi_1' = P_{11}Q_1 \qquad \text{and} \qquad \Phi_2' = P_{21}Q_1$$

The potentials at any point in space are completely specified by the net charges on the conductors and linearity is ensured, just as in the case of a single conductor. In particular, this is true of the potentials, Φ_1' and Φ_2', of the conductors themselves. The P's depend only on the forms and relative configuration of the conductors. If the charge density, and therefore Q, is increased by a factor α, the potentials increase similarly.

2. Analogously, consider the case where $Q_1 = 0$, but conductor 2 is given a charge Q_2. We have

$$\Phi_1'' = P_{12}Q_2 \qquad \text{and} \qquad \Phi_2'' = P_{22}Q_2$$

3. Now, invoking superposition, we know that the sum $\Phi = \Phi' + \Phi''$, of the two solutions above provide the solution to the problem where Q_1 and Q_2 are specified simultaneously. That solution is the same as Eqs. (6.12) and (6.13).

This result can be easily generalized to the case where N conductors are present—namely, Eq. (6.9). Equation (6.10) is also expected to be linear because one can arrive at it by solving Eq. (6.9) for the charge Q_j as a function of the various potentials of the conductors.

The electric energy of N conductors can now be written in terms of the P_{ij} or C_{ij}. Substituting Eq. (6.9) or (6.10) into Eq. (6.8) gives the following expressions for U:

$$U = \frac{1}{2}\sum_{j=1}^{N}\sum_{k=1}^{N} P_{jk}Q_jQ_k \tag{6.14}$$

$$U = \frac{1}{2}\sum_{j=1}^{N}\sum_{k=1}^{N} C_{jk}\Phi_j\Phi_k \tag{6.15}$$

We now consider some properties of P_{ij}. The energy given by Eq. (6.14) can be used to show that the "off-diagonal" elements of the coefficients of potential are equal

$$P_{ij} = P_{ji} \tag{6.16}$$

Consider the increase in electrostatic energy of a system of N charged conductors when an infinitesimal charge dQ_i is added to the ith conductor. If only charge Q_i is thus changed, the change in U will be given by

$$dU = \frac{\partial U}{\partial Q_i}\, dQ_i$$

Taking the differential of Eq. (6.14) gives:

$$dU = \frac{1}{2}\sum_j \sum_k P_{jk}\left(\frac{\partial Q_j}{\partial Q_i} Q_k + \frac{\partial Q_k}{\partial Q_i} Q_j\right) dQ_i. \tag{6.17}$$

Since

$$\frac{\partial Q_j}{\partial Q_i} = \frac{\partial Q_k}{\partial Q_i} = 0 \qquad \text{if } j \neq i \neq k \quad \text{and 1 if } i = j = k$$

then the sums over j in the first term, and over k in the second term of dU in Eq. (6.17) include only those terms where $j = i$ and $k = i$, respectively. Thus we obtain

$$dU = \frac{1}{2}\left(\sum_k P_{ik} Q_k \, dQ_i + \sum_j P_{ji} Q_j \, dQ_i\right) \tag{6.18}$$

Replacing k by j in the first sum, we can write Eq. (6.18) as

$$dU = \frac{1}{2}\sum_{j=1}^{N} (P_{ij} + P_{ji}) Q_j \, dQ_i \tag{6.19}$$

However, another expression for dU is simply $\Phi_i \, dQ_i$, by the definition of potential. Therefore

$$dU = \Phi_i \, dQ_i = \sum_{j=1}^{N} P_{ij} Q_j \, dQ_i \tag{6.20}$$

Comparing Eqs. (6.19) and (6.20), we see that equality is possible only if

$$\frac{1}{2}(P_{ij} + P_{ji}) = P_{ij} \qquad \text{or} \qquad P_{ij} = P_{ji}$$

Another property of P_{ij} is $P_{ii} > 0$. This property implies the obvious statement that a positive charge produces a positive potential. Finally, we show that the diagonal coefficient P_{ii} is larger or equal to the off-diagonal coefficient P_{ij} for any j; that is,

$$P_{ii} \geq P_{ij} \tag{6.21}$$

Consider a system of two conductors, conductor 1 bears a positive charge Q_1 and conductor 2 is neutral. There are two distinct configurations of the system: In one configuration, conductor 1 encloses conductor 2; in the other configuration, the conductors are exterior to each other. In the first configuration the potential of conductor 2, Φ_2, is the same as that of conductor 1, Φ_1. Using Eq. (6.9) we find that $\Phi_1 = P_{11} Q_1$ and $\Phi_2 = P_{12} Q_1$, and therefore $P_{11} = P_{12}$. In the other configuration, on the other hand, $\Phi_1 > \Phi_2$. This can be realized since all lines of force impinging on conductor 2 have to be traced back to conductor 2 as a result of the fact that conductor 2 is neutral. Thus this configuration gives $P_{11} > P_{22}$. Therefore both configurations imply that $P_{11} \geq P_{22}$. This result can be easily generalized to the case involving N conductors, hence establishing Eq. (6.21).

Because the coefficients C_{ij} are related to the coefficients P_{ij} through Eqs. (6.9) and (6.10), then their properties follow from those of P_{ij}. In fact, one can show that

$$C_{ij} = C_{ji}, \; C_{ii} > 0, \quad \text{and } C_{ij} \leq 0 \qquad \text{for } i \neq j$$

Example 6.3 The Coefficients of Potential—Three Identical Spheres

This example shows how the method of coefficients of potential can be used to solve some electrostatic problems. Consider three initially isolated and uncharged equal conducting spheres placed with their centers at the vertices of an equilateral triangle as shown in Fig. 6.1. Conductor 1 is now charged to potential V and isolated, and it is found to have a charge Q_1 on it. Conductor 2 is then charged to potential V and isolated, and it is found to have a charge Q_2 on it. Conductor 3 is finally charged to potential V and isolated.

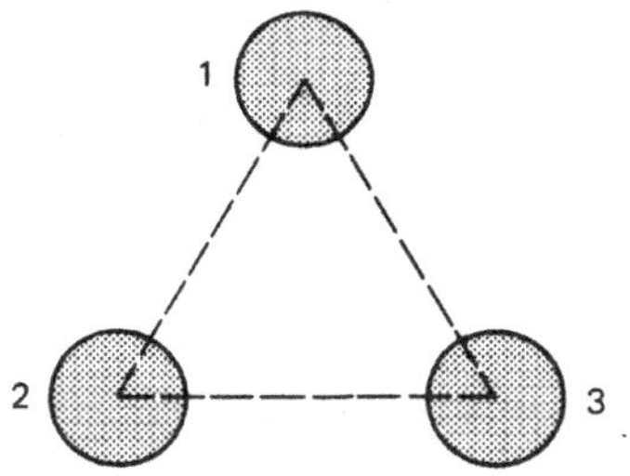

Figure 6.1 Three identical conducting spheres placed at the vertices of an equilateral triangle.

This given information about conductors 1 and 2 can now be shown to be sufficient to yield information about the status of conductor 3, such as the total charge on it, and to determine the sizes and distances between the spheres.

The coefficients of potential P_{ij} where $i, j = 1$ to 3 are not all distinct. Since the spheres are identical, then $P_{11} = P_{22} = P_{33}$; and because of the symmetrical positioning of the spheres, then $P_{12} = P_{13} = P_{23}$. Moreover, using the property $P_{ij} = P_{ji}$, Eq. (6.16), then there are only two distinct coefficients: P_{11} and P_{12}.

The potential of the conductors is related to the charges on them by Eq. (6.9). In the first step, the charges on conductors 2 and 3 are zeros, and therefore the potential on conductor 1 is related to the charges simply by

$$V = P_{11}Q_1 \tag{6.22}$$

In the second step, the charges on conductors 1, 2, and 3 are Q_1, Q_2 and zero. The potential on conductor 2 is given to be V. Therefore the potential on conductor 2 is related to the charges as follows:

$$V = P_{11}Q_2 + P_{12}Q_1 \tag{6.23}$$

Finally, the given potential V of conductor 3 is related to the charges by

$$V = P_{11}Q_3 + P_{12}(Q_1 + Q_2) \tag{6.24}$$

where Q_3 is the charge on conductor 3, an unknown quantity at this point.

Equations (6.22) to (6.24) can now be solved for P_{11}, P_{12} and Q_3. Equation (6.22) gives $P_{11} = V/Q_1$, which upon substitution into Eq. (6.23) gives

$$P_{12} = \frac{V}{Q_1}\left(1 - \frac{Q_2}{Q_1}\right)$$

Substituting for P_{11} and P_{12} in Eq. (6.24) yields $Q_3 = Q_2^2/Q_1$. The determination of the size and the distance between the spheres will be left as an exercise.

Example 6.4 Coefficients of Potential of Concentric Spheres

This example shows how one determines the coefficients of potential of a given system of conductors of known geometry. We consider the spherical capacitor which was treated in Example 2.12*a* and was shown in Fig. 2.20. The potential between the spheres was determined from the relation $\Phi = -\int \mathbf{E}\cdot d\mathbf{r}$, where the electric field was determined from Gauss' law, with the result

$$\Phi(\mathbf{r}) = \frac{1}{4\pi\varepsilon_0}\left(\frac{Q_1}{r} + \frac{Q_1 + Q_2}{R_2} - \frac{Q_1}{R_3}\right) \qquad R_1 < r < R_3$$

The potential of the inner sphere, Φ_1, can be easily determined from this expression by taking $r = R_1$; hence

$$\Phi_1 = \frac{1}{4\pi\varepsilon_0}\left(\frac{1}{R_1} + \frac{1}{R_2} - \frac{1}{R_3}\right)Q_1 + \frac{1}{4\pi\varepsilon_0}\frac{Q_2}{R_2} \tag{6.25}$$

Similarly, one can determine the potential at the outer sphere, Φ_2 by taking $r = R_3$; that is,

$$\Phi_2 = \frac{1}{4\pi\varepsilon_0}\frac{Q_1}{R_2} + \frac{1}{4\pi\varepsilon_0}\frac{Q_2}{R_2} \tag{6.26}$$

Comparing Eqs. (6.25) and (6.26) with the equation defining the coefficients of potential, Eq. (6.9), gives

$$P_{11} = \frac{1}{4\pi\varepsilon_0}\left(\frac{1}{R_1} + \frac{1}{R_2} - \frac{1}{R_3}\right)$$

$$P_{12} = P_{21} = \frac{1}{4\pi\varepsilon_0}\frac{1}{R_2}$$

$$P_{22} = \frac{1}{4\pi\varepsilon_0}\frac{1}{R_2}$$

It appears that $P_{12} = P_{22}$ and thus only two distinct coefficients of potential are needed to describe this system. The equality between P_{12} and P_{22} is a direct result of the fact that one of the spheres is enclosed by the other. A special case of this geometry is when the outer shell is very thin such that $R_2 \simeq R_3$; in that case,

$$P_{11} = \frac{1}{4\pi\varepsilon_0}\frac{1}{R_1}, \quad P_{22} = P_{12} = P_{21} = \frac{1}{4\pi\varepsilon_0}\frac{1}{R_2}$$

6.4 Capacitors

In this section we consider a special configuration of conductors wherein the charges on a pair of conductors are $\pm Q_i$ and hence the total charge of the pair is zero. Such a pair, called a *capacitor*, exhibits a shielding effect. In this shielding effect, the presence of other conductors affect the potentials of each conductor of the pair by the same amount and hence leave their potential difference unaffected.

6.4.1 Capacitance of an Isolated Conductor

We will first consider a single isolated conductor. Such a conductor can be considered to be part of a capacitor whose other conductor has a radius that extends to infinity. In other words, there exists capacitance to "ground" or "earth"—that is, to the external world. From Eq. (6.11), the potential on the surface of an isolated conductor is $\Phi = PQ$. Solving for Q in terms of Φ gives

$$Q = P^{-1}\Phi \equiv C\Phi \tag{6.27}$$

The geometrical coefficient C is called the *capacitance* of the conductor. Clearly it gives the total amount of charge an isolated conductor can carry when at a potential of 1 volt. It measures the "capacity" of a conductor to hold charge.

As an application of Eq. (6.27) we determine the capacitance of a single isolated conducting sphere. First we take the sphere to be placed in vacuum, and then we consider the case where the sphere is enclosed by a dielectric shell. In the first case, we simply note that the potential function Φ is given by

$$\Phi(\mathbf{r}) = \frac{1}{4\pi\varepsilon_0}\frac{Q}{r} \qquad r \geq R$$

and

$$\Phi(\mathbf{r}) = \frac{1}{4\pi\varepsilon_0}\frac{Q}{R} \qquad r \leq R$$

From the latter equation, and from the definition of capacitance given by Eq. (6.27), we see that

$$C = 4\pi\varepsilon_0 R \tag{6.28}$$

When the sphere is surrounded by a dielectric shell of radius R_2 and dielectric constant K, as was considered earlier in Fig. 4.7, the potential of the sphere with respect to a reference at infinity is given by Eq. (4.54). The capacitance can then be easily determined

$$C^{-1} = \frac{1}{4\pi\varepsilon_0 R_2}\left(1 - \frac{1}{K}\right) + \frac{1}{4\pi\varepsilon R_1} \tag{6.29}$$

where R_1 is the radius of the sphere and $\varepsilon = \varepsilon_0 K$.

The unit of capacitance in the SI system is the coulomb per volt, called the farad. Because $4\pi\varepsilon_0 \approx \frac{1}{9} \times 10^{-9}$, we see that the capacitance of most ordinary isolated conductors is of the order of 10^{-9} farad. The earth itself, considered as an isolated conductor, has a capacitance of "only" about 10^{-3} farad. Most laboratory sized objects have capacitances of the order of 10^{-9} farad or smaller. Hence most commonly encountered capacitances are conveniently expressed in picofarads or pF (10^{-12} F), in nanofarads or nF (10^{-9} F), or in microfarads or μF (10^{-6} F).

6.4.2 The Two-Conductor Capacitor

The next simplest capacitor beyond the isolated one-conductor system is the isolated system of two conductors. Let (Φ_i and Q_i) be the potential and the charge on the ith conductor, where $i = 1$ and 2. An important special case of the preceding relations is one where $Q_1 = -Q_2$, so that the total charge of the two conductor system is zero. For this system,

$$\Phi_1 - \Phi_2 = (P_{11} - P_{12} - P_{21} + P_{22})Q_1 \tag{6.30}$$

Thus the potential difference between the conductors is proportional to the charge (on either one) of the conductors. This relation is more conventionally written

$$Q_1 = C\Phi_{12} \tag{6.31}$$

where $\Phi_{12} = \Phi_1 - \Phi_2$ and $C = (P_{11} - P_{12} - P_{21} + P_{22})^{-1}$ is an intrinsically positive geometrical coefficient called the capacitance of the two-conductor system. A device consisting of two isolated conductors and used to store charge is called a

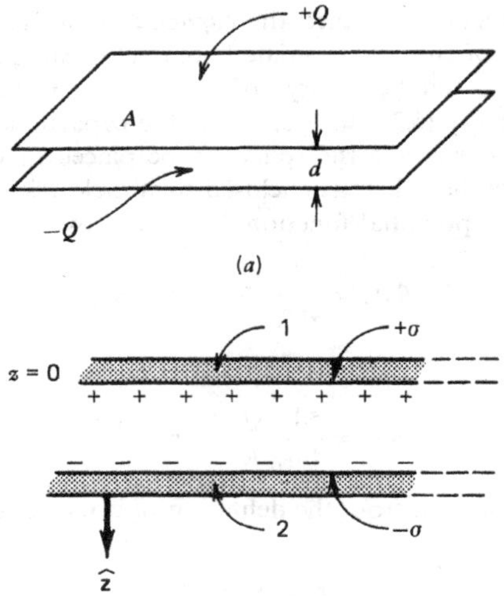

Figure 6.2 Calculation of the capacitance of a parallel-plate capacitor. (*a*) Geometry of the capacitor. (*b*) A side view showing the charge distribution.

capacitor (formerly "condenser"). It is useful to think of the capacitor as being charged by transferring charge between the originally neutral conductors.

We shall find the capacitance of two identical parallel conducting plates of area A separated by a distance d, as shown in Fig. 6.2*a*. We shall assume the plates large enough so that end effects may be legitimately neglected and the plates therefore considered as infinite in lateral extent (see Example 4.4).

Assume that each plate is given a charge of magnitude Q, but with opposite signs. By symmetry, the distribution of charge on each plate will be the same, and consequently the electric field will exist only in the space between the plates. (Why?) Moreover, this electric field will be normal to the plates and approximately constant, and will be related to the surface charge density on the inner plate surfaces by $\varepsilon_0 E = \sigma$, where $\sigma = Q/A$ is the surface charge density. This result can be derived using Gauss' law (see Example 4.4) or direct integration [see Eqs. (4.27) and (4.28) in Example 4.1].

In order to find the potential difference between the plates, we evaluate the line integral $\int \mathbf{E} \cdot d\mathbf{r}$. Using the notation of Fig. 6.2*b* and substituting for $E = \sigma/\varepsilon_0$, we obtain

$$\Phi_{12} \equiv \Phi_1 - \Phi_2 = -\int_2^1 \mathbf{E} \cdot d\mathbf{r} = -\int_2^1 \frac{\sigma}{\varepsilon_0} dz = \frac{\sigma}{\varepsilon_0} d \tag{6.32}$$

Referring to our definition of capacitance in Eq. (6.31), the capacitance of the two plates is

$$C = \frac{\varepsilon_0 A}{d} \tag{6.33}$$

We should note that if the capacitor had been filled with a dielectric material of permittivity ε, then the capacitance would have been $\varepsilon A/d$. (Show this.)

The capacitance per unit area and unit length of the conductors, C_s and C_l, are often used. These are defined as follows:

$$C_s \equiv \frac{\sigma}{\Phi_{12}} \qquad C_l \equiv \frac{\lambda}{\Phi_{12}} \tag{6.34}$$

where λ is the charge density per unit length.

These expressions for capacitance will be in error for actual plates of finite size. For the latter case, the capacitance is increased because near the edges of the plates the charge density is increased. For a circular parallel-plate capacitor of radius R, the capacity is increased by approximately 10 percent when the ratio d/R equals $\frac{1}{20}$.

Example 6.5 Capacitance of a Cylindrical Capacitor

Suppose we have the long coaxial line shown in Fig. 6.3. The inner conductor of the line has a radius a, and the outer conductor has an inner radius b. In this geometry, the field will be

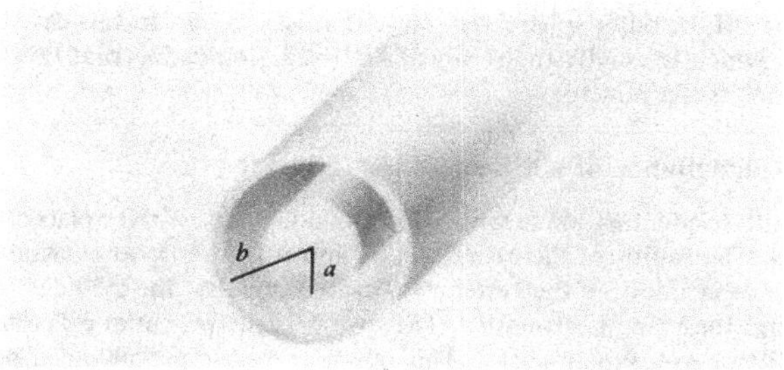

Figure 6.3 Coaxial cylindrical capacitor.

radial from the axis of the line if we are far removed from the ends of the line. By Gauss' law we find that the electric field is given by*

$$\mathbf{E} = 0 \qquad \text{for } \rho < a \text{ and } \rho > b$$

$$\mathbf{E} = \frac{\lambda}{2\pi\varepsilon_0 \rho}\hat{\boldsymbol{\rho}} \qquad a < \rho < b$$

where λ represents the charge per unit length (per meter) on the inner conductor. An equal and opposite charge is present on the inner surface of the outer conductor. The electric field is zero for $\rho > b$ because of our assumption that the total charge on the line is zero. With the knowledge of $\mathbf{E}$, we can now find the potential difference $\Phi(a) - \Phi(b)$. We have

$$\Phi(a) - \Phi(b) = \int_{\rho=a}^{\rho=b} \mathbf{E} \cdot d\mathbf{r} = \int_a^b E\, d\rho = \frac{\lambda}{2\pi\varepsilon_0}\int_a^b \frac{d\rho}{\rho} = \frac{\lambda}{2\pi\varepsilon_0}\ln\frac{b}{a} \tag{6.35}$$

Consequently, from the definition of capacitance per unit length given in Eq. (6.34), and from Eq. (6.35) we have

$$C = \frac{2\pi\varepsilon_0}{\ln(b/a)} = \frac{10^{-9}}{18}\frac{1}{\ln(b/a)} \text{ farads per meter} \tag{6.36}$$

---

* Components tangential to the surfaces—that is, E_ϕ and E_z—must be zero since they are zero at the surfaces, and they are constant otherwise.

Example 6.6 Concentric Spherical Conductors

Consider a concentric spherical capacitor with R_1 and R_2 the radii of the inner and outer spheres, and Q and $-Q$ are the charges on them respectively. This configuration was treated earlier in Example 2.12*a*, where it was shown that the potential for the region $R_1 \leq r \leq R_2$ is given by

$$\Phi(r) = \frac{Q}{4\pi\varepsilon_0 r} - \frac{Q}{4\pi\varepsilon_0 R_2}$$

It therefore follows that

$$\Phi(R_1) - \Phi(R_2) = \frac{Q}{4\pi\varepsilon_0 R_1} - \frac{Q}{4\pi\varepsilon_0 R_2}$$

Therefore

$$C = \frac{4\pi\varepsilon_0 R_1 R_2}{R_2 - R_1} \tag{6.37}$$

Let us take the size of the outer sphere very large that is $R_2 \to \infty$. In this case the capacitance becomes $4\pi\varepsilon_0 R_1$, which is exactly the result of Eq. (6.28), indicating that this limit is equivalent to a single isolated conductor.

Example 6.7 Capacitance of a Plane and a Cylinder

Consider an infinite conducting plane with a long, conducting cylinder placed parallel to it as shown in Fig. 6.4. The radius of the cylinder is R and its center is at a distance x_0 from the plane. To find the capacitance of the system we first assume that the cylinder has the charge λ per unit length and the plane is grounded. Then we find the potential difference between the cylinder and the plane as a function of λ. This boundary value problem can be solved by the method of images, which was discussed in Section 3.5.3. In this method the potential between the conductors is given by a line charge placed inside the cylinder at the diameter normal to the plane and carrying the charge λ per unit length, and an image line charge placed behind the plane at an equidistance from the first and with the charge $-\lambda$ per unit length.

The distance of the line charges from the plane are to be chosen such that an equipotential surface exactly coincides with the cylindrical surface. Using Eq. (3.129) and (3.130),

$$x_0 = \frac{m^2 + 1}{m^2 - 1}a \qquad R = \frac{2ma}{m^2 - 1}$$

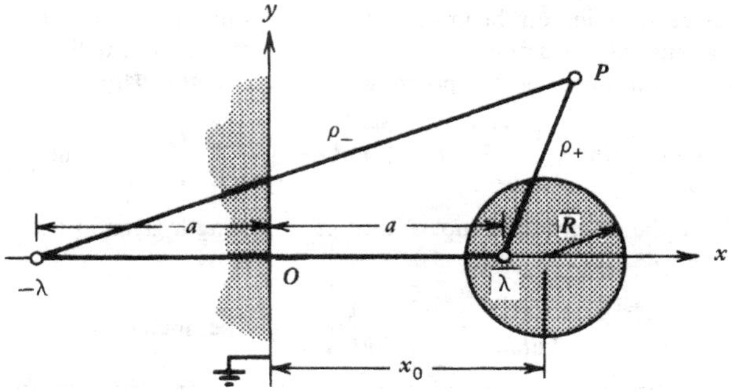

Figure 6.4 Calculation of the capacitance of a conducting cylinder in the vicinity of a conducting parallel plate using the method of images.

These two relations are solved for m, where $m = (\rho_-/\rho_+)$:

$$m = \frac{x_0}{R} + \left[\left(\frac{x_0}{R}\right)^2 - 1\right]^{1/2} \tag{6.38}$$

and hence the potential of the cylinder is given by Eq. (3.131) with m given by this value.

$$\Phi_c = \frac{\lambda}{2\pi\varepsilon_0} \ln m = \frac{\lambda}{2\pi\varepsilon_0} \ln\left\{\frac{x_0}{R} + \left[\left(\frac{x_0}{R}\right)^2 - 1\right]^{1/2}\right\} = \frac{\lambda}{2\pi\varepsilon_0} \cosh^{-1}\frac{x_0}{R} \tag{6.39}$$

The capacitance of the system per unit length is then given by

$$C = \frac{\lambda}{\Phi_c} = \frac{2\pi\varepsilon_0}{\cosh^{-1}(x_0/R)} \tag{6.40}$$

6.4.3 Combinations of Capacitors

The symbol used to represent a capacitor is —||—, that is, an object suggesting a parallel-plate capacitor. Consider now the situation where several capacitors are connected to each other. There are two simple configurations which are easy to treat, in which we consider the capacitors to be "in series" or "in parallel."

In the series case, it is assumed that the conductors are arranged such that the lines of force terminating on any conductor originate on at most one other conductor. Consider Fig. 6.5*a*, showing four capacitors of capacitances C_1, C_2, C_3, and C_4 connected in series. The "end" conductors are attached to terminals A and B, and kept at a potential difference $\Phi_{AB} \equiv \Phi_A - \Phi_B$. The *conservation of charge* requires that each capacitor acquire the same charge Q. Thus $\Phi_{AB} = \Delta\Phi_1 + \Delta\Phi_2 + \Delta\Phi_3 + \Delta\Phi_4 = Q/C_1 + Q/C_2 + Q/C_3 + Q/C_4$, where $\Delta\Phi_i$ is the potential difference across the ith capacitor. This result can now be written in terms of an equivalent single capacitor of capacitance C. That is $\Phi_{AB} = Q/C$, where $1/C = 1/C_1 + 1/C_2 + 1/C_3 + 1/C_4$. In the general case of N capacitors, we have

$$\frac{1}{C} = \sum_{j=1}^{N} \frac{1}{C_j} \tag{6.41}$$

The case of parallel capacitors is sketched in Fig. 6.5*b*, showing four capacitors so arranged that each has the same potential difference Φ_{AB}, and charges Q_1, Q_2, Q_3, and Q_4. Again the conductors are assumed to be arranged so that lines of force terminating (or originating) on any one conductor originate from (or terminate at) a single other conductor. In other words, the electric field of a pair of conductors is

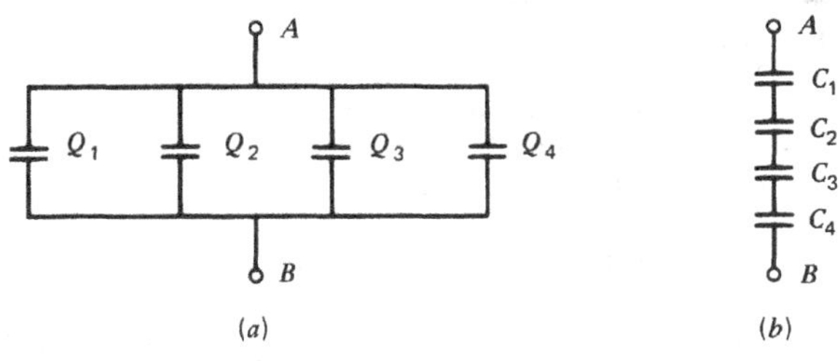

Figure 6.5 Combinations of capacitors. (*a*) Series connection. (*b*) Parallel connection.

completely contained between this pair. The total charge on the four capacitors is $Q = Q_1 + Q_2 + Q_3 + Q_4 = C_1\,\Delta\Phi_1 + C_2\,\Delta\Phi_2 + C_3\,\Delta\Phi_3 + C_4\,\Delta\Phi_4$. But each of the potential differences is equal to Φ_{AB}; thus $Q = (C_1 + C_2 + C_3 + C_4)\Phi_{AB}$. In terms of an equivalent single capacitor of capacitance C, we write $Q/\Phi_{AB} = C = C_1 + C_2 + C_3 + C_4$. In the general case of N capacitors we have, for parallel connection,

$$C = \sum_{j=1}^{N} C_j \tag{6.42}$$

Finally we note that capacitors in parallel give a capacitance greater than that of any single one of them, whereas capacitors in series have a capacitance less than that of any single one.

6.4.4 Energy Storage in Capacitors

One of the primary uses of capacitors in electrostatic applications is as receptacle of electrostatic energy. In order to charge up a capacitor, work must be done to move the charge against the electrostatic forces present. Because electrostatic forces are conservative, the work done in the charging process is available subsequently as the electrostatic energy of the capacitor. Work can be done at the expense of this energy when the capacitor is discharged.

If we have a single conductor at potential Φ (relative to infinity), an amount of work $\Phi\,dq$ is required to bring a charge dq up to the conductor surface (from infinity). To charge the conductor from a condition of zero charge to a total charge Q therefore requires an amount of work

$$W = \int_0^Q dW = \int_0^Q \Phi\,dq = \int_0^Q \frac{q}{C}\,dq = \frac{1}{2}\frac{Q^2}{C} = \frac{1}{2}CV^2 \tag{6.43}$$

where C is the capacitance of the conductor, and V is the potential of the conductor when having a charge Q.

Similarly, if we wish to charge a two-conductor capacitor, we start with the capacitor uncharged, and charge it by removing charge from one conductor and placing it on the other. The work necessary to charge the conductors to a final potential difference $\Phi_{12} \equiv V$ and corresponding charge Q:

$$W = \int_0^Q dW = \int_0^Q \Phi_{12}\,dq = \int_0^Q \frac{q}{C}\,dq = \frac{1}{2}\frac{Q^2}{C} = \frac{1}{2}CV^2 \tag{6.44}$$

where $C = Q/\Phi_{12}$ is the capacitance of the system.

The result in this case is identical in form to the preceding case of a single conductor, as is expected, since as one of the conductors moves off to infinity, we are left with the case of a single conductor.

Example 6.8 Electrostatic Energy of Two Charged Spheres

Two identical conducting spheres 1 and 2 of radius R carry charges q_1 and q_2, respectively, and the distance between their centers is $r \gg R$, as shown in Fig. 6.6. The electrostatic energy of the system can be calculated using Eq. (6.14); that is,

$$U = \frac{1}{2}\sum P_{ij}q_iq_j = \frac{1}{2}P_{11}(q_1^2 + q_2^2) + P_{12}q_1q_2 \tag{6.45}$$

where P_{11} and P_{12} are the only distinct coefficients of potential of the system. If charge q_2 is zero, then the potential of sphere 1 is $\Phi_1 = P_{11}q_1 = q_1/4\pi\varepsilon_0 R$; thus $P_{11} = 1/4\pi\varepsilon_0 R$. Also the potential at distance r from the sphere is $\Phi_2 = P_{12}q_1 = q_1/4\pi\varepsilon_0 r$. Thus $P_{12} = 1/4\pi\varepsilon_0 r$.

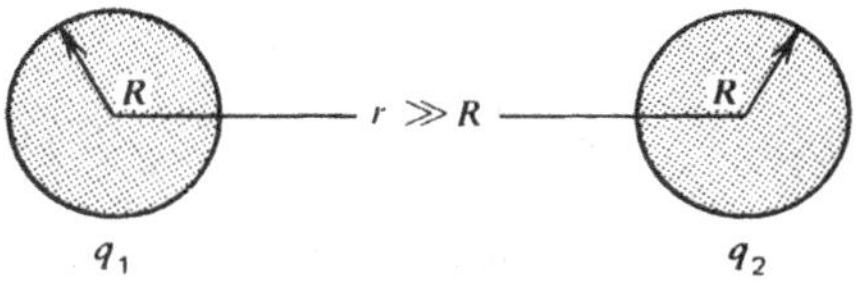

Figure 6.6 Two identical, small, charged conducting spheres at a large distance from each other.

Therefore

$$U = \frac{1}{8\pi\varepsilon_0 R}(q_1^2 + q_2^2) + \frac{q_1 q_2}{4\pi\varepsilon_0 r} \tag{6.46}$$

Let us calculate the change in electrostatic energy when a thin wire is used to connect the spheres together electrically. After connection, each sphere carries a charge $\frac{1}{2}(q_1 + q_2)$. Replacing both q_1 and q_2 in Eq. (6.46) with $\frac{1}{2}(q_1 + q_2)$ gives the energy of the new situation:

$$U' = \frac{1}{4\pi\varepsilon_0 R}\left(\frac{q_1 + q_2}{2}\right)^2 + \frac{1}{4\pi\varepsilon_0 r}\left(\frac{q_1 + q_2}{2}\right)^2$$

The change in the electrostatic energy, $\Delta U = U' - U$, is then equal to

$$\Delta U = \frac{1}{16\pi\varepsilon_0}(q_1 - q_2)^2 \frac{R - r}{Rr} < 0$$

indicating a reduction in the electrostatic energy.

Example 6.9 Expansion of a Spherical Capacitor

The inner sphere of a spherical capacitor, shown in Fig. 6.7, has a radius R_1; the outer concentric shell is very thin and has a radius R_2. The inner sphere is kept at a constant potential V by a battery. The outer shell is insulated and has a charge q_2.

The potentials Φ_1 and Φ_2 of the inner and outer conductors, respectively, are related to the charges on the conductors by Eqs. (6.12) and (6.13), that is,

$$\Phi_1 = P_{11}q_1 + P_{12}q_2$$

$$\Phi_2 = P_{22}q_2 + P_{12}q_1$$

where q_1 (unknown at this point) is the charge on the inner sphere. Using the results of Example 6.4 we find that:

$$P_{11} = \frac{1}{4\pi\varepsilon_0}\frac{1}{R_1} \qquad P_{22} = \frac{1}{4\pi\varepsilon_0}\frac{1}{R_2} \qquad P_{12} = P_{22}$$

Substituting these in the above relations yields the following expressions for q_1 and Φ_2.

$$q_1 = 4\pi\varepsilon_0 R_1 V - \frac{R_1 q_2}{R_2} \quad \text{and} \quad \Phi_2 = \frac{1}{4\pi\varepsilon_0}\left(\frac{q_1 + q_2}{R_2}\right) \tag{6.47}$$

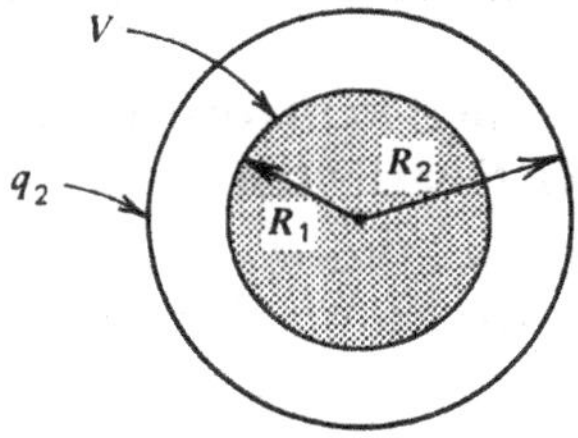

Figure 6.7 A spherical capacitor whose inner electrode undergoes expansion.

Using Eqs. (6.47) and (6.3) the electrostatic energy of the system is

$$U_1 = \frac{1}{2}\sum_{i=1}^{2} q_i\Phi_i = \frac{1}{2}\left[q_1 V + \frac{q_2}{4\pi\varepsilon_0}\left(\frac{q_1 + q_2}{R_2}\right)\right] \tag{6.48}$$

Suppose that the inner sphere now expands to radius R_3. As a result of the expansion, the electric energy changes and some mechanical work is performed by the attractive forces between the spheres. We first calculate the change in the electric energy. The charge at the inner sphere q_3 is determined by replacing R_1 by R_3 in Eq. (6.47); hence

$$q_3 = 4\pi\varepsilon_0 R_3 V - \frac{R_3 q_2}{R_2} \tag{6.49}$$

Replacing q_1 by q_3 in Eq. (6.48) gives the new electric energy of the system:

$$U_2 = \frac{1}{2}\left[q_3 V + \left(\frac{q_2}{4\pi\varepsilon_0}\right)\left(\frac{q_3 + q_2}{R_2}\right)\right] \tag{6.50}$$

Thus the change in electric energy $\Delta U_e = U_2 - U_1$ is

$$\Delta U_e = \frac{1}{2}(q_3 - q_1)\left[V + \frac{q_2}{4\pi\varepsilon_0 R_2}\right] = 2\pi\varepsilon_0\left[V^2 - \left(\frac{q_2}{4\pi\varepsilon_0 R_2}\right)^2\right](R_3 - R_1) \tag{6.51}$$

This result indicates that the electric energy increases upon expansion and decreases upon contraction of the inner sphere.

Let us now calculate the mechanical work done by the attractive forces. From Eqs. (6.47) and (6.49), the charge on the inner sphere q_r as a function of r is $q_r = q_1 r/R_1$, and hence the charge density σ_r is $q_r/4\pi r^2 = q_1/4\pi R_1 r$. From Eq. (4.116) the outward force F_r acting on the sphere is given by

$$F_r = \frac{1}{2\varepsilon_0}\int_S \sigma^2\, da$$

where S is the surface of the sphere. Hence

$$F_r = \frac{1}{2}\frac{q_1^2}{4\pi\varepsilon_0 R_1^2}$$

which is independent of r. The mechanical work done when the sphere expands from R_1 to R_3 is equal to

$$\Delta U_m = \int_{R_1}^{R_3} F_r\, dr = \frac{q_1^2}{8\pi\varepsilon_0 R_1^2}(R_3 - R_1) = 2\pi\varepsilon_0\left(V - \frac{q_2}{4\pi\varepsilon_0 R_2}\right)^2 (R_3 - R_1) \tag{6.52}$$

Thus the energy supplied by the battery, ΔU, is the sum of Eqs. (6.51) and (6.52); that is,

$$\Delta U = \Delta U_e + \Delta U_m = 4\pi\varepsilon_0(R_3 - R_1)V\left(V - \frac{q_2}{4\pi\varepsilon_0 R_2}\right) \tag{6.53}$$

6.5 Electrostatic Energy: An Alternative Expression in Terms of the Field Distribution

Let us assume that we have a particular charge distribution everywhere characterized by a charge density ρ in vacuum. We then know that the electrostatic energy for this distribution is given as

$$U = \frac{1}{2}\int_V \rho(\mathbf{r})\Phi(\mathbf{r})dv \tag{6.5}$$

We have indicated that the integration is over any volume V finite in extent containing all the charge in the distribution. We shall consider it a sphere with its center in the charge distribution. We can just as well choose V to be all of space; the contribution to the integral will be zero wherever $\rho = 0$. Substituting for $\rho(\mathbf{r})$ from Gauss' law ($\nabla \cdot \mathbf{E} = \rho/\varepsilon_0$), we have

$$U = \frac{\varepsilon_0}{2} \int_V \Phi(\nabla \cdot \mathbf{E}) dv \tag{6.54}$$

The form of this equation can be changed using the vector identity $\nabla \cdot (\Phi \mathbf{E}) = \nabla\Phi \cdot \mathbf{E} + \Phi(\nabla \cdot \mathbf{E})$, whence

$$U = \frac{\varepsilon_0}{2} \int_V \nabla \cdot (\Phi \mathbf{E}) dv - \frac{\varepsilon_0}{2} \int_V \nabla\Phi \cdot \mathbf{E}\, dv \tag{6.55}$$

Applying the divergence theorem to the first integral and using $E = -\nabla\Phi$ in the second integral, we get

$$U = \frac{\varepsilon_0}{2} \oint_S (\Phi \mathbf{E}) \cdot d\mathbf{a} + \frac{\varepsilon_0}{2} \int_V \mathbf{E} \cdot \mathbf{E}\, dv, \tag{6.56}$$

where S is a closed surface enclosing the volume V. Now, if the charge distribution is localized in space, and if surface S is considered as a spherical surface very far removed from the charge distribution,* then $|\mathbf{E}|$ on this surface will in good approximation be constant and proportional to $|\hat{\mathbf{r}}/r^2|$, and Φ will be proportional to $1/r$. Consequently,

$$\oint_S (\Phi \mathbf{E}) \cdot d\mathbf{a} \approx \oint \frac{Q^2}{r^3} \hat{\mathbf{r}} \cdot d\mathbf{a} = \frac{Q^2}{r^3} \cdot 4\pi r^2 = \frac{4\pi Q^2}{r}$$

and as the spherical surface recedes to infinity, the surface integral becomes arbitrarily small, and may consequently be neglected. Thus Eq. (6.56) becomes

$$U = \frac{\varepsilon_0}{2} \int E^2\, dv \tag{6.57}$$

where the volume of integration must now include all space. If we assigned to each volume of space dv an energy $u\, dv$, we would say that the total energy in space would be just

$$U = \int u\, dv \tag{6.58}$$

On this basis we would be led to the energy density expression:

$$u = \frac{\varepsilon_0}{2} E^2 \tag{6.59}$$

If the charge density was embedded in space of permittivity ε (dielectric constant $K \neq 1$), then one can easily show that Eqs. (6.57) and (6.59) become

$$U = \frac{1}{2} \int \mathbf{E} \cdot \mathbf{D}\, dv = \frac{1}{2} \varepsilon \int E^2\, dv \tag{6.60}$$

$$u = \frac{1}{2} \mathbf{E} \cdot \mathbf{D} = \frac{1}{2} \varepsilon_0 E^2 + \frac{1}{2} \mathbf{E} \cdot \mathbf{P} = \frac{1}{2} \varepsilon E^2 \tag{6.61}$$

* At "infinity," the charge distribution looks like a point charge of magnitude $\int_V \rho\, dv$.

We can call the results given in Eq. (6.59) or Eq. (6.61) the electro(static) field *energy density*. In this interpretation we would say that the energy associated with a charge distribution was really stored in the electric field of this distribution. This interpretation will be especially useful in nonstatic situations (to be discussed in later chapters), where the possibility of electromagnetic radiation exists. Energy may then indeed be transferred through space, and this energy will be associated with the electromagnetic fields that exist there. In the static situation, this interpretation can lead to ambiguities, and perhaps it is best to think of Eq. (6.57) as merely another way to calculate U.

We calculate the energy associated with a uniformly charged sphere, employing Eq. (6.57). In order to use this formula we need to know what **E** is everywhere in space (due to the charged sphere). Referring to Eqs. (2.52) and (2.53) in Example 2.12*b*, we have

$$U = \frac{\varepsilon_0}{2} \int_{\text{inside}} \left(\frac{Qr}{4\pi\varepsilon_0 R^3} \right)^2 dv + \frac{\varepsilon_0}{2} \int_{\text{outside}} \left(\frac{Q}{4\pi\varepsilon_0 r^2} \right)^2 dv$$

Taking $dv = 4\pi r^2 \, dr$ we get

$$U = \frac{\varepsilon_0}{2} \int_{r=0}^{R} \frac{Q^2 r^2}{(4\pi\varepsilon_0)^2 R^6} \cdot 4\pi r^2 \, dr + \frac{\varepsilon_0}{2} \int_{r=R}^{\infty} \frac{Q^2}{(4\pi\varepsilon_0)^2 r^4} \cdot 4\pi r^2 \, dr$$

which, upon integration, gives

$$U = \frac{Q^2}{2} \cdot \frac{1}{4\pi\varepsilon_0} \left(\frac{1}{5R} + \frac{1}{R} \right) = \frac{3}{5} \frac{Q^2}{4\pi\varepsilon_0 R}$$

verifying the result previously obtained in Example 6.1.

Example 6.10 The Calculation of Capacitance Using the Energy Relation

One direct use of the energy formulas is in calculating the capacitance of a two-conductor system. For such a capacitor, we know from Eq. (6.44) that

$$U = \frac{1}{2} C V^2 = \frac{1}{2} \frac{Q^2}{C}$$

Consequently,

$$C = \frac{2U}{V^2} = \frac{1}{2} \frac{Q^2}{U} \tag{6.62}$$

A knowledge of U can be used directly in calculating C. Below we use this definition to calculate the capacitance of a parallel-plate and a spherical capacitor.

Consider the parallel-plate capacitor shown in Fig. 6.8*a*. The area of each plate is A and the separation is d. The two plates are kept at V potential difference. The electric field produced by the plates was determined previously (see Example 4.4); that is, it is zero outside and V/d inside. Therefore, for energy density, Eq. (6.59) gives

$$u = \frac{1}{2} \varepsilon_0 E^2 = \frac{1}{2} \varepsilon_0 \frac{V^2}{d^2}$$

inside and zero outside the capacitor. The total electrostatic energy can be determined by integrating the energy density over the volume of the capacitor. Since u is a constant, then

$$U = uAd = \frac{1}{2} \varepsilon_0 A \frac{V^2}{d}.$$

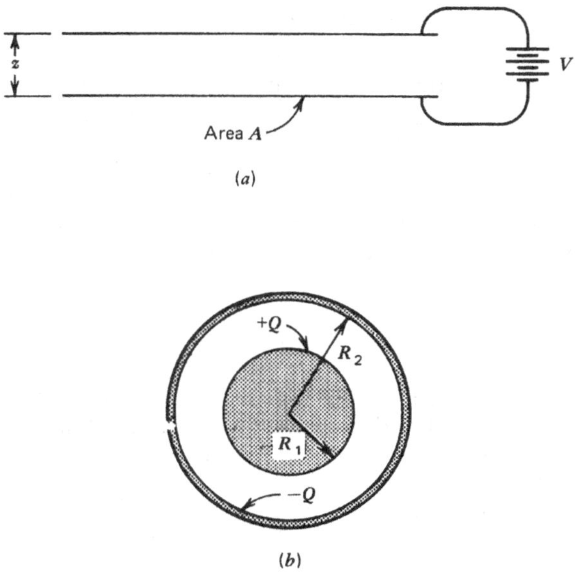

Figure 6.8 Calculation of capacitance using energy considerations. (*a*) Parallel-plate capacitor. (*b*) Spherical capacitor.

Equating the total energy to $\frac{1}{2}CV^2$ gives

$$C = \frac{\varepsilon_0 A}{d}$$

Consider the spherical capacitor shown in Fig. 6.8*b* as another example. Here, if the inner conductor has a charge Q and the outer conductor a charge $-Q$, the field outside the outer sphere is zero. The field between the spheres ($R_1 < r < R_2$) can be determined, using Gauss' law, as $E = Q/4\pi\varepsilon_0 r^2$. Therefore

$$u = \frac{1}{2}\varepsilon_0 E^2 = \frac{1}{2}\varepsilon_0 \frac{Q^2}{(4\pi\varepsilon_0)^2 r^4}$$

and hence the total energy is

$$U = \int u\, dv = \int \frac{1}{2}\varepsilon_0 \frac{Q^2}{(4\pi\varepsilon_0)^2}\cdot\frac{dv}{r^4}$$

Substituting $dv = 4\pi r^2\, dr$ and integrating, we obtain

$$U = \int_{R_1}^{R_2} \frac{1}{2}\frac{Q^2}{4\pi\varepsilon_0}\cdot\frac{dr}{r^2} = \frac{Q^2}{4\pi\varepsilon_0}\cdot\frac{1}{2}\left(\frac{1}{R_1} - \frac{1}{R_2}\right)$$

Using the second form of Eq. (6.62), we find that

$$\frac{1}{C} = \frac{1}{4\pi\varepsilon_0}\left(\frac{1}{R_2} - \frac{1}{R_1}\right)$$

which is exactly the result arrived at in Example 6.6.

6.6 Self-Energies and Interaction Energies

By the self-energy $U^{(S)}$ of a charge distribution, one means the work required to assemble this charge distribution when it is outside the influence of any other charges external to the charge distribution. Thus, for a charge distribution characterized by the charge density ρ, $U^{(S)} = \frac{1}{2}\int \rho\Phi\, dv$, where Φ is the potential generated by the charge distribution itself.

In contrast to this situation, one is sometimes interested in knowing how much work is required to place one distribution of charges in the field of another (distinct and disjoint) set of charges. Directly from the definition of electric potential, the work required is given by

$$U^{(\text{int})} = \int \Phi^{(e)}\, dq$$

where $\Phi^{(e)}$ denotes the potential function characteristic of the "external" charge distribution, and is assumed to be completely independent of the dq distribution by which it is multiplied. Note that the factor of $\frac{1}{2}$ is missing from this letter expression.

These results may be seen to emerge formally if we write the total electrostatic energy to two charge distribution as

$$U = \frac{1}{2}\varepsilon_0 \int E^2\, dv = \frac{1}{2}\varepsilon_0 \int [\mathbf{E}_1 + \mathbf{E}_2]^2\, dv \tag{6.63}$$

where $\mathbf{E}_1$ and $\mathbf{E}_2$ denote the electric fields of those charge distributions separately. Expanding the integrand of Eq. (6.63) we find

$$U = \frac{1}{2}\varepsilon_0 \int E_1^2\, dv + \frac{1}{2}\varepsilon_0 \int E_2^2\, dv + \varepsilon_0 \int \mathbf{E}_1 \cdot \mathbf{E}_2\, dv$$

It is now apparent that the first two terms on the right-hand side of this equation represent the self-energies of the two distributions, whereas the third term is evidently the energy of interaction of these charge distributions, $U^{(\text{int})}$.

The latter statement is proved directly as follows: Substituting $-\nabla\Phi_1$ for $\mathbf{E}_1$ in the third term, we get

$$U^{(\text{int})} = \varepsilon_0 \int \mathbf{E}_1 \cdot \mathbf{E}_2\, dv = -\varepsilon_0 \int (\nabla\Phi_1 \cdot \mathbf{E}_2) dv$$

Using Eq. (1.57) yields

$$U^{(\text{int})} = -\varepsilon_0 \int \nabla \cdot (\Phi_1 \mathbf{E}_2) dv + \varepsilon_0 \int \Phi_1 (\nabla \cdot \mathbf{E}_2) dv$$

The first integral is zero if neither charge distribution is infinite in extent. (Why?) Thus

$$U^{(\text{int})} = \int \rho_2 \Phi_1\, dv = \int \Phi_1\, dq$$

We may associate $\rho_2\, dv$ with dq and Φ_1 with $\Phi^{(e)}$, and thus the third term of Eq. (6.63) is indeed the interaction energy. Also one can easily show that

$$U^{(\text{int})} = \varepsilon_0 \int \mathbf{E}_1 \cdot \mathbf{E}_2\, dv = \int \rho_1 \Phi_2\, dv$$

so that

$$U^{(\mathrm{int})} = \int \Phi_1 \, dq_2 = \int \Phi_2 \, dq_1 \tag{6.64}$$

The interaction energy between two charge systems generally may also take the form:

$$U^{(\mathrm{int})} = \frac{1}{2}\left(\int \Phi_1 \, dq_2 + \int \Phi_2 \, dq_1\right) \tag{6.65}$$

The latter expression appears automatically in the expansion of Eq. (6.4), $U = \frac{1}{2}\int \Phi \, dq$, where $\Phi = \Phi_1 + \Phi_2$ and $dq = dq_1 + dq_2$. Thus

$$U = \frac{1}{2}\int \Phi_1 \, dq_1 + \frac{1}{2}\int \Phi_2 \, dq_2 + \frac{1}{2}\left[\int \Phi_1 \, dq_2 + \int \Phi_2 \, dq_1\right]$$

or

$$U \equiv U^{(S_1)} + U^{(S_2)} + U^{(\mathrm{int})}$$

Note that for two point charges, the self-energies are indeterminate (or infinite). The interaction energy is just

$$q_2\Phi_1 = q_2 \cdot \frac{q_1}{4\pi\varepsilon_0 r_{12}} = q_1\Phi_2$$

Often, one is interested in the force experienced by a charge distribution placed in some fixed external field $\mathbf{E}^{(e)}$—that is, the interaction force between the charge distribution and the field $\mathbf{E}^{(e)}$. Clearly, a small displacement of each part of the charge distribution by the same amount will change the interaction energy of the whole system only. (By "system" is meant the sources of $\mathbf{E}^{(e)}$ as well as the distribution acted on by $\mathbf{E}^{(e)}$.) Therefore, in calculating the forces we can use $U^{(\mathrm{int})}$ rather than the total U (see the following section).

6.7 Forces and Torques Using the Electrostatic Energy

We have observed that we must do work to assemble a charge distribution against the (conservative) electrostatic forces. Because the forces are conservative, we have been able to define a potential energy for the charge distribution that depends only upon the particular charge configuration we construct. It does not, for example, depend on the particular paths by which we bring up the charge elements from infinity.

Thus an isolated system of charges possesses an electrostatic energy that can change only if the charge configuration itself is changed—as, for example, by changing the location of the charges or by adding charge. We consider only static situations where the charges are stationary, so there is no heating or cooling of the charge distribution. Therefore, if mechanical work is done on the charge distribution by some external agent, the work done, by definition, must increase the energy of the charge distribution; that is,

$$dW = dU \tag{6.66}$$

If we imagine this work, dW, to be done mechanically in such a way that the external applied forces are always in *equilibrium* with the electrostatic forces during the work process, then dW is just the negative of the work done by these electrostatic forces, which we label $dW^{(\text{mech})}$. In such a case, if the system of charges has no interaction with the outside (i.e., it is isolated) and the other charge of the system remains constant, we have

$$dW^{(\text{mech})} = -dU \qquad Q = \text{constant} \tag{6.67}$$

Suppose, for example, that $dW^{(\text{mech})} = F_\xi \, d\xi$; that is, suppose we have a system of conductors, one of which is moved through a slight displacement $d\boldsymbol{\xi}$ under the influence of the electrostatic force $\mathbf{F}$. Since $\mathbf{F} \cdot d\boldsymbol{\xi} = F_\xi \, d\xi$, then $F_\xi \, d\xi = -dU$, where $Q = constant$ on all conductors and dU is the change in electrostatic energy as the conductor is moved through $d\boldsymbol{\xi}$. We therefore have

$$F_\xi = -\left.\frac{dU}{d\xi}\right|_Q \tag{6.68}$$

giving the force F_ξ acting on the conductor in the $\hat{\boldsymbol{\xi}}$ direction. The vertical bar with its subscript Q is inserted to emphasize the constancy of Q on all conductors in taking this derivative.

In some other applications Q will not be constant, and one should know what to do in this case. Suppose, e.g., that the *potentials* of all conductors are kept constant as the configuration of the system is changed. These potentials can in general only be maintained constant if the charges on the conductors are changed. But the charges can be changed only if some agent external to the conductors themselves provides the charge. Batteries will do the job, for example. In this case, then, if one of the conductors is permitted a virtual displacement and an increment of mechanical work $dW^{(\text{mech})}$ is done by the electrostatic forces in the process, additional work must be done to maintain all the conductors at a constant potential. Charge is supplied at the potential of the various conductors. It is now the sum of these effects that must be equal to the change in electrostatic energy. We have (keeping all potential constant)

$$-dW^{(\text{mech})} + dW^{(B)} = +dU \tag{6.69}$$

In words, the *total* work done by *external* agents $[-dW^{(\text{mech})} + dW^{(B)}]$ equals the increase in electrostatic energy. If the change in the charge on conductor m is called dQ_m, then

$$dW^{(B)} = \sum_m \Phi_m \, dQ_m$$

But $U = \frac{1}{2}\Sigma\Phi_m Q_m$, and so $dU = \frac{1}{2}\Sigma\Phi_m \, dQ_m$ at constant potential. Therefore, the energy balance equation above may be written

$$-dW^{(\text{mech})} = -dW^{(B)} + dU = -\sum \Phi_m \, dQ_m + \frac{1}{2}\sum \Phi_m \, dQ_m = -\frac{1}{2}\sum \Phi_m \, dQ_m$$

or

$$dW^{(\text{mech})} = dU|_\Phi \tag{6.70}$$

The work done by the batteries in keeping all the conductors at a constant potential is equal to twice the change in the electrostatic energy. The result of Eq. (6.70) leads to the expression:

$$F_\xi = \left.\frac{dU}{d\xi}\right|_\Phi \tag{6.71}$$

for the force on a conductor in the "ξ direction," when all potentials are fixed.

Finally, it should be noted that in our discussion of forces acting on conductors (via energy arguments), the expression for mechanical work $-F_\xi\, d\xi$ does not necessarily imply (force)·(displacement). It can also represent (torque)·(rotation). Consequently, in the formulas derived,

$$F_\xi = -\left.\frac{\partial U}{\partial \xi}\right|_Q \quad \text{(a)}$$
$$F_\xi = +\left.\frac{\partial U}{\partial \xi}\right|_\Phi \quad \text{(b)} \qquad (6.72)$$

if ξ is a displacement (or rotation), then F_ξ represents a force (or torque) in the direction of ξ.

Example 6.11 Force on Plates of Parallel Capacitor—Constant Charge

A simple example of the preceding discussion is afforded by the parallel-plate capacitor. Let us find the force acting on the top plate of the isolated (Q is constant) capacitor of Fig. 6.9*a*. We have seen that the electrostatic energy is given by $U = \frac{1}{2}Q^2/C$. The only geometrical quantity in these equations is the capacitance C, where $C = \varepsilon_0 A/z$. In using the formula for force, we utilize the fact that Q is constant but the potential difference is not, implying that the energy expression $Q^2/2C$ is the appropriate one; that is,

$$F_z = -\left.\frac{dU}{dz}\right|_Q = -\frac{1}{2}Q^2\frac{d}{dz}\left(\frac{1}{C}\right) = -\frac{1}{2}Q^2\frac{1}{\varepsilon_0 A} \qquad (6.73)$$

It is constructive to obtain this result without using the formula for F_z directly. For this purpose, consider a small (virtual) displacement dz of the upper plate of the capacitor while keeping Q constant (the system isolated). The electrostatic forces acting on the plate do an amount of work $F_z\, dz$ in this displacement. Therefore the electrostatic energy must decrease. It decreases by the amount $\frac{1}{2}Q^2/C_f - \frac{1}{2}Q^2/C_i$, where C_f and C_i are the final and initial capacitances. Since $-dW^{(\text{mech})} = +dU$, then

$$-F_z\, dz = U(z+dz) - U(z) = \frac{1}{2}Q^2\left(\frac{1}{C(z+dz)} - \frac{1}{C(z)}\right)$$

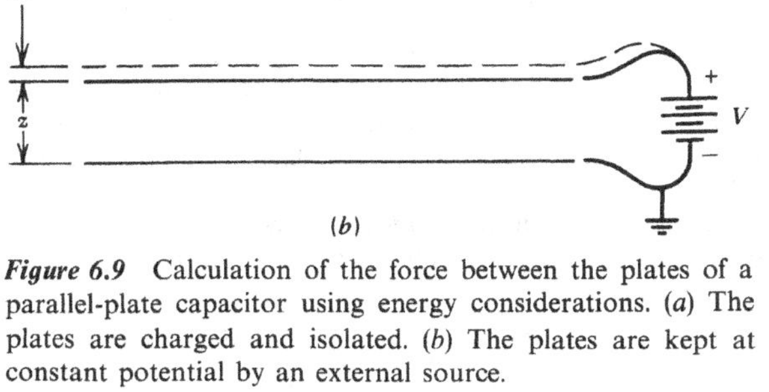

Figure 6.9 Calculation of the force between the plates of a parallel-plate capacitor using energy considerations. (*a*) The plates are charged and isolated. (*b*) The plates are kept at constant potential by an external source.

Substituting $C = \varepsilon_0 A/z$ gives

$$-F_z\,dz = \frac{1}{2}Q^2\left(\frac{z + dz}{\varepsilon_0 A} - \frac{z}{\varepsilon_0 A}\right) = \frac{1}{2}\frac{Q^2}{\varepsilon_0 A}dz$$

Therefore,

$$F_z = -\frac{1}{2}\frac{Q^2}{\varepsilon_0 A}$$

which is exactly the result of Eq. (6.73).

Example 6.12 Force on Plate of Parallel-Plate Capacitor—Constant Voltage

A simple illustration of the arguments presented above again comes from the parallel-plate capacitor. In Fig. 6.9*b*, since the potential is kept constant, we may express $U = \frac{1}{2}CV^2$. Therefore,

$$F_z = \left.\frac{dU}{dz}\right|_V = \frac{1}{2}V^2\frac{dC}{dz} = \frac{1}{2}V^2\left(\frac{-\varepsilon_0 A}{z^2}\right)$$

or

$$F_z = -\frac{1}{2}\frac{V^2C^2}{\varepsilon_0 A} = -\frac{1}{2}\frac{Q^2}{\varepsilon_0 A}$$

which shows that the force is the same as in the case of constant Q.

Again let us do this problem from first principles.

1. The mechanical work done (by external agents) in displacing the top plate by dz is $-F_z\,dz$, the minus sign arising from the fact that the *net* force during the displacement is zero. F_z is the electrostatic force.

2. The charge on the plates of the capacitor was originally $Q_i = VC_i$. After the displacement, the charge is $Q_f = VC_f$, so a charge of magnitude $Q_f - Q_i = V(C_f - C_i) \equiv V\,dC$ was supplied during the displacement. This charge was supplied to the upper plate at potential V, so an amount of work was done on the system equal to $(Q_f - Q_i)V = V^2\,dC$.

3. The sum of the work done in steps 1 and 2 above equals dU; that is,

$$dU = \frac{1}{2}V^2\,dC = -F_z\,dz + V^2\,dC$$

or

$$F_z = +\frac{1}{2}V^2\frac{dC}{dz}$$

as derived above.

Example 6.13 Force Exerted by a Capacitor on a Dielectric Slab

In this example we analyze the force acting on dielectric materials by charged conductors using the energy method. Consider Fig. 6.10, which shows a parallel-plate capacitor with plate separation d and dimensions a and l. A dielectric slab of permittivity ε, thickness d, and dimensions a and l is partially inserted between the plates as shown. The plates are kept at a potential difference $\Delta\Phi$.

To calculate the force exerted on the slab we find first the capacitance of the system as a function of x. For a given x, the system can be viewed as two capacitors connected in parallel. Using Eq. (6.33) one writes

$$C_1 = \frac{\varepsilon a x}{d} \qquad \text{and} \qquad C_2 = \frac{\varepsilon_0 a(l - x)}{d}$$

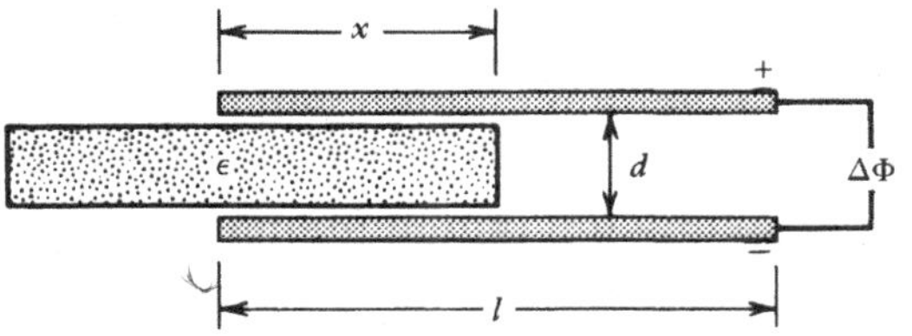

Figure 6.10 Calculation of the force on a dielectric slab partially inserted between the plates of a capacitor.

where C_1 is the capacitance of the part of the capacitor filled with the dielectric and C_2 is the capacitance of the rest of it. The total capacitance is then

$$C = C_1 + C_2 = \frac{a}{d}[(\varepsilon - \varepsilon_0)x + \varepsilon_0 l]$$

The electrostatic energy of the system is then written, using Eq. (6.44).

$$U(x) = \frac{1}{2}CV^2 = \frac{1}{2}\frac{a}{d}[(\varepsilon - \varepsilon_0)x + \varepsilon_0 l](\Delta\Phi)^2$$

The dependence of U on x can now be utilized to determine the force experienced by the slab. Using Eq. (6.72) we find

$$\mathbf{F} = \left.\frac{\partial U(x)}{\partial x}\right|_{\Delta\Phi} \hat{\mathbf{x}} = \frac{1}{2}\frac{a}{d}(\Delta\Phi)^2(\varepsilon - \varepsilon_0)\hat{\mathbf{x}}$$

indicating that the force is attractive.

6.8 Summary

To assemble a system of N point charges from a condition where all the charges are not interacting, work must be done by an external source against the Coulomb forces between them. This work will be stored in the system as electrostatic energy of the system U:

$$U = \sum_{j=1}^{N}\sum_{i<j}^{N} \frac{q_i q_j}{4\pi\varepsilon_0 r_{ij}} = \frac{1}{2}\sum_{j=1}^{N} q_j \Phi(j) \tag{6.1),(6.3}$$

where r_{ij} is the distance between charges q_i and q_j, and $\Phi(j)$ is the potential at the location of q_j due to the other charges. For a continuous charge distribution ρ embedded in a linear dielectric material, the potential energy becomes

$$U = \frac{1}{2}\int \rho(\mathbf{r})\Phi(\mathbf{r})dv \tag{6.5}$$

When the charges reside on conductors whose surfaces are equipotentials, this energy becomes

$$U = \frac{1}{2}\sum_{j}^{N} Q_j \Phi_j \tag{6.8}$$

The potential of each conductor is expressible in terms of the charges on all the conductors by the linear expansion

$$\Phi_j = \sum_{k=1}^{N} P_{jk} Q_k \tag{6.9}$$

where P_{jk} are geometrical parameters independent of the charges and the potential and are called the coefficients of potential. The inverse relation

$$Q_j = \sum_{k=1}^{N} C_{jk}\Phi_k \tag{6.10}$$

is also a linear relation with C_{jk} called the coefficients of induction or capacitance. The coefficients have the following properties

$$P_{ij} = P_{ji} \qquad P_{ii} \geq P_{ij} \qquad C_{ij} = C_{ji} \qquad C_{ii} > 0 \tag{6.16}$$

$$C_{ij} \leq 0 \qquad \text{for } i \neq j \tag{6.21}$$

For a two-conductor capacitor with charges $\pm Q$ on the conductors we have

$$Q = C\Phi_{12} \qquad \text{and} \qquad U = \frac{1}{2}Q\Phi_{12} \tag{6.31), (6.44}$$

where C is the capacitance of the system and Φ_{12} is the potential difference between the conductors. For example the capacitance of a parallel-plate capacitor of area A, separation d, and filling material of permittivity ε is

$$C = \frac{\varepsilon A}{d} \tag{6.33}$$

The electrostatic energy of a charge distribution can alternatively be written in terms of the electric field produced by the distribution. This is convenient since it allows the introduction of an energy density; however, this energy density exists over all space since the electric field is a long-range field. In terms of E, the energy density u for a medium of permittivity ε is

$$u = \frac{1}{2}\varepsilon E^2 = \frac{1}{2}\mathbf{E}\cdot\mathbf{D} = \frac{1}{2}\frac{D^2}{\varepsilon} \tag{6.61}$$

The total energy is the integral of u over all space:

$$U = \int u\, dv \tag{6.58}$$

Forces between the charge elements of a charge distribution can be conveniently determined from the knowledge of the electrostatic energy. There are two cases to consider: The system is isolated, with constant charge on each conductor; and the system is not isolated, but instead the potential on each conductor is kept constant by external sources such as batteries. In the isolated case the force F on an element is the negative gradient of the electrostatic energy, or

$$F_\xi = -\frac{\partial U}{\partial \xi} \qquad \text{constant charge} \tag{6.68}$$

In the case of constant potential we have

$$F_\xi = +\frac{\partial U}{\partial \xi} \qquad \text{constant potential} \tag{6.71}$$

Problems

6.1 Assuming that the electric charge Q of an atomic nucleus is uniformly distributed inside a sphere of radius R, determine the electrostatic energy using $\frac{1}{2}\int \rho\Phi\, dv$.

6.2 A volume charge distribution is distributed throughout space in such a way that the electrostatic potential at a distance r from the origin is given by $\Phi(r) = Ae^{-\alpha r^3}$, where A and α are constants. (a) Find the density of the corresponding charge distribution. (b) Calculate the electrostatic energy $\frac{1}{2}\int \rho\Phi\, dv$ of the distribution.

6.3 Three identical spheres of radius a are placed at the corners of an equilateral triangle with side $l \gg a$. Each sphere carries a charge q. One of the spheres is now grounded until equilibrium is reached. The same procedure is repeated for the other two spheres. Determine the charge on each sphere at the end of the process.

6.4 Consider two equal, fixed, insulated conducting spheres S_1 and S_2. Initially S_1 has a charge q_1 and potential V while S_2 is uncharged. It is found that the spheres attract each other with a force F. Then S_2 is raised to potential V by placing a charge q_2 on it. It is then found that the spheres repel each other with the force F. Sphere S_1 is now grounded. (a) Find the charge induced on it, and (b) show that the spheres attract each other with a force $q_2(2q_1^2 - q_2^2)F/q_1^3$.

6.5 Four small identical spheres of radius a are placed at the corners of a square of side l, where $l \gg a$. Sphere 1 carries a charge q. Sphere 1 is then connected, using a thin wire, in turn to spheres 2, 3, and 4 until equilibrium is reached in each of the operations. (a) What are the coefficients P_{ii}, where $i = 1, 2, 3$, and 4? (b) Find the coefficients P_{ij} $(i \neq j)$. (c) Find the charge on each of the spheres at the end of the operation.

6.6 (a) A spherical capacitor consists of two concentric, spherical shells of radii a and b, with $b > a$. Find its capacitance. (b) If the radii of the capacitor differ by a small amount d, where $d \ll a$, show that the expression for the capacitance reduces to that for a parallel-plate capacitor having the same surface area.

6.7 A conducting sphere of radius a is surrounded by an isolated, thick, spherical conducting shell of inner radius b and outer radius c. The thick, outer shell is isolated and initially uncharged. A charge $+Q$ is placed at the inner sphere. (a) Determine the electric field in all regions. (b) What is the potential difference between $r = \infty$ and $r = a$? (c) What is the capacitance of the system? (d) Show that the calculated capacitance is equivalent to two capacitors connected in series. What is the capacitance of each capacitor?

6.8 A parallel-plate capacitor has plates of area A, separated by a distance d. A potential difference of V is applied between the plates, after which they are isolated. (a) What is the energy stored in the capacitor? (b) An uncharged sheet of metal of thickness a is placed between the plates and parallel to them. Find its new capacitance. How much work is done by electric forces during the insertion of the metal sheet? (c) What is the potential difference between the capacitor plates after the sheet has been inserted?

6.9 A spherical capacitor has its spheres assembled in a nonconcentric way. The departure from concentricity is very small. Determine the correction to the capacitance that is due to this departure from concentricity (see Example 3.5).

6.10 A metal sphere of radius R in an infinite dielectric medium of permittivity ε bears a charge Q. (a) Determine the work that had to be done to charge it using the definition of work in terms of the charge and potential. (b) Use the fields $\mathbf{E}$ and $\mathbf{D}$ produced by the sphere to calculate the energy stored in the electric field. How does this result compare with (a)? (c) If the sphere expands to radius R', what would be the change in the electric energy? (d) Account for the energy change in (c).

6.11 Use the concept of coefficient of potential to solve the following problems. (a) Two spherical conductors are located in vacuum, with a distance d between them. One of the spheres is of radius R and is grounded. The other has a very small radius and bears a charge q (can be looked at as a point charge). What is the charge induced on the large sphere? (b) In part (a) the sphere is neutral and insulated. What is its potential? (c) Compare these results with the results of the method of images.

6.12 Repeat Problem 6.1 using $U = \frac{1}{2}\int \varepsilon_0 E^2\,dv$ and compare results.

6.13 Show that $\frac{1}{2}\varepsilon_0 \int E^2\,dv$ gives the same result as part (b) of Problem 6.2, where E is the electric field of the distribution.

6.14 Two identical spherical capacitors with inner and outer radii r_1 and r_2 are insulated and placed such that the distance between them is very large. Charges q_1 and q_2 are placed on the inner spheres. Determine the change in the energy of the system when the outer spheres are joined by a thin wire.

6.15 Consider a large wedge capacitor defined by the plates $\phi = 0$ and $\phi = \pi/6$, which are insulated and kept at zero and $-V_0$ volts, respectively. Given that the voltage in the capacitor is given by $\Phi = -(6\phi/\pi)V_0$ (see Example 3.3). (a) Calculate the energy density between the plates. (b) Calculate the energy stored between the plates for $0.1 \text{ m} \leq \rho \leq 0.6 \text{ m}$ and $0 \leq z \leq 1$ m and $V_0 = -10$ volts.

6.16 Two point charges q_1 and q_2 are separated by a distance d. (a) Calculate the energy stored in this system using Eq. (6.2). (b) Repeat using Eq. (6.57). Explain the difference in sign between the results of (a) and (b).

6.17 Calculate the energy stored in the volume bounded by $0 \leq x, y, z \leq 1$ m due to the potential $V = 3x^2 + 4y^2$.

6.18 The potential due to a spherical conducting shell of radius R with center at the origin is $V = V_0$ for $r \leq R$ and $V_0 R/r$ for $r \geq R$. (a) Determine the stored electric energy in this system using Eq. (6.8). (b) Repeat using Eq. (6.57) and compare.

6.19 Consider the two large metallic plates forming a wedge capacitor of Problem 3.2 in Fig. 3.31. (a) Determine the potential inside the capacitor. (b) What is the charge density and the total charge residing on the plates? (c) Calculate the capacitance of the system. (d) Determine the torque between the plates.

6.20 Determine the total force per unit area acting on the dielectric slab of Problem 4.6.

6.21 An insulating rod of length l and negligible polarizability has two small conducting spheres of radius $a \ll l$ attached to its two ends. The rod is attached at its center such that it can rotate freely about its center. It is placed in a uniform external electric field E_0. (a) What are the equilibrium orientations of rod? Which of them are stable, neutral, or unstable? (b) Calculate the work needed to align the rod with field starting from a position normal to it?

6.22 An electric dipole of moment p is located a distance d from an infinite conducting plane. It is inclined at an angle θ with the normal to the plane. Referring to the results of Example 3.13, determine the work necessary to remove the dipole to an infinite distance above the conducting plane.

6.23 (a) How much work is necessary to move a charge q from infinity to a distance r from the center of a conducting sphere (radius b) grounded by a resistanceless wire? (b) Will a current flow in the wire as a result of this operation? (c) If the sphere were isolated from ground and had a charge $+Q$ on it, what would have been the work necessary to bring the charge to its above location? (d) Compare the results in (a) and (c) and explain the differences, if any.

6.24 Determine the energy density stored in the uniformly polarized cylinder of Problem 4.19.

SEVEN

STEADY CURRENTS

So far in our study of electricity we have dealt with situations that were completely static; that is, all charge densities considered were independent of time. We have assumed that what we have studied applies to the real world in situations where macroscopic charge densities are constant in time. An important implication here, which we have not yet discussed, is that macroscopic time-independent charge distribution can arise from time-dependent microscopic charge distributions. (Electrons tend to dance around.) Evidently, if we use coarse enough measuring instruments, nature performs a time and space average in such a way that the averaged, microscopic, time-dependent charge distributions appear truly static from a macroscopic point of view; moreover, the time and space averages of the microscopic electromagnetic fields (produced by the microscopic charge elements) appear identical with those that would be produced from the macroscopic, static charge distributions. The static macroscopic fields are related to the macroscopic densities via Eqs. (2.33) and (2.37); that is

$$\nabla \times \mathbf{E} = 0 \qquad \text{and} \qquad \nabla \cdot \mathbf{E} = \frac{\rho}{\varepsilon_0}$$

If the microscopic charge distributions discussed are time-independent,* these classical equations also describe microscopic fields.

In this chapter we will briefly discuss a situation where although charges are in motion, the condition is static insofar as calculations of electric fields are concerned. The chief ingredient here is that macroscopically all charge densities are constant in time, and any currents that exist also remain constant in time. In such a case, the equations of electrostatics remain valid.

* A detailed discussion of when we may consider the charge distribution to be microscopically time-dependent from the point of view of the equations above requires the consideration of quantum mechanics. If the charges are in "stationary states" quantum mechanically, then charge distributions may be considered static.

7.1 Definition of Electric Current

A conducting medium is one in which charges are free to move. A conducting medium is also one that contains a great number of mobile charges. Different species of mobile charges may coexist in a given medium (electrons, "holes," positive ions, and so on). Focus on a particular species in the medium that carries a charge q. The motion of members of this species are assumed to have an average velocity $\langle \mathbf{v} \rangle$. Thus, if there are n members of this species per unit volume, and N in a small volume, then

$$\langle \mathbf{v} \rangle \equiv \frac{1}{N} \sum_{j=1}^{N} \mathbf{v}_j$$

where $\mathbf{v}_j$ is the velocity of a particular (j) member of the species. Thus $\langle \mathbf{v} \rangle$ represents the *drift velocity* of the species, and is assumed to be a macroscopically smooth function of position in the medium. The number of charges of this species that in a time dt cross an area $d\mathbf{a}$ is given by

$$dI = \frac{dQ}{dt} = qn\langle \mathbf{v} \rangle \cdot d\mathbf{a} \tag{7.1}$$

Since $\rho = nq$ is the amount of charge due to this species per unit volume—that is, the charge density associated with this species—then

$$dI \equiv \rho \langle \mathbf{v} \rangle \cdot d\mathbf{a} \tag{7.2}$$

If now there exist a number of species $\{i = 1, 2, \ldots\}$, there will exist a current through $d\mathbf{a}$ due to each one, the total current being the sum over Eq. (7.2); that is,

$$dI = \left[\sum_{\substack{\text{all} \\ \text{species}}} \rho_i \langle \mathbf{v}_i \rangle \right] \cdot d\mathbf{a} \tag{7.3}$$

The term in square brackets of Eq. (7.3) now conveniently defines the *current density* $\mathbf{J}$:

$$\mathbf{J} \equiv \sum_{\substack{\text{species} \\ i}} \rho_i \langle \mathbf{v}_i \rangle \tag{7.4}$$

so that

$$dI = \mathbf{J} \cdot d\mathbf{a} \tag{7.5}$$

Clearly, if only one species of charge carrier exists in a conducting medium, then

$$\mathbf{J} = \rho \langle \mathbf{v} \rangle = nq \langle \mathbf{v} \rangle \tag{7.6}$$

The utility of the idea of a current density is analogous to that of the charge density. It is a (vector point) function that is assumed to vary smoothly (on a macroscopic scale) in a conducting medium. In terms of it, the current through any surface is given by

$$I = \int_S \mathbf{J} \cdot d\mathbf{a} \tag{7.7}$$

The units of $\mathbf{J}$ in the SI system of units are coulombs per [(meter)^2·second] $\equiv$ amperes/meter^2; that is, $\mathrm{C \cdot m^{-2} \cdot s^{-1} \equiv A \cdot m^{-2}}$.

In discussing electric current, one often distinguishes among different types: *convection current*, *conduction current*, *polarization current*, *displacement current*, and so forth. (In the next chapter we shall also encounter the "magnetization current".) A convection current is one wherein a material appears to move "en masse," containing and carrying along with it whatever net charge is associated with it. It depends directly upon the motion of the observer, for if the observer were to move along with the moving charged material, the current density would appear to be zero. Thus an insulating belt onto which is "sprayed" a positive charge, when moving, would give rise to a convection current. A moving charged mass of gaseous ions, as in a particle accelerator, would constitute a convection current. In a convection current, there thus appears to be movement of mass relative to an observer at rest whose associated charge produces a current.

A conduction current usually denotes movement of charge carriers through a neutral medium, as electrons through a wire, or ions through a solution. Here there may be no apparent mass movement. The crucial distinction is that the conduction current is independent of the motion of the observer because of the relative motion of the positive and negative charges in the medium. Thus, if the drift velocity of electrons is $\mathbf{v}$, an observer moving also with velocity $\mathbf{v}$ will not consider the electron drift as a current, but that observer will detect an opposite drift velocity for the positive charges that he or she will then perceive as a current. Since there are equal densities of positive and negative charges in this case, the current will be independent of the velocity of the observer (for nonrelativistic velocities).

A polarization current is associated with the movement of polarization charge as a medium is being polarized. Steady polarization currents do not exist because the charge movement is bounded. It may be easily seen dimensionally that the current density associated with polarization current is, in amperes,

$$\mathbf{J}_P = \frac{\partial \mathbf{P}}{\partial t} \tag{7.8}$$

where $\mathbf{P}$ is the polarization in coulombs/meter^2 and time t is in seconds. Referring to Eq. 4.16, we see that a polarization charge $\delta\mathbf{P}$ is caused when a charge distribution of density ρ_+ is displaced by $\delta\mathbf{s}$ relative to the negative charge density distribution of ρ_-. Therefore

$$\frac{\delta \mathbf{P}}{\delta t} = \rho_+ \frac{\delta \mathbf{s}}{\delta t} \equiv \rho_+ \langle \mathbf{v} \rangle_P \equiv \mathbf{J}_P$$

Finally, there is the displacement current. It is not really associated with the movement of charge, but it is a construct required in electromagnetic theory in order to maintain a consistent theory. We shall discuss it in greater detail in Chapter 14. The current density associated with the displacement current is

$$\mathbf{J}_E \equiv \varepsilon_0 \frac{\partial \mathbf{E}}{\partial t}.$$

If we deal with static fields, then $\mathbf{J}_E = 0$ because $\partial \mathbf{E}/\partial t = 0$.

Example 7.1 Drift Velocities in Wires

We here acquaint ourselves with a typical magnitude of $\langle \mathbf{v} \rangle$ in a conducting wire. We shall see that generally for currents in metallic wires, it is very small compared to the actual velocities of the electrons in the metal. Consider a wire with a square cross section of area 10^{-6} m^2 (1 mm × 1 mm) carrying a current of one ampere (1 A). If the current density in the wire is uniform and along the wire, then $J \equiv |\mathbf{J}| = 1\ \text{A}/10^{-6}\ \text{m}^2 = 10^6\ \text{A/m}^2$. Now in the wire, the current is typically carried by electrons having a charge $e \approx 1.6 \times 10^{-19}$ C. There is approximately one such electron for each atom of the metal wire, distributed in the wire so that the actual charge density is zero. Consequently, Eq. (7.6) is relevant: $n \approx 10^{22}$ electrons per cubic centimeter $= 10^{28}$ electrons per cubic meter, and $\rho = ne \simeq 10^{28} \times 1.6 \times 10^{-19} \simeq 1.6 \times 10^9\ \text{C/m}^3$. Therefore

$$|\langle \mathbf{v} \rangle| = \frac{J}{\rho} \cong \frac{10^6}{1.6 \times 10^9} \cong 6 \times 10^{-4}\ \text{m/s}$$

This is a rather small speed, compared to the actual electron speed, which may be estimated from the formula

$$\frac{3}{2} kT = \frac{1}{2} m_e v^2$$

where k is Boltzmann's constant, m_e is the electron mass and T is the absolute temperature in kelvins.* We find that, at room temperature, $v \simeq 10^5$ m/s, hence $|\langle \mathbf{v} \rangle| \ll v$. The actual electron velocities are essentially randomly directed in space, so their average is almost zero.

7.2 The Continuity Equation: Local Conservation of Charge

Local conservation of charge means that if a net charge leaves any volume of space V whose closed surface is S, then the charge within the volume must be decreased accordingly. Now $\oint_S \mathbf{J} \cdot d\mathbf{a}$ is the charge per unit time leaving V through surface S, if $d\mathbf{a}$ is taken outward from V. This is a current, dQ'/dt, through S. Consequently, the charge inside V is increasing at a rate $dQ'/dt \equiv -dQ/dt$ (i.e., it is decreasing if $dQ/dt > 0$) if charge conservation is to hold the quantity, where Q represents the total net charge in V at any time, and may be represented as $\int_V \rho\, dv$. We may thus write

$$\oint_S \mathbf{J} \cdot d\mathbf{a} = -\frac{dQ}{dt} = -\frac{d}{dt} \int_V \rho\, dv = + \int -\frac{\partial \rho}{\partial t}\, dv$$

The divergence theorem gives $\oint_S \mathbf{J} \cdot d\mathbf{a} = \int_V (\nabla \cdot \mathbf{J}) dv$, and so

$$\int_V \left(\nabla \cdot \mathbf{J} + \frac{\partial \rho}{\partial t} \right) dv = 0$$

Our argument holds for *all possible volumes*, V. Therefore, the integrand must itself be identically zero; that is,

$$\nabla \cdot \mathbf{J} = -\frac{\partial \rho}{\partial t} \tag{7.9}$$

This is known as the *continuity equation* and is a basic equation of electromagnetism, relating the charge density at a point in space to the current density at that point. It assumes that charge can be neither created nor destroyed.

* Actually, the average electron temperature in a conducting wire turns out to be much higher than room temperature. For example, in copper a Fermi-Dirac theory predicts a temperature of 80,000 K, thus giving average speeds 14 to 15 times what we estimated above. See F. Reif, *Fundamentals of Statistical and Thermal Physics* (New York: McGraw-Hill, 1965), McGraw-Hill.

In the presence of dielectric materials, the current density $\mathbf{J}$ is the sum of the external current density $\mathbf{J}_f$ and the current density produced by the motion of the bound changes $\mathbf{J}_p$. Similarly the charge density ρ is the sum of ρ_f and ρ_p, that is

$$\mathbf{J} = \mathbf{J}_f + \mathbf{J}_p \qquad \rho = \rho_f + \rho_p \tag{7.10}$$

As we noted in Chapter 4, it is customary to call the external sources free sources, hence, we use the subscript f in both $\mathbf{J}_f$ and ρ_f. The continuity equation as given by Eq. (7.9) now becomes

$$\nabla \cdot \mathbf{J}_f + \frac{\partial \rho_f}{\partial t} + \nabla \cdot \mathbf{J}_p + \frac{\partial \rho_p}{\partial t} = 0$$

But from Eq. (7.8) we have $\mathbf{J}_p = \partial \mathbf{P}/\partial t$, and from Eq. (4.13) we have $\rho_p = -\nabla \cdot \mathbf{P}$; thus

$$\nabla \cdot \mathbf{J}_p + \frac{\partial \rho_p}{\partial t} = \nabla \cdot \frac{\partial \mathbf{P}}{\partial t} + \frac{\partial \rho_p}{\partial t}$$

$$= \frac{\partial}{\partial t} \nabla \cdot \mathbf{P} + \frac{\partial \rho_p}{\partial t} = 0.$$

Hence, the continuity relation given by Eq. (7.9) is equivalent to

$$\nabla \cdot \mathbf{J}_f + \frac{\partial \rho_f}{\partial t} = 0 \qquad \text{and} \qquad \nabla \cdot \mathbf{J}_p + \frac{\partial \rho_p}{\partial t} = 0 \tag{7.11}$$

If $\partial \rho/\partial t = 0$ everywhere, all charge densities are fixed in time, and

$$\nabla \cdot \mathbf{J} = 0 \qquad \text{steady current} \tag{7.12}$$

This result is the condition of *steady* currents. Clearly, if $\nabla \cdot \mathbf{J} = 0$, then $\oint_S \mathbf{J} \cdot d\mathbf{a}$ is zero for all possible (closed) surfaces S. This means that the net current entering (or leaving) any closed surface is zero. Charge thereby cannot accumulate anywhere, so the "lines of current density" are continuous.

7.3 Ohm's Law

In many, but not all, conducting media there is a simple *linear* relationship between the potential difference of two boundaries of the medium and the (constant) current flowing between them, usually expressed as Ohm's law:

$$V = RI$$

where R is a constant, of dimensionality volts per ampere $\equiv$ ohms. In "ohmic" (linear) materials, R depends on the composition of the material and the geometrical shape between the "electrodes" (equipotential boundaries). The quantity V is the "voltage drop" between electrodes and the current I flows in the direction of the voltage drop. In a purely electrostatic environment, the voltage drop is simply the drop in potential. Alternatively, one occasionally writes the proportionality expressed above as

$$I = \frac{1}{R} V = GV,$$

where G is called the *conductance*.

Now, it may be seen that if everywhere in a material the current density is proportional to the electric field, then one obtains Ohm's law as stated above. In fact, the relation

$$\mathbf{J} = \sigma_c \mathbf{E} \tag{7.13}$$

is itself often called (the *differential* form of) Ohm's law, where σ_c is called the electrical *conductivity* of the material. Its reciprocal $1/\sigma_c \equiv \rho_c$ is called the *resistivity*. If a conducting medium is "ohmic" [if Eq. (7.13) is satisfied], has a well-defined constant cross-sectional area A through which current flows, and has a well-defined constant length l (like a wire with no kinks) then the current density $\mathbf{J}$ (and $\mathbf{E}$) is constant in the medium (see Example 7.2), and the current flowing is just $I = JA = \sigma_c EA = \sigma_c(V/l)A$. Therefore $V = (l/\sigma_c A)I$, and the resistance is

$$R = \frac{l}{\sigma_c A} \tag{7.14}$$

The electrical conductivity of a substance depends upon its atomic structure and its thermodynamic state. It is a function of the temperature of the substance, as well as its density, purity, and so forth. For certain crystalline materials, the relationship of current density to electric field is not a simple proportionality representable by a single number, σ_c (a scalar), but is rather a tensor relationship, for it depends upon the direction in which $\mathbf{E}$ is applied. Also, especially in gases, σ_c may itself depend upon E, so that the relationship between J and E is not a proportional (linear) one. Such a nonlinear dependence shows up in most substances at high enough values of E.

Despite all of these caveats, under normal conditions most liquids and many solids are characterizable by electrical conductivities (which may be found in handbooks). The difference in the conductivities of different substances may be enormous. A metallic conductor has a conductivity of about (10^5 to 10^8) Ω^{-1} per meter, whereas an insulator—a very poor conductor—has a conductivity that is less by a factor of perhaps of 10^{22}. Conductivities of certain common substances are listed in Table 7.1. So-called *superconductors* have essentially an infinite conductivity, but usually only at temperatures within a few degrees of absolute zero. In metals the conductivity generally decreases with increasing temperature—in fact, the resistivity changes by approximately 0.4 percent per degree Celsius at room temperatures. In *semiconductors*, like germanium and silicon, and in most solid dielectrics the opposite is true; that is, the conductivity increases with increasing temperature. The differences in behavior are related to the way in which charge transport occurs in these substances and is a proper concern of solid-state physics.

It is of interest to know the time required for the charge that might be present inside a conductor to be neutralized and appear at the conductor surface. This characteristic time is called a *relaxation time* [see Eq. (7.17)]. In periods of time commensurate with the relaxation time a nonzero charge density may persist inside a conductor. In periods of time much longer than the relaxation time the charge that may have been present initially will have been neutralized. (We are assuming that all electric fields not emanating from the charge of the conductor itself are constant in time or varying with time characteristics much longer than the relaxation time.) Thus, given a conductor that is placed in an external electrostatic field at time zero, a time of the order of relaxation time will pass before charge appears at the conductor surface so as to neutralize the external field at all points interior to the conductor. In a conductor,

$\mathbf{J} = \sigma_c \mathbf{E}$. Assuming σ_c to be a constant in the conductor medium, then the continuity equation becomes

$$\sigma_c \nabla \cdot \mathbf{E} = -\frac{\partial \rho}{\partial t} \tag{7.15}$$

Table 7.1 **Table of Conductivities at Approximate Room Temperature**

Material	Conductivity[a]	Material	Conductivity[a]
Silver	6.2×10^7	H_2O	2×10^{-4}
Copper	5.8×10^7	Marble	10^{-5}
Pure iron	1.0×10^7	Wood	10^{-9}
Steel	0.2×10^7	Glass	10^{-11}
Mercury	10^6	Oil	10^{-14}
Carbon	10^4	Polyethylene	10^{-15}
Silicon	10^{-2}	Fused quartz	10^{-17}
Alcohol	3×10^{-4}	True vacuum	?

[a] Conductivities are expressed in $(\text{ohm-meters})^{-1}$, or $(\Omega \cdot \text{m})^{-1}$.

Therefore, since $\nabla \cdot \mathbf{E} = \rho/\varepsilon_0$, Eq. (7.15) gives the simple relation

$$\frac{\sigma_c}{\varepsilon_0} \rho = -\frac{\partial \rho}{\partial t} \tag{7.16}$$

This first-order linear differential equation has the solution

$$\rho(t) = \rho(0) e^{-t/(\varepsilon_0/\sigma_c)} \tag{7.17}$$

where ε_0/σ_c has the dimensions of time, and is called the *charge relaxation time*. In such a time, the charge density decays to e^{-1} of its original value. Even for a relatively poor metallic conductor like iron, $\varepsilon_0/\sigma_c \simeq 10^{-18}$ second. (For water, $\varepsilon_0/\sigma_c \approx 10^{-8}$ second; for glass, $\varepsilon_0/\sigma_c \approx 2$ seconds.) Therefore, for metallic conductors, unless charge is continuously injected into the conductor, the charge density is zero even when the most rapidly changing electric fields (frequencies of the order of 10^{12} Hz) exist around the conductor.

7.4 Steady Currents

7.4.1 Equations Governing J

The current vector density $\mathbf{J}$ constitutes a vector field that may vary (macroscopically) from point to point in a material. In a superconductor, $\mathbf{J}$ can be finite only if $\mathbf{E} = 0$ there. In a perfect insulator, $\mathbf{J}$ is, by definition, zero. For the present discussion, we shall assume that $\mathbf{J} = \sigma_c \mathbf{E}$ in a given medium. $\mathbf{J}$ being a vector point function, it will be completely and uniquely determined if its divergence and curl are specified within a region and if appropriate boundary conditions at the surfaces of that region are specified.

For electrostatic fields, whether or not steady currents flow, Eqs. (2.33) and (2.37) hold; that is, $\nabla \cdot \mathbf{E} = \rho/\varepsilon_0$ and $\nabla \times \mathbf{E} = 0$. The electric field is still conservative, Gauss's law is still valid, and we can attempt to find $\mathbf{E}$ just as we did formerly. Having found $\mathbf{E}$, it is a simple matter to find $\mathbf{J}$ using Ohm's law.

We will consider only ohmic materials. The equations of steady currents inside such media can now be derived. In fact the equations for **E** may be transcribed into equation for **J**. Substituting $\mathbf{E} = \mathbf{J}/\sigma_c$ in the curl and divergence equations of **E** gives

$$\nabla \times \mathbf{E} = 0 \quad \rightarrow \quad \nabla \times \frac{\mathbf{J}}{\sigma_c} = 0 \tag{7.18}$$

$$\nabla \cdot \mathbf{E} = \frac{\rho}{\varepsilon_0} \quad \rightarrow \quad \nabla \cdot \frac{\mathbf{J}}{\sigma_c} = \frac{\rho}{\varepsilon_0} \tag{7.19}$$

The integral equations corresponding to Eqs. (7.18) and (7.19) are

$$\oint \mathbf{E} \cdot d\mathbf{r} = 0 \quad \rightarrow \quad \oint \frac{\mathbf{J}}{\sigma_c} \cdot d\mathbf{r} = 0 \tag{7.20}$$

and

$$\oint \mathbf{E} \cdot d\mathbf{a} = \frac{Q}{\varepsilon_0} \quad \rightarrow \quad \oint_S \frac{\mathbf{J}}{\sigma_c} \cdot d\mathbf{a} = \frac{Q}{\varepsilon_0} \tag{7.21}$$

where Q is the net charge inside a volume V whose (closed) surface is S. If σ_c is constant,

$$\nabla \cdot \frac{\mathbf{J}}{\sigma_c} = \frac{1}{\sigma_c} (\nabla \cdot \mathbf{J}) = 0$$

implying that $\rho = 0$ inside the medium. This is consistent with our discussion of relaxation times. [If σ_c is itself a function of position, the continuity requirement, $\nabla \cdot \mathbf{J} = 0$, means that $\nabla \cdot (\mathbf{J}/\sigma_c) = \mathbf{J} \cdot \nabla(1/\sigma_c) = \rho/\varepsilon_0$. This indicates that in general the charge density will be nonzero in such a medium.]

In this chapter we will consider only media with constant conductivities—that is, homogeneous materials. For σ_c a constant, Eqs. (7.18) to (7.21) for **J** inside the medium then become

$$\left.\begin{aligned} \nabla \times \mathbf{J} &= 0 \\ \oint \mathbf{J} \cdot d\mathbf{r} &= 0 \end{aligned}\right\} \quad \text{and} \quad \left.\begin{aligned} \nabla \cdot \mathbf{J} &= 0 \\ \oint \mathbf{J} \cdot d\mathbf{a} &= 0 \end{aligned}\right\} \tag{7.22}$$

The surface integral $\oint \mathbf{J} \cdot d\mathbf{a} = 0$ might be called Gauss' law of currents.

7.4.2 The Boundary Conditions

If surfaces or line integrals cross boundaries of different media, then Eqs. (7.20) and (7.21) are especially relevant. Since in the case of steady currents $\nabla \cdot \mathbf{J} = 0$, then $\oint \mathbf{J} \cdot d\mathbf{a} = 0$. Using the same procedure used previously in Chapter 4 for the determination of the boundary conditions satisfied by the displacement vector **D**, we find that

$$\oint \mathbf{J} \cdot d\mathbf{a} = 0 \quad \rightarrow \quad J_{1n} = J_{2n} \tag{7.23}$$

where J_{in} is the normal component of the current density in the ith medium. The boundary condition on the tangential components of **J** follows directly from the condition on the tangential components of the electric field. That is, the condition $\mathbf{E}_{1t} = \mathbf{E}_{2t}$, along with $\mathbf{J} = \sigma_c \mathbf{E}$, implies that

$$\frac{\mathbf{J}_{1t}}{\sigma_{c1}} = \frac{\mathbf{J}_{2t}}{\sigma_{c2}} \tag{7.24}$$

where $\mathbf{J}_{it}$ is the tangential component of the current density in the ith medium. The same result can also be derived from the line integral of $\mathbf{J}$, Eq. (7.20), by using the same procedure used previously in Chapter 4 for the determination of the boundary conditions satisfied by $\mathbf{E}$. That is,

$$\oint \frac{\mathbf{J}}{\sigma_c} \cdot d\mathbf{r} = 0 \quad \rightarrow \quad \frac{\mathbf{J}_{1t}}{\sigma_{c1}} = \frac{\mathbf{J}_{2t}}{\sigma_{c2}} \tag{7.24}$$

Let us examine the boundary condition given by Eq. (7.23) more closely in relation to the dielectric properties of the conducting materials. Since $\mathbf{J} = \sigma_c \mathbf{E}$, then Eq. (7.23) requires that the normal components of the electric field at the interface to be related by

$$\sigma_{c1} E_{1n} = \sigma_{c2} E_{2n} \tag{7.25}$$

If the two media have permittivities ε_1 and ε_2, then Gauss' law in dielectrics requires that the displacement vectors in the media be related by $D_{1n} - D_{2n} = \sigma_f$, or, in terms of the corresponding electric fields,

$$\varepsilon_1 E_{1n} - \varepsilon_2 E_{2n} = \sigma_f \tag{7.26}$$

where σ_f is the free surface charge density at the interface. Equations (7.25) and (7.26) of course have to be consistent if the fields are to be physical. In other words, there is a restriction on the magnitude of σ_f that can be stationed at the interface. Solving for σ_f shows that it depends on the permittivities and conductivities of the media and on the magnitude of the current, according to the relation:

$$\sigma_f = J_n \left(\frac{\varepsilon_2}{\sigma_{c2}} - \frac{\varepsilon_1}{\sigma_{c1}} \right) \tag{7.27}$$

It is apparent from Eq. (7.27) that one should be very careful in dealing with boundary value problems that involve dielectrics with conductive properties (see Example 7.4). The problem should be treated from the point of view of currents, and the correct boundary conditions are those of currents [see Eqs. (7.23 and 7.24)].

Let us now consider some special cases of Eq. (7.27). No free charge exists at the interface if $\varepsilon_2/\varepsilon_1 = \sigma_{c2}/\sigma_{c1}$ and conversely. If this is not the case (as is generally true) a free charge density must exist there. In the case where one medium is a much better conductor than the other ($\sigma_{c2} \ll \sigma_{c1}$), the boundary conditions become

$$J_{2n} = J_{1n} \qquad \text{and} \qquad \mathbf{J}_{2t} = \frac{\sigma_{c2}}{\sigma_{c1}} \mathbf{J}_{1t}$$

so that if $\mathbf{J}_1$ at the interface is finite, $\mathbf{J}_{2t} \to 0$ and $J_2 = J_{2n}$. This implies that the electric field is almost perpendicular to the better conductor outside this conductor and that the normal component of field in medium 2 is much larger than it is in medium 1. An electrode is usually considered to be a conductor whose conductivity is much greater than that of the medium in which it is immersed, so that the electric field is very small inside the electrode and the electric field outside it is perpendicular to the electrode surface.

Example 7.2 Carbon Resistors

Consider the resistor in Fig. 7.1. Between two parallel aluminum plates is sandwiched a block of graphite of thickness t. The surrounding medium is air. The conductivities of aluminum, graphite and air are $\sigma_c \simeq 3.5 \times 10^7$, 10^4 and 10^{-12} $(\Omega \cdot \text{m})^{-1}$ respectively. Since the conductivity of air is so low, essentially no current flows there, so if a potential difference V is applied between the plates, current will flow solely through the graphite. Now we note the following facts.

1. The aluminum plates (electrodes) are essentially equipotential regions, and the conductivity of aluminum is so large compared to that of carbon that **E** is perpendicular to the plates at the surfaces.
2. At the air-graphite surfaces **J** is parallel to the surface because air is essentially a nonconductor, and J_n in air is zero. It follows that **E** also is tangential to these surfaces just inside the medium.

These conditions dictate that the **E** field in the carbon block is everywhere constant and points from the higher equipotential electrode to the lower. The reason is that, with such a field, the boundary conditions 1 and 2 are satisfied, and the equations $\nabla \cdot \mathbf{E} = \rho/\varepsilon_0 = 0$ and $\nabla \times \mathbf{E} = 0$ are everywhere satisfied in the block. Uniqueness of the solution specifies the constant field solution.

It is interesting to note that without the carbon block the field would not be uniform between the electrodes. If the area of the plates were small, edge effects would be severe. In order for the field to become uniform, when the carbon block was inserted the charge flowed to the surfaces of the block initially. Since the charge is confined to the block, a charge density was built up at the surface just adequate to modify the internal electric field and render it constant everywhere.

Similar arguments show that the current density in ordinary wires is uniform and the electric field is constant.* However, some charge does exist on the surface of these wires. Since the potential difference from one end of the wire to the other is ordinarily small, the charge on the wires is correspondingly small.

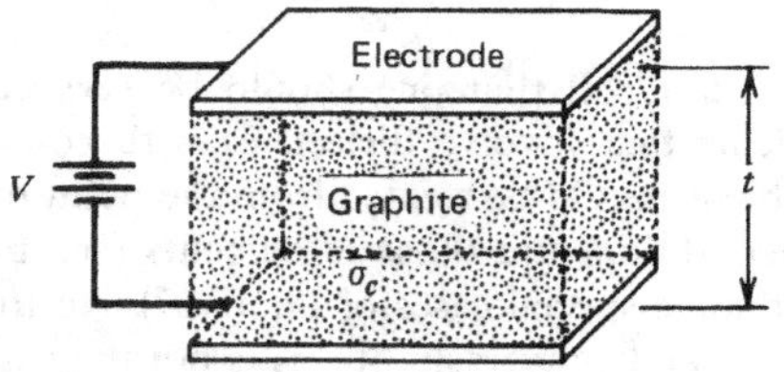

Figure 7.1 Schematic diagram of a carbon resistor.

Example 7.3 Gauss' Law of Currents—Sphere in Conducting Medium

Suppose a steady current I enters a sphere that is surrounded by an effectively infinite medium of conductivity σ_c and permittivity ε_0. The current I enters through a thin wire and, after a steady state is established, leaves through the medium as shown in Fig. 7.2. The conductivity of the sphere is very large compared to σ_c.

* This is a practical result that assumes that the current flow is unaffected by the magnetic fields present in the wire. If the conductivity depended upon the magnetic field, the current density would not be constant.

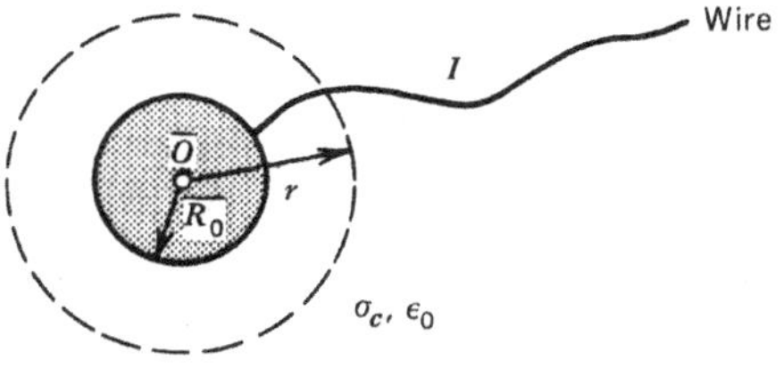

Figure 7.2 Application of Gauss' law to current flow from a highly conducting sphere placed in a conducting medium.

The sphere is here effectively an equipotential region. If we ignore the asymmetry caused by the wire.† we expect that a charge will build up symmetrically on the sphere, giving rise to a spherically symmetric electric field and current density distribution. We have for a spherical surface at $r > R$, $\oint \mathbf{J} \cdot d\mathbf{a} = \int_w \mathbf{J} \cdot d\mathbf{a} + \int_o \mathbf{J} \cdot d\mathbf{a} = 0$, where the w and o labels refer to the wire and the spherical surface, respectively. Therefore

$$\int_o \mathbf{J} \cdot d\mathbf{a} = -\int_w \mathbf{J} \cdot d\mathbf{a} = I$$

Because of the spherical symmetry, $\mathbf{J}$ is radial and depends only on r, and therefore

$$\int_o \mathbf{J} \cdot d\mathbf{a} = 4\pi r^2 J = I \qquad \text{or} \qquad \mathbf{J}(\mathbf{r}) = \frac{I}{4\pi r^2}\,\hat{\mathbf{r}}$$

The electric field in the medium can be calculated using $\mathbf{E}(\mathbf{r}) = \mathbf{J}(\mathbf{r})/\sigma_c$. Therefore

$$\mathbf{E}(\mathbf{r}) = \frac{I}{4\pi\sigma_c r^2}\,\hat{\mathbf{r}} \tag{7.28}$$

Inside the sphere, however, the electric field vanishes because it is highly conducting. The potential of the sphere, V, as a function of the current can be calculated using Eqs. (2.42) and (7.28):

$$V = -\int_\infty^{R_0} \mathbf{E} \cdot d\mathbf{r} = \frac{I}{4\pi\sigma_c R_0} \tag{7.29}$$

The resistance of the sphere-medium system, $R = V/I$, is

$$R = \frac{1}{4\pi\sigma_c R_0} \tag{7.30}$$

The capacitance of a conducting sphere of radius R_0 placed in vacuum was calculated in Section 6.4.1. Taking ε_0 to be the permittivity of the conducting medium surrounding the sphere, then the capacitance is given by Eq. (6.28) as $C = 4\pi\varepsilon_0 R_0$. It is instructive, however, to rederive the capacitance of the system in terms of our present example. Since inside the sphere, $\mathbf{E} = 0$, then the charge density on the sphere is just $\sigma = \varepsilon_0 E(r \to R_0)$. Using Eq. (7.28) gives

$$\sigma = \frac{\varepsilon_0 I}{4\pi\sigma_c R_0^2} \tag{7.31}$$

† Presumed thin enough that a negligible charge lies on it and that it occupies a negligible volume.

The total charge on the sphere is $Q = 4\pi R_0^2\sigma = \varepsilon_0 I/\sigma_c$. The potential of the sphere can be written in terms of total charge:

$$V \equiv -\int_{\infty}^{R_0} \mathbf{E}\cdot d\mathbf{r} = \frac{I}{4\pi\sigma_c R_0} = \frac{Q}{4\pi\varepsilon_0 R_0} \tag{7.32}$$

The capacitance of the sphere is $Q/V = 4\pi\varepsilon_0 R_0$, which is just Eq. (6.28). Moreover, the product of the resistance and the capacitance of the system is

$$RC = \frac{1}{4\pi\sigma_c R_0}\cdot 4\pi\varepsilon_0 R_0 = \frac{\varepsilon_0}{\sigma_c} \tag{7.33}$$

The previous example brings out a special relationship between the capacitance and resistance between two electrodes separated by a medium of conductivity σ_c and dielectric permittivity ε_0—namely, Eq. (7.33). In fact, this relation is a general result. Consider two electrodes kept at potential difference V and carrying charges $\pm Q_f$ as shown in Fig. 7.3*a*. The permittivity and conductivity of the medium between them, ε and σ_c, are constants and characterize the medium at all points of the surface S. We assume that the only free charge inside S is what resides on the electrode. Using the definitions of the resistance and the capacitance of the system ($R = V/I$ and $C = Q_f/V$, respectively, where I is the total current between the electrodes), we find that

$$RC = \frac{V}{I}\cdot\frac{Q_f}{V} = \frac{Q_f}{I} \tag{7.34}$$

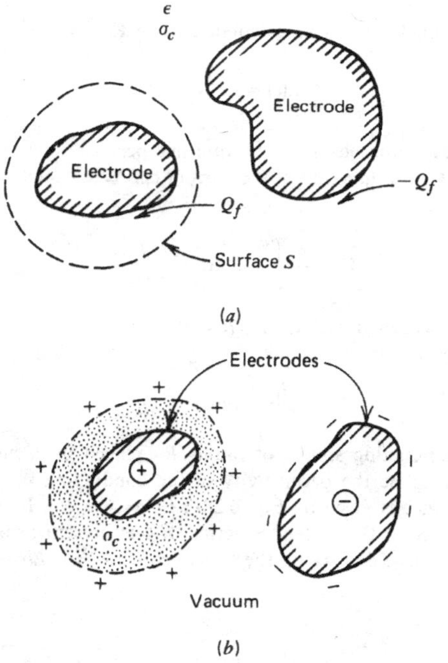

Figure 7.3 Two electrodes separated by a conducting medium. (*a*) Uniform material. (*b*) Nonuniform material.

Using Gauss's law [Eq. (4.40)] and the definition of the current in terms of **J** [Eq. (7.7)], we get:

$$RC = \frac{Q_f}{I} = \frac{\oint_S \mathbf{D} \cdot d\mathbf{a}}{\oint_S \mathbf{J} \cdot d\mathbf{a}} \tag{7.35}$$

Using $\mathbf{D} = \varepsilon \mathbf{E}$ and $\mathbf{J} = \sigma_c \mathbf{E}$ in this ratio, the general result—is

$$RC = \frac{\varepsilon}{\sigma_c} \tag{7.36}$$

Sometimes it happens that the conducting medium that lies between the electrodes is not uniform, such as that shown in Fig. 7.3*b*. In this figure, for example, a vacuum surrounds one electrode and a medium of finite conductivity surrounds the other. The charge will in general collect at the interface of the two media, so that Q_f may not represent the charge that defines the capacitance. In this case, under steady-state current conditions, most of the free charge will have migrated to the vacuum interface, so it would make little sense to describe the capacitance as between the original electrodes. (See the following two examples.)

Example 7.4 Parallel Plates with Nonuniform Media

Consider the parallel-plate configuration shown in Fig. 7.4. Between the two electrodes, 1 and 2, are two media with constants ε_1, σ_{c1} and ε_2, σ_{c2}, respectively. Let us find the resistance of this system.

Assume that the potential difference between the plates is V. The fields E_1 and E_2 and the currents J_1 and J_2 are expected to be normal to the plates and the interface, and to be constant in each material. Then using $V = -\int \mathbf{E} \cdot d\mathbf{r}$, we find that

$$V = E_1 t_1 + E_2 t_2$$

Now, one is tempted to carry on by saying that $D_1 = D_2$. But this is wrong because in general free charge σ_f will collect at the interface between the media [see Eq. (7.26) and the discussion following it]. Rather, we must say that $\sigma_f = D_2 - D_1 = \varepsilon_2 E_2 - \varepsilon_1 E_1$. Also, since σ_f exists at the interface, the correct, normal, boundary condition on **J** gives Eq. (7.23); in other words,

$$J_1 = J_2 \qquad \text{or} \qquad \sigma_{c1} E_1 = \sigma_{c2} E_2$$

This equation and the equation for V yield the values of E_1 and E_2, as follows:

$$E_1 = \frac{\sigma_{2c} V}{\sigma_{c2} t_1 + \sigma_{c1} t_2} \qquad E_2 = \frac{\sigma_{1c} V}{\sigma_{c2} t_1 + \sigma_{c1} t_2}$$

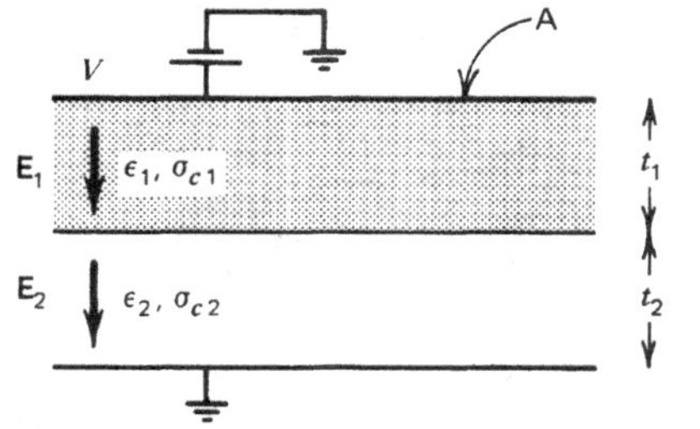

Figure 7.4 Parallel-plate capacitor filled with a nonuniform conducting medium.

It is seen that $\sigma_{1c} \to 0$, then $E_2 \to 0$, and $E_1 \to V/t_1$, the value we would obtain for a capacitor of plate separation t_1.

The free charge at the interface is derived from $\sigma_f = \varepsilon_2 E_2 - \varepsilon_1 E_1$; that is,

$$\sigma_f = \frac{\varepsilon_2 \sigma_{c1} - \varepsilon_1 \sigma_{c2}}{\sigma_{c2} t_1 + \sigma_{c1} t_2} V$$

The free charge densities at the two plates are unequal by just this amount.

The resistance per square meter is a constant and is given by Eq. (7.12), or

$$R \equiv \frac{V}{JA} = \frac{V}{\sigma_{c1} E_1 A} = \frac{V}{\sigma_{c2} E_2 A} = \frac{\sigma_{c2} t_1 + \sigma_{c1} t_2}{\sigma_{c2} \sigma_{c1} A}$$

In terms of resistivities, one will recognize this configuration as equivalent to two resistors in series.

Example 7.5 A Sphere Partially Immersed in a Nonuniform Conducting Medium—Gauss' Law

This example deals with a simplified model of grounding an electric circuit using a spherical electrode. A system is grounded by using a perfectly conducting sphere of radius a with half of the sphere in contact with the ground, as shown in Fig. 7.5. The layer of earth of radius b that is in immediate contact with the sphere has a conductivity σ_{c2}, and the rest of the ground has a conductivity σ_{c1}. Assuming that there is a current I flowing from the sphere to the ground, then the current density in region 2 is given by Eq. (7.7); that is,

$$\int_{S_2} \mathbf{J} \cdot \hat{\mathbf{n}}\, da = I$$

where S_2 is a hemisphere with radius r. Because of the spherical symmetry, the current density is radial, and therefore

$$\mathbf{J}_2 = \frac{I}{2\pi r^2} \hat{\mathbf{r}}$$

where $\hat{\mathbf{r}}$ is a unit vector in the radial direction. The corresponding electric field in this region is equal to $\mathbf{E}_2 = \mathbf{J}_2/\sigma_{c2}$, or

$$\mathbf{E}_2 = \frac{I}{2\pi \sigma_{c2} r^2} \hat{\mathbf{r}}$$

The current density and the electric field in region 1 can be similarly determined using Gauss' and Ohm's laws, as follows:

$$\mathbf{J}_1 = \frac{I}{2\pi r^2} \hat{\mathbf{r}} \quad \text{and} \quad \mathbf{E}_1 = \frac{I}{2\pi \sigma_{c1} r^2} \hat{\mathbf{r}}$$

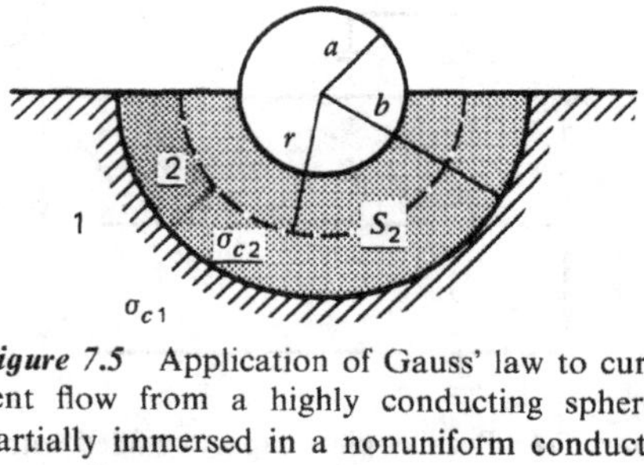

Figure 7.5 Application of Gauss' law to current flow from a highly conducting sphere partially immersed in a nonuniform conducting medium.

The pontential V of the sphere can now be determined using Eq. (2.42); that is,

$$V = -\int_{\infty}^{a} \mathbf{E} \cdot d\mathbf{r} = -\int_{\infty}^{b} \frac{I}{2\pi\sigma_{c1} r^2}\, dr - \int_{b}^{a} \frac{I}{2\pi\sigma_{c2} r^2}\, dr$$

Thus

$$V = \frac{I}{2\pi\sigma_{c1}}\left(\frac{1}{b}\right) + \frac{I}{2\pi\sigma_{c2}}\left(\frac{1}{a} - \frac{1}{b}\right)$$

The resistance R between the sphere and ground is equal to V/I; therefore,

$$R = \frac{1}{2\pi\sigma_{c1} b} + \frac{1}{2\pi\sigma_{c2}}\left(\frac{1}{a} - \frac{1}{b}\right)$$

Example 7.6 Calculation of R using $RC = \varepsilon/\sigma_c$-Coaxial Line

Suppose a coaxial line, as shown in Fig. 7.6, has a material of permittivity ε and conductivity σ_c, the potential difference between the electrodes (indicated by radii ρ_1 and ρ_2) is V, and the charge per unit length on the inner electrode is λ. Using Gauss law we find that the electric field between the electrodes is $\mathbf{E} = (\lambda/2\pi\varepsilon\rho)\hat{\boldsymbol{\rho}}$. The potential difference between ρ_1 and ρ_2 is now calculated using Eq. (2.42):

$$V = \frac{\lambda}{2\pi\varepsilon} \ln \frac{\rho_2}{\rho_1}$$

The capacitance of the line per unit length C_l is λ/V, and therefore $RC = \varepsilon/\sigma_c$ gives

$$R = \frac{\varepsilon}{\sigma_c} \cdot \frac{1}{C} = \frac{1}{2\pi\sigma_c l} \cdot \frac{1}{\ln(\rho_2/\rho_1)}$$

where l is the length of the line. This is the (leakage) resistance of the line.

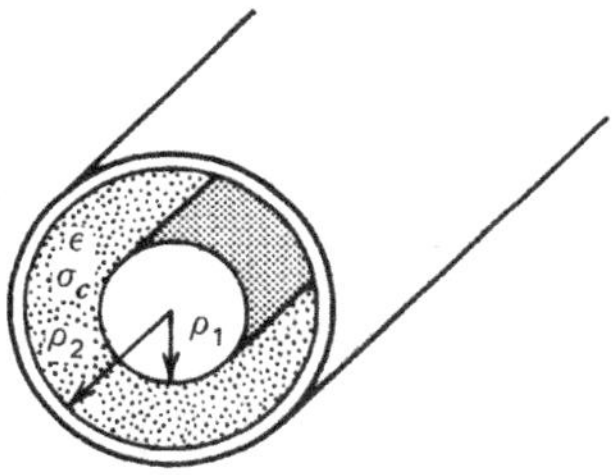

Figure 7.6 Coaxial line filled with a conducting material.

7.4.3 Boundary Value Problems

We will now show that boundary value problems of steady currents in conducting media can be described in an analogous way to problems in electrostatics. Under steady-state conditions, the rate of change of the charge distribution with respect to time vanishes, and therefore the continuity equation that expresses the law of conservation of charge reduces to

$$\nabla \cdot \mathbf{J} = 0 \tag{7.12}$$

For ohmic media, the current density is proportional to the total electric field—that is, the sum of the electric field **E** and the external electric field $\mathbf{E}_e$, which may include nonelectrical effects (electromotive forces—see Section 7.8). That is,

$$\mathbf{J} = \sigma_c(\mathbf{E} + \mathbf{E}_e) \tag{7.37}$$

The current density **J** and the electric field **E** can be described by a scalar potential according to Eq. (2.36): $\mathbf{E} = -\nabla\Phi$. Substituting Eqs. (7.37) and (2.36) into Eq. (7.12) gives:

$$\nabla\cdot(\sigma_c\nabla\Phi) = \nabla\cdot(\sigma_c\mathbf{E}_e) = \sigma_c\nabla^2\Phi + (\nabla\sigma_c)\cdot\nabla\Phi \tag{7.38}$$

For a homogeneous medium, and in the absence of external sources, Eq. (7.38) reduces to Laplace's equation:

$$\nabla^2\Phi = 0 \qquad \text{linear material, no external emf.} \tag{7.39}$$

The techniques developed previously in Chapter 4 for the solution of this equation can be used to solve the current problem. As we encountered before, the appropriate solution of Laplace's equation is determined by the boundary conditions. The appropriate boundary conditions satisfied by **J** are [see (7.23) and (7.24)] as follows:

$$J_{1n} = J_{2n} \qquad \frac{\mathbf{J}_{1t}}{\sigma_{c1}} = \frac{\mathbf{J}_{2t}}{\sigma_{c2}}$$

The condition on the tangential components of **J** implies that the tangential components of **E** are continuous at the boundary, and hence the scalar potentials are also continuous at the boundary. Therefore, the boundary conditions on the potential are given as follows:

$$\begin{gathered}\Phi_1 = \Phi_2 \\ \sigma_{c1}\frac{\partial\Phi_1}{\partial n} = \sigma_{c2}\frac{\partial\Phi_2}{\partial n}\end{gathered} \tag{7.40}$$

In the presence of external current sources, $\nabla\cdot\mathbf{E}_e$ in Eq. (7.38) is not zero, and thus the boundary conditions become

$$\begin{gathered}\Phi_1 = \Phi_2 \\ \sigma_{c2}\frac{\partial\Phi_2}{\partial n} - \sigma_{c1}\frac{\partial\Phi_1}{\partial n} = (\mathbf{J}_{e1} - \mathbf{J}_{e2})\cdot\hat{\mathbf{n}}\end{gathered} \tag{7.41}$$

where $\mathbf{J}_e = \sigma_c\mathbf{E}_e$ is the external current produced by the external electromotive force, and $\hat{\mathbf{n}}$ is a unit vector normal to the interface and pointing away from medium 1.

Thus it is apparent from Eqs. (7.40) to (7.41) that there is a close correspondence between this current problem and the analogous electrostatic problems considered in Chapter 4. The solution of the current problem may therefore be obtained by solving the corresponding electrostatic problem with the following replacements made:

$$\begin{gathered}\varepsilon \to \sigma_c, \quad \mathbf{D} \to \mathbf{J} = -\sigma_c\nabla\Phi \\ \frac{\rho}{\varepsilon} \to -\nabla\cdot\mathbf{J}_e \qquad \text{and} \qquad \frac{\sigma}{\varepsilon} \to (\mathbf{J}_{e1} - \mathbf{J}_{e2})\cdot\hat{\mathbf{n}}\end{gathered} \tag{7.42}$$

Example 7.7 Spherical Boundary—A Sphere with Angular Potential Distribution

Let us consider a homogeneous, isotropic sphere with radius R, conductivity σ_c, and a surface kept at a potential $V_0 \cos\theta$, where θ is the angle measured with respect to an axis through the center of the sphere—say, the z axis—as shown in Fig. 7.7. The conductivity of the material surrounding the sphere is taken to be zero.

The steady-state current situation implies that $\nabla \cdot \mathbf{J} = 0$ and hence for a homogeneous material gives $\sigma_c \nabla \cdot \mathbf{E} = 0$. In the absence of external electromotive forces, $\mathbf{E}_e = 0$ and $\mathbf{J}_e = 0$, therefore, the electrostatic potential satisfies Laplace's equation $\nabla^2 \Phi = 0$.

Because the potential on the surface of the sphere depends on θ, then the potential inside the sphere can be represented by the solution of Laplace's equation in two dimensions, r and θ. From Eq. (3.28) we get

$$\Phi(r, \theta) = \sum_{n=0}^{\infty} [A_n r^n + B_n r^{-(n+1)}] P_n(\cos\theta) \tag{3.28}$$

The constants A_n and B_n can now be evaluated by applying the boundary conditions. The potential at $r = 0$ should be finite, and therefore $B_n = 0$ for all n. The condition on the surface of the sphere gives

$$V_0 \cos\theta = \sum_{n=0}^{\infty} A_n R^n P_n(\cos\theta) \tag{7.43}$$

Equating coefficients of $P_n(\cos\theta)$ on both sides gives $A_1 = V_0/R$ and $A_n = 0$ for $n \neq 1$. Thus the potential, electric field, and the current density are, respectively,

$$\Phi(r, \theta) = \frac{V_0}{R} r \cos\theta \qquad \mathbf{E}(r, \theta) = -\frac{V_0}{R} \hat{\mathbf{z}} \qquad \mathbf{J}(r, \theta) = -\sigma_c \frac{V_0}{R} \hat{\mathbf{z}} \tag{7.44}$$

These results indicate that the current density inside the sphere is uniform and it is along the negative z axis, and of magnitude proportional to the amplitude of the voltage V_0 and inversely proportional to the radius of the sphere. The current density outside the sphere is zero since the conductivity of the material surrounding the sphere is zero.

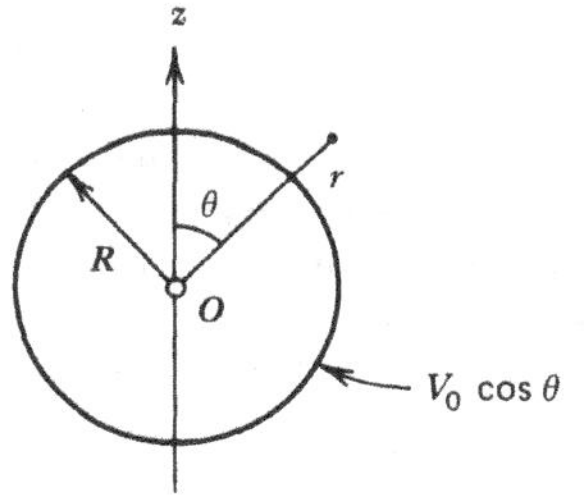

Figure 7.7 A partially conducting sphere whose surface is kept at an angle-dependent potential.

Example 7.8 Cylindrical Boundary—A Circular Cavity in a Plane Conductor

A conducting plate of conductivity σ_c, length d, width b, and thickness t has a small hole of radius $\eta \ll d, b$ at its center. Two opposite sides are kept at V_0 and $-V_0$ potentials as shown in Fig. 7.8.

The potential of the plate satisfies Laplace's equation because the plate is homogeneous and there are no external sources of electromotive force (emf). In the absence of the hole, the current distribution $\mathbf{J}$ is expected to be uniform with a direction along the x axis, and the corresponding potential Φ is expected to depend on x only. Laplace's equation in a single

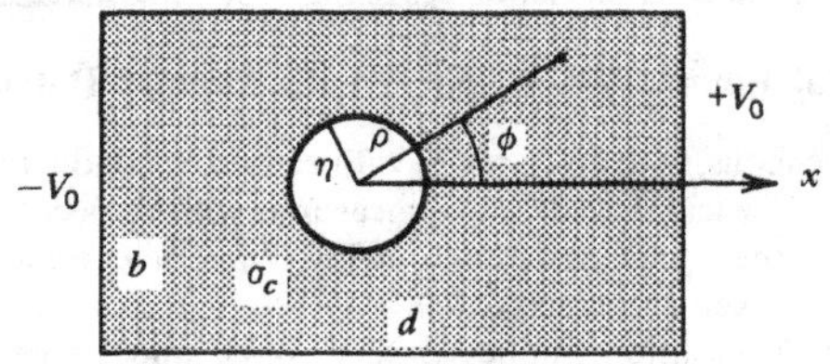

Figure 7.8 Current flow in a thin metallic plate with a small circular cavity.

cartesian variable gives $d^2\Phi/dx^2 = 0$, which yields $\Phi = Ax + B$. The constants A and B are evaluated from the conditions on the potential at $x = \pm d/2$. This yields $A = (2V_0/d)$ and $B = 0$; thus

$$\Phi = \frac{2V_0}{d} x \tag{7.45}$$

The current $\mathbf{J} = -\sigma_c \nabla \Phi$ is

$$\mathbf{J} = -\frac{2V_0}{d} \sigma_c \hat{\mathbf{x}} \tag{7.46}$$

The total current I in the plate is determined by integrating the current density $\mathbf{J}$ over the cross section of the plate, as follows:

$$I = \int \mathbf{J} \cdot \hat{\mathbf{n}}\, da = -\frac{2V_0 \sigma_c tb}{d}$$

and thus the resistance R of the plate is

$$R = -\frac{2V_0}{I} = \frac{d}{\sigma_c tb} = \frac{d}{\sigma_c A} \tag{7.47}$$

where $A = tb$.

In the presence of the hole, the potential becomes also dependent on the angle ϕ measured from the x axis, and therefore it becomes dependent on two cylindrical coordinates ρ and ϕ, where ρ is the distance from the center of the hole. The most general solution of Laplace's equation in two dimensions in cylindrical geometry is given in Eq. (3.65). We note, however, that only a subset of the most general solution will actually contribute. Close to the ends of the plate, the solution can be essentially taken to be that of the plate without the hole. This is true because the hole radius is much smaller than the plates length. Thus, as $\rho \to d/2$,

$$\Phi(\rho, \phi) \approx \frac{2V_0}{d} x = \frac{2V_0 \rho}{d} \cos\phi \tag{7.48}$$

This property implies that the solution should not include the sine terms and should include only $\cos\phi$ terms. In addition, terms involving $\ln \rho$ factors will not contribute; hence

$$\Phi(\rho, \phi) = A_0 + A_1 \rho \cos\phi + A_1' \frac{\cos\phi}{\rho} \qquad \rho > \eta \tag{7.49}$$

Using the boundary condition of Eq. (7.48) gives $A_0 = 0$, and $A_1 = 2V_0/d$. The constant A_1' can now be determined from the boundary conditions at $\rho = \eta$. Since the inside of the hole is not conducting, then J_n for $\rho < \eta$ is zero. Therefore, Eq. (7.23) gives $J_n = 0$ for $\rho = \eta$; that is, $-\sigma_c(\partial\Phi/\partial\rho) = 0$, which yields $A_1' = -A_1\eta^2$. Thus the potential in the plate is

$$\Phi(\rho, \phi) = \frac{2V_0 \rho}{d} \cos\phi - \frac{2V_0 \eta^2}{d} \frac{\cos\phi}{\rho} \tag{7.50}$$

The first term on the right-hand side of this equation is the potential we obtained in the absence of the hole. The second term is an angular dependence produced by the hole. Note that the edges of the plates are not exactly equipotential surfaces because η/d is finite; it is a consequence of the approximate nature of the condition given by Eq. (7.48).

Example 7.9 Analogy with Electrostatics—A Conducting Sphere Placed in a Uniform Current

Consider a sphere of radius R and conductivity σ_{c1} placed in an initially uniform current with density $\mathbf{J} = J_0\hat{\mathbf{z}}$. The medium surrounding the sphere is of conductivity σ_{c2}, as shown in Fig. 7.9*a*.

This problem is analogous to the dielectric problem treated in Example 4.8, where a dielectric sphere is placed in an external electric field, as shown in Fig. 7.9*b*. The potentials inside and outside the sphere Φ_1 and Φ_2 in the dielectric case are as follows:

$$\Phi_1(r, \theta) = V_0 - \frac{3\varepsilon_2 E_0}{\varepsilon_1 + 2\varepsilon_2} r \cos\theta \qquad r < R \tag{4.84}$$

$$\Phi_2(r, \theta) = V_0 - E_0 r \cos\theta + \frac{\varepsilon_1 - \varepsilon_2}{\varepsilon_1 + 2\varepsilon_2} E_0 R^3 \frac{\cos\theta}{r^2} r > R \tag{4.85}$$

The potentials and hence the current distributions of the current problem can now be determined from these expressions using the transformations given in Eq. (7.42). Replacing ε_i by σ_{ci} and $\mathbf{D}$ by $\mathbf{J}$ (that is, $\varepsilon_2 E_0$ by J_0), we get

$$\Phi_1(r, \theta) = V_0 - \frac{3J_0}{\sigma_{c1} + 2\sigma_{c2}} r \cos\theta \tag{7.51}$$

$$\Phi_2(r, \theta) = V_0 - \frac{J_0}{\sigma_{c2}} r \cos\theta + \frac{(\sigma_{c1}/\sigma_{c2}) - 1}{\sigma_{c1} + 2\sigma_{c2}} \frac{J_0 R^3}{r^2} \cos\theta \tag{7.52}$$

The determination of the fields and the current distribution will be left as an exercise.

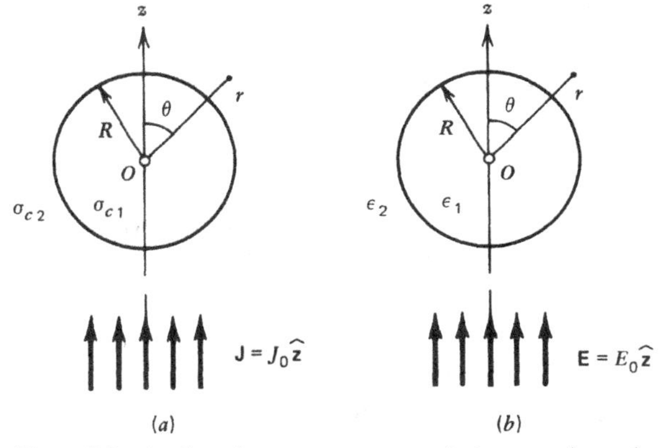

Figure 7.9 Analogy between current and electrostatic problems. (*a*) A conducting sphere in a uniform current. (*b*) A dielectric sphere in a uniform electric field.

Example 7.10 Boundary Current Problem with External Nonelectrical Sources

This example involves external emf and thus brings out the application of the boundary conditions of Eq. (7.41). Consider three wires of conductivities σ_{c1}, σ_{c2}, and σ_{c3}, of permittivities ε_1, ε_2, and ε_3, and of lengths l_1, l_2, and l_3, respectively, connected together in series in the

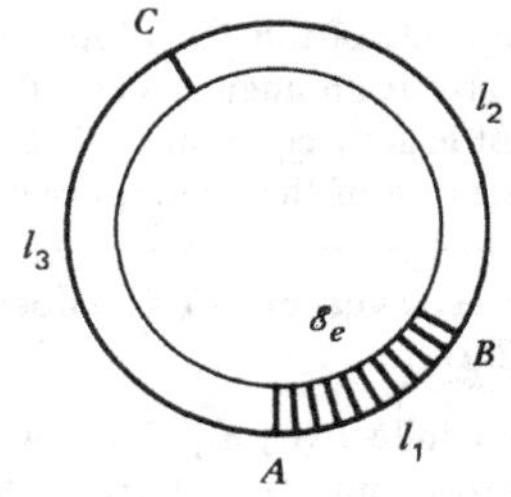

Figure 7.10 Three conducting wires connected together, through one of which is uniformly distributed an external source of emf.

shape of a ring as shown in Fig. 7.10. The wires have equal circular cross sections of radius R such that $R \ll l_i$. The wire of length l_1 has an external (electromotive) source, $\mathscr{E}_e$, uniformly distributed through it which can be represented in terms of a uniform external electric field E_e

$$\mathscr{E}_e = E_e l_1 \tag{7.53}$$

Because E_e is uniform, then the right-hand side of Eq. (7.38) is zero, and therefore the potential everywhere in the ring satisfies Laplace's equation. Moreover, because the cross sections of the wires are small, then the electric field is expected to be along the wires. Therefore, the potential will depend only on the distance along the wire, and consequently the electric fields (or current densities) take on constant values in the wires.

Let us take J_i to represent the current density in the ith wire. We next use the boundary conditions given by Eq. (7.41).

1. At point C, the normal components of the currents are continuous since there are no external electromotive forces. Therefore

$$J_2 = J_3 = J$$

2. At point B, the current densities are related as follows:

$$J_2 - J_1 = J - J_1 = J_e \tag{7.54}$$

Since $J_e = \sigma_{c1} E_e$, using $\mathscr{E}_e = E_e l_1$ we find that $J_e = (\sigma_{c1}/l_1)\mathscr{E}_e$, which, upon substitution in Eq. (7.54), gives

$$J - J_1 = \frac{\sigma_{c1}\mathscr{E}_e}{l_1} \tag{7.55}$$

3. The last boundary condition is the continuity of the potential. For example, the potential difference around the closed loop should be zero. Therefore using Eq. (2.42) gives $E_1 l_1 + E_2 l_2 + E_3 l_3 = 0$. Using $E_i = J_i/\sigma_{ci}$ in this relation gives

$$\frac{J_1 l_1}{\sigma_{c1}} + J\left(\frac{l_2}{\sigma_{c2}} + \frac{l_3}{\sigma_{c3}}\right) = 0 \tag{7.56}$$

The equations relating J and J_1 can now be solved simultaneously, as follows:

$$J_1 = -\alpha\sigma_{c1}(\sigma_{c2} l_3 + \sigma_{c3} l_2)\varepsilon_e \qquad \text{and} \qquad J = \alpha l_1 \sigma_{c2}\sigma_{c3}\mathscr{E}_e$$

where

$$\alpha = \frac{\sigma_{c1}}{l_1(\sigma_{c1}\sigma_{c2} l_3 + \sigma_{c1}\sigma_{c3} l_2 + \sigma_{c2}\sigma_{c3} l_1)}$$

The calculation of the electric fields in the wires and the charge densities located at the separation boundaries of the wires will be left as an exercise. We also note that the above results can be arrived at using Ohm's law. Do it.

7.5 The Coefficients of Resistance

As discussed in Section 7.4, Steady Currents, the methods of electrostatics can be used to solve current problems. The correspondence between the two cases can be extended further by finding a corresponding formulation to the method of coefficients of potential encountered in the electrostatic case. This can be realized since in the case of a perfect conductor ($\sigma_c \to \infty$) embedded in a conducting medium of finite conductivity, the potential on the surface of the conductor is constant.

In the electrostatic case, the potentials of the conductors were related to the charges on them. The corresponding quantities in the case of a current problem are the potentials of the electrodes and the currents leaving them. Thus one can write

$$\Phi_i = \sum_j R_{ij} I_j \tag{7.57}$$

where Φ_i is the voltage of the ith electrode, I_j is the current leaving the jth electrode, and R_{ij} is the coefficient of resistance. The coefficients R_{ij} are independent of the potentials and the currents; they are completely determined by the geometry of the electrodes and the conductivity of the material surrounding them. These coefficients are analogous to the coefficients of potential. Below we give an example of this method.

Example 7.11 Calculation of Resistance Using the Coefficients of Resistance

Consider a grounded circuit consisting of two perfectly conducting spheres, as shown in Fig. 7.11. The radii of the spheres are a_1 and a_2 and the distance between them is l, where $l \gg a_1$, a_2. One-half of each sphere is immersed in a ground of conductivity σ_c and forms a good contact. Assume that I_1 and I_2 are currents leaving spheres 1 and 2, respectively, and the corresponding voltages on the spheres are Φ_1 and Φ_2. These voltages are related to the currents through the coefficients of resistance as given by Eq. (7.57); that is,

$$\Phi_1 = R_{11} I_1 + R_{12} I_2 \tag{7.58}$$

$$\Phi_2 = R_{22} I_2 + R_{21} I_1 \tag{7.59}$$

where R_{11}, R_{22}, and $R_{12} = R_{21}$ are constants to be evaluated from the geometry. If $I_2 = 0$, then $\Phi_1 = R_{11} I_1$. From Example 7.5 we find that $\Phi_1 = I_1/2\pi\sigma_c a_1$, and thus $R_{11} = 1/2\pi\sigma_c a_1$. On the other hand, if $I_1 = 0$ and $I_2 \neq 0$, then, again from Example 7.5, $\Phi_2 = I_2/2\pi\sigma_c a_2$, and thus $R_{22} = 1/2\pi\sigma_c a_2$. To calculate R_{12}, one refers again to Example 7.5. If $I_2 = 0$, the potential at distance $l \gg a_1$, a_2 from the sphere is $\Phi_2(l) = I_1/2\pi\sigma_c l$, which gives $R_{12} = 1/2\pi\sigma_c l$. Substituting the values of the coefficients determined above in Eqs. (7.58) and (7.59), we get

$$\Phi_1 = \frac{1}{2\pi\sigma_c a_1} I_1 + \frac{1}{2\pi\sigma_c l} I_2 \tag{7.60}$$

$$\Phi_2 = \frac{1}{2\pi\sigma_c a_2} I_2 + \frac{1}{2\pi\sigma_c l} I_1 \tag{7.61}$$

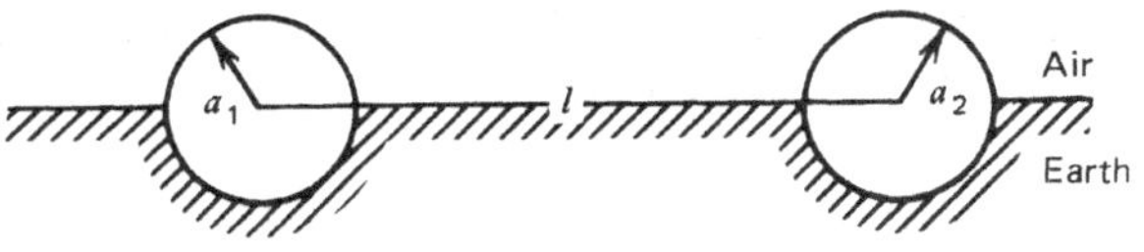

Figure 7.11 Calculation of the resistance between two small, highly conducting spheres partially immersed in ground.

In the case where $I_1 = I = -I_2$ (that is, the current leaving sphere 1 enters sphere 2), then Eqs. (7.60) and (7.61) give

$$\Phi_1 - \Phi_2 = \left(\frac{1}{2\pi\sigma_c a_1} + \frac{1}{2\pi\sigma_c a_2}\right)I - \frac{1}{\pi\sigma_c l}I$$

The resistance R of the system can now be determined as $R = (\Phi_1 - \Phi_2)/I$, which gives

$$R = \frac{1}{2\pi\sigma_c}\left(\frac{1}{a_1} + \frac{1}{a_2} - \frac{2}{l}\right)$$

7.6 The Method of Images for Currents

The method of images was shown to be very powerful for the solution of electrostatic problems that otherwise would be very difficult to solve via the use of expansions in terms of zonal, cylindrical, or cartesian harmonics. Such problems involve, for example, a conducting cylinder near a large conducting plane or near another conducting cylinder.

The same difficulty also arises in the case of steady currents. For example, in the case of a pair of cylindrical electrodes or a pair composed of one cylindrical electrode and one plane electrode, the methods developed so far are not very useful. On the other hand, the method of images is very convenient and powerful in obtaining solutions for these problems. The analogy with the electrostatic image case is obvious in view of the analogy developed in Section 7.4. Therefore, more details of the method can be best introduced through examples.

Example 7.12 Calculation of Resistance Using the Method of Images— A Cylindrical Electrode Parallel to a Plane Electrode

Consider a very long, highly conducting cylindrical electrode of radius R placed parallel to, and with its center at a distance x_0 from, a highly conducting, infinite-plane plate. The half space containing the cylinder is filled with a medium of conductivity σ_c. The cylinder is maintained at a potential Φ_c relative to the plate.

Because there are no external electromotive sources and because the medium is of uniform conductivity, Eq. (7.38) reduces to Laplace's equation. The problem can then be solved as an electrostatic one (see Example 6.7) by assuming the medium to be a dielectric with permittivity ε_0, for example. Assuming the cylinder has the charge λ per unit length, one finds that its potential relative to the plane is given by Eq. (6.39) and the capacitance per unit length of the system is given by Eq. (6.40); that is,

$$C_l = \frac{2\pi\varepsilon_0}{\cosh^{-1}(x_0/R)} \tag{6.40}$$

Using Eq. (7.36), which relates the capacitance of the electrostatic case to the resistance of the corresponding current case ($RC = \varepsilon_0/\sigma_c$), we get

$$R = \frac{1}{2\pi\sigma_c l}\cosh^{-1}\frac{x_0}{R} \tag{7.62}$$

The current between the electrodes may now be easily determined from the relation $I = \Phi_c/R$.

Example 7.13 Calculation of Resistance Using the Method of Images—Two Cylindrical Electrodes

This example deals with the current distribution between two cylindrical electrodes. Consider a very large plate of thickness t and conductivity σ_c. Two highly conducting disks of thickness t and radius R are implanted in the plate with distance Δ between their centers, as shown in Fig. 7.12. A potential difference $\Delta\Phi$ is imposed between the two electrodes.

The method of images of electrostatics along with Eq. (7.36) can be used to solve for the resistance and the current between the cylinders. Let us first consider the case where the cylinders are surrounded by a dielectric material of permittivity ε_0. Let us also assume that the cylinders have the charges λ and $-\lambda$ per unit length. The electrostatic problem can be solved with the help of the discussion in Section 3.5.3. Since $R_1 = R_2 = R$, then $x_{02} = x_{01} = x_0$, and $\Delta = 2x_0$. The values of m_1 and m_2 can be shown to be

$$\begin{pmatrix} m_1 \\ m_2 \end{pmatrix} = \frac{x_0}{R} \pm \sqrt{\left(\frac{x_0}{R}\right)^2 - 1}$$

The potential difference between the electrodes $\Delta\Phi = \Phi_1 - \Phi_2$ is

$$\Delta\Phi = \frac{\lambda}{2\pi\varepsilon_0} \ln \frac{m_1}{m_2}$$

Substituting for m_1 and m_2 gives

$$\Delta\Phi = \frac{\lambda}{2\pi\varepsilon_0} \ln m_1^2 = \frac{\lambda}{\pi\varepsilon_0} \ln m_1$$

which can be written as

$$\Delta\Phi = \frac{\lambda}{\pi\varepsilon_0} \cosh^{-1} \frac{x_0}{R}$$

The resistance, therefore, can be determined from Eqs. (6.34) and (7.36); that is,

$$R = \frac{1}{\pi\sigma_c l} \cosh^{-1} \frac{x_0}{R} \tag{7.63}$$

which is just twice that of a cylinder and a plane.

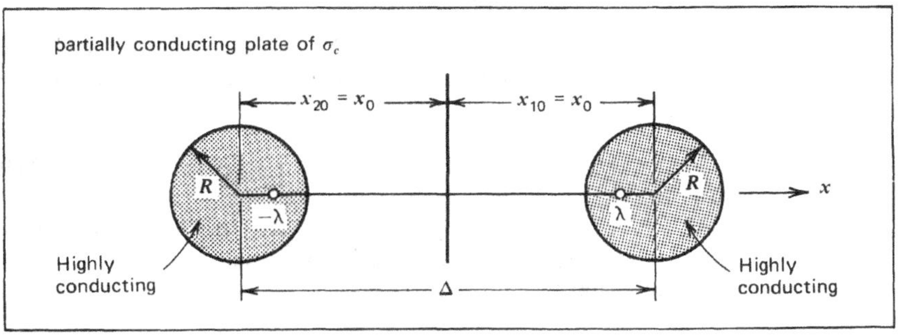

Figure 7.12 Calculation of the resistance between two cylindrical electrodes implanted in a large partially conducting plate using the method of images.

7.7 Microscopic Origin of Conduction

In this section we take on the problem of explaining the concept of conductivity from a microscopic point of view. Specifically, we would like to establish the linear behavior expressed by Ohm's law given in Eq. (7.13).

Consider a dilute ensemble of identical particles each having charge q and mass m. These particles are taken to interact with another system of particles with the interaction causing them to scatter with mean time τ between collisions. The time τ is often called the *collision time* or the *relaxation time* of the particles.

When an electric field (which, say, is along the z axis) is applied to the system, then the charges experience forces according to Coulomb's law. Therefore, the equation of motion of a charge between collisions is

$$m\frac{dv_z}{dt} = qE \tag{7.64}$$

where t is measured from the instance right after the last collision. The solution of this equation of motion when the field is time independent is

$$v_z = \frac{qE}{m}t + v_z(0) \tag{7.65}$$

which indicates that along the field the particle acquires a velocity that is linear in time. The velocity $v_z(0)$ is an initial velocity. Because of collisions, the particle will not continue to accelerate since they tend to interrupt the period of acceleration. As a result this period is cut short to $\langle t\rangle = \tau$. In fact, the probability theory tells us that the probability that a particle, after surviving without collisions for time t, will suffer a collision in the time interval between t and $t + dt$ is given by*

$$P(t)dt = \frac{1}{\tau}\mathbf{e}^{-t/\tau}\,dt \tag{7.66}$$

Therefore the average speed is given by

$$\langle v_z\rangle = \int_0^\infty v_z P(t)dt \tag{7.67}$$

Substituting Eq. (7.66) in Eq. (7.76), and integrating, we get

$$\langle v_z\rangle = \frac{qE}{m}\tau + \langle v_z(0)\rangle \tag{7.68}$$

The average initial velocity $\langle v_z(0)\rangle$ can be taken to be zero especially if the charges collide with particles of considerably larger masses, and especially when $v_z(0)$ has random directions. Thus

$$\langle v_z\rangle = \frac{qE}{m}\tau \tag{7.69}$$

The effect of collisions can alternatively be incorporated in the motion of the particles by including a velocity-dependent force in the equation of motion of the particles (damping force). That is,

$$m\frac{dv_z}{dt} = qE - \frac{m}{\tau}v_z \tag{7.70}$$

* See F. Reif, *Fundamentals of Statistical and Thermal Physics* (New York: McGraw-Hill, 1965).

where $(m/\tau)v_z$ opposes the action of the electric field. It becomes more important at higher velocities, with the result of limiting the velocity that the particle can acquire. The steady-state solution—that is, when the velocity of the particle ceases to vary as a function of time—is easily determined from Eq. (7.70) by taking $dv_z/dt = 0$; then $v_z = qE\tau/m$, which is the result arrived at in Eq. (7.69).

If the density of the charges is n, then the current density can be calculated from Eqs. (7.6) and (7.69) as

$$\mathbf{J} = \frac{nq^2\tau\mathbf{E}}{m} \tag{7.71}$$

which indicates that the relation between **J** and **E** is linear, as given by Ohm's law. The proportionality constant is just the conductivity, as given by Eq. (7.13),

$$\sigma_c = \frac{nq^2\tau}{m} \tag{7.72}$$

This result can be easily generalized to the case where many types of charges are present. The result is

$$\sigma_c = \sum_{i=1}^{p} \frac{n_i q_i^2 \tau_i}{m_i} \tag{7.73}$$

Where the summation is taken over the number of the different types of the charges, p.

7.8 Joule Heating and Batteries

The fact that charge carriers are not accelerated in an ohmic medium (i.e., that the drift velocity is constant—see Eq. 7.69) means that energy must be dissipated in the medium. The atomic constituents of the medium scatters the charge carriers and, in so doing, are given kinetic energy, which appears as a heating of the medium. The average kinetic energies of the charge carriers is to a first approximation constant in the medium, so that the energy injected into the medium all appears as heat. Consider a volume element of a conducting medium dv, which is not necessarily ohmic, where the applied electric field is **E**. The work done by **E** in moving a charge dq through a displacement $d\mathbf{l}$ is just $dq\,\mathbf{E}\cdot d\mathbf{l}$. Since $dq = \rho\, dv$ (by definition), we can write this work as

$$d^2W = (\rho\, dv)\mathbf{E}\cdot d\mathbf{l}$$

If the charge is displaced by $d\mathbf{l}$ in a time dt, $d\mathbf{l} = \langle\mathbf{v}\rangle dt$, where $\langle\mathbf{v}\rangle$ is the average (drift) velocity of the charge; then

$$d^2W = \rho(\langle\mathbf{v}\rangle\cdot\mathbf{E})dv\, dt$$

Using the relation $\mathbf{J} = \rho\langle\mathbf{v}\rangle$, we have

$$d^2W = \mathbf{J}\cdot\mathbf{E}\, dv\, dt$$

The power, or work done by **E** per unit time in sustaining the current, is then

$$\frac{d^2W}{dt} = \mathbf{J}\cdot\mathbf{E}\, dv$$

and is proportional to the volume element dv. This work contributes to the heating in dv. In a finite volume V the power generated, P, is consequently

$$P = \frac{dW}{dt} = \int_V \mathbf{J} \cdot \mathbf{E}\, dv \tag{7.74}$$

The power volume density, dP/dv, is given by

$$\frac{dP}{dv} = \mathbf{J} \cdot \mathbf{E} \tag{7.75}$$

and hence the scalar product, $\mathbf{J} \cdot \mathbf{E}$, might be called the "power density." If the medium is ohmic, $\mathbf{J} = \sigma_c \mathbf{E}$, and we can also write

$$P = \frac{dW}{dt} = \int_V \rho_c J^2\, dv = \int_V \sigma_c E^2\, dv$$

where $\rho_c = 1/\sigma_c$. Now consider a conducting medium, with opposite faces A and B equipotential regions. If a charge $dq = (dq/dt)dt$ is transported from A to B in a time dt, then the work performed by the electric field $\mathbf{E}$ is given by

$$dW = \frac{dW}{dt}\, dt = dq \int_A^B \mathbf{E} \cdot d\mathbf{r} \equiv dq\, V = IV\, dt$$

where $I = dq/dt$. The power generated in the medium with passage of the current I is therefore

$$P = \frac{dW}{dt} = IV \tag{7.76}$$

In the static situation, where $\mathbf{E}$ is a conservative field, $V = \Phi(A) - \Phi(B)$. Since $V = RI$ in an ohmic medium, we have also

$$P = IV = RI^2$$

as the power dissipation for this region of the medium.

It is now apparent that since energy is constantly dissipated, in an electric circuit containing resistive elements, this energy must be replenished in equal amounts by some energy source if constant currents are to be maintained. Considering a simple circuit consisting of a loop, one realizes that an electrostatic field alone does not provide a source of energy for any unit charge that traverses the loop, since $\oint \mathbf{E} \cdot d\mathbf{r} = 0$ for conservative fields.

A mechanism that provides for the possibility of maintaining currents in a dissipative (resistive) medium is called an *emf*, or *electromotive force*, because it provides "motive" force to move the charge. It is the energy source that maintains the currents.

A common example of a source of steady-state emf ("direct current ≡ dc") is the ordinary voltaic battery found in automobiles, flashlight batteries, or standard cells used with potentiometers. The mechanism whereby energy is made available to produce electric currents in such batteries has a chemical origin. Chemical reactions (atomic or molecular transformations) occur in which chemical energy is released and is available to do the work required to produce a charge separation. The charge separation in turn produces an electric potential difference between points in space between which charge can be made to move (as in wires).

A different kind of battery is the solar battery, shown in Fig. 7.13. Here rays of sunlight fall on a sensitive metal surface, which consequently emits electrons (via the

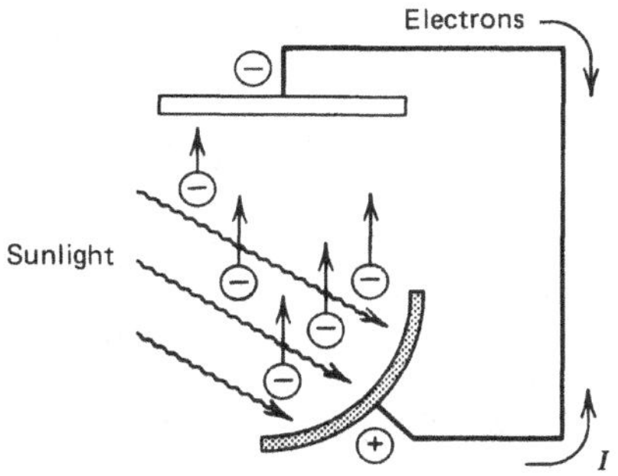

Figure 7.13 Solar battery.

photoelectric effect). These electrons are collected at another nearby metal surface. Thus, a charge separation is affected that can thereby produce an electric current. The light meters on cameras operate on this principle.

Yet another kind of battery with promising applications is the nuclear battery (see Example 7.14). Here a radioactive source is placed at one terminal, and the charged radiation emitted is collected at another terminal. The action is similar to the solar battery except that the source of energy here is nuclear rather than electromagnetic (sunlight).

The common characteristic of all sources of emf is their ability to effect a charge separation. This separation must be accomplished against electrostatic forces created by the charge separation. Ultimately, if charge continues to build up, the electrostatic forces will become large enough to prevent further charge separation. When this occurs, current ceases. The electromotive or impressed force that causes a charge q to move against the electrostatic fields will be denoted as $q\mathbf{E}^N$, where q is the value of the charge. The field $\mathbf{E}^N$ thus constitutes a kind of force field, which, however, typically exists only in a very restricted region of space, such as inside a battery. If $\mathbf{E}^q$ represents the electric field produced by the charge separation processes, and is static in nature, then the work done in moving a unit charge around a loop through the source of emf is

$$\mathscr{E} = \oint_C (\mathbf{E}^q + \mathbf{E}^N)\cdot d\mathbf{l}$$

since $\oint \mathbf{E}^q \cdot d\mathbf{l} = 0$ if $\mathbf{E}$ is static and thus conservative; then

$$\mathscr{E} = \oint_C \mathbf{E}^N \cdot d\mathbf{l}$$

Consequently, if work is done, $\mathbf{E}^N$ must be nonconservative in the region containing the loop. Since, usually, $\mathbf{E}^N \neq 0$ only "inside" the source of emf, meaning that the source is localized, one writes

$$\mathscr{E} = \oint_C \mathbf{E}^N \cdot d\mathbf{l} = \int_A^B \mathbf{E}^N \cdot d\mathbf{l} \tag{7.77}$$

where A and B are points at the terminals of the source (only one is assumed here) and $\mathscr{E}$ is called the emf of the loop C. The exact nature of $\mathbf{E}^N$ may be difficult to envision, but it is a well-defined quantity for any loop equal to the net work that must be done to carry a unit charge around that loop. Multiple sources of energy may exist in the loop. If the terminals A and B are insulated from each other, a stable situation will exist in the source such that no current flow will occur from A to B. This means that

$$\int_A^B \mathbf{E}^N \cdot d\mathbf{l} + \int_A^B \mathbf{E}^q \cdot d\mathbf{l} = 0 \tag{7.78}$$

In other words, the work done by both $\mathbf{E}^N$ and $\mathbf{E}^q$ in transporting charge from A to B is zero. The conservative forces in effect oppose the nonconservative forces, so that current flow ceases. No work is done in the source, thus we write

$$\int_A^B \mathbf{E}^N \cdot d\mathbf{l} - [\Phi(B) - \Phi(A)] = 0$$

and hence we may write Eq. (7.78) as

$$\mathscr{E} = \Phi(B) - \Phi(A) \qquad \text{(no current)} \tag{7.79}$$

Consider now the case where connection is made from B to A outside the source so that current can flow in a continuous loop. It is clear that if there is resistance to the current flow inside the source, then current can flow only if

$$\int_A^B (\mathbf{E}^N + \mathbf{E}^q) \cdot d\mathbf{l} \equiv \int_A^B \mathbf{E} \cdot d\mathbf{l} > 0$$

That is, there must be over the length of path C a motive force, $\mathbf{E} \equiv \mathbf{E}^N + \mathbf{E}^q$, moving the charge from A to B so as to overcome the resistance. If the medium in which the charge moves in the source of emf is ohmic ($\mathbf{J} = \sigma_c \mathbf{E}$),

$$V_{AB} \equiv \int_A^B \mathbf{E} \cdot d\mathbf{l} = rI$$

where r is called the internal resistance. Thus, when current flows through the source we have

$$\mathscr{E} - rI = \Phi(B) - \Phi(A) \tag{7.80}$$

The representation of such an ohmic source is thus a "pure" emf, $\mathscr{E}$, in series with the "internal resistance" r (see Fig. 7.14). In good sources of emf, r will be small compared to the resistance of the circuit to which it is attached. As a consequence, the work performed by the source of emf ($\mathbf{E}^N$) is largely delivered as energy to the rest of the circuit and is not dissipated as heat inside the source of emf itself.

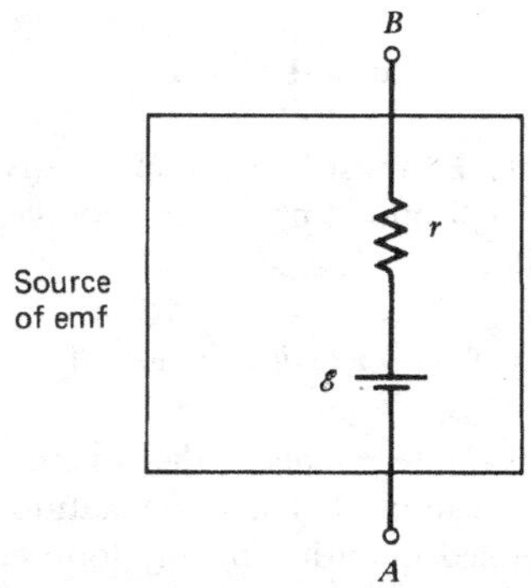

Figure 7.14 The standard representations of a source of emf.

The power delivered by the source of emf when an external circuit extracts a current I from the emf is just $\mathscr{E}I$. If the external circuit contains a total resistance R, the total power dissipation in R and r is just $(R + r)I^2$; that is, $\mathscr{E}I = RI^2 + rI^2$, or

$$\mathscr{E} = RI + rI \tag{7.81}$$

With the above representation of the ohmic source, we see that this equation becomes $\mathscr{E} = RI + rI = \Phi(B) - \Phi(A) + rI$ as was encountered in Eq. (7.80).

We note that the internal resistance is in principle easy to determine simply by measuring the potential difference $\Phi(B) - \Phi(A)$ without and with current flow from the source, as described by Eqs. (7.79) and (7.80), respectively.

*Example 7.14 Nuclear Battery—Current Source

Consider the nuclear battery sketched in Fig. 7.15. We assume that the radioactive source emits alpha particles (helium nuclei that have a charge of $+2e$) having kinetic energies of 5.0 MeV. These α particles are collimated into a beam and pass to a metal collector electrode at B. If switch S were open and no charge could leak off the collector, the potential of B relative to the box (ground) would rise to a value of 2.5×10^6 V. At this point, no further charge would be collected, because the initial kinetic energy of the α's would just equal the work done against the electrostatic field in moving to B from the box. At a higher positive potential, the α's would be deflected back to the box.

The nonconservative field in this battery, $\mathbf{E}^N$, must be ascribed to a nuclear force and it is zero everywhere except inside the nucleus. It is the force that kicks the α's out of the nuclei of the radioactive material. (As it does so, two electrons are released in the material.) The work done by the nuclear forces is clearly equal to the kinetic energy imparted on the α particles. Thus

$$\frac{1}{2}m_\alpha v^2 = 2e\int \mathbf{E}^N \cdot d\mathbf{r} = 5 \times 10^6 \text{ eV}$$

Consider the case when S is open and equilibrium is achieved. In the equilibrium situation, the α's reach the collector at B with zero kinetic energy. Thus

$$2e\{\Phi(B) - \Phi(A)\} = \frac{1}{2}m_\alpha v^2$$

From

$$\int_A^B \mathbf{E}^N \cdot d\mathbf{r} = \Phi(B) - \Phi(A)$$

it follows that

$$\mathscr{E} = \Phi(B) - \Phi(A)$$

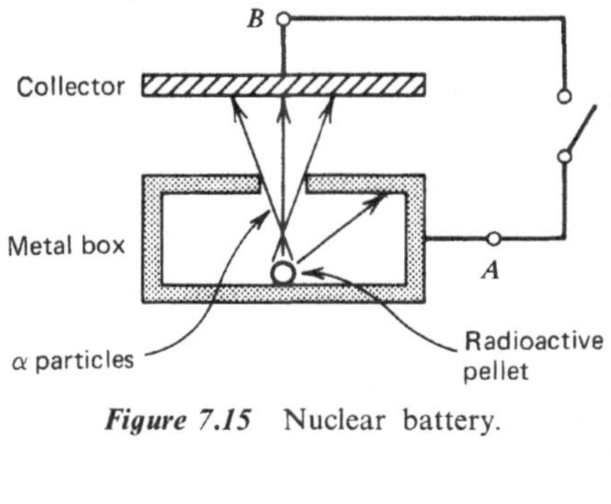

Figure 7.15 Nuclear battery.

Thus, $\mathscr{E} = 2.5 \times 10^6$ V. This is the open-circuit potential difference at the terminals of the source.

Now, suppose S is closed. A current I_0 will flow through R such that $\Phi(B) - \Phi(A) = RI_0$ in an equilibrium situation. But now I_0 must just be equal to $2e$ times the number of α's each second moving from A to B, and $\Phi(B) - \Phi(A)$ is no longer equal to $\mathscr{E}$. For example, if 10^6 α particles per second pass from A to B, then $I_0 = 1.6 \times 10^{-19} \times 2 \times 10^6 = 3.2 \times 10^{-13}$ A. If R is 10^7 Ω, then $RI_0 = \Phi(B) - \Phi(A) = 3 \times 10^{-5}$ V. We might describe the situation by saying that our source of emf has an internal resistance r, such that $\mathscr{E} - rI_0 = RI_0$. Then, for example, $r \approx 10^{19}$ Ω.*

This type of emf is known as a current source, because the current $I = I_0$ is essentially independent of R, and r is essentially infinite compared to any practical R's one may use.

7.9 Kirchhoff's Laws and Resistive Networks

We consider in this section problems involving electromotive sources connected to various combinations of external loads in the form of loops. Also, we include cases where the loops include many of these sources, as shown in Fig. 7.16.

Network problems are analyzed by means of two rules, called Kirchhoff's rules:

1. The first rule states that the algebraic sum of the currents flowing towards a junction is zero. This is a statement expressing the conservation of charge; that is if there is no piling of charge at a given junction, then the rate at which charge enters the junction should be equal to the rate at which it leaves the junction. Therefore

$$\Sigma I = 0 \tag{7.82}$$

2. The second rule states that the sum of the electromotive forces and the voltage drops across all resistances (including the internal resistances of the sources) in a closed loop is zero. That is,

$$\sum_{i=1}^{N} \mathscr{E}_i - r_i I_i - \sum_{k=1}^{M} (RI)_k = 0 \quad \text{or} \quad \sum V = 0 \tag{7.83}$$

The procedure for applying Kirchhoff's rules to a current circuit can now be outlined: First, we assume a direction and a magnitude to the currents going through all elements of the circuit. Second, we choose a direction to go around the different loops. Third, a convention for the sign of the potential drops across resistances and the sign of the electromotive force as we go around the loops must be chosen. The emf is taken as positive if we go through the source from negative to positive. The voltage drop across a resistance is taken as positive if we go across it in the same direction as the assigned current, and negative if we go in the opposite direction to the current.

With this procedure one may find that the number of equations derived is larger than the number of unknown currents. Such a case arises because not all the loops are independent. In fact, one has to make sure that none of the independent loops is missed.

The resulting equations are algebraic and therefore can be easily solved for the unknowns. A negative current means that the correct direction is the opposite of what we have assumed.

* In this example, r is not a well-defined resistance, in fact, since it depends upon the current drawn. We do *not* have an ohmic medium here.

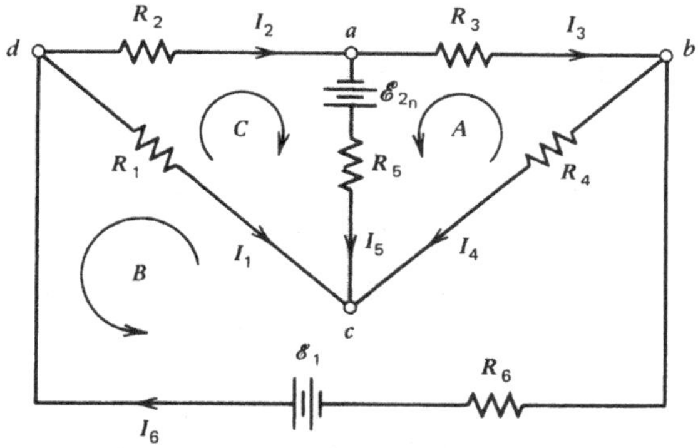

Figure 7.16 Application of Kirchhoff's laws to a multiloop dc circuit.

In the course of analyzing electric networks, one encounters series and parallel connections of resistors. When a number of resistors of resistances $R_1, R_2, \ldots, R_n$ are connected in series (that is, the current through each of them is the same), then it can be easily shown that they are equivalent to a single resistor of resistance R, as follows:

$$R = R_1 + R_2 + \cdots + R_n \tag{7.84}$$

On the other hand, when these resistors are connected in parallel (that is, the voltage across each one of them is the same), then they can be easily shown to be equivalent to a single resistor of resistance R; that is,

$$\frac{1}{R} = \frac{1}{R_1} + \frac{1}{R_2} + \cdots + \frac{1}{R_n} \tag{7.85}$$

Let us now analyze the circuit of Fig. 7.16. The assumed currents and their directions and the sense of going around the loops are labeled in the figure. Kirchhoff's first rule gives the following relations between the currents at the various junctions.

$$\begin{aligned} I_2 &= I_3 + I_5 && \text{at } a \\ I_3 &= I_6 - I_4 && \text{at } b \\ I_4 &= I_1 + I_5 && \text{at } c \\ I_6 &= I_1 + I_2 && \text{at } d \end{aligned}$$

Another set of equations can now be determined using Kirchhoff's second rule, as follows:

$$\begin{aligned} -\mathscr{E}_2 - I_3R_3 + I_4R_4 + I_5R_5 &= 0 && \text{in loop } A \\ -\mathscr{E}_1 - I_1R_1 - I_4R_4 - I_6R_6 &= 0 && \text{in loop } B \\ \mathscr{E}_2 + I_5R_5 + I_2R_2 - I_1R_1 &= 0 && \text{in loop } C \end{aligned}$$

The above equations are algebraic and hence can be easily solved for the unknown currents. We will leave such determination as an exercise.

7.10 Summary

In a conducting medium the current density $\mathbf{J}$ and the current I through any surfaces are defined as follows:

$$\mathbf{J} = \sum_i \rho_i \langle \mathbf{v}_i \rangle \tag{7.4}$$

$$I = \int_S \mathbf{J} \cdot d\mathbf{a} \tag{7.7}$$

where ρ_i is the charge density of species i, and $\langle \mathbf{v}_i \rangle$ is the average dreft velocity of this species. The total charge Q crossing the surface S is related to I as follows:

$$I = \frac{dQ}{dt}$$

A basic equation of electromagnetism is the continuity equation, which expresses the concept of conservation of charge. It relates the charge density at a point in space to the current density at that point, or

$$\nabla \cdot \mathbf{J} + \frac{\partial \rho}{\partial t} = 0 \tag{7.9}$$

For steady-state applications, where ρ ceases to vary with time, the continuity equation becomes

$$\nabla \cdot \mathbf{J} = 0 \qquad \text{(steady state)} \tag{7.12}$$

The density of conduction current in a medium under the influence of an electric field $\mathbf{E}$ is given by

$$\mathbf{J} = \sigma_c \mathbf{E} = \frac{\mathbf{E}}{\rho_c} \tag{7.13}$$

where σ_c is the conductivity of the medium, and $\rho_c = 1/\sigma_c$ is the resistivity of the medium. For linear (ohmic) media, σ_c is a constant (independent of $\mathbf{E}$). This relation is often called the differential form of Ohm's law. The integral form of this law is the familiar relation

$$V = IR$$

The resistance of a wire of length l, cross section A, and conductivity σ_c is

$$R = \frac{l}{\sigma_c A} \tag{7.14}$$

The distribution of steady currents in a conducting medium can be described in an analogous way to problems in electrostatics in dielectric media. The equation $\nabla \cdot \mathbf{J} = 0$ implies, for linear material, that $\sigma_c \nabla \cdot \mathbf{E} = 0$. Taking $\mathbf{E} = -\nabla \Phi$, then

$$\nabla \cdot \mathbf{J} = 0 \rightarrow \nabla^2 \Phi = 0 \qquad \text{(linear material, no external emf)} \tag{7.39}$$

When the space considered has more than one material, then Laplace's equation can be solved in each material separately, and the solutions (potentials, fields, and currents) are then matched at the interfaces of the materials. In the absence of external current sources at the interfaces produced by electromotive sources, then using $\nabla \cdot \mathbf{J} = 0$ and $\nabla \times \mathbf{E} = \nabla \times \mathbf{J}/\sigma_c = 0$ we find that the boundary conditions are

$$J_{1n} = J_{2n} \qquad \text{and} \qquad \frac{\mathbf{J}_{1t}}{\sigma_{c1}} = \frac{\mathbf{J}_{2t}}{\sigma_{c2}} \tag{7.23),(7.24}$$

where n and t stand for components that are, respectively, normal and tangent to the interface. These two conditions are equivalent to the following two conditions on Φ.

$$\sigma_{c1} \frac{\partial \Phi_1}{\partial n} = \sigma_{c2} \frac{\partial \Phi_2}{\partial n} \qquad \text{and} \qquad \Phi_1 = \Phi_2 \tag{7.40}$$

Thus it is apparent that there is a close correspondence between current problems and electrostatic problems. The solution of the current problem may therefore be obtained by solving the corresponding electrostatic problem with the following replacements made:

$$\varepsilon \to \sigma_c \qquad \text{and} \qquad \mathbf{D} \to \mathbf{J} \tag{7.42}$$

Other electrostatic techniques such as the method of images and the method of coefficients of resistance (analogous to coefficients of potential and capacitance), are also useful here. Also this correspondence implies that the capacitance C and resistance R between two conductors embedded in an infinite medium of ε and σ_c satisfy the following simple relation

$$RC = \frac{\varepsilon}{\sigma_c} \tag{7.36}$$

The differential form of Ohm's law and hence its integral form can be derived by considering the microscopic response of a dilute conducting medium to an external electric field. The motion of the charges are governed by the acceleration qE/m and by a linear deceleration due to collisions with other particles: $v/m\tau$ where q, m, and v are the charge, mass, and velocity of the charges and τ is a time constant for ohmic media that gives the mean time between two collisions. It is a measure of how frequent these retarding collisions. This treatment gives

$$\mathbf{J} = \frac{nq^2\tau}{m}\mathbf{E} \qquad \text{or} \qquad \sigma_c = \frac{nq^2\tau}{m} \tag{7.71),(7.72}$$

where n is the number density of the charges. This indicates that charges in an ohmic medium are not accelerated by an external electric field.

Because of the absence of acceleration of charges in an ohmic medium, then energy must be dissipated in the medium. The change of power per unit volume is as follows:

$$\frac{dP}{dv} = \mathbf{J} \cdot \mathbf{E} \tag{7.75}$$

Electric circuits are analyzed by Kirchhoff's two laws: The summation of all currents at any junction is zero, and the summation of all voltage drops in any loop is zero:

$$\Sigma I = 0 \qquad \text{and} \qquad \Sigma V = 0 \tag{7.82),(7.83}$$

These two laws embody the two basic laws of steady currents: $\nabla \cdot \mathbf{J} = 0$ and $\nabla \times \mathbf{E} = 0$.

Problems

7.1 A current of 10 A flows through a wire with a cross section of 2 mm^2. If the density of charge carriers in the wire is $10^{21}/\text{cm}^3$, determine the average drift velocity of the electrons.

7.2 The current distribution in a given three-dimensional conductor of conductivity σ_c is such that the electric field strength and therefore the current density are constant on an equipotential circuit. In this case one can show that the resistance of the conductor is given by $R = \int dl/\sigma_c A$, where $d\mathbf{l}$ is normal to the equipotential surface of area S. Using this result find the resistance for a spherical capacitor of inner and outer radii a and b filled with a homogeneous medium of conductivity σ_c.

7.3 Use the result of Problem 7.2 to determine the resistance of a spherical capacitor of inner and outer radii a and b, filled to a radius c with a material of conductivity σ_{c1}, and from c to b with a material of conductivity σ_{c2}.

7.4 Use the result of Problem 7.2 to determine the resistance of a cylindrical capacitor, of inner and outer radii a and b and length l, that is filled with a conducting material of conductivity σ_c.

7.5 Consider a parallel-plate capacitor that is filled with a partially conducting material of dielectric constant K and conductivity σ_c. The capacitor is charged with an initial charge Q_0. (a) Determine the charge on the plates as a function of time. (b) Calculate the total Joule heat produced and show that is equal to the initially stored electrostatic energy. (c) If $K = 4.3$ and $\sigma_c = 10^{-13}$ $(\Omega \cdot \text{m})^{-1}$, calculate the time constant for the discharging of the capacitor.

7.6 The current flow lines make angles θ_1 and θ_2 with the normal to the separation boundary between two conducting media of conductivities σ_{c1} and σ_{c2}. Derive the law of refraction of the current flow lines.

7.7 A block of conducting material is in the form of a cube of side a as shown in Fig. 7.17. Its conductivity is not uniform and at any point in the block is $\alpha(a + x)$, where α is a constant. Assume that current flows only along the x axis from face S to the opposite face S'. (a) Taking the electrostatic potential Φ to depend only on x, use Eq. (7.38) to show that

$$\alpha(a + x)\frac{d^2\Phi}{dx^2} + \alpha\frac{d\Phi}{dx} = 0$$

(b) Solve the equation satisfied by Φ and show that the potential difference between S and S' is $\Delta\Phi = A \log 2$ where A is a constant. (c) Using the solution for Φ, determine the total current between S and S', and hence show that the resistance between S and S' is $(\ln 2)/\alpha a^2$.

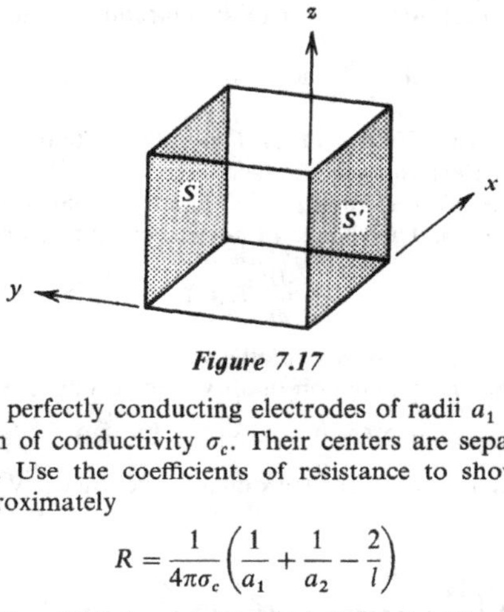

Figure 7.17

7.8 Two small, spherical, perfectly conducting electrodes of radii a_1 and a_2 are embedded in an infinite medium of conductivity σ_c. Their centers are separated by a distance l such that $l \gg a_1, a_2$. Use the coefficients of resistance to show that the resistance between them is approximately

$$R = \frac{1}{4\pi\sigma_c}\left(\frac{1}{a_1} + \frac{1}{a_2} - \frac{2}{l}\right)$$

7.9 If the medium in Problem 7.8 is bounded by an infinite-plane interface as shown in Fig. 7.18, use the method of images by introducing image currents to show that

$$R' = R + \frac{1}{4\pi\sigma_c}\left(\frac{1}{b} - \frac{1}{\sqrt{a^2 + b^2}}\right)$$

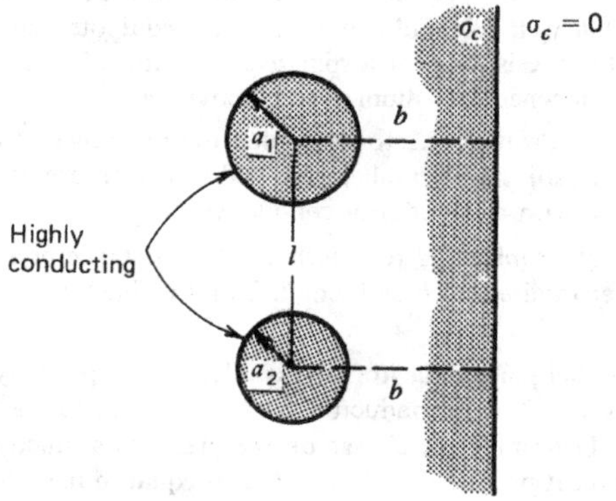

Figure 7.18

7.10 A capacitor of arbitrary form is filled with a homogeneous dielectric of permittivity ε. If it is known that when the capacitor is filled with a homogeneous conductor of conductivity σ_c its dc resistance is R, then determine its capacitance.

7.11 The space between two parallel conducting plates of area $A = 0.05\ \text{m}^2$, and separation 0.2 cm is filled with a lossy dielectric for which $K = 8$ and $\sigma_c = 0.8 \times 10^3\ (\Omega\cdot\text{m})^{-1}$. Calculate the total rms (root-mean-square) current when a voltage $V = 10 \sin \omega t$, where $\omega = 10^7$ radian/s is applied across the plates.

7.12 The conductivity σ_c of copper at room temperature is $0.59 \times 10^8\ (\Omega\cdot\text{m})^{-1}$ and the density of mobile electrons is $10^{21}/\text{cm}^3$. Find the relaxation time τ for electrons in copper.

7.13 A system of electrodes is characterized by the coefficients of resistance R_{ik}. Determine the amount of heat Q generated per second in the space between the electrodes in terms of the currents I_k leaving the electrodes.

7.14 A circuit for measuring resistance in a Wheatsone bridge is shown in Fig. 7.16, with $\mathscr{E}_2 = 0$. Consider the case where $R_5 = R_6 = 0$. Show that for the current I_5 to vanish (which can be monitored via a galvanometer), the condition $R_1/R_2 = R_4/R_3$ must be satisfied. This condition allows one of the resistances to be measured if the other three are known.

7.15 Determine the current I_2 when the Wheatsone bridge of Problem 7.14 is off balance. Show that $S = cR_3(\partial I_2/\partial R_3)$, where c is the deflection of the galvanometer per unit current, and S is the sensitivity of the bridge (neglect the resistance of the galvanometer):

$$S = \frac{c\mathscr{E}_1}{R_1 + R_2 + R_3 + R_4}$$

7.16 The superposition theorem does not apply to power. Superposition is a property that depends on the linearity of the quantity in question, and power is a quadratic rather than a linear quantity. (a) Calculate the power delivered by source $\mathscr{E}_1$ in the circuit of Fig. 7.19 with $\mathscr{E}_2$ dead. (b) Calculate the power delivered by $\mathscr{E}_2$ with $\mathscr{E}_1$ dead. (c) Calculate the power delivered to the circuit in the presence of both and show that the sum of the results in (a) and (b) does not give the power absorbed in the circuit.

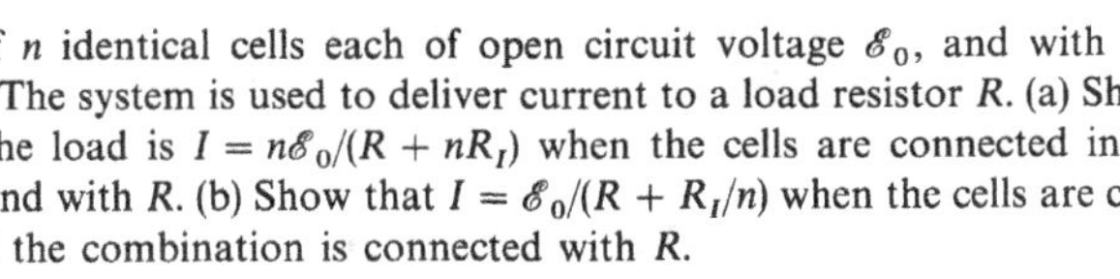

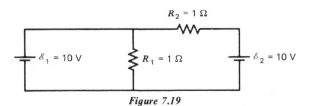

Figure 7.19

7.17 A system of n identical cells each of open circuit voltage $\mathscr{E}_0$, and with internal resistance R_I. The system is used to deliver current to a load resistor R. (a) Show that the current in the load is $I = n\mathscr{E}_0/(R + nR_I)$ when the cells are connected in series with each other and with R. (b) Show that $I = \mathscr{E}_0/(R + R_I/n)$ when the cells are connected in parallel and the combination is connected with R.

7.18 A square net is made of a wire of uniform cross section. It consists of n^2 identical square cells, and the resistance of one side of each of these cells is r. If a current enters at one of the corners and leaves at the opposite corner, find the resistance of the entire net for $n = 2$, 3, and 4. (*Hint*: The symmetry of the circuit can be used to reduce the number of circulating currents; for example, in the case of $n = 3$, the number of circulating currents may be reduced to three.)

EIGHT

MAGNETISM OF STEADY CURRENTS

8.1 The Lorentz Force

An electric field **E** is essentially defined by the relation $\mathbf{F}_e = q\mathbf{E}$, where $\mathbf{F}_e$ is the force exerted by the electric field. It can be determined in any reference system by determining the force on a point charge q stationary in that reference system.* It is natural that in electrostatics we considered the charges to be at rest. If we remove this restriction, such that q is allowed to move with nonzero velocity **v** (in the given reference system), we adhere to the equation $\mathbf{F}_e = q\mathbf{E}$, but we find by experiment that another force exists, given by

$$\mathbf{F}_m = q\mathbf{v} \times \mathbf{B} \tag{8.1}$$

This is a *magnetic force*, commonly referred to as the *Lorentz force*. The quantity **B** in Eq. (8.1) is sometimes called the *magnetic induction*: we shall call it the magnetic field or simply the **B** field. Its existence is inferred from observations that establish (in the absence of any electric field) the existence of a force (1) proportional to the magnitudes of q and **v**, (2) perpendicular to **v** and to another direction, $\pm\hat{\mathbf{B}}$, and (3) such that the magnitude F_m varies as the sine of the angle between **v** and $\hat{\mathbf{B}}$.

As we noted above, the Lorentz force is perpendicular to the plane defined by **v** and **B**. As a consequence, the work done by $\mathbf{F}_m$ on q (in a time dt) as q moves through a displacement $d\mathbf{r} \equiv \mathbf{v}\,dt$ is zero: $dW = \mathbf{F}_m \cdot d\mathbf{r} = \mathbf{F}_m \cdot \mathbf{v}\,dt = 0$. That is, magnetic forces alone cannot do work on charged particles.

Given that vector fields **B** and **E** coexist, the *electromagnetic force* on a point charge q is given by

$$\mathbf{F} = q(\mathbf{E} + \mathbf{v} \times \mathbf{B}) \tag{8.2}$$

* We ignore gravitational forces (which are much weaker than electromagnetic forces) and the other fundamental forces that can appear in the vicinity of atomic nuclei. (The latter are "short-range" forces, which can be ignored if q is distant from such nuclei. This is the usual case.)

It is this force that must be used in determining the motion of charged particles. The study of charged-particle behavior in electromagnetic fields forms the subject of *electrodynamics*. Depending on whether the equations of motion are those of classical or quantum physics, the modifiers *classical* electrodynamics or *quantum* electrodynamics (abbreviated QED) are appended.

As here defined in Eqs. (8.1) and (8.2), **B** has units of electric field divided by velocity. The following equivalent dimensional relations hold, where N, C, V, A, m, s, kg, Wb, and T stand for newton, coulomb, volt, ampere, meter, second, kilogram, weber, and tesla respectively.

$$B = \frac{\text{N/C}}{\text{m/s}} = \frac{\text{V}\cdot\text{s}}{\text{m}^2}$$

or

$$B = \frac{\text{kg}}{\text{C}\cdot\text{s}} = \frac{\text{kg}}{\text{A}\cdot\text{s}^2} = \frac{\text{Wb}}{\text{m}^2} = \text{T}$$

The weber and tesla are both derived units.

8.2 Forces on Current Distribution—Motion in Crossed Fields

We had $\mathbf{F} = q\mathbf{v} \times \mathbf{B}$ (we drop the subscript m), where q was a point charge, or a charge whose physical dimensions are small enough so that **B** is constant over them. Consider now a volume of space, V, in which the mobile scalar charge density $\rho(\mathbf{r})$ is everywhere defined. Assume also that in V there exists a vector field, $\mathbf{v}(\mathbf{r})$, which gives the velocity with which $\rho(\mathbf{r})$ is moving.* Then the Lorentz force given by Eq. (8.1) acting upon the mobile charge contained in a volume element dv at position r is simply

$$d\mathbf{F} = dq(\mathbf{r})[\mathbf{v}(\mathbf{r}) \times \mathbf{B}(\mathbf{r})] = \rho(\mathbf{r})dv[\mathbf{v}(\mathbf{r}) \times \mathbf{B}(\mathbf{r})].$$

By definition, $\rho(\mathbf{r})\mathbf{v}(\mathbf{r}) \equiv \mathbf{J}(\mathbf{r})$ [see Eq. (7.6)], so we may write

$$d\mathbf{F}(\mathbf{r}) = [\mathbf{J}(\mathbf{r}) \times B(\mathbf{r})]dv \tag{8.3}$$

The Lorentz force on the volume V† is then given by

$$\mathbf{F} = \int_V \mathbf{J}(\mathbf{r}) \times \mathbf{B}(\mathbf{r})dv \tag{8.4}$$

Analogously, if there exist scalar charge density fields on surfaces or curves in space given by σ and λ, respectively, then there can exist Lorentz forces on these surfaces and curves. Again, by definition,

$$\sigma\mathbf{v} \equiv \mathbf{K} \qquad \text{and} \qquad \lambda\mathbf{v} \equiv \mathbf{I} \tag{8.5}$$

where **K** is a surface current density, and **I** is a linear or *filamentary current* density. Clearly, such concepts have significance only from a macroscopic point of view. The current density **J** is more fundamental, since it is expected to have significance even on a microscropic level. In any case, we obtain for the Lorentz force expression

$$d\mathbf{F} = \mathbf{K} \times \mathbf{B}\, da \qquad \text{or} \qquad \mathbf{F} = \int_S [\mathbf{K}(\mathbf{r}) \times \mathbf{B}(\mathbf{r})]da \tag{8.6}$$

* By this we mean that whatever charge is in the volume element dv at **r** has an average velocity $\mathbf{v}(\mathbf{r})$.

† That is, the force on the charge of V.

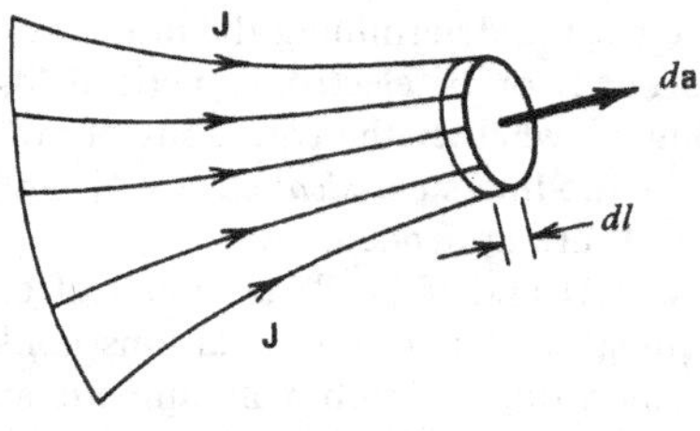

Figure 8.1 Tubular region in space whose side surfaces are formed by streamlines of **J**.

for surface currents and

$$d\mathbf{F} = \mathbf{I} \times \mathbf{B}\, dl \qquad \text{or} \qquad \mathbf{F} = \int_C \mathbf{I} \times \mathbf{B}\, dl \tag{8.7}$$

for filamentary currents. If $d\mathbf{l}$ denotes a filamentary element of length in the direction of **I**, we can write the force also as

$$\mathbf{F} = \int I\, d\mathbf{l} \times \mathbf{B}(\mathbf{r}) \tag{8.8}$$

Finally, a current distribution specified in terms of all three types of the current densities discussed above will be a sum of Eqs. (8.4), (8.6), and (8.7).

We now show an important relation in the case of tubular currents or filamentary currents. Let us consider a current density **J**, restricted to a tubular region in space whose side surfaces are formed by streamlines of **J** (refer to Fig. 8.1), such that the total current flowing past a cross-sectional area of the tube S is given by $\int_S \mathbf{J} \cdot d\mathbf{a} = \int_S J\, da \equiv I$. The relation we want to prove for this type of current is:

$$\int_V \mathbf{J}\, dv * [\quad] = \int_C I\, d\mathbf{l} * [\quad] \tag{8.9}$$

where the brackets may respresent any field defined in V, either vector or scalar, and $*$ represents a binary operation such as a cross operation ($\times$) or dot operation ($\cdot$). For example, one may be interested in showing $\int \mathbf{J}\, dv \times \mathbf{B} = \int I\, d\mathbf{l} \times \mathbf{B}$. We can write the left-hand side of Eq. (8.9) as $\int_C \int_S \{\mathbf{J}\, da\, dl * [\quad]\}$ where da is an area element of S, and dl is an element of length in the direction of **J** and hence perpendicular to da. If $[\quad]dl$ varies insignificantly over the area S (or if we represent by $[\quad]dl$ the average of $[\quad]dl$ over the area S), then we can factor the above integral to obtain

$$\int_C \int_S \{\mathbf{J}\, da\, dl * [\quad]\} = \int_C \left\{ \int_S \mathbf{J}\, da \right\} * [\quad] dl \equiv \int_C \mathbf{I} * [\quad] dl \equiv \int_C I\, d\mathbf{l} * [\quad]$$

In the last step, we have assumed that $\mathbf{I}\, dl = I\, d\mathbf{l}$; that is, the "direction" of the current is assumed to be in the "direction" of the element of length dl. The conditions listed above under which these operations are valid we shall call the filamentary conditions, or the *filamentary approximation*.

An immediate illustration of the utility of this transformation is in the calculation of the force acting on a current-carrying wire placed in an external magnetic field **B**. If **B** denotes the average value of the field over a cross-sectional area of the wire and

J is the current density in the wire, we note that the bracket and the star in Eq. (8.9) are given by [ ] ≡ **B**, * ≡ ×, and therefore Eq. (8.4) transforms as follows:

$$\mathbf{F} = \int \{\mathbf{J}\, dv \times \mathbf{B}\} = \int I\, d\mathbf{l} \times \mathbf{B}$$

for the force on the wire as in Eq. (8.8). Under steady-state conditions, the current I is constant along the wire and can be factored from beneath the integral sign:

$$\mathbf{F} = I \int d\mathbf{l} \times \mathbf{B} \qquad \text{(constant current)}$$

If, furthermore, **B** were constant along the path of integration $\mathbf{F} = I\{\int d\mathbf{l}\} \times \mathbf{B}$. If $\int d\mathbf{l}$ approximates a closed loop, then $\int d\mathbf{l} = \oint d\mathbf{l} = 0$. The net force on a closed filamentary current loop in a constant magnetic field is zero.

It is perhaps not obvious that one can equate the force on the moving charge in a wire to the force on the wire itself. That such an equation can be made results because in an equilibrium situation the force acting on the moving charge is transmitted to the atoms of the wire, the charge not being able to leave the wire.

Example 8.1 Charged Particle in a Constant Magnetic Field

Let us consider the motion of a particle of mass m and charge q in a magnetic field **B**. Moreover, let us take the case where the electric field in the region is zero. Since $\mathbf{F} = q(\mathbf{v} \times \mathbf{B})$, Newton's second law tells us that

$$m\frac{d\mathbf{v}}{dt} = q(\mathbf{v} \times \mathbf{B}) \tag{8.10}$$

If we take the scalar product of both sides of this equation with **v**, we obtain $m\mathbf{v} \cdot d\mathbf{v}/dt = q\mathbf{v} \cdot (\mathbf{v} \times \mathbf{B})$. Since $\mathbf{v} \cdot (\mathbf{v} \times \mathbf{B}) = (\mathbf{v} \times \mathbf{v}) \cdot \mathbf{B} = 0$, then

$$m\mathbf{v} \cdot \frac{d\mathbf{v}}{dt} = \frac{1}{2} m \frac{d}{dt}(v^2) = \frac{d}{dt}\left(\frac{1}{2} mv^2\right) = 0$$

The kinetic energy of the particle is then constant in time, consistent with the fact that **F** is always perpendicular to the motion.

If **B** is a constant field, it is not difficult to show (show it) that the trajectory of the particle is a combination of uniform circular motion, and motion in a straight line. This motion when $\mathbf{B} = B\hat{\mathbf{z}}$ may be described by the equations (see Fig. 8.2)

$$x - x_0 = \frac{v_\perp}{\omega} \sin \omega(t - t_0) \qquad y - y_0 = \frac{v_\perp}{\omega} \cos \omega(t - t_0) \quad z - z_0 = v_\parallel(t_0 - t_0) \tag{8.11}$$

where $v_\perp$ and $v_\parallel$ are the projections of the velocity normal to and along the field respectively, and $\omega = qB/m$ is called the *cyclotron frequency*. Thus, the projection of the trajectory on a plane normal to **B** (the x-y plane) is a circle centered on the point (x_0, y_0), with radius $R = v_\perp/\omega = mv_\perp/qB$.

Often, one has the situation where $v_\parallel = 0$, and the particle moves in uniform circular motion of angular frequency ω. In this case the momentum of the particle is given by

$$p = mv = qBR \tag{8.12}$$

This is a relationship often used when one employs magnetic fields (as, e.g., in magnetic spectrometers) to determine a particle's momentum. The product BR is often called the "magnetic rigidity."

It is interesting to note an important practical property of the **B** field displayed here: that the **B** field can be used to contain a distribution of charged particles without changing their

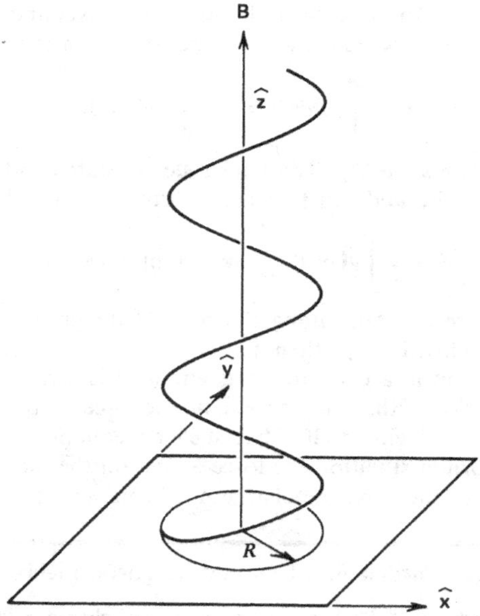

Figure 8.2 Trajectory of a charge in a constant **B** field showing a combination of uniform circular motion and motion in a straight line. (Helix motion)

distribution of energies. Thus, for any particle that leaves a point on a line parallel to a constant **B** field, **B** will return it to that line at a later time given by the period of its projected circular orbit; that is, $T = 2\pi R/v = 2\pi m/qB$. This time is independent of the particle velocity (nonrelativistically). All particles that have the same component of velocity, $v_{\parallel}$, parallel to **B** will thus return to the line at the same point. Such a property is used in devices called magnetic lenses for focusing beams of charged particles.

Example 8.2 A Charge Particle in Crossed Fields

Assume that there exists in a reference system designated O the mutually perpendicular ("crossed") constant fields

$$\mathbf{B} = B\hat{\mathbf{z}} \qquad \mathbf{E} = E\hat{\mathbf{y}} \tag{8.13}$$

and assume that there is a charge q moving with a velocity $\mathbf{v} = v\hat{\mathbf{x}}$ in this region. Using Eq. (8.2), one can show that q will feel no force if $v = E/B$; that is,

$$\mathbf{F} = q(\mathbf{E} + \mathbf{v} \times \mathbf{B}) = q\left[E\hat{\mathbf{y}} + \left(\frac{E}{B}\hat{\mathbf{x}}\right) \times B\hat{\mathbf{z}}\right] \equiv 0$$

Now suppose a charge q is stationary at time $t = 0$ in such crossed fields. What is its subsequent motion? This can be determined using the above result. We study the motion from the point of view of a reference frame, O', moving with velocity $\mathbf{v} = (E/B)\hat{\mathbf{x}}$. Thus consider a charge q' and an observer moving with velocity $\mathbf{v} = (E/B)\hat{\mathbf{x}}$.

The magnetic force as observed in the reference frame of the moving particle, O', is zero, because the velocity of the particle is by definition zero there. Since the total force at the particle is zero, then the observer must conclude that the electric force $q\mathbf{E}$ and hence the electric field must also be zero. Only a magnetic field may exist, and it is uniform. As a consequence, if the charge q' is moving in frame O', its motion in it will be governed by only a magnetic field, and therefore its orbit will be circular. Now to the observer with a velocity $\mathbf{v}$, the charge q at rest in O appears to move with a velocity $-\mathbf{v}$, and so moves in circles! The motion is described by equations of the form:

$$x' = R \cos \omega t \qquad y' = R \sin \omega t \qquad x'^2 + y'^2 = R^2 \tag{8.14}$$

As a result, motion in O is a superposition of circular motion plus linear motion—i.e., cycloidal motion (see Fig. 8.3). In the O system (x, y, z), the equations of motion are obtained from the transformation formulas* $x' = x - vt$, $y' = y$. Thus, the equations of motion are

$$x - vt = R \cos \omega t \qquad y = R \sin \omega t \tag{8.15}$$

These are parametric equations of a cycloid.

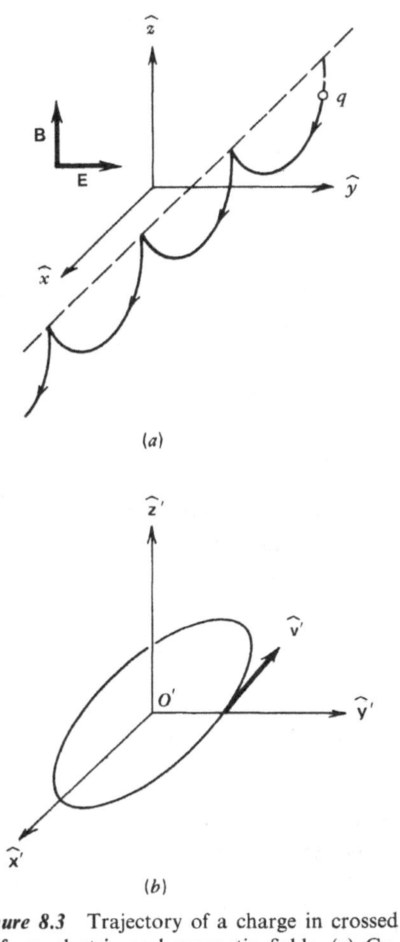

Figure 8.3 Trajectory of a charge in crossed uniform electric and magnetic fields. (*a*) Cycloidal motion observed in the laboratory frame. (*b*) Circular motion observed in a reference frame moving with velocity $v = E/B$ in the direction of $\mathbf{E} \times \mathbf{B}$.

* We assume $v \ll c$ here, so we can use the Galilean transformation formulas (see Chapter 17.)

8.3 The Sources of B

Thus far, we have postulated the existence of a **B** field without describing either its sources or its properties. It is now appropriate to give such a description. In so doing we shall assume that the sources of the magnetostatic field are steady electric currents. The justification of this assumption has evolved from many experimental studies made on the properties of magnetic forces, and we shall not detail these studies here. Suffice it to say that magnetic forces appear to coexist with electric currents, and that even the magnetic forces associated with permanent magnets, where the existence of currents is not obvious, can be explained by a representation of the magnet as a current distribution (see Chapter 9).

Assume that the magnetostatic **B** is a well-defined vector field in space; it ought to be characterized by its curl and divergence. In general we could write

$$\nabla \cdot \mathbf{B} = 4\pi k'_M \rho_M \tag{8.16}$$

$$\nabla \times \mathbf{B} = 4\pi k_M \mathbf{g} \tag{8.17}$$

where k'_M and k_M are constants of proportionality having to do with choices of units, and ρ_M and **g** are source densities independent of time that themselves represent scalar and vector fields. In analogy to electrostatics, where $\nabla \cdot \mathbf{E} = 4\pi k_E \rho \equiv \rho/\varepsilon_0$, ρ_M is called the magnetic charge density. If the (vector) field **g** were zero everywhere and ρ_M were nonzero, then the **B** field would be in exact analogy with the **E** field and would possess the same properties:

$$\mathbf{B}(\mathbf{r}) = k'_M \int \frac{dq_M(\mathbf{r}')}{|\mathbf{r} - \mathbf{r}'|^3} (\mathbf{r} - \mathbf{r}') \tag{8.18}$$

In fact, the early development of magnetism used such formulas for the calculation of **B** in regard to free space, and it still proves a useful calculational device (see Section 9.2). However, we now believe* that there are no such physical entities as magnetic charge, so that, in truth, $\rho_M \equiv 0$ everywhere, and in general **g** is equal to **J**, the electric current density. Thus, we believe the true equations of the static magnetic field to be

$$\nabla \cdot \mathbf{B} = 0 \tag{8.19}$$

$$\nabla \times \mathbf{B} = \mu_0 \mathbf{J} \tag{8.20}$$

where $\mu_0 = 4\pi k_M \equiv 4\pi \times 10^{-7}$ N/A^2 is a constant so chosen as to make the unit of current a convenient one (the ampere).† Equation (8.20) asserts the curl of **B** is linearly related to the electric current density.

In a sense Eq. (8.20) thus relates magnetism to electricity, since **J** represents a flow of electric charge, obeying the equation of continuity $\nabla \cdot \mathbf{J} = -\partial\rho/\partial t$. Indeed, Eq. (8.20) is telling us that we have a static condition for the electric as well as the magnetic fields, for if we take the divergence of both sides of Eq. (8.20) and use the vector identity [Eq. (1.66)] we obtain

$$\nabla \cdot (\nabla \times \mathbf{B}) \equiv 0 \equiv \mu_0 \nabla \cdot \mathbf{J}$$

*There is no experimental evidence for magnetic monopoles (charges) but some theories allow for their existence. Experiments are still being done to search for them.

†This is shown in Example 8.4.

implying

$$\nabla \cdot \mathbf{J} = 0 \qquad \text{and} \qquad \frac{\partial \rho}{\partial t} = 0$$

The charge density ρ is independent of time everywhere—the condition of electrostatics.

We shall assume in the following that if the current distribution is known everywhere, then a unique solution to the magnetostatic equation exists, and that if **J** is everywhere zero, so is **B**. This amounts to saying that any constant **B** field independent of **J** has no physical meaning, although it satisfies Eqs. (8.19) and (8.20). If the sources of **B** (that is, **J**) are localized, it implies that the **B** field at infinity must tend to zero.

We stress that although Eqs. (8.19) and (8.20) are the equations of the magnetic field of steady currents, and in general will be modified if the currents change in time, the equation $\nabla \cdot \mathbf{B} = 0$ will not have to be modified. There exists no evidence under any conditions, for magnetic charges.*

8.4 Integral Equations of Magnetostatics and Ampere's Law

That the divergence of **B** is everywhere zero means that there are no sources of magnetic flux. That is,

$$\oint_S \mathbf{B} \cdot d\mathbf{a} = 0 \tag{8.21}$$

This can be shown by integrating $\nabla \cdot \mathbf{B} = 0$ over a volume V and using the divergence theorem, $\oint_S \mathbf{B} \cdot d\mathbf{a} = \int_V \nabla \cdot \mathbf{B}\, dv = 0$. This result shows the flux of **B** through any closed surface S is zero, so that the streamlines of **B** ("magnetic lines of force") are everywhere continuous. There is no magnetic charge on which these streamlines can originate or terminate. Equation (8.21) might be called Gauss' law of magnetism.

From the equation $\nabla \times \mathbf{B} = \mu_0 \mathbf{J}$, we derive the second fundamental integral relationship, known as *Ampere's (circuital) law*: Through any possible orientable surface S, one can perform the scalar surface integral of this equation:

$$\int_S (\nabla \times \mathbf{B}) \cdot d\mathbf{a} = \mu_0 \int_S \mathbf{J} \cdot d\mathbf{a}$$

But, by Stokes' theorem,

$$\int_S (\nabla \times \mathbf{B}) \cdot d\mathbf{a} = \oint_C \mathbf{B} \cdot d\mathbf{r}$$

so that

$$\oint_C \mathbf{B} \cdot d\mathbf{r} = \mu_0 \int_S \mathbf{J} \cdot d\mathbf{a} \equiv \mu_0 I \tag{8.22}$$

* In truth, if magnetic charges (called "poles") always occurred as inseparable positive and negative pairs (dipoles) infinitesimally displaced from each other, each dipole producing a dipole **B** field as in electrostatics, the relation $\nabla \cdot \mathbf{B} = 0$ might still be valid (just as in dielectrics, where $\nabla \cdot \mathbf{E} = 0$ if $\rho_f = 0$ and K = constant everywhere). However, the behavior in and around matter indicates that no such poles exist, since the consequences of a magnetic polarization charge are not observed there. (Indeed, they are contradicted.) Therefore, we assume that magnetic poles are fictitious. Nonetheless, this fiction is sometimes useful in calculating magnetic fields.

which is Ampere's law. The sense of rotation of C is related to the direction chosen for $d\mathbf{a}$, a vector normal to S at a point of S, as the sense of rotation of a right-handed screw is related to its motion along its axis. The current I is the net current flowing through S. Note, however, that for a given closed curve C there exist infinitely many choices of S for which to evaluate I, the only criterion on S being that its periphery be C. In applying Eq. (8.22) one chooses S so as to make calculations easiest.

Example 8.3 Integral Equations of Magnetostatics—Long, Straight Wire

We consider a wire that is "infinitely" long—that is, long enough that end effects are negligible. We assume that a current I flows in the wire with a constant current density, so $J = I/\pi R^2$. We note that $\mathbf{B}$ can be a function of ρ alone, not of ϕ or z, due to symmetry considerations. Now we use these symmetry considerations and the integral equations of magnetostatics [Eqs. (8.21) and (8.22)] to show that $\mathbf{B} = B_\phi \hat{\boldsymbol{\phi}}$; that is, $\mathbf{B}$ lies in the "ϕ" direction. We apply Eq. (8.21) to the pillbox shown in Fig. 8.4*a*:

$$\oint \mathbf{B} \cdot d\mathbf{a} = 0 \rightarrow B_\rho 2\pi\rho l = 0$$

which yields $B_\rho = 0$. We next apply Eq. (8.22) to the loop *abcd* of Fig. 8.4*b*, and use the fact that the current going through the loop $I = 0$; that is,

$$\oint \mathbf{B} \cdot d\mathbf{r} = 0 = \mathbf{B} \cdot \mathbf{ab} + \mathbf{B} \cdot \mathbf{cd} = B_z(\rho_2) - B_z(\rho_1),$$

which yields $B_z =$ constant. We see that B_z, if it exists, must be a constant. It seems reasonable that the constant is zero.*

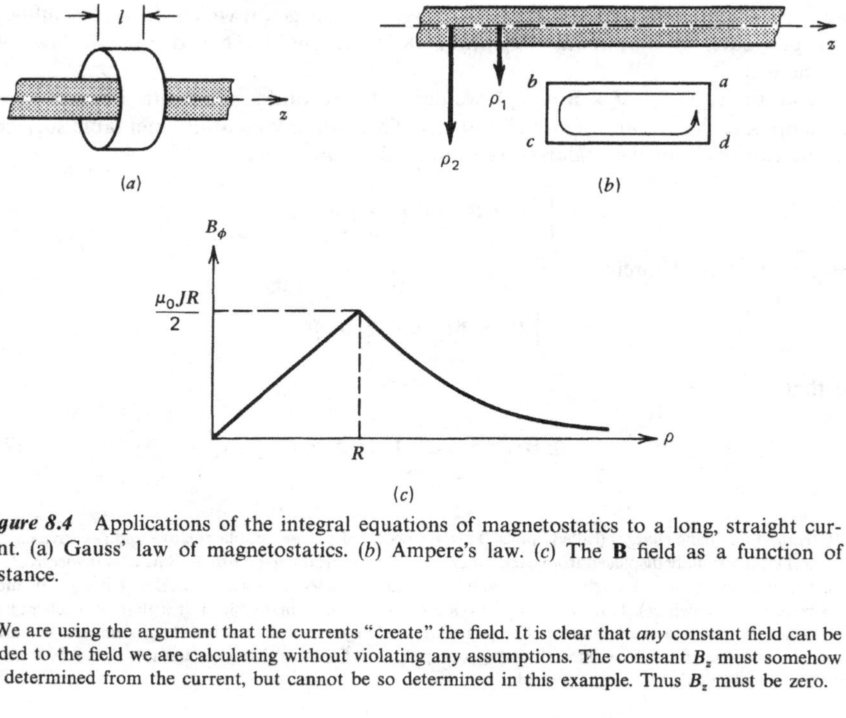

Figure 8.4 Applications of the integral equations of magnetostatics to a long, straight current. (a) Gauss' law of magnetostatics. (*b*) Ampere's law. (*c*) The $\mathbf{B}$ field as a function of distance.

* We are using the argument that the currents "create" the field. It is clear that *any* constant field can be added to the field we are calculating without violating any assumptions. The constant B_z must somehow be determined from the current, but cannot be so determined in this example. Thus B_z must be zero.

Now, applying Ampere's law to the curve C, which is a circle whose center is at the axis of the wire and whose plane is normal to the wire:

$$\oint_C \mathbf{B} \cdot d\mathbf{r} = \int_0^{2\pi} B_\phi \rho \, d\phi = 2\pi\rho B_\phi = \mu_0 I_S \tag{8.23}$$

For $\rho \geq R$, $I_S = I$, and Eq. (8.23) gives

$$B = B_\phi = \frac{\mu_0 I}{2\pi\rho} \qquad \rho \geq R \tag{8.24}$$

If the curve C was a similar circular path of radius $\rho < R$, then $I_S = \pi\rho^2 J$ and Eq. (8.23) gives

$$B = B_\phi = \frac{\mu_0 J \rho}{2} \qquad \rho \leq R \tag{8.25}$$

A plot of $B_\phi = B$ versus ρ, the distance from the wire axis is shown in Fig. 8.4*c*.

Note that the same arguments may be applied to any current distribution that is a function of the cylindrical coordinate ρ only and lies in the z direction $\mathbf{J} = J_z(\rho)\hat{\mathbf{z}}$. Again one obtains, from Ampere's law,

$$2\pi\rho B_\phi = \mu_0 I(\rho) \equiv \mu_0 \int_0^\rho J_z(\rho') 2\pi\rho' \, d\rho'$$

so that

$$B_\phi = \frac{\mu_0}{\rho} \int_0^\rho J_z(\rho')\rho' \, d\rho' \tag{8.26}$$

To get an idea of the magnitude of $\mathbf{B}$ for "laboratory" currents, insert the values $I = 1$ A, $\rho = 1$ mm $= 10^{-3}$ m in Eq. (8.24); B then equals 2×10^{-4} tesla. The laboratory unit of field strength is thus more conveniently given as 10^{-4} tesla $\equiv 1$ gauss. The gauss, in fact, turns out to be the cgs (gaussian) unit of field strength. See Appendix I.

The above example illustrates the fact that for filamentary currents the streamlines of $\mathbf{B}$ tend to circle around the filaments in the right-hand sense. With the thumb pointing along the current flow, $\mathbf{B}$ curls in the directions of the fingers of the right hand.

Example 8.4 The Force Between Current-Carrying Wires

If there are two long, filamentary, parallel wires I_z and I'_z, the force between them is given by Eq. (8.7), where $\mathbf{B}$ is the field produced by I_z at the position of an element $d\mathbf{l}'$ of the wire carrying current I'_z. With the coordinate directions indicated in Fig. 8.5, and using the field produced by a wire, Eq. 8.24, we find

$$d\mathbf{F} = I'_z \, dl'\hat{\mathbf{z}} \times \mathbf{B} = I'_z \, dl'\hat{\mathbf{z}} \times \left(\frac{\mu_0 I_z}{2\pi d} \hat{\boldsymbol{\phi}} \right)$$

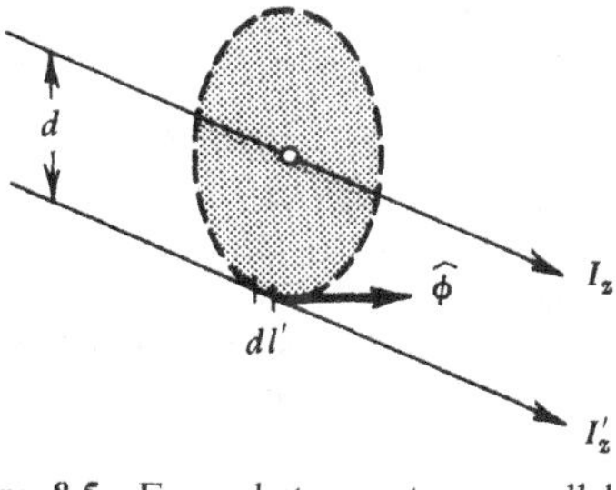

Figure 8.5 Force between two parallel filamentary currents.

or

$$d\mathbf{F} = \mu_0 \frac{I_z I'_z}{2\pi d} dl'(\hat{z} \times \hat{\boldsymbol{\phi}}) \tag{8.27}$$

The force on element dl' is thus directed toward or away from the wire carrying current I_z depending on whether $I_z I'_z \gtrless 0$, respectively. Thus, wires carrying antiparallel currents repel each other. The force per meter of the wires can be easily determined from Eq. (8.27); that is,

$$\frac{dF}{dl'} = \mu_0 \frac{I_z I'_z}{2\pi d} \tag{8.28}$$

In an instrument like the "current balance" shown in Fig. 8.6, $I = I'$. The unit of current can be chosen arbitrarily by picking an appropriate value for μ_0 to use in Eq. (8.20). (The other quantities are purely mechanical in nature.) In the SI system, μ_0 is chosen to be $4\pi \times 10^{-7}$ in magnitude. The resulting unit of current, the ampere, then can be used to define the coulomb, and hence the constant ε_0 of electrostatics. It just "happens" that $\mu_0\varepsilon_0 = 1/c^2$, where c is the speed of light.

To get some idea of the strength of the magnetic force, note that if $I = I' = 1$ ampere and $d = 1$ cm, then $dF/dl' = 2 \times 10^{-5}$ N/m. A copper wire whose cross section is 1 mm^2 has a weight per millimeter of about 8×10^{-5} N.

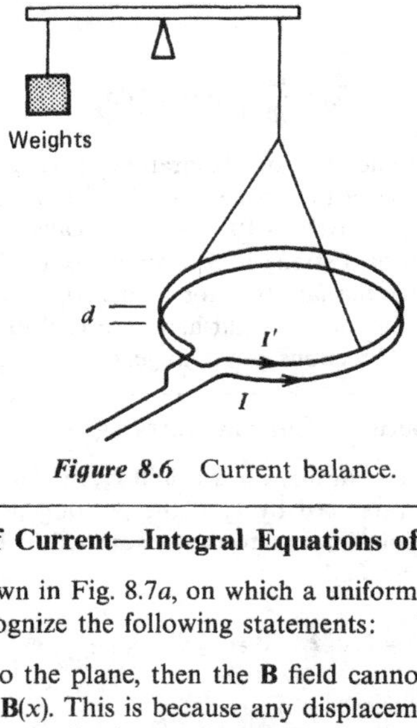

Figure 8.6 Current balance.

Example 8.5 A sheet of Current—Integral Equations of Magnetostatics

Consider a large plane, shown in Fig. 8.7*a*, on which a uniform surface current density flows: $\mathbf{K} = K\hat{z}$(A/m). We now recognize the following statements:

1. If $\hat{x}$ is perpendicular to the plane, then the **B** field cannot depend on the (z, y) coordinates orthogonal to x: $\mathbf{B} = \mathbf{B}(x)$. This is because any displacement of a field point parallel to the plane leaves the current distribution unchanged in relation to the field point.

2. The symmetry of the problem implies some relations between the field components:

$$B_x(x) = -B_x(-x) \qquad B_y(x) = -B_y(-x) \qquad B_z(x) = B_z(-x) \tag{8.29}$$

In words, upon moving from points (x, y, z) to $(-x, y, z)$, the x and y components are reversed in direction and the z component remains unchanged.

We can now apply the integral relations given in Eqs. (8.21) and (8.22) to find all the components of **B**. First, we apply Eq. (8.21) to the gaussian pillbox of Fig. 8.7*b*. Note that the contribution to the flux from the sides of the pillbox normal to the plane is zero because **B** is a function of x only. Then

$$0 = \int_S \mathbf{B} \cdot d\mathbf{a} = [\mathbf{B}(x) \cdot \hat{\mathbf{x}} + \mathbf{B}(-x) \cdot (-\hat{\mathbf{x}})]A$$

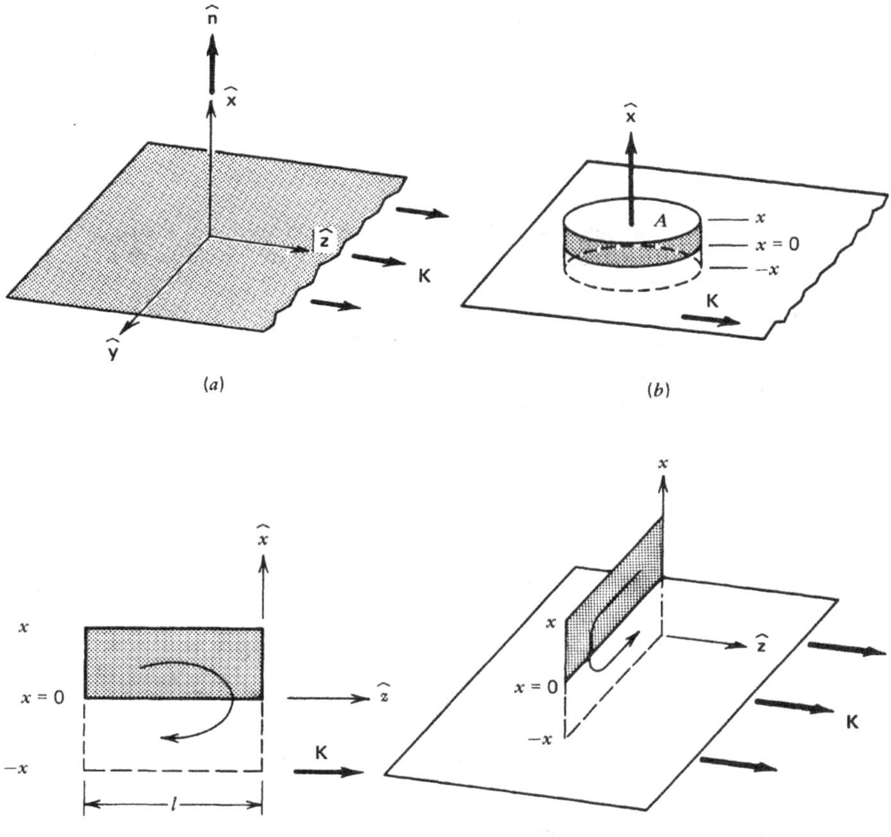

Figure 8.7 Application of the equations of magnetostatics to a current sheet to determine **B**. (*a*) Current sheet along the z direction. (*b*) Gauss's law of magnetostatics to determine the x component. (*c*) Ampere's law to determine the z component. (*d*) Ampere's law to determine the y component.

Therefore, the x component B_x is constant: $B_x(-x) = B_x(x)$. However, in light of the symmetry condition, Eq. (8.29), $B_x(x) = -B_x(-x)$. These two relations can simultaneously be valid only if $B_x = 0$. Next, consider the rectangular loop in the $x - z$ plane, shown in Fig. 8.7*c*. Application of Ampere's law and the realization that the current going through the loop is zero yields $\oint \mathbf{B} \cdot d\mathbf{r} = 0$. Since the line integrals of the other sides cancel identically, then

$$\oint \mathbf{B} \cdot d\mathbf{r} = [(\mathbf{B}(x) \cdot \hat{\mathbf{z}}) + (\mathbf{B}(-x) \cdot (-\hat{\mathbf{z}}))]l = 0$$

which gives $B_z(x) = B_z(-x)$, and the z component of **B** is constant. This constant turns out to be zero for reasons analogous to those applied in Example 8.3: $B_z = 0$.

Finally, for a rectangular loop in the $x - y$ plane, shown in Fig. 8.7*d*, application of Ampere's law yields

$$[\mathbf{B}(x) \cdot \hat{\mathbf{y}} + \mathbf{B}(-x) \cdot (-\hat{\mathbf{y}})]l = \mu_0 I$$

$$B_y(x) - B_y(-x) = \mu_0 K$$

Using the symmetry relation of Eq. (8.29) gives $2B_y(x) = \mu_0 K$, or

$$B_y(x) = \frac{\mu_0 K}{2} \tag{8.30}$$

Since $B_x = B_z = 0$, we can summarize the result in vector form as follows:

$$\mathbf{B} = \frac{\mu_0}{2}(\mathbf{K} \times \hat{\mathbf{n}}) \tag{8.31}$$

The field has constant magnitude on each side of the current sheet and a direction along $\mathbf{K} \times \hat{\mathbf{n}}$, where $\hat{\mathbf{n}}$ is the perpendicular unit vector from the plane to the point where $\mathbf{B}$ is calculated.

Example 8.6 Superposition of Filamentary Currents—Current Sheet

Another way to find the magnetic field due to the infinite plane current sheet of Fig. 8.7*a* is to regard the plane as a superposition of straight filamentary currents, each of magnitude $d\mathbf{I} = \mathbf{K}\,dy$. In Fig. 8.8 we have paired off filamentary currents, symmetrically dispersed about the field point. From Fig. 8.8 it can be seen that the resultant field of such pair is given by

$$d\mathbf{B} = \frac{\mu_0 K\,dy}{2\pi}\,\frac{\hat{\boldsymbol{\phi}}_1 + \hat{\boldsymbol{\phi}}_2}{\xi}$$

where $\hat{\boldsymbol{\phi}}_1$ and $\hat{\boldsymbol{\phi}}_2$ are unit vectors in the ϕ direction with respect to origins at each of the currents of the pair under consideration. The sum of the unit vectors $\hat{\boldsymbol{\phi}}_1 + \hat{\boldsymbol{\phi}}_2$ can be written as $2\hat{\mathbf{y}} \cos\phi$. Therefore

$$d\mathbf{B} = \frac{\mu_0 K}{2\pi}\,\frac{2\cos\phi}{\xi}\,\hat{\mathbf{y}}\,dy$$

Summing over all such pairs and noticing that $dy = \xi\,d\phi/\cos\phi$, we obtain

$$\mathbf{B}(x) = \int d\mathbf{B}(x) = \frac{\mu_0 K}{\pi}\,\hat{\mathbf{y}} \int_0^{\pi/2} d\phi = \frac{\mu_0 K}{2}\,\hat{\mathbf{y}} \tag{8.32}$$

as was found in the previous example.

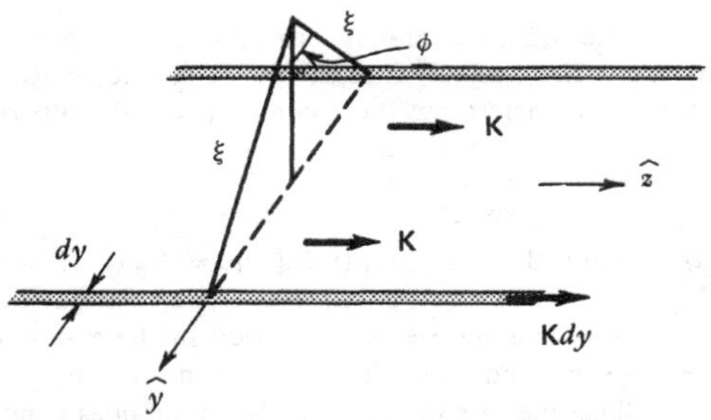

Figure 8.8 Determination of the field of a current sheet of Fig. 8.7*a* by the superposition of filamentary currents.

Example 8.7 Toroid—Ampere's Law

Suppose we have closely spaced filamentary windings wound around the doughnut-shaped object in Fig. 8.9. Such a configuration is known as a toroid. We take the z axis to be normal to the circular cross section of the toroid (along the axis).

If the current distribution has rotational symmetry about the z axis, the $\mathbf{B}$ field cannot depend upon the angle ϕ. Consequently, on the circular paths shown, one inside, C_i, and one

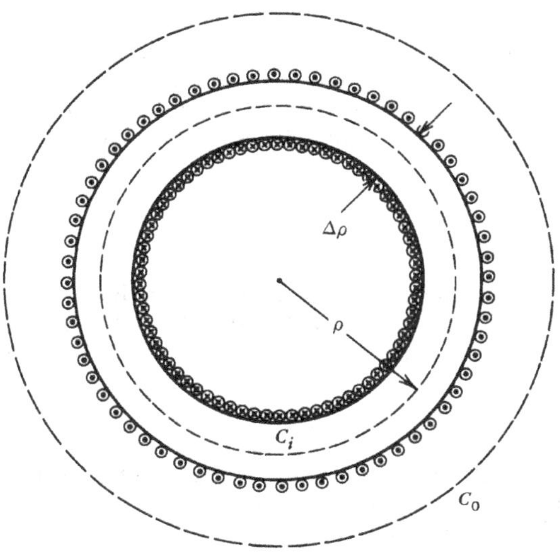

Figure 8.9 A toroid showing two amperean loops C_i and C_0.

outside the toroid, C_o, the line integrals have the form $\oint_C \mathbf{B}\cdot d\mathbf{r} = 2\pi\rho B_\phi$, where ρ is the distance from the z axis. If there are N total turns on the toroid, and the current through the wire is I, then Ampere's law for curve C_i gives

$$B_\phi = \mu_0 \frac{NI}{2\pi\rho} \tag{8.33}$$

For curve C_o, $B_\phi = 0$, because the net current through any surface whose periphery is C_o is zero.

If the "thickness" of the toroid, $\Delta\rho$, is much smaller than its mean radius R, we may write Eq. (8.33) for the field B_ϕ everywhere inside the toroid as

$$B_\phi = \mu_0 \frac{NI}{2\pi\rho} = \mu_0 n_l I \tag{8.34}$$

where n_l is the number of turns per unit length (meter) of the toroid. In this approximation, B_ϕ is constant inside the toroid. Finally, it can be argued from symmetry considerations involving the turns that $B_\rho = 0$ and $B_z = 0$, so $B = B_\phi$. (This is true if the turns essentially form a sheet of current everywhere perpendicular to $\hat{\boldsymbol{\phi}}$.)

Example 8.8 The Long Solenoid

As the radius R of a toroid goes to infinity but the dimensions of its cross section remain constant, the toroid approximates an endless solenoid. Therefore, from Eq. (8.34), the field inside such a solenoid is given by

$$B = \mu_0 n_l I \tag{8.35}$$

and the field outside the solenoid is zero. These are good approximations near the center of a long solenoid, where end effects may be neglected. Note that $n_l I$ is the current flowing per unit length on the "surface" of the solenoid, so we could also write

$$B = \mu_0 K \tag{8.36}$$

The field is constant and independent of the cross-sectional dimensions of the solenoid. Thus the solenoid can be deformed so that it approximates two current sheets, and it is then to be expected that its field is $\mu_0 K$, which is the field of two current sheets.

A Remark

In none of the foregoing examples have we used Ampere's law alone to find the **B** field. Two conditions are necessary, corresponding to specifications of the curl and divergence of **B**. Ampere's law is equivalent to specification of the curl only, so we have also had to invoke the flux continuity condition, $\oint \mathbf{B}\cdot d\mathbf{a} = 0$.

8.5 The Vector Potential

We now desire explicit methods for calculating **B** given the currents. To this end we turn to Eqs. (8.19) and (8.20)—that is, $\nabla\cdot\mathbf{B} = 0$ and $\nabla\times\mathbf{B} = \mu_0\mathbf{J}$—and use a result of vector calculus that says that if $\nabla\cdot\mathbf{B} = 0$ everywhere, there exists some vector field **A** such that

$$\mathbf{B} = \nabla \times \mathbf{A} \tag{8.37}$$

We shall demonstrate this result by actually constructing the new vector field **A**, which is then called the *vector potential* of **B**.

The converse of this result—namely, that if $\mathbf{B} = \nabla\times\mathbf{A}$, then $\nabla\cdot\mathbf{B} = 0$—is a consequence of the vector identity $\nabla\cdot\nabla\times\mathbf{f} \equiv 0$, where **f** is any (well-behaved) vector point function. We note that the vector **A** is not uniquely defined via Eq. (8.37). For example, a field $\mathbf{A}' = \mathbf{A} + \nabla\psi$ also satisfies $\mathbf{B} = \nabla\times\mathbf{A}'$, where ψ is any scalar field, since

$$\nabla\times\mathbf{A}' = \nabla\times\mathbf{A} + \nabla\times(\nabla\psi) = \nabla\times\mathbf{A}$$

In fact, **A** will be uniquely specified only if its divergence as well as its curl is given. Thus, to uniquely define **A** let us choose*

$$\nabla\cdot\mathbf{A} = 0 \tag{8.38}$$

Then, from $\nabla\times\mathbf{B} = \mu_0\mathbf{J}$, we obtain

$$\nabla\times(\nabla\times\mathbf{A}) = \mu_0\mathbf{J}$$

Using the vector expansion $\nabla\times(\nabla\times\mathbf{A}) = \nabla(\nabla\cdot\mathbf{A}) - (\nabla\cdot\nabla)\mathbf{A}$, which is valid in cartesian coordinates,† we obtain

$$\nabla^2\mathbf{A} \equiv (\nabla\cdot\nabla)\mathbf{A} = -\mu_0\mathbf{J} \tag{8.39}$$

which means that the cartesian components of **A** satisfy the following equations:

$$\nabla^2 A_x = -\mu_0 J_x \qquad \nabla^2 A_y = -\mu_0 J_y \qquad \nabla^2 A_z = -\mu_0 J_z \tag{8.40}$$

These are scalar equations; each of these component equations has precisely the form of Poisson's equation $\nabla^2\Phi = -\rho/\varepsilon_0$. The solution for $\mathbf{A} \equiv (A_x, A_y, A_z)$ is then

$$A_i(\mathbf{r}) = \frac{\mu_0}{4\pi}\int\frac{J_i(\mathbf{r}')dv'}{\xi} \tag{8.41}$$

* This choice is referred to as the Coulomb gauge.

† More generally, $\nabla\times(\nabla\times\mathbf{A}) = \nabla(\nabla\cdot\mathbf{A}) - \nabla\cdot(\nabla\mathbf{A})$, where $\nabla\mathbf{A}$ is a second-rank tensor or dyadic (see Example 1.3). In cartesian coordinates, $\nabla\cdot(\nabla\mathbf{A}) = (\nabla\cdot\nabla)\mathbf{A} = \nabla^2\mathbf{A}$, where ∇^2 is the Laplacian operator. (See Eq. 1.67.)

where $\xi = |\mathbf{r} - \mathbf{r}'|$ and $[A_i(\infty) = 0]$; and i stands for x, y, and z. More succinctly, we can write $[\mathbf{A}(\infty) \equiv \mathbf{0}]$:

$$\mathbf{A}(\mathbf{r}) = \frac{\mu_0}{4\pi} \int \frac{\mathbf{J}(\mathbf{r}')dv'}{\xi} \tag{8.42}$$

Observe that the contribution of $\mathbf{A}$ to a point $P(\mathbf{r})$ from the current density $\mathbf{J}(\mathbf{r}')$ (in a volume element dv' located at $\mathbf{r}'$) is parallel to $\mathbf{J}(\mathbf{r}')$.

$$d\mathbf{A}(\mathbf{r}) = \frac{\mu_0}{4\pi} \cdot \frac{\mathbf{J}(\mathbf{r}')dv'}{\xi} \tag{8.43}$$

In the case of a filamentary current I,

$$d\mathbf{A}(\mathbf{r}) = \frac{\mu_0}{4\pi} I \frac{d\mathbf{l}'}{\xi}$$

where $d\mathbf{l}'$ is a differential length along the wire. Summing up over all volume elements of the filament, we obtain

$$\mathbf{A}(\mathbf{r}) = \frac{\mu_0}{4\pi} \int \frac{d\mathbf{l}'}{\xi} I(\mathbf{r}') \tag{8.44}$$

which defines the filamentary approximation. We have now shown that a vector potential exists. One may now verify directly that the expression given in Eqs. (8.42) and (8.44) satisfy $\nabla \cdot \mathbf{A} = 0$.

We now reintroduce the concept of *magnetic flux* and derive a very useful relation between it and the vector potential. The magnetic flux F passing through a surface S is defined as

$$F = \int_S \mathbf{B} \cdot \hat{\mathbf{n}}\, da \tag{8.45}$$

This equation reduces to Eq. (8.21) when S is taken to be a closed surface; that is, F vanishes. The flux can be rewritten in terms of the vector potential $\mathbf{A}$ by substituting $\mathbf{B} = \nabla \times \mathbf{A}$ and using Stokes' theorem:

$$F = \int_S \nabla \times \mathbf{A} \cdot \hat{\mathbf{n}}\, da = \oint_C \mathbf{A} \cdot d\mathbf{l} \tag{8.46}$$

where C is a closed loop bounding the surface S.

This definition of flux in terms of $\mathbf{A}$ can be used to find the behavior of the tangential components of $\mathbf{A}$ when crossing an interface of two regions of space. Consider Fig. 8.10, which shows two regions 1 and 2 with a common interface, which has a surface current $\mathbf{K}$ with the vector potentials in the regions are $\mathbf{A}_1$ and $\mathbf{A}_2$. We apply Eq. (8.46) to the rectangular path $ABCD$, whose sides $AB = DC = l$ are taken small. In addition, $BC = AD$ are taken to be very small such that the area of the rectangle is vanishingly small thus making F vanish (for finite $\mathbf{B}$), and their contribution to the line integral is also vanishingly small. Thus

$$\mathbf{A}_{1t} = \mathbf{A}_{2t} \tag{8.47}$$

which establishes the continuity of the tangential component of the vector potential irrespective of the presence of $\mathbf{K}$.

The boundary condition on the normal component of the vector potential in passing through regions of currents can be determined from $\nabla \times \mathbf{B} = \mu_0 \mathbf{J}$. However, it is more convenient to use the equivalent boundary condition on $\mathbf{B}$ instead. Such a

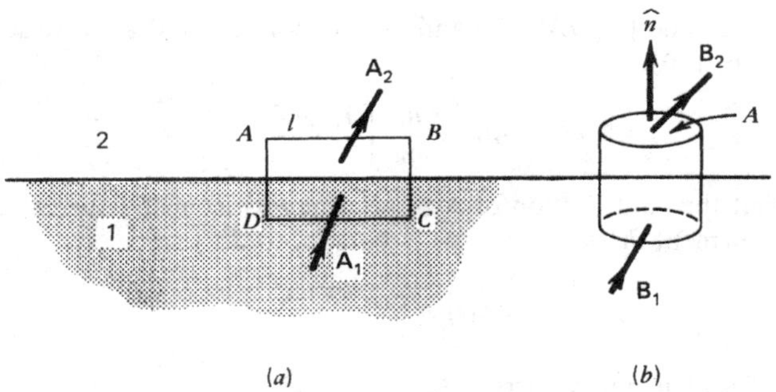

Figure 8.10 (*a*) Application of the integral form of $\mathbf{B} = \nabla \times \mathbf{A}$ to a rectangle at the interface of two regions to determine boundary conditions on the vector potential. (*b*) Application of the integral form of $\nabla \cdot \mathbf{B} = 0$ to a pillbox at the same interface to find bound conditions on **B** field.

boundary condition can be determined by applying Ampere's law to the same rectangle shown in Fig. 8.10. Using similar arguments on the size of the rectangle, one can easily show that

$$\mathbf{B}_{2t} - \mathbf{B}_{1t} = \mu_0 \mathbf{K} \times \hat{\mathbf{n}} \tag{8.48}$$

where $\mathbf{B}_{it}$ is the tangential component of the field at the interface, **K** is the surface current density, and $\hat{\mathbf{n}}$ a unit vector normal to the interface and pointing away from material 1. For a current sheet such that **K** is along the z axis, and $\hat{\mathbf{n}}$ is along the x axis, Eq. (8.48) reduces to

$$B_{2y} - B_{1y} = \mu_0 K$$

which is the result arrived at in Example 8.5.

Now we show that the continuity of the tangential components of the vector potential is equivalent to the continuity of the normal component of **B**. When the flux is evaluated over a closed surface, Eq. (8.45) gives the same result obtained in Eq. (8.21); that is, $\oint \mathbf{B} \cdot d\mathbf{a} = 0$. (This result is a direct consequence of $\nabla \cdot \mathbf{B} = 0$.) Evaluating the surface integral over a pillbox partially immersed in both regions and of vanishingly small height as shown in Fig. 8.10, one can show that

$$B_{1n} = B_{2n} \tag{8.49}$$

Since the continuity of the tangential component of **A** was established using the definition of the flux, then such a condition is equivalent to the continuity of the normal component of **B**.

The vector potential **A** will be of primary use in calculating other quantities. As a simple analytical means of calculating **B**, its use is limited. The following example, however, describes one case where it is used to calculate a magnetic field.

Example 8.9 Vector Potential and B Field for a Filamentary Current

We choose cylindrical coordinates with the z axis along the wire to determine the vector potential just above the center of a long straight filamentary wire of length $2L$ (shown in Fig. 8.11). We employ the filamentary approximation for the wire and note that, from Eq. (8.44), **A** must be in the z direction. Thus

$$A = A_z = \frac{\mu_0}{4\pi} \int_{-L}^{L} \frac{I\,dz}{\sqrt{z^2 + \rho^2}} = \frac{\mu_0 I}{2\pi} \int_0^L \frac{dz}{\sqrt{z^2 + \rho^2}}$$

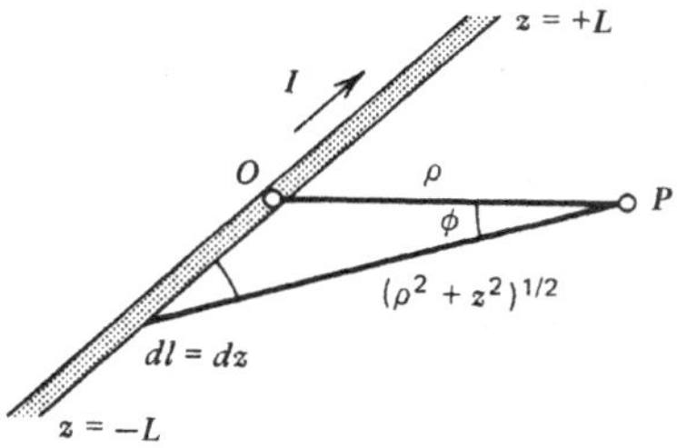

Figure 8.11 A finite, straight filamentary current. Determination of the vector potential and from which **B** may be calculated.

which can be evaluated to give the result:

$$A_z = \frac{\mu_0 I}{2\pi}[\ln(z + \sqrt{z^2 + \rho^2})]_0^L = \frac{\mu_0 I}{2\pi}\ln\left[\frac{L}{\rho}\left(1 + \sqrt{1 + \frac{\rho^2}{L^2}}\right)\right]$$

In the limit $\rho \ll L$, this result reduces to

$$A_z = \frac{\mu_0 I}{2\pi}\ln\left(\frac{2L}{\rho}\right) \qquad \frac{L}{\rho} \gg 1. \tag{8.50}$$

As $L/\rho \to \infty$, the value of $A \to \infty$. However, if we calculate $\mathbf{B} = \nabla \times \mathbf{A} = -\hat{\boldsymbol{\phi}}\,\partial A_z/\partial\rho$ we obtain the result

$$\mathbf{B} = -\hat{\boldsymbol{\phi}}\frac{\mu_0 I}{2\pi}\left(\frac{1}{L + \sqrt{L^2 + \rho^2}} \cdot \frac{\rho}{\sqrt{L^2 + \rho^2}} - \frac{1}{\rho}\right) \tag{8.51}$$

and if $L/\rho \gg 1$, this equation gives the finite result obtained previously for the field outside of an infinitely long straight wire: $\mathbf{B} = \hat{\boldsymbol{\phi}}\mu_0 I/2\pi\rho$.

It will be observed that as $L/\rho \to \infty$, A_z takes the form

$$A_z = \frac{-\mu_0 I}{2\pi}\ln\rho + \text{constant} \tag{8.52}$$

since upon differentiation to calculate **B**, the constant disappears. That is, as long as A_z has this form, one obtains the value of **B** given in Eq. (8.24). The constant is used to determine the zero of A. Thus, the reason for the infinity in A as $L \to \infty$ is that we have tried to set $A(\infty) = 0$. This is not feasible when the current distribution itself extends to infinity, a situation reminiscent of the electrostatic potential of an infinitely long straight wire, uniformly charged with a charge per meter of λ. In that case we have, from Example 3.1,

$$\Phi(\rho) = -\frac{\lambda}{2\pi\varepsilon_0}\ln\rho + \text{constant}$$

Formulas (8.42) or (8.44) for $\mathbf{A}(\mathbf{r})$ can thus only be safely used to calculate $\mathbf{A}$ for current distributions that are essentially localized inside a finite volume V (that is, $\mathbf{J}(\mathbf{r}')$ must fall off faster than $1/r$ as $r \to \infty$). The results of this example can be used to determine the potential of two parallel currents (see Problem 8.7).

Example 8.10 Magnetic Vector Potential Due to a Nonfilamentary Current

We consider an infinitely long cylindrical conductor of radius ρ_0 with a constant current I flowing in. Taking the z axis along the axis of the conductor, then the cartesian components of the magnetic vector potential satisfy Eqs. (8.40):

$$\nabla^2 A_x = \nabla^2 A_y = 0 \qquad \text{and} \qquad \nabla^2 A_z = -\mu_0 J \tag{8.53}$$

where $J = I/\pi\rho_0^2$ for $\rho \le \rho_0$ and $J = 0$ for $\rho > \rho_0$.

Because there are no current sources in the x and y directions, A_x and A_y may be taken to be zero. Moreover, we take the component A_z to depend only on the distance from the axis of the conductor. Thus A_z satisfies the equation

$$\frac{1}{\rho}\frac{d}{d\rho}\left(\rho\frac{dA_z}{d\rho}\right) = -\mu_0 J \tag{8.54}$$

which can be easily integrated twice to give:

$$A_z = -\frac{1}{4\pi}\mu_0 I\left(\frac{\rho}{\rho_0}\right)^2 + C_1 \ln \rho + C \qquad \rho \le \rho_0 \tag{8.55}$$

$$A_z = D_1 \ln \rho + D \qquad \rho > \rho_0 \tag{8.56}$$

where C, C_1, D, and D_1 are constants.

The boundary conditions on the potential and the corresponding fields can now be used to evaluate these constants. Since the potential inside the conductor must be finite because the current is uniformly distributed (no filamentary currents at the axis), then C_1 must be taken to be zero. Moreover, the potentials should give the same value at $\rho = \rho_0$ as required by $\mathbf{A}_{1t} = \mathbf{A}_{2t}$, thus giving the following relationship among D_1, D, and C:

$$-\frac{1}{4\pi}\mu_0 I + C = D_1 \ln \rho_0 + D \tag{8.57}$$

The constant D_1 can now be evaluated. Taking $\nabla \times \mathbf{A}$ in the region $\rho > \rho_0$ gives

$$\mathbf{B} = -\frac{D_1}{\rho}\hat{\boldsymbol{\phi}} \tag{8.58}$$

and applying Ampere's law to a circle of radius $\rho > \rho_0$, with its center at the axis of the current, and its plane normal to the axis of the current gives

$$\oint \mathbf{B}\cdot d\mathbf{l} = -\int_0^{2\pi} \frac{D_1}{\rho}\rho\, d\phi = \mu_0 I$$

This equation yields $D_1 = -(\mu_0 I/2\pi)$. Substituting for D_1 in Eq. (8.57) gives

$$D = -\frac{1}{4\pi}\mu_0 I + \frac{\mu_0 I}{2\pi}\ln \rho_0 + C$$

Thus the potentials and the fields are

$$A_z = -\frac{1}{4\pi}\mu_0 I\left(\frac{\rho}{\rho_0}\right)^2 + C \qquad \mathbf{B} = \frac{\mu_0 I}{2\pi}\frac{\rho}{\rho_0^2}\hat{\boldsymbol{\phi}} \qquad \rho < \rho_0$$

$$A_z = -\frac{\mu_0 I}{4\pi}\left(1 + 2\ln\frac{\rho}{\rho_0}\right) + C \qquad \mathbf{B} = \frac{\mu_0 I}{2\pi}\frac{1}{\rho}\hat{\boldsymbol{\phi}} \qquad \rho > \rho_0$$

The above magnetic potential is determined to within an arbitrary constant C; however, the field is uniquely determined, indicating that there is no arbitrariness in evaluating measurable quantities like the forces on charge distribution.

Example 8.11 Determination of the Magnetic Vector Potential Using $\nabla \times \mathbf{A} = \mathbf{B}$

In cases of high symmetry where it is easy to calculate the magnetic field, as in cases where Ampere's law is applicable, then the magnetic vector potential can be calculated from the relation $\nabla \times \mathbf{A} = \mathbf{B}$. For example, we calculate the vector potential produced by the current sheet examined in Example 8.5. Taking $\mathbf{B}$ from Eq. (8.32), we find

$$\nabla \times \mathbf{A} = \pm\frac{1}{2}\mu_0 K\hat{\mathbf{y}} \tag{8.59}$$

where the + and − signs apply to $x < 0$ and $x > 0$ regions, respectively. Because the current is along the z axis, we expect $\mathbf{A}$ to be also along the z axis. Therefore, Eq. (8.59) reduces to

$$\frac{dA_z}{dx} = \pm\frac{1}{2}\mu_0 K \tag{8.60}$$

which can be easily integrated, as follows:

$$A_z = \pm\frac{1}{2}\mu_0 K(x - x_0) \tag{8.61}$$

where x_0 is a constant which defines the position at which the potential vanishes.

8.6 The Biot-Savart Law

We shall now find a direct procedure for calculating $\mathbf{B}$ from a known current distribution. This is simply achieved by employing the relation $\mathbf{B} = \nabla \times \mathbf{A}$, where $\mathbf{A}$ is the vector potential of the current distribution written as in Eq. (8.42),

$$\mathbf{B}(\mathbf{r}) = \nabla \times \left[\frac{\mu_0}{4\pi}\int_V \frac{\mathbf{J}(\mathbf{r}')dv'}{\xi}\right]$$

We assume $\mathbf{J}$ to be nonzero only in the volume V. The ∇ operator is understood only to operate on functions of $\mathbf{r}(x, y, z)$. The volume V over which integration occurs is independent of the point $\mathbf{r}$ where $\mathbf{B}$ is to be evaluated, and so the curl operation can be taken under the integral sign:

$$\mathbf{B}(\mathbf{r}) = \frac{\mu_0}{4\pi}\int_V \nabla \times \left[\frac{\mathbf{J}(\mathbf{r}')}{\xi}\right]dv'$$

Using the vector identity given by Eq. (1.58) yields

$$\nabla \times \left[\frac{\mathbf{J}(\mathbf{r}')}{\xi}\right] = \frac{1}{\xi}[\nabla \times \mathbf{J}(\mathbf{r}')] + \nabla\left(\frac{1}{\xi}\right) \times \mathbf{J}(\mathbf{r}')$$

Since $\mathbf{J}$ is independent of $\mathbf{r}(x, y, z)$, then

$$\nabla \times \left[\frac{\mathbf{J}(\mathbf{r}')}{\xi}\right] = \nabla\left(\frac{1}{\xi}\right) \times \mathbf{J}(\mathbf{r}')$$

Therefore,

$$\mathbf{B}(\mathbf{r}) = \frac{\mu_0}{4\pi}\int \mathbf{J}(\mathbf{r}') \times \frac{\boldsymbol{\xi}}{\xi^3}\,dv' \tag{8.62}$$

To each element of dv' there may be ascribed a contribution $d\mathbf{B}$ to the $\mathbf{B}$ field of

$$d\mathbf{B}(\mathbf{r}) = \frac{\mu_0}{4\pi}\frac{\mathbf{J}(\mathbf{r}') \times \boldsymbol{\xi}}{\xi^3}\,dv' \tag{8.63}$$

which shows that $d\mathbf{B}$ is normal to the plane of $\boldsymbol{\xi}$ and $\mathbf{J}$.

If we have filamentary currents, like currents in thin wires, we can use the filamentary approximation of Eq. (8.9), as long as $(\boldsymbol{\xi}/\xi^3)$ does not vary appreciably over the cross section of the filament:

$$\mathbf{B}(\mathbf{r}) = \frac{\mu_0}{4\pi}\int dl\,\mathbf{I} \times \frac{\boldsymbol{\xi}}{\xi^3} = \frac{\mu_0}{4\pi}\int I\,d\mathbf{l} \times \frac{\boldsymbol{\xi}}{\xi^3} \tag{8.64}$$

Each element of current, $I\,d\mathbf{l}$, may be interpreted as contributing to the total field an amount

$$d\mathbf{B}(\mathbf{r}) = \frac{\mu_0}{4\pi} I \frac{d\mathbf{l} \times \boldsymbol{\xi}}{\xi^3}. \tag{8.65}$$

Equations (8.62) to (8.65) are all variously referred to as "Biot-Savart" formulas. They are especially useful in calculating **B** directly from the currents, as will be illustrated in the examples below. They indicate that the **B** field for a current element falls off as $1/\xi^2$ as did the **E** (electrostatic) field for the charge element. This is in consonance with the fact that the vector potential **A** varies as $1/\xi$, like the electric potential Φ.

Example 8.12 Biot-Savart Law—Field on Axis of Circular Loop

Current flows, as shown in Fig. 8.12, around a circular loop of radius R. We observe that as $d\mathbf{l}$ is summed up around the loop, the only component of **B** that emerges lies along the z axis; that is, $\mathbf{B} = B\hat{\mathbf{z}}$. Using the filamentary approximation given in Eq. (8.65),

$$dB_z = \left[\frac{\mu_0}{4\pi} \frac{I\,d\mathbf{l} \times \boldsymbol{\xi}}{\xi^3}\right]\cdot\hat{\mathbf{z}} \tag{8.66}$$

where $d\mathbf{l} = R\,d\phi\,\hat{\boldsymbol{\phi}}$, $\xi = z\hat{\mathbf{z}} - R\hat{\boldsymbol{\rho}}$. Note that $(d\mathbf{l} \times \boldsymbol{\xi})\cdot\hat{\mathbf{z}} = R^2\,d\phi$; hence

$$dB_z = \frac{\mu_0}{4\pi} \frac{IR^2\,d\phi}{\xi^3}$$

Since ξ remains constant as we integrate around the loop, we obtain

$$B = B_z = \frac{\mu_0 IR^2}{2\xi^3} \tag{8.67}$$

The field at the center of the circle is thus $\mu_0 I/2R$. It is not easy to find **B** off the axis, however, the field can be calculated near the axis using $\nabla\cdot\mathbf{B} = 0$. Note that the vector potential of the loop on the axis is zero (see Problem 8.17).

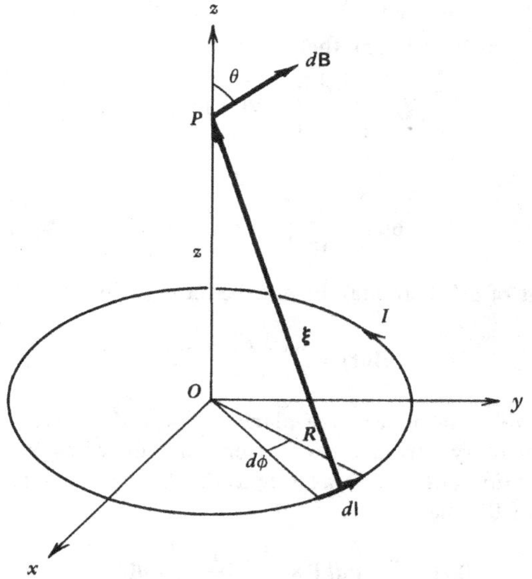

Figure 8.12 A current loop.

Example 8.13 Field on the Axis of a Circular Solenoid

Let us calculate the field on the axis of the solenoid in Fig. 8.13. We assume the solenoid to be constituted of adjacent current filamentary loops, closely packed and of negligible pitch. The current distribution is thus akin to a cylindrical sheet of current. If there are n_l such loops per unit length and the current per loop is I, the current per unit length, K, of the equivalent current sheet is $K = n_l I$. The direction of K is indicated in the figure. In a length dz, the current is $K\,dz$. The contribution to B on the axis of the solenoid of a loop carrying this current is, from Example 8.12,

$$dB = dB_z = \frac{\mu_0 K\,dz}{2} \cdot \frac{R^2}{\xi^3}$$

Noting that $R/z = \tan\theta$, $dz = -R\,d\theta/\sin^2\theta$, and $\xi = R/\sin\theta$,

$$dB = -\frac{\mu_0 K}{2} \frac{R\,d\theta}{\sin^2\theta} \cdot \frac{\sin^3\theta}{R} = -\frac{\mu_0 K}{2} \sin\theta\,d\theta$$

Therefore,

$$B = \int_{\theta_2}^{\theta_1} dB = \frac{\mu_0 K}{2}[\cos\theta_1 - \cos\theta_2] \tag{8.68}$$

For a point P at the center of the solenoid,

$$\theta_2 = \pi - \theta_1 \qquad \cos\theta_2 = -\cos\theta_1 \qquad B_{\text{center}} = \mu_0 K \cos\theta_1 \tag{8.69}$$

If $\theta_1 \to 0$—that is, if the solenoid is very long compared to its radius—then $B = \mu_0 K$, which is a result obtained previously using Ampere's law (see Example 8.8). Thus the field at the center of a finite solenoid is less than for an infinite solenoid. In order to have $\cos\theta_1 = 0.99$, the diameter $2R$ of the solenoid should be about 1/7 of its length.

As an idea of the magnitudes involved, note that if $I = 1\ A$ and $n_l = 10/\text{cm} = 1000/\text{m}$, then $K = n_l I = 1000$ A/m, B for the infinitely long case becomes $4\pi \times 10^{-4}$ tesla (or, in gaussian units, 4π gauss).

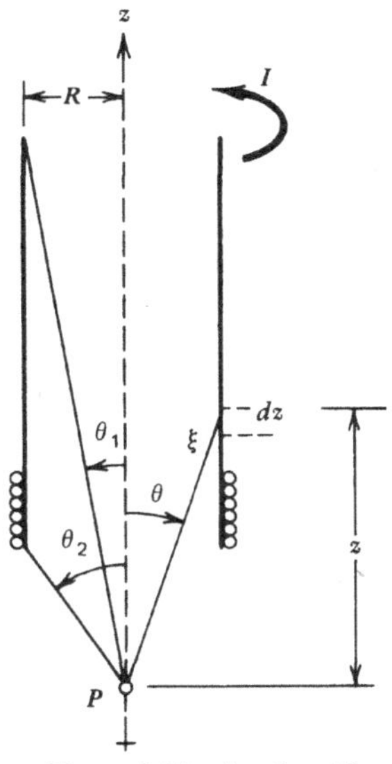

Figure 8.13 A solenoid.

If the solenoid is semiinfinite (i.e., if $\theta_1 \to 0$), then $B \to \mu_0 K[1 - \cos\theta_2]/2$. If $\theta_2 = \pi/2$ (at edge of solenoid), $B \to \mu_0 K/2$. The field is just one-half the value of that in the infinite solenoid.

Example 8.14 Biot-Savart Law—Field of a Straight Wire Segment

Many circuits may be considered to be composed of straight wire segments. The **B** field of these segments may easily be computed as follows. Consider Fig. 8.14. The field from each

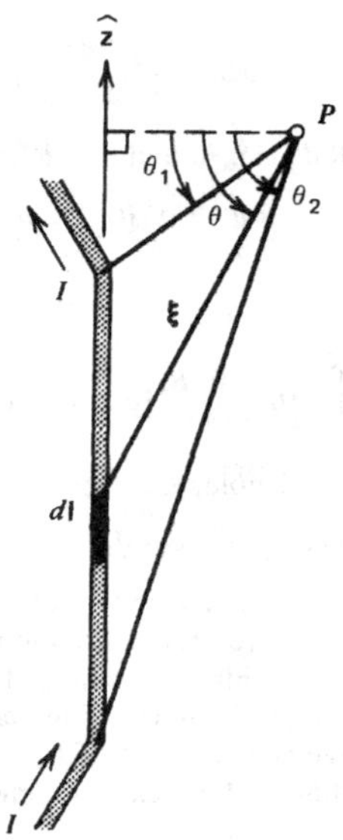

Figure 8.14 The **B** field of a straight wire segment, calculated by means of the Biot-Savart law.

elementary segment $d\mathbf{l}$ will be in the same direction at a given point, P. This direction is "around" the z axis (the direction of $\hat{\mathbf{z}} \times \boldsymbol{\xi} = \hat{\boldsymbol{\phi}}$). From Eq. (8.65),

$$d\mathbf{B} = \frac{\mu_0 I}{4\pi} \frac{d\mathbf{l} \times \boldsymbol{\xi}}{\xi^3}$$

where $d\mathbf{l} = dz'\hat{\mathbf{z}}$ and $\boldsymbol{\xi} = -z'\hat{\mathbf{z}} + \rho\hat{\boldsymbol{\rho}}$. Note that $d\mathbf{l} \times \boldsymbol{\xi} = \rho\, dz'\, \hat{\boldsymbol{\phi}}$; thus

$$d\mathbf{B} = \frac{\mu_0 I}{4\pi} \frac{\rho\, dz'}{\xi^3} \hat{\boldsymbol{\phi}}$$

Noting that $\xi = \rho/\cos\theta$ and $z' = -\rho\tan\theta$,

$$d\mathbf{B} = -\frac{\mu_0 I}{4\pi} \frac{\rho^2\, d\theta}{\cos^2\theta} \frac{\cos^3\theta}{\rho^3} \hat{\boldsymbol{\phi}} = -\frac{\mu_0 I}{4\pi\rho} \cos\theta\, d\theta\, \hat{\boldsymbol{\phi}}$$

Therefore

$$\mathbf{B} = \int_{\theta_2}^{\theta_1} d\mathbf{B} = \frac{\mu_0 I}{4\pi\rho} \int_{\theta_1}^{\theta_2} \cos\theta\, d\theta\, \hat{\boldsymbol{\phi}} = \frac{\mu_0 I}{4\pi\rho} (\sin\theta_2 - \sin\theta_1)\hat{\boldsymbol{\phi}} \tag{8.70}$$

Note that, for a line segment of infinite length, $\theta_1 \to -\pi/2$ and $\theta_2 \to +\pi/2$, so $B \to \mu_0 I/2\pi\rho$, as found previously using Ampere's law in Example 8.3. We note that Eq. (8.70) may be used to determine the field of multisegment circuit (see Problem 8.13).

Example 8.15 B Field on the Axis of a Spinning Charged Disk

A uniformly charged thin disk of charge density σ, radius R, and thickness $t < R$ rotates with an angular velocity $\boldsymbol{\omega}$ about the z axis of symmetry, as shown in Fig. 8.15.

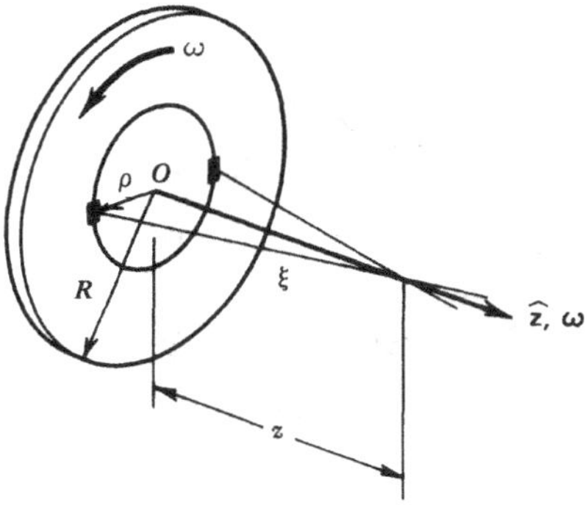

Figure 8.15 Spinning charged disk.

The current density in the disk is given by $\rho\mathbf{v} = \sigma\omega\rho\hat{\boldsymbol{\phi}}$, where ρ is the distance from the axis of the disk. A ring-shaped portion of the disk of radial thickness $d\rho$ thus constitutes a current ring of current $dI = \sigma\omega\rho(d\rho)$.

The **B** field on the z axis due to this current ring was given in Eq. (8.67):

$$dB = \frac{\mu_0}{2} \cdot \frac{(dI)\rho^2}{\xi^3}$$

Substituting for dI, writing ξ in terms of ρ and z, and integrating from $\rho = 0$ to $\rho = R$, we get

$$B = B_z = C \int_0^R \frac{\rho^3 \, d\rho}{(\rho^2 + z^2)^{3/2}}$$

where $C \equiv [\mu_0 \sigma\omega/2]$. With the substitution $u^2 = \rho^2 + z^2$, we find

$$B = C \int_{|z|}^{\sqrt{z^2+R^2}} \left(1 - \frac{z^2}{u^2}\right) du = C\left[\frac{R^2 + 2z^2}{\sqrt{R^2 + z^2}} - 2|z|\right] \tag{8.71}$$

This result can now be used to determine the field due to a spinning charged sphere (see Problem 8.16).

8.7 The Magnetic Scalar Potential

We have seen that because $\nabla \times \mathbf{B} = \mu_0 \mathbf{J}$ for steady currents, **B** is not a conservative field. Therefore it makes no sense to introduce a *magnetic scalar potential* in the same sense as we introduced an electrostatic potential. Nonetheless, there usually are certain regions of space where $\nabla \times \mathbf{B} = 0$. For these regions we are permitted to introduce a scalar potential function just so long as the space is "simply connected." A simply connected region is one for which any closed curve constructed therein

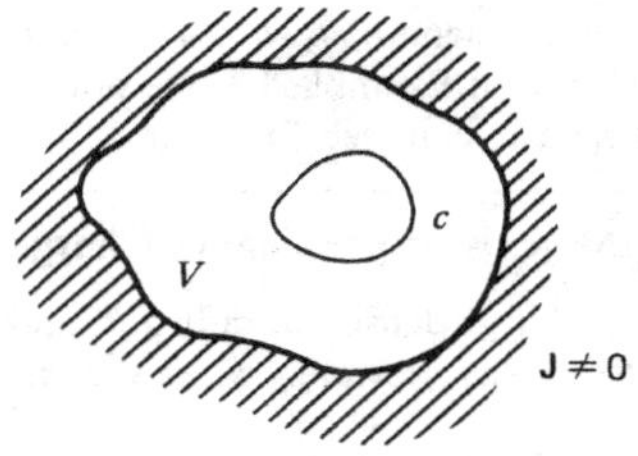

(a)

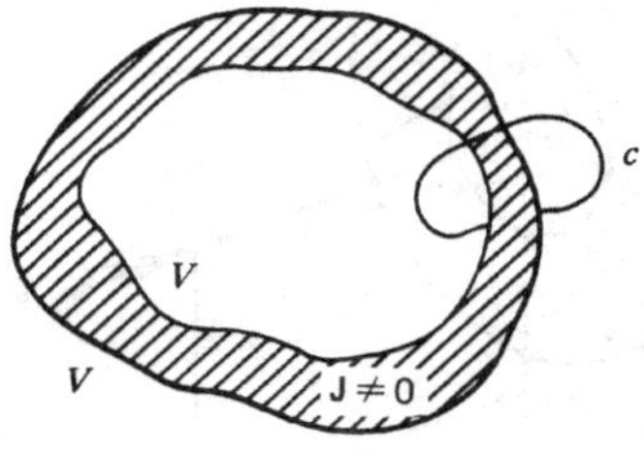

(b)

Figure 8.16 Simply connected regions. (*a*) Because any loop c in V can be shrunk to zero without crossing the currents, then V is a simply connected region. (*b*) The volume V is not a simply connected region.

can be shrunk down continuously to a point without the curve leaving the region, as in Fig. 8.16. The relevance of simple connectedness to magnetism is that the sources of magnetic fields are current loops, which by their nature render space not simply connected. If the current loops are finite in size, however, it is possible to imagine them as spanned or enclosed by surfaces that leave the rest of space simply connected.

Assuming then that there exists a simply connected volume V, where $\nabla \times \mathbf{B} = 0$, we define a magnetic scalar potential function Φ_m in V, such that

$$\mathbf{B} = -\mu_0 \nabla \Phi_m \tag{8.72}$$

Substituting this in $\nabla \cdot \mathbf{B} = 0$ gives

$$\nabla^2 \Phi_m = 0 \tag{8.73}$$

This is just Laplace's equation again. We can find unique solutions for it if, on the boundaries of V, certain boundary conditions are satisfied (e.g., the normal components of **B** must be continuous across a boundary surface, and the discontinuity of the tangential components is proportional to the surface current density, see Eqs. 8.48–8.49). The techniques to be used for finding **B** in simply connected regions in which there are no currents, therefore, are just those that are used in electrostatics.

Let us now attempt to find an explicit expression for Φ_m in terms of the current distribution outside V by use of the Biot-Savart law and Eq. (8.72). Consider a single current loop shown in Fig. 8.17*a*. The space outside this loop is not simply connected, but we can make it so by erecting an imaginary surface to span the loop.

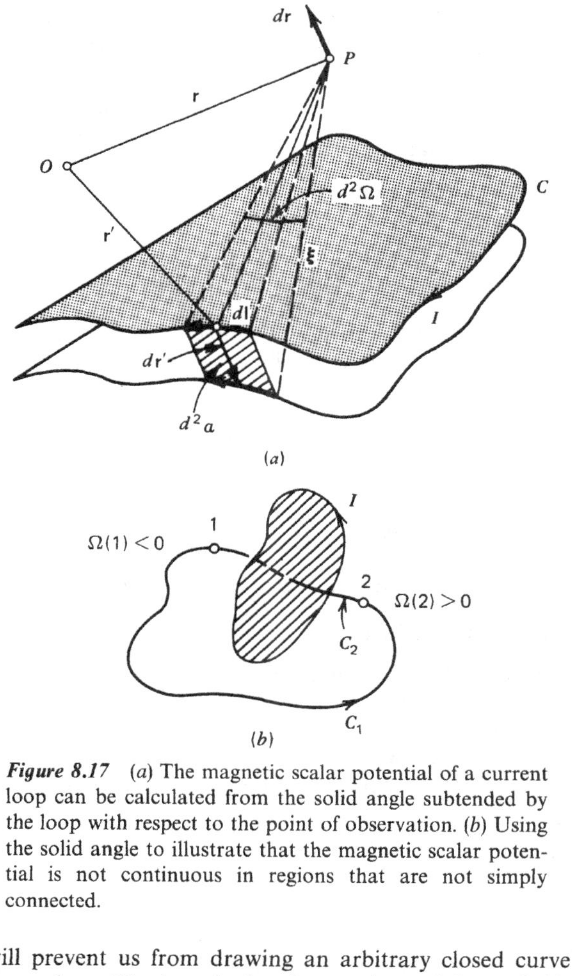

Figure 8.17 (*a*) The magnetic scalar potential of a current loop can be calculated from the solid angle subtended by the loop with respect to the point of observation. (*b*) Using the solid angle to illustrate that the magnetic scalar potential is not continuous in regions that are not simply connected.

This surface will prevent us from drawing an arbitrary closed curve around and through the current loop. The loop is therefore at the periphery of this surface. Any such surface will suffice.

Now, if Φ_m is to be a potential function in V, then

$$d\Phi_m \equiv \nabla\Phi_m \cdot d\mathbf{r} = -\frac{\mathbf{B}\cdot d\mathbf{r}}{\mu_0} \tag{8.74}$$

where $d\mathbf{r}$ is an infinitesimal displacement of the field point P, which brings about a change in potential $d\Phi_m$. But, from Eq. (8.64), the magnetic field for a filamentary current loop C is

$$\mathbf{B} = \frac{\mu_0 I}{4\pi}\oint_C \frac{d\mathbf{l} \times \boldsymbol{\xi}}{\xi^3} \tag{8.64}$$

Therefore,

$$d\Phi_m = \frac{-I}{4\pi}\left[\oint_C \frac{d\mathbf{l} \times \boldsymbol{\xi}}{\xi^3}\right]\cdot d\mathbf{r} = \frac{-I}{4\pi}\oint_C \left[\frac{d\mathbf{l} \times \boldsymbol{\xi}}{\xi^3}\cdot d\mathbf{r}\right]$$

Now note that a change $d\Phi_m$, affected by a displacement of P by $d\mathbf{r}$, is the same as the change in Φ_m affected by displacing the loop rigidly by $d\mathbf{r}' = -d\mathbf{r}$. Thus, writing $-d\mathbf{r} = d\mathbf{r}'$ in the above integral, we have

$$d\Phi_m = \frac{I}{4\pi}\oint_C \frac{d\mathbf{l} \times \boldsymbol{\xi}}{m^3}\cdot d\mathbf{r}' = \frac{I}{4\pi}\oint_C \frac{d\mathbf{r}' \times d\mathbf{l}}{\xi^3}\cdot \boldsymbol{\xi}$$

We observe that $d\mathbf{r}' \times d\mathbf{l}$ is an element of area, $d^2\mathbf{a}'$. Consequently the integrand of the integral is an element of solid angle $d^2\Omega$; that is,

$$d\Phi_m = \frac{I}{4\pi}\oint_C \frac{d^2\mathbf{a}'\cdot\boldsymbol{\xi}}{\xi^3} = \frac{I}{4\pi}\oint_C d^2\Omega$$

Taking the line integral around C then yields the *change* in solid angle, as observed at P, when the whole current loop is rigidly displaced by $d\mathbf{r}'$; that is,

$$d\Phi_m = \frac{I}{4\pi}\oint_C d^2\Omega = \frac{I}{4\pi}\, d\Omega$$

or*

$$\Phi_m \equiv \frac{I}{4\pi}\Omega$$

$$\mathbf{B} = -\mu_0 \nabla\Phi_m = -\frac{\mu_0 I}{4\pi}\nabla\Omega \tag{8.75}$$

Two comments are pertinent here:

1. $\nabla\Omega$, the gradient of the solid angle Ω, is obtained by taking derivatives with respect to the coordinates $\mathbf{r}(x, y, z)$ of point P, since Ω will be a function of these coordinates.
2. Ω is reckoned positive when $\int_S d\mathbf{a}' \equiv \mathbf{A}$ points toward the point P, or, in other words, when the current I rotates in a counterclockwise direction as observed from P.

We emphasize that in computing

$$\mu_0[\Phi_m(2) - \Phi_m(1)] = -\int_1^2 \mathbf{B}\cdot d\mathbf{r}$$

along a path C, we must not pass through the borders of our simply connected region. Thus, for the paths C_1 and C_2 in Fig. 8.17*b*, different results (differing by $\mu_0 I$) obtain for the integral because either C_1 or C_2 will penetrate any imaginary surface constructed to keep the space simply connected. It is for analogous physical reasons that in general we cannot use the equation $\nabla^2\Phi_m = 0$ to solve magnetic-field problems in regions where $\mathbf{J} \neq 0$.

Example 8.16 The Magnetic Scalar Potential of a Long Filamentary Current

In Example 8.9 the magnetic vector potential of a long filamentary current was calculated. In this example we show that the magnetic field of such a current can also be calculated from a scalar potential. Away from the current, the scalar potential Φ_m satisfies Laplace's equation. Because of the symmetry of the problem, the potential is expected to depend on only one coordinate.

* Note that $\Phi_m(\infty) = 0$. The constant of integration is thus chosen to be zero.

The solution of Laplace's equation in cylindrical coordinates was used to solve for the electric potential and electric field of a long, charged wire. The potential was taken to depend on a single variable, namely, ρ. Such a dependence results in a radial electric field. In the magnetic case the nature of the magnetic field is quite different, and it is expected to be in the $\hat{\boldsymbol{\phi}}$ direction. Therefore it is natural to take the magnetic scalar potential to depend on ϕ rather than ρ. Thus

$$\frac{d^2\Phi_m}{d\phi^2} = 0 \tag{8.76}$$

which has the solution

$$\Phi_m = C\phi \tag{8.77}$$

The corresponding magnetic field is derived from $\mathbf{B} = -\mu_0 \nabla\Phi_m$:

$$\mathbf{B} = -\frac{\mu_0}{\rho}\frac{d\Phi}{d\phi}\hat{\boldsymbol{\phi}} = -\mu_0\frac{C}{\rho}\hat{\boldsymbol{\phi}}$$

Ampere's law can be used to evaluate the constant C. We take a circular loop normal to the wire with a radius ρ and the center located at the wire. Then $\oint \mathbf{B}\cdot d\mathbf{l} = \mu_0 I$ gives $C = -I/2\pi$. Thus

$$\Phi_m = -\frac{I}{2\pi}\phi \qquad \text{and} \qquad \mathbf{B} = \frac{\mu_0 I}{2\pi\rho}\hat{\boldsymbol{\phi}} \tag{8.78}$$

It is to be noted that this scalar potential is not one of the cylindrical harmonics derived in Eq. (3.65). Those cylindrical harmonics were constructed by requiring the solution of Laplace's equation $\nabla^2\Phi = 0$ to be single-valued; that is, $\Phi(\phi = 0, \rho) = \Phi(\phi = 2\pi, \rho)$. This requirement is not invoked in the present case, since the magnetic potential is not single-valued. Whereas the cylindrical harmonics correspond to the cases of nonzero separation constant of the radial and the angular parts of Laplace's equation [Eq. (3.60)], the magnetic potential corresponds to the case where the separation constant is zero. Taking $K = 0$ in Eqs. (3.61) and (3.64) gives

$$\frac{d^2 Y}{d\phi^2} = 0 \qquad \text{and} \qquad \frac{d}{d\rho}\left(\rho\frac{dR}{d\rho}\right) = 0 \tag{8.79}$$

which have $Y = C_1\phi$, and $R = C_2$ and $R = C_3 \ln\rho$ solutions respectively. Hence the potential, $\Phi_m = RY$, can be represented by the sum of two terms; that is,

$$\Phi_m = C\phi + D\phi\ln\rho \tag{8.80}$$

where C_1, C_2, C_3, C, and D are constants. The second term in Eq. (8.80) introduces magnetic fields that are not physical; they depend on ϕ and $\ln\rho$ and have radial components. Therefore D is set to zero, and consequently the potential becomes a function of a single coordinate, as was arrived at in Eq. (8.77).

Example 8.17 Determination of the Magnetic Scalar Potential Using $\mathbf{B} = -\mu_0\nabla\Phi_m$

In cases of some degree of symmetry in which the magnetic field can be easily determined, the relation $\mathbf{B} = -\mu_0 \nabla\Phi_m$ can be used to determine the magnetic scalar potential. We consider as an example a long filamentary current, and write, from Eq. (8.74), $\mu_0\Phi_m = -\int \mathbf{B}\cdot d\mathbf{r}$. Substituting for $\mathbf{B}$ from Eq. (8.24), which was determined from Ampere's law, we get

$$\begin{aligned}\Phi_m &= -\frac{I}{2\pi}\int\frac{\hat{\boldsymbol{\phi}}}{\rho}\cdot d\mathbf{r} = -\frac{I}{2\pi}\int\frac{\hat{\boldsymbol{\phi}}}{\rho}\cdot(d\rho\,\hat{\boldsymbol{\rho}} + \rho\,d\phi\,\hat{\boldsymbol{\phi}} + dz\,\hat{\mathbf{z}}) \\ &= -\frac{I}{2\pi}\int_0^\phi d\phi = -\frac{I}{2\pi}\phi\end{aligned}$$

which is exactly the result arrived at in Eq. (8.78).

Example 8.18 The Magnetic Scalar Potential on the Axis of a Circular Loop

We calculate $\Phi_m(P)$ on the z axis of a filamentary circular current loop. In order to determine Ω, we imagine the loop to be the periphery of a spherical surface centered at P, as shown in Fig. 8.18. Then $\Omega =$ (area of spherical cap)/ξ^2, or

$$\Omega = \frac{1}{\xi^2}\int_0^\theta\int_0^{2\pi} \xi^2 \sin\theta'\, d\theta'\, d\phi' = 2\pi(1 - \cos\theta)$$

Therefore

$$\Phi_m(P) = \frac{I}{2}(1 - \cos\theta) = \frac{I}{2}\left(1 - \frac{z}{(z^2 + R^2)^{1/2}}\right) \tag{8.81}$$

Let us now calculate $B_z = -\mu_0(\partial\Phi_m/\partial z)$. It is to be noted that we cannot calculate the other derivatives of Φ_m because our expression for Φ_m is true only on the z axis. However, by symmetry, these other derivatives are zero. Thus

$$B = B_z = -\frac{\mu_0 I}{2}\frac{d}{dz}\left[1 - \frac{z}{(z^2 + R^2)^{1/2}}\right] = \frac{\mu_0 I}{2}\frac{R^2}{(R^2 + z^2)^{3/2}} = \frac{\mu_0 I}{2}\frac{R^2}{\xi^3}$$

which is the same result obtained before in Example 8.12. At points that are not located on the axis of the loop, the potential and the field cannot, as mentioned above, be obtained in a closed simple form. However, we can show that expansions for the radial and angular components of the magnetic field, B_r and B_θ, can be derived using the expression on the axis and with the help of Laplace's equation. We consider the region $z < R$, and expand the potential given in Eq. 8.81 in powers of z/R. The result is

$$\Phi_m = \frac{I}{2}\left[1 - \frac{z}{R} + \frac{1}{2}\left(\frac{z}{R}\right)^3 - \frac{3}{8}\left(\frac{z}{R}\right)^5 + \frac{5}{16}\left(\frac{z}{R}\right)^7 - \cdots\right] \tag{8.82}$$

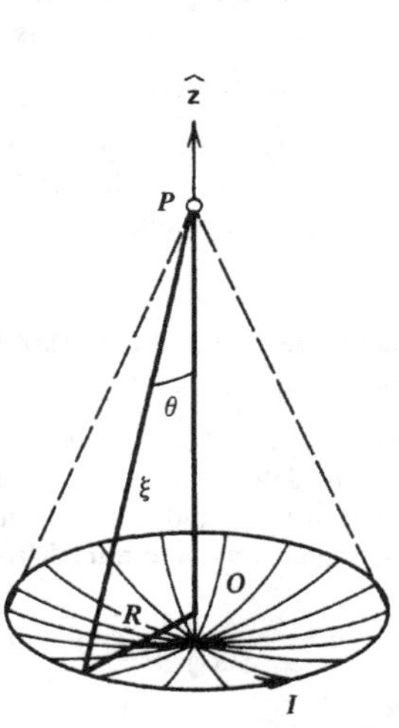

Figure 8.18 Magnetic scalar potential of a current loop in terms of solid-angle considerations.

On the other hand, the potential can be represented by the spherical-zone solution of Laplace's equation since the potential is expected not to depend on ϕ. Therefore, for the $r < R$ region,

$$\Phi_m = \sum_{n=0}^{\infty} A_n r^n P_n(\cos\theta) \tag{8.83}$$

At the z axis—i.e., $\theta = 0$—$P_n = 1$ for all n: Thus Eq. (8.83) is reduced to:

$$\Phi_m = A_0\left(1 + \frac{A_1}{A_0}z + \frac{A_2}{A_0}z^2 + \frac{A_3}{A_0}z^3 + \frac{A_4}{A_0}z^4 + \frac{A_5}{A_0}z^5 + \cdots\right) \tag{8.84}$$

Comparing Eqs. (8.82) and (8.84), we can find these constants and hence the potential. The result is

$$\Phi_m = \frac{I}{2}\left\{1 - \frac{r\,(\cos\theta)}{R} + \frac{1}{2}\left(\frac{r}{R}\right)^3 P_3\,(\cos\theta) - \frac{3}{8}\left(\frac{r}{R}\right)^5 P_5(\cos\theta)\right.$$
$$\left. + \frac{5}{16}\left(\frac{r}{R}\right)^7 P_7\,(\cos\theta) + \cdots\right\} \tag{8.85}$$

The corresponding magnetic field components can now be easily obtained by taking the gradient of Φ_m:

$$B_r = \frac{\mu_0 I}{2R}\left[P_1(\cos\theta) - \frac{3}{2}\left(\frac{r}{R}\right)^2 P_3(\cos\theta) + \cdots\right]$$

$$B_\theta = \frac{\mu_0 I}{2R}\left[-\sin\theta + \frac{3}{4}\left(\frac{r}{R}\right)^2 (5\cos^2\theta - 1)\sin\theta + \cdots\right]$$

8.8 Magnetic Effects of a Small Current Loop

8.8.1 The Scalar Potential

Consider a small current loop located at $\mathbf{r}'$, of area $\Delta\mathbf{a}'$ and current I. If the current loop is small enough and lies in a plane, then we may use Eq. (8.75) to represent the potential Φ_m at $\mathbf{r}$, calling it $\Delta\Phi_m$, as follows:

$$\Delta\Phi_m = \frac{I}{4\pi}\,\Delta\Omega = \frac{I}{4\pi}\,\frac{\Delta\mathbf{a}'\cdot\boldsymbol{\xi}}{\xi^3} \tag{8.86}$$

where $\Delta\mathbf{a}'$ is perpendicular to the plane of the loop. This expression is reminiscent of the electric dipole potential [Eq. (2.44)]:

$$\Delta\Phi_e = \frac{1}{4\pi\varepsilon_0}\,\frac{\Delta\mathbf{p}\cdot\boldsymbol{\xi}}{\xi^3}$$

where $\Delta\mathbf{p}$ is the dipole moment. Thus, aside from the numerical constants, the potentials, and thus the fields, of small currents loops and dipoles are identical. For this reason, the quantity

$$I\,\Delta\mathbf{a}' \equiv \Delta\mathbf{m} \tag{8.87}$$

is called the *magnetic dipole moment* of the loop. Thus

$$\Delta\Phi_m = \frac{1}{4\pi}\,\frac{\Delta\mathbf{m}\cdot\boldsymbol{\xi}}{\xi^3} \tag{8.88}$$

and

$$\Delta\mathbf{B} = -\mu_0 \nabla(\Delta\Phi_m) = \frac{\mu_0}{4\pi}\frac{1}{\xi^3}[3(\Delta\mathbf{m}\cdot\hat{\boldsymbol{\xi}})\hat{\boldsymbol{\xi}} - \Delta\mathbf{m}] \tag{8.89}$$

just as the case of the field of an electric dipole (see Example 2.9):

$$\Delta\mathbf{E} = \frac{1}{4\pi\varepsilon_0}\frac{1}{\xi^3}[3(\Delta\mathbf{p}\cdot\hat{\boldsymbol{\xi}})\hat{\boldsymbol{\xi}} - \Delta\mathbf{p}].$$

If we had an area $d\mathbf{a}'$ and the polarization per unit area were P_s, then the dipole moment of the area, $d\mathbf{p}$, would equal $P_s\, d\mathbf{a}'$. The quantity analogous to P_s in the magnetic case is the current I. The current I is thus the "magnetization per unit area" of the loop $d\mathbf{a}'$ and the magnetic phenomena of small current loops at points distant from the loops (thus the loops are small) are indistinguishable from the magnetic phenomena of "magnetic dipoles."

8.8.2 Magnetic Moments

Even if a filamentary current loop does not lie in a plane, its magnetic moment may be defined as $I\mathbf{S}$, where $\mathbf{S} \equiv \int_S d\mathbf{a}$. Consider Fig. 8.19. One may imagine the surface S spanning this loop to be composed of small planar surface elements $d\mathbf{a}'$, around each of which there flows a filamentary current I, so that the current internal to the actual loop is zero (the currents of adjacent elementary loops canceling). Then, from Eq. (8.86), the potential for each elementary loop is $d\Phi_m = (4\pi)^{-1} I\, d\mathbf{a}'\cdot\boldsymbol{\xi}/\xi^3$, and consequently

$$\Phi_m = \frac{I}{4\pi}\int_S \frac{d\mathbf{a}\cdot\boldsymbol{\xi}}{\xi^3} \tag{8.90}$$

Note that this formula is identical to Eq. (8.75) since $\Omega \equiv \int_S d\mathbf{a}'\cdot\boldsymbol{\xi}/\xi^3$. Also note that $I\, d\mathbf{a}'$ is the magnetic dipole moment associated with the area $d\mathbf{a}'$; that is, $d\mathbf{m} = I\, d\mathbf{a}'$. In our simply connected region, it is thus seen that the field due to a filamentary current loop may be ascribed to fictitious magnetic (di)poles lying on any surface of the loop whose periphery is the loop itself. The area density of these fictitious

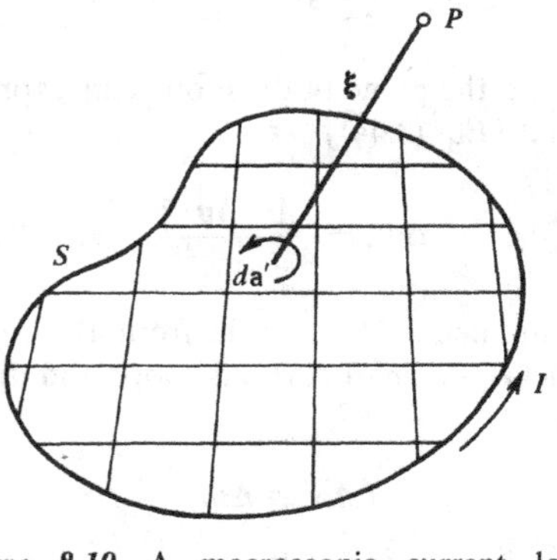

Figure 8.19 A macroscopic current loop constructed from elemental magnetic loops (dipoles).

dipoles is I. They are aligned along $d\mathbf{a}'$. When $\xi \gg$ largest dimension of S, then $\boldsymbol{\xi}/\xi^3$ can be taken outside the integral; that is,

$$\Phi_m \approx 4\pi\left[I\int_S d\mathbf{a}'\right]\cdot\frac{\boldsymbol{\xi}}{\xi^3} \equiv 4\pi\frac{\mathbf{m}\cdot\boldsymbol{\xi}}{\xi^3} \tag{8.91}$$

where $\mathbf{m} \equiv I\mathbf{S}$ is the dipole moment of the loop.

The magnetic dipole moment of a current loop has been defined above as $I\mathbf{S}$. If a current distribution is considered as consisting of many individual filamentary current loops, each of magnetic moment $I_j\mathbf{S}_j$, the magnetic moment of the sum is just the vector sum $\Sigma_j I_j \mathbf{S}_j$.

We now derive general expressions for the magnetic moment of filamentary and volume current distributions and discrete charges in motion. Taking an origin at the apex of the conical surface, S_j, we see that we may express the element of area as

$$d\mathbf{a} = \frac{1}{2}(\mathbf{r}' \times d\mathbf{l}')$$

Thus, from Eq. (8.87),

$$\mathbf{m} = \oint I\frac{1}{2}(\mathbf{r}' \times d\mathbf{l}') = \frac{1}{2}\oint \mathbf{r}' \times I\, d\mathbf{l}' \tag{8.92}$$

If we make the transformation from a filamentary current distribution to a volume current-density distribution by the identification $\Sigma_j \int I\, d\mathbf{l} * [\ \ ] \to \int_V \mathbf{J}\, dv * [\ \ ]$, whereby the sum Σ_j is meant of summation over all the filamentary current tubes of the total current distribution, then the following expression for the magnetic dipole moment is obtained.

$$\mathbf{m} = \frac{1}{2}\int \mathbf{r}' \times \mathbf{J}(\mathbf{r}')dv' \tag{8.93}$$

$$d\mathbf{m} = \frac{1}{2}\mathbf{r}' \times \mathbf{J}(\mathbf{r}')dv' \tag{8.94}$$

where $\mathbf{r}'$ is the displacement of element dv' from an arbitrarily chosen origin.

It is interesting to remark that the all-encompassing expression for the magnetic moment of a distribution of charge elements $dq(\mathbf{r})$ moving at $\mathbf{r}'$ with average velocity $\mathbf{v}(\mathbf{r}')$ is given by

$$\mathbf{m} = \frac{1}{2}\int \mathbf{r}' \times \mathbf{v}(\mathbf{r}')dq(\mathbf{r}') \tag{8.95}$$

where the sum is over all the charge elements of the distribution and $dq(\mathbf{r}')$ and $\mathbf{v}(\mathbf{r}')$ are assumed to be scalar and vector fields, respectively. For example, if one thinks of an atom as a positive stationary nucleus around which electrons are moving, then in general the atom will possess a magnetic moment. For example, if the hydrogen atom is represented by an electronic charge, $-e$, moving in circular motion at radius a_0 around a proton of charge $+e$, then Eq. (8.95) yields for the magnetic moment the single term

$$\mathbf{m} = -\frac{1}{2}ea_0\hat{\mathbf{r}} \times \mathbf{v}$$

Example 8.19 Magnetic Dipole Moment of a Charged Spinning Disk and a Solenoid

If one surface of the spinning disk has a constant surface charge density σ, then when the disk spins about its axis with angular frequency ω, each circular ring of the disk at distance ρ consists of a current $dI = \sigma v\, d\rho = \sigma\omega\rho\, d\rho$, and a magnetic moment $dm = dI(\pi\rho^2) = \pi\sigma\omega\rho^3\, d\rho$. The total magnetic moment of the disk is thus

$$\mathbf{m} = \int d\mathbf{m} = \omega\pi\sigma \int_0^R \rho^3\, d\rho = \frac{\pi\sigma\omega R^4}{4}$$

In this problem, we have used the vector property of magnetic moments; that is, the magnetic dipole moment of an object is the sum of the magnetic moments of its constituents.

Let us calculate the magnetic dipole moment of a straight circular solenoid of length L and radius R that has n turns per meter and a current I per turn. Regarding the turns as plane filamentary loops, each loop has a dipole moment equal to $I\mathbf{S} = IA\hat{\mathbf{z}}$ with $A = \pi R^2$. The dipole moment of the solenoid, which consists of nL loops, is then

$$\mathbf{m} = \sum_j I_j \mathbf{S}_j = nLIA\hat{\mathbf{z}}$$

Alternatively, the solenoid may be regarded as a circular sheet of current whose surface current density $\mathbf{K}$ equals $nI\hat{\boldsymbol{\phi}} = K\hat{\boldsymbol{\phi}}$, then Eq. (8.93) assumes an integral over the surface of the solenoid of the form

$$\mathbf{m} = \frac{1}{2}\int [\mathbf{r}' \times \mathbf{K}(\mathbf{r}')]da'$$

where we have assumed the current to flow in a vanishing thickness at the surface of the solenoid. Taking the origin of our (cylindrical) coordinate system on the z axis so that $\mathbf{r}' = z'\hat{\mathbf{z}} + R\hat{\boldsymbol{\rho}}$, we have

$$\mathbf{r}' \times \mathbf{K}(\mathbf{r}') = (z'\hat{\mathbf{z}} + R\hat{\boldsymbol{\rho}}) \times K\hat{\boldsymbol{\phi}} = -\hat{\boldsymbol{\rho}}z'K + \hat{\mathbf{z}}RK$$

In the integration, the $\hat{\boldsymbol{\rho}}$ term vanishes by symmetry, leaving the result

$$\mathbf{m} = \frac{1}{2}\hat{\mathbf{z}}RK \int_S da' = \frac{1}{2}\hat{\mathbf{z}}RK(2\pi RL) = \pi R^2 KL\hat{\mathbf{z}} = nLIA\hat{\mathbf{z}}$$

as before.

One motive for characterizing the solenoid by its dipole moment is that at distances far from the solenoid (compared to its dimensions), its $\mathbf{B}$ field is the same as that of a dipole (of moment $\mathbf{m}$).

8.8.3 The Vector Potential of a Small Current Loop

In the preceding discussion we have used the concept of a magnetic scalar potential to introduce the notion of a magnetic dipole moment. In the present subsection we show that with the use of the vector potential the concept of magnetic dipole moment plays a similar role. To this end, we consider the vector potential of a small, filamentary current loop, "small" meaning that we are interested in calculating the vector potential only at distances large compared to the dimensions of the loop.

As indicated by the notation of Fig. 8.20 and Eq. (8.44),

$$\mathbf{A} = \frac{\mu_0}{4\pi}\oint_C \frac{I\, d\mathbf{l}'}{\xi}$$

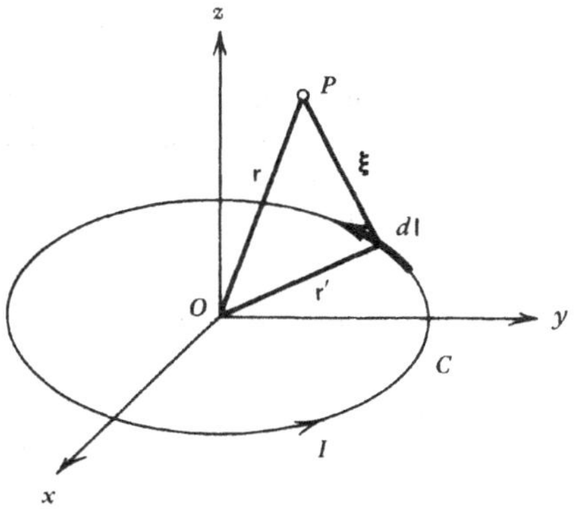

Figure 8.20 Vector potential of a small current loop.

The integral is transformed via the corollary of Stokes' theorem [Eq. (1.74)]:

$$\oint_C \Phi \, d\mathbf{l} = \int_S d\mathbf{a} \times \nabla\Phi$$

so that (see the notation of Fig. 8.19)

$$\mathbf{A} = \frac{\mu_0 I}{4\pi} \int_S d\mathbf{a}' \times \nabla'\left(\frac{1}{\xi}\right) = \frac{\mu_0 I}{4\pi} \int_S d\mathbf{a}' \times \frac{\boldsymbol{\xi}}{\xi^3}$$

Since [from Eq. (8.87)] $I\, d\mathbf{a}' \equiv d\mathbf{m}$, we have the result

$$\mathbf{A} = \frac{\mu_0}{4\pi} \int d\mathbf{m} \times \frac{\boldsymbol{\xi}}{\xi^3} \tag{8.96}$$

where the integral indicates a summation over all the dipole elements $d\mathbf{m} = I\, d\mathbf{a}'$. A "small" current loop now means that the term $\boldsymbol{\xi}/\xi^3$ remains essentially constant for all $d\mathbf{m}$, so it can be factored from the integral, yielding

$$\mathbf{A} = \frac{\mu_0}{4\pi}\left[\int d\mathbf{m}\right] \times \frac{\boldsymbol{\xi}}{\xi^3} \equiv \frac{\mu_0}{4\pi}\left(\mathbf{m} \times \frac{\boldsymbol{\xi}}{\xi^3}\right) \tag{8.97}$$

where $\mathbf{m}$ is the magnetic dipole moment of the small loop and $\boldsymbol{\xi}$ is now the displacement of the field point of P from the loop. That we have the right to call this field a dipole field only becomes apparent when we calculate $\mathbf{B} = \nabla \times \mathbf{A}$. For purposes of calculation, it is convenient to use a spherical coordinate system with the origin at the dipole, so that $(r, \theta, \phi) \to (\xi, \theta, \phi)$; then we get the two forms

$$\mathbf{B}(\xi, \theta, \phi) = \nabla \times \mathbf{A}(\xi, \theta) = \frac{\mu_0}{4\pi}\frac{m}{\xi^3}\{2\cos\theta\hat{\boldsymbol{\xi}} + \sin\theta\hat{\boldsymbol{\theta}}\}$$

$$\mathbf{B} = \frac{\mu_0}{4\pi\xi^3}[3(\mathbf{m}\cdot\hat{\boldsymbol{\xi}})\hat{\boldsymbol{\xi}} - \mathbf{m}] \tag{8.98}$$

which are precisely the forms of the electric dipole field [Eqs. (2.46) to (2.47)], which we also got from the magnetic scalar potential [see Eq. (8.89)]. As was done in connection with the scalar potential Φ_m (which we could now define from this latter expression), we can find the **B** field for an arbitrary loop by considering it as a mesh of infinitesimally small loops, each having a dipole moment $d\mathbf{m} = I\, d\mathbf{a}'$, and summing over all these constituent dipoles. In fact, Eq. (8.96) represents just such a summation. To obtain the resultant **B** from any distribution of dipoles (i.e., from any distribution of loops) we need only calculate $\mathbf{B} = \nabla \times \mathbf{A}$ from this equation.

Equation (8.97) is valid for any current distribution whose dimensions become small compared to the distance from any point inside the distribution to the field point. It is not restricted to current loops. To show this, consider a current distribution whose current density **J** is given at any point inside a finite volume V and is nonzero only inside V. The vector potential is given by (Eq. 8.42), as follows:

$$\mathbf{A}(\mathbf{r}) = \frac{\mu_0}{4\pi} \int_V \frac{\mathbf{J}(\mathbf{r}')dv'}{\xi} \tag{8.42}$$

If $\xi = |\mathbf{r} - \mathbf{r}'|$ is larger than the greatest linear dimension of V, we can expand the function $1/\xi$ using the binomial theorem, just as was done in the electrostatic multipole expansion in Chapter 2:

$$\mathbf{A}(\mathbf{r}) = \frac{\mu_0}{4\pi}\frac{1}{r}\int \mathbf{J}(\mathbf{r}')dv' + \frac{1}{r^3}\int (\mathbf{r}\cdot\mathbf{r}')\mathbf{J}\,dv' + \frac{1}{2r^5}\int [3(\mathbf{r}\cdot\mathbf{r}')^2 - \mathbf{r}'^2]\mathbf{J}(\mathbf{r}')dv' + \cdots$$

Since $r'/r < 1$, successive terms of this expansion become progressively weaker, so if $r'/r \ll 1$, only the first nonvanishing term need be considered. Here a departure from the case of the electrostatic multipole expansion becomes evident, because the monopole term is *always* zero *if* the currents are steady:

$$\int \mathbf{J}(\mathbf{r}')dv' = 0 \qquad \text{if } \nabla'\cdot\mathbf{J}(\mathbf{r}') = 0 \tag{8.99}$$

Example 8.20 Proof of $\int \mathbf{J}\,dv = 0$

We use cartesian coordinates $\{\hat{\mathbf{x}}_i\}$ and the implied summation convention to show the truth of the above assertion. Writing $\mathbf{J} = \hat{\mathbf{x}}_i(\hat{\mathbf{x}}_i\cdot\mathbf{J})$ and noting that $\hat{\mathbf{x}}_i = \nabla' x_i'$, we have

$$\mathbf{J} = \hat{\mathbf{x}}_i(\nabla' x_i')\cdot\mathbf{J} = \hat{\mathbf{x}}_i\{\nabla'\cdot(x_i'\mathbf{J}) - x_i'\nabla'\cdot\mathbf{J}\} = \hat{\mathbf{x}}_i\nabla'\cdot(x_i'\mathbf{J}).$$

Therefore

$$\int_V \mathbf{J}(\mathbf{r}')dv' = \hat{\mathbf{x}}_i\int_V \nabla'\cdot(x_i'\mathbf{J})dv' = \hat{\mathbf{x}}_i\int_S x_i'\mathbf{J}(\mathbf{r}')\cdot d\mathbf{a}'$$

where V is any volume that encloses all the current. Therefore, S can be chosen to lie outside all the current; that is, $\mathbf{J}(\mathbf{r}_S') = 0$. Therefore, $\int_V \mathbf{J}(\mathbf{r}')dv' = 0$.

Note that since streamlines of **J** are continuous (nonending), any steady current distribution can be thought of as made up of a set of filamentary current tubes. Hence, for any volume integral over **J**, one can make the substitution

$$\int \mathbf{J}\,dv' * [\quad] \rightarrow \sum_i \oint \mathbf{I}_i\,dl_i * [\quad]. \tag{8.100}$$

The volume distribution is conceived of as a (possibly infinite) set of filamentary loops. Thus, the integral of **J** over V becomes a sum of closed line integrals each of the form $\oint I\,d\mathbf{l} = I\oint d\mathbf{l} = 0$. Clearly, the volume integral must then be zero.

The second or dipole term in the above expansion thus assumes the primary significance. It can be shown to have the same form as Eq. (8.97). In fact, one can show (see Problem 8.26) that

$$\int_V (\mathbf{r}\cdot\mathbf{r}')\mathbf{J}(\mathbf{r}')dv' = \mathbf{m}\times\mathbf{r} \tag{8.101}$$

where **m** was previously defined in Eq. (8.93)

$$\mathbf{m} \equiv \frac{1}{2}\int \mathbf{r}' \times \mathbf{J}(\mathbf{r}')dv' \tag{8.93}$$

It is now clear that far enough away from any current distribution whose dipole moment is **m**, the vector potential is given by Eq. (8.97). In addition, since any current distribution can be decomposed into an infinite sum of vanishingly small components each with its own dipole moment $d\mathbf{m}$, then as long as (the observation) point $P(\mathbf{r})$ is some finite distance from all current elements, we can use Eqs. 8.94 and 8.96 for the potential. It will be observed that successive terms of this expansion differ by a factor of the order of R'/r, where R' is a linear dimension characteristic of the current distribution. Therefore, the dominant term of the distribution, when $r \gg R'$, will be the first non vanishing term. If the dipole term vanishes, then the dominant contribution in the expansion of Eq. 8.42 would be the third contribution which involves terms of the order R'^2/r^2. These terms will not be discussed in this book.

8.8.4 Localized Current Distribution in an External Magnetic Field

If a current distribution is placed in an external magnetic field **B**, it experiences a force whose general expression is given by Eq. (8.4). Because many applications involve steady current distributions that are localized in small regions of space ("small" being relative to the scale of length of interest to the observer), we will derive an approximate expression of the general force appropriate for this special case of localized distributions. Let us take the external magnetic field **B** to vary slowly over the region of the current and assert that $\nabla \times \mathbf{B} = 0$, and $\nabla \cdot \mathbf{B} = 0$ in this region. Because of the slow variation of **B**, we utilize a Taylor's series expansion to write the following approximate expression for **B**:

$$\mathbf{B}(\mathbf{r}') = \mathbf{B}(0) + \mathbf{r}' \cdot \nabla\mathbf{B}(0) + \cdots \tag{8.102}$$

where ∇ depends on $\mathbf{r}'$ and operate only on **B**, and the origin of the coordinate system used is chosen suitably within or very close to the distribution. The force given by Eq. (8.4) then becomes

$$\mathbf{F} = \mathbf{B}(0) \times \int \mathbf{J}(\mathbf{r}')dv' + \int \mathbf{J}(\mathbf{r}') \times [\mathbf{r}' \cdot \nabla\mathbf{B}(0)]dv' + \cdots \tag{8.103}$$

The first integral vanishes since the volume integral of **J** vanishes for steady currents [this fact was proved in Example 8.20]. Thus the lowest-order contribution to **F** involves the gradient of **B**. The expression for the force will be manipulated further in order to transform it to a form that involves the magnetic moment of the distribution $\mathbf{m} = \int \mathbf{r}' \times \mathbf{J}(\mathbf{r}')dv'$, previously defined in Eq. (8.93). To accomplish this we use the vector identity given in Eq. (1.59), and utilize the fact that $\nabla \times \mathbf{B} = 0$ and the fact that the gradient operator operates only on **B**

$$\mathbf{F} = \int \mathbf{J}(\mathbf{r}') \times [\nabla(\mathbf{r}' \cdot \mathbf{B}(0))]dv' \tag{8.104}$$

Also one can write $\mathbf{J}(\mathbf{r}') \times [\nabla(\mathbf{r}' \cdot \mathbf{B}(0))] = -\nabla \times [\mathbf{J}(\mathbf{r}' \cdot \mathbf{B})]$, and hence

$$\mathbf{F} = -\nabla \times \int \mathbf{J}(\mathbf{r}' \cdot \mathbf{B})dv' \tag{8.105}$$

Finally one can show that

$$\int \mathbf{J}(\mathbf{r}'\cdot\mathbf{B})dv' = \mathbf{B} \times \int \mathbf{r}' \times \mathbf{J}(\mathbf{r}')dv' = -\mathbf{B} \times \mathbf{m}$$

Thus

$$\mathbf{F} = \nabla \times (\mathbf{B} \times \mathbf{m}) \tag{8.106}$$

This expression of the force can also be written in two other useful forms, which follow from the properties $\nabla\cdot\mathbf{B} = 0$ and $\nabla \times \mathbf{B} = 0$:

$$\mathbf{F} = \nabla \times (\mathbf{B} \times \mathbf{m}) = (\mathbf{m}\cdot\nabla)\mathbf{B} = \nabla(\mathbf{m}\cdot\mathbf{B}) \tag{8.107}$$

The total torque on the localized current distribution is determined by summing over the torques $d\boldsymbol{\tau}$ on the various elements of the distribution:

$$d\boldsymbol{\tau} = \mathbf{r}' \times d\mathbf{F} = \mathbf{r}' \times (\mathbf{J} \times \mathbf{B})dv'$$

Thus (8.108)

$$\boldsymbol{\tau} = \int \mathbf{r}' \times (\mathbf{J} \times \mathbf{B})dv'$$

Using the triple vector product relation, the torque is written as

$$\boldsymbol{\tau} = \int (\mathbf{r}'\cdot\mathbf{B})\mathbf{J}\,dv' - \int (\mathbf{r}'\cdot\mathbf{J})\mathbf{B}\,dv'$$

The second integral can be shown to vanish for a localized current distribution, as follows: Writing

$$\mathbf{r}'\cdot\mathbf{J} = \frac{1}{2}\nabla\cdot(r'^2\mathbf{J}) - r'^2\nabla\cdot\mathbf{J} = \frac{1}{2}\nabla\cdot(r'^2\mathbf{J})$$

and transforming the volume integral to a surface integral, one can easily convince oneself of this fact. The first integral is the same one considered in Eq. (8.105); thus

$$\boldsymbol{\tau} = \mathbf{m} \times \mathbf{B} \tag{8.109}$$

Now we calculate the mechanical work needed to be done to place a localized current distribution of magnetic moment **m** in a uniform magnetic field **B**, with **m** making an angle θ with **B**. To simplify the calculation we imagine that the distribution is brought into the field with its moment pointing along the field. After it is in place, the distribution is then rotated to its final position.

Since the total force on the loop is zero in a uniform magnetic field, mechanical work is done only in the rotation step. The principle of virtual work says that the mechanical torque is the rate of change of work with angle; that is, $dW = \tau\, d\theta$. According to Eq. (8.109), the torque exerted by the magnetic field is restoring $\tau = -mB\sin\theta$. Thus

$$dW^{(\text{mech})} = -mB\sin\theta\, d\theta$$
$$W^{(\text{mech})} = mB\cos\theta + \text{constant}$$

Taking $W^{(\text{mech})}$ to be zero at $\theta = \pi/2$ requires the constant to be zero; thus

$$W^{(\text{mech})} = \mathbf{m}\cdot\mathbf{B} \tag{8.110}$$

This work is stored in the system as energy in the dipole-field system

$$U = -W^{(\text{mech})} = -\mathbf{m}\cdot\mathbf{B} \tag{8.111}$$

This expression is analogous to the case of an electric dipole $\mathbf{p}$ placed in an electric field

$$U = -\mathbf{p}\cdot\mathbf{E}$$

We should note that while $-\mathbf{p}\cdot\mathbf{E}$ is the true total electrostatic energy of the dipole, $-\mathbf{m}\cdot\mathbf{B}$ is only part of the total magnetic energy of the dipole. In the magnetic case we will find in later chapters that the current in the distribution is affected by the magnetic field, and hence we have to take into account the energy required to maintain the current in the distribution (see Faraday's law in Chapters 11 and 12). We should, however, note that the result $-\mathbf{m}\cdot\mathbf{B}$ can still be used to find the forces on localized steady currents.

8.9 Summary

In the presence of an electric field $\mathbf{E}$ and a magnetic field $\mathbf{B}$, the electromagnetic force on a charge q moving with velocity $\mathbf{v}$ is

$$\mathbf{F} = q(\mathbf{E} + \mathbf{v}\times\mathbf{B}) \tag{8.2}$$

The force $q\mathbf{v}\times\mathbf{B}$ is commonly referred to as the Lorentz force. The Lorentz force on the charge of volume V is given in terms of the current density $\mathbf{J}$ in the volume V

$$\mathbf{F} = \int_V \mathbf{J}\times\mathbf{B}\,dv \tag{8.4}$$

The condition under which the operation

$$\mathbf{J}\,dv \to I\,d\mathbf{l} \tag{8.9}$$

is valid is called the filamentary condition or approximation. Thus the Lorentz force on a filamentary current is

$$\mathbf{F} = \int I\,d\mathbf{l}\times\mathbf{B}$$

The basic equations of magnetostatics where the currents are steady ($\nabla\cdot\mathbf{J} = 0$) are

$$\nabla\cdot\mathbf{B} = 0 \qquad \text{and} \qquad \nabla\times\mathbf{B} = \mu_0\mathbf{J} \tag{8.19),(8.20}$$

where $\mu_0/4\pi = 10^{-7}\ \text{N/A}^2$. The relation $\nabla\cdot\mathbf{B} = 0$ implies that there are no magnetic monopoles. The curl relation is called the differential form of Ampere's law. The integral form of Ampere's law can be derived by integrating the differential one over a surface S of perimeter C and then applying Stokes' theorem:

$$\oint_C \mathbf{B}\cdot d\mathbf{r} = \mu_0 I \tag{8.22}$$

where $I = \int_S \mathbf{J}\cdot d\mathbf{a}$ is the total current through S.

The nonexistence of monopoles—$\nabla\cdot\mathbf{B} = 0$—implies that $\mathbf{B}$ can be written as a curl of a vector potential $\mathbf{A}$; that is,

$$\mathbf{B} = \nabla\times\mathbf{A} \tag{8.37}$$

When this is substituted in $\nabla\times\mathbf{B} = \mu_0\mathbf{J}$, and using $\nabla\cdot\mathbf{A} = 0$ (Coulomb gauge), we get

$$\nabla^2\mathbf{A} = -\mu_0\mathbf{J} \tag{8.39}$$

which has the solution

$$\mathbf{A} = \frac{\mu_0}{4\pi} \int_V \frac{\mathbf{J}\, dv}{|\mathbf{r} - \mathbf{r}'|} \tag{8.42}$$

Taking $\nabla \times \mathbf{A}$ explicitly, we arrive at the Biot-Savart law

$$\mathbf{B} = \frac{\mu_0}{4\pi} \int \frac{\mathbf{J} \times (\mathbf{r} - \mathbf{r}')}{|\mathbf{r} - \mathbf{r}'|^3}\, dv \tag{8.62}$$

or for a filamentary current

$$d\mathbf{B} = \frac{\mu_0}{4\pi} I \frac{d\mathbf{l} \times (\mathbf{r} - \mathbf{r}')}{|\mathbf{r} - \mathbf{r}'|^3}. \tag{8.65}$$

The differential equation for **A** can be solved in cases of cylindrical symmetry. If more than one region in space separated by current distributions are considered, then boundary conditions on **A** and **B** are needed to match the solution at the interfaces. These are derivable from the basic differential equations of **B**:

$$\mathbf{A}_{1t} = \mathbf{A}_{2t} \qquad \text{or} \qquad B_{1n} = B_{2n} \tag{8.47}$$

$$(\mathbf{B}_2 - \mathbf{B}_1)_t = \mu_0 \mathbf{K} \times \hat{\mathbf{n}} \tag{8.48}$$

where t and n stand for tangential and normal to the interface and **K** is the surface current density.

In regions where $\mathbf{J} = 0$, the curl of **B** vanishes and hence one can introduce a magnetic scalar potential Φ_m in analogy with electrostatics

$$\mathbf{B} = -\mu_0 \nabla \Phi_m \tag{8.72}$$

Substituting this in $\nabla \cdot \mathbf{B} = 0$ shows that Φ_m satisfies Laplace's equation

$$\nabla^2 \Phi_m = 0 \tag{8.73}$$

The magnetic scalar potential for a current loop I at a point of observation P is

$$\Phi_m = \frac{I}{4\pi} \Omega \tag{8.75}$$

where Ω is the solid angle of the loop with respect to an origin at the point of observation.

The magnetic scalar potential of a current loop of area **a** and current I at large distances compared to the dimension of the loop is

$$\Phi_m = \frac{1}{4\pi} \frac{\mathbf{m} \cdot \boldsymbol{\xi}}{\xi^3} \tag{8.88}$$

where $\boldsymbol{\xi} = \mathbf{r} - \mathbf{r}'$ and $\mathbf{m} = I\mathbf{a}$ is the magnetic moment of the loop. This potential has a similar form to that of an electric dipole, and hence it is referred to as the field of a magnetic dipole of moment **m**. The **B** field of the loop is a dipole field:

$$\mathbf{B} = \frac{\mu_0}{4\pi \xi^3} [3(\mathbf{m} \cdot \boldsymbol{\xi})\hat{\boldsymbol{\xi}} - \mathbf{m}] \tag{8.89}$$

The magnetic moment of a current distribution is

$$d\mathbf{m} = \frac{1}{2} \mathbf{r} \times \mathbf{J}\, dv \tag{8.94}$$

The vector potential of a magnetic dipole is

$$\mathbf{A} = \frac{\mu_0}{4\pi} \frac{\mathbf{m} \times \boldsymbol{\xi}}{\xi^3} \tag{8.97}$$

If a magnetic dipole is placed in an external magnetic field, then it will exhibit the following force, torque, and energy.

$$\mathbf{F} = \nabla \times (\mathbf{B} \times \mathbf{m}) = (\mathbf{m} \cdot \nabla)\mathbf{B} = \nabla(\mathbf{m} \cdot \mathbf{B}) \tag{8.107}$$

$$\boldsymbol{\tau} = \mathbf{m} \times \mathbf{B} \tag{8.109}$$

$$U = -\mathbf{m} \cdot \mathbf{B} \tag{8.111}$$

Problems

8.1 Write down the equation of motion for a particle of mass m and charge q in a region where $\mathbf{E}$ and $\mathbf{B}$ are nonzero. What is the magnetic field needed to confine an electron with velocity $v = 10^5$ m/s to an orbit of 10^{-10} m radius?

8.2 A long cylinder of radius ρ_0 and axis along the z axis carries a current of density $\mathbf{J} = e^{-2\rho}\hat{\mathbf{z}}$. Determine the magnetic field everywhere.

8.3 A filamentary current $\mathbf{I} = I_0\hat{\mathbf{z}}$ is at a distance h just above and parallel to the symmetry axis of a current sheet of width W and density $\mathbf{K} = K_0\hat{\mathbf{z}}$. Determine the force per unit length on the filamentary current. Find the force when W becomes very large.

8.4 A cylindrical conductor along the z axis of radius $\rho_0 = 10$ cm produces a magnetic field

$$\mathbf{B} = \frac{100\mu_0}{\rho}\left(\frac{4R^2}{\pi^2}\sin\frac{\pi\rho}{2R} - \frac{2\rho R}{\pi}\cos\frac{\pi\rho}{2R}\right)\hat{\boldsymbol{\phi}}\,(T) \qquad \rho < \rho_0$$

(a) Determine the current density and the total current in the conductor. (For the latter, Ampere's law may be used.) (b) Determine $\mathbf{B}$ outside the conductor.

8.5 Show explicitly that $\nabla \cdot \mathbf{B} = 0$ near a long, straight wire that carries a current I. Use either cartesian or cylindrical coordinates. What is $\nabla \times \mathbf{B}$ near this wire?

8.6 A rigid triangular loop carrying current I_1 is in the plane of a long wire carrying current I_2, as shown in Fig. 8.21. Calculate the force $\mathbf{F}$ on the diagonal side by I_2.

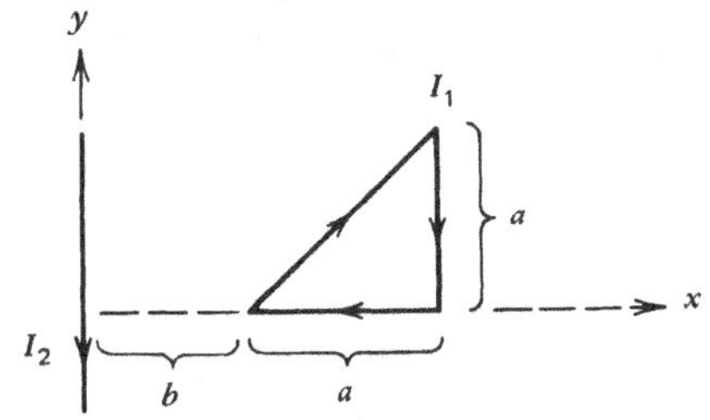

Figure 8.21

8.7 Show that the vector potential due to two parallel line currents flowing in opposite direction is $\mathbf{A} = \mu_0 I\hat{\mathbf{z}}\ln(\rho_2/\rho_1)/2\pi$, where ρ_1 and ρ_2 are the distances from the observation point to the wires and $\hat{\mathbf{z}}$ is a unit vector parallel to the wires.

8.8 Show that the following are all possible vector potentials of the uniform field, $\mathbf{B} = B\hat{\mathbf{z}}$: $\mathbf{A}_1 = -By\hat{\mathbf{x}}$, $\mathbf{A}_2 = Bx\hat{\mathbf{y}}$, $\mathbf{A}_3 = -\frac{1}{2}\mathbf{r} \times \mathbf{B}$. For which of these is $\nabla \cdot \mathbf{A} = 0$? Show that $\mathbf{A}_1 - \mathbf{A}_2$ is the gradient of a function, $\nabla\psi$. Plot $\mathbf{A}_1$, $\mathbf{A}_2$, and $\mathbf{A}_1 - \mathbf{A}_2$ in the $x - y$ plane.

8.9 Given the two circuits shown in Fig. 8.22: a very long, straight wire and a rectangular loop which lie in the same plane. A current I_1 flows in the long wire. (a) Calculate the magnetic field $\mathbf{B}$ produced by the long wire at a distance ρ from it. (b) Calculate the magnetic flux through the rectangular loop. (c) Determine the vector potential difference between ρ_2 and ρ_1. (d) If a current I_2 flows clockwise in the rectangular circuit, find the forces on the segments ab and bc.

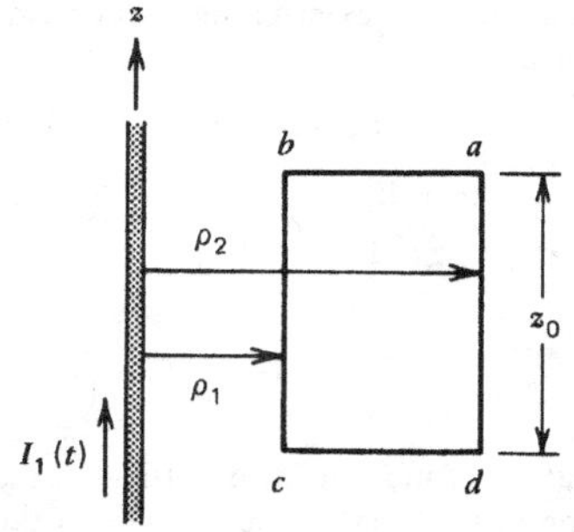

Figure 8.22

8.10 Consider a very long solenoid with n turns per unit length and a current I. The z axis is taken along its axis, as shown in Fig. 8.23. (a) Determine the magnetic field inside the solenoid, and the flux through the rectangular curve C. (b) Assume the vector potential $\mathbf{A}$ to be along the y axis and independent of y. Determine $\mathbf{A}$ inside the solenoid for the case where $\mathbf{A}$ is zero at the axis of the solenoid, using $\oint_C \mathbf{A} \cdot d\mathbf{l} = F$. (c) Repeat part b but with the assumption that the vector potential is along the x axis and independent of x. (d) Write a linear combination of the potentials in parts b and c that gives the same magnetic field and satisfies $\nabla \cdot \mathbf{A} = 0$. (e) Show that the sketches in Fig. 8.23*b*, 8.23*c*, and 8.23*d* represent the potential of parts b, c, and d.

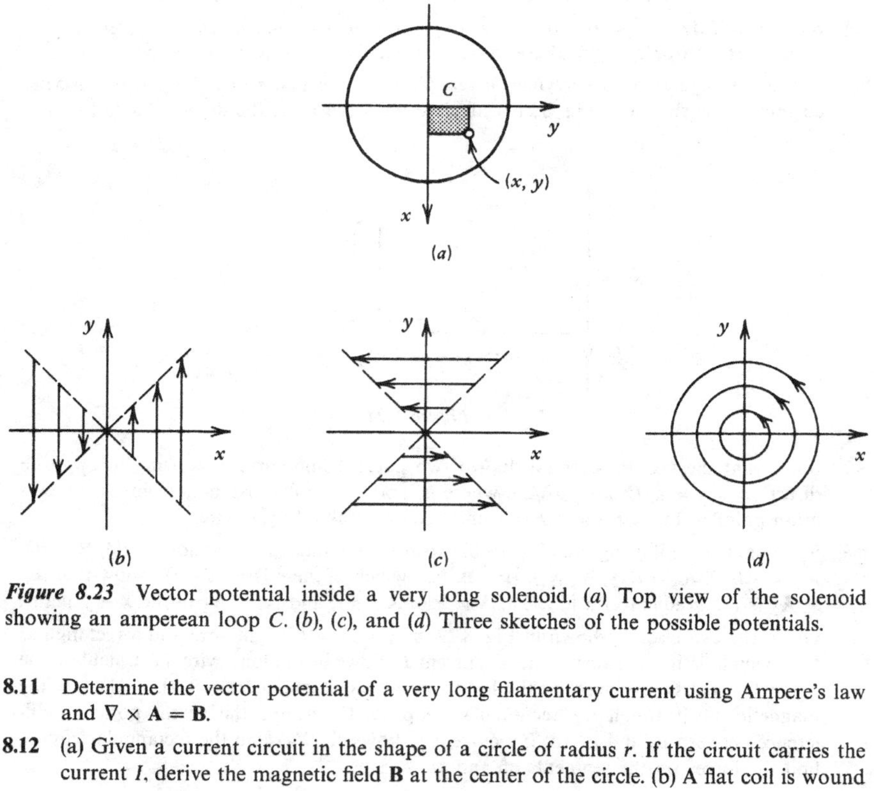

Figure 8.23 Vector potential inside a very long solenoid. (*a*) Top view of the solenoid showing an amperean loop C. (*b*), (*c*), and (*d*) Three sketches of the possible potentials.

8.11 Determine the vector potential of a very long filamentary current using Ampere's law and $\nabla \times \mathbf{A} = \mathbf{B}$.

8.12 (a) Given a current circuit in the shape of a circle of radius r. If the circuit carries the current I, derive the magnetic field $\mathbf{B}$ at the center of the circle. (b) A flat coil is wound so that it contains a very large uniform number of turns per unit distance along its radius. The inside and outside radii of the coil are a and b, respectively. It carries a current I, and has N turns per unit distance. Find the **B** field at the center of the coil.

8.13 Calculate the **B** field at the center of a wire square of side a with current I flowing through it.

8.14 A very long wire is bent 180° around a wooden cyclinder of radius b. If the wire carries current I, what is the **B** field at the center of the cylinder in the plane of the wire?

8.15 A large number N of closely spaced turns of fine wire are wound in a single layer upon the surface of a wooden sphere such that they completely cover it. The planes of the turns are perpendicular to an axis of the sphere. If the current in the wire is I, determine the magnetic field at the center of the sphere.

8.16 Determine the **B** field at the center of a sphere of radius R, uniform volume charge distribution ρ, and rotating about one of its diameters with an angular velocity $\boldsymbol{\omega}$.

8.17 (*a*) Determine the magnetic field near the axis of the current-carrying loop of Example 8.12 using $\nabla \cdot \mathbf{B} = 0$ and the field on its axis. (*b*) Show that the vector potential on the axis of the loop is zero.

8.18 A frequently used source of a reasonably uniform magnetic field is the Helmholtz coil. The coil consists of two circular coils of the same radius a with a common axis z, separated by a distance equal to the coil radius and carrying the current I. Show that, at the midpoint on the common axis, dB/dz and d^2B/dz^2 equal zero. Find B at the midpoint.

8.19 Consider the magnetic scalar potential in empty space $\Phi = B_0(z + xz/b)\mu_0$, where B_0 and b are constants. (a) Show that this is a reasonable potential from which a static magnetic field can exist and find it. (b) If an atom whose nucleus is stationary at the origin, has its electron in a circular orbit of radius a in the $x - y$ plane, find the force exerted by the field of part (a) on the atom.

8.20 A small current circular loop of radius a and current I lies in the $x - y$ plane with its center at the origin. Show that the vector potential at large distances is $A_r = A_\theta = 0$ and $A_\theta = (\pi\mu_0 a^2 I \sin\theta)/4\pi r^2$.

8.21 An N-turn, thin, circular coil of radius r and current I lies in the $z = 0$ plane. The current is in the $\hat{\boldsymbol{\phi}}$ direction and there is an external uniform magnetic field $\mathbf{B} = (\hat{\mathbf{x}} + \hat{\mathbf{y}})B_0/\sqrt{2}$. (a) Find the magnetic moment of the coil. (b) Find the force acting on the coil. (c) Find the torque acting on the coil.

8.22 In a triangular loop of wire in which there is a current of 6 A, a magnetic field $B = 1.1\ \text{Wb/m}^2$ is uniform over the triangle and parallel to the side AC as shown in Fig. 8.24. (a) Find the magnitude and direction of the force acting on each side. (b)

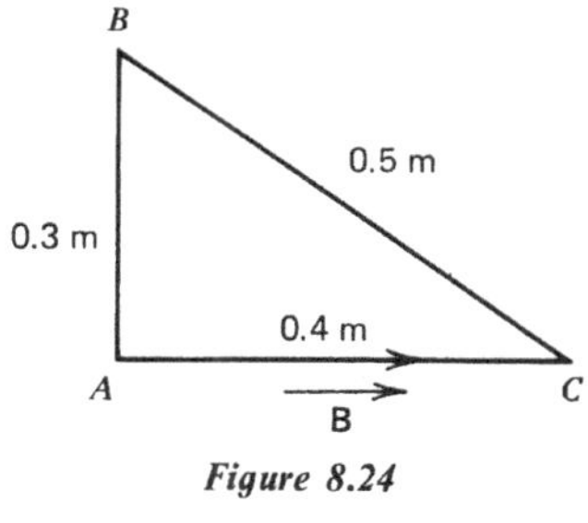

Figure 8.24

Calculate the dipole moment of the loop and the magnitude and direction of the torque acting on it.

8.23 A circular loop of wire with radius $a = 1$ centimeter and center at the origin is bent so that half lies in the $y - z$ plane and half lies in the $x - y$ plane. A current $I = 2$ A flows in the wire. (a) What is the magnetic moment of this loop? (b) What is the magnetic field **B** at $(x, y, z) = (3, 4, 0)$ meters from the origin.

8.24 An electron is moving in a circular orbit of radius 3.5×10^{-11} m in the presence of a uniform magnetic field $\mathbf{B} = 4 \times 10^{-2}$ T. If the electron experiences a torque of magnitude 7.85×10^{-26} N·m, determine the electron's angular velocity and magnetic moment.

8.25 (a) Determine the magnetic moment of a sphere of radius R, uniform volume charge distribution ρ, and rotating about one of its diameter. (b) Repeat for the case of surface charge distribution σ.

8.26 Prove Eq. (8.101).

NINE

FORMAL THEORY OF MAGNETISM AND MATTER

In Chapter 8 it was indicated that current loops produce dipole fields at large distances compared to their dimensions and may be characterized conveniently in terms of magnetic dipole moments. It is now a conceptually simple step to describe how the presence of matter affects and produces magnetic fields because we may regard the matter simply as a collection of atomic or molecular dipoles, each with its own dipole moment. It is, in fact, well established that such atoms or molecules possess magnetic dipole moments, and thus we tend to conceive of them as small circulation electronic currents, the electrons of the atoms being located in orbits around the atomic nuclei. The magnitude of these magnetic moments are usually specified in terms of the quantity of $e\hbar/2m_e$, called the Bohr magneton, and the magnitude of the atomic or molecular dipole moment, if it exists, is of this order of magnitude, where $\hbar$ = Planck's constant/2π, m_e = electron mass, and e = electron charge.

9.1 Magnetization

Classically, an electron rotating about a nucleus at a distance r and with a speed v would be equivalent to a current $I = e \times$ frequency of rotation $= ev/2\pi r$. Since the area of the (planar) orbit is πr^2, the magnetic moment is

$$m = \frac{ev}{2\pi r} \cdot \pi r^2 = \frac{evr}{2} \tag{9.1}$$

The orbital angular momentum of this electron is $m_e \mathbf{v} \times \mathbf{r} = \mathbf{L}$, and thus the magnetic moment is given by $eL/2m_e$. From quantum mechanics it is known that $L = \hbar\sqrt{l(l+1)}$, where l is a positive integer. Therefore the semiclassical picture predicts a magnetic moment per electron of magnitude as follows:

$$m = \frac{e\hbar}{2m_e}\sqrt{l(l+1)} \tag{9.2}$$

An atom may contain many electrons, but in general their magnetic moments tend to cancel each other, and only the "unpaired" electrons finally contribute to the atomic moment. In any case, atoms are found to have magnetic moments whose magnitude, if nonzero, is of the order $e\hbar/2m_e$ (called Bohr Magneton)—that is, 0.9×10^{-23} J/T (joules per tesla). It is also interesting that the effective currents are of the order of

$$\frac{ev}{2\pi r} = \frac{e\hbar}{2\pi m_e r^2} = 0.19 \times 10^{-2}\ \mathrm{A}$$

that is, of the order of 2 mA.

In each macroscopically small element of matter there are many atoms, and the dipole moment of such an element is simply the vector sum of the moments of the atomic constituents. Because of the large numbers of atomic dipoles per macroscopic element of matter, the dipole moments of contiguous macroscopic elements will vary smoothly, and we shall assume that these are characterized by a field vector, **M**, called the *magnetization*, defined so that in any macroscopically small volume element, dv, there is a dipole moment

$$d\mathbf{m} = \mathbf{M}\, dv \tag{9.3}$$

This relation indicates that **M** is the dipole moment per unit volume (density of magnetic dipoles, similar to **P** for the electric case) and hence will be a smooth function of position in the matter.

Previously [in Eq. (8.93)] we saw that

$$\mathbf{m} = \frac{1}{2}\int_V \mathbf{r}' \times \mathbf{J}(\mathbf{r}')dv'$$

for a volume V, where a current density **J** was defined. In such a case, Eq. (9.3) implies that **M** is the integrand of the integral:

$$\mathbf{M} = \frac{1}{2}[\mathbf{r}' \times \mathbf{J}(r')] \tag{9.4}$$

Thus, if we know the current distribution inside matter (a macroscopic current distribution), we can find **M** from this formula. It is therefore to be expected that if indeed a nonzero **M** exists in matter, there will be macroscopic currents associated with it (which we call the *magnetization currents*). To see what these currents are, we consider Fig. 9.1*a*, which shows macroscopic blocks of matter of volume ΔV stacked on each other. The magnetization at the center of each block is in general $\mathbf{M} = M_x\hat{\mathbf{x}} + M_y\hat{\mathbf{y}} + M_z\hat{\mathbf{z}}$, and thus each of these is equivalent to a macroscopic magnetic dipole $\mathbf{M}\,\Delta V = M_x\,\Delta V\hat{\mathbf{x}} + M_y\,\Delta V\hat{\mathbf{y}} + M_z\,\Delta V\hat{\mathbf{z}}$. Each block may then be regarded as consisting of current loops flowing in three mutually perpendicular planes (see Fig. 9.1*b*).* These currents will be labeled as I_{yz}, I_{zx}, and I_{xy}. Thus, using $d\mathbf{m} = I\, d\mathbf{a}$,

$$M_x\,\Delta V = I_{yz}\,\Delta y\,\Delta z \qquad M_y\,\Delta V = I_{zx}\,\Delta z\,\Delta x \qquad M_z\,\Delta V = I_{xy}\,\Delta x\,\Delta y$$

If each block has the same dipole moment (constant **M**), their equivalent current loops (I_{yz}, I_{zx}, I_{xy}) will be the same and no current will exist inside the material because of the current cancellation from contiguous blocks. If **M** is not constant, this cancellation will not be complete and a nonzero current will exist. Consider Fig. 9-1*c*, which shows the front view of blocks 1 and 2 with currents $I_{yz}(1)$ and $I_{yz}(2)$,

* This assumption will be seen to be consistent with our final result.

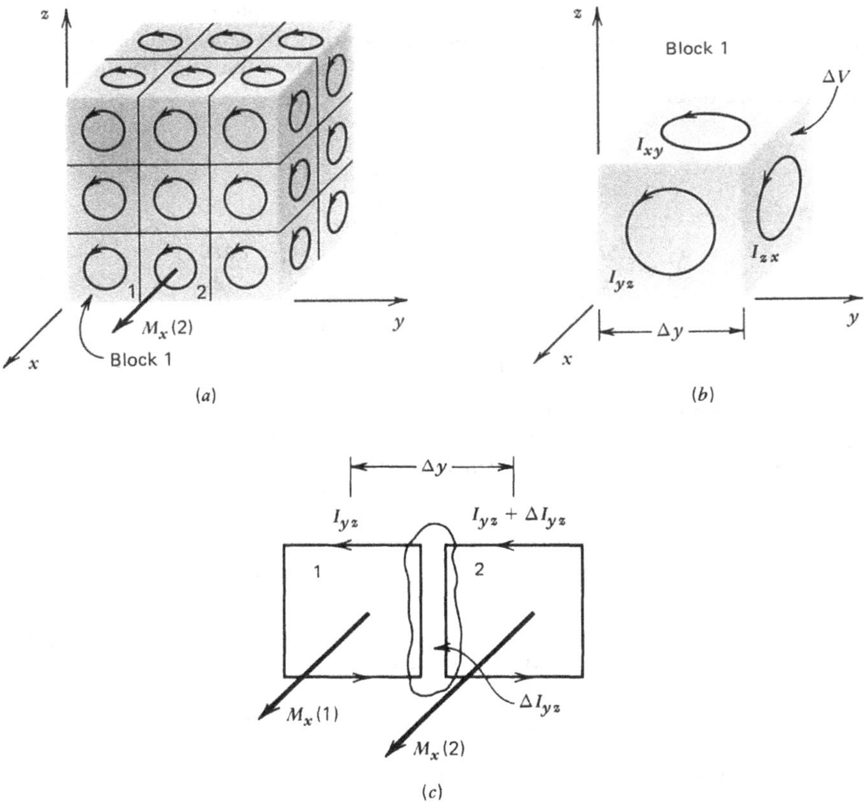

Figure 9.1 Derivation of the magnetization currents in terms of the magnetization **M**. (*a*) A state of macroscopic blocks of matter. (*b*) Enlargement of the block labeled 1 showing the three-dimensional currents flowing in it. (*c*) Front view of the two blocks 1 and 2.

and corresponding magnetization $M_x(1)$ and $M_x(2)$, respectively. We take $I_{yz}(2) = I_{yz}(1) + \Delta I_{yz}$. Thus

$$M_x(1)\Delta V = I_{yz}(1)\Delta y\,\Delta z \qquad \text{and} \qquad M_x(2)\Delta V = [I_{yz}(1) + \Delta I_{yz}]\Delta y\,\Delta z$$

Therefore,

$$[M_x(1) - M_x(2)]\Delta V = -\Delta I_{yz}\,\Delta y\,\Delta z$$

Taking $M_x(2) = M_x(1) + (\partial M_x/\partial y)\Delta y$ gives

$$-\frac{\partial M_x}{\partial y}\Delta y\,\Delta V = -\Delta I_{yz}\,\Delta y\,\Delta z$$

But ΔI_{yz} is the current flowing "between" blocks 1 and 2 and is seen to equal the current flowing in the $-z$ direction; that is, $-\Delta I_{yz} = I_z \equiv J_z\,\Delta x\,\Delta y$. As a result,

$$-\frac{\partial M_x}{\partial y}\Delta V \equiv J_z\,\Delta x\,\Delta y\,\Delta z \qquad \text{or} \qquad -\frac{\partial M_x}{\partial y} = J_z$$

Now it will be noted that there may exist another component to J_z due to the variation of **M** in the x direction. One finds, in precisely the same fashion as above, that this component of J_z is given by $+\partial M_y/\partial x$. Therefore the total component in the z direction of the current density is

$$J_z = \frac{\partial M_y}{\partial x} - \frac{\partial M_x}{\partial y}$$

Similarly,

$$J_x = \frac{\partial M_z}{\partial y} - \frac{\partial M_y}{\partial z} \quad \text{and} \quad J_y = \frac{\partial M_x}{\partial z} - \frac{\partial M_z}{\partial x}$$

The three components can be combined into one relation:

$$\mathbf{J}_m = \nabla \times \mathbf{M} \tag{9.5}$$

which we call the *magnetization current density.*

Finally, let us note that even if $\mathbf{M}$ = constant, and so $\mathbf{J}_m = 0$, currents will exist on the surfaces of the blocks, where cancellation is absent. Thus, on the top surface of the large block there will be currents due to M_x and M_y. In fact, considering (the top surface of) a single block,

$$\frac{-I_{yz}\hat{\mathbf{y}}}{\Delta x} = K_y\hat{\mathbf{y}} \quad \text{and} \quad \frac{I_{zx}\hat{\mathbf{x}}}{\Delta y} = K_x\hat{\mathbf{x}}$$

Therefore, the total surface current in the $x - y$ plane is $\mathbf{K}_{xy} = K_x\hat{\mathbf{x}} + K_y\hat{\mathbf{y}}$. Using the expressions $M_x\,\Delta V = I_{yz}\,\Delta y\,\Delta z, \ldots$, we get

$$\mathbf{K}_{xy} = \hat{\mathbf{x}}\left(\frac{I_{zx}}{\Delta y}\right) - \hat{\mathbf{y}}\left(\frac{I_{yz}}{\Delta x}\right) = \hat{\mathbf{x}}M_y - \hat{\mathbf{y}}M_x \quad \text{or} \quad \mathbf{K}_{xy} = \mathbf{M} \times \hat{\mathbf{z}}$$

The same result is true for any surface,* so in general we may write

$$\mathbf{K}_m = \mathbf{M} \times \hat{\mathbf{n}} \tag{9.6}$$

where $\hat{\mathbf{n}}$ is the outward normal to the block. In conclusion, we see that a magnetization $\mathbf{M}$ in a material is completely equivalent to a macroscopic current distribution in and on the material.

We should note that the divergence of $\mathbf{J}_m$ is identically zero, that is

$$\nabla \cdot \mathbf{J}_m = \nabla \cdot (\nabla \times \mathbf{M}) = 0$$

Hence this current will not affect the continuity relations as given by Eqs. (7.9) and (7.11).

Example 9.1 Magnetization Currents—Uniformly Magnetized Cylinder (a Bar Magnet)

Assume that we have a solid cylinder magnetized uniformly along its axis with a magnetization $\mathbf{M} = M_0\hat{\mathbf{z}}$. Then $\mathbf{J}_m = 0$, and $\mathbf{K}_m = M_0\hat{\mathbf{z}} \times \hat{\boldsymbol{\rho}} = M_0\hat{\boldsymbol{\phi}}$ on the curved sides of the cylinder. At the ends, $\mathbf{K}_m = 0$. Clearly, this gives a field of a solenoid whose surface current density equals M_0. At distances far from the axis, the field is that of a dipole of dipole moment $\mathbf{m} = \mathbf{M}V$, where V is the volume of the cylinder. Note that if the cylinder were hollowed out, the field inside would be reduced by virtue of the fact that $\mathbf{K}_m$ from the inner surface would be in the $-\hat{\boldsymbol{\phi}}$ direction.

* The only significance to z here is that it is normal to the surface.

9.2 The Vector and Scalar Potentials of a Magnetized Material

In the previous section we have shown that the magnetization of a material is completely equivalent to a macroscopic current distribution in and on the material; these are given by Eqs. (9.5) and (9.6), respectively. Consequently, one may calculate the magnetic effects directly from the currents—e.g., using the methods developed in Chapter 8. Thus, for the contribution to the vector potential from matter with magnetization **M**, one has, from Eq. (8.42),

$$\mathbf{A}_m = \frac{\mu_0}{4\pi}\left[\int_V \frac{\mathbf{J}_m\, dv'}{\xi} + \int_S \frac{\mathbf{K}_m\, da'}{\xi}\right] \tag{9.7}$$

and for the **B** field, one has, from (Eq. 8.62),

$$\mathbf{B}_m = \frac{\mu_0}{4\pi}\left[\int_V \frac{\mathbf{J}_m \times \boldsymbol{\xi}}{\xi^3}\, dv' + \int_S \frac{\mathbf{K}_m \times \boldsymbol{\xi}}{\xi^3}\, da'\right] \tag{9.8}$$

where V is the volume of the region where $\mathbf{M} \neq 0$, S includes all surfaces where $\mathbf{K}_m$ is defined, and $\boldsymbol{\xi} = \mathbf{r} - \mathbf{r}'$.

It will be observed that Eq. (9.7) above must be consistent with Eq. (8.96) of the previous chapter, where $\mathbf{A}_m$ was given (using $d\mathbf{m} = \mathbf{M}\, dv'$) as

$$\mathbf{A}_m = \frac{\mu_0}{4\pi}\int_V \frac{\mathbf{M} \times \boldsymbol{\xi}}{\xi^3}\, dv' \tag{9.9}$$

This consistency is proved in the following development: Since

$$\nabla'\left(\frac{1}{\xi}\right) = +\frac{\boldsymbol{\xi}}{\xi^3}$$

then the integral of Eq. (9.9) transforms as follows:

$$A_m = \frac{\mu_0}{4\pi}\int_V \frac{\mathbf{M} \times \boldsymbol{\xi}}{\xi^3}\, dv' = \frac{\mu_0}{4\pi}\int_V \mathbf{M} \times \nabla'\left(\frac{1}{\xi}\right) dv'$$

We apply the integral to a volume V' that is enclosed by V so that discontinuities in **M** at the surface could be avoided, and we use the vector identity, Eq. (1.58), to write

$$\nabla' \times \left(\frac{\mathbf{M}}{\xi}\right) = \nabla'\left(\frac{1}{\xi}\right) \times \mathbf{M} + \frac{\nabla' \times \mathbf{M}}{\xi}$$

Thus

$$\mathbf{A}_m = \frac{\mu_0}{4\pi}\lim_{V' \to V}\int_{V'} \mathbf{M} \times \nabla'\left(\frac{1}{\xi}\right) dv' = \lim_{V' \to V}\int_{V'}\left[-\nabla' \times \left(\frac{\mathbf{M}}{\xi}\right) + \frac{\nabla' \times \mathbf{M}}{\xi}\right] dv'$$

By use of Eq. (1.70), the potential becomes

$$\mathbf{A}_m = \frac{\mu_0}{4\pi}\lim_{V' \to V}\left[\int_{V'} \frac{\nabla' \times \mathbf{M}}{\xi}\, dv' + \int_{S'} \frac{\mathbf{M} \times d\mathbf{a}'}{\xi}\right]$$

or

$$\mathbf{A}_m = \frac{\mu_0}{4\pi}\left(\int_V \frac{\mathbf{J}_m\, dv'}{\xi} + \int_S \frac{\mathbf{K}_m\, da'}{\xi}\right)$$

We now discuss the magnetic properties of matter from the point of view of the magnetic scalar potential. It was shown in Chapter 8 that the magnetic field due to some real current distributions can be derived from a scalar potential according to the relation $\mathbf{B} = -\mu_0 \nabla \Phi_m$. We now show that the fields produced by magnetized materials can also be derived from a scalar potential. Taking the curl of Eq. (9.9) with respect to $\mathbf{r}$ and noting that this operation does not affect functions which depend on $\mathbf{r}'$, we get

$$\mathbf{B}_m = \frac{\mu_0}{4\pi} \int \nabla \times \left(\frac{\mathbf{M}(\mathbf{r}') \times \boldsymbol{\xi})}{\xi^3} \right) dv' \tag{9.10}$$

This expression can be rearranged by expanding the triple cross product of the integrand according to Eq. (1.61); that is,

$$\nabla \times \left(\frac{\mathbf{M} \times \boldsymbol{\xi}}{\xi^3} \right) = \mathbf{M}\, \nabla \cdot \frac{\boldsymbol{\xi}}{\xi^3} - (\mathbf{M} \cdot \nabla) \frac{\boldsymbol{\xi}}{\xi^3} \tag{9.11}$$

where differentiation of $\mathbf{M}(\mathbf{r}')$ with respect to $\mathbf{r}$ is taken to be zero. Equation (9.11) can be further recast into a more useful form. We write $\nabla \cdot (\boldsymbol{\xi}/\xi^3) = 4\pi\delta(\boldsymbol{\xi})$, as was shown in Eq. (1.81). Moreover, using the triple cross product, we write the second term of the right-hand side of Eq. (9.11) as

$$(\mathbf{M} \cdot \nabla) \frac{\boldsymbol{\xi}}{\xi^3} = \nabla \left[\mathbf{M} \cdot \frac{\boldsymbol{\xi}}{\xi^3} \right] - \mathbf{M} \times \nabla \times \frac{\boldsymbol{\xi}}{\xi^3}$$

The last term involves

$$\nabla \times \frac{\boldsymbol{\xi}}{\xi^3} = \nabla \times \left(\nabla \frac{1}{\xi} \right) \equiv 0$$

Thus

$$(\mathbf{M} \cdot \nabla) \frac{\boldsymbol{\xi}}{\xi^3} = \nabla \left(\mathbf{M} \cdot \frac{\boldsymbol{\xi}}{\xi^3} \right)$$

Therefore Eq. (9.10) becomes

$$\mathbf{B}_m(\mathbf{r}) = -\frac{\mu_0}{4\pi} \int \nabla \left[\mathbf{M}(\mathbf{r}') \cdot \frac{\boldsymbol{\xi}}{\xi^3} \right] dv' + \mu_0 \int \mathbf{M}(\mathbf{r}')\delta(\boldsymbol{\xi})\, dv' \tag{9.12}$$

The gradient in Eq. (9.12) can be taken outside the integral since it is with respect to $\mathbf{r}$, and the integration of $\mathbf{M}(\mathbf{r}')\delta(\boldsymbol{\xi})$ over the volume gives $\mu_0 \mathbf{M}(\mathbf{r})$. Thus

$$\mathbf{B}_m(\mathbf{r}) = -\mu_0 \nabla \Phi_m(\mathbf{r}) + \mu_0 \mathbf{M}(\mathbf{r}) \tag{9.13}$$

where

$$\Phi_m(\mathbf{r}) = \frac{1}{4\pi} \int \mathbf{M}(\mathbf{r}') \cdot \frac{\boldsymbol{\xi}}{\xi^3}\, dv' \tag{9.14}$$

is the scalar magnetic potential produced by the magnetized material. Since $\mathbf{M}(\mathbf{r}')dv' = d\mathbf{m}(\mathbf{r}')$ is a differential of a dipole moment, then the scalar potential is just the sum of dipole fields [Eq. (8.90)]. The second term of Eq. (9.13) is μ_0 times the local magnetization of the magnetic material. Outside the magnetic material $\mathbf{M}(\mathbf{r}) = 0$, and thus

$$\mathbf{B}_m(\mathbf{r}) = -\mu_0 \nabla \Phi_m(\mathbf{r}) \qquad \text{(outside material)} \tag{9.15}$$

The similarity between this result and the relation between the electric field and the electric potential, Eq. (2.36), is clear. Therefore the manipulations used in writing the electrostatic potential in terms of the volume and surface polarization charge densities ρ_p and σ_p can be used in the magnetized material case. The result is

$$\Phi_m(\mathbf{r}) = \frac{1}{4\pi}\int \frac{\rho_m}{\xi}\,dv' + \frac{1}{4\pi}\int \frac{\sigma_m}{\xi}\,da' \tag{9.16}$$

and, from (Eq. 9.13),

$$\mathbf{B}_m(\mathbf{r}) = \frac{\mu_0}{4\pi}\int \frac{\rho_m \boldsymbol{\xi}}{\xi^3}\,dv' + \frac{\mu_0}{4\pi}\int \frac{\sigma_m \boldsymbol{\xi}}{\xi^3}\,da' + \mu_0 M(\mathbf{r}) \tag{9.17}$$

where

$$\rho_m = -\nabla\cdot\mathbf{M} \qquad \text{and} \qquad \sigma_m = \mathbf{M}\cdot\hat{\mathbf{n}} \tag{9.18}$$

The quantity $-\nabla\cdot\mathbf{M}$ has the character of a magnetic charge density, or *magnetic pole density*. If $\mathbf{M}$ is discontinuous at a boundary, such that $\mathbf{M} = 0$ on one side (side 2) of the boundary and finite on the other, then there will be a surface pole density given by $\sigma_m = \mathbf{M}\cdot\hat{\mathbf{n}}$ called the *surface density* or *magnetic pole strength.*

It is interesting to observe that the total pole strength associated with any piece of magnetized material is zero. Thus, if the material has a volume V, then the total pole strength q_m is calculated by integrating the densities of Eq. (9.18); that is

$$q_m = -\int_V \nabla\cdot\mathbf{M}\,dv + \int_S \mathbf{M}\cdot\hat{\mathbf{n}}\,da$$

Since the divergence theorem gives $\int_S \mathbf{M}\cdot\hat{\mathbf{n}}\,da = \int_V \nabla\cdot\mathbf{M}\,dv$, then q_m vanishes.

The magnetic moment of the material can be calculated from the poles just as in electrostatics [see Eq. (4.15)]:

$$\mathbf{m} = -\int_V \mathbf{r}'(\nabla\cdot\mathbf{M})dv' + \int_S (\mathbf{M}\cdot\hat{\mathbf{n}})\mathbf{r}'\,da' \tag{9.19}$$

This expression for $\mathbf{m}$ can now be shown to be equivalent to the simple form: $\mathbf{m} = \int \mathbf{M}\,dv'$, which was introduced in Eq. (9.3). Applying Eq. (1.72),

$$\oint_S \mathbf{B}(\mathbf{A}\cdot d\mathbf{a}) = \int_V [\mathbf{B}(\nabla\cdot\mathbf{A}) + (\mathbf{A}\cdot\nabla)\mathbf{B}]dv$$

gives $\mathbf{m} = \int (\mathbf{M}\cdot\nabla')\mathbf{r}'\,dv'$. Noting that $(\mathbf{M}\cdot\nabla)\mathbf{r} = \mathbf{M}$, then $\mathbf{m} = \int \mathbf{M}\,dv'$, the result expected from the definition of $\mathbf{M}$.

Example 9.2 Magnetization Pole and Current Densities— Uniformly Magnetized Sphere

In this example we consider the magnetization of a sphere to be $\mathbf{M} = M_0\hat{\mathbf{z}}$. The fields produced will have cylindrical symmetry, so if we use spherical coordinates, $\mathbf{B}$ will be a function of (r, θ) alone.

The magnetization current densities will be surface currents (see Fig. 9.2); that is,

$$\mathbf{J}_m = 0 \qquad \mathbf{K}_m = \mathbf{M}\times\hat{\mathbf{r}} = M_0 \sin\theta\,\hat{\boldsymbol{\phi}} \tag{9.20}$$

Rather than calculating $\mathbf{B}$ directly from this current distribution, we calculate first the vector potential. Thus Eqs. (8.42) and (9.20) give

$$\mathbf{A} = \frac{\mu_0}{4\pi}\int \frac{M_0 \sin\theta'\,\hat{\boldsymbol{\phi}}}{\xi}\,da'$$

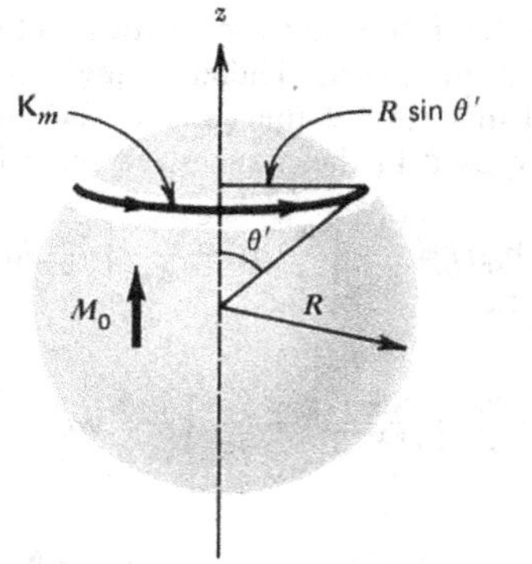

Figure 9.2 Uniformly magnetized sphere.

Clearly $\mathbf{A} = \hat{\boldsymbol{\phi}} A_\phi$ by the cylindrical symmetry, and thus

$$A_\phi = \frac{\mu_0}{4\pi} \int \frac{M_0 \sin \theta'}{\xi} \hat{\boldsymbol{\phi}}(\mathbf{r}') \cdot \hat{\boldsymbol{\phi}}(\mathbf{r}) R^2 \, d\Omega' \tag{9.21}$$

This integral can be integrated, but not with elementary methods. The result is*

$$\begin{aligned} A_\phi &= \frac{\mu_0 M_0}{3} \frac{R^3}{r^2} \sin \theta \qquad r > R \\ A_\phi &= \frac{\mu_0 M_0}{3} r \sin \theta \qquad r < R. \end{aligned} \tag{9.22}$$

Taking the curl of the vector potential gives

$$\mathbf{B} = \frac{2\mu_0 M_0}{3} (\cos \theta \, \hat{\mathbf{r}} - \sin \theta \, \hat{\boldsymbol{\theta}}) = \frac{2\mu_0 M_0}{3} \hat{\mathbf{z}} \qquad r < R \tag{9.23}$$

$$\mathbf{B} = \frac{\mu_0 M_0 R^3}{3} \frac{1}{r^3} (2 \cos \theta \, \hat{\mathbf{r}} + \sin \theta \, \hat{\boldsymbol{\theta}}) \qquad r > R \tag{9.24}$$

We recognize the latter as a dipole field with a dipole moment $M_0 V = M_0(4\pi R^3/3)$. The field inside the sphere is constant, of magnitude $\frac{2}{3}\mu_0 M_0$. In fact, this could have been proved by rather simple arguments.†

Let us now find the same results using the concept of the magnetic scalar potential. In this case, $\mathbf{B}$ is produced by a surface pole density $\sigma_m = M_0 \cos \theta$. Then, the problem henceforth involves the analogous surface electric charge density of $\sigma_0 \cos \theta$. It has been shown for that problem (see Examples 2.17 and 3.6) that the external potential (for $r > R$) is a dipole potential [Eq. (3.44)]. Thus, one can write along the same lines

$$\Phi_m = \frac{M_0 R^3}{3} \frac{1}{r^2} \cos \theta \qquad r > R \tag{9.25}$$

The internal potential can also be written, in analogy to the electric case [Eq. (3.43)],

$$\Phi_m = \frac{M_0}{3} r \cos \theta \qquad r < R \tag{9.26}$$

Using $\mathbf{B} = -\mu_0 \nabla \Phi_m$ outside the sphere and $\mathbf{B} = -\mu_0 \nabla \Phi_m + \mu_0 \mathbf{M}$ inside the sphere, we get the same field derived above using the magnetization current method.

*See, for example, J. D. Jackson, *Classical Electrodynamics*, 2d ed. (New York: Wiley, 1975), p. 197.

† That is, a simple integration would show that, along the z axis, B = constant, whence it could be argued that inside the sphere in general, B = constant. (How?)

Example 9.3 Uniformly Magnetized Cylinder

In this example we consider again a solid cylinder with a uniform magnetization $\mathbf{M} = M_0\hat{\mathbf{z}}$ along its axis, and with no free currents. We have seen that a surface current $K_m = \hat{\boldsymbol{\phi}}M_0$ due to the magnetization flows around the curved parts of the cylinder and truly gives rise to the macroscopic **B** field everywhere (see Example 9.1). The flow of current has the form of a solenoidal current. Therefore, the magnetic field can be calculated using the results of Example 8.13. Using the notations of Fig. 8.13 and replacing K by M_0 in Eq. (8.68), we get

$$B = B_z = \frac{\mu_0 M_0}{2}(\cos\theta_1 - \cos\theta_2) \tag{9.27}$$

This problem can also be solved using the magnetic scalar potential. The magnetic poles occur only at the ends of the cylinder (that is, $\nabla\cdot\mathbf{M} = 0$ everywhere, $\sigma_m = \pm M_0$). The magnetic scalar potential of the ends of the cylinder can be calculated in the same fashion as the case of two electrically charged disks with uniform densities. In Example 2.11, the electric potential was calculated for a uniformly charged disk, and hence replacing ε_0 by $1/\mu_0$ and σ_0 by M_0 in Eq. (2.49) gives the magnetic scalar potential. The total magnetic potential can then be determined by superposition of the potential produced by both ends of the cylinder. The magnetic fields inside and outside the cylinder are then determined using Eqs. (9.13) and (9.15), respectively. Figure 9.3 shows the similarity of field lines of the uniformly magnetized cylinder to those of a solenoid.

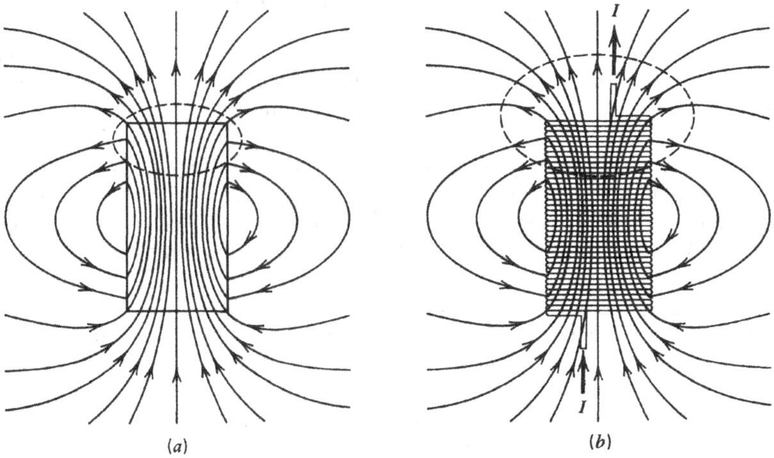

Figure 9.3 Field lines of (a) uniformly magnetized cylinder along its axis and (b) a solenoid showing the similarity between them.

As an experimental note, it is worthwhile to mention that a "strong" magnetic **B** field produced by magnetization alone is of the order of 10,000 gauss or 1 Wb/m^2, or 1 T. If we let the cylinder being discussed be very long, so that $\theta_1 \to -\pi$ and $\theta_2 \to \pi$, we find that $B = \mu_0 K_m$ inside the cylinder. Thus, the equivalent surface current density has the magnitude $(10^7/4\pi)$ A/m. It would take a formidable power supply to maintain such a steady current in wires alone, and generally would involve considerable heat dissipation, whereas no dissipation is associated with the magnetization currents when **M** is constant. The atomic currents require no external energy to be maintained.

9.3 The Equations of Macroscopic Magnetostatics

We have seen that the basic equations of steady-current magnetism are $\nabla\cdot\mathbf{B} = 0$ and $\nabla\times\mathbf{B} = \mu_0\mathbf{J}$. We now write these equations in a different form by regarding the **B** and **J** vectors as representing macroscopic fields (that is, we have taken space and time averages of both sides of the equation.) Moreover, the current density **J** is split into two parts—one representing free (generally conduction) currents $\mathbf{J}_f$ and the other representing magnetization currents $\mathbf{J}_m$:

$$\mathbf{J} = \mathbf{J}_f + \mathbf{J}_m = \mathbf{J}_f + \nabla \times \mathbf{M} \tag{9.28}$$

Then we have $\nabla \times \mathbf{B} = \mu_0 \mathbf{J}_f + \mu_0 \nabla \times \mathbf{M}$, or $\nabla \times (\mathbf{B} - \mu_0 \mathbf{M}) = \mu_0 \mathbf{J}_f$. We shall now define a new vector field, the *magnetic intensity* **H**:

$$\mathbf{H} = \frac{\mathbf{B}}{\mu_0} - \mathbf{M} \tag{9.29}$$

Then, we may write for the equations of macroscopic magnetism:

$$\nabla \cdot \mathbf{B} = 0 \qquad \nabla \times \mathbf{H} = \mathbf{J}_f \tag{9.30}$$

In these equations we are implicitly assuming that **B**, **H**, **M**, and $\mathbf{J}_f$ are everywhere continuous vector functions so that the derivatives have meaning. In dealing with regions where discontinuities occur, it is frequently useful to employ the analogous integral equations

$$\oint_S \mathbf{B} \cdot d\mathbf{a} = 0 \qquad \text{and} \qquad \oint_C \mathbf{H} \cdot d\mathbf{r} = I_f \tag{9.31}$$

where I_f is the current through the closed curve C, the positive sense of I_f being determined via the sense of traversal of the line integral by the right-hand rule.

It is legitimate to ask the questions, "Why introduce **H**?" and "What physical meaning does it have?" The answer to these questions will only become evident in the sequel. However, one may observe that the differential equation of **H** involves only the conventional current density—i.e., external currents. Also, if in a region $J_f = 0$, then the fact that $\nabla \times \mathbf{H} = 0$ will permit us to define a scalar potential function for **H** there. In fact, when Eqs. (9.13) and (9.29) are compared, we find that

$$\mathbf{H} = -\nabla \Phi_m \tag{9.32}$$

This will ease the calculational burden of computing the fields. Thus, ultimately **H** is introduced because in some situations it is easier to calculate **H** directly and then **B** rather than to calculate **B** directly.

In the general case where free currents, $\mathbf{J}_f$, exist in magnetic materials, the total **H** field can be written as the following sum.

$$\mathbf{H} = \frac{1}{4\pi} \int \frac{\mathbf{J}_f \times \boldsymbol{\xi}}{\xi^3} \, dv' - \nabla \Phi_m \tag{9.33}$$

9.4 The Magnetic Constitutive Relations

In the preceding section the magnetic intensity **H** was related to the magnetic field **B** and the magnetization **M** by the relation $\mathbf{B} = \mu_0 \mathbf{H} + \mu_0 \mathbf{M}$. This relation can be cast in another useful form, by relating **M** to **H** and hence **B** to **H** directly. The degree of response of a material to an external field **H** depends on the microscopic structure of the material. We will defer the nature of these responses to the following chapter. The response falls into a number of classes. The first class involves what is called *linear materials*, where the degree of magnetization is proportional to **H**; that is,

$$\mathbf{M} = \chi \mathbf{H} \tag{9.34}$$

where χ is dimensionless, does not depend on **H**, and is called the *magnetic susceptibility*. In the case of isotropic materials, χ does not depend on the direction of the external field; that is, it is a scalar quantity. If the material is anistropic, then the direction of **M** is not necessarily along the direction of the field, and becomes a tensor of rank 2:

$$\mathbf{M}_i = \sum \chi_{ij} H_j = \chi_{ix} H_x + \chi_{iy} H_y + \chi_{iz} H_z$$

where $i = x$, y, and z. Such anistropic materials will not be discussed any further in this book. Linear isotropic materials are further classified into two categories: Materials with $\chi < 0$ are called *diamagnetic*, and materials with $\chi > 0$ are called *paramagnetic*. Diamagnetic materials cause a weakening in **B** when placed in an external field, while paramagnetic materials cause a strengthening of **B**. The magnetic susceptibilities of some paramagnetic and diamagnetic materials are given in Table 9.1.

We should also note that most references on physical data list the *mass* (*molar*) *susceptibility* χ_{mass} (χ_{molar}) instead of χ. These are defined as

$$\chi_{\text{mass}} = \frac{\chi}{d} \qquad \text{and} \qquad \chi_{\text{molar}} = \frac{\chi A}{d} \tag{9.35}$$

where d is the mass density of the material and A is its molecular weight.

Linear magnetic materials exhibit an interesting effect when placed in a magnetic field. It is found that paramagnetic materials get attracted to magnetic fields, whereas diamagnetic materials get repelled. In Chapter 4 we found, contrary to this effect, that the corresponding electrical effect always causes dielectrics to be attracted to electric fields. These two types of magnetic materials can be easily tested by using the magnet shown in Fig. 9.4. The magnet produces a much stronger field near the pointed pole than near the flat one. A small piece of the material is

***Table 9.1* Magnetic Susceptibility of Some Materials at Room Temperature**

Material	χ
Paramagnetic materials	
Aluminum	2.1×10^{-5}
Sodium	0.84×10^{-5}
Titanium	18.0×10^{-5}
Tungsten	7.6×10^{-5}
Gadolinium chloride ($GdCl_3$)	603.0×10^{-5}
Oxygen (1 atm)	193.5×10^{-8}
Magnesium	1.2×10^{-5}
Diamagnetic materials	
Carbon dioxide (1 atm)	-1.19×10^{-8}
Hydrogen (1 atm)	-0.22×10^{-8}
Nitrogen (1 atm)	-0.67×10^{-8}
Bismuth	-1.64×10^{-5}
Copper	-0.98×10^{-5}
Diamond	-2.2×10^{-5}
Gold	-3.5×10^{-5}
Mercury	-2.8×10^{-5}
Silver	-2.4×10^{-5}

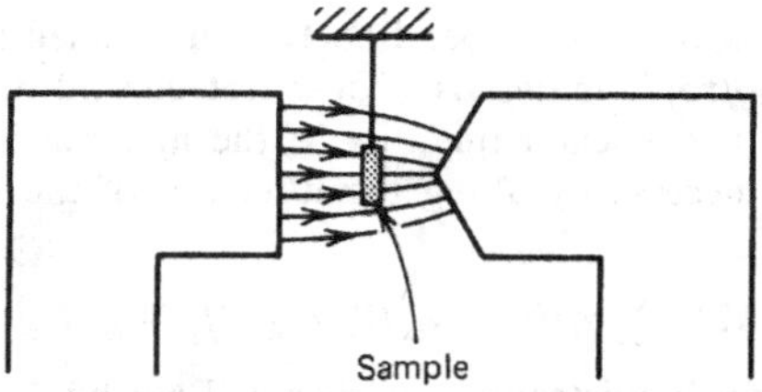

Figure 9.4 Determination of the type of magnetic materials (diamagnetic or paramagnetic) by insertion of a sample in a nonuniform magnetic field.

suspended between the poles, after which the magnet is turned on and the displacement of the sample from the vertical direction is noted. For example, when a sample of bismuth is used, the fact that it gets repelled by the pointed pole, indicates that it is a diamagnetic material. On the other hand, when an aluminum sample is used, the force is attractive; hence it is a paramagnetic material.

Substituting Eq. (9.34) into Eq. (9.29), we get

$$\mathbf{B} = \mu_0(1 + \chi)\mathbf{H} = \mu\mathbf{H} \tag{9.36}$$

where μ is called the *magnetic permeability*. Again it is a constant, and has the dimensions of μ_0. The magnetic permeability when measured in units of μ_0 yields a scalar quantity K_m, which is called the *relative permeability*; that is,

$$K_m = \frac{\mu}{\mu_0} = 1 + \chi \tag{9.37}$$

There are some materials that respond to external magnetic fields in a *nonlinear* fashion; that is, μ becomes dependent on $\mathbf{H}$. These materials are called *ferromagnetic*; they exhibit a high degree of magnetization compared to the paramagnetic materials. In certain cases K_m can reach 1.5×10^5, which is almost five orders of magnitude larger than the largest K_m of paramagnetic materials. These materials also exhibit an irreversible phenomenon called *hysteresis*. These properties make possible a number of important applications in technology, which include making permanent magnets and building transformers and motors.

The nonlinear magnetic permeability $\mu(H)$ of different materials can be either tabulated or plotted as a function of $\mathbf{H}$. In the literature, however, this information is not usually given directly; instead, the dependence of $\mathbf{B}$ on $\mathbf{H}$ is given. The dependence can be determined experimentally in the following way. A specimen of a material, say iron, is placed in an external magnetic field $B_0 = \mu_0 H_0$. The magnetic field B inside the specimen is then measured. The value of B will exceed B_0 as mentioned above because of the presence of the specimen. Thus $B = B_0 + B_m$, where the difference $B_m = B - B_0$ is the magnetic field produced by the specimen. Plotting B_m as a function of H_0 gives what is called the *magnetization curve* of the material studied.

Figure 9.5 shows a Rowland ring, which can be used to perform such a measurement. The iron specimen is made in the form of a ring whose radius r is made much larger than its thickness t. A toroidal coil is then wound around it, with N_1 turns per unit length. The external field B_0 is produced by setting up a current I_1 in the coil, which in the absence of the iron core is equal to $\mu_0 N_1 I_1$. The magnetic field $\mathbf{B}$ is measured by a secondary coil shown also in the figure.

Figure 9.6*a* shows the magnetization curve of iron showing the relation between $\mathbf{H}$ and $\mathbf{B}$ measured using the Rowland ring. Starting from the unmagnetized state

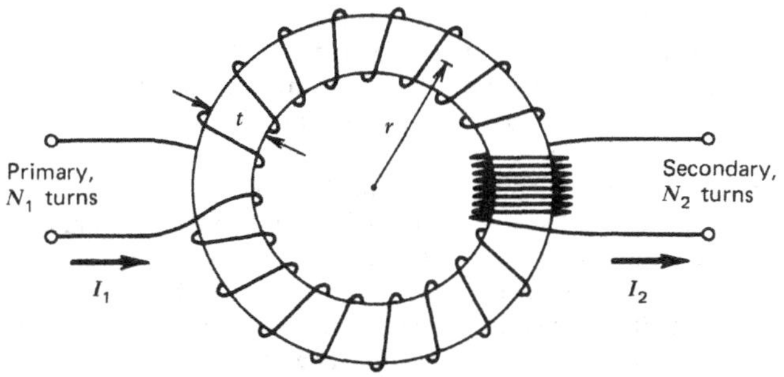

Figure 9.5 Rowland ring for the measurement of the magnetization curves of ferromagnetic materials.

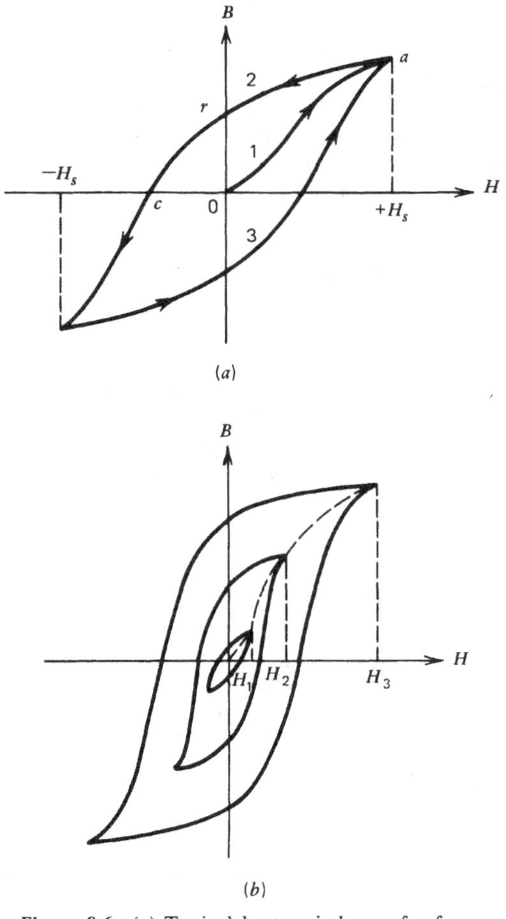

Figure 9.6 (*a*) Typical hysteresis loop of a ferromagnetic material. (*b*) A number of minor hysteresis loops, along with the major loop of a ferromagnetic material.

(that is, from point 0), **B** increases with increasing **H** along curve 1. The initial increase is very steep; however, at higher values of **H**, the iron saturates, and the result is a leveling off of the magnetization (saturation magnetization M_s). In this high-H regime beyond H_s, the increase in B is only due to the increase in H, since M is not increasing any further.

The next operation after reaching the saturation point, a, is reducing the current, and hence H, in the toroid back to zero. It is found that the magnetization curve does not fall back along curve 1; it falls back along curve 2. Curve 2 shows that even when H is zero, B is not zero; that is, the specimen has become permanently magnetized. The B field at r is called *retentivity* or *remanence*. The current in the coil is now reversed, and increased in magnitude. The B-H relationship follows curve 2 until the medium saturates in the reverse direction. The magnitude of H at point c is called the *coercive force* or *coercivity* of the material. As the current is reduced back to zero, the relationship traces curve 3. Again, even when **H** is zero, the specimen has a residual negative **B** field. When the current is now reversed and increased, the relationship continues to follow curve 3 until the saturation point is reached and the magnetization curve closes.

This property of magnetization curves wherein they do not retrace themselves is typical of ferromagnetic materials. The shape of the loop depends on the nature of the specimen and on the maximum value of H reached. When the material is subjected to a maximum value of H where saturation is achieved, as in Fig. 9.6*a*, the loop ceases to change as the material is subjected to even higher fields. Figure 9.6*b* shows the dependence of the shape of the loop on the maximum value of H reached for values below the saturation field.

The relative permeability of a ferromagnetic material as a function of **H** can be calculated from the B-H curve using the relation

$$K_m = \frac{1}{\mu_0}\frac{B}{H}$$

Figure 9.7 shows K_m versus H for silicon steel; it shows extreme nonlinearity of the response at $H \leq 100$ A/m and a saturation effect at $H = 300$ A/m. Figure 9.8 gives the hysteresis curves of some high permeability and some permanent magnetic materials. Table 9.2 gives properties of ferromagnetic materials at room temperature.

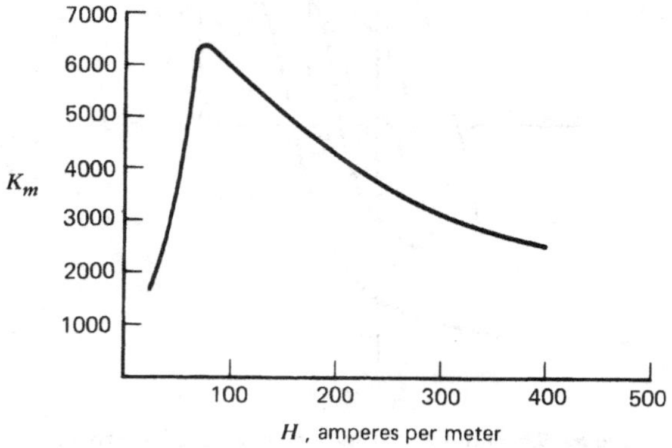

Figure 9.7 Relative permeability of a ferromagnetic material.

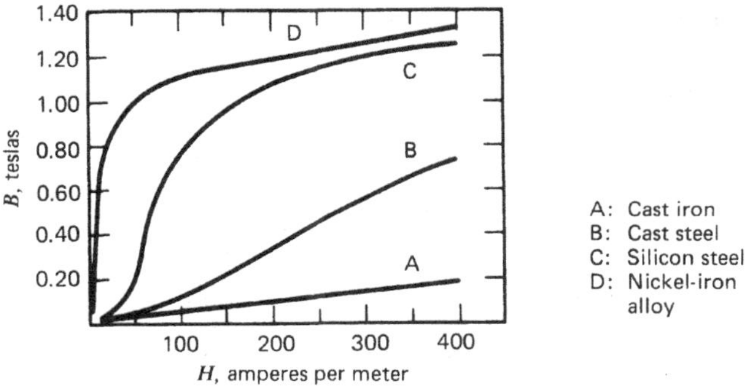

(a)

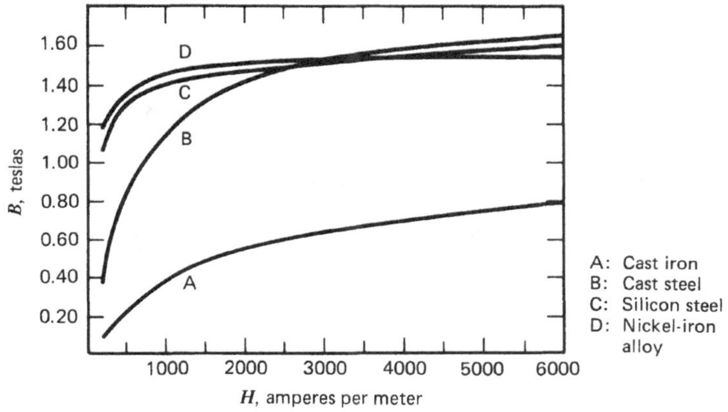

(b)

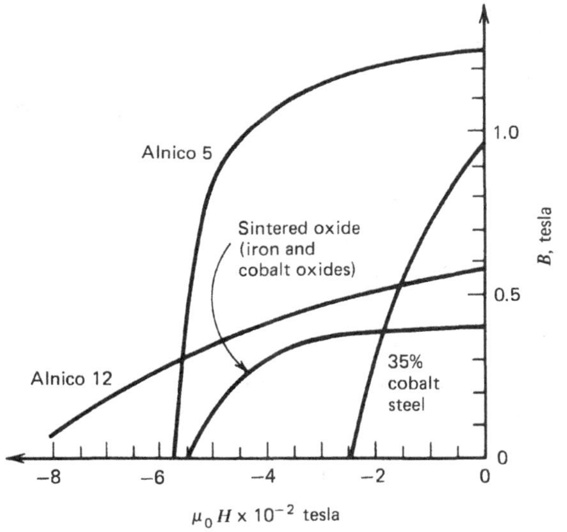

(c)

Figure 9.8 Hysteresis curves of some ferromagnetic materials. (*a*) and (*b*) High permeability materials, (*c*) Permanent magnetic materials.

Table 9.2 **Data for Ferromagnetic Materials**

High-Permeability Materials

Material	Percent Composition	Maximum Relative Permeability	Saturation Flux Density B_{sat} (Wb/m^2)	Coercive Force H_c (A/m)
Cold rolled steel	98.5 Fe	2,000	2.10	145
Iron	99.9 Fe	5,000	2.15	80
Mu metal	18 Fe, 75 Ni, 2 Cr, 5 Cu	100,000	0.65	4
Purified iron	99.95 Fe	180,000	2.15	4
78 Permalloy	21.2 Fe, 78.5 Ni, 0.3 Mn	100,000	1.07	4
Supermalloy	15.7 Fe, 79 Ni, 5 Mo, 0.3 Mn	800,000	0.80	0.16

Permanent-Magnet Materials

Material	Percent Composition	Remanent Flux Density B_r (W/m^2)	Coercive Force H_c (A/m)
Alnico II (sintered)	64.5 Fe, 10 Al, 17 Ni, 2.5 Co, 6 Cu	0.69	41,600
Alnico V	53 Fe, 8 Al, 14 Ni, 24 Co, 3 Cu	1.25	44,000
Carbon steel	98.1 Fe, 1 Mn, 0.9 C	1.0	4,000
Platinum-cobalt	77 Pt, 23 Co	0.45	208,000
Remalloy	71 Fe, 17 Mo, 12 Co	1.05	20,000
Tungsten steel	94 Fe, 5 W, 0.3 Mn, 0.7 C	1.03	5,600

9.5 Boundary Value Problems

9.5.1 The Potential Equations

In Sections 9.2 and 9.3, the Biot-Savart law, Coulomb's law, and Ampere's law were used to determine the magnetic field in a nonferromagnetic medium—that is, a linear medium. In this section we present a boundary value method. The macroscopic equations of magnetism are

$$\nabla \cdot \mathbf{B} = 0 \qquad \text{and} \qquad \nabla \times \mathbf{H} = \mathbf{J}_f \tag{9.30}$$

The boundary value method can be subdivided into two methods: boundary value via the scalar potential and boundary value via the vector potential. We will treat both methods. In the scalar potential method one writes $\nabla \cdot \mathbf{B} = 0$ as follows:

$$\nabla \cdot \mathbf{B} = \nabla \cdot \mu_0(\mathbf{H} + \mathbf{M}) = 0 \qquad \text{or} \qquad \nabla \cdot \mathbf{H} = -\nabla \cdot \mathbf{M}$$

Thus, in terms of **H**, the macroscopic equations become

$$\nabla \cdot \mathbf{H} = -\nabla \cdot \mathbf{M} \qquad \text{and} \qquad \nabla \times \mathbf{H} = \mathbf{J}_f \tag{9.38}$$

In regions of space where $\mathbf{J}_f = 0$, Eq. (9.38) becomes

$$\nabla \cdot \mathbf{H} = -\nabla \cdot \mathbf{M} \qquad \text{and} \qquad \nabla \times \mathbf{H} = 0 \tag{9.39}$$

Because the curl of **H** is zero, then **H** can be written in terms of the gradient of a scalar potential, as was introduced previously in Eq. (9.32).

$$\mathbf{H} = -\nabla \Phi_m \tag{9.32}$$

Taking the divergence of this equation and substituting for $\nabla \cdot \mathbf{H}$ from Eq., (9.39), we get

$$\nabla^2 \Phi_m = \nabla \cdot \mathbf{M} = -\rho_m \tag{9.40}$$

which is analogous to Poisson's equation of electrostatics. However, contrary to the electrostatic case, and as we previously noted in Section 8.7, the magnetic scalar potential is valid only in simply connected regions. If the region is not simply connected the potential would not be single valued (see Example 8.16).

Poisson's equation of magnetostatics, Eq. (9.40), reduces to Laplace's equation when $\rho_m = -\nabla \cdot \mathbf{M}$ vanishes:

$$\nabla^2 \Phi_m = 0 \tag{9.41}$$

The vanishing of ρ_m can take place in a number of cases: (1) $\mathbf{M} = 0$, that is, when one is dealing with currents placed in vacuum, as discussed in Section 8.7; (2) linear magnetic materials with uniform magnetization, where $\mathbf{M}$ = a constant vector; and (3) a material with nonuniform magnetization, but where $\nabla \cdot \mathbf{M}$ is zero (see Example 9.6).

In the vector potential method, $\mathbf{A}$ is defined by $\mathbf{B} = \nabla \times \mathbf{A}$, which automatically makes $\nabla \cdot \mathbf{B} = 0$. The curl equation $\nabla \times \mathbf{H} = \mathbf{J}_f$ gives $\nabla \times \mathbf{B} = \mu \mathbf{J}_f$ in all regions in which the magnetic material is homogeneous. Thus

$$\nabla \times (\nabla \times \mathbf{A}) = \mu \mathbf{J}_f$$

Expanding the triple vector product gives

$$\nabla^2 \mathbf{A} - \nabla(\nabla \cdot \mathbf{A}) = -\mu \mathbf{J}_f. \tag{9.42}$$

Again, as we did in Section 8.5, we choose $\nabla \cdot \mathbf{A} = 0$, which is called the Coulomb's gauge. Thus

$$\nabla^2 \mathbf{A} = -\mu \mathbf{J}_f \tag{9.43}$$

The different components A_x, A_y, and A_z satisfy precisely the form of Poisson's equation:

$$\nabla^2 A_i = -\mu J_{fi} \tag{9.44}$$

where $i = x$, y, and z.

Laplace's equation [Eq. (9.41)] can be solved using the techniques already developed for the electrical case (Chapters 3 and 4). Along with Laplace's equation, one needs a set of boundary conditions satisfied by the scalar potential and the fields in order to determine them uniquely.

Similarly, in regions where $\mathbf{J}_f = 0$ the cartesian components of the vector potential satisfy Laplace's equation. The boundary conditions on the vector potential and the fields are also to be used to determine the solution uniquely.

In the cases where the sources in Eq. (9.40) and (9.43) are not zero, the solution of each of these equations consists of two components: a particular solution of Poisson's equation and a solution of Laplace's equation. The total solution is then made to satisfy the boundary conditions. Examples utilizing those methods will be given after we discuss the boundary conditions on the fields and the potentials.

9.5.2 The Boundary Conditions on the Fields and the Potentials

We now turn to the discussion of the boundary conditions to be satisfied by the fields and the potentials in the presence of magnetic materials. These conditions are

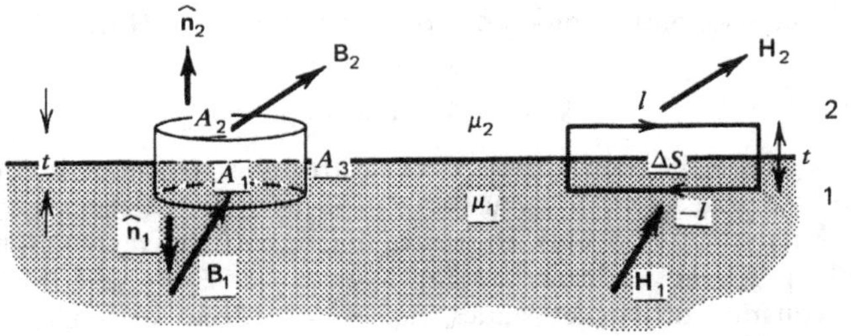

Figure 9.9 Obtaining the boundary conditions on the fields **B** and **H** at the interface between two magnetic materials by applying Gauss' law to the pillbox and Ampere's law to the rectangle.

derived from the macroscopic equations $\nabla \cdot \mathbf{B} = 0$ and $\nabla \times \mathbf{H} = \mathbf{J}_f$; they prescribe relations describing the change undergone by them in passing an interface between two media of different magnetic properties.

The relation $\nabla \cdot \mathbf{B} = 0$ can now be integrated over a volume of a pillbox that is partially immersed in material 1, as shown in Fig. 9.9. The pillbox has a thickness t, flat areas A_1 and A_2, curved surface A_3, volume V, and $\hat{\mathbf{n}}_1$, $\hat{\mathbf{n}}_2$, and $\hat{\mathbf{n}}_3$ unit vectors normal to A_1, A_2, and A_3, respectively. Using the divergence theorem, the volume integral $\int \nabla \cdot \mathbf{B}\, dv = 0$ can be transformed into a surface integral $\int_S \mathbf{B} \cdot \hat{\mathbf{n}}\, da = 0$, where S is the surface of the pillbox, and $\hat{\mathbf{n}}$ is a unit vector normal to the surface. Thus

$$\mathbf{B}_1 \cdot \hat{\mathbf{n}}_1 A_1 + \mathbf{B}_2 \cdot \hat{\mathbf{n}}_2 A_2 + \int_{A_3} \mathbf{B} \cdot \hat{\mathbf{n}}\, da = 0$$

In order to find the change in the field in passing through the interface, we take t to be very small; therefore A_3 vanishes, causing the last contribution of the surface integral to vanish, $-\hat{\mathbf{n}}_1$ to become equal to $\hat{\mathbf{n}}_2 = \hat{\mathbf{n}}$, and $A_1 = A_2 = A$. Thus

$$(\mathbf{B}_2 - \mathbf{B}_1) \cdot \hat{\mathbf{n}} = 0 \qquad \text{or} \qquad B_{1n} = B_{2n} \tag{9.45}$$

That is, the components of the field normal to the interface are continuous across the interface.

A condition on the tangential component of the field intensity **H** can now be derived by integrating $\nabla \times \mathbf{H} = \mathbf{J}_f$ over a surface area, ΔS, bounded by the rectangular curve C shown in Fig. 9.9. The rectangle, which has a length l and width t is partially immersed in medium 1. Thus

$$\int_{\Delta S} \nabla \times \mathbf{H} \cdot d\mathbf{a} = \int_{\Delta S} \mathbf{J}_f \cdot d\mathbf{a}$$

Using Stokes' theorem, the left-hand side of this equation can be transformed to a line integral over the curve C. Thus

$$\oint_C \mathbf{H} \cdot dl = \int_{\Delta S} \mathbf{J}_f \cdot d\mathbf{a}$$

To determine the change in H in passing through the interface we take the width of the rectangle t to be very small; hence

$$(\mathbf{H}_2 - \mathbf{H}_1) \cdot \mathbf{l} = \lim_{t \to 0} \mathbf{J}_f \cdot \Delta\mathbf{S} = K_f l$$

where K_f is the surface current density normal to the plane of the rectangle. This result can also be written in terms of the total surface current density $\mathbf{K}_f$ as

$$(\mathbf{H}_2 = \mathbf{H}_1)_t = \mathbf{K}_f \times \hat{\mathbf{n}} \qquad \text{or} \qquad \hat{\mathbf{n}} \times (\mathbf{H}_2 - \mathbf{H}_1) = \mathbf{K}_f \tag{9.46}$$

where $\hat{\mathbf{n}}$ is a unit vector normal to the interface. Thus the tangential components of H are discontinuous when a surface current density exists on the interface; the discontinuity is the surface current density normal to the field component. The components are consequently continuous when K_f is zero; that is,

$$\mathbf{H}_{1t} = \mathbf{H}_{2t} \tag{9.47}$$

The boundary conditions of the scalar and vector potentials may be deducted from the boundary conditions of **B** and **H**. Since $\mathbf{H}_m = -\nabla\Phi_m$, then

$$\Phi_m = -\int \mathbf{H}_m \cdot d\mathbf{r}. \tag{9.48}$$

The potential difference between two closely located points can be written as $\Delta\Phi_m = -\mathbf{H}_m \cdot \mathbf{l}$, where $\mathbf{l}$ is the separation between the points, and $\mathbf{H}_m$ was taken to be finite and fairly constant across the points. Because $\mathbf{H}_{1t} = \mathbf{H}_{2t}$, across an interface that has *no* surface currents, then Eq. (9.48) implies that the scalar potential is continuous in passing that interface; that is,

$$\Phi_{m1} = \Phi_{m2} \tag{9.49}$$

This continuity of the potential is not independent of the above conditions on the fields, and in fact it is equivalent to the condition on the tangential components of **H**.

The boundary condition on the vector potential can be derived from the boundary condition on **B**. Since $\mathbf{B} = \nabla \times \mathbf{A}$, then

$$\int_{\Delta S} \mathbf{B} \cdot \hat{\mathbf{n}}\, da = \int_{\Delta S} \nabla \times \mathbf{A} \cdot \hat{\mathbf{n}}\, da = F$$

The area ΔS is the rectangular loop shown in Fig. 9.9, and F is the magnetic flux passing through it. Using Stokes' theorem, the above equation can be written as follows:

$$\oint_C \mathbf{A} \cdot d\mathbf{l} = F \tag{9.50}$$

To determine the change in **A** just across the interface we take the width of the rectangle to be very small. In this case the area of the loop vanishes, and hence the flux through it vanishes. Therefore

$$\oint_C \mathbf{A} \cdot d\mathbf{l} = \mathbf{A}_2 \cdot \mathbf{l} - \mathbf{A}_1 \cdot \mathbf{l} = 0$$

Thus the tangential components of the vector potential are continuous:

$$\mathbf{A}_{1t} = \mathbf{A}_{2t} \tag{9.51}$$

As in the case of the scalar potential, this condition on the tangential component of the vector potential is not independent of the above conditions. It is precisely equivalent to the continuity of the normal component of the **B** field.

We would now like to show an important property of **B**—namely, that its flux is continuous everywhere. Let us construct imaginary *magnetic field lines* in a region of space (the direction of the lines at any point is the same as the direction of the

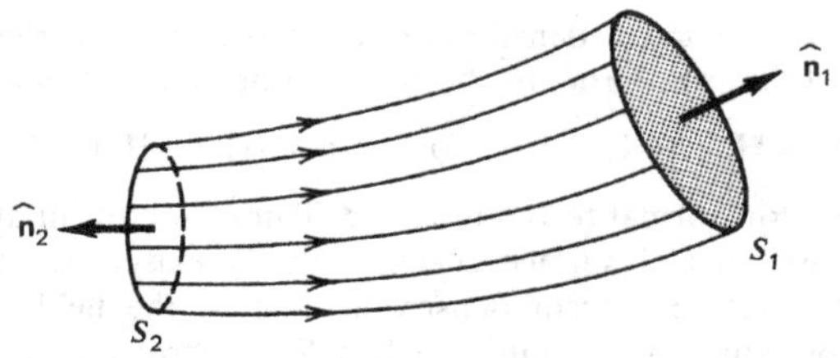

Figure 9.10 A tube of **B** field flux.

magnetic field at the point). Next we consider a volume whose side of area S is bounded by the field lines and whose end surfaces are S_1 and S_2, as shown in Fig. 9.10. Integrating $\nabla \cdot \mathbf{B} = 0$ over this volume and using the divergence theorem to transform the volume integral to a surface integral over the surface of the volume, we get

$$\oint \mathbf{B} \cdot \hat{\mathbf{n}}\, da = 0 = \int_{S_1} \mathbf{B}_1 \cdot \hat{\mathbf{n}}_1\, da - \int_{S_2} \mathbf{B}_2 \cdot (\hat{\mathbf{n}}_2) da \tag{9.52}$$

where the lateral surface did not contribute because B is tangential to it. Thus, Eq. (9.52) establishes flux continuity; that is,

$$F(S_1) = F(S_2) \tag{9.53}$$

where $F(S_i)$ is the magnetic flux passing through surface S_i.

Finally, we would like to note that the flux of the magnetic intensity H, F_H, is not continuous. This can be easily realized since $\nabla \cdot \mathbf{B} = 0$ implies $\nabla \cdot \mathbf{H} = -\nabla \cdot \mathbf{M}$, where $\mathbf{M}$ is the magnetization; and therefore the above procedure leading to Eq. (9.53) gives

$$F_H(S_2) - F_H(S_1) = -\int \nabla \cdot \mathbf{M}\, dv = \int \rho_m\, dv \tag{9.54}$$

Example 9.4 A Filamentary Current at a Plane Interface—Scalar and Vector Potential

An infinite, filamentary, straight wire carrying a current I lies along the z axis on the plane interface between two media with magnetic permeabilities μ_1 and μ_2, respectively. At points away from the wire the magnetic scalar potential satisfies Laplace's equation: $\nabla^2 \Phi = 0$, (we dropped the sublabel m). It was shown in Example 8.16 that the magnetic scalar potential of a current wire in vacuum is of the form $\Phi = C\phi$ and that it is not single-valued. Thus the potentials Φ_1 and Φ_2 in regions 1 and 2 are written as

$$\Phi_1 = C_1 \phi \tag{9.55}$$

$$\Phi_2 = C_2 \phi \tag{9.56}$$

where C_1 and C_2 are constants to be evaluated from the boundary conditions.

One boundary condition is the fact that the total current is I. This can be utilized by using Ampere's law on a circle of radius ρ and center at the wire. The magnetic intensities $\mathbf{H}_1$ and $\mathbf{H}_2$ are first evaluated from Φ_1 and Φ_2, respectively, using the relation $\mathbf{H} = -\nabla\Phi$.

$$\mathbf{H}_1 = -\frac{C_1}{\rho}\hat{\boldsymbol{\phi}} \qquad \mathbf{H}_2 = -\frac{C_2}{\rho}\hat{\boldsymbol{\phi}} \tag{9.57}$$

Ampere's law then gives

$$C_1 + C_2 = -\frac{I}{\pi} \tag{9.58}$$

Another relationship between C_1 and C_2 can now be determined from the continuity of the normal components of the magnetic field at the interface, $B_{1n} = B_{2n}$:

$$\mu_1 C_1 = \mu_2 C_2 \tag{9.59}$$

Equations (9.58) and (9.59) give

$$C_1 = -\frac{I\mu_2}{\pi(\mu_1 + \mu_2)} \qquad C_2 = -\frac{I\mu_1}{\pi(\mu_1 + \mu_2)} \tag{9.60}$$

Thus

$$\Phi_1 = -\frac{I\mu_2}{\pi(\mu_1 + \mu_2)}\phi \qquad \Phi_2 = -\frac{I\mu_1}{\pi(\mu_1 + \mu_2)}\phi \tag{9.61}$$

$$H_1 = \frac{I\mu_2}{\pi(\mu_1 + \mu_2)\rho}\hat{\boldsymbol{\phi}} \qquad H_2 = \frac{I\mu_1}{\pi(\mu_1 + \mu_2)\rho}\hat{\boldsymbol{\phi}} \tag{9.62}$$

In the absence of the materials—that is when the wire is in vacuum ($\mu_1 = \mu_2 = \mu_0$)—the fields reduce to

$$\mathbf{H}_1 = \mathbf{H}_2 = \mathbf{H}_0 = \frac{I}{2\pi\rho}\hat{\boldsymbol{\phi}}.$$

The fields can now be written in terms of $\mathbf{H}_0$, as follows:

$$\mathbf{H}_1 = \frac{2\mu_2}{\mu_1 + \mu_2}\mathbf{H}_0 \qquad \text{and} \qquad \mathbf{H}_2 = \frac{2\mu_1}{\mu_1 + \mu_2}\mathbf{H}_0 \tag{9.63}$$

This result is in fact true for any circuit lying on the plane interface provided $\mathbf{H}_0$ is the field of the circuit in vacuum.

We would now like to treat the same problem using the vector potential method. Since the current is along the z axis, one can take $A_x = A_y = 0$, and thus A_z is the only nonzero component, and at points away from the current it satisfies Laplace's equation: $\nabla^2 A_z = 0$. Moreover, because of the cylindrical symmetry and the fact that the wire is very long, A_z will not depend on ϕ and z; therefore, Laplace's equation reduces to

$$\frac{1}{\rho}\frac{d}{d\rho}\left(\rho\frac{dA_z}{d\rho}\right) = 0 \tag{9.64}$$

which can be easily integrated to give

$$\mathbf{A} = (C \ln \rho + C')\hat{\mathbf{z}} \tag{9.65}$$

where C and C' are constants to be evaluated from the boundary conditions. It is apparent that A_z is single-valued since $A_z(\phi) = A_z(\phi + 2\pi)$, which is in contrast to the magnetic scalar potential discussed above. The potentials $\mathbf{A}_1$ and $\mathbf{A}_2$ in regions 1 and 2, respectively, are written as follows:

$$\mathbf{A}_1 = C_1\hat{\mathbf{z}} \ln \rho \tag{9.66}$$

$$\mathbf{A}_2 = C_2\hat{\mathbf{z}} \ln \rho \tag{9.67}$$

where the constant term in Eq. (9.65) was taken to be zero.

The continuity of the tangential components of the potential at the boundary, Eq. (9.51), requires $C_1 = C_2 = C$. The constant C is now evaluated using Ampere's law: $\oint \mathbf{H}\cdot d\mathbf{l} = I_f$ where the integral is taken over a circle normal to the wire, of radius ρ and with its center at the wire. Taking $H = (1/\mu)\nabla \times \mathbf{A}$, we find

$$\oint \mathbf{H}\cdot d\mathbf{l} = \pi C\left(\frac{1}{\mu_1} + \frac{1}{\mu_2}\right) = I$$

which gives

$$C = \frac{I}{\pi} \frac{\mu_1 \mu_2}{\mu_1 + \mu_2}$$

Hence

$$\mathbf{A} = \frac{I}{\pi} \frac{\mu_1 \mu_2}{\mu_1 + \mu_2} \ln \rho \, \hat{\mathbf{z}} \tag{9.68}$$

It is now straightforward to show that this vector potential gives precisely the fields we got using the scalar potential [Eq. (9.62)].

Example 9.5 A Magnetic Sphere in a Uniform External Magnetic Field

We consider a sphere of radius R, made of a linear magnetic material of permeability μ_1 and embedded in a medium of permeability μ_2. The sphere is placed in a magnetic field $\mathbf{H}_0$ which is initially uniform and pointing along the z direction, as shown in Fig. 9.11.

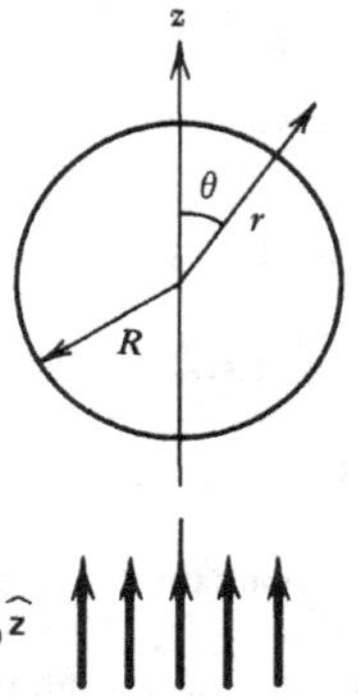

Figure 9.11 A magnetic sphere in an initially uniform magnetic field.

Since there are no external currents and the materials are linear, the magnetic scalar potential satisfies Laplace's equation. The total spherical symmetry of the problem is broken by the presence of the external field; however, because the field is uniform and pointing along the z direction, there is still symmetry about the ϕ direction, thus making the potential dependent on r and θ only. The potentials Φ_1 and Φ_2 inside and outside the sphere can be written as a linear combination of the zonal harmomics [see Eq. (3.28)]. Because at distances far away from the sphere the magnetic field is $H_0\hat{\mathbf{z}}$, the corresponding potential $\Phi_2 = -\int \mathbf{H} \cdot d\mathbf{r}$ is $-H_0 z = -H_0 r \cos 0$. This implies that only a subset of the zonal harmonics will contribute; therefore we only retain terms up to $P_1(\cos\theta)$, and write:

$$\Phi_1 = A_1 r \cos\theta + \frac{A_2}{r^2}\cos\theta \qquad r < R \tag{9.69}$$

$$\Phi_2 = C_1 r \cos\theta + \frac{C_2}{r^2}\cos\theta \qquad r > R \tag{9.70}$$

Note that the potentials are single-valued; that is, $\Phi_i(r, \theta) = \Phi_i(r, \theta + 2\pi)$. Also, it is to be noted that the $1/r$ term was dropped because it implies the existence of magnetic monopoles.

The constant A_2 is now taken zero because the potential inside the sphere should be finite at the origin, and the constant C_1 is taken equal to $-H_0$ since $\Phi_2 = -H_0 r \cos\theta$ as r becomes very large. Thus Φ_1 and Φ_2 become

$$\Phi_1 = A_1 r \cos\theta \qquad r < R \tag{9.71}$$

$$\Phi_2 = -H_0 r \cos\theta + \frac{C_2}{r^2}\cos\theta \qquad r > R \tag{9.72}$$

We need two more equations relating A_1 and C_2 in order to determine the potentials. These relations are now deduced from the boundary condition on the surface of the sphere, $r = R$. The continuity of the tangential components of **H** on the surface is equivalent to the continuity of the potentials on the surface. Thus

$$A_1 R = -H_0 R + \frac{C_2}{R^2} \tag{9.73}$$

The second equation is determined from the continuity of the normal components of **B**. Since

$$B_n = \mu H_n = -\mu \frac{\partial \Phi}{\partial r}$$

we find from Eqs. (9.71) and (9.72) that

$$\mu_1 A_1 = -\mu_2 H_0 - 2\frac{\mu_2 C_2}{R^3} \tag{9.74}$$

Solving the above equations relating A_1 and C_2 simultaneously gives

$$A_1 = -\frac{3\mu_2 H_0}{\mu_1 + 2\mu_2} \qquad C_2 = \mu_2 H_0 R^3 \frac{(\mu_1/\mu_2 - 1)}{\mu_1 + 2\mu_2}.$$

Substituting for A_1 and C_2 in Eqs. (9.71) and (9.72) and using $B = -\mu \nabla\Phi$, we obtain

$$\mathbf{B}_1 = \frac{3\mu_2 H_0}{1 + 2\mu_2/\mu_1}\hat{\mathbf{z}} \tag{9.75}$$

$$\mathbf{B}_2 = \mu_2 H_0 \hat{\mathbf{z}} + \mu_2 \frac{\mu_1/\mu_2 - 1}{\mu_1/\mu_2 + 2} H_0 \left(\frac{R}{r}\right)^3 [2\hat{\mathbf{r}}\cos\theta + \hat{\boldsymbol{\theta}}\sin\theta] \tag{9.76}$$

Note the resemblance to the case of a dielectric sphere in an external electric field [Eqs. (4.86) and (4.87)].

Example 9.6 A Cylinder with a Nonuniform Magnetization

We consider a boundary value problem where the magnetization is nonuniform. A long cylinder of radius R is magnetized with the magnetization given by $M = (p\rho \sin 2\phi + q\rho \cos\phi)\hat{\boldsymbol{\rho}} + (\frac{1}{2}p\rho\cos 2\phi - q\rho\sin\phi)\hat{\boldsymbol{\phi}}$, where p and q are constants. It is easy to show that $\nabla \cdot \mathbf{M} = 0$; thus the magnetic scalar potential Φ satisfies Laplace's equation.

Because the cylinder is very long, Φ will not depend on z. The fact that **M** depends on ϕ, however, makes Φ dependent on both ρ and ϕ. Equation (3.65) gives the solution for the potential as a function of ρ and ϕ, in terms of cylindrical harmonics. The solution Φ_1 and Φ_2 for the inside and the surrounding regions of the cylinder, respectively, however, are described by only a few terms of the infinite expansion. To determine these contributing terms we consider the boundary conditions:

1. Φ_1 does not include terms with $\ln\rho$ and ρ^{-n} dependence since it should be finite at $\rho = 0$.
2. Φ_2 does not include terms with ρ^n dependence since it should be finite as $\rho \to \infty$.
3. Φ_2 does not include the $\ln\rho$ term since there are no monopoles.
4. $\Phi_1 = \Phi_2$ at the cylinder's surface—that is, at $\rho = \rho_0$.
5. The normal components of **B** are continuous at $\rho = \rho_0$.

Thus we write

$$\Phi_1(\rho, \phi) = A_0 + \sum_{n=1}^{\infty} (A_n \cos n\phi + C_n \sin n\phi)\rho^n \qquad \rho < \rho_0 \tag{9.77}$$

$$\Phi_2(\rho, \phi) = \sum_{n=1}^{\infty} (A'_n \cos n\phi + C'_n \sin n\phi)\rho^{-n} \qquad \rho > \rho_0 \tag{9.78}$$

The continuity of the potentials at $\rho = \rho_0$, $\Phi_1(\rho_0, \phi) = \Phi_2(\rho_0, \phi)$, gives $A_0 = 0$ and the following relations between the constants

$$A_n = \frac{A'_n}{\rho_0^{2n}} \qquad C_n = \frac{C'_n}{\rho_0^{2n}} \qquad \text{for } n = 1, 2, \ldots \tag{9.79}$$

We now consider condition number 5: Since $\mathbf{B} = \mu_0 \mathbf{H} + \mu_0 \mathbf{M}$, the continuity of the normal components of $\mathbf{B}$ implies that

$$\mathbf{H}_1 \cdot \hat{\mathbf{n}} + \mathbf{M} \cdot \hat{\mathbf{n}} = \mathbf{H}_2 \cdot \hat{\mathbf{n}} \tag{9.80}$$

The normal component of $\mathbf{M}$ at the surface of the cylinder is

$$M_n = \mathbf{M} \cdot \hat{\boldsymbol{\rho}} = p\rho_0 \sin 2\phi + q\rho_0 \cos \phi \tag{9.81}$$

The normal component of $\mathbf{H}$ at the surface of the cylinder is calculated from using the relation $H_n = -\partial\Phi/\partial\rho$ at $\rho = \rho_0$. Substituting for the normal components of $\mathbf{H}$ and $\mathbf{M}$ in Eq. (9.80) and equating coefficients of $\sin n\phi$ and coefficients of $\cos n\phi$, we get

$$\begin{aligned} -A_1 + q\rho_0 &= \frac{A'_1}{\rho_0^2} \qquad \text{and} \qquad -A_n = \frac{A'_n}{\rho_0^{2n}} \qquad \text{for } n \neq 1 \\ -2C_2\rho_0 + p\rho_0 &= \frac{2C'_2}{\rho_0^3} \qquad \text{and} \qquad -C_n = \frac{C'_n}{\rho_0^{2n}} \qquad \text{for } n \neq 2 \end{aligned} \tag{9.82}$$

Solving Eqs. (9.79) and (9.82) simultaneously yields the following nonzero constants.

$$A_1 = \frac{q\rho_0}{2} \qquad A'_1 = \frac{q\rho_0^3}{2} \qquad C_2 = \frac{p}{4} \qquad C'_2 = \frac{p}{4}\rho_0^4$$

Hence the potentials take the following expressions:

$$\Phi_1(\rho, \phi) = \frac{1}{2} q\rho_0\rho \cos \phi + \frac{1}{4} p\rho^2 \sin 2\phi \tag{9.83}$$

$$\Phi_2(\rho, \phi) = \frac{1}{2} q\rho_0^3 \frac{\cos \phi}{\rho} + \frac{p}{4} \rho_0^4 \frac{\sin 2\phi}{\rho^2} \tag{9.84}$$

It is apparent from these potentials that only the harmonics $\cos \phi$ and $\sin 2\phi$ contribute to the potential. This dependence in fact results from the fact that the angular dependence of M_n involves only these harmonics. Therefore, we could have started with a much smaller number of terms in Eqs. (9.77) and (9.78); that is, we could have retained only the terms that have $\cos \phi$ and $\sin 2\phi$ dependence.

Also, it is to be noted that this magnetization problem is analogous to a problem in electrostatics (see Example 3.10) where a long cylinder has a surface charge distribution of the form

$$\sigma = \sigma_1 \sin 2\phi + \sigma_2 \cos \phi \tag{9.85}$$

The analogy becomes clear when we assign:

$$p\rho_0 = \sigma_1 \qquad \text{and} \qquad q\rho_0 = \sigma_2$$

Example 9.7 The Fields Due to a Uniformly Magnetized Sphere

In this example we consider a sphere of radius R that has a uniform magnetization $M_0\hat{\mathbf{z}}$ along the z axis. This problem was analyzed in Example 9.2 using both the magnetization pole and current densities. Here we use the boundary value method. Since $\mathbf{M}$ is constant, then $\nabla \cdot \mathbf{M} = 0$, and thus the scalar magnetic potential satisfies Laplace's equation. The fact that $\mathbf{M}$ is along the z axis makes the potential dependent on θ as well as on r. Because

$\mathbf{M}\cdot\hat{\mathbf{n}} = M_0 \cos\theta$, we expect the angular dependence of the potential to involve only $\cos\theta$; thus we write:

$$\Phi_1 = A_1 r \cos\theta + \frac{C_1 \cos\theta}{r^2} \qquad r < R \tag{9.86}$$

$$\Phi_2 = A_1' r \cos\theta + \frac{C_1' \cos\theta}{r^2} \qquad r > R \tag{9.87}$$

We now apply the boundary conditions in order to evaluate A_1, A_1', C_1, and C_1'. We take $C_1 = 0$ because Φ_1 should be finite at $r = 0$, and take $A_1' = 0$ because Φ_2 should vanish at large distances from the sphere. At the surface of the sphere the potential is continuous: $\Phi_1(R, \theta) = \Phi_2(R, \theta)$; thus

$$A_1 R = \frac{C_1'}{R^2} \tag{9.88}$$

The last boundary condition is the continuity of the normal component of $\mathbf{B}$ at the surface of the sphere:

$$-\mu_0 \frac{\partial \Phi_1}{\partial r} + \mu_0 M_0 \cos\theta = -\mu_0 \frac{\partial \Phi_2}{\partial r} \qquad \text{at } r = R \tag{9.89}$$

Substituting for Φ_1 and Φ_2 from Eqs. (9.86) and (9.87) and equating the coefficients of $\cos\theta$, we get:

$$-A_1 + M_0 = 2\frac{C_1'}{R^3} \tag{9.90}$$

Solving Eqs. (9.88) and (9.90) simultaneously yields:

$$A_1 = \frac{1}{3} M_0 \qquad \text{and} \qquad C_1' = \frac{1}{3} M_0 R^3$$

Hence

$$\Phi_1(r, \theta) = \frac{1}{3} M_0 r \cos\theta, \tag{9.91}$$

$$\Phi_2(r, \theta) = \frac{1}{3} M_0 \frac{R^3}{r^2} \cos\theta \tag{9.92}$$

The external potential produced by the sphere is a dipole potential of the form

$$\Phi_2 = \frac{\mathbf{m}\cdot\mathbf{r}}{4\pi r^3} = \frac{m\cos\theta}{4\pi r^2} \tag{9.93}$$

where $m = (4\pi/3)M_0 R^3 = M_0 V$ is the magnetic dipole moment and V is the volume of the sphere. The internal potential $\Phi_1 = \frac{1}{3}M_0 z$, on the other hand, is linear with z and thus corresponds to a uniform magnetic field. The corresponding magnetic fields are

$$\mathbf{H}_1 = -\frac{1}{3} M_0 \hat{\mathbf{z}} \tag{9.94}$$

$$\mathbf{H}_2 = \frac{1}{3} M_0 \left(\frac{R}{r}\right)^3 (2\hat{\mathbf{r}} \cos\theta + \hat{\boldsymbol{\theta}} \sin\theta] \tag{9.95}$$

Figure 9.12 gives the lines of $\mathbf{H}$ and the lines of $\mathbf{B}$. The lines of $\mathbf{B}$ are continuous (closed curves) where as those of $\mathbf{H}$ are discontinuous; they originate from the effective magnetic pole surface charges located at the surface of the sphere.

We would like to say a few more words about the internal fields. It is important to note that $\mathbf{H}_1$ points along the $-z$ direction; that is, it opposes the magnetization $M_0\hat{\mathbf{z}}$; hence it

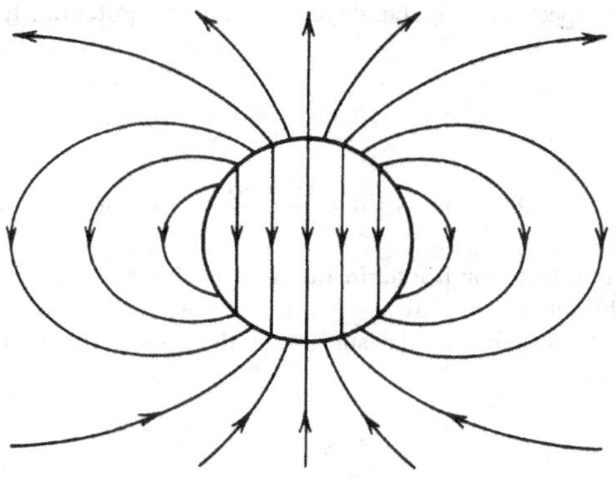

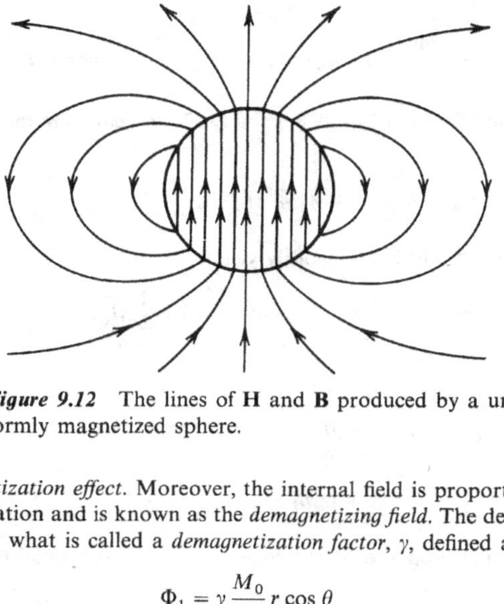

Figure 9.12 The lines of **H** and **B** produced by a uniformly magnetized sphere.

results in a *demagnetization effect*. Moreover, the internal field is proportional to the magnitude of the mangetization and is known as the *demagnetizing field*. The demagnetization effect may be described by what is called a *demagnetization factor*, γ, defined as follows:

$$\Phi_1 = \gamma \frac{M_0}{4\pi} r \cos \theta \tag{9.96}$$

Therefore, in this case, it is $4\pi/3$. This value pertains only to a sphere; other geometries have other values. For example, a similar procedure will yield $\gamma = 2\pi$ for an infinite cylinder whose axis is perpendicular to the direction of magnetization. Also the value of γ for a large, flat sheet magnetized normal to its surface is approximately 4π. This large value for γ is due to the fact that most of the lines of **H** passing from one pole on one face to the other pole on the other must pass through the sheet.

---

*Example 9.8 Vector Potential Boundary Problem—Angular Current Distribution

This example involves a current distribution that depends on angles, and consequently the vector potential produced by it will depend on more than one coordinate. Let us consider a sphere of radius R that has a charge uniformly distributed over its surface. The sphere is spinning with a constant angular velocity ω about an axis going through its center (see Example 9.2).

We describe this geometry with spherical polar coordinates with the z axis taken along the axis of rotation. The density of the surface current produced as a result of spinning is $\mathbf{K} = \sigma\mathbf{v}$, where $\sigma = q/4\pi R^2$ is the surface charge density, and $\mathbf{v} = \omega R \sin\theta\hat{\boldsymbol{\phi}}$ is the velocity of the charge element at angle θ, and $\hat{\boldsymbol{\phi}}$ is a unit vector in the ϕ direction. Thus

$$\mathbf{K} = \frac{q\omega}{4\pi R}\sin\theta\hat{\boldsymbol{\phi}} \tag{9.97}$$

The vector potential outside and inside the sphere satisfy Laplace's equation. Due to the symmetry of the geometry, $\mathbf{A}$ is taken in the ϕ direction and independent of ϕ, but it is of course dependent on r and θ. Away from the currents (that is, away from the surface of the sphere) Eq. (9.43) gives the following for the ϕ component of $\mathbf{A}$

$$(\nabla^2\mathbf{A})_\phi = \nabla^2 A_\phi - \frac{A_\phi}{r^2\sin^2\theta} = 0 \tag{9.98}$$

or

$$\frac{1}{r^2}\frac{\partial}{\partial r}\left(r^2\frac{\partial A_\phi}{\partial r}\right) + \frac{1}{r^2\sin\theta}\frac{\partial}{\partial\theta}\left(\sin\theta\frac{\partial A_\phi}{\partial\theta}\right) + \frac{A_\phi}{r^2\sin^2\theta} = 0 \tag{9.99}$$

Because the current density is proportional to $\sin\theta$, it is reasonable to take the angular dependence of A_ϕ to be simply $\sin\theta$ and hence write A_ϕ as a product of a radial function and $\sin\theta$; that is,

$$A_\phi = F(r)\sin\theta \tag{9.100}$$

Substituting this in Eq. (9.99) gives the following equation for $F(r)$.

$$\frac{1}{r^2}\frac{d}{dr}\left(r^2\frac{dF}{dr}\right) - \frac{2F}{r^2} = 0. \tag{9.101}$$

We take a solution for $F(r)$ of the form r^n, which upon substitution gives an equation for n:

$$n(n+1) - 2 = 0 \tag{9.102}$$

which gives $n = 1$ or -2. Therefore, there are two solutions for A_ϕ:

$$A_\phi = \frac{\sin\theta}{r^2} \quad \text{and} \quad r\sin\theta \tag{9.103}$$

Because the potential is expected to be finite at the center of the sphere and not to blow up outside the sphere, we choose $r\sin\theta$ for the internal region, and $(\sin\theta)/r^2$ for the surrounding region. Thus

$$\mathbf{A}_1 = C_1 r\sin\theta\,\hat{\boldsymbol{\phi}} \qquad r < R \tag{9.104}$$

$$\mathbf{A}_2 = C_2\frac{\sin\theta}{r^2}\hat{\boldsymbol{\phi}} \qquad r > R \tag{9.105}$$

At $r = R$, the potentials from both regions are equal according to (Eq. 9.51). Thus

$$C_2 = C_1 R^3 \tag{9.106}$$

Another relation between C_1 and C_2 can now be determined from the condition relating the tangential components of $\mathbf{H}$ [Eq. (9.46)]. Using $\mathbf{H} = (1/\mu_0)\nabla\times\mathbf{A}$ gives

$$\mathbf{H}_1 = \frac{2C_1}{\mu_0}(\cos\theta\,\hat{\mathbf{r}} - \sin\theta\,\hat{\boldsymbol{\theta}}) \qquad \mathbf{H}_2 = \frac{C_2}{r^3\mu_0}(2\cos\theta\,\hat{\mathbf{r}} + \sin\theta\,\hat{\boldsymbol{\theta}})$$

Thus

$$\hat{\mathbf{r}}\times(\mathbf{H}_2 - \mathbf{H}_1)|_{r=R} = \frac{q\omega}{4\pi R}\sin\theta\,\hat{\boldsymbol{\phi}}$$

Substituting for $\mathbf{H}_1$ and $\mathbf{H}_2$ gives

$$\frac{1}{\mu_0}\left(2C_1 \sin\theta + \frac{C_2}{R^3}\sin\theta\right)\hat{\boldsymbol{\phi}} = \frac{q\omega \sin\theta}{4\pi R}\hat{\boldsymbol{\phi}}$$

or

$$2C_1 + \frac{C_2}{R^3} = \mu_0 \frac{q\omega}{4\pi R} \tag{9.107}$$

Equations (9.106) and (9.107) are now solved simultaneously for C_1 and C_2; that is,

$$C_1 = \frac{\mu_0 q\omega}{12\pi R} \qquad \text{and} \qquad C_2 = \frac{\mu_0 q\omega R^2}{12\pi}$$

Hence

$$A_1 = \frac{\mu_0 q\omega}{12\pi R} r\sin\theta \qquad \text{and} \qquad A_2 = \frac{\mu_0 q\omega R^2}{12\pi r^2}\sin\theta \tag{9.108}$$

It is to be noted that the potential produced by the spinning sphere is similar to the potential produced by a magnetized sphere (see Example 9.2). In fact, the two results are identical if M_0 is taken equal to $q\omega/4\pi R$.

*9.6 Method of Images for Magnetic Interfaces

The method of images used for solving electrostatic problems can be used for the solution of some magnetostatic problems. For a description of the method see Chapters 3 and 4. The method is very powerful in solving some magnetic problems that contain free currents and prescribed boundary conditions in nonsymmetric configurations. Such problems are not easily solvable using the techniques developed so far in this chapter. Below are some examples of this method.

Example 9.9 A Filamentary Current Parallel to an Interface

In Example 9.4 we considered a filamentary current placed in the plane interface of two magnetic materials. Because of the symmetry, the fields were functions of the distance from the filament only and were determined by solving the radial Laplace's equation in the media in one dimension and applying the boundary conditions. When the current is not located in the plane interface, however, the symmetry is broken and the problem cannot be solved by the same method.

Consider a very long wire that carries a current I, embedded in a semiinfinite magnetic material of permeability μ_1 as shown in Fig. 9.13. The wire is parallel to and at a distance $x = d$ from a plane interface separating another semi-infinite magnetic material of permeability μ_2. According to the method of images, the field in region 1 is given by the current I and an image current I_1 (unknown for now) located at distance $x = -d$ just under the current I. The field in region 2, on the other hand, is given by a current I_2 (unknown for now) located at the same position of current I. Thus the vector potentials produced by the wire can be easily written, using (Eq. 8.52), as follows:

$$\mathbf{A}_1 = -\frac{\mu_1 I}{2\pi}\hat{\mathbf{z}}\ln\rho_1 - \frac{\mu_1 I_1}{2\pi}\hat{\mathbf{z}}\ln\rho_2 \qquad x > 0 \tag{9.109}$$

$$\mathbf{A}_2 = -\frac{\mu_2 I_2}{2\pi}\hat{\mathbf{z}}\ln\rho_1 \qquad x < 0 \tag{9.110}$$

where $\hat{\mathbf{z}}$ is unit vector along the z axis, the values

$$\rho_1 = [(x-d)^2 + y^2]^{1/2} \qquad \text{and} \qquad \rho_2 = [(x+d)^2 + y^2]^{1/2} \tag{9.111}$$

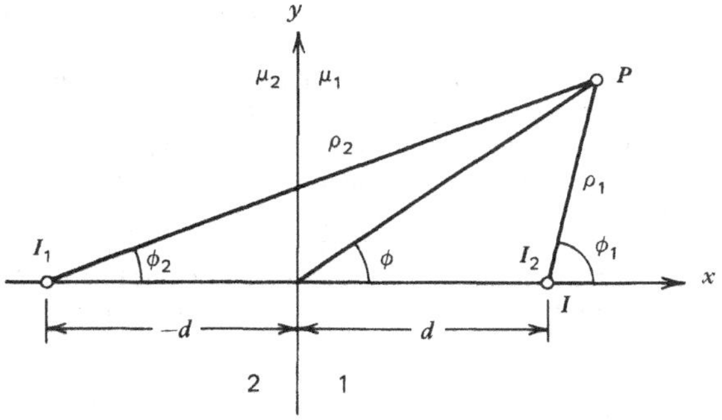

Figure 9.13 Application of the method of images to a filamentary straight current parallel to the interface of two magnetic media.

are defined in the figure, and I_1 and I_2 are constants to be evaluated from the boundary conditions.

At the interface the tangential component of **A** is continuous; that is, $\mathbf{A}_{1t} = \mathbf{A}_{2t}$. Since $\rho_1 = \rho_2 = (d^2 + y^2)^{1/2}$ in the plane of the interface, we get

$$\mu_1(I + I_1) = \mu_2 I_2 \tag{9.112}$$

The second boundary condition is the continuity of the tangential component of the magnetic field **H**. Taking the curl of $\mathbf{A}_1/\mu_1$ and $\mathbf{A}_2/\mu_2$ and equating the y components at the interface (there is no z component), we get

$$I - I_1 = I_2 \tag{9.113}$$

Solving Eqs. (9.112) and (9.113) simultaneously yields

$$I_1 = \frac{\mu_2 - \mu_1}{\mu_1 + \mu_2} I \qquad \text{and} \qquad I_2 = \frac{2\mu_1}{\mu_1 + \mu_2} I \tag{9.114}$$

Note that I_2 has the same direction as I, whereas I_1 is in the same direction only when $\mu_2 > \mu_1$.

It is instructive to apply the magnetic scalar potential method to this problem. The scalar potentials in both regions are

$$\Phi_1 = -\frac{I}{2\pi}\phi_1 - \frac{I_1}{2\pi}\phi_2 \qquad \text{and} \qquad \Phi_2 = -\frac{I_2}{2\pi}\phi_1 \tag{9.115}$$

where the angles ϕ_1 and ϕ_2 are shown in Fig. 9.13 and are given as follows:

$$\phi_1 = \tan^{-1}\left(\frac{y}{x - d}\right) \qquad \text{and} \qquad \phi_2 = \tan^{-1}\left(\frac{y}{x + d}\right) \tag{9.116}$$

We now use the boundary conditions to evaluate I_1 and I_2. One condition requires the normal component of **B** to be continuous at the interface. The magnetic field is calculated from the scalar potential using the relation $\mathbf{B} = -\mu\,\nabla\Phi$. Thus

$$\mathbf{B}_1 = \frac{\mu_1 I}{2\pi\rho_1}[-\sin\phi_1\,\hat{\mathbf{x}} + \cos\phi_1\,\hat{\mathbf{y}}] + \frac{\mu_1 I_1}{2\pi\rho_2}[-\sin\phi_2\,\hat{\mathbf{x}} + \cos\phi_2\,\hat{\mathbf{y}}] \tag{9.117}$$

$$\mathbf{B}_2 = \frac{\mu_2 I_2}{2\pi\rho_1}[-\sin\phi_1\,\hat{\mathbf{x}} + \cos\phi_1\,\hat{\mathbf{y}}] \tag{9.118}$$

Note that at the interface $x = 0$, ρ_1 becomes equal to ρ_2, and $\phi_1 + \phi_2$ becomes equal to π. Equating the normal components of **B**—that is, the components along the x axis—gives

$\mu_1(I + I_1) = \mu_2 I_2$, which is exactly the result of Eq. (9.112) arrived at by using the vector potential method.

The second condition requires that the tangential components of H be continuous. Using $\mathbf{H} = \mathbf{B}/\mu$ and taking the y components of $\mathbf{B}$ from Eqs. (9.117) and (9.118) gives $I - I_1 = I_2$, which is precisely what we arrived at in Eq. (9.113) using the vector potential method.

Both methods therefore give exactly the same image currents. Also it is easy to show that the fields derived from the vector potentials of Eqs. (9.109) and (9.110) are identical to those of Eqs. (9.117) and (9.118), which were derived from the scalar potential.

We would like to note that the scalar potentials of Eqs. (9.115) and (9.116) are not continuous at the interface. Substituting $\phi_2 = \pi - \phi_1$ and I_1 and I_2 from Eqs. (9.114) into Eq. (9.115) gives

$$\Phi_1(0, y) = -\frac{\mu_2 I}{\pi(\mu_1 + \mu_2)}\phi_1 - \frac{(\mu_2 - \mu_1)I}{2(\mu_1 + \mu_2)}$$

and

$$\Phi_2(0, y) = \frac{-\mu_1 I}{(\mu_1 + \mu_2)}\phi_1$$

This result reflects the fact that the region considered is not simply connected and hence the scalar potential in this region is not single-valued, as we discussed above.

Example 9.10 A Magnet Near a Plane Interface of Magnetic Materials

A magnet of magnetic moment **m** is placed at distance z_0 at an angle θ from the surface of a semiinfinite slab of magnetic material whose relative permeability is $\mu \gg \mu_0$, as shown in Fig. 9.14.

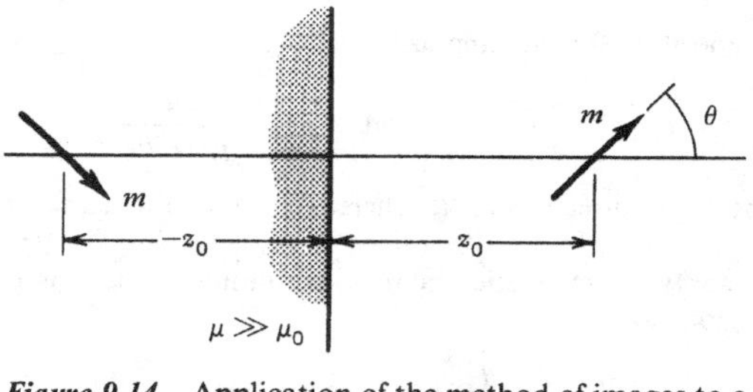

Figure 9.14 Application of the method of images to a magnetic dipole near the interface of a ferromagnetic material.

The interaction between the magnet and the slab can be analyzed using the method of images. In fact, if the magnet is represented by an effective length l and a pole strength q, then the problem becomes completely analogous to the problem of an electric dipole placed near a conducting plane (see Example 3.13). The magnetic slab can be replaced by an image magnet placed at distance $-z_0$ at an angle $-\theta$ from the interface, as shown in Fig. 9.14. The force and the torque experienced by the magnet can be evaluated using the same procedure followed in Example 3.13; thus we leave the details as an exercise. Specifically, one finds that

$$F = \frac{3\mu_0}{64\pi}\frac{m^2}{z_0^4}(1 + \cos^2\theta) \tag{9.119}$$

$$\tau = \frac{\mu_0}{64\pi}\frac{m^2 \sin 2\theta}{z_0^3} \tag{9.120}$$

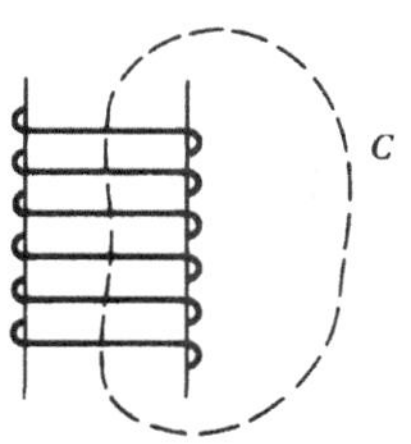

Figure 9.15 A solenoid with a line of **B** is shown and labeled as a loop C.

*9.7 Magnetic Circuits

So far in this chapter and the previous we have analyzed various problems involving current distributions embedded in infinite media. The fields were found to extend over all space. In this section we consider some problems where these fields are confined to well-defined paths. The situation is different from the free-space case to the degree that it embodies an entirely different subject matter called *magnetic circuits*, which are characterized by a close analogy to current circuits.

Consider as an example the long air-core coil shown in Fig. 9.15, which was treated in Example 8.8. Applying Ampere's law to the path C gives

$$\oint_C \mathbf{H} \cdot d\mathbf{l} = NI \qquad (9.121)$$

where N is the number of turns and I is the current in the coil. Outside the coil the field lines are widely spread, and hence the field is weak. Thus the flux is effectively confined to the internal region with the field given by

$$H \approx \frac{NI}{l} \qquad (9.122)$$

where l is the length of the coil.

When ferromagnetic materials are present in space, most of the magnetic flux can be channeled through them. This is so because they have very large relative permeabilities; hence the magnetic field inside these materials for a given H is much larger than the corresponding field in free space. Figure 9.16 shows the idea of a magnetic circuit. A coil of N turns is distributed over a small section of an iron ring. Although the coil does not cover all the ring, it is found that the flux produced by the coil just follows the ring, with very negligible flux outside of it.

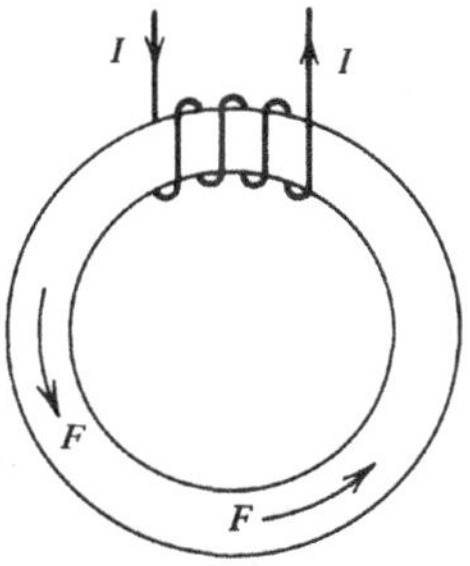

Figure 9.16 An iron ring with a current coil distributed over a small section of it constitues a magnetic circuit.

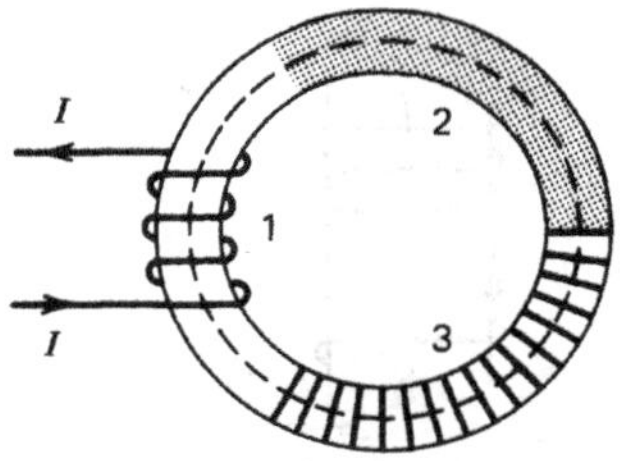

Figure 9.17 A magnetic circular core composed of several materials.

In the magnetic-circuit point of view, the external flux is neglected and thus the fields are assumed to be totally confined to the core. Therefore, Eq. (9.53) implies that the magnetic flux is continuous throughout the circuit. Moreover, the flux is assumed to be uniformly distributed over the cross section of the core.

Let us now analyze in detail a magnetic circuit. Consider an N-turn coil distributed over a section of a ferromagnetic core. The core is composed of several materials—three, for example, in Fig. 9.17. Applying Ampere's law around a path taken in the center of the core gives

$$\oint \mathbf{H}\cdot d\mathbf{l} = H_1 l_1 + H_2 l_2 + H_3 l_3 = NI \tag{9.123}$$

where l_i and H_i are the length and the magnetic intensity in the ith section. Equation (9.123) is closely analogous to electric circuits. Applying Kirchhoff's law, Eq. (7.83), around a single loop containing three resistors and an electromagnetic source $\mathscr{E}$ gives $\mathscr{E} = V_1 + V_2 + V_3$, where V_i is a voltage drop across the ith resistor. This comparison suggests that NI in Eq. (9.123) can be viewed as a *magnetomotive force*, $\mathscr{M}$, and $H_i l_i$ as *amperage drop*. The analogy can be developed further by using the magnetic flux. Using $H_i = B_i/\mu_i$ and $F_i = B_i A_i$, where A_i and μ_i are the cross section and the permeability, respectively of the ith section of the core, transforms Eq. (9.123) to

$$\frac{F_1 l_1}{\mu_1 A_1} + \frac{F_2 l_2}{\mu_2 A_2} + \frac{F_3 l_3}{\mu_3 A_3} = NI \tag{9.124}$$

Since we are dealing with a magnetic circuit where the flux is assumed to be confined to the core, then $F_1 = F_2 = F_3 = F$, and thus

$$F(\mathscr{R}_1 + \mathscr{R}_2 + \mathscr{R}_3) = F\mathscr{R} = \mathscr{M} \tag{9.125}$$

where

$$\mathscr{R}_i = \frac{l_i}{\mu_i A_i} \tag{9.126}$$

is the *reluctance* of the ith section and $\mathscr{R}$ is the sum of the individual reluctances. The analogy with electric circuits is now more evident. The relation in Eq. (9.125) is similar to what we encountered in a series current circuit where the current and the electromotive force are related by $\mathscr{E} = RI$, where R is the total resistance of the circuit. Moreover, the expression for the reluctance of the iron core in terms of l, μ,

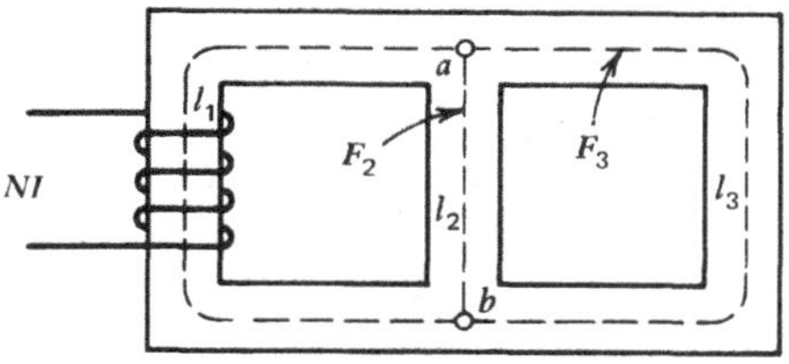

Figure 9.18 A multiloop magnetic circuit.

and A is very analogous to the expression for the resistance of a wire length l, cross-sectional area A, and conductivity σ_c; that is, $R = l/\sigma_c A$.

A word of caution needs to be mentioned now concerning the analogy between the electrical and magnetic cases. The permeability must be known for each of the materials before one can calculate the reluctance of the circuit and hence the fields in the circuit. However, B or H must be known in each material before the permeability can be calculated. It is clear that iterative procedures or graphical methods will have to be used to solve simultaneously the problems, as we will show later in this section. This is in contrast to the electrical case, where the conductivities, and hence the resistances, of the elements are independent of the current (for linear materials).

So far only reluctances in series were considered. We will now discuss parallel magnetic circuits. Consider Fig. 9.18. The magnetomotive force $\mathscr{M} = NI$ sets up a magnetic flux F_1 in the circuit. At junction a the flux encounters two parallel paths 2 and 3 where part of it, F_2, takes path 2 and the rest, F_3, takes path 3. The equivalent reluctance of the two paths is $\mathscr{R}_1\mathscr{R}_2/(\mathscr{R}_1 + \mathscr{R}_2)$ where $\mathscr{R}_1$ and $\mathscr{R}_2$ are the reluctances of the individual paths, and the ratio F_2/F_1 is $\mathscr{R}_1/\mathscr{R}_2$, with the result that the flux prefers passing through the lower reluctance.

Because the reluctance is inversely proportional to the permeability, cores made from ferromagnetic materials have lower reluctances than those made from nonferromagnetic materials, since ferromagnetic permeabilities are about 10^3 to 10^4 μ_0. This enforces the accuracy of the assumption that most of the flux goes through the ferromagnetic core, with only negligible amounts leaking into free space.

It is common to have *air gaps* in magnetic circuits in order to have access to the magnetic field. The gap size is chosen as small as possible in order to minimize its reluctance. The area of the gap is not completely defined because the field fringes out at the gap, thus making the effective cross section of the gap larger than its iron-core face. A first-order correction to the area of a rectangular gap is given by

$$A = A_0\left(1 + \frac{l_g}{d_1}\right)\left(1 + \frac{l_g}{d_2}\right) \tag{9.127}$$

where d_1 and d_2 are the geometrical dimensions of the gap, $A_0 = d_1 d_2$, l_g is the gap length, and l_g/d_1, $l_g/d_2 \approx 10^{-1}$.

Example 9.11 A Cast-Iron Core with an Air Gap

Consider a cast-iron magnetic core with $NI = 1000$ A, as shown in Fig. 9.19. It has an area $A_c = 4$ cm^2 and a length $l_c = 44$ cm. The length of the air gap is $l_g = 0.2$ cm and its effective area is $A_g = 4.8$ cm^2. Applying Ampere's law to the circuit gives

$$H_c l_c + H_g l_g = NI \tag{9.128}$$

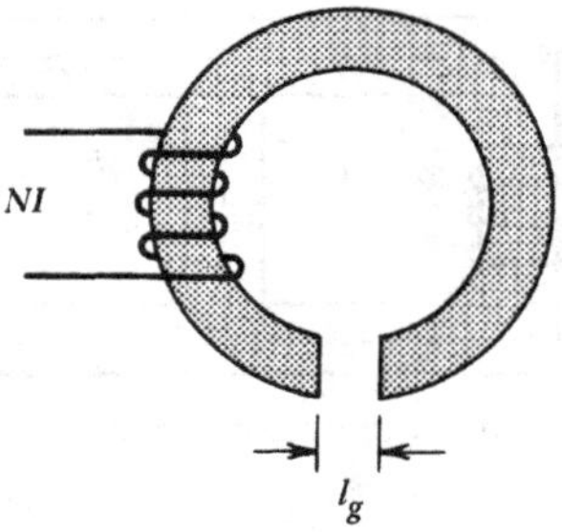

Figure 9.19 A magnetic circuit with an air gap.

where H_c and H_g are the magnetic fields in the core and the gap, respectively. The magnetic flux F is continuous throughout the circuit; thus we take $F/\mu_0 A_g$ for H_g, which gives

$$H_c l_c + \frac{F l_g}{\mu_0 A_g} = NI \tag{9.129}$$

Again, because the flux in the different sections of the circuit is conserved, we take $F = B_c A_c$; hence Eq. (9.129) becomes:

$$B_c = -\frac{\mu_0 l_c}{l_g}\frac{A_g}{A_c} H_c + \frac{\mu_0 A_g}{A_c l_g} NI \tag{9.130}$$

which is a linear relationship between B_c and H_c of the cast-iron core. The slope of the line is proportional to the ratio of the cross section of the gap to that of the cast-iron core and also proportional to the ratio of the length of the cast-iron core to the length of the gap. The intercept of the line is proportional to the magnetomotive force NI. In most cases $A_g/A_c \approx 1$, so the slope effectively becomes proportional to l_c/l_g and the intercept becomes proportional to NI/l_g.

To solve for the flux and the fields in the gap and in the cast-iron core, the above-derived relation between B_c and H_c, and the B-H magnetization curve of the iron core are solved simultaneously. We plot Eq. (9.130) using the given dimensions and quantities of the above circuit along the magnetization curve, as shown in Fig. 9.20. The intersection of these two curves gives the operating point of the magnet: $0.4T$ for the B field in the core, and hence the field in the gap $B_g = (B_c A_c/A_g) = 0.33T$.

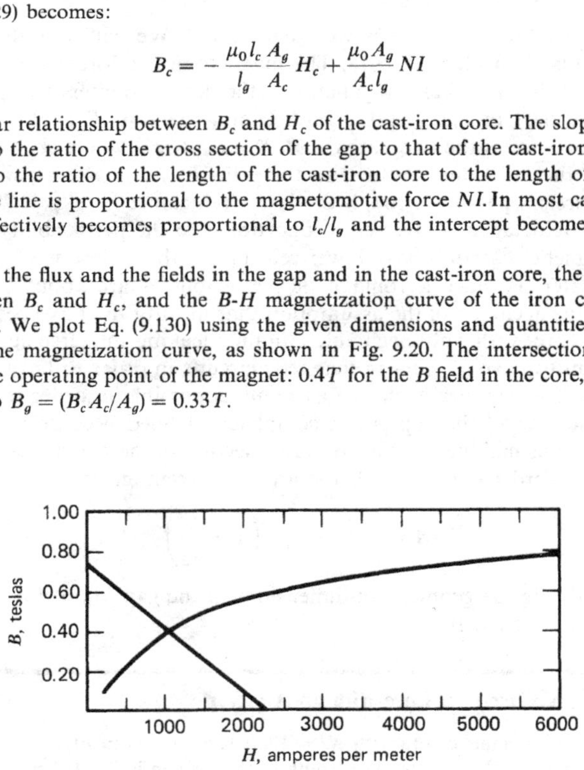

Figure 9.20 Graphical solution for the operating point of magnetic circuit with an air gap.

Example 9.12 A Magnetic Circuit with a Permanent Magnet

In the previous example the magnetomotive force in the circuit was produced by current windings. Magnetomotive forces can also be supplied by permanent magnets. Consider the magnetic circuit shown in Fig. 9.21, which is made up of a permanent magnet of length l_m, two soft-iron side arms of total length l_s, and an air gap of length l_g. The cross-sectional areas of the magnet, the arms, and the gap are A_m, A_s, and A_g, respectively. Moreover, the arms have an N-turn current winding and a current I.

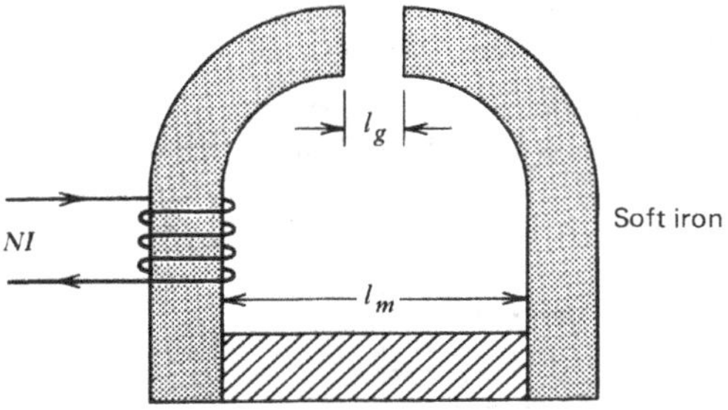

Figure 9.21 A magnetic circuit that includes a current coil and a permanent magnet.

Applying Ampere's law to the circuit gives

$$H_m l_m + H_s l_s + H_g l_g = NI \tag{9.131}$$

where H_m, H_s, and H_g are the magnetic fields in the magnet, arms, and the gap, respectively. Using the constancy of the flux in the circuit we write

$$H_g = \frac{B_m A_m}{\mu_0 A_g} \tag{9.132}$$

and therefore Eq. (9.131) becomes

$$H_m l_m + H_s l_s + \frac{B_m A_m}{\mu_0 A_g} l_g = NI \tag{9.133}$$

We should note that the relation between B_m and H_m as given in this equation is not linear because of the presence of the $H_s l_s$ term. Again, using the constancy of the flux gives

$$H_s = \frac{B_m A_m}{\mu_s A_s} \tag{9.134}$$

which clearly shows that it is a nonlinear equation since μ_s of soft iron is a nonlinear function of the fields, $\mu_s(H_m)$. The same relation can also be written in terms of the total reluctance of the circuit $\mathscr{R}$ as follows:

$$B_m = -\frac{l_m}{\mathscr{R} A_m} H_m + \frac{NI}{\mathscr{R} A_m} \tag{9.135}$$

where

$$\mathscr{R} = \frac{l_g}{\mu_0 A_g} + \frac{l_s}{\mu_s A_s}. \tag{9.136}$$

To determine the fields, Eq. (9.135) and the hysteresis curve must be solved simultaneously. The complication, of course, arises because the circuit contains two nonlinear media: the magnet and soft iron. In cases where the gap length is not too small, one can have

$$\frac{l_s}{\mu_s A_s} \ll \frac{l_g}{\mu_0 A_g}$$

In this case the effect of the arms can be neglected; hence Eq. (9.133) or Eq. (9.135) becomes

$$B_m = -\mu_0 \frac{l_m A_g}{l_g A_m} H_m + \frac{\mu_0 A_g}{A_m l_g} NI \tag{9.137}$$

The fields can now be easily determined by solving this linear equation with the hysteresis curve of the magnet, as was done in the previous example. Situations where two nonlinear materials that are equally important in a magnetic circuit are discussed in the following example.

Example 9.13 Two Nonlinear Core Sections

We now take up the case where the circuit consists of two nonlinear materials. The circuit shown in Fig. 9.22 consists of parts 1 and 2, which are of lengths l_1 and l_2 and cross-sectional

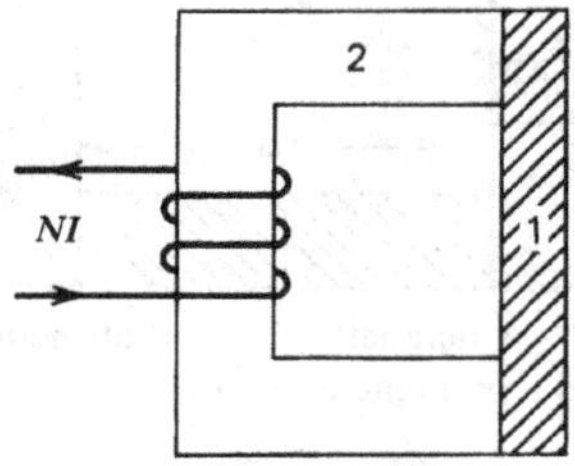

Figure 9.22 A magnetic circuit composed of two ferromagnetic materials.

areas A_1 and A_2, respectively. Moreover, an N-turn coil carrying a current I is wound on the coil. Applying Ampere's law gives

$$H_1 l_1 + H_2 l_2 = NI \qquad \text{or} \qquad H_2 = \frac{NI}{l_2} - \frac{H_1 l_1}{l_2}$$

where H_1 and H_2 are the magnetic fields in parts 1 and 2, respectively. The continuity of flux in the circuit gives an equation relating B_1 and B_2; that is, $B_2 = B_1 A_1 / A_2$. These two equations relating the pairs (B_2, H_2) and (B_1, H_1) can be solved graphically with the aid of the magnetization curve of material 1. The generated points (B_2, H_2) are then plotted simultaneously with the magnetization curve of material 2. The intersection of the latter curves gives the operating point of the circuit.

Example 9.14 Magnetic Shielding

The fact that most of the magnetic flux can be channeled through ferromagnetic materials as a result of their very large relative permeability (small reluctance) makes possible the concept of magnetic shielding. Magnetic shielding is of considerable practical importance since low field regions are often required in experiments or for electronic devices to work reliably.

To see this effect let us consider a very long cylindrical shell of magnetic material, with relative permeability K_m and internal and external radii a and b, respectively. It is placed in a uniform magnetic field H_0, with its axis normal to the field. The relative permeability of the material in the cavity and outside the shell is 1.

We try the magnetic scalar potentials

$$\Phi_0 = -H_0 \rho \cos\phi + \frac{A\cos\phi}{\rho}$$

$$\Phi_s = B\rho\cos\phi + \frac{C\cos\phi}{\rho}$$

$$\Phi_i = E\rho\cos\phi$$

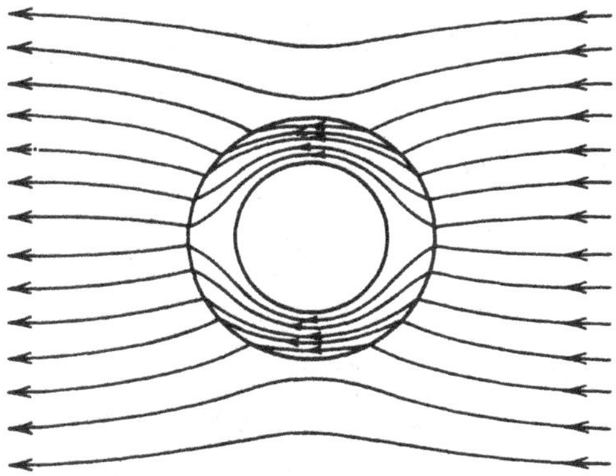

Figure 9.23 The lines of **B** when a ferromagnetic cylindrical shell is placed in an initially uniform magnetic field with its axis normal to the field.

for the regions outside, within, and inside the cylindrical shell, respectively. Taking Φ and B_n to be continuous on all boundaries, we can find all of the magnetic scalar potentials, and hence all of the fields. For example, we have

$$\Phi_m = -2H_0 a^2(K_m - 1)\left\{\rho \cos\phi + \frac{b^2}{\rho}\left(\frac{K_m - 1}{K_m + 1}\right)\cos\phi\right\}\{a^2(K_m + 1)^2 - b^2(K_m - 1)^2\}^{-1}$$

The magnetic field inside the cylinder for example is

$$\mathbf{B} = \frac{4B_0 K_m a^2}{a^2(K_m + 1)^2 - b^2(K_m - 1)^2}$$

This result illustrates the phenomenon of shielding. We see that the inner field is proportional to K_m^{-1} as $K_m \gg 1$. Therefore a cylindrical shield made of high-permeability material with K_m of 10^3 to 10^5 induces a great reduction in the field inside it even with a relatively thin shell. Figure 9.23 shows a schematic of the lines of **B**, showing their tendency to pass through the permeable medium if possible. See Problems 9.9 and 9.10 for further details of shielding.

9.8 Summary

The magnetization **M** of a macroscopic piece of material is defined as the density of atomic magnetic dipoles:

$$\mathbf{M} = \frac{d\mathbf{m}}{dv} \tag{9.3}$$

The magnetic properties of a macroscopic piece of material of magnetization **M** can be calculated by replacing all of the atomic dipoles by effective volume and surface currents $\mathbf{J}_m$ and $\mathbf{K}_m$ (called magnetization currents); hence

$$\mathbf{J}_m = \nabla \times \mathbf{M} \qquad \mathbf{K}_m = \mathbf{M} \times \hat{\mathbf{n}} \qquad (9.5),(9.6)$$

where $\hat{\mathbf{n}}$ is a unit vector normal to the surface of the material. The vector potential of such a material is

$$\mathbf{A}_m = \frac{\mu_0}{4\pi}\int \frac{\mathbf{J}_m\, dv'}{|\mathbf{r} - \mathbf{r}'|} + \frac{\mu_0}{4\pi}\int \frac{\mathbf{K}_m\, da'}{|\mathbf{r} - \mathbf{r}'|} \tag{9.7}$$

and hence $\mathbf{B} = \nabla \times \mathbf{A}_m$.

Alternatively, the magnetic properties of the material can be calculated by replacing the atomic dipoles by an effective volume and surface magnetic pole densities ρ_m and σ_m; that is,

$$\rho_m = -\nabla\cdot\mathbf{M} \qquad \sigma_m = \mathbf{M}\cdot\hat{\mathbf{n}} \tag{9.18}$$

The magnetic scalar potential Φ_m and hence the magnetic field take the expressions

$$\Phi_m = \frac{1}{4\pi}\int\frac{\rho_m\,dv'}{|\mathbf{r}-\mathbf{r}'|} + \frac{1}{4\pi}\int\frac{\sigma_m\,da'}{|\mathbf{r}-\mathbf{r}'|} \tag{9.16}$$

$$\mathbf{B}_m(\mathbf{r}) = -\mu_0\,\nabla\Phi_m + \mu_0\mathbf{M}(\mathbf{r}) \tag{9.13}$$

It is convenient to define the field **H** such that

$$\mathbf{B}(\mathbf{r}) = \mu_0\mathbf{H} + \mu_0\mathbf{M} \tag{9.29}$$

$$\mathbf{H} = -\nabla\Phi_m \tag{9.32}$$

In the presence of external currents $\mathbf{J}_f$, then

$$\mathbf{H} = \frac{1}{4\pi}\int\frac{\mathbf{J}_f\times(\mathbf{r}-\mathbf{r}')}{|\mathbf{r}-\mathbf{r}'|^3}\,dv' - \nabla\Phi_m \tag{9.33}$$

The differential equations of magnetostatics in magnetic materials are $\nabla\times\mathbf{B} = \mu_0\mathbf{J}$, and $\nabla\cdot\mathbf{B} = 0$, where $\mathbf{J} = \mathbf{J}_f + \mathbf{J}_m$ is the total current including the external current $\mathbf{J}_f$ and the magnetization current $\mathbf{J}_m$. Using **H**, these equations become

$$\nabla\times\mathbf{H} = \mathbf{J}_f \qquad \nabla\cdot\mathbf{B} = 0 \tag{9.30}$$

Ampere's integral law in the presence of magnetic material becomes

$$\oint_C \mathbf{H}\cdot d\mathbf{l} = I_f \tag{9.31}$$

where I_f is the total free current that goes through C.

The response of a material to an external field **H** depends on the microscopic structure of the material. Here we classify materials according to their macroscopic response

$$\mathbf{M} = \chi(\mathbf{H})\mathbf{H} \tag{9.34}$$

where χ is called the magnetic susceptibility. If χ is independent of **H** (magnitude and direction) and independent of space, the material is said to be linear (simple). The material is further classified as diamagnetic or paramagnetic if $\chi < 0$ or $\chi > 0$, respectively. For linear materials $\mathbf{M} = \chi\mathbf{H}$, and hence

$$\mathbf{B} = \mu_0(1+\chi)\mathbf{H} = \mu\mathbf{H} = \mu_0 K_m\mathbf{H} \tag{9.36}$$

where μ is the permeability of the material, and K_m is its relative permeability.

Ferromagnetic materials are nonlinear materials whose μ is a nonlinear function of H; they exhibit a high degree of magnetization compared to paramagnetic materials, with K_m in certain cases reaching about 10^5. They also exhibit the irreversible phenomenon of hysteresis.

When a given space is composed of regions of different magnetic properties, then the fields can be determined using boundary value techniques. In a given region of permeability μ, the equations $\nabla\times\mathbf{A} = \mathbf{B}$ and $\nabla\times\mathbf{H} = \mathbf{J}_f$, and hence $\nabla\times\mathbf{B} = \mu\mathbf{J}_f$, combine to give (taking $\nabla\cdot\mathbf{A} = 0$)

$$\nabla^2\mathbf{A} = -\mu\mathbf{J}_f \tag{9.43}$$

On the other hand, in regions away from $\mathbf{J}_f$ we have $\mathbf{H} = -\nabla\Phi_m$, which, when substituted in $\nabla\cdot\mathbf{B} = 0$, gives

$$\nabla^2\Phi_m = \nabla\cdot\mathbf{M} = -\rho_m \tag{9.40}$$

The equation for Φ_m (or **A**) can be solved in the different regions independently. The solutions then are matched at the boundaries according to the following rules

$$B_{1n} = B_{2n} \quad \text{or} \quad \mathbf{A}_{1t} = \mathbf{A}_{2t} \qquad (9.45),(9.51)$$

$$\mathbf{H}_{1t} = \mathbf{H}_{2t} \quad \text{or} \quad \Phi_{m1} = \Phi_{m2} \quad (\mathbf{K}_f = 0 \text{ on boundary}) \qquad (9.47),(9.49)$$

$$(\mathbf{H}_2 - \mathbf{H}_1)_t = \mathbf{K}_f \times \hat{\mathbf{n}} \qquad (9.46)$$

where t means tangent to the boundary and $\hat{\mathbf{n}}$ is a unit vector normal to the boundary.

The method of images is useful in solving some magnetic problems that contain free currents and prescribed boundary conditions.

The magnetic field of current distributions in the presence of low-permeability materials extends over large distances. In the presence of ferromagnetic materials of high μ, the field can be mostly trapped in these materials and hence confine it to well-defined paths, thus creating what are called magnetic circuits, which exhibit a close analogy to current circuits. In this analogy we have the quantities (NI, F, $\mathscr{R}$) parallel to (emf, I, R) of current circuits, where $\mathscr{R}$ is called the reluctance, given as follows:

$$\mathscr{R} = \frac{l}{\mu A} \qquad (9.126)$$

in analogy with the resistance of a circuit $R = l/\sigma_c A$. Additions of reluctances obey the same rules as additions of resistances.

Problems

9.1 A sphere of radius r is magnetized, with the magnetization given by $\mathbf{M} = (a_1 y^2 + b_1)\hat{\mathbf{y}} + a_2 x^2 \hat{\mathbf{x}}$, where x and y are measured using a coordinate system with origin at the center of the sphere. (a) Calculate the pole densities. (b) Calculate the magnetization current.

9.2 A hemicylinder of radius R and length L is magnetized in the y direction. The magnetization is $\mathbf{M} = M_0 \hat{\mathbf{y}}$, with M_0 a constant (see Fig. 9.24). (a) Find the volume magnetic pole density. Find the surface magnetic pole density on all the surfaces. (b) Find by integration the net poles on the hemicylinder. (c) Explain why your answer to (b) is expected for physical reasons. (d) Find by integration the dipole moment relative to the origin O. Calculate the dipole moment also by using the definition of the magnetization **M**. Do the two answers agree?

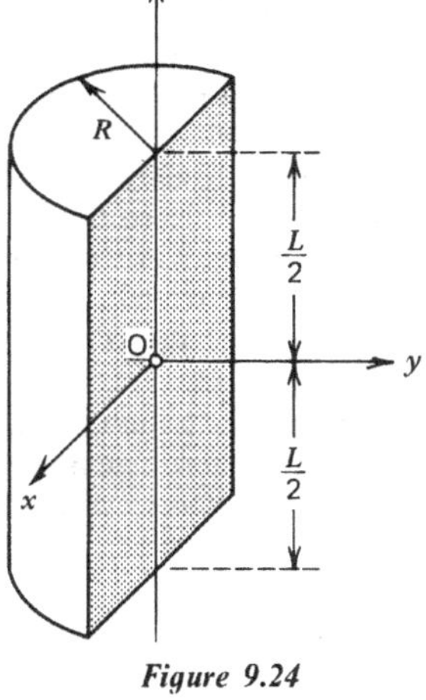

Figure 9.24

9.3 A disk of a magnetic material has a radius R and thickness $T \ll R$. It is uniformly magnetized, and the magnetization, $\mathbf{M} = M_0\hat{\mathbf{x}}$, is in the plane of the disk (x-y plane) (Fig. 9.25). (a) Determine the magnetic pole densities. (b) Determine the magnetization current densities. (c) Determine **B** and **H** at the center of the disk.

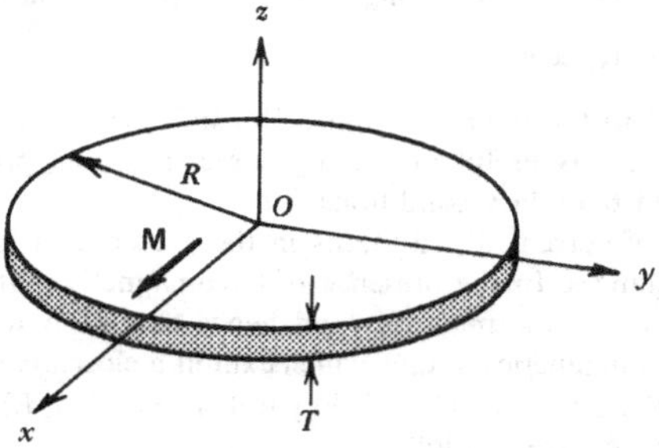

Figure 9.25

9.4 Determine the potential produced by a cylinder uniformly magnetized along its axis (Example 9.3) using the method of magnetic pole density.

9.5 A sphere of radius R is magnetized with the magnetization given by $\mathbf{M} = M_1(r)\hat{\mathbf{r}} + \mathbf{M}_0$, where $M_1(r)$ is any function of $r = |\mathbf{r}|$, and $\mathbf{M}_0$ is a constant vector. Show that (a) the external field produced by the sphere is independent of M_1, and (b) the magnetic scalar potential inside the sphere is

$$\Phi_m(r) = 4\pi \int_r^R M_1(r)dr + \frac{4\pi}{3}\mathbf{r}\cdot\mathbf{M}_0$$

9.6 A circuit is lying on the plane interface between two media with magnetic permeabilities μ_1 and μ_2. If the field produced by the circuit in the absence of the magnetic materials is $\mathbf{H}_0$, show that the fields in the magnetic materials are given by Eq. (9.63).

9.7 An infinitely long cylindrical tube of inner radius ρ_1 and outer radius ρ_2 carries a uniformly distributed axial current I. (a) Determine the magnetic field everywhere using Ampere's law. (b) Determine the vector potential everywhere by solving Eq. (9.43), and then determine the magnetic field. (c) Compare the results of (a) and (b).

9.8 A uniformly magnetized sphere with magnetization **M** is placed in a uniform magnetic field **H**. Determine the torque acting on the sphere.

9.9 A spherical shell of a magnetic material of permeability μ_1 and internal and external radii R_1 and R_2, respectively, is placed in a uniform magnetic field H_0. The permeability of the material in the cavity and in the surrounding region is μ_2. Determine the magnetic field in the cavity. Discuss the case $\mu_1 \gg \mu_2$ (see Example 9.14).

9.10 A very long cylindrical shell of magnetic material of permeability μ, and of internal and external radii a and b, respectively, is placed in a uniform magnetic field H_0 with its axis normal to the field. The permeability of the material in the cavity and outside the shell is μ_0. (a) Determine the scalar potentials and the fields in all regions. (b) Determine the equation of the lines of the magnetic field **B** within the material of the cylinder.

9.11 Determine the magnetic field produced by an infinitely long magnetic cylinder of radius ρ_0 and permeability μ_1. The cylinder is uniformly magnetized, with its magnetization M_0 being normal to the axis of the cylinder, and immersed in a material of permeability μ_2.

9.12 An infinitely long magnetic wire of radius ρ_0 and permeability μ_1 is placed in a uniform magnetic field H_0 with its axis normal to the field. The wire carries a current I, and the permeability of the material surrounding the wire is μ_2. Determine the magnetic field inside and outside the wire.

9.13 A very large ferromagnetic material of uniform magnetization **M** has small cavities in it. There also exists in the material a magnetic field **H**, which is in the direction of **M**. Determine **B** in (a) a spherical cavity of radius r, (b) a very long cylindrical cavity of small radius ρ, whose axis is parallel to **M**, and (c) a very long cylindrical cavity of radius ρ and axis normal to the magnetization.

9.14 Two magnetic media 1 and 2 of permeabilities μ_1 and μ_2, respectively, are separated by a plane interface. The field B_1 just below the interface in medium 1 makes an angle θ_1, and the field B_2 just above the interface in medium 2 makes an angle θ_2. The angles are measured with respect to the normal to the interface. Show that $\mu_1 \tan\theta_2 = \mu_2 \tan\theta_1$.

9.15 Determine the magnetic moment of the current distribution produced by the spinning, charged sphere of Example 9.8 by (a) using $\mathbf{m} = \int \mathbf{r} \times \mathbf{J}\, dv$, (b) using $\mathbf{m} = \int \mathbf{M}\, dv$, and (c) using the potentials produced by it [Eq. (9.108)].

9.16 A wire is placed at a distance d parallel to the plane face of a semiinfinite medium of permeability μ. If a current I flows in the wire, show that the wire is attracted to the material and determine the force.

9.17 Show that the force acting on a unit length of the current of Example 9.9 is $dF/dl = (\mu_0/4\pi d)I^2$ if $\mu_2 \gg \mu_1 = \mu_0$.

9.18 A straight wire carrying a current I is parallel to the axis of an infinitely long cylinder of radius a and at a distance $b > a$ from it. The permeability of the cylinder is $\mu_1 \gg \mu_2$, where μ_2 is the permeability of the material surrounding the cylinder. Find the force per unit length of the wire.

9.19 Solve Problem 9.18 for the case where $a > b$ and $\mu_2 \gg \mu_1$ (a wire inside a cylindrical cavity).

9.20 Prove Eqs. (9.119) and (9.120).

9.21 Two very long cylindrical magnets are placed next to each other, as shown in Fig. 9.26. Each magnet is of length L and diameter ρ_0, and has uniform magnetization **M** along the axis. The magnets are separated by a small gap of lengths $l_g \ll L, \rho_0$. Determine the magnetic fields **H** and **B** in the gap.

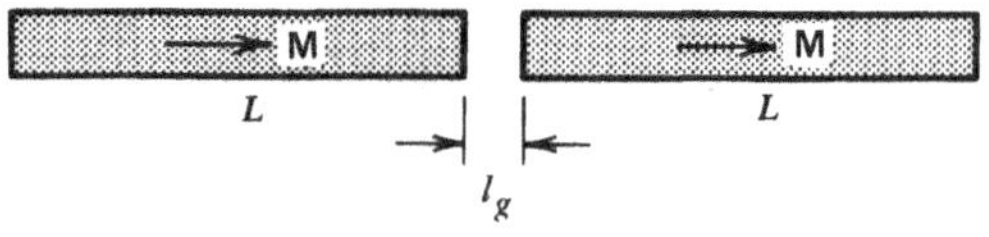

Figure 9.26

9.22 Consider the magnetic circuit shown in Fig. 9.18. The circuit is made of cast iron. The circuit dimensions are $l_2 = 0.04$ m, $l_1 = l_3 = 0.1$ m, and the circuit cross section is 1.5 $\times 10^{-4}$ m^2. Determine the flux in the circuit if $NI = 5.25 \times 10^2$ A.

9.23 Consider the magnetic circuit given in Fig. 9.21 with $I = 0$. The hysteresis curve of the permanent magnet (Alnico 5) is given in Fig. 9.8c. The arms are made of soft iron whose hysteresis is negligible compared with that of the permanent magnet. The dimensions of the circuit are: length of gap $l_g = 1$ cm, area of the gap $A_g = 2$ cm^2, length of the magnet $l_m = 10$ cm, and cross section of the magnet $A_m = 2$ cm^2. Determine the fields in the gap and in the magnet.

9.24 Consider the magnetic circuit given in Fig. 9.22 of Example 9.13 where material 1 is cast steel and material 2 is a nickel-iron alloy. The dimensions of the circuit are $l_1 = 0.08$ m, $l_s = 0.1$ m, $A_1 = 3 \times 10^{-4}$ m^2, and $A_s = 2.25 \times 10^{-4}$ m^2. Determine the flux and the fields in the two materials when $NI = 40$ A.

9.25 Consider two coaxial toroids of a ferromagnetic core, of mean radius R and thickness $t \ll R$ and separated by $2d$. The magnetic field inside each of them is B and outside is zero. (a) Determine the vector potential at the axis due to one of them. (b) Determine the magnetic vector potential at the axis of the toroids. (*Hint*: Use $\nabla \times \mathbf{A} = \mathbf{B}$, in analogy with $\nabla \times \mathbf{B} = \mu_0 \mathbf{J}$, along with the result of a current loop).

9.26 The hysteresis curves of certain special ferrites have nearly rectangular loops, as shown in Fig. 9.27, thus enabling them to be used as memory elements. Consider a toroid of

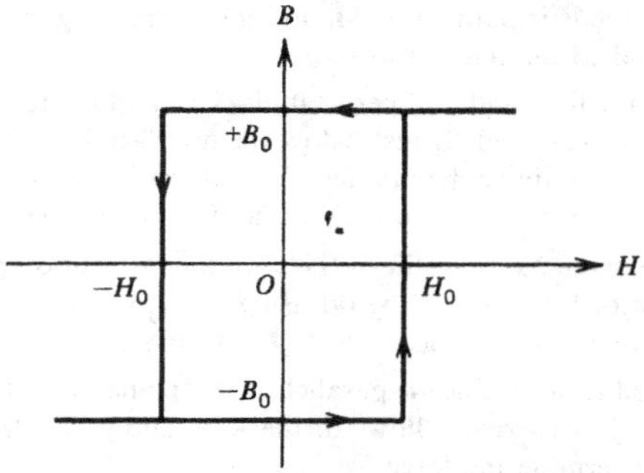

Figure 9.27

this material of inner and outer radii a and b, thickness c and a current-carrying coil of N_0 turns, and current I. (a) If $I = 1$ ampere, determine the minimum N_0 required to saturate it. (b) What is the magnetization when the current is reduced to zero?

TEN

THE MICROSCOPIC THEORY OF MAGNETISM

In Chapter 9 we discussed the macroscopic properties of magnetic materials, but deferred the discussion of their microscopic nature. In this chapter we take on the microscopic aspect of magnetization by analyzing the response of individual atoms or molecules to external magnetic fields. The objective here is not to present a full discussion of this subject, but as we dealt with the microscopic theory of dielectrics in Chapter 5, we will discuss simple ideas that describe the qualitative nature of the different classes of magnetic materials and give an order of magnitude of their permeabilities.

Part of the treatment requires the introduction and utilization of two quantum mechanical properties of the atoms; however, the actual treatment is carried out classically.

10.1 The Interaction of Atoms and Molecules with Magnetic Fields

We now consider the interaction of individual atoms or molecules with magnetic fields. Unlike the electrical effects of matter, where the atomic charge interacts with electric fields whether it is in motion or static, only charges in motion interact with magnetic fields. Since atomic electrons orbit around the nuclei, they produce a magnetic moment at large distances (rotational or orbital magnetic moments) (see Example 8.15, which deals with a rotating ring of charge) and hence interact with external fields via this moment; as discribed in Section 8.8.4. This picture is approximate, of course, since the electronic orbits are not actually well defined according to quantum mechanics. The correct solution of this problem can be found in quantum mechanics books; it involves the solution of the Schrödinger wave equation of the electron in the presence of a magnetic field. Nevertheless, we will go ahead with the foregoing simplified model using classical ideas and be content with an order of magnitude of the effects.

We should mention that the applicability of classical models is close to reality in some physical cases. For example, in a plasma or in regions where the number of free electrons is large, the behavior of electrons is governed by classical mechanics. In fact, the early work on the nature of magnetic materials was based on classical ideas and surprisingly led to intelligent guesses at the behavior of these materials.

Charges in matter can also interact with external magnetic fields via their *spin*, which is an *intrinsic* quantum mechanical effect that can be viewed classically from the point of view of a charge distribution spinning about itself. Such a classical analog predicts a magnetic moment (see Example 8.19). Both the electrons and the nuclei have spin or intrinsic magnetic moment; hence they interact with external magnetic fields.

The total magnetic moment of an electron is the sum of its rotational and intrinsic magnetic moments. Based on this total moment, various atoms are classified into two categories: atoms that have net permanent magnetic moments (magnetic), and atoms that have no net permanent magnetic moments (nonmagnetic). In the latter case, the rotational and the intrinsic moments cancel.

As we will see in Section 10.2, when a nonmagnetic atom is placed in an external magnetic field, an induced current is produced in the atom. The direction of this additional current is such that the induced magnetic field opposes the external field in accordance with Lenz' law (to be introduced in Chapter 11). Thus, as a result of the interaction, the atom acquires an induced magnetic moment, which is in opposite direction to the external field. Such an atom is called a diamagnetic atom. We should note that this effect occurs in all atoms whether they are nonmagnetic or magnetic. However, in the case of magnetic atoms other effects may occur.

As we will see in Section 10.3, the permanent moment of a magnetic atom experiences some degree of alignment along the external field (analogous to the alignment of electric dipoles in an electric field). This alignment effect is the origin of paramagnetism. Moreover, in some substances called ferromagnetic materials there are internal forces (called *exchange forces*) between the magnetic moments of different atoms; they are pure quantum mechanical forces that result in a great degree of alignment.

We now consider the question: Which atoms of the periodical table are magnetic? Atoms that have "unfilled shells" have magnetic moments. In these atoms the number of electrons is odd, and it is the electron in the unfilled shell that gives the atom a spin and hence a magnetic moment. Electrons in "filled shells" are paired; every two electrons have their spins in opposite directions such that the sum of their magnetic moments vanishes. Consequently, atoms with filled shells have no magnetic moments.

The compounds formed from magnetic atoms are not necessarily magnetic. Compounds whose atomic constituents have unfilled *outer* shells are not necessarily magnetic because the outer-shell electrons of opposite spins get coupled, thus resulting in cancellation of the total moment. Such coupling, however, is absent in compounds whose atomic constituents have unfilled *inner* shells. The latter class of atoms include the transition elements and the rare earth elements.

10.2 The Origin of Diamagnetism—Induced Dipole Moments

The origin of diamagnetism can be described using the following simple model of the electronic motion around the nucleus. The electron of charge e is taken to rotate around the nucleus in a circular orbit of radius ρ and frequency ω (see Fig. 10.1), which constitutes a current loop with a current equal to $-e\omega/2\pi$ and a magnetic

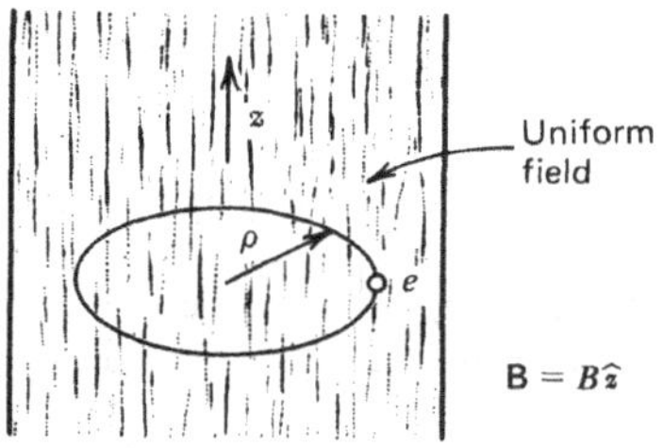

Figure 10.1 An electron of circular orbit in a uniform magnetic field that is normal to the orbit.

moment equal to the product of the current and the area of the orbit and a direction normal to the plane of the motion:

$$\mathbf{m} = -\frac{e\omega}{2}\rho^2\hat{\mathbf{z}} \tag{10.1}$$

where $\hat{\mathbf{z}}$ is a unit vector normal to the plane of the motion. The moment can be directly related to the angular momentum of the motion $\mathbf{L} = m_e\omega\rho^2\hat{\mathbf{z}}$, where m_e is the mass of the electron, as follows:

$$\mathbf{m} = -\frac{e}{2m_e}\mathbf{L} \tag{10.2}$$

A magnetic field $\mathbf{B}$ whose direction is normal to the plane of the orbit—that is, along the z axis—is now switched on the atom. The act of switching induces an electric field in the atom according to Faraday's law (to be discussed in detail in Chapter 11):

$$\oint_C \mathbf{E}\cdot d\mathbf{l} = -\frac{dF}{dt} \tag{10.3}$$

where $\mathbf{E}$ is the induced electric field, C is the orbit of the electron, and F is the magnetic flux of the external field traversing the orbit. Taking the tangential component of the electric field, E_t, constant for a given ρ in Eq. (10.3) gives

$$2\pi\rho E_t = -\frac{d}{dt}(\pi\rho^2 B) = -\pi\rho^2\frac{dB}{dt}$$

or

$$E_t = -\frac{1}{2}\rho\frac{dB}{dt} \tag{10.4}$$

The induced electric field E_t produces a torque on the electron equal to $\tau = -eE_t\rho$, which when equated to the rate of change of the angular momentum of the electron (Newton's law for rotation) gives

$$\tau = \frac{dL}{dt} \tag{10.5}$$

or

$$\frac{dL}{dt} = -\frac{1}{2}e\rho^2\frac{dB}{dt} \tag{10.6}$$

which gives a change δL in the angular momentum resulting from a change in the magnetic field ΔB as follows:

$$\delta L = -\frac{1}{2} e\rho^2 \, \Delta B \tag{10.7}$$

Thus Eq. (10.7) shows that a buildup of a magnetic field **B** causes a change in the angular momentum of the electron, $\Delta\mathbf{L}$, and hence a change in the magnetic moment governed by Eq. (10.2); that is,

$$\Delta\mathbf{L} = -\frac{1}{2} e\rho^2 \mathbf{B} \tag{10.8}$$

$$\Delta\mathbf{m} = -\frac{1}{4}\frac{e^2\rho^2}{m_e}\mathbf{B} \tag{10.9}$$

We can now calculate the induced magnetization of an ensemble of molecules placed in an external magnetic field. If each individual molecule has n electrons, each with an orbit of radius r_i making an angle θ_i with the magnetic field, then the change in the magnetic moment of the molecule is

$$\Delta\mathbf{m} = -\frac{1}{4}\frac{e^2}{m_e}\mathbf{B}\sum_{i=1}^{n} r_i^2 \cos^2\theta_i \tag{10.10}$$

The magnetization **M** of a material of N molecules per unit volume can be determined from Eqs. (9.3) and (10.10) as $\mathbf{M} = N\,\Delta\mathbf{m}$. Therefore

$$\mathbf{M} = -\frac{1}{4}\frac{e^2}{m_e}N\mathbf{B}\sum_{i=1}^{n} r_i^2 \cos^2\theta_i \tag{10.11}$$

and thus the macroscopic diamagnetic susceptibility of the medium, $\chi = \mathbf{M}/\mathbf{H}$, is

$$\chi = -\frac{1}{4}\frac{\mu_0 e^2}{m_e}N\sum_{i=1}^{n} r_i^2 \cos^2\theta_i \tag{10.12}$$

This expression can be alternatively written in terms of an average of the radius. Since $r^2\cos^2\theta = \rho^2 = x^2 + y^2$, $\langle\rho^2\rangle = (2/3)\langle r^2\rangle$; therefore

$$\chi = -\frac{1}{6}\frac{\mu_0 e^2}{m_e}N\langle r^2\rangle \tag{10.13}$$

10.3 Paramagnetism—Permanent Moments

As mentioned above, when the atoms being investigated are magnetic (that is, when they have permanent magnetic moments), the field causes some degree of alignment of the moment along its direction in addition to simultaneously inducing a magnetic moment (diamagnetic effect). Since the diamagnetic effect has been discussed, we will now turn to the alignment effect.

According to quantum mechanics, the motion of atomic electrons and the orientation of their spins under the influence of an external magnetic field are quantized. In other words, the total magnetic moment of the atom has a discrete set of orientations relative to the field. From a classical point of view, on the other hand, one ignores these restrictions on the direction and assumes that all orientations are possible. Although we will discuss the problem from the classical point of view, we will later give an example where the discreteness is used and compare both results.

Also we will defer the discussion of the dependence of the magnetic moment on the spin to the following section.

In a macroscopic sample of a magnetic material, some thermal effects take place. The thermal energy of the molecules, which results in collisions among them, tends to randomize the molecular dipole orientation. In fact, in the absence of an external field the vector sum of the dipole moments of all the molecules vanishes. The magnetic field, on the other hand, exerts torques on the individual dipoles and tend to align them. The degree of alignment for a particular sample, therefore, depends both on the strength of the field and on the temperature.

The classical treatment of paramagnetism is very similar to that of the alignment of electric dipoles of molecules placed in an external electric field (discussed in Chapter 5), which was derived quantitatively using statistical methods. Consider an assembly of N magnetic molecules per unit volume, each of magnetic moment m and at a temperature T. Classically, each dipole moment can make an arbitrary angle θ with respect to a given direction such as the z axis. In the absence of an external magnetic field, the probability that the dipole will be between angles θ and $\theta + d\theta$ is proportional to $2\pi \sin\theta \, d\theta$, which is the solid angle $d\Omega$ subtended by this range of angle. This probability leads to a zero average of the dipoles. When the field is present and is taken along the z axis, this probability becomes also proportional to the Boltzmann distribution $e^{-U/kT}$, where U is the magnetic energy of the dipole when it is making an angle θ with the magnetic field, k is the Boltzmann constant, and T is the absolute temperature. The Boltzmann factor introduces the dependence of the probability on the field and on the temperature in a quantitative way.

In a magnetic field $\mathbf{B}$, the magnetic energy of an atomic dipole is $U = -\mathbf{m}\cdot\mathbf{B} = -mB\cos\theta$, where $\mathbf{B}$ is along the z axis and θ is the angle between the moment and the z axis. The same procedure used in Section 5.2.2 gives the average of the dipole moment, $\langle\mathbf{m}\rangle$, and hence the magnetization, $\mathbf{M}$, in terms of the Langevin function:

$$\langle\mathbf{m}\rangle = m\left[\coth\eta - \frac{1}{\eta}\right]\hat{\mathbf{z}} \tag{10.14}$$

$$\mathbf{M} = \langle\mathbf{m}\rangle N \tag{10.15}$$

where

$$\eta = \frac{mB}{kT} \tag{10.16}$$

For most cases of paramagnetic materials, $\eta \ll 1$ at ordinary temperatures even for B as large as 10^4 gauss (1 tesla); therefore, $\mathbf{M}$ can be approximated by the lowest order in η;

$$\mathbf{M} = \frac{Nm^2\mathbf{B}}{3kT} \tag{10.17}$$

It is apparent that the magnetization at high temperatures is proportional to the magnetic field that yields a constant (independent of the field) magnetic susceptibility:

$$\chi = \frac{\mathbf{M}}{\mathbf{H}} = \frac{Nm^2\mu_0}{3kT} \tag{10.18}$$

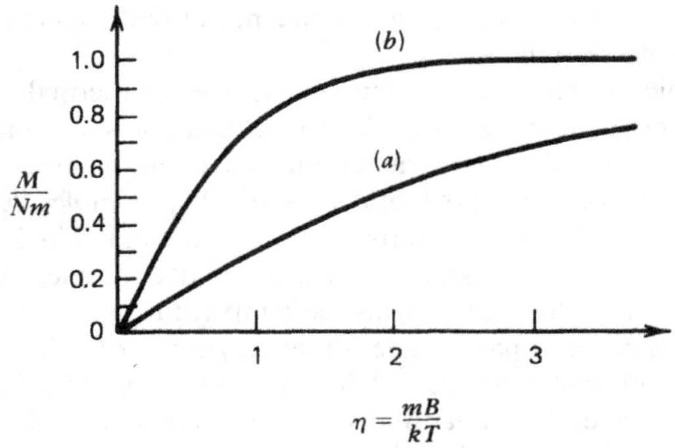

Figure 10.2 The magnetization of an ensemble of paramagnetic material placed in a magnetic field **B** as a function of $\eta = mB/KT$. (*a*) Classical calculation with no restriction on the direction of the dipole. (*b*) Quantum mechanical calculation with direction restricted.

The fact that χ is proportional to T^{-1} is known as Curie's law.

At lower temperatures, such that $\eta \gg 1$, **M** becomes

$$\mathbf{M} = Nm\hat{\mathbf{z}} \tag{10.19}$$

which is independent of magnetic field. In this case **M** takes its maximum value; that is, it reaches *saturation.* For intermediate temperatures, the dependence of **M** on **B** is nonlinear. The complete dependence is shown in Fig. 10.2*a.*

Example 10.1 The Magnetization of a Quantum System

We would like to consider paramagnetism from the point of view of quantum mechanics. For atoms with a magnetic moment due to a spin of $\frac{1}{2}$, there are only two discrete orientations of the moment **m** with respect to the external field, **B**: parallel or antiparallel corresponding to angles $\theta = 0$ and π, as shown in Fig. 10.3. The corresponding magnetic energies of these

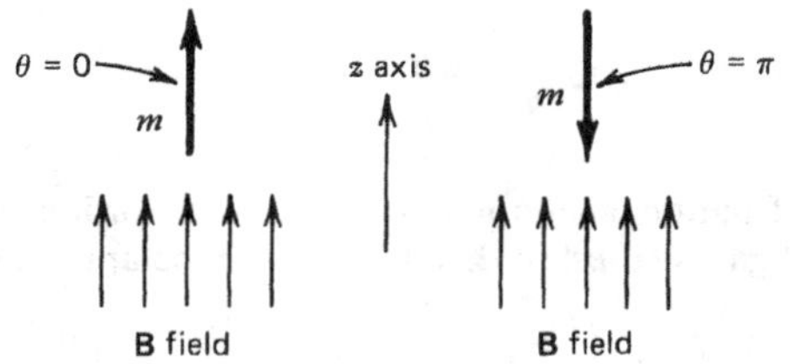

Figure 10.3 A magnetic atom of spin $\frac{1}{2}$ in an external **B** field showing two possible configurations along and opposite to the field.

orientations are $-mB$ and mB. The probability of finding the moments in the parallel and antiparallel orientations are Ce^{η} and $Ce^{-\eta}$ respectively, where C is a normalization constant; hence the mean magnetic moment is

$$\langle \mathbf{m} \rangle = m \frac{e^{\eta} - e^{-\eta}}{e^{\eta} + e^{-\eta}} \hat{\mathbf{z}} = m\hat{\mathbf{z}} \tanh \eta \tag{10.20}$$

where $\eta = mB/kT$ was defined in Eq. (10.16), and $\hat{\mathbf{z}}$ is a unit vector along the field. The magnetization of the ensemble is then given by:

$$\mathbf{M} = Nm\hat{\mathbf{z}} \tanh \eta \tag{10.21}$$

Figure 10.2*b* shows a plot of M/Nm as a function of η. When η is very large, $\tanh \eta = 1$; therefore M/Nm approaches unity, indicating saturation where the moments are all in the spin up position.

For most materials at ordinary temperatures and fields, $\eta \ll 1$; hence $\tanh \eta = \eta$ and the magnetization is

$$\mathbf{M} = \frac{Nm^2\mathbf{B}}{kT} \tag{10.22}$$

indicating a linear dependence on $\mathbf{B}$ and on T^{-1} just as was found using the classical treatment. There seems, however, a discrepancy in the magnitude of the effect. The quantum result appears to predict that the magnetization is a factor of 3 larger than the classical prediction. In actuality this discrepancy is not genuine since the definitions of the magnetic moments in both treatments differ. In fact they are related as follows:*

$$m_c^2 = 3m_q^2 \tag{10.23}$$

where m_c and m_q are the classical and quantum mechanical magnetic dipole moments of the atom, respectively, thus clearing the discrepancy.

Example 10.2 Susceptibility of A Free-Electron Gas

Consider a system that consists of particles of charge e and mass m_e. Each particle moves in a circular orbit. The system is placed in a magnetic field and taken to be in a state of statistical equilibrium. We would like to calculate the magnetic susceptibility of the gas. Classically, the total magnetic susceptibility is the sum of the paramagnetic and diamagnetic susceptibilities given in Eqs. (10.13) and (10.18):

$$\chi = \frac{Nm^2}{3kT}\mu_0 - \frac{Ne^2}{6m_e}\mu_0 r^2 \tag{10.24}$$

The kinetic energy of the electron E_k, in terms of its angular momentum L, is $E_k = L^2/2m_e r^2$. Thus m^2 can be written as:

$$m^2 = \frac{e^2}{4m_e^2}L^2 = \frac{e^2 E_K r^2}{2m_e} \tag{10.25}$$

In the case of statistical equilibrium, the average kinetic energy $\langle E_K \rangle$ can be found from the theorem of equipartition of energy. Because the electrons are rotating in planes, they have two degrees of freedom; hence their average kinetic energy is equal to kT. As a result, the average magnetic moment is

$$m^2 = \frac{e^2 kT}{2m_e} r^2 \tag{10.26}$$

Substituting this result in Eq. (10.24) gives $\chi = 0$.

This result indicates that the magnetic susceptibility, and hence the magnetic moment of a charged body that obeys classical statistics, is zero. It is to be noted that if quantum effects are included, which are important in the case of atomic electrons (bound electrons), a finite magnetic moment would exist. These effects include the spin angular momentum and quantization of the electronic orbits.

* See R. Feynman, R. Leighton, and M. Sands, *The Feynman Lectures on Physics*, (Reading, Mass.: Addison-Wesley, 1966), Vol. 2, page 35.9.

10.4 Ferromagnetism

10.4.1 Spin-Spin (Exchange) Interaction

In accounting for the diamagnetic property of matter, atoms were not required to have permanent magnetic dipole moments or to have any interaction among themselves. The paramagnetic property of matter, however, can be accounted for only by a quantum mechanical property—a permanent magnetic moment produced by the *intrinsic spin, S*, accompanied by an interaction among themselves—that is, interaction through collisions that tend to randomize the directions of the moments. As we previously stated, the spin is an intrinsic angular momentum, analogous to the orbital angular momentum, that can only have discrete orientation in a magnetic field (see Example 10.1). Associated with the atom's spin is a magnetic moment, which is related to it by

$$\mathbf{m} = g\frac{e\hbar}{2m_e}\mathbf{S} = g\beta'\mathbf{S} = \beta\mathbf{S} \tag{10.27}$$

where $\beta' = e\hbar/2m_e$ is the Bohr magneton, $\hbar$ is Planck's constant divided by 2π, and g is a constant called the g factor, which has a value of 2.001 (≈ 2). Equation (10.27) is analogous to Eq. (10.2), which relates the orbital magnetic moment to the orbital angular momentum.

Another magnetic property of matter is the ferromagnetic effect exhibited by some materials whose macroscopic properties were discussed in Section 9.4, The Magnetic Constitutive Relation's. It turns out that these above principles are not sufficient to account for this effect, which point towards the fact that there must be other types of interactions between the atoms of ferromagnetic materials. One type of interaction that may readily come to mind (but does not, however, account for the effect) is the dipole-dipole interaction; it is caused by the magnetic field produced by one of the atomic dipoles at the position of another. This effect will be considered in further detail in Example 10.3. Also see Example 5.3 for the corresponding effect in the electrical property of matter.

The effect that accounts for the ferromagnetic property is a quantum mechanical effect resulting from the Pauli exclusion principle; it is called *the spin-spin (exchange) interaction.** According to the Pauli exclusion principle, electrons cannot occupy the same total state; for example, if two electrons on neighboring atoms occupy the same orbital state, then their spins have to be antiparallel. Conversely, if the electrons have parallel spins, then they have to be in different orbital states. Because electrons that occupy the same orbital state can get closer to each other more than those that are only allowed to be in different orbital states, the electrostatic energy of the two electrons depends on the configuration of their spins.

The quantum mechanical exchange interaction energy U' of a pair of atomic electrons is written in terms of their spins, and hence in terms of their magnetic moments, as follows:†

$$U' = -2\alpha\mathbf{S}_j\cdot\mathbf{S}_k = \frac{-2\alpha}{\beta^2}\mathbf{m}_j\cdot\mathbf{m}_k \tag{10.28}$$

which can be approximated by the much simpler expression

$$U' = -2\alpha S_{jz}S_{kz} = \frac{-2\alpha}{\beta^2}m_{jz}m_{kz} \tag{10.29}$$

* W. Heisenberg, *Zeitschrift für Physik*, vol. 49, p. 619, 1928.

† See the discussion of the interaction of two classical magnetic dipoles in Examples 10.3 and 12.2.

where $(\mathbf{S}_j, S_{jz})$ and $(\mathbf{S}_k, S_{kz})$ are the spin and its z component of the jth and the kth atom, respectively, and $\alpha > 0$, a quantity that depends on the distance between the interacting atoms, measures the strength of the interaction; specifically, it falls rapidly with increasing distance.

We should note that the quantum mechanical interaction energy between the spins (moments) of atoms j and k implicitly defines a field H'_k along the z axis produced, say, by atom k at the site of atom j. Comparing Eq. (10.29) with $U' = -\mu_0 m_{jz} H'_k$ gives

$$H'_k = \frac{2\alpha}{\mu_0 \beta^2} m_{kz} \tag{10.30}$$

With the introduction of the quantum mechanical permanent magnetic moments [Eq. (10.27)] and their quantum mechanical interactions [Eq. (10.29)] and fields [Eq. (10.30)], the rest of the treatment of the various effects are done classically.

10.4.2 The Molecular Field

To analyze the ferromagnetic effect, one needs to calculate the field at the site of a particular atom caused by the external sources and by all the magnetic atoms in the medium. This field is called the *molecular field* or the *local field*, H. We will first set aside the spin-spin interactions and hence calculate the classical local field. Although we already mentioned that this field does not account for the permanent magnetization of ferromagnetic materials (sometimes called *spontaneous magnetization*), it will shed some light on the nature of the effect. After all, it was the breakdown of the classical theory that led to the development of the quantum theory.

The classical local field of a magnetized material of magnetization $\mathbf{M}$ can be determined using the procedure introduced in Section 5.1 for the calculation of the local electric field produced by the electric dipole moments of neighboring atoms. The method used assumes that such a field at a particular atom is the same as would be found in a small spherical hole in the material, whose presence is assumed not to affect the rest of the material. The field H [see (Eqs. 4.34) and (5.2)] is therefore given by

$$\mathbf{H} = \mathbf{H}_0 + \frac{1}{3}\mathbf{M} \tag{10.31}$$

where $\mathbf{H}_0$ is the macroscopic field in the material. We should note that the corresponding electric molecular field of a polarized dielectric material, Eq. (5.2), was sufficient to account for the ferroelectric property of dielectrics. In the magnetic case, however, Eq. (10.31) does not explain ferromagnetism, as we will see later.

We now turn to the calculation of H in a ferromagnetic material placed in an external field $\mathbf{H}_0$ taken along the z axis when the spin-spin interactions are included. It is to be noted that the general quantum mechanical problem is so difficult that up to the present time it has not been solved exactly. However, some approximate methods have been developed. Here we will briefly describe the *molecular-field method* developed by Weiss.

In the Weiss molecular-field approximation the attention is focused on a particular central jth atom. Because the spin-spin interaction falls rapidly with distance, only a few neighboring atoms interact appreciably with this central atom. As an approximation, the effect of these neighboring atoms is replaced by their average interaction. Thus one can write, using Eq. (8.111), $\mathbf{U} = -\mathbf{m}\cdot\mathbf{B}$, or

$$E = -\mu_0 m_{jz}(H_0 + H') = -\mu_0 m_{jz} H \tag{10.32}$$

where H' is the field produced along the z axis by n neighboring atoms; that is,

$$H' = 2\alpha\left(\sum_{k=1}^{n} m_{kz}\right)\frac{1}{\mu_0\beta^2} \tag{10.33}$$

and H is the molecular field at the site of jth atom. The number of contributing atoms n is not known at this stage and is expected to be a function of the strength of the exchange interaction (namely, α) and hence of the bulk properties of the ferromagnetic material. In any case, we take it at this stage as a parameter in the theory (see Problem 10.8).

The response of the jth atom to the total field $H_0 + H'$ can be calculated using the procedure followed in Section 10.3, Paramagnetism. However, because of the quantum nature of this effect, it is more accurate to use the procedure discussed in Example 10.1, which imposes quantum mechanical restrictions on the direction of the magnetic moment with respect to the direction of the magnetic field. Nonetheless, we will follow the classical treatment here.* The average $\langle m_{jz}\rangle$ is given by the Langevin function [see Eq. (10.14)]:

$$\langle m_{jz}\rangle = m\left(\coth\eta - \frac{1}{\eta}\right) = mB(\eta) \tag{10.34}$$

where

$$\eta = \frac{\mu_0 mH}{kT} = \frac{\mu_0 m(H_0 + H')}{kT} \tag{10.35}$$

In the above derivation we concentrated on the effect of neighboring atoms on a given central jth atom. Because the atoms are identical, the result would be the same if one repeated the derivation with any of the neighboring atoms taken as the central atom. Therefore, we conclude that the average magnetic moment given in Eq. (10.34) is the same for all neighboring atoms, or

$$\langle m_{kz}\rangle = mB(\eta) \tag{10.36}$$

where $k = 1, 2, \ldots, n$. The average molecular field at a given central atom can now be determined in terms of the Langevin function by substituting Eq. (10.36) into Eq. (10.33); that is,

$$H' = \frac{2\alpha n}{\mu_0\beta^2}\langle m_z\rangle = \frac{2\alpha nm}{\mu_0\beta^2}B(\eta) \tag{10.37}$$

* A quantum mechanical treatment gives

$$\langle m_{jz}\rangle = \beta S B_S(\eta')$$

where

$$B_S(\eta') = \frac{1}{S}\left[\left(S + \frac{1}{2}\right)\coth\left(S + \frac{1}{2}\right)\eta' - \frac{1}{2}\coth\frac{1}{2}\eta'\right]$$

is called the Brillouin function, and

$$\eta' = \frac{\mu_0\beta H}{kT}$$

In fact, for large S the Brillouin function is essentially equivalent to the Langevin function. (Quantum mechanics approaches classical mechanics for large quantum numbers.)

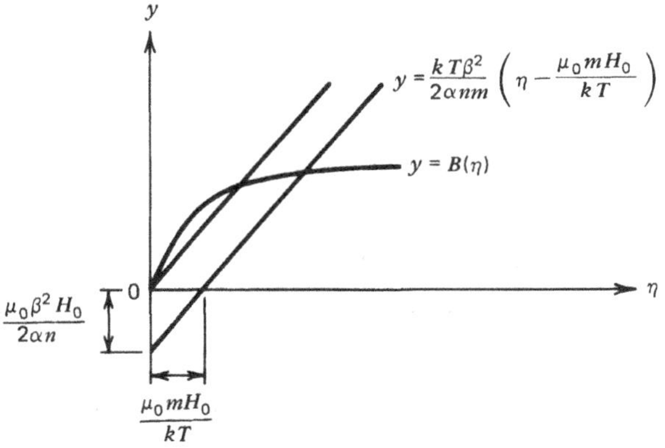

Figure 10.4 A self-consistent graphical solution of the ferromagnetic effect using the Langevin function. The magnetization in the presence of an external field and in its absence (spontaneous magnetization) are determined.

Since $H = H_0 + H'$, then this result relates H to the atomic properties of the sample. This relation is nonlinear and therefore cannot be solved analytically; however, a graphical solution can be obtained by parameterizing it, via the introduction of the function y,

$$y = B(\eta). \tag{10.38}$$

$$y = \frac{kT\beta^2}{2\alpha nm^2}\left(\eta - \frac{\mu_0 m H_0}{kT}\right) \tag{10.39}$$

Equation (10.37) can now be solved by plotting y as a function η for both Eqs. (10.38) and (10.39), as shown in Fig. 10.4. The intersection of the curves gives the value of η and hence the value of H'.

10.4.3 Spontaneous Magnetization

The magnetization of the medium can now be evaluated using Eq. (10.36). If N is the density of the magnetic atoms in the medium, the magnetization M will be

$$M = N\langle m_{jz}\rangle = NmB(\eta) \tag{10.40}$$

Now we can answer the question whether a magnetization M can exist even in the absence of the external magnetic field H_0. Such magnetization, if it exists, is called *spontaneous magnetization*, and the materials that exhibit this property are called ferromagnetic materials. To answer the question we reexamine the graphical solution of Eqs. (10.38) and (10.39). For $H_0 = 0$, Eq. (10.38) becomes $y = (kT\beta^2/2\alpha nm)\eta$, which is the equation of a straight line passing through the origin (see Fig. 10.4). Since the function $B(\eta)$ passes through the origin too, a solution exists at the origin—namely, $\eta = 0$. This solution, however, is a trivial solution. Another solution may exist at $\eta \neq 0$ if the initial slope of $B(\eta)$ at $\eta = 0$ is larger than the slope of the line. To investigate this possibility we focus on the limit $\eta \ll 1$. For $\eta \ll 1$, Eq. (10.34) gives

$$B(\eta) \approx \frac{1}{3}\eta \qquad \eta \ll 1 \tag{10.41}$$

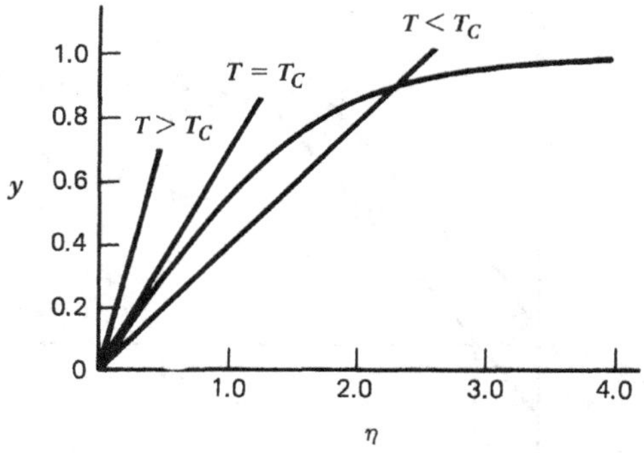

Figure 10.5 The solution for the spontaneous magnetization using the Langevin function as a function of temperature showing the Curie temperature T_C.

which is a linear function of slope $\frac{1}{3}$. The slope of the line, on the other hand, is $kT\beta^2/2\alpha nm^2$; therefore the condition for nontrivial solution is

$$\frac{1}{3} \geq \frac{kT\beta^2}{2\alpha nm^2} \tag{10.42}$$

This result may be viewed as a condition on the temperature of a magnetic material of given atomic properties that allows the occurrence of spontaneous magnetization. The limiting case where the slopes are equal defines what is called the *Curie temperature*, T_C, an experimentally measurable quantity below which the magnetization is nonzero (see Fig. 10.5). Taking the equality sign in Eq. (10.42) gives:

$$T_C = \frac{2m^2}{3k\beta^2}\,\alpha n \tag{10.43}$$

As T decreases, the spontaneous magnetization, which is determined by the intersection of the line with the Langevin function, increases until it reaches its maximum value (saturation limit), $M_S = Nm$, which is an experimentally measurable quantity. Figure 10.6 shows a comparison between theory and experiment for the spontaneous magnetization M as a function of T/T_C for the case $S = \frac{1}{2}$.

It is instructive to recast the quantum mechanical expression of the molecular field $H = H_0 + H'$ in the form of the classical result, given in Eq. (10.31). The field H' can be easily written in terms of the magnetization M of the medium by use of Eqs. (10.37) and (10.40); that is,

$$H' = \frac{2\alpha n}{\mu_0 \beta^2 N} M = \gamma M \tag{10.44}$$

Hence

$$H = H_0 + \gamma M$$

where

$$\gamma = \frac{2\alpha n}{\mu_0 \beta^2 N} \tag{10.45}$$

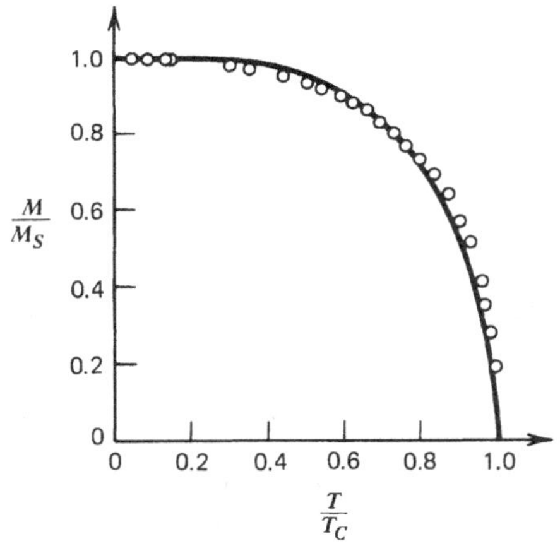

Figure 10.6 The spontaneous magnetization of a ferromagnetic material as a function of temperature, with T_C being the Curie temperature, along with the experimental measurements, shown as circles.

The quantity γ can also be written more conveniently in terms of the measurable properties of the material (Curie temperature). Using Eq. (10.43), one finds

$$\gamma = \frac{3kT_C}{m^2 N\mu_0} = \frac{3kT_C N\mu_0}{\mu_0^2 M_S^2} \tag{10.46}$$

where $M_S = Nm$ is the saturation magnetization. We take iron as an example; its experimental saturation magnetization $\mu_0 M_S = 2.15$ T, and its experimental Curie temperature is 1043 K. Thus, $\gamma = 995$, which is much larger than the classical prediction of $\frac{1}{3}$, and hence shows the need for the spin-spin interaction to account for ferromagnetism.

10.4.4 The Magnetic Susceptibility of Ferromagnetic Materials Above the Curie Temperature—The Curie-Weiss Law

Above the Curie temperature, ferromagnetic materials do not exhibit spontaneous magnetizations. However, because they have permanent magnetic moments, they exhibit paramagnetic properties in the presence of external magnetic fields. The paramagnetic susceptibility calculated above for nonferromagnetic material [Eq. (10.18)], however, will turn out to be not accurate in the case of ferromagnetic materials. We will show below that the departure from Eq. (10.18) is caused by their ferromagnetic properties.

Consider Eqs. (10.38) and (10.39). Above T_C, η can be taken small, so one can represent $B(\eta)$ by Eq. (10.41); therefore these equations give

$$\frac{kT\beta^2}{2\alpha nm^2}\left(\eta_0 - \frac{\mu_0 mH_0}{kT}\right) = \frac{1}{3}\eta_0 \tag{10.47}$$

or, in terms of the Curie temperature,

$$kT\left(\eta_0 - \frac{\mu_0 m H_0}{kT}\right) = kT_C\eta_0 \tag{10.48}$$

which can be easily solved for η_0:

$$\eta_0 = \frac{\mu_0 m H_0}{k(T - T_C)} \tag{10.49}$$

The magnetization of the medium can now be calculated using (Eq. 10.40). Taking η_0 to be small gives

$$M = \frac{1}{3} N m \eta_0 \tag{10.50}$$

which, when η_0 is written in terms of H_0 as in Eq. (10.49), gives

$$M = \frac{\mu_0 N m^2 H_0}{3k(T - T_C)}. \tag{10.51}$$

Thus, the magnetic susceptibility $\chi = M/H_0$ is

$$\chi = \frac{\mu_0 m^2 N}{3k(T - T_C)} = \frac{C}{T - T_C} \tag{10.52}$$

This is the *Curie-Weiss law*, and C is called the Curie constant. Table 10.1 gives the Curie temperatures of ferromagnetic elements.

Table 10.1 **Curie Temperatures of Ferromagnetic Elements**

Element	T_C, in kelvins
Fe	1043
Co	1393
Ni	631
Gd	289
Dy	105

Example 10.3 Dipole-dipole Interaction and Ferromagnetism

As remarked above, the dipole-dipole interaction between magnetic atoms is not sufficient to explain ferromagnetism. Let us concentrate on a central atom of magnetic moment **m** in a ferromagnetic sample and consider its interaction with its nearest n atoms. The magnetic field produced by a magnetic dipole moment **m** is given by Eq. (8.98):

$$\mathbf{B} = \frac{\mu_0}{4\pi}\left[-\frac{\mathbf{m}}{r^3} + \frac{3(\mathbf{m}\cdot\mathbf{r})\mathbf{r}}{r^5}\right]$$

where **r** is the distance from the dipole to the point of observation. The interaction energy of an atom of magnetic moment $\mathbf{m}_i$ placed in this field is $-\mathbf{m}_i \cdot \mathbf{B}$ [see (Eq. 8.111)]. Therefore the total interaction energy of n nearest atoms with the central atom is

$$U = \frac{\mu_0}{4\pi}\left[\sum_{i=1}^{n} \frac{\mathbf{m}\cdot\mathbf{m}_i}{r_i^3} - \frac{3(m\cdot\mathbf{r}_i)(\mathbf{m}_i\cdot\mathbf{r}_i)}{r_i^5}\right]$$

where $\mathbf{r}_i$ is the distance from the central atom to the ith nearest atom.

Because we are interested in an order of magnitude of this interaction we make the following approximations: $r_i \approx r$ for all i, $\mathbf{m}\cdot\mathbf{m}_i - 3(\mathbf{m}\cdot\hat{\mathbf{r}}_i)(\mathbf{m}_i\cdot\hat{\mathbf{r}}_i) \approx m^2$ for all i, where $\hat{\mathbf{r}}_i$ is a unit vector along $\mathbf{r}_i$. Thus

$$U = \frac{\mu_0}{4\pi}\frac{nm^3}{r^3}$$

An order of magnitude can now be evaluated by taking $n = 15$, $m = 10^{-23}$ joules per tesla (J/T), and $r = 2$ Å; thus $U = 1.5 \times 10^{-23}$ joules. The ratio U/KT is of the order of unity at ~ 1 K; therefore this type of interaction cannot produce ferromagnetic effects in the region below 1000 K, where metallic iron is ferromagnetic.

10.4.5 Ferromagnetic Domains

It is observed that ferromagnetic materials do not always exhibit the ferromagnetic properties at temperatures below the Curie temperatures. In order to explain this phenomenon, Weiss* advanced in 1907 the hypothesis that the bulk of the material may be divided into *domains*, as shown schematically in Fig. 10.7 for a single crystal and a polycrystal. Each domain is taken to obey the above theory; that is, to be spontaneously magnetized at temperatures below the Curie point, with a random direction of its resultant magnetization. The degree of magnetization of each domain, however, is governed by the temperature of the material. The overall magnetization of the material is the vector sum of all the magnetizations of the various domains. With the domains randomly magnetized, the bulk magnetization of the specimen would be zero. On the other hand, when the domains are magnetized in the same direction, they reinforce each other, thus giving a large magnetic field. Such reinforcing can be achieved by the action of a weak external magnetic field.

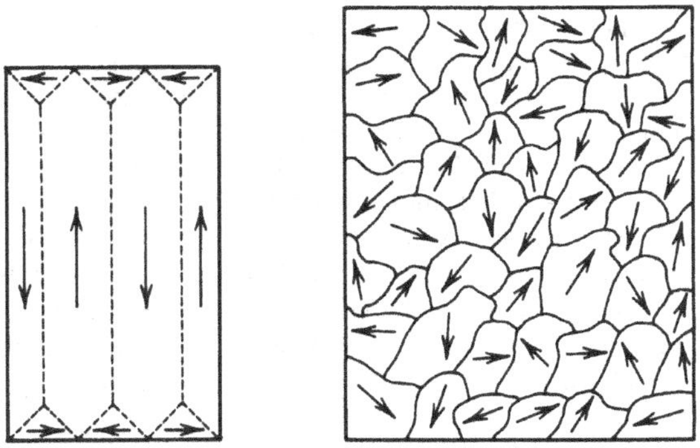

(*a*) (*b*)

Figure 10.7 Ferromagnetic domain structure. (*a*) Single crystal, showing domain walls dotted. (*b*) Polycrystal, showing crystal walls solid, and with domains not shown. The arrows indicate the direction of magnetization of individual crystals.

* P. Weiss, *Journal de Physique*, vol. 6, p. 667, 1907.

An experimental technique developed by Bitter* provided evidence supporting the domain hypothesis. In this technique, a colloidal suspension of a fine ferromagnetic powder is prepared. The domain structure of a ferromagnetic crystal is studied by placing a drop of this colloidal suspension on the surface of the crystal. When a photomicrograph of the surface is investigated, it is observed that the particles gather at some boundaries (*domain boundaries*). The gathering occurs because adjacent domains are magnetized in different directions, thus resulting in strong local fields at interfaces of the domains. With this method, the actual movements of the *walls* and the sizes of the domains are observed under the influence of an external magnetic field.

The formation of multidomain structure of a ferromagnetic specimen is usually energetically favored. There are three types of magnetic energy that enter into consideration. We will see in chapter twelve that a magnetized crystal has stored magnetic energy associated with its magnetization; hence this energy will depend on the domain structure. This is so because if the crystal can arrange itself into domains with opposite magnetization, the overall magnetization will be reduced, resulting in the stored magnetic energy. Another type of magnetic energy is the magnetic energy needed to set up and maintain the domain boundaries. Such energy is needed because the exchange forces favor parallel and oppose antiparallel orientations of the magnetization (see Section 10.4.6). In addition to these considerations, most crystals have two axes, where establishing magnetization is easier along one axis (easy axis) than along the other axis (hard axis). The difference in energy between these two cases is called the anistropy energy. The number and shapes of the domains that a perfect ferromagnetic crystal may tend to form is dictated by the above energy considerations, in such a way as to minimize the total magnetic energy of the crystal.

10.4.6 Antiferromagnetism and Ferrimagnetism (Ferrite)

Equations (10.28) and (10.29) give the exchange interaction energy between two magnetic atoms. For ferromagnetic materials, $\alpha > 0$; that is, parallel spins have lower magnetic energy than antiparallel spins. In certain materials such as chromium and manganese, α is negative; that is, antiparallel spins have lower magnetic energy than parallel spins. These materials, which are called *antiferromagnetic*, have a tendency for antiparallel spin alignment. This alignment property is also temperature-dependent; below a certain temperature, which is also called the Curie temperature, the spins become aligned in an alternation configuration, as shown in Fig. 10.8*a*. Above this temperature the spin directions are random.

This alternating spin configuration in chromium can be proved by scattering some particles, which themselves have spins such as neutrons, from a crystal of

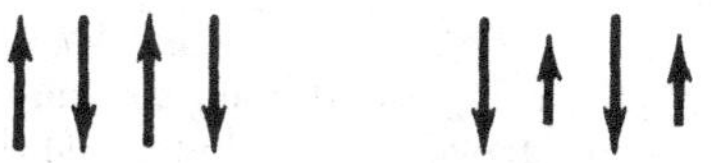

(*a*) (*b*)

Figure 10.8 Schematic representation of atomic spins in (*a*) antiferromagnetic and (*b*) ferromagnetic material.

* F. H. Bitter, *Physical Review*, vol. 41, p. 507, 1932.

chromium. The degree of scattering depends on whether the neutron spin is parallel or antiparallel to the scatterer. Thus, the interference pattern of the scattered neutrons changes when the spin state changes from a random state (above the Curie temperature) to an alternate state (below the Curie temperature).

Another type of antiferromagnetic materials has a combination of the ferromagnetic and antiferromagnetic alignment, and is called a *ferrimagnet*, or simply a *ferrite*. This is possible in the case of $ZnOFe_2O_3$ because it has two magnetic atoms (metal atoms zinc and iron) arranged in a spiral structure. The various iron spins line up together, and the various zinc spins also line up together. The lowest energy of the whole system together is the situation where the resultant iron spin is antiparallel to the resultant zinc spin. If the opposed spins have unequal moments, as in Fig. 10.8*b*, the antiferromagnetic substance is called a ferrimagnetic material or simply a ferrite. The mineral magnetite (Fe_3O_4) is an example of a ferrite which has been known since ancient times. Other examples of ferrimagnets (ferrites) include the oxides in which Zn in $ZnFe_2O_3$ is replaced by Co, Ni, Mn, Cu, Mg, Cd, or divalent iron.

An important property of ferrites is their large resistivity (up to $10^4\ \Omega\cdot\text{m}$). Such large resistivity makes them poor conductors of electricity—a property that is valuable in high-frequency applications. When very good conductors are used in these applications, eddy currents become very effective in causing large heat losses (see Example 11.3). Therefore, the use of ferrites instead of good conductors in these applications helps to minimize these losses.

10.5 Summary

Atoms are classified into two categories: atoms which have net permanent magnetic dipole moments (magnetic) resulting from spin, and atoms which have no net permanent magnetic moments (nonmagnetic). When a nonmagnetic atom is placed in an external magnetic field normal to its orbit, the field induces a magnetic moment opposing the field itself (Lenz' law)

$$\Delta\mathbf{m} = -\frac{1}{4}\frac{e^2\rho^2}{m_e}\mathbf{B} \tag{10.9}$$

where ρ and m_e are the radius of the orbit and mass of the electron. For an ensemble of atoms of number density N placed in the magnetic field, this effect results in the induced susceptibility

$$\chi = -\frac{1}{6}\frac{\mu_0 e^2}{m_e}N\langle r^2\rangle \tag{10.13}$$

where an average over the orientation of the orbit was made.

When magnetic atoms of permanent moment m are placed in a magnetic field $\mathbf{B} = B\hat{\mathbf{z}}$, they exhibit an additional susceptibility resulting from the tendency of the field to align permanent dipoles along its direction, just as in the electrical case. This tendency is opposed by the randomizing collisions with other atoms. Just as in the electrical case, this orientational susceptibility, at a given absolute temperature T, is given by the Langevin function

$$\langle\mathbf{m}\rangle = m\left(\coth\eta - \frac{1}{\eta}\right)\hat{\mathbf{z}} \tag{10.14}$$

where $\eta = mB/kT$, and $\langle\mathbf{m}\rangle$ is the average component of $\mathbf{m}$ along the field. At high temperature, we get the Curie law

$$\langle m_z\rangle \approx \frac{m^2 B}{3kT} \quad \text{and} \quad \chi = \frac{\mu_0 m^2 N}{3kT} \tag{10.18}$$

where N is the number density of the ensemble. At very low temperatures and/or large B, the field may overcome the effect of randomization and result in complete alignment (saturation). The magnetization of the ensemble then approaches

$$\mathbf{M} = Nm\hat{\mathbf{z}} \tag{10.19}$$

Ferromagnetic atoms are magnetic atoms that can have an additional interaction among themselves that works in favor of alignment in addition to the above randomizing interaction. This additional interaction is a quantum mechanical effect arising from what is called spin-spin interaction. The effect of this interaction effectively results as an additional magnetic field at the site of every atom due to the n neighboring identical atoms. This field is effectively equal to

$$\mathbf{H}' = \frac{2\alpha n}{\mu_0 \beta^2} \langle m_z \rangle \hat{\mathbf{z}}$$

where $\langle m_z \rangle$ is the average moment of the identical atoms along the external field, $\beta = e\hbar/m_e$, and α is a constant that depends on the type of material. Thus the total field at the site of each indentical atom is given by the sum of this field and the external field H_0; that is,

$$H = H_0 + H'$$

Under the influence of this field, the magnetic moment of one of the identical atoms will have a degree of alignment just as the case of paramagnetic atoms. Hence $\langle m_z \rangle$ is given by the Langevin function

$$\langle m_z \rangle = m\left[\coth \eta - \frac{1}{\eta}\right] \tag{10.14}$$

where $\eta = \mu_0 mH/kT$. Because H itself depends on $\langle m_z \rangle$ through H', then the question that remains to be answered whether there is a nontrivial physical solution of the Langevin relation for $\langle m_z \rangle$.

If such a solution exists, then one would like to know how does $\langle m_z \rangle$ depend on T, compared to $\langle m_z \rangle$, which one can get in the absence of the spin-spin interaction (compare to the paramagnetic effect), and finally whether there will be a solution even when the external magnetic field H_0 vanishes.

Graphical as well as numerical examinations of the above condition indicates that there exists a nontrivial solution, and the solution exists even when $H_0 = 0$; that is, the material can exhibit spontaneous alignment (magnetization) under the influence of this additional interaction. The spontaneous magnetization, however, has a temperature threshold, called the Curie temperature T_C, above which it is ineffective.

$$T_C = \frac{2m^2 \alpha n}{3k\beta^2} \tag{10.43}$$

At $T < T_C$ the interaction can cause a sizable alignment (magnetization M), which explains the phenomenon of ferromagnetism. The magnetic field H' can be easily cast in terms of such magnetization M, as follows:

$$H' = \gamma M \tag{10.44}$$

where γ comes out to be nearly 1000 for iron as an example. This is much larger than what the classical relationship gives $H' = \frac{1}{3}M$. At $T > T_C$, the material behaves as a paramagnetic material with

$$\chi \approx \frac{\mu_0 m^2 N}{3k(T - T_C)} \tag{10.52}$$

Because of domain structure, even below the Curie temperature, a macroscopic ensemble of a ferromagnetic material may not show a permanent magnetic moment.

Problems

10.1 Consider an atomic electron in circular motion of radius ρ and frequency ω. The atom is placed in a magnetic field **B** normal to the orbit of the electron. (a) Determine the magnetic moment of the atom in the absence of the field. (b) Determine the magnetic moment in the presence of the field. (c) If the magnetic susceptibility of an ensemble of density 2.8×10^{28} atoms per cubic meter is -10^{-6}, determine the atomic radius.

10.2 An atom has two electrons traveling in opposite directions in the same circular orbit of radius ρ and frequency ω. The atom is placed in an external magnetic field **B** normal to the orbit of the electrons. Determine the magnetic moment of the atom in the absence and in the presence of the field.

10.3 A solid contains weakly interacting paramagnetic atoms of spin $\frac{1}{2}$, and magnetic moment $m = 9.27 \times 10^{-24}$ J/T. To what temperature does one need to cool the sample in the presence of a field of 3×10^4 G to have 75 percent of the atoms with their spins parallel to the field? (See Example 10.1.)

10.4 Paraffin contains many protons, each of which has a spin $\frac{1}{2}$ and a magnetic moment $m = 1.4 \times 10^{-26}$ J/T (about a factor of a thousand smaller than the moment of atomic electrons). To what temperature does one need to cool the sample in the presence of a field of 3 T to have 75 percent of the protons with their spins parallel to the external field?

10.5 A bar of metallic iron is of length 10 cm, cross section 1 cm^2, and density 10^{23} atoms per cubic centimeter. The magnetic moment of each atom is 1.8×10^{-19} A$\cdot$cm^2. (a) Assume that ferromagnetism does not play a role in iron; that is, if iron were paramagnetic, determine the susceptibility of iron at 300 K. (b) What is the dipole moment of the bar when immersed in a field of 10^3 G, if it has the susceptibility found in part (a)? (c) What would be the magnetization and dipole moment of the bar if all the atoms were aligned in one domain (saturation) as a result of the ferromagnetic interaction?

10.6 The inverse of the magnetic susceptibility of gadolinium metal above its Curie temperature was found to be equal to 5.82×10^3 cm^3/g at 600 K and 1.35×10^4 cm^3/g at 1000 K. Determine the Curie temperature. Plot $1/\chi$ as a function of temperature.

10.7 Nickel has a density of 9.1×10^{28} atoms per cubic meter, an atomic magnetic moment of 9.28×10^{-24} A$\cdot$m^2, and a Curie temperature of 631 K. Determine the saturation magnetization and γ for nickel.

10.8 Determine the product of α (the strength of exchange interaction) and n (number of contributing atoms to this interaction) from the macroscopic properties of nickel. (Thus, if one is given α, n can be estimated.)

ELEVEN
INDUCTION

Until now we have limited our considerations to static phenomena. We have considered electric and magnetic fields that did not change in time. In this chapter we shall discuss situations where variations of charge and current distributions, and hence electric and magnetic fields, are permitted in time. We shall see that once such time variations are allowed, we no longer have two separate subjects, electricity and magnetism, but one subject called electromagnetism. The phenomena and applications of this subject are infinitely more diverse than the disjoint subjects of electrostatics and magnetostatics.

Once we allow for the possibility for time-varying charge densities, several complications arise. Most of these are related to having a consistent interpretation of physical phenomena that are observed from different *inertial reference frames* of motion—that is, frames that have no acceleration with respect to each other. Thus, as a simple example, whether a point charge in the presence of a magnetic field experiences what we interpret to be a *Lorentz* force or not depends upon whether or not the charge is moving in our reference frame. Because we believe that all inertial reference frames are equivalent with respect to the validity of physical laws, it turns out that what we call an electric field in one reference frame must be interpreted as a magnetic field in another. Thus, the requirements of the special theory of relativity lead naturally to a kind of coupling of electric and magnetic fields, which is manifest in the basic electromagnetic equations for the field vectors. As another aspect of these relativity requirements it will be seen that the impossibility of physical signals propagating across space with infinite speed requires changes in the electromagnetic equations to account for this phenomenon. In our studies of statics such considerations were unimportant. In the dynamic real world, however, they assume a primary importance. For these consideration see Chapter 14.

As our starting point we shall assume that Gauss' law has a more general validity than was implied in our study of electrostatics. Many experiments have shown this law to be true at any point in time even when the charge distribution and electric fields are changing in time. Thus, we shall assume that the equations

$$\nabla \cdot \mathbf{E} = \frac{\rho}{\varepsilon_0} \qquad \text{and} \qquad \oint_S \mathbf{E} \cdot d\mathbf{a} = \frac{Q(t)}{\varepsilon_0} \tag{11.1}$$

are true at any instant in time, even though $\mathbf{E}$, ρ, and Q may all be functions of the time variable t. The surface integral is always to be understood as being evaluated at some definite instant in time.

Analogously, the equations for the magnetic field vector $\mathbf{B}$

$$\nabla \cdot \mathbf{B} = 0 \qquad \text{and} \qquad \oint_S \mathbf{B} \cdot d\mathbf{a} = 0 \tag{11.2}$$

are also thought to be valid even when $\mathbf{B}$ is a function of time. Equations (11.1) and (11.2) are two of the keystones of the theory of electromagnetism and will henceforth be assumed to be most generally valid.

11.1 Faraday's Law

Equations (11.1) and (11.2) have the same forms as encountered in statics. On the other hand, a departure from the behavior observed in statics is implicit in what is known as *Faraday's law*. Consider an open surface S, bounded by a curve C and located in a region in which there is magnetic field $\mathbf{B}$. We assume that S, C, and $\mathbf{B}$ *in general* all depend on time. The flux $F(t)$ through S is then given by the surface integral

$$F = \int_S \mathbf{B} \cdot d\mathbf{a}$$

Faraday's law states that an *electomotive force* $\mathscr{E}$ is generated around the curve C, and $\mathscr{E}$ is proportional to the time rate of change of F. That is,

$$\mathscr{E} = -\frac{dF}{dt} \tag{11.3}$$

If the electric field in space is denoted by $\mathbf{E}$, then, by definition, the *induced emf* or the electromotive force around C is

$$\mathscr{E} = \oint_C \mathbf{E} \cdot d\mathbf{r} \tag{11.4}$$

From Eqs. (11.3) and (11.4) we get

$$\mathscr{E} = \oint_C \mathbf{E} \cdot d\mathbf{r} = -\frac{d}{dt}\left(\int_S \mathbf{B} \cdot d\mathbf{a}\right) \tag{11.5}$$

We now consider a curve C that is fixed in our reference system—and, correspondingly, a fixed surface S and defer the general case to Section 11.2. Thus for now neither S nor C depends on t, and as a result we can move d/dt inside the integral. Therefore Eq. (11.5) becomes

$$\oint_C \mathbf{E} \cdot d\mathbf{r} = -\int_S \frac{\partial \mathbf{B}}{\partial t} \cdot d\mathbf{a} \tag{11.6}$$

where partial derivatives are used inside the integral because $\mathbf{B}$ is a function of position and time. Using Stokes' theorem, the left-hand side of Eq. (11.6) can be changed to a surface integral. Thus

$$\int_S (\nabla \times \mathbf{E}) \cdot d\mathbf{a} = -\int_S \frac{\partial \mathbf{B}}{\partial t} \cdot d\mathbf{a}$$

Because this is true for any arbitrary surface S, we may equate the integrands of these expressions, as follows:

$$\nabla \times \mathbf{E} = -\frac{\partial \mathbf{B}}{\partial t} \tag{11.7}$$

This is the (*local or differential*) equivalent of Faraday's (integral) law and thus is itself to be considered as one of the fundamental equations of electromagnetism, and it will be assumed to apply at every point in space.

We observe that the electric field $\mathbf{E}$ is now not in general a conservative field. This follows directly from Eq. (11.4), which asserts that, in general, $\oint \mathbf{E}\cdot d\mathbf{r} \neq 0$. We thus observe that the sources of $\mathbf{E}$ are not only charges (from $\nabla\cdot\mathbf{E} = \rho/\varepsilon_0$), but $\partial\mathbf{B}/\partial t$ fields (ultimately arising from nonsteady, time-varying currents). In general, neither the curl nor divergence of $\mathbf{E}$ is zero. We should, however, note that $\mathbf{E}$ is conservative in the special case when $\mathbf{B}$ is constant in time.

The minus sign in Eq. (11.3) of Eq. (11.7) embodies *Lenz' law*, which ensures that perpetual-motion machines will be impossible to construct. Thus, if we have a curve C that lies in a conductor, the induced emf $\mathscr{E}$ will produce currents in the conductors that tend to oppose the change in flux through the curve. If the minus sign were absent, this would imply that the induced currents could increase the flux change, possibly giving rise to a runaway, regenerative behavior. Lenz' law ensures that "negative feedback" is present, implying a tendency to stability, rather than "positive feedback," which could result in the creation of infinitely intense $\mathbf{B}$ fields. Thus, Lenz' law is the law of the status quo: the systems to which it applies will always resist change.

We shall now show how $\mathbf{E}$ can be decomposed into two parts, of which one is conservative and the other is not. The latter is called *solenoidal* (meaning its divergence is zero.) Since the divergence of $\mathbf{B}$ is zero, then it follows that $\mathbf{B}$ can be written as $\nabla\times\mathbf{A}$. Substituting this result in $\nabla\times\mathbf{E} = -\partial\mathbf{B}/\partial t$ gives

$$\nabla \times \mathbf{E} = -\frac{\partial}{\partial t}(\nabla \times \mathbf{A}) = \nabla \times \left(-\frac{\partial \mathbf{A}}{\partial t}\right)$$

that is,

$$\nabla \times \left(\mathbf{E} + \frac{\partial \mathbf{A}}{\partial t}\right) = 0 \tag{11.8}$$

If the curl of a vector field is zero in a region, it can be written as the gradient of a scalar field there [see Eq. (1.52); hence Eq. (11.8) gives

$$\mathbf{E} + \frac{\partial \mathbf{A}}{\partial t} = -\nabla\Phi \tag{11.9}$$

where Φ is a scalar field. Because $\nabla\Phi$ is a conservative vector field, and $-\partial\mathbf{A}/\partial t$ satisfies

$$\nabla \times \left(-\frac{\partial \mathbf{A}}{\partial t}\right) = -\frac{\partial \mathbf{B}}{\partial t} \neq 0$$

we have established the proposed decomposition of $\mathbf{E}$:

$$\mathbf{E} = -\frac{\partial \mathbf{A}}{\partial t} - \nabla\Phi \tag{11.10}$$

The quantity Φ is a scalar function, but its relationship to the potential function introduced in electrostatics is perhaps not yet clear. Therefore we write $\mathbf{E} = \mathbf{E}^i + \mathbf{E}^q$, where

$$\mathbf{E}^i = -\frac{\partial \mathbf{A}}{\partial t} \qquad \mathbf{E}^q = -\nabla\Phi \tag{11.11}$$

The field $\mathbf{E}^i$ is the induced field; it is nonconservative (solenoidal). The field $\mathbf{E}^q$, however, is emanating from charges; it is *nonsolenoidal* and conservative. Thus

$$\nabla \times \mathbf{E}^i = -\frac{\partial \mathbf{B}}{\partial t} \qquad \nabla \times \mathbf{E}^q = 0 \tag{11.12}$$

We now determine an integral relationship between the induced field and the vector potential associated with the magnetic field. Integrating the curl equation of $\mathbf{E}^i$ over an area, S, substituting $\mathbf{B} = \nabla \times \mathbf{A}$, and using Stokes' theorem, we get

$$\oint_C \mathbf{E}^i \cdot d\mathbf{r} = -\frac{d}{dt}\oint_C \mathbf{A} \cdot d\mathbf{r} \tag{11.13}$$

Example 11.1 Pulsating Solenoid

Consider a long solenoid of radius R and of n_l closely packed turns of wire per meter, as shown in Fig. 11.1. Let current $I(t)$ flow in the wire. For a solenoid of length $\gg$ diameter, $\mathbf{B}$ is

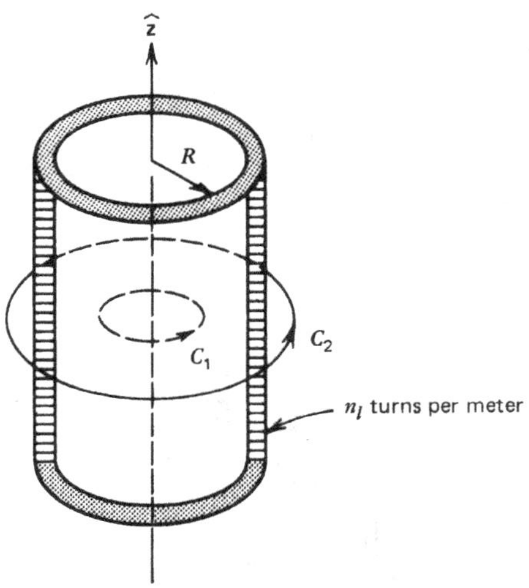

Figure 11.1 Pulsating solenoid illustrating Faraday's law.

negligible outside the solenoid, and is approximately uniform inside. For path C_1, of radius ρ less than R, Eq. (11.5) gives

$$\mathscr{E} = \oint_{C_1} \mathbf{E} \cdot d\mathbf{r} = -\frac{d}{dt}(B\pi\rho^2)$$

where $B\pi\rho^2$ is the "flux-linking" curve C_1, and $\mathbf{E}$ is the tangential component of the electric field on C_1. Since $\mathbf{E}$ is not dependent on the rotational angle ϕ by symmetry considerations, and $B = \mu_0 n_l I$ for the interior solenoidal field, we have

$$\mathscr{E} = -\pi\rho^2\frac{\partial B}{\partial t} = -\pi\rho^2\mu_0 n_l \frac{dI}{dt} \tag{11.14}$$

and

$$E_\phi = \frac{\mathscr{E}}{2\pi\rho} = -\frac{\mu_0 n_l \rho}{2}\frac{dI}{dt} \qquad \rho < R \tag{11.15}$$

The induced emf and field for $\rho > R$ can be calculated in a similar way. (We leave this as an exercise.)

Example 11.2 A Wire and a Rectangular Loop

Consider a long straight wire whose current is given by $I(t)\hat{\mathbf{z}}$, as shown in Fig. 11.2. The emf induced around a rectangular curve C of area S in a plane containing the wire can be calculated from $\mathscr{E} = -dF/dt$. The flux F through C is

$$F = \int_S \mathbf{B}\cdot d\mathbf{a} = \int_S \mathbf{B}\cdot(\hat{\mathbf{z}} \times \hat{\boldsymbol{\rho}})da = \int_S B_\phi \, da$$

But $da = z_0 \, d\rho$ and B_ϕ is given by Eq. (8.24); that is,

$$B_\phi = \frac{1}{2\pi\rho}\mu_0 I(t)$$

Therefore,

$$F = \frac{\mu_0}{2\pi}\left(I(t)\int_{\rho=\rho_1}^{\rho_2} \frac{1}{\rho} z_0 \, d\rho\right) = \frac{\mu_0 z_0}{2\pi}\left(\ln\frac{\rho_2}{\rho_1}\right)I(t) \tag{11.16}$$

The induced $\mathscr{E}$ is then

$$\mathscr{E} = \frac{-\mu_0 z_0}{2\pi}\left(\ln\frac{\rho_2}{\rho_1}\right)\frac{dI}{dt}$$

If a wire is placed coincident with C, and $dI/dt > 0$, a current will flow around C in a counterclockwise direction.

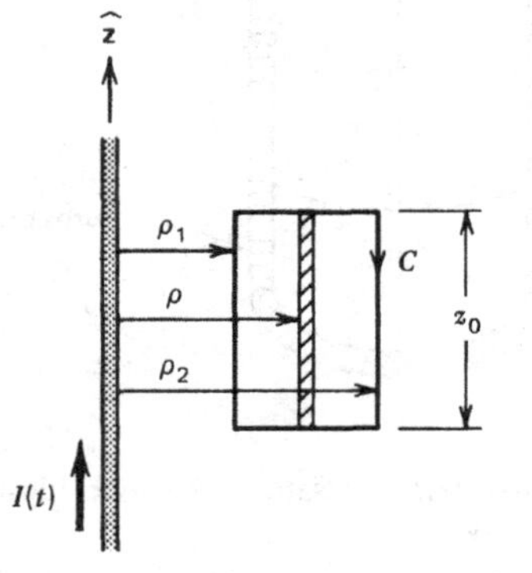

Figure 11.2 A current-carrying wire in the plane of a rectangular wire.

Example 11.3 Eddy Currents—Induction Heating

It is known that when **B** changes in time, currents are induced in the conducting materials near the region where $\mathbf{B} \neq 0$. Such currents are called *eddy currents*. In electromagnetic machinery they are often undesirable because they absorb energy and heat up the conducting materials through $\rho_c J^2$ losses (Joule losses). In order to eliminate the eddy currents, conductors are often laminated. The effect of such lamination is illustrated in Fig. 11.3. In Fig. 11.3*a*

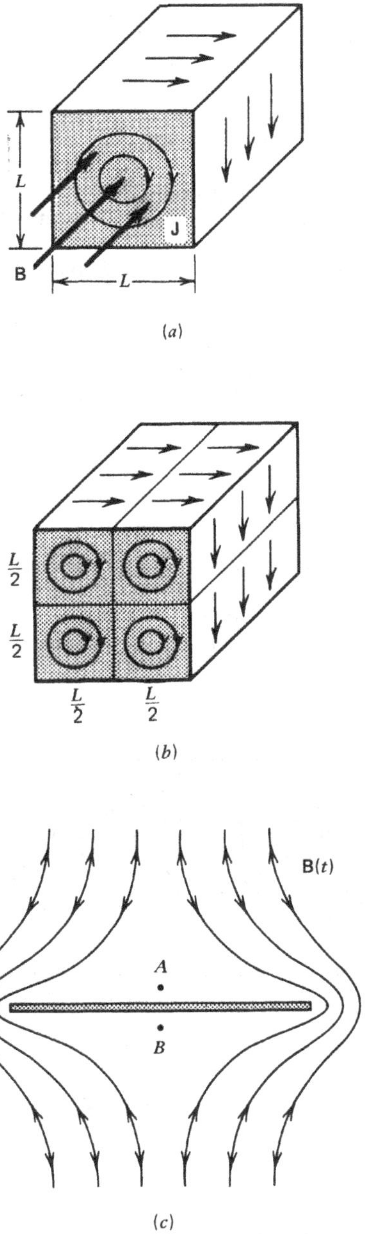

Figure 11.3 Lamination of a conductor in order to minimize the effect of eddy currents. (*a*) A block without lamination in a changing **B** field. (*b*) A block laminated into four sections. (*c*) The reduction of the change **B** field near a conducting plate due to eddy currents in the plate.

we show a block of a conductor in a changing field, $B(t)$. The circulating current in such a block will be proportional to L^2 (see Problem 11.3) and thus the power dissipation, which is proportional to the square of the current, will be proportional to L^4. If we now subdivide the block in Fig. 11.13*a* into four sections, as shown in Fig. 11.3*b*, such that the current cannot flow from one section into another, then the current in each section will be proportional to $(L/2)^2$. The power dissipation in each section will thus be proportional to $(L/2)^4$, and the total power dissipation in all four blocks will be proportional to $4(L/2)^4 = L^4/4$. This is less by a factor of 4 than we had originally. The general result is that if a cross section (perpendicular to the **B** field direction) of the conducting material is laminated by subdividing it into N parts, then power losses due to eddy currents are diminished by a factor proportional to $1/N$.

The effect of eddy currents near a conducting material is to reduce the magnitude of the changing **B** field, because the eddy currents are so set up as to oppose the change in the inducing field **B**. Such an effect is shown in Fig. 11.3*c*. [If the conductor had infinite conductivity (as, e.g., in a superconductor) the **B** field in and near the conductor would be reduced to zero thereby.] This type of effect may be used to shield a region from changing magnetic fields. It does not, however, provide shielding against constant magnetic fields.

Finally, eddy current "induction" heating may be employed beneficially to heat an object without disturbing it otherwise. Thus, food may be uniformly cooked in an induction field by being placed in a region of intense alternating **B** fields.

Example 11.4 The Betatron*

The operation of a betatron, an accelerator of electrons colloquially known as an "ausserordentlichhochgeswindigkeitselectronentwickeldenschwerarbeitsbeigollitron"† is a prime example of the application of Faraday's law. The idea of a betatron (see Fig. 11.4) is to accelerate charged particles by means of an induction field and to contain the particles in space by using that same changing magnetic field.

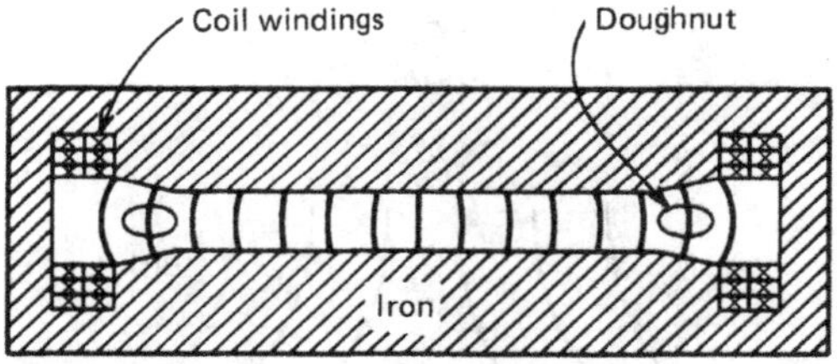

Figure 11.4 Schematic diagram of cross section of a betatron used to accelerate and confine charged particles by means of a changing magnetic field.

The particles are to be contained in an evacuated ring-shaped object known as a "doughnut." If the average radius of the circular particle trajectory is R, its charge is q, and the frequency of revolution is ω, we require that the change of its momentum to be equal to the centripetal force

$$\left|\frac{d\mathbf{p}}{dt}\right| = p\omega$$

where **p** is the momentum of the charge. Equating the centripetal force to the Lorentz force ($qR\omega B$) gives

$$p = BqR \tag{11.17}$$

*Here at the University of Illinois, where the betatron was developed, there were two active betatrons whose electron beams were used in nuclear physics research.

†Copyright, E. C. Hill (notes on electromagnetism).

where B is the magnetic field. Thus the momentum of the particles, of charges q, is directly related to the field B. Now, if B is made to increase in time in the space through the hole of the doughnut, there will be an induced electromotive force

$$\mathscr{E} = -\frac{dF}{dt} = -\frac{d}{dt}\langle B\rangle A$$

where $\langle B\rangle$ is the average field inside the circle of radius R. The corresponding induced field $E^i = \mathscr{E}/2\pi R$ exerts a force on the charge such that

$$\frac{dp}{dt} = qE^i = q\frac{\mathscr{E}}{2\pi R} = \frac{q}{2\pi R}A\left[-\frac{d\langle B\rangle}{dt}\right]$$

An integration of this equation yields

$$p = p_0 - \frac{qA\langle B\rangle}{2\pi R}$$

where p_0 is the initial momentum of the particle (when $\langle B\rangle \approx 0$) and may be taken to be zero. Then

$$p = \frac{qA\langle B\rangle}{2\pi R} \tag{11.18}$$

We must now require the two expressions for the momentum of the particle: $p = BqR$ given by Eq. (11.17) and that given by Eq. (11.18) to be equal. Thus

$$BqR = \frac{qA\langle B\rangle}{2\pi R}$$

or, if $A = \pi R^2$,

$$B = \frac{1}{2}\langle B\rangle$$

Thus, the "guide field," B, must equal one-half the average induction field, $\langle B\rangle$, if the particle is to be maintained on a constant radius as it is accelerated. A typical (small) betatron might have a radius R of 0.5 m and a maximum field $\langle B\rangle_{\max}$ of 0.5 T. The energy of electrons that would be produced by such a machine would be given by $pc \approx eR\langle B\rangle c/2$ (relativistic) or about 35 MeV.*

11.2 Motional EMF

Thus far, we have considered only the consequence that changing magnetic fields may induce electric fields in the space about us in fixed and rigid loops. Now we consider the case where the loops may change their size, orientation, position, or shape. Let us consider first the question of what happens if a conductor moves with respect to our assumed stationary reference frame. A hint of the answer is obtained by considering an observer O' to move along with the conductor. Consider a current circuit that is stationary in the laboratory frame O and carries a steady-state current. A conductor is moving in the magnetic field of the circuit, $\mathbf{B}$, with a constant velocity $\mathbf{v}$. From the point of O' the sources move in the opposite direction. Therefore, O' will notice a varying magnetic field, and consequently an induced electric field, E^i. This electric field will exert a force qE^i on q and hence cause it to move in the conductor. We at rest, however, see no induced electric field because for us the magnetic field is unvarying. We know, however, that the free charge q in the

*The *relativistic energy* of a particle of mass m_0 is given by $\sqrt{(pc)^2 + (m_0c^2)^2}$, which is approximated by pc if $pc \gg m_0c^2$.

moving conductor experiences a Lorentz force **F** by virtue of the motion of the conductor through the magnetic field: $\mathbf{F} = q(\mathbf{v} \times \mathbf{B})$. If $v \ll c$, as we shall assume here, both we and the observer who moves with the conductor should observe the same phenomenon of moving charge and the same acceleration of this charge. Only the explanations of the source of the forces that cause the charge motion will differ between us. Thus

$$\mathbf{E}^i = \mathbf{v} \times \mathbf{B} \tag{11.19}$$

If, in addition to the force associated with the magnetic fields there are also electric fields produced by charges, in the rest frame of reference **E**, then (nonrelativistically) the electric field observed by one moving with the conductor is $\mathbf{E}' = \mathbf{E} + \mathbf{E}^i$; that is,

$$\mathbf{E}' = \mathbf{E} + \mathbf{v} \times \mathbf{B} \qquad v/c \ll 1 \tag{11.20}$$

The use of the term "nonrelativistic" is a hint that the effects being discussed here may only be fully understood with the help of the special theory of relativity (discussed in Chapter 17). It is to be noted that Eqs. (11.19) and (11.20) have not been proved. With their use, however, we can now find expressions for the emf of a loop (closed curve), each part of which may be moving through our space.

An emf for a loop is defined at a specific instant in time, t. It is the work that would be required to move an (imaginary) unit point charge around that loop at that instant. It is thus the energy delivered to a unit point charge imagined to proceed around the loop under the conditions prevailing at time t. Assuming that the only forces acting are from electric and magnetic fields, we have the following expressions for the emf of a loop lying in a conductor.

1. In a rest system O, with measurable fields $\mathbf{E}(t)$ and $\mathbf{B}(t)$, and where a point of the loop C instantaneously has a velocity $\mathbf{v}(\mathbf{r}, t)$, the emf around C is given by

$$\mathscr{E}(t) = \oint_C [\mathbf{E} + \mathbf{v} \times \mathbf{B}] \cdot d\mathbf{r} \tag{11.21}$$

where all quantities are evaluated at points on the loop at time t. Note that **E** in general has components $\mathbf{E}^q$ and $\mathbf{E}^i$ governed by Eqs. (11.11) and (11.12).

2. In the reference frame O' of a loop which is moving rigidly with a velocity **v** with respect to the rest system above, the emf $\mathscr{E}'$ is given effectively by

$$\mathscr{E}'(t) = \oint_C \mathbf{E}' \cdot d\mathbf{r} \tag{11.22}$$

where $\mathbf{E}'$ is measured by an observer in the moving frame O'. The values measured are related to the values of **E** and **B** of the rest frame O by Eq. (11.20).

We will now take on Faraday's law from the point of view of the motional electromotive force. As was stated above, we will consider loops changing their size, orientation, position, or shape in a magnetic field. For simplicity we assume that the magnetic field is constant in time—since, when it varies, we know how to calculate the induced electric fields (see the previous section). Example 11.11 discusses a situation where both are occurring simultaneously.

Consider loop C to be a conducting filament that moves in a static magnetic field $\mathbf{B}(\mathbf{r})$, such that it assumes the shapes $C(t)$ and $C(t + dt)$ at times t and $t + dt$, respectively, as shown in Fig. 11.5. Because of this motion an emf is induced, according to (Eq. 11.21), as follows:

$$\mathscr{E} = \oint_{C(t)} (\mathbf{v} \times \mathbf{B}) \cdot d\mathbf{l}$$

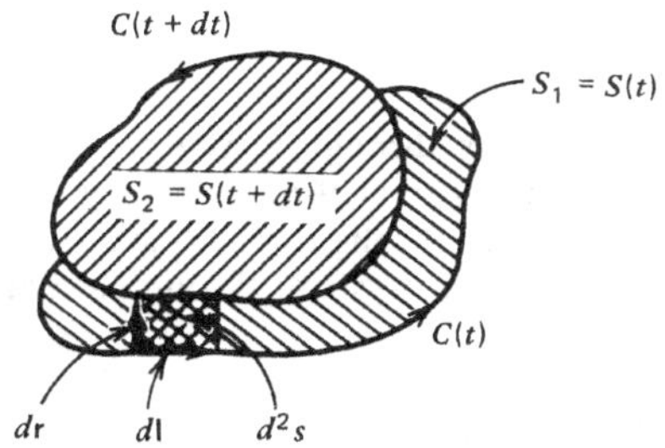

Figure 11.5 A conducting filament in the form of a closed loop in motion in a static magnetic field showing two positions at t and $t + dt$.

where $\mathbf{v}$ is the velocity of the element of C, $d\mathbf{l}$. Since $\mathbf{v} \times \mathbf{B} \cdot d\mathbf{l} = d\mathbf{l} \cdot \mathbf{v} \times \mathbf{B} = (d\mathbf{l} \times \mathbf{v}) \cdot \mathbf{B}$, then

$$\mathscr{E}(t) = \oint_{C(t)} (d\mathbf{l} \times \mathbf{v}) \cdot \mathbf{B}$$

Taking $\mathbf{v} = d\mathbf{r}/dt$, where $d\mathbf{r}$ is the displacement of the element $d\mathbf{l}$ in a time dt, we get

$$\mathscr{E} = \oint_{C(t)} \left(d\mathbf{l} \times \frac{d\mathbf{r}}{dt} \right) \cdot \mathbf{B}$$

Now, $d\mathbf{l} \times d\mathbf{r}$ is the element of area $d^2\mathbf{a}$, swept out in a time dt by the d**l**, and $\mathbf{B} \cdot d^2\mathbf{a}$ is the flux through this element of area. Thus

$$\mathscr{E}(t) = \frac{1}{dt} \oint (d^2\mathbf{a}) \cdot \mathbf{B} \equiv \frac{1}{dt} (-dF)$$

where $-dF$ is the flux of $\mathbf{B}$ through $\oint d^2\mathbf{a}$, the total area swept out by C in a time dt. Note that we have chosen the flux to be positive when it points in the direction $-d^2\mathbf{a} = d\mathbf{r} \times d\mathbf{l}$. With this definition, we have

$$\mathscr{E} = -\frac{dF}{dt} \tag{11.23}$$

Now we show that dF is actually the change in flux through loop C between times t and $t + dt$. Considering the closed surface composed of $S(t)$, $S(t + dt)$, and $\oint d^2\mathbf{a}$ we know that, evaluated at any time, the total flux through these surfaces is zero: $\oint \mathbf{B} \cdot d\mathbf{a} = 0$ [see Eq. (9.52)]. But

$$\oint \mathbf{B} \cdot d\mathbf{a} = \int_{S(t+dt)} \mathbf{B} \cdot d\mathbf{a} - \int_{S(t)} \mathbf{B} \cdot d\mathbf{a} - dF$$

[The minus sign in the second integral over $S(t)$ is required because of the sense chosen for $C(t)$.] Thus

$$0 = F(t + dt) - F(t) - dF$$

that is,

$$dF = F(t + dt) - F(t)$$

Thus dF is shown to be the change in flux through loop C between times t and $t + dt$.

We may summarize these results by stating that dF/dt in Eq. (11.23) may represent

1. The change in magnetic flux through a loop fixed in space (i.e., in our reference system) due to variation of **B** in time, as given by Eq. (11.3).
2. The change in magnetic flux through a well-defined conducting loop which moves (relative to our reference system) through a magnetic **B** field constant in time.
3. The flux "swept out" by a conducting loop as it changes its dimensions in the presence of a **B** field constant in time.
4. A linear combination of items 1 and 2 or 1 and 3 above.

It is well to emphasize that the law expressed in Eq. (11.23) is useful for moving media when such media consist of well-defined conducting filamentary loops. It cannot be used without ambiguity when such loops are not clearly defined. (Example 11.7 illustrates this fact.) It is nonetheless remarkable that two different kinds of physical phenomena can be described by the one law: $\mathscr{E} = -dF/dt$.

Example 11.5 Conducting Bar Moving Through a Constant Magnetic Field

We consider a conducting bar moving with velocity $v \ll c$ normal to its axis, through a constant magnetic field. This situation is illustrated in the Fig. 11.6. From the rest frame,

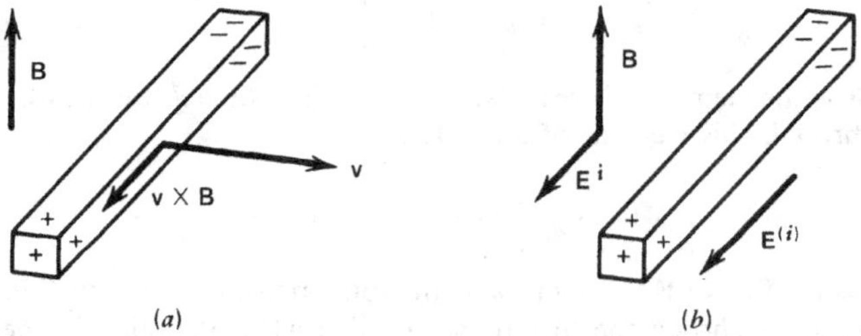

Figure 11.6 A conducting bar moving with velocity $v \ll c$ normal to its axis through a constant magnetic field: (*a*) as seen from the rest system; (*b*) as seen from the bar.

when motion starts, charges under the influence of the Lorentz force and the forces constraining the charges to remain in the bar will move until equilibrium is established, at which time no further charge movement with respect to the bar is observed. In this latter condition, it must then be true that on any free point charge q inside the bar the total force given by Eq. (11.20) must vanish; that is,

$$q(\mathbf{v} \times \mathbf{B}) + q\mathbf{E} = 0$$

where **E** is the field due to the equilibrium charge distribution on the bar. Thus, inside the bar,

$$\mathbf{E} = -(\mathbf{v} \times \mathbf{B})$$

meaning a uniform electric field exists inside the bar. Outside the bar the electric field produced by the charges will not be uniform.

From the observer moving with the bar, an electric field $\mathbf{E}^i$ is observed everywhere in space having the constant value

$$\mathbf{E}^i = \mathbf{v} \times \mathbf{B}$$

(The observer might not know that his or her velocity is $\mathbf{v}$, but this expression is what he or she finds anyway.) The bar, being a conductor, cannot sustain an internal electric field. Therefore, a redistribution of charge occurs tending to cancel the field $\mathbf{E}^i$, so ultimately the total electric field inside the bar is zero: $\mathbf{E}'$ (inside bar) $= 0 = \mathbf{E} + \mathbf{E}^i$. Inside the bar then, $\mathbf{E} = -\mathbf{E}^i = -(\mathbf{v} \times \mathbf{B})$, as above.

One should carefully note the different interpretations given by observers in the rest and moving frames of reference to the physical phenomena observed, namely, the occurrence of charges appearing on the surface of the conducting bar. This is one thing that both observers agree upon. They do not agree upon the values of electric fields existing (except for $\mathbf{E}^q$, which both agree upon nonrelativistically). Nonrelativistically, both see the same $\mathbf{B}$ field.

Example 11.6 Conducting Bar Moving on Stationary Tracks Through B Field

We now extend the content of Example 11.5 above by making the bar move at constant velocity $\mathbf{v}$ on stationary conducting tracks, again in a constant $\mathbf{B}$ field, as shown in Fig. 11.7. Now in an equilibrium situation, we observe a steady current flow around the loop formed by the tracks and the bar.

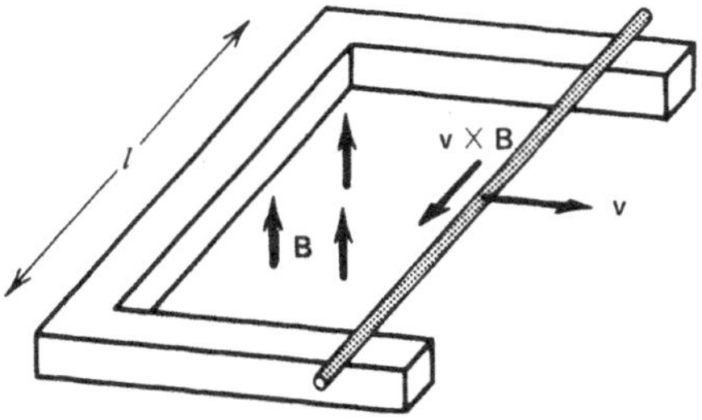

Figure 11.7 A conducting bar moving at constant velocity on stationary conducting tracks through a constant magnetic field.

We interpret this result as follows. We know that the charges in the bar move with its velocity $\mathbf{v}$, and so feel a Lorentz force $q(\mathbf{v} \times \mathbf{B})$, which, in the configuration shown, at any instant is directed along the bar, as shown. In addition, there will also be forces due to charges appearing on the bar that keep the charges inside the conductor. These forces, for example, must annul the effect of the force that $\mathbf{B}$ exerts on the charges moving along the bar. (The latter force is in a direction perpendicular to the bar.) If we calculate the emf around the loop at any instant, we thus find

$$\mathscr{E} = \int_b (\mathbf{v} \times \mathbf{B}) \cdot d\mathbf{r} + \int_b \mathbf{E} \cdot d\mathbf{r} + \int_t \mathbf{E} \cdot d\mathbf{r} + \int_{b+t} (\mathbf{u} \times \mathbf{B}) \cdot d\mathbf{r}$$

where b and t stand for the bar and the tracks, respectively, and $\mathbf{u}$ is the drift velocity of the charges in the conductor and is parallel to the sides of the conductor. Thus, $\mathbf{u} \times \mathbf{B} \cdot d\mathbf{r} = 0$, and the last integral may be neglected. The second and third integrals, when lumped together, give $\oint \mathbf{E}^q \cdot d\mathbf{r}$, which is zero by the conservative nature of $\mathbf{E}^q$. Consequently,

$$\mathscr{E} = \int_b (\mathbf{v} \times \mathbf{B}) \cdot d\mathbf{r} = vBl$$

If the total resistance of the loop is R, then a current I will flow, of magnitude $\mathscr{E}/R = vBl/R$.

Let us next analyze the result from the vantage point of one moving with the bar. This observer sees everywhere the electric field $\mathbf{E}' = \mathbf{E} + \mathbf{E}^i = \mathbf{E} + \mathbf{v} \times \mathbf{B}$, and knows that inside the bar

$$\int \mathbf{E}' \cdot d\mathbf{r} = \int \mathbf{E} \cdot d\mathbf{r} + \int (\mathbf{v} \times \mathbf{B}) \cdot d\mathbf{r} = 0$$

However, the moving observer sees the U-shaped section of the loop moving with velocity $\mathbf{v}' = -\mathbf{v}$. Thus just as the rest observer found an emf vBl because of the moving bar, so our moving observer will find an emf $|vBl|$ due to the moving loop. The sense of current flow will be the same as was found above. Again, consistent results are obtained for the two observers.

Example 11.7 Faraday Disk—Induction Generators (Motors)

The example of the Faraday disk shown in Fig. 11.8 is basic to an understanding of induction generators (or motors). Here a conducting disk is made to rotate with a constant angular frequency ω about its central axis. If the disk is placed in a (uniform) magnetic field $\mathbf{B}$ that is perpendicular to its plane, then the Lorentz force on the charge carriers in the disk will cause

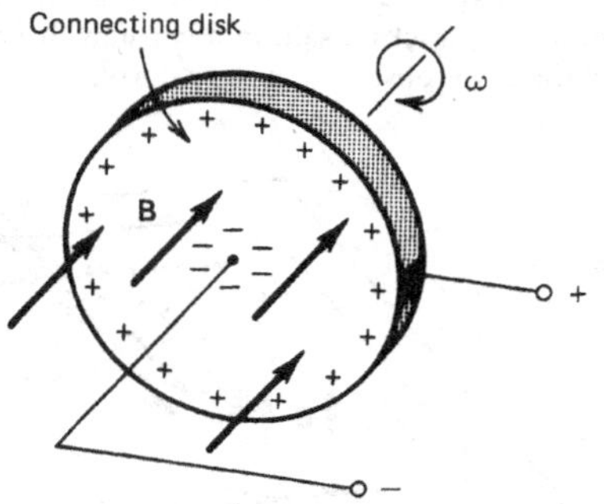

Figure 11.8 Schematic representation of Faraday's disk, which is basic to the understanding of induction generators (or motors).

a charge separation to occur. After equilibrium is attained, the electric field produced by the charge separation will balance the Lorentz fields at all points in the disk. Thus, $\mathbf{E} + \mathbf{v} \times \mathbf{B} = 0$. Since $\mathbf{v} = \boldsymbol{\omega} \times \boldsymbol{\rho} = \hat{\boldsymbol{\phi}}\omega\rho$, we have $\mathbf{E} = -\omega\rho\hat{\boldsymbol{\phi}} \times \mathbf{B}$, or $\mathbf{E} = -\omega B\boldsymbol{\rho}$ inside the disk. As shown in Fig. 11.8, positive charge accumulates in the outer edge of the disk and negative charge toward the center. Thus, a constant potential difference exists between the rim of the disk and its axis

$$\Delta\Phi = \Phi(R) - \Phi(0) = -\int_0^R \mathbf{E}\cdot d\mathbf{r}$$

which, upon using the above expression for $\mathbf{E}$, gives

$$\Delta\Phi = -\int_0^R -\omega B\boldsymbol{\rho}\cdot d\boldsymbol{\rho} = \frac{\omega BR^2}{2}$$

If one now attaches a wire between the + and − terminals, a current will flow. The emf associated with the Faraday disk is given by Eq. (11.21); that is,

$$\mathscr{E} = \int_C \mathbf{E}\cdot d\mathbf{r} + \int_C \mathbf{v} \times \mathbf{B}\cdot d\mathbf{r} = 0 + \frac{\omega BR^2}{2}$$

where the loop C passes through the disk along a radius and back outside the disk. This shows that the induced electromotive force is identical to the potential difference. A disk spinning at 1000 Hz, with a radius of 0.2 m, placed in a field of 1000 G would give rise to an emf of approximately 12 V.

If, instead of mechanically rotating the disk, we caused a current to flow from the axis to the rim, the disk would be set into rotation. The Faraday disk generator would then become a Faraday disk motor.

It should be observed that in calculating the emf associated with the Faraday disk, we have not employed the relation $\mathscr{E} = -dF/dt$, because we do not have here a case of a loop moving through a magnetic field, and thus there seems to be no well-defined area to discuss when calculating dF/dt.

11.3 Application of Faraday's Law to Circuits: Coefficients of Inductance

Electric circuits always consist of closed loops. Thus, if we are to account for the electrical behavior of such circuits, we must include the effects associated with Faraday's law, which, for closed loops, means that we must take all induced emf's into account.

Suppose, for example, that we have a loop in a changing magnetic field. This field must itself be produced by currents presumably flowing in other loops. A very simple situation exists if we consider only two loops, C_1 and C_2, with currents I_1 and I_2 flowing in these loops. Consider surface S_1 associated with loop C_1. If the flux of **B** through S_1 changes for any reason, then an emf will be induced about C_1. This flux, F_1, may change either because the magnetic field acting at points of S_1 is changing or because of loop C_1, and so the surface S_1 is changing. Moreover, it may be split into two parts: the flux F_{12} through C_1 due to the current I_2 around C_2, and the flux F_{11}, through C_1 due to the current I_1;

$$F_1 = F_{11} + F_{12} \tag{11.24}$$

Similarly, for the total flux linking C_2,

$$F_2 = F_{21} + F_{22} \tag{11.25}$$

Now F_{11} and F_{21} are proportional to I_1, and F_{12} and F_{22} are proportional to I_2, because the magnetic fields are directly proportional to the currents. We write this proportionality as follows:

$$F_{11} = L_{11}I_1 \qquad F_{21} = L_{21}I_1 \qquad F_{22} = L_{22}I_2 \qquad F_{12} = L_{12}I_2 \tag{11.26}$$

where the L's are called "coefficients of inductance." In particular, L_{11} and L_{22} are coefficients of self-inductance and L_{12} and L_{21} coefficients of mutual inductance. We remark that the self-inductances depend only upon the geometrical properties of the individual loops, whereas the mutual inductances depend upon the geometrical properties of both loops or circuits. Below, we discuss the properties of these coefficients. We will discuss the mutual inductance before we discuss the self-inductance.

11.3.1 Mutual Inductance

We now prove an important property of the mutual inductance: $L_{12} = L_{21} \equiv M$. Consider Fig. 11.9. The flux F_{12} is written in terms of the magnetic field produced by loop C_2 at the site of loop C_1, or $\mathbf{B}_{12}$, as follows:

$$F_{12} = \int_{S_1} \mathbf{B}_{12} \cdot d\mathbf{a}_1 \tag{11.27}$$

Using Eq. (8.46) one can write F_{12} in terms of the corresponding vector potential, $\mathbf{A}_{12}$:

$$F_{12} = \oint_{C_1} \mathbf{A}_{12} \cdot d\mathbf{l}_1 \tag{11.28}$$

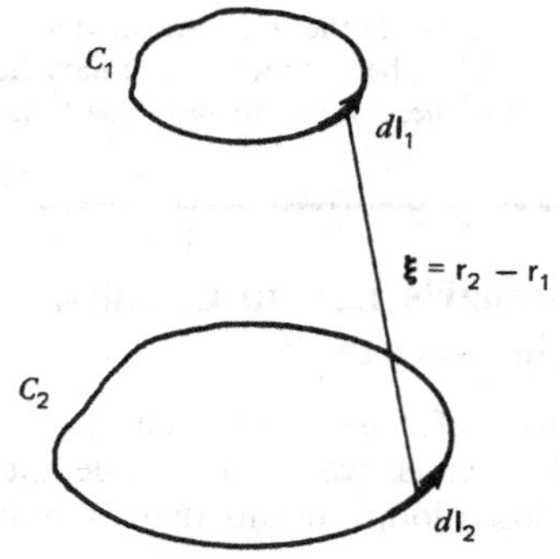

Figure 11.9 A schematic diagram of two loops to aid in the derivation of the mutual inductance (Neumann's formula) between any two loops.

We next write the vector potential in terms of the parameters of the circuits by using Eq. (8.44), as follows:

$$F_{12} = \oint_{C_1} \left[\frac{\mu_0}{4\pi} \oint_{C_2} \frac{I_2\, d\mathbf{l}_2}{|\mathbf{r}_1 - \mathbf{r}_2|} \right] \cdot d\mathbf{l}_1$$

or

$$F_{12} = \frac{\mu_0}{4\pi} I_2 \oint_{C_1} \oint_{C_2} \frac{d\mathbf{l}_2 \cdot d\mathbf{l}_1}{|\mathbf{r}_1 - \mathbf{r}_2|} \equiv L_{12} I_2$$

Therefore,

$$L_{12} = \frac{\mu_0}{4\pi} \oint_{C_1} \oint_{C_2} \frac{d\mathbf{l}_2 \cdot d\mathbf{l}_1}{|\mathbf{r}_1 - \mathbf{r}_2|} \qquad |\mathbf{r}_1 - \mathbf{r}_2| \neq 0$$

Similarly,

$$L_{21} = \frac{\mu_0}{4\pi} \oint_{C_2} \oint_{C_1} \frac{d\mathbf{l}_1 \cdot d\mathbf{l}_2}{|\mathbf{r}_2 - \mathbf{r}_1|}$$

and since $|\mathbf{r}_1 - \mathbf{r}_2| = |\mathbf{r}_2 - \mathbf{r}_1|$, and the order of integration may be interchanged [the integrals are well-behaved (continuous and bounded)], $L_{12} = L_{21} = M$. The inductance M is simply called the *mutual inductance* of the loops, and it may in fact be calculated via the formula obtained, called *Neumann's formula*:

$$L_{12} = L_{21} = M = \frac{\mu_0}{4\pi} \oint_{C_1} \oint_{C_2} \frac{d\mathbf{l}_1 \cdot d\mathbf{l}_2}{|\mathbf{r}_1 - \mathbf{r}_2|} \tag{11.29}$$

Using Faraday's law, one can write the emf's $\mathscr{E}_1$ and $\mathscr{E}_2$ for the two loops in terms of these inductance coefficients. We write

$$\mathscr{E}_1 = -\frac{dF_{11}}{dt} - \frac{dF_{12}}{dt} = \mathscr{E}_{11} + \mathscr{E}_{12} \tag{11.30}$$

and

$$\mathscr{E}_2 = -\frac{dF_{21}}{dt} - \frac{dF_{22}}{dt} = \mathscr{E}_{21} + \mathscr{E}_{22} \tag{11.31}$$

where $\mathscr{E}_{ij}$ is defined as the emf induced in loop i due to a current in loop j. Using Eq. (11.26) to write F_{ij} explicitly in terms of the currents, and the inductances of loops that are both rigid and fixed gives

$$\mathscr{E}_1 = -L_1 \frac{dI_1}{dt} - M \frac{dI_2}{dt} \qquad \mathscr{E}_2 = -M \frac{dI_1}{dt} - L_2 \frac{dI_2}{dt} \tag{11.32}$$

where we have contracted the subscripts so that $L_{11} \equiv L_1$, $L_{22} \equiv L_2$, and $L_{12} \equiv L_{21} = M$. It is seen that the unit of inductance is volts × seconds/amperes, which is called the "henry" (abbreviated H).

For fixed geometry, the relations $\mathscr{E}_{11} = -L_1\, dI_1/dt$, $\mathscr{E}_{12} = -M\, dI_2/dt$, and so on can be used to define the inductance coefficients. The minus signs here simply indicate that the induced emf's oppose the change in currents.

Calculations of inductance are complex for all but a few simple geometrical arrangements. We shall defer until a later time calculations of self-inductance coefficients. Here we shall simply illustrate methods used in calculating the mutual inductance coefficients, methods that follow directly from the definitions:

$$M \equiv \frac{F_{12}}{I_2} = \frac{F_{21}}{I_1} \qquad \text{or} \qquad M = \frac{-\mathscr{E}_{12}}{dI_2/dt} = \frac{-\mathscr{E}_{21}}{dI_2/dt} \tag{11.33}$$

Many common configurations have mutual inductance coefficients listed in handbooks of electrical engineering or physics.

Example 11.8 Mutual Inductance of a Loop Around a Toroid or Solenoid

Consider a toroid of cross-sectional area S_2 and n_l windings per meter, as shown in Fig. 11.10. We assume that it has an essentially constant field B_{12} in its core. If we have a loop of cross-sectional area S_1 encircling the toroid, then the flux through the loop due to the toroid is the flux in the toroid itself. Thus, using Eqs. (11.27) and (8.35), we get

$$F_{12} \approx B_{12} S_2 = \mu_0 n_l I_2 S_2$$

The mutual inductance can now be calculated using F_{12} and Eq. (11.33), as follows:

$$M \equiv \frac{F_{12}}{I_2} = \mu_0 n_l S_2 \tag{11.34}$$

It is interesting that M does not here depend upon the geometry of loop 1 so long as it encompasses the toroid. Thus a very long, straight wire on the axis of the toroid has the same mutual inductance. (In fact, such a wire would in practice have a "return" path and so would constitute a loop.)

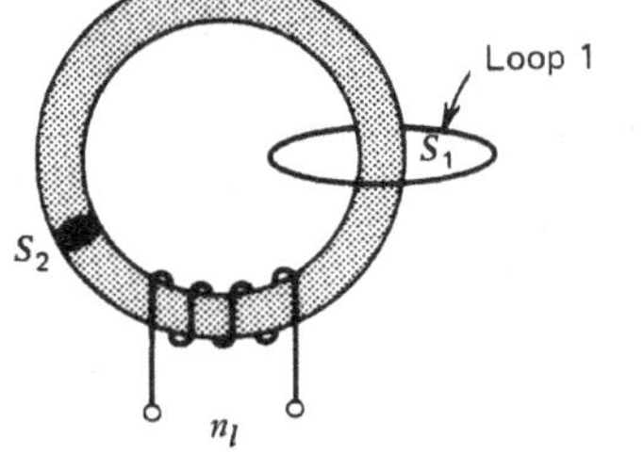

Figure 11.10 A loop around a toroid of a ferromagnetic core.

Note here that the toroid is not a single loop of current. Yet we still employ the definitions given, since circuit 2 creates a flux that "links" loop 1. It is as if we calculated the mutual inductances between the loop 1 and all the individual loops constituting the toroid, and then added them together. (We neglect here the pitch of the toroid windings.) Similarly, if we have N loops in circuit 1 rather than one loop, we can find the mutual inductance between the "circuits" by adding up the mutual inductances between all the pairs of loops in each circuit. Thus, if circuit 1 has N loops, each loop having the same mutual inductance with the toroid circuit, we find

$$M_{(\text{total})} = NM_{(\text{one loop})} = \mu_0 n_l N S_2 \tag{11.35}$$

for the mutual inductance between the circuits.

One should also note that though $M = L_{12} = L_{21}$, it would have been quite difficult to have calculated L_{21}. We have chosen the simpler mutual inductance coefficient to calculate, knowing both are equal.

If we have a solenoid (see Fig. 11.11) whose length is much greater than its diameter, then for loops placed near the center of the solenoid, the results of the toroid are valid. Thus, for a single loop encircling the solenoid near its center, $M = \mu_0 n_l S_2$, where S_2 is the cross-sectional area of the solenoid.

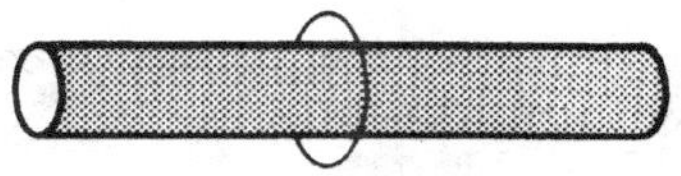

Figure 11.11 A loop around a long solenoid.

Note, however, that if the loop were placed at one end of the solenoid, where the magnetic field is only approximately $\frac{1}{2}\mu_0 n_l I_2$ (see Example 8.13), then the mutual inductance also would be one-half the value given above. Thus if one has two long solenoids, one surrounding the other, and of the same approximate length, the mutual inductance will lie between the values $\mu_0 n_l N S_2$ and $\mu_0 n_l N S_2/2$, where N is the number of turns on the outer solenoid.

Similar arrangements are frequently used to "couple" circuits together magnetically—i.e., inductively. (See the later discussion of transformers given in Chapter 13.) Note also that if the magnetic flux could be increased by insertion of a ferromagnetic material, the coefficient of inductance would be correspondingly increased.

Example 11.9 Mutual Inductance of Two Circular, Coaxial Loops

One might think that the simple geometry of two single loop coils (see Figs. 11.12*a* and 11.12*b*) whose planes are parallel and whose axes are coincident would be amenable to a mutual-inductance calculation. As we shall see, however, this is not the case. Nonetheless, because it illustrates the use of the Neumann formula, we shall consider this problem in detail.

We take the radii of loops 1 and 2 to be R_1 and R_2 and the distance between their centers to be h. We also take the currents flowing in them to be I_1 and I_2, respectively. In order to apply Neumann's formula, we choose a cylindrical coordinate system whose origin is located at the center of loop 1 and whose axis is along the axis of the loops. Consider two differential elements $d\mathbf{l}_1$ and $d\mathbf{l}_2$ of circuits 1 and 2 and coordinates $(R_1, \phi_1, 0)$ and (R_2, ϕ_2, h), respectively. The elements can be written as

$$d\mathbf{l}_1 = R_1\, d\phi_1 \hat{\boldsymbol{\phi}}_1 \qquad d\mathbf{l}_2 = R_2\, d\phi_2\, \hat{\boldsymbol{\phi}}_2 \qquad d\mathbf{l}_1 \cdot d\mathbf{l}_2 = R_1 R_2\, d\phi_1\, d\phi_2\, \hat{\boldsymbol{\phi}}_1 \cdot \hat{\boldsymbol{\phi}}_2$$

The distance between the elements $|\mathbf{r}_2 - \mathbf{r}_1|$ is written in terms of their coordinates as

$$|\mathbf{r}_2 - \mathbf{r}_1| = |h\hat{\mathbf{z}} + R_2\hat{\boldsymbol{\rho}}_2 - R_1\hat{\boldsymbol{\rho}}_1| = [(h^2 + R_1^2 + R_2^2 - 2R_1R_2\hat{\boldsymbol{\rho}}_1 \cdot \hat{\boldsymbol{\rho}}_2)]^{1/2}$$

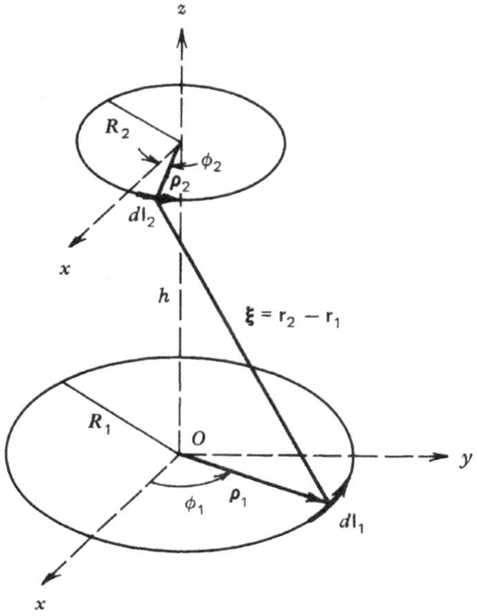

(*a*)

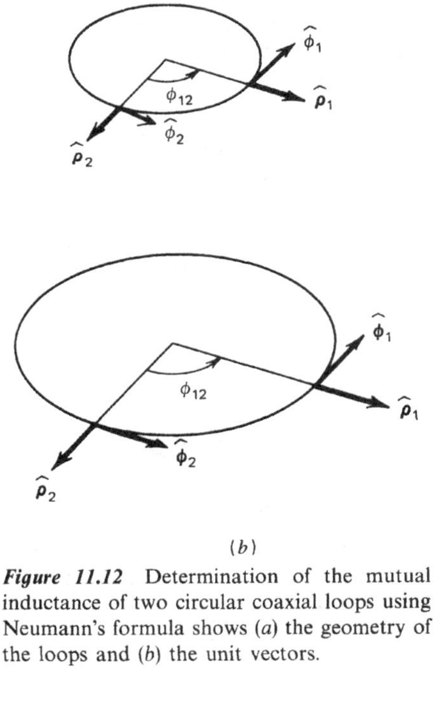

(*b*)

Figure 11.12 Determination of the mutual inductance of two circular coaxial loops using Neumann's formula shows (*a*) the geometry of the loops and (*b*) the unit vectors.

Now $\hat{\boldsymbol{\phi}}_1 \cdot \hat{\boldsymbol{\phi}}_2 = \cos\phi_{12} = \hat{\boldsymbol{\rho}}_1 \cdot \hat{\boldsymbol{\rho}}_2$, where $\phi_{12} = \phi_1 - \phi_2$ (see Fig. 11.12*b*). Therefore, substituting these expressions in Neumann's formula [Eq. (11.29)] gives

$$M = \frac{\mu_0}{4\pi} R_1 R_2 \int_0^{2\pi}\int_0^{2\pi} \frac{d\phi_2\, d\phi_{12} \cos\phi_{12}}{\sqrt{h^2 + R_1^2 + R_2^2 - 2R_1R_2\cos\phi_{12}}} \tag{11.36}$$

where we have changed our variables of integration from (ϕ_1, ϕ_2) to (ϕ_2, ϕ_{12}). The integration over ϕ_2 gives just 2π, so

$$M = \frac{\mu_0 R_1 R_2}{2} \int_0^{2\pi} \frac{d\phi_{12} \cos\phi_{12}}{\sqrt{h^2 + R_1^2 + R_2^2 - 2R_1R_2\cos\phi_{12}}}. \tag{11.37}$$

This integral may be written in terms of elliptic integrals:

$$K(k) \equiv \int_0^{\pi/2} \frac{d\phi}{\sqrt{1 - k^2\sin^2\phi}} \qquad \text{("first kind")}$$

$$E(k) \equiv \int_0^{\pi/2} d\phi\sqrt{1 - k^2\sin^2\phi} \qquad \text{("second kind")} \tag{11.38}$$

where $k^2 = 4R_1R_2/(h^2 + (R_1 + R_2)^2)$. In terms of these integrals, which are functions of k, the mutual inductance is:

$$M = \mu_0\sqrt{R_1R_2}\left[-\left(\frac{2}{k}E + \frac{2}{k} - k\right)\right]K(k) \tag{11.39}$$

11.3.2 Self-Inductance—Inductances in Series and in Parallel

In attempting to calculate the coefficient of self-inductance, $L_{11} = F_{11}/I$, where I is the current through a filamentary loop, a difficulty is encountered—namely, that F_{11} becomes infinite for the case when a finite current is supposed to pass through the filament. The reason is that the magnetic field very close to the filament varies inversely with the distance from the filament, and so becomes infinite as that distance approaches zero.

Actually, wires of finite surface area or cross section carry the currents around the loops, so these infinities do not have physical significance. The magnetic fields produced are everywhere finite. The only possible trouble in this case is to decide on the area through which to calculate the flux. Often, the so-called "external" self-inductance is calculated. In this case, the area chosen is external to the wire—for example, in the shaded region of Fig. 11.13. This will usually underestimate the self-inductance by a small factor (if the wire's transverse dimensions are small compared with the dimensions of the external area). In fact, if the current is assumed to flow

Figure 11.13 The "external" self-inductance of a loop is often calculated by calculating the flux through the area external to the wire (shaded area) to avoid possible divergences at the wire itself.

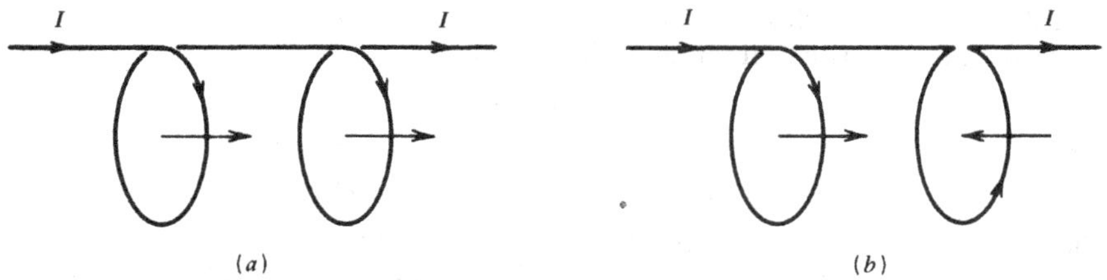

Figure 11.14 Two current loops connected in series. (*a*) Fluxes enforce each other and (*b*) fluxes opposite to each other.

on the surface of the wire, the external inductance can equal the inductance since the interior **B** field can then be zero (if the wire is thin and "kinkless").

The exact calculation of self-inductance may therefore be difficult. It is often calculated more simply and unambiguously using energy considerations than by employing its definition in terms of flux and will thus be largely deferred until such considerations are made (see Chapter 12). However, there are some simple considerations that merit discussion, and we shall describe these here. These include the connection of inductances in series and in parallel.

Consider first the "self-inductance" of *loops in series*. A schematic diagram of two loops is shown in Fig. 11.14 and of the corresponding conventional circuit is shown in Fig. 11.15a. (We ignore the flux through the loops due to the straight sections between the loops, etc.) Therefore, the fluxes through them are given by Eqs. (11.24) and (11.25).

$$F_1 = F_{11} + F_{12} \qquad F_2 = F_{21} + F_{22}$$

The flux F_1 through loop 1 has two components, one due to the current flowing around loop 1 and the other due to the current flowing around loop 2. The latter

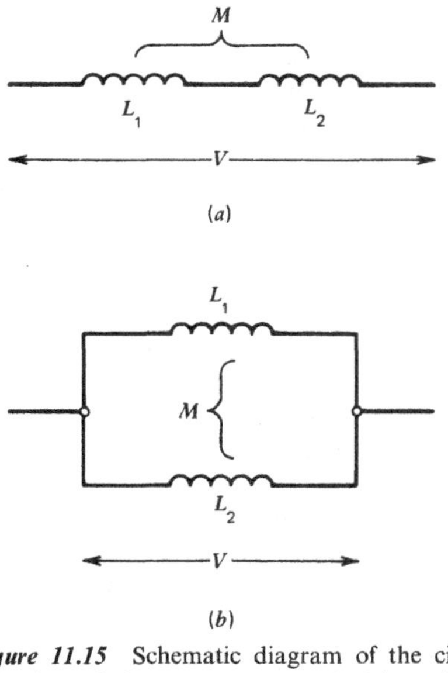

Figure 11.15 Schematic diagram of the circuit of two inductances connected (*a*) in series and (*b*) in parallel.

contributes a "mutual flux," F_{21} to F_1. Using Eq. (11.33), the various components of the fluxes can be written in terms of the currents:

$$F_{11} \equiv L_1 I \qquad F_{12} = \pm MI = F_{21} \qquad F_{22} = L_2 I$$

Thus

$$F_1 = (L_1 \pm M)I \qquad F_2 = (L_2 \pm M)I$$

The plus sign is used when the currents traverse the loops "in the same sense," that is, when the "mutual flux" reinforces the "self-flux" as shown in Fig. 11.14*a*. Otherwise, the minus sign is appropriate as in Fig. 11.14*b*. Now, the induced emf's around these two loops, $\mathscr{E}_1$ and $\mathscr{E}_2$, are given by Eq. (11.3)

$$\mathscr{E}_1 = -\frac{dF_1}{dt} \qquad \mathscr{E}_2 = -\frac{dF_2}{dt}$$

and the total emf is given by

$$\mathscr{E} = \mathscr{E}_1 + \mathscr{E}_2 = -\frac{d}{dt}(F_1 + F_2)$$

We may thus define the "total flux linkage" for the two loops as $F_1 + F_2$, and define a self-inductance L for the whole circuit (of two loops) by

$$F = \sum_{i=1,2} F_i \equiv LI = (L_1 + L_2 \pm 2M)I \tag{11.40}$$

$$L = L_1 + L_2 \pm 2M \tag{11.41}$$

In fact, for N loops, one uses a similar procedure such that

$$F = \sum_{i=1}^{N} F_i \equiv LI \tag{11.42}$$

Then, for fixed geometry,

$$\mathscr{E} = -L\frac{dI}{dt} \tag{11.43}$$

as we had for a single loop. It must only be remembered that the fluxes are calculated through areas whose normals, or positive sides, are defined by the direction of current flow in a right-handed-screw sense.

If one considers two loops, it is clear that the mutual flux linking the loops can *only* be less than (or at best equal to) the self-fluxes of the two loops. Thus $F_{12} \le F_{22}$, $F_{21} \le F_{11}$, and so forth. Writing $F_{12} = k_2 F_{22}$, $F_{21} = k_1 F_{11}$, where $k_1, k_2 \le 1$, we have $MI_2 = k_2 L_2 I_2$ and $MI_1 = k_1 L_1 I_1$. Multiplying MI_1 by MI_2 gives $M^2 I_1 I_2 = k_1 k_2 L_1 L_2 I_1 I_2$, implying that

$$M = \sqrt{k_1 k_2 L_1 L_2} \quad \text{or} \quad M = k\sqrt{L_1 L_2} \qquad -1 < k < 1 \tag{11.44}$$

The purely geometrical constant is known as the "coefficient of coupling" of the loops. If k can be varied by varying the geometry, then a variable inductance can be constructed. (In fact, this was a popular way of tuning resonant circuits in the early days of radio.)

If the inductors are now connected in *parallel* as shown in Fig. 11.15*b*, then one expects the effective inductance of the system to be different from that of the series connection [Eq. (11.41)]. One can show in this case (see Problem 11.18) that

$$L = \frac{L_1 L_2 - M^2}{L_1 + L_2 \pm 2M} \tag{11.45}$$

where the sign of M depends on the way in which the inductors are connected.

Example 11.10 Self-Inductance of a Solenoid or Toroid

A solenoid consists of many closely spaced loops. If it is very long, so that end effects are negligible, we know that the average flux through each of its loops is given by $F_j = \mu_0 n_l S I$, where S is the cross-sectional area of a loop, n_l gives the number of turns per meter, and I is the current passing through it. The sum of all the fluxes going through the loops is

$$F = \sum_{j=1}^{N} F_j = \mu_0 n_l S N I$$

The self-inductance of the solenoid is now calculated using Eq. (11.42). Thus

$$L = \mu_0 n_l N S$$

The result of a toroid in which the magnetic field is essentially constant, is exactly the same.

Since $n_l \approx N/l$, it may be noted that L is proportional to N^2. The reason is simply that there is mutual flux between the different loops which multiplies the average flux in a single loop by a factor N. Since there are N loops, the result is proportional to N^2. If there were no mutual flux, the inductance would be proportional to N alone.

Example 11.11 Superposition of Time Variation and Motional EMF

In this example we discuss a conducting loop of area A moving in a magnetic field that itself is varying with time. Consider a region in which the magnetic field $\mathbf{B} = B(t)\hat{\mathbf{z}}$. A planar rectangular conducting loop is in the x-y plane with its center at the origin. At $t = 0$, the loop is set rotating about the x axis with angular velocity ω. We will now calculate the induced emf using two methods:

$$\mathscr{E} = -\frac{dF}{dt} \qquad \text{and} \qquad \mathscr{E} = -\int \frac{\partial \mathbf{B}}{\partial t} \cdot d\mathbf{a} + \oint (\mathbf{v} \times \mathbf{B}) \cdot d\mathbf{l}$$

In the first method, one calculates the flux passing through the loop at time t: $F = B(t) A \cos \omega t$. Thus

$$\mathscr{E} = -\frac{d}{dt}[A B(t) \cos \omega t]$$

or

$$\mathscr{E} = -(A \cos \omega t)\frac{dB}{dt} + BA\omega \sin \omega t$$

In the second method, we have to evaluate two integrals. The first one results from the time variation of the B field:

$$\mathscr{E}_1 = \int -\frac{\partial \mathbf{B}}{\partial t} \cdot d\mathbf{a} = -\int \frac{dB}{dt} \hat{\mathbf{z}} \cdot \hat{\mathbf{n}}\, da = -A \cos \omega t \frac{dB}{dt}$$

The second contribution to $\mathscr{E}$ is the motional emf. The velocity of a point on the loop is $\mathbf{v} = r\omega\hat{\mathbf{n}} = y\omega\hat{\mathbf{n}}/\cos \omega t$. Thus

$$\mathbf{v} \times \mathbf{B} = y\omega B \hat{\mathbf{n}} \times \frac{\hat{\mathbf{z}}}{\cos \omega t} = -y\omega B \tan \omega t\, \hat{\mathbf{x}}$$

Therefore

$$\mathscr{E}_2 = \oint (\mathbf{v} \times \mathbf{B}) \cdot d\mathbf{l} = -\omega B \tan \omega t \oint y\hat{\mathbf{x}} \cdot d\mathbf{l}$$

Since $\nabla \times (y\hat{\mathbf{x}}) = -\hat{\mathbf{z}}$, then Stokes' theorem gives

$$\oint y\hat{\mathbf{x}} \cdot d\mathbf{l} = \int_A \nabla \times (y\hat{\mathbf{x}}) \cdot d\mathbf{a} = -\int \hat{\mathbf{z}} \cdot \hat{\mathbf{n}}\, da = -A \cos \omega t$$

and thus $\mathscr{E}_2 = BA\omega \sin \omega t$. It is apparent that $\mathscr{E}_1$ and $\mathscr{E}_2$ are exactly the same as the two terms arrived at in the first method.

11.4 Summary

The independent subjects of electrostatics and magnetostatics are coupled together via the time variation of the magnetic field. Faraday's experimental law states that the rate of change with respect to time of the magnetic flux through an area S produces an induced emf, and hence an induced electric field, at the perimeter of the area C; that is,

$$\mathscr{E} = \oint_C \mathbf{E} \cdot d\mathbf{l} = -\frac{d}{dt}\int_S \mathbf{B} \cdot d\mathbf{a} = -\frac{dF}{dt} \tag{11.3)–(11.5}$$

The minus sign is a statement of Lenz' law. This integral law for a stationary rigid area S can be changed to the differential law

$$\nabla \times \mathbf{E} = -\frac{\partial \mathbf{B}}{\partial t} \tag{11.7}$$

which shows that in the presence of a varying magnetic field, the total electric field is not a conservative field. This differential equation along with the two divergence equations are three of Maxwell's equations which always hold. Substituting $\mathbf{B} = \nabla \times \mathbf{A}$ in Ampere's differential law gives

$$\nabla \times \mathbf{E} = -\frac{\partial}{\partial t} \nabla \times \mathbf{A} \qquad \text{or} \qquad \nabla \times \left[\mathbf{E} + \frac{\partial \mathbf{A}}{\partial t}\right] = 0$$

which implies that

$$\mathbf{E} = -\nabla\Phi - \frac{\partial \mathbf{A}}{\partial t} \tag{11.10}$$

The contribution $\mathbf{E}^q = -\nabla\Phi$ is a conservative field where as $\mathbf{E}^i = -\partial\mathbf{A}/\partial t$ is the nonconservative.

The magnetic flux through a nonrigid or/and nonstationary area S can vary with time even if the magnetic field is not a function of time. Faraday's law is still applicable here, and the induced emf, called the motional emf, is

$$\mathscr{E} = -\frac{dF}{dt} \tag{11.23}$$

In order to account for the electrical behavior of electric circuits that always consists of closed loops, we must include the effects associated with Faraday's law and hence for all induced emf. It is convenient for this purpose to relate the flux through each loop directly in terms of the current through all of them:

$$F_i = \sum_{j=1}^{N} L_{ij} I_j$$

The coefficients L_{ij} are geometrical coefficients that are independent of the currents and the fluxes. For $j = i$, $L_{ii} = L_i$ depends on the self-geometry of the individual loops and is called the coefficient of self-inductance, and the coefficient $L_{ij} = M_{ij}$ for $i \neq j$ depends on the relative geometry of a pair and is called coefficient of mutual inductance. In terms of these coefficients, Faraday's law becomes

$$\mathscr{E}_i = -\frac{dF_i}{dt} = -\sum_j L_{ij}\frac{dI_j}{dt}$$

The mutual inductance of two loops C_1 and C_2 is given by Neumann's formula

$$L_{12} = M = \frac{\mu_0}{4\pi}\oint_{C_1}\oint_{C_2} \frac{d\mathbf{l}_1 \cdot d\mathbf{l}_2}{|\mathbf{r}_1 - \mathbf{r}_2|} \tag{11.29}$$

where $\mathbf{r}_1 = \mathbf{r}_2$ is the distance between elements $d\mathbf{l}_1$ and $d\mathbf{l}_2$ of the two loops. In fact, this formula shows an important property of the coefficients, $L_{12} = L_{21}$.

When two loops of L_1, L_2, mutual inductance M, and negligible resistance are connected in series and in parallel we have, for the effective inductance L_{eff},

$$L_{\text{eff}} = L_1 + L_2 \pm 2M \qquad \text{(series)} \tag{11.41}$$

$$L_{\text{eff}} = \frac{L_1 L_2 - M^2}{L_1 + L_2 \pm 2M} \qquad \text{(parallel)} \tag{11.45}$$

If M is negligible, then this shows that L_1 and L_2 combine like resistors. In general, $M = k\sqrt{L_1 L_2}$ where $|k| \leq 1$ is the coefficient of coupling.

Problems

11.1 The magnetic field in the region $\rho \leq \rho_0 = 0.1$ m is increasing at a rate of $0.1\hat{\mathbf{z}}$ T/s. Determine the magnitude of the electric field at any radius ρ. Plot E as a function of ρ.

11.2 A long solenoid has 100 turns per centimeter and a diameter of 3.0 cm. A thin coil having 20 turns and a diameter of 2 cm is placed inside the solenoid such that their axes are parallel. The current in the solenoid $I = 3t + 2t^2$ where I is in amperes and t in seconds. Determine the induced emf in the coil. What is the instantaneous current in the coil at $t = 2$ s if its resistance is 0.15 Ω? Neglect the flux produced by the induced current.

11.3 A conducting disk of radius a, thickness δ, and conductivity σ_c is placed in a cylindrically symmetric $\mathbf{B}$ field: $\mathbf{B} = B_0(t)\hat{\mathbf{z}}$ for $0 \leq \rho \leq R$, and $\mathbf{B} = 0$ for $\rho > R$ $(R < a)$ such that its axis is along the z axis, and its center is at the origin. (a) Determine the vector potential $\mathbf{A}$ associated with $\mathbf{B}$ in all regions. (b) Determine the induced electric field in all regions. (c) Determine the current density inside the disk. (d) Show that the total power dissipated in the disk is

$$P = \frac{\pi\delta\sigma_c}{8} R^4 \left(\frac{dB_0}{dt}\right)^2 \left(1 + 4\ln\frac{a}{R}\right)$$

(If the field completely covered the disk, the logarithmic term would be zero, and the power would be proportional to the square of the disk area and to the square of dB_0/dt. This is a general property of such "eddy" currents. See Example 11.3).

11.4 A semicircular piece of wire, shown in Fig. 11.16, travels with constant velocity $\mathbf{v} = v_0\hat{\mathbf{x}}$ in a constant magnetic field $\mathbf{B} = B_0\hat{\mathbf{z}}$. What is the emf induced between the ends of the wire?

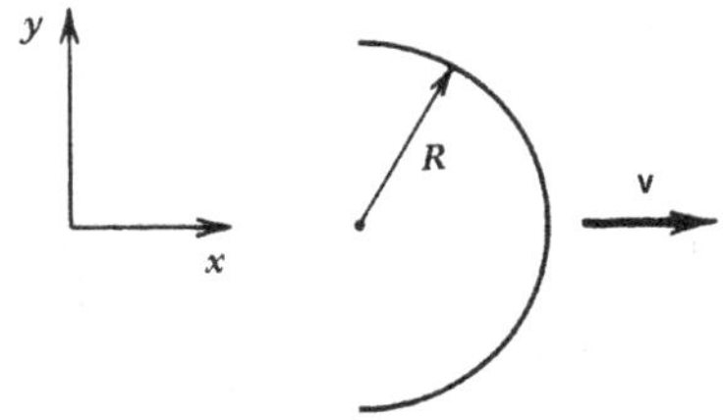

Figure 11.16

11.5 A rectangular wire of width a and length b is moving parallel to its width at a uniform speed v. A very long wire, carrying current I, stationary, coplanar with the rectangle, and parallel to its length is at distance l from the nearest long side. (a) Determine the emf induced in the rectangle. (b) Determine the mutual inductance of the circuits.

11.6 Two thin vertical conductors of length 50 cm are parallel to the z axis and at distances $\rho_1 = 3$ cm, and $\rho_2 = 5$ cm from it. A resistive wire connects the two tops and the two bottoms, thus making a rectangular loop of resistance $R = 0.2\,\Omega$. The loop rotates about the z axis at $\omega = 2\pi \times 500$ rad/min (keeping the same geometry), in a nonuniform magnetic field: $\mathbf{B}_1 = 0.25\hat{\boldsymbol{\rho}}T$ at ρ_1 and $B_2 = 0.8\hat{\boldsymbol{\rho}}T$ at ρ_2. (a) Determine the current in the loop. (b) Determine the electric power generated in the loop. (c) Determine the rate at which mechanical work is done on the loop, and compare with the result in (b).

11.7 Consider the bar moving on the stationary conducting tracks shown in Fig. 11.7. We take the B field here to be normal to the plane of the tracks and to depend on time: $B = 0.3 \sin \omega t(T)$, where $\omega = 10^4$ rad/s. (a) Determine the induced voltage when the bar is stationary and the loop area is 25 cm^2. (b) Repeat for the case where the bar is moving with uniform speed $v = 1.5 \times 10^4$ cm/s and at an instant of time when the area is 25 cm^2. Take the area to be zero at $t = 0$. Take $l = 5$ cm.

11.8 Two infinitely long, parallel, fixed wires are located at $x = \pm a$ and are parallel to the y axis. The wires carry constant currents $\pm I$, respectively. Another infinitely long parallel wire is moving with its instantaneous position at distances r_1 and r_2 from the wires at $x = a$ and $x = -a$, respectively. (a) Use the Neumann formula to show that the mutual inductance between the circuit composed of the two fixed wires and the circuit composed of a unit length of the moving wire is $L = \mu_0 \log(r_2/r_1)/2\pi$. (b) Calculate the induced emf per unit length in the moving wire when it is at the origin if its velocity there is $\mathbf{v} = v_0\hat{\mathbf{x}}$.

11.9 A loop of wire of area A_1, with its center at the origin and carrying a constant current I_1, is rotated with constant angular velocity ω about the z axis. At $t = 0$, the loop is in the y-z plane. Find the current induced in a stationary loop at a large distance r on the y axis. The second loop is fixed in the y-z plane and has area A_2 and resistance R.

11.10 Determine the inductance of an N turn toroid of square cross section of inner and outer radii ρ_1 and ρ_2 and thickness a.

11.11 A toroidal coil has N turns, a mean radius b, and a circular cross section of radius a. Show that its inductance per unit length is $L = \mu_0 N^2(b - \sqrt{b^2 - a^2})$.

11.12 The space between two coaxial conducting cylindrical shells of inner and outer radii a and b is filled with a material of permeability μ. Find the self-inductance of the system per unit length.

11.13 A conducting wire of radius a is surrounded by a thin coaxial conducting cylindrical shell of radius. The wire and the shell have magnetic permeability μ_0. The space between them is filled with a material of permeability μ. Determine the self-inductance per unit length of the line.

11.14 Consider a very long cylindrical conductor of radius a parallel to a very large ground plane. The distance between them is d. Show that the mutual inductance per unit length is $L/l = \mu_0 \cosh^{-1}(d/a)/2\pi$ and $\mu_0 \ln(2d/a)/2\pi$ if $d \gg a$.

11.15 Consider two identical parallel conductors of radius a and separation d. (a) Show that the inductance per unit length is $L/l = \mu_0 \cosh^{-1}(d/2a)/\pi$. (b) Calculate the inductance per unit length when $d = 25$ feet and $a = 0.8$ inch.

11.16 A circular loop of radius a, resistance R, and inductance L is rotating in a magnetic field $H = H_0 \sin \omega t$, about a diameter that is normal to H. (a) Determine the current in the loop. (b) Find the retarding torque. (c) Calculate the average power necessary to maintain the rotation.

11.17 Determine the mutual inductance of the two loops of Example 11.10 when the two loops (a) are far away from each other ($h \gg R_1, R_2$) and (b) have about the same radius and are close together ($R_1 \approx R_2 \approx R$ and $h \ll R$).

11.18 Two inductors L_1 and L_2 with a mutual inductance M and negligible resistance are connected in parallel. If the total current through them is I, determine (a) the voltage across them, and (b) the effective inductance of the system. See Eq. 11.45.

TWELVE

MAGNETIC ENERGY

In Chapter 6 we calculated the electric energy stored in charge distributions and in polarized dielectric materials. The same energy was shown to be alternatively expressible in terms of energy stored in the corresponding electric field.

In this chapter we consider the magnetic case: We calculate energy stored in current distributions and in magnetized materials. We will also show that this energy can be accounted for in terms of the energy stored in the corresponding magnetic field. Moreover, we will show that the energy formulation is very convenient for calculating forces and torques between different elements of the current distributions.

12.1 A Current Loop Immersed in a Linear Magnetic Material

Let us consider a closed loop. Between $t = 0$ and $t = t$ a current I is established in the loop with the aid of an external source. In establishing such currents (fields) the external source does electric work that gets stored as magnetic energy. If the material the loop is immersed in is a linear magnetic material and there are no losses to heat in the loop, this energy may be recovered when the current is switched off.

The energy stored can be calculated with the aid of Faraday's electromotive force law [see Eq. (11.3)]. We assume that the circuit has negligible resistance such that we can neglect energy losses to heat (I^2R). The rate of *electric work*, dW/dt, in the circuit caused by the induced electromotive force is

$$\frac{dW}{dt} = \mathscr{E} I \tag{12.1}$$

where $\mathscr{E}$ and I are the induced instantaneous emf and current in the loop. But, from Eq. (11.3), $\mathscr{E} = -dF/dt$, where F is the flux passing through the loop; therefore

$$dW = -I\,dF \tag{12.2}$$

the corresponding rate of work done by external sources (batteries, for example) is $dW^{(b)}/dt = -\mathscr{E}I$; hence

$$dW^{(b)} = I\,dF \tag{12.3}$$

Alternatively, dW can be written in terms of L and I, where L is the self-inductance. From Eq. (11.43), $\mathscr{E} = -L(dI/dt)$; thus Eq. (12.1) becomes

$$\frac{dW}{dt} = -LI\frac{dI}{dt}$$

or, in differential form,

$$dW = -LI\,dI \tag{12.4}$$

This can be easily integrated from 0 to the final current I: $W = -\frac{1}{2}LI^2$. Thus the magnetic energy stored in the system, $U = -W$ is

$$U = \frac{1}{2}LI^2 \tag{12.5}$$

It is useful to express this energy in terms of the total flux passing through the loop. For a single loop $L = dF/dI$, which is equal to F/I since F is linear with the current; thus, Eq. (12.5) becomes

$$U = \frac{1}{2}IF \tag{12.6}$$

12.2 *N* Loops Immersed in a Linear Magnetic Medium

In this section we generalize the above results for a single loop to N coupled loops. Again, we assume the resistances of the given loops to be negligible, and thus the energy losses to heat can be neglected. The currents in the circuits are raised from zero to some final values. At a given instant of time the induced electromotive force in the mth circuit, $\mathscr{E}_m$ is given by

$$\mathscr{E}_m = -\sum_{k \neq m}^{N} M_{mk}\frac{dI_k}{dt} - L_m\frac{dI_m}{dt} \tag{12.7}$$

where M_{mk} is the mutual inductance between the mth and kth circuits and I_m and I_k are the currents in them, respectively. Thus the rate of work by the induced emf in the mth circuit is

$$\frac{dW_m}{dt} = -\sum_{k \neq m}^{N} M_{mk}I_m\frac{dI_k}{dt} - L_mI_m\frac{dI_m}{dt} \tag{12.8}$$

or

$$dW_m = -\sum_{k \neq m}^{N} M_{mk}I_m\,dI_k - L_mI_m\,dI_m \tag{12.9}$$

The total differential work done in all circuits is calculated by summing Eq. (12.9) over m, as follows:

$$dW = -\sum_{\substack{m,k=1 \\ m \neq k}}^{N} M_{mk}I_m\,dI_k - \sum_{m}^{N} L_mI_m\,dI_m \tag{12.10}$$

We now write $M_{mk} I_m \, dI_k + M_{km} I_k \, dI_m = M_{mk} \, d(I_k I_m)$, where $m < k$. Thus, Eq. (12.10) becomes

$$dW = -\sum_{\substack{m,k \\ m<k}}^{N} M_{mk} \, d(I_m I_k) - \sum_{m}^{N} L_m I_m \, dI_m \tag{12.11}$$

which can be easily integrated from 0 to I_m and I_k:

$$W = -\sum_{\substack{m,k \\ m<k}}^{N} M_{mk} I_m I_k - \frac{1}{2} \sum_{m}^{N} L_m I_m^2 \tag{12.12}$$

Therefore, the total magnetic energy stored in the system $U = -W$ is

$$U = \sum_{\substack{m,k \\ m<k}}^{N} M_{mk} I_m I_k + \frac{1}{2} \sum_{m}^{N} L_m I_m^2 \tag{12.13}$$

To express the energy of the N circuits in terms of the flux passing through them, we first rewrite Eq. (12.13) as follows:

$$U = \frac{1}{2} \sum_{m \neq k}^{N} M_{mk} I_m I_k + \frac{1}{2} \sum_{m}^{N} L_m I_m^2$$

or

$$U = \frac{1}{2} \sum_{m,k}^{N} M_{mk} I_m I_k \tag{12.14}$$

where $M_{mk} = L_m$ for $m = k$. The inductance M_{mk} is defined in terms of the flux passing through the mth circuit and caused by the kth circuit, F_{mk}; that is, $M_{mk} = dF_{mk}/dI_k$, which gives for linear systems $M_{mk} = F_{mk}/I_k$. The total flux passing through the mth loop; therefore

$$F_m = \sum_{k=1}^{N} F_{mk} = \sum_{k=1}^{N} M_{mk} I_k \tag{12.15}$$

Substituting this result in Eq. (12.14) gives

$$U = \frac{1}{2} \sum_{m=1}^{N} I_m F_m \tag{12.16}$$

For a single circuit, Eq. (12.16) gives the previous result of Eq. (12.6).

12.3 Energy Stored in a Magnetic Field in the Presence of Linear Materials

In the previous sections the magnetic energy of current circuits was expressed in terms of the currents, self-inductances, and mutual inductances of the circuits. Also, it was expressed in terms of the currents and the total magnetic fluxes passing through the circuits. In this section we give yet another way of expressing the energy, by expressing it in terms of the associated magnetic field.

In order to do this we use a more general concept of a "current circuit," which is not necessarily defined by wires but can be defined by a line of current (line of force of the corresponding electric field). The magnetic energy given by Eq. (12.16) will now be transformed to an expression in terms of the associated magnetic field and the permeability of the medium. We can express the flux passing through the circuits in terms of the local magnetic field via Eq. (8.45); however, for the transformation, it is more convenient to express it in terms of the local magnetic vector potential **A**, given by Eq. (8.46). Upon substitution of Eq. (8.46) in Eq. (12.16), we get:

$$U = \frac{1}{2}\sum_m I_m \oint_{C_m} \mathbf{A}\cdot d\mathbf{l}_m = \frac{1}{2}\sum_m \oint_{C_m} I_m \mathbf{A}\cdot d\mathbf{l}_m \tag{12.17}$$

where C_m is the boundary of the mth "circuit" and $d\mathbf{l}_m$ is a differential displacement along C_m. The product $I_m\,d\mathbf{l}_m$ can be changed to $\mathbf{J}\,dv$ and the sum over the different line integrals can be changed to an integration over the volume

$$\sum_m \oint_{C_m} = \int_V$$

That is,

$$U = \frac{1}{2}\int_V \mathbf{J}\cdot\mathbf{A}\,dv \tag{12.18}$$

The current density can now be written in terms of the magnetic intensity $\mathbf{J} = \nabla \times \mathbf{H}$; that is,

$$U = \frac{1}{2}\int_V (\nabla \times \mathbf{H})\cdot\mathbf{A}\,dv \tag{12.19}$$

Using the vector identity given by Eq. (1.60), $\nabla\cdot(\mathbf{A}\times\mathbf{H}) = (\nabla\times\mathbf{A})\cdot\mathbf{H} - (\nabla\times\mathbf{H})\cdot\mathbf{A}$, Eq. (12.19) becomes

$$U = \frac{1}{2}\int_V \mathbf{H}\cdot(\nabla\times\mathbf{A})dv - \frac{1}{2}\int_V \nabla\cdot(\mathbf{A}\times\mathbf{H})dv \tag{12.20}$$

The second integral of this equation becomes a surface integral when the divergence theorem is applied to it. Moreover, using $\mathbf{B} = \nabla\times\mathbf{A}$, U becomes

$$U = \frac{1}{2}\int_V \mathbf{H}\cdot\mathbf{B}\,dv - \frac{1}{2}\oint_S (\mathbf{A}\times\mathbf{H})\cdot d\mathbf{a} \tag{12.21}$$

where S is the surface that encloses the volume V. We would like now to choose S and hence V such that the surface integral vanishes. If the current distributions are bounded—that is, if they do not extend to infinity—it is possible to take V very large and hence S becomes very large, so all points on it are at very large distances from the distributions. The dominating radial dependence of the vector potential and the magnetic fields at the surface can then be taken of the form $1/r$ and $1/r^2$, respectively, where r is the distance from the points at the surface to the points in the current distributions. The other radial dependences, if any, fall off as r^{-m}, where m is an integer ≥ 3 in the case of the H and > 2 in the case of the potential. In any case the integrand $\mathbf{A}\times\mathbf{H}$ falls off as $1/r^3$ or faster and because the area element da goes as r^2, the surface integral vanishes as r goes to infinity. Therefore, Eq. (12.21) becomes

$$U = \frac{1}{2}\int \mathbf{H}\cdot\mathbf{B}\,dv \qquad \text{(all space)} \tag{12.22}$$

where V is now the volume of all space (infinite volume). For linear isotropic media of permeability μ such as being considered in this section we have $\mathbf{B} = \mu\mathbf{H}$ and

$$u = \frac{1}{2}\mu H^2 = \frac{1}{2}\frac{B^2}{\mu} \tag{12.23}$$

for the energy density which depends on the final values of the fields.

It is instructive to express the energy density in terms of $\mathbf{H}$ and the magnetization $\mathbf{M}$. Substituting $\mathbf{B} = \mu_0\mathbf{H} + \mu_0\mathbf{M}$ in Eq. (12.22), we get

$$u = \frac{1}{2}\mu_0 H^2 + \frac{1}{2}\mu_0\mathbf{H}\cdot\mathbf{M} \tag{12.24}$$

The first term of u is interpreted as the energy density of the magnetic field in vacuum. The second term is therefore interpreted as the energy density stored in the material itself. This form of writing the energy density will be very useful in the case of nonlinear materials (to be discussed in the next section).

Example 12.1 The Magnetic Energy of Two Filamentary Currents

The *total* magnetic energy of two long filamentary currents can be evaluated using Eq. (12.18): $U = \frac{1}{2}\int \mathbf{J}\cdot\mathbf{A}\,dv$ where $\mathbf{A}$ is the total vector potential and the volume includes both currents. The *interaction* energy $U^{(\text{int})}$, on the other hand, is given by (using the filamentary approximation)

$$U^{(\text{int})} = \int I_2\mathbf{A}_1\cdot d\mathbf{l}_2 = \int I_1\mathbf{A}_2\cdot d\mathbf{l}_1$$

(See the discussion in Section 6.6.) Consider the currents to be along the z axis—one located at the origin and the other at $\rho = \rho_0$. The vector potential at $\rho = \rho_0$ due to the current at $\rho = 0$ is

$$\mathbf{A} = \frac{\mu_0 I_1\hat{\mathbf{z}}}{2\pi}\ln\rho_0$$

Thus

$$U = \frac{\mu_0 l I_1 I_2}{2\pi}\ln\rho_0$$

where l is the length of a section of the second filament. Thus, the interaction energy per unit length, is

$$u = \frac{1}{2\pi}\mu_0 I_1 I_2\ln\rho_0 + C \tag{12.25}$$

where C is a constant.

Example 12.2 The Magnetic Energy of a Magnetic Dipole in a Uniform B Field

Current Loop. In Section 8.8.4 we discussed the forces, torques, and energy of a current loop in an external magnetic field. It was remarked that the result $U = -\mathbf{m}\cdot\mathbf{B}$ is only a part of the magnetic energy in a uniform B that is associated with the torque exerted by $\mathbf{B}$ on the distribution. The other part, which we did not calculate, is associated with the work done by the external source that maintains the current in the loop when it is introduced in the field.

In this example we derive the total result using the methods developed in this chapter. Consider a long solenoid of n_l turns per unit length, carrying current I, whose axis is along the z axis as shown in Fig. 12.1. A small loop, carrying a *constant* current I_1, and of an area A is placed inside the solenoid.

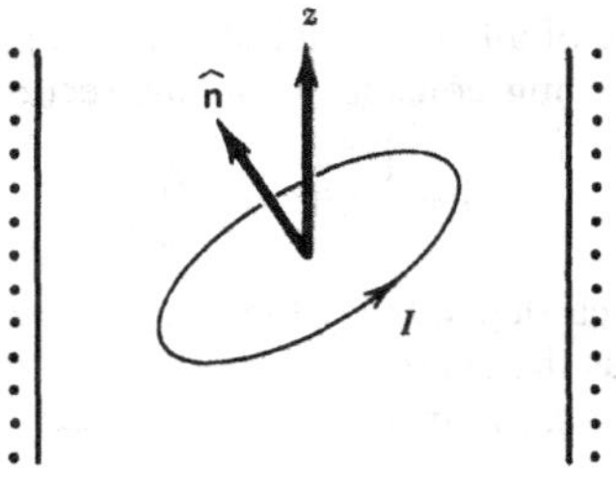

Figure 12.1 Magnetic energy of a current-carrying loop inside a long solenoid.

The magnetic field produced by the solenoid is uniform and parallel to its axis; it is equal to $\mathbf{B} = \mu_0 n_l I \hat{\mathbf{z}}$ (see Example 8.8). The magnetic flux passing through the loop is $F_1 = \mu_0 n_l I A|\hat{\mathbf{z}}\cdot\hat{\mathbf{n}}|$, where $\hat{\mathbf{n}}$ is a unit vector normal to the loop.

The mutual inductance of the system is determined by taking the derivative of F with respect to I:

$$M = \frac{dF}{dI} = \mu_0 n_l A|\hat{\mathbf{z}}\cdot\hat{\mathbf{n}}|$$

Using Eq. (12.14), the interaction energy of the loop with the solenoid is

$$U = MI_1 I = \mu_0 n_l A I_1 I|\hat{\mathbf{z}}\cdot\hat{\mathbf{n}}|$$

which can be rewritten in terms of the magnetic dipole moment of the loop, $\mathbf{m} = I_1 A\hat{\mathbf{n}}$, and the magnetic field $\mathbf{B}$:

$$U = +\mathbf{m}\cdot\mathbf{B} \qquad \text{(current loop)} \tag{12.26}$$

This result shows that the total magnetic energy when the current is *maintained* is the negative of the energy associated with the torque, $W^{(m)} = -\mathbf{m}\cdot\mathbf{B}$, which tells us that the batteries maintaining the current must have done the work

$$W^{(b)} = 2\mathbf{m}\cdot\mathbf{B}$$

This can be understood as follows. When a loop connected to a constant current source I is introduced in a magnetic field and the magnetic flux increases in it by dF, the battery, according to Eq. (12.2), does the work

$$W^{(b)} = I\,dF$$

if it is to maintain the current I. The magnetic energy $U = \frac{1}{2}IF$, on the other hand, changes due to the increase in the flux according to Eq. (12.6) by the amount

$$dU = \frac{1}{2}I\,dF$$

Thus

$$W^{(b)} = 2\,dU$$

But

$$-W^{(\text{mech})} + W^{(b)} = dU$$

which gives $W^{(b)} = 2\mathbf{m}\cdot\mathbf{B}$.

Permanent Atomic Dipole (or Very Small Magnet). We now consider a magnetic dipole moment that is associated with the spin of a magnetic atom. Since there is no real current in this case, then the interaction energy of this dipole with a uniform external magnetic field is associated only with the torque exerted on the dipole. Thus the magnetic energy in this case is

$$U = -\mathbf{m}\cdot\mathbf{B} \qquad \text{(atomic permanent dipole)} \tag{12.27}$$

Example 12.3 The Magnetic Energy of Two Magnetic Dipoles

In this example we consider the magnetic interaction of two permanent atomic magnetic dipoles (or two very small magnets). We take the dipoles sufficiently small compared to the distance between them that the variation of the magnetic field produced by one of them over the other can be neglected. Thus the interaction energy is: $U = -\mathbf{m}_1 \cdot \mathbf{B}_2$, where $\mathbf{m}_1$ is the magnetic moment of the first dipole and $\mathbf{B}_2$ is the magnetic field produced by the second dipole at the site of the first.

The magnetic field produced by the second dipole of moment $\mathbf{m}_2$ is given by Eq. (8.98); that is,

$$\mathbf{B}_2 = \frac{\mu_0}{4\pi}\left[-\frac{\mathbf{m}_2}{r^3} + 3\frac{(\mathbf{m}_2 \cdot \mathbf{r})}{r^5}\mathbf{r}\right]$$

where $\mathbf{r}$ is the distance of the point of observation to the second dipole. Taking the point of observation at the position of the first one gives

$$U = \frac{\mu_0}{4\pi}\left[\frac{\mathbf{m}_1 \cdot \mathbf{m}_2}{r^3} - 3\frac{(\mathbf{m}_1 \cdot \mathbf{r})(\mathbf{m}_2 \cdot \mathbf{r})}{r^5}\right] \tag{12.28}$$

12.4 Magnetic Energy in Nonlinear Materials

In linear systems such as the cases considered so far in this chapter the work needed to establish a magnetic field depends only on the final value of the magnetic field. This implies that these systems are reversible, meaning that the energy consumed in establishing the magnetic system can be recovered as the field is switched off. In this section we take on the case of nonlinear materials where this is not true because hysteresis plays an important role. The irreversible changes in the domain configurations that are responsible for the hysteresis cause energy losses in the form of heat.

Consider a circuit in the form of a solenoid that has N current turns and negligible resistance, and completely filled with a ferromagnetic material, as shown in Fig. 12.2. Because of the large air gap, the flux lines will not be confined to a single, well-defined path. Each flux line, however, may be viewed as a magnetic circuit, and all of them are connected in parallel. Each line C_k is characterized by length $\mathbf{l}_k$ and area A_k, which may be a function of $\mathbf{l}_k$. The work done by an *external source* $\Delta W^{(b)}$ in a time interval Δt at time t and current I is given by means of Eq. (12.3); that is,

$$\Delta W^{(b)} = -\sum_k \mathscr{E}_k I \,\Delta t \tag{12.29}$$

where $\mathscr{E}_k$ is the induced emf associated with C_k. Now one substitutes for $\mathscr{E}_k$ in terms of the flux I_k along the curve C_k, or $\mathscr{E}_k = -dF_k/dt$; therefore Eq. (12.29) becomes

$$\Delta W^{(b)} = \sum_k I \,\Delta F_k \tag{12.30}$$

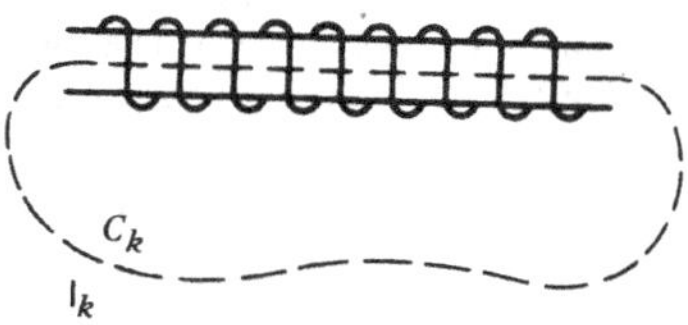

Figure 12.2 A solenoid with a ferromagnetic core, showing one of the flux lines. Each flux line may be viewed as a magnetic circuit, and all of them are connected in parallel.

The magnetic intensity **H** at time t can be related to I by using Ampere's law $\oint_{C_k} \mathbf{H} \cdot d\mathbf{l}_k = NI$. Therefore Eq. (12.30) becomes

$$\Delta W^{(b)} = \frac{1}{N} \sum_k \oint_{C_k} \mathbf{H} \cdot d\mathbf{l}_k \, \Delta F_k \tag{12.31}$$

Since the magnetic flux is continuous along each circuit [see Eq. (9.53)], we write $\Delta F_k = NA_k \, \Delta B$, where A_k is the cross section of the circuit at the interval $d\mathbf{l}_k$. Therefore

$$\Delta W^{(b)} = \sum_k \oint_{C_k} \mathbf{H} \cdot d\mathbf{l}_k A_k \, \Delta B. \tag{12.32}$$

Because $\mathbf{l}_k$ is along the flux line, the quantity $d\mathbf{l}_k \, \Delta B$ can be written as $dl_k \, \Delta\mathbf{B}$. Moreover, we can replace

$$\sum_k \oint_{C_k} A_k \, dl_k$$

by $\int dv$, where v is the volume containing all flux lines. Therefore

$$\Delta W^{(b)} = \int_v \mathbf{H} \cdot \Delta\mathbf{B} \, dv \tag{12.33}$$

The magnetic energy density in an increment $d\mathbf{B}$ per unit volume is then given by

$$dW^{(b)} = \mathbf{H} \cdot d\mathbf{B} \tag{12.34}$$

This expression represents the energy required to change the magnetic field from **B** to $\mathbf{B} + d\mathbf{B}$. Writing $\mathbf{B} = \mu_0 \mathbf{H} + \mu_0 \mathbf{M}$ in this expression gives

$$dW^{(b)} = \mu_0 H \, dH + \mu_0 \mathbf{H} \cdot d\mathbf{M} \tag{12.35}$$

which indicates that $dW^{(b)}$ represents the work necessary to establish the magnetic field from **H** to $\mathbf{H} + d\mathbf{H}$ and to magnetize the material from **M** to $\mathbf{M} + d\mathbf{M}$. The total energy required to establish the field from 0 to H_0 and **M** from 0 to $\mathbf{M}(H_0)$ per unit volume of the sample is determined by integrating Eq. (12.35) or Eq. (12.34):

$$W^{(b)} = \int_0^{B_0} \mathbf{H} \cdot d\mathbf{B} = \frac{1}{2} \mu_0 H_0^2 + \mu_0 \int_0^{H_0} \mathbf{H} \cdot d\mathbf{M} \tag{12.36}$$

Example 12.4 Hysteresis Loss

Consider a ferromagnetic material whose hysteresis curve is shown in Fig. 12.3. The material is initially in a magnetic field $-\mathbf{H}_{\max}$. The magnitude of the field is gradually decreased to zero, reversed in polarity, and gradually increased to $H_{\max}$. The opposite of this procedure is then performed such that the field is returned to its initial condition.

Since the system is returned to its initial magnetic state (that is, to the same H and B values), the overall change in its magnetic energy is zero. The total work done by the external source per unit volume of the material, however, is not zero; it is equal to

$$W^{(b)} = \oint_C \mathbf{H} \cdot d\mathbf{B} = \oint_C \mathbf{H} \cdot d\mathbf{M} \tag{12.37}$$

where C is the hysteresis loop. The integral can be easily calculated; it is the area enclosed by the hysteresis loop. It is now apparent that it takes more energy to produce the magnetization than is returned when the magnetization is reduced. The energy lost is called the hysteresis loss; it goes into heat.

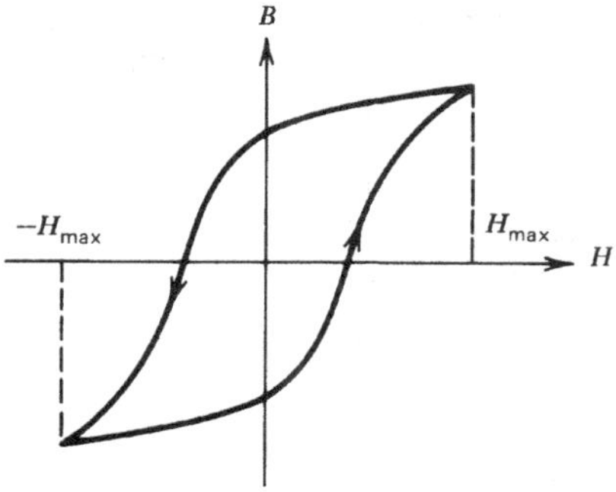

Figure 12.3 A hysteresis curve of a ferromagnetic material.

Example 12.5 Magnetic Cooling

Because we have seen in Section 12.3 and in this one that it is possible to do work on a magnetic sample by varying the magnetic field applied to it, it is possible to change the temperature of the sample (heat it or cool it). This principle has been used to provide a means for attaining very low temperatures. The magnetic cooling method can be easily described as follows. The sample is first placed in thermal contact with liquid helium at about 1 K. The contact is achieved by heat conduction via a low pressure of helium gas. The helium liquid at this temperature constitutes a heat bath at temperature T_i. In a second step, a magnetic field of intensity H_i is switched on. As a result, the sample becomes magnetized and work is done. Because the sample is in thermal contact with the heat bath, it gives the produced heat to the bath; hence it remains at temperature T_i. The helium gas, which maintains the thermal contact with the heat bath, is now removed, resulting in the thermal isolation of the sample. Finally, the magnetic field is reduced quasi-statically to a final value H_f, which results in a reduction in the temperature of the sample to $T_f < T_i$. This *adiabatic demagnetization* process can cool the sample to as low as 0.01 K. In fact, further elaborations of this process have produced temperatures as low as 10^{-6} K.

12.5 Forces and Torques Using the Magnetostatic Energy

We have found in the previous sections that work must be done against the induced electromotive forces in order to establish some given currents in a set of rigid, fixed current circuits. This work has to be supplied by the external source being used such as batteries. We have also previously found that current circuits produce magnetic fields and hence exert forces on each other. These magnetic forces become more involved if the rigid circuits were allowed to move during the process. As a result of the motion, mechanical work would be done by these forces. In this section we show how the magnetic-energy formalism discussed in the previous sections can be conveniently used in determining such forces.

Consider a system of rigid circuits. Let one circuit make a virtual rigid displacement $d\mathbf{r}$ while the currents in all of the circuits are kept constant by the external sources. Moreover, consider only static situations where the circuits are stationary (no kinetic energies are involved) such that there is no heating or cooling of the circuits. As a result of the virtual displacement, a mechanical work $dW^{(\mathrm{mech})}$ is done by the magnetic forces, and an electrical work $dW^{(b)}$ is done by the batteries against the induced electromotive forces to maintain the currents in the circuits. The sum of these two effects is equal to the change in the magnetic energy of the system:

$$-dW^{(\text{mech})} + dW^{(b)} = dU \tag{12.38}$$

If the change in the flux passing through circuit i is called dF_i, then using Eq. (12.3) we write $dW^{(b)} = \Sigma_i I_i\, dF_i$. On the other hand, the corresponding change in the magnetic energy is given by Eq. (12.16); that is,

$$dU = d\left(\frac{1}{2}\sum_i I_i F_i\right) = \frac{1}{2}\sum_i I_i\, dF_i \tag{12.39}$$

which indicates that $dW^{(b)} = 2\, dU$. Thus Eq. (12.38) becomes

$$dW^{(\text{mech})} = dU \qquad \text{(constant currents)} \tag{12.40}$$

The mechanical work $dW^{(\text{mech})}$ can now be written in terms of the magnetic force $\mathbf{F}$ acting on the circuit in question, as follows:

$$dW^{(\text{mech})} = \mathbf{F}\cdot d\mathbf{r} \tag{12.41}$$

Thus, Eqs. (12.40) and (12.41) give

$$dU = \mathbf{F}\cdot d\mathbf{r} \tag{12.42}$$

which implies the existence of a magnetic energy U such that

$$\mathbf{F} = \nabla U \qquad \text{(constant currents)} \tag{12.43}$$

The various components of $\mathbf{F}$ follow from this result:

$$F_\xi = \left.\frac{\partial U}{\partial \xi}\right|_I \tag{12.44}$$

where F_ξ is the force acting on the circuit in the $\boldsymbol{\xi}$ direction, and the vertical bar with its subscript I is inserted to emphasize the constancy of I in all the circuits in taking this derivative.

Another physical situation arises when the circuits are isolated from the external sources (batteries). In this case, a virtual rigid movement of one of the circuits results in a change in the currents in all of them. However, according to Lenz' law, the amount of change in the induced current due to the induced emf is such that the magnetic flux passing through the circuits stays the same. Therefore, we take $dW^{(b)} = 0$ in Eq. (12.38); thus

$$dU = -dW^{(\text{mech})} \qquad \text{(constant flux } F\text{)} \tag{12.45}$$

It is interesting to note that in this case the change in the magnetostatic energy is the negative of the mechanical work, whereas in the case of constant currents [Eq. (12.40)] the change is equal to the mechanical work. Again, in this case, $dW^{(\text{mech})}$ can be written in terms of the magnetic force $\mathbf{F}$ acting on the circuit and in terms of the differential displacement $d\mathbf{r}$; that is, $dW^{(\text{mech})} = \mathbf{F}\cdot d\mathbf{r}$. Thus

$$\mathbf{F} = -\nabla U \qquad \text{(constant flux)} \tag{12.46}$$

and the various components of $\mathbf{F}$ are

$$F_\xi = -\left.\frac{\partial U}{\partial \xi}\right|_{\text{flux } F} \tag{12.47}$$

where F_ξ is the component along the $\boldsymbol{\xi}$ direction, and the differentiation is carried out keeping the flux in all the circuits constant.

Finally, the same procedure followed above can be used to find the magnetic torques acting on the circuits. If, instead of a virtual translation, the circuit is allowed to have a rigid virtual rotation $d\theta$, then

$$\tau_\theta = \frac{\partial U}{\partial \theta}\bigg|_I \tag{12.48}$$

gives the torque in the increasing θ direction in the case of constant currents, and

$$\tau_\theta = -\frac{\partial U}{\partial \theta}\bigg|_F \tag{12.49}$$

gives the torque in the increasing θ direction in the case of constant fluxes.

Example 12.6 Force Exerted by a Solenoid on a Magnetic Slab

This example is analogous to the problem of finding the force exerted on a dielectric slab partially inserted in the field of a capacitor (Example 6.13). Consider a solenoid of cross-sectional area A, N turns, and length l. The solenoid is connected to an external source that sets up a constant current I through it. A rod of a magnetic material of constant permeability μ and cross-sectional area A is partially inserted in the solenoid while keeping their axes parallel to each other, as shown in Fig. 12.4.

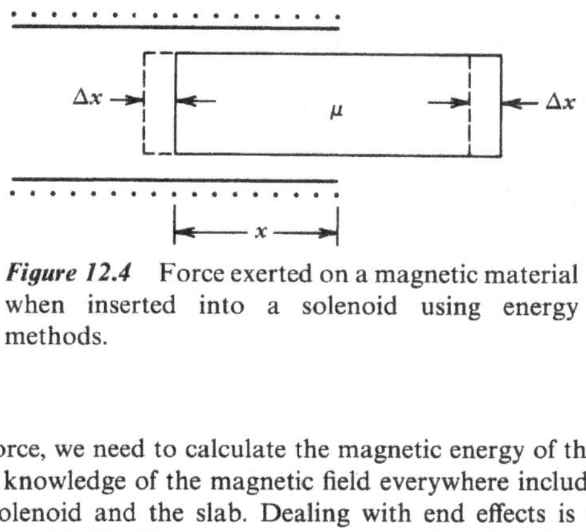

Figure 12.4 Force exerted on a magnetic material when inserted into a solenoid using energy methods.

To calculate the force, we need to calculate the magnetic energy of the system as a function of x. This requires a knowledge of the magnetic field everywhere including the neighborhood of the ends of the solenoid and the slab. Dealing with end effects is not simple since they involve very complicated fields. However, an approximation that simplifies the solution tremendously can be used. We assume that when the slab is slightly moved by a distance Δx from its position, the structure of the field remains the same, the only difference being that a Δx of the slab is effectively transferred from the very outer region to the region well inside the solenoid. With this approximation, the energy difference of the two configurations as a function of Δx can easily be determined:

$$\Delta U = U(x + \Delta x) - U(x) \approx \frac{1}{2}(\mu - \mu_0) \int_V H^2 \, dv \tag{12.50}$$

where $V = A\,\Delta x$ is the change in volume caused by the displacement, and $H = NI/l$ is the magnetic intensity inside the solenoid. The integration can be easily carried out since H is constant; thus, Eq. (12.50) becomes

$$\Delta U \approx \frac{1}{2}(\mu - \mu_0)\left(\frac{NI}{l}\right)^2 A\,\Delta x \tag{12.51}$$

The force on the slab is then equal to

$$F \approx \left.\frac{\Delta U}{\Delta x}\right|_I \approx \frac{1}{2}(\mu - \mu_0)\left(\frac{NI}{l}\right)A \tag{12.52}$$

***Example 12.7 The Force Between a Magnet and a Magnetic Slab—Magnetic Circuits**

Consider the magnetic circuit shown in Fig. 12.5. It consists of a U-shaped arm of a magnetic material of high constant permeability μ and is wound with N turns of wire that carries a constant current I. The cross section of the arm is rectangular, with an area A. A bar of the same material and the same cross section is placed against its poles, thus making the total length of the circuit l.

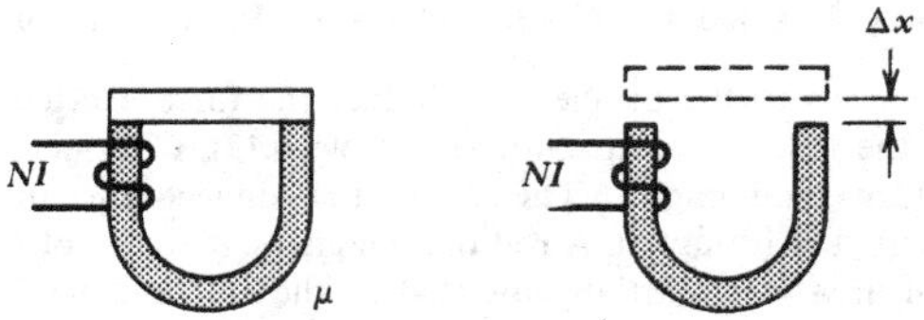

Figure 12.5 Force on a magnetic slab by a U-shape magnet using energy methods.

To determine the force between the bar and U arm, we make a virtual displacement of the bar from the poles. The magnetic field in the various parts of the circuit can be found using Ampere's law and the principle of continuity of flux. We apply Ampere's law to the circuit before the virtual displacement is made $\oint \mathbf{H}_0 \cdot d\mathbf{l} = NI$. Thus

$$H_0 = \frac{NI}{l} \tag{12.53}$$

After the displacement of the bar by an amount $\Delta x \ll l$, Ampere's law gives

$$H_m l + 2H_g \Delta x = NI \tag{12.54}$$

where H_m and H_g are the magnetic fields in the material and in the gap, respectively. The continuity of flux gives another relation between H_m and H_g; that is,

$$\mu H_m = \mu_0 H_g \tag{12.55}$$

Equations (12.54) and (12.55) are now solved simultaneously for H_m and H_g; that is,

$$H_m = \frac{NI}{l + 2\dfrac{\mu}{\mu_0}\Delta x} = H_0\left(1 - 2\frac{\mu}{\mu_0}\frac{\Delta x}{l}\right) \tag{12.56}$$

$$H_g = \frac{\mu}{\mu_0} H_m \tag{12.57}$$

The change in the magnetic energy ΔU is

$$\Delta U = \frac{1}{2}\int_{V_m} \mu H_m^2\, dv + \frac{1}{2}\int_{V_g} \mu_0 H_g^2\, dv - \frac{1}{2}\int_{V_m} \mu H_0^2\, dv \tag{12.58}$$

where V_m and V_g are the volumes of the magnetic material and the gap, respectively. Substituting H_m and H_g from Eqs. (12.56) and (12.57), respectively, gives

$$\Delta U = -\frac{\mu^2}{\mu_0} H_0^2 A\, \Delta x \tag{12.59}$$

Thus the force acting on the bar is

$$F = \frac{\Delta U}{\Delta x}\bigg|_I = -\frac{\mu^2}{\mu_0} H_0^2 A$$

Example 12.8 Force Between a Wire and a Circular Loop Using the Energy Method

In this example we use the energy method to calculate the force between the two current-carrying circuits shown in Fig. 12.6. A current I_1 flows in a circular loop of radius R. An infinite wire carrying a current I_2 is in the plane of the loop and at a distance $d > R$ from the center of the loop.

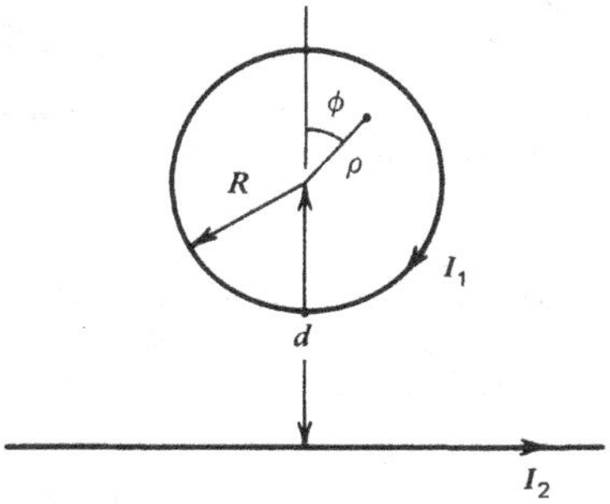

Figure 12.6 Force between a current-carrying wire and a current-carrying loop placed in the same plane.

Since we are interested in the force between the wires, we will calculate only the interaction magnetic energy $U = I_1 F_1$, where F_1 is the flux passing through the loop due to the field of the wire. The flux $F_1 = \int \mathbf{B}_2 \cdot \hat{\mathbf{n}}\, da$, where $\hat{\mathbf{n}}$ is a unit vector normal to the loop and $\mathbf{B}_2$ is the field produced by the wire can be calculated as follows: Consider a differential area $\rho\, d\rho\, d\phi$ located at distance ρ from the center of the loop and hence at a distance $d + \rho \cos\phi$ from the wire and at an angle ϕ with respect to the diameter of the loop normal to the wire. Therefore

$$F_1 = \frac{\mu_0}{2\pi} I_2 \int_0^R \int_0^{2\pi} \frac{\rho\, d\rho\, d\phi}{d + \rho \cos\phi} \tag{12.60}$$

Noting that

$$\int_0^{2\pi} \frac{d\phi}{1 + \beta \cos\phi} = \frac{2\pi}{\sqrt{1 - \beta^2}} \qquad -1 < \beta < 1 \tag{12.61}$$

then

$$F_1 = \mu_0 I_2 \int_0^R \frac{\rho\, d\rho}{\sqrt{d^2 - \rho^2}} \tag{12.62}$$

Integrating over ρ gives

$$F_1 = \mu_0 I_2 (d^2 - \rho^2)^{1/2}\Big|_0^R = \mu_0 I_2[(d^2 - R^2)^{1/2} - d] \tag{12.63}$$

The interaction magnetic energy is therefore

$$U = \mu_0 I_1 I_2[(d^2 - R^2)^{1/2} - d] \tag{12.64}$$

The force between the wire and the loop can now be calculated from U using Eq. (12.44), as follows:

$$\text{Force} = \frac{\partial U}{\partial d}\bigg|_{I_1, I_2} = \mu_0 I_1 I_2 \left[\left(1 - \frac{R^2}{d^2}\right)^{-1/2} - 1\right] \tag{12.65}$$

Example 12.9 Calculation of Self-Inductance Using Energy Considerations

The energy formalism will be used in this example to calculate the self-inductance of two infinitely long wires, shown in Fig. 12.7. The wires, of radii ρ_1 and ρ_2, carry currents I and

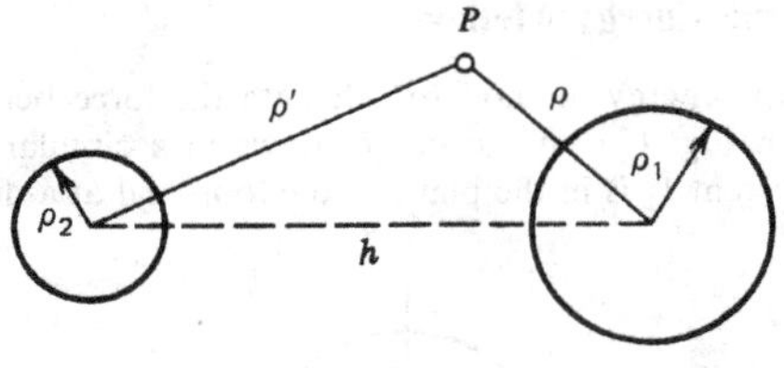

Figure 12.7 Self-inductance of two infinitely long parallel wires using magnetic-energy methods.

$-I$, respectively, and their centers are a distance h apart. The magnetic energy U of the system can be calculated using Eq. (12.18): $U = \frac{1}{2}\int \mathbf{J}\cdot\mathbf{A}\,dv$. The vector potential of a straight wire of finite diameter was obtained in Example 8.10. For wire 1, the vector potential is

$$\mathbf{A}_1 = -\frac{\mu_0}{4\pi} I\left(\frac{\rho}{\rho_1}\right)^2 \hat{\mathbf{z}} \qquad \rho < \rho_1 \tag{12.66}$$

$$\mathbf{A}_1 = -\frac{\mu_0 I}{4\pi}\left(1 + 2\ln\frac{\rho}{\rho_1}\right)\hat{\mathbf{z}} \qquad \rho > \rho_1 \tag{12.67}$$

For wire 2, the vector potential is

$$\mathbf{A}_2 = \frac{\mu_0}{4\pi} I\left(\frac{\rho'}{\rho_2}\right)^2 \hat{\mathbf{z}} \qquad \rho < \rho_2 \tag{12.68}$$

$$\mathbf{A}_2 = \frac{\mu_0 I}{4\pi}\left(1 + 2\ln\frac{\rho'}{\rho_2}\right)\hat{\mathbf{z}} \qquad \rho > \rho_2$$

Taking $J_1 = I/\pi\rho_1^2$ and $J_2 = -I/\pi\rho_2^2$, then U takes the form

$$U = \frac{I}{2\pi\rho_1^2}\int_{S_1}(\mathbf{A}_1 + \mathbf{A}_2)\cdot d\mathbf{a}_1\,dz - \frac{I}{2\pi\rho_2^2}\int_{S_2}(\mathbf{A}_1 + \mathbf{A}_2)\cdot d\mathbf{a}_2\,dz \tag{12.69}$$

where S_1 and S_2 are the cross-sectional areas of wires 1 and 2, respectively. Since the potentials are independent of z and are in the direction of z, then the energy per unit length $u = U/l$ is

$$u = \frac{U}{l} = \frac{I}{2\pi\rho_1^2}\int_{S_1}(A_1 + A_2)da_1 - \frac{1}{2\pi\rho_2^2}\int_{S_2}(A_1 + A_2)da_2 \tag{12.70}$$

The integration gives

$$u = \frac{\mu_0 I^2}{16\pi}\left[1 + 2\ln\left(\frac{h^2}{\rho_1\rho_2}\right)\right] \tag{12.71}$$

Using the definition of the self-inductance per unit length, Eq. (12.5), we find that

$$L = \frac{\mu_0}{8\pi}\left[1 + 2\ln\left(\frac{h^2}{\rho_1\rho_2}\right)\right] \tag{12.72}$$

12.6 Summary

To establish a current I in a loop of inductance L, negligible resistance, and immersed in a linear magnetic material requires an external source, such as a battery, to do work $dW^{(b)}$; then

$$dW^{(b)} = I\,dF \tag{12.3}$$

where dF is the change in flux through the loop. The corresponding work done by the induced emf is just the negative of this work, or

$$dW = -I\,dF \tag{12.2}$$

Taking $dF = L\,dI$ gives

$$dW = -LI\,dI \qquad \text{or} \qquad W = -\frac{1}{2}LI^2$$

The magnetic energy stored in the system is therefore

$$U = -W = W^{(b)} = \frac{1}{2}LI^2 = \frac{1}{2}IF \tag{12.5), (12.6}$$

This result can be generalized to a case of N loops of self-inductances and mutual inductances L_m and M_{mk}, as follows:

$$U = \frac{1}{2}\sum_{m,k}^{N} M_{mk}I_mI_k = \frac{1}{2}\sum_{m=1}^{N} I_mF_m \tag{12.14), (12.6}$$

where $M_{mk} = L_m$ for $m = k$, and F_m is the total flux through the mth loop due to all currents.

Alternatively, the energy of the loops can be written in terms of the magnetic field or the vector potential produced by the loops:

$$U = \frac{1}{2}\int \mathbf{H}\cdot\mathbf{B}\,dv \tag{12.22}$$

$$U = \frac{1}{2}\int I\,d\mathbf{l}\cdot\mathbf{A} = \frac{1}{2}\int \mathbf{J}\cdot\mathbf{A}\,dv \tag{12.18}$$

where the integration over $d\mathbf{l}$ includes a summation over the various loops, and the expression in terms of $\mathbf{J}$ describes a more general current distribution, which is not necessarily filamentary. Thus one can define a density of magnetic energy

$$u = \frac{1}{2}\mathbf{H}\cdot\mathbf{B} = \frac{1}{2}\frac{B^2}{\mu} = \frac{1}{2}\mu H^2 \tag{12.23}$$

$$= \frac{1}{2}\mu_0 H^2 + \frac{1}{2}\mu_0\mathbf{H}\cdot\mathbf{M} \tag{12.24}$$

In the presence of nonlinear magnetic materials, the work necessary to be done by the external source to establish the magnetic field (H_0, B_0) is

$$W^{(b)} = \int_0^{B_0}\mathbf{H}\cdot d\mathbf{B} = \frac{1}{2}\mu_0 H^2 + \mu_0\int_0^{H_0}\mathbf{H}\cdot d\mathbf{M} \tag{12.36}$$

The last term is a hysteresis loss.

In analogy with the electrical case, we can find forces acting on the various elements of a magnetic system from its magnetic energy. For isolated systems, flux stays constant by Lenz' law (if there are no losses due to Joule heating), and hence

$$F_\xi = -\frac{\partial U}{\partial \xi} \qquad \text{(constant flux)} \tag{12.47}$$

For a system whose currents are maintained at certain values by regulated external sources, we have

$$F_\xi = \frac{\partial U}{\partial \xi} \qquad \text{(constant currents)} \tag{12.44}$$

If the ξ represent a rotation, the F_ξ represent a torque.

Problems

12.1 A coil has an inductance of $L = 5$ H and a resistance of 20 Ω. An emf of $\mathscr{E} = 100$ V is applied. (a) What energy is stored in it after the current has built up to its maximum value $\mathscr{E}/R$? (b) What is the flux in the coil?

12.2 A coaxial conductor of inner radius a and outer radius b carries a current $\pm I$. (a) Determine the B field between the conductors and the magnetic energy per unit length. (b) Determine the inductance per unit length.

12.3 A cylindrical conductor of radius a carries a current I, which is distributed uniformly across its cross section. (a) Determine the B field and the stored magnetic energy per unit length in the conductor. (b) Determine the self-inductance (internal) per unit length.

12.4 A toroidal coil of N turns is wound on a nonmagnetic form of square cross section of side a. The inner and outer radii of the toroid are ρ_1 and ρ_2, and it carries a current I. (a) Determine the magnetic field in the toroid using Ampere's law. (b) Determine the stored magnetic energy. (c) Determine the self-inductance of the toroid.

12.5 An inductance is formed of 100 turns of wire wrapped around a closed iron loop 20 cm in length and a cross section of 1 × 1 cm. A 60-Hz alternating current is passed through the coil. The iron goes through the hysteresis loop once each cycle. (a) Find the approximate power loss due to hysteresis. (b) Use the expression for $\oint \mathbf{H} \cdot d\mathbf{l}$ to obtain the peak current in the coil from information given on the hysteresis plot shown in Fig. 12.8. (c) What is the self-inductance of the inductor under the conditions of the problem? (d) If twice the current were passed through the coil, how would this affect the inductance (qualitative)?

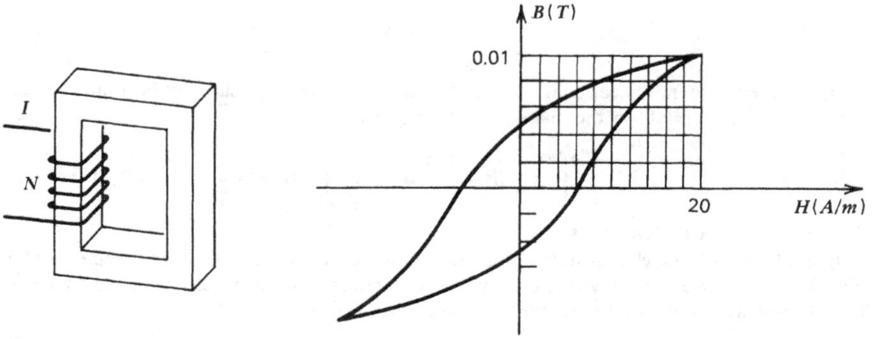

Figure 12.8 A inductor formed from a closed rectangular iron loop with a coil wrapped around it, along with its hysteresis loop.

12.6 The magnetic field B between the poles of an electromagnet is uniform and is held at a constant value $\mathbf{B}_0$ even in the absence or in the presence of magnetic materials (case of constant flux). A thin, paramagnetic rod of susceptibility χ_m is partially inserted in the field, as shown in Fig. 12.9. Determine the force on the rod.

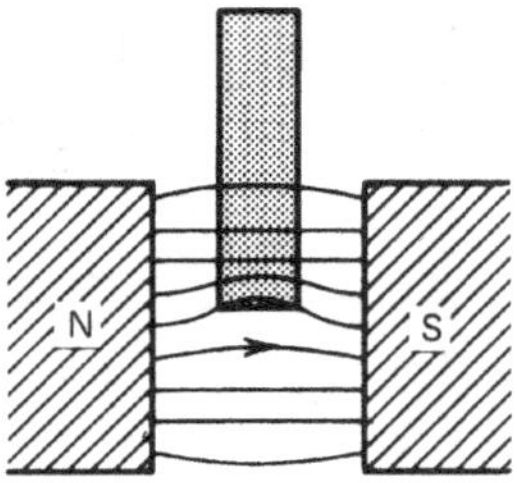

Figure 12.9 The force on a paramagnetic rod when inserted in a region of uniform B field that is held constant between the pole faces of an electromagnet.

12.7 The current I in a very long solenoid of N turns, length l, and radius R is kept constant by a battery. (a) Find the magnetic energy and hence the force on one turn of the winding per unit length of the circumference. (b) Repeat assuming that the flux remains constant instead of the current, and the system is isolated (using superconducting windings).

12.8 Two small coplanar magnets of moments $2\,m_0$ and $3\,m_0$, where m_0 is in $\mathrm{A \cdot m^2}$ are free to turn about their fixed centers. The line joining their centers has a length d and is perpendicular to an external uniform field H. Calculate the energy of the system. Show that a position of equilibrium is one in which their axes are in the direction of H.

12.9 Consider two parallel, infinitely long, straight-line conductors with currents I_1 and I_2 are placed at a distance R from each other. (a) Determine the vector potential of current I_1. (b) Determine the interaction energy between the two currents. (c) Use the interaction energy to find the force between the wires.

12.10 Consider a long solenoid of N turns, radius R, and length L_0 carrying current I_0. The turns are closely and uniformly spaced and have negligible resistance. The ends of the solenoid are connected by a conductor that has negligible resistance and inductance, and the system is isolated (flux remains constant). (a) Determine the magnetic field, the magnetic-energy density, and the total magnetic energy inside the solenoid. (b) How does the current I_0 vary if the length of the solenoid is changed to L_1 while keeping the radius unchanged? (c) Determine the tension in the solenoid—that is, the force required to stretch the solenoid. Neglect the stiffness of the wire.

12.11 Two thin rings of radii a and b are placed such that their planes are normal to the line joining their centers. The rings carry currents I_1 and I_2, and the distance between their centers is $l \gg a, b$. Determine the force between the rings.

12.12 A rod of paramagnetic material of permeability μ and uniform cross section A is placed in a nonuniform magnetic field between the poles of a magnet similar to the arrangement of Fig. 12.10. If the field at the bottom of the rod is H_1 and at the top is H_2, determine the vertical force on the rod.

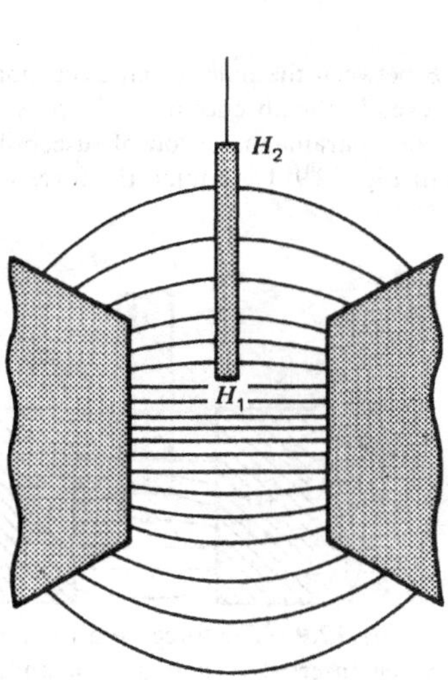

Figure 12.10 The force on a paramagnetic rod when inserted in a nonuniform H field.

12.13 Consider two rigid circuits carrying constant currents I_1 and I_2. The force between these circuits can be calculated using the Biot-Savart law. Show that this force can be derived from the magnetic energy of interaction $U = I_1 I_2 M$, according to the definition $\mathbf{F}_2 = -\mathbf{F}_1 = \nabla_2 U$, where M is the mutual inductance given by Neumann's formula, and ∇_2 operates on the coordinates of the second circuit.

12.14 An electromagnet is constructed as shown in Fig. 12.11. The total length of iron in the two parts is L, the cross-sectional area is A, and the permeability of the iron is μ. A current I through N turns activates the magnet. The two halves of the magnet are separated by a distance x, where $x \ll L$. Find the force of attraction between the two halves of the magnet.

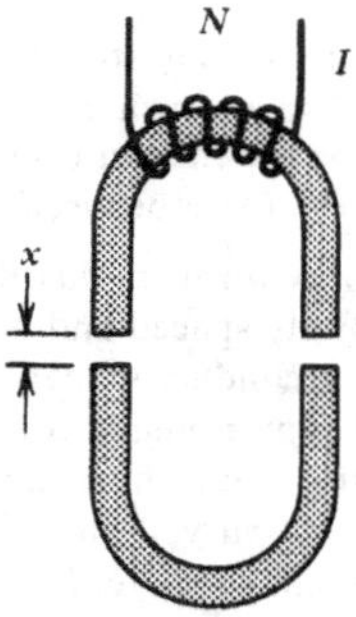

Figure 12.11

THIRTEEN

CIRCUITS WITH NONSTEADY CURRENTS

In Chapter 7 we treated electric circuits that contained only resistors. Resistors are passive elements that dissipate energy as heat or in some other fashion. In this chapter we consider electric circuits that contain elements that store energy, such as inductors and capacitors. For example, an inductor is a device that stores energy by virtue of a current passing through it, and a capacitor stores energy by virtue of a voltage across it. These elements have quite different voltage, current, and charge relationships: For a resistor, the current is proportional to the voltage; for an inductor, the voltage is proportional to the rate of change of the current; and for a capacitor, the voltage is proportional to the charge or the integral of the current.

The methods that were applicable to the purely resistive circuits—namely, Kirchhoff's laws—are also applicable to the more general circuits containing capacitors and inductors. In the more general case, however, Kirchhoff's equations are integral differential equations rather than algebraic equations.

The electromotive sources that can supply energy to the circuits may include any time dependence in principle, but the most commonly used are *step functions*, *impulses*, and *sinusoidal functions*. A step function is a suddenly applied constant source; it represents a single jump. On the other hand, an impulse is a very large pulse of voltage with a very short time duration. A sinusoidal source is a periodic source such as a cosine or a sine function of time.

In these three source cases the response of the circuit can be readily found analytically. The response to other types of excitations are difficult to be directly determined analytically. However, because the elements in the circuits being discussed are linear, and hence the principle of superposition is applicable, solutions to an arbitrary source can be represented by a sum of responses to steps, impulses, or sinusoidal sources. This is possible since an arbitrary source can be decomposed into summations of steps, impulses, or sinusoidal sources.

If the decomposition of the source results in step functions or impulses, then the problem is solved in the *time domain*. If, on the other hand, the decomposition results in sinusoidal functions, then it is solved in the *frequency domain*. It is to be noted that the time domain is a *transient* response (that is, the time dependence eventually decays to zero), whereas each wave in the frequency domain has a *steady-state solution.*

13.1 Definition of Quasi-Static Circuits

When energy-storing elements are present in the circuit, the currents and charge densities, and consequently the magnetic and electric fields, vary with time. In Chapter 11 the equations of electrostatic and magnetostatics that describe only static distributions were modified to include the variations of the magnetic field via Faraday's law:

$$\nabla \cdot \mathbf{E} = \frac{\rho}{\varepsilon_0} \qquad \nabla \cdot \mathbf{B} = 0$$

$$\nabla \times \mathbf{E} = -\frac{\partial \mathbf{B}}{\partial t} \qquad \nabla \times \mathbf{B} = \mu_0 \mathbf{J}$$

where the total electric field $\mathbf{E} = \mathbf{E}^q + \mathbf{E}^i$ [see Eq. (11.11)]. These resulting equations are called *quasi-static* equations because they still do not account for all the time variations. In Chapter 14 they will be modified further with the incorporation of the time variation of the electric field. It will be found that the time variations of **E** and **B** result in the production of waves that carry energy away from the conductors. As a result of this propagation, the analysis of circuits becomes very complicated. However, there exist physical conditions under which the time variation of the electric field, and hence the propagation, can be neglected. When these conditions are met, circuits can then be analyzed using the quasi-static equations. The above equations are sometimes referred to as "quasi-static" inasmuch as they are valid only if the fields (and consequently currents, and charge densities) do not vary "too rapidly" in time. What "too rapidly" implies will be specified more precisely later. Suffice it to say here that with the combinations of wires and materials used to construct "circuits" of physical dimensions whose distances are of order d, the time parameters τ characterizing fluctuations of the fields should satisfy

$$\tau \gg \frac{d}{c} \tag{13.1}$$

For sinusoidal signals, τ is chosen as the period. If L is the length of a segment of wire, we may disregard propagation effects without undue error if

$$L \ll cT \equiv \frac{c}{f} \tag{13.2}$$

Thus, Table 13.1 indicates that for frequencies associated with power lines (60 Hz), transmission-line effects are only important for distances of the order of 10^6 meters (or close to a thousand miles), whereas at 10^9 Hz we must account for these effects in the wires of our laboratory circuits.

It is apparent from the study of these equations that they will not be truly consistent with our circuit equations. In particular, the fact that $\nabla \times \mathbf{B}$ is taken to equal $\mu_0 \mathbf{J}$ implies that $\nabla \cdot \mathbf{J} = 0$, which implies further that all current filaments are continuous. However, we know this is not true when a wire carrying current terminates in a capacitor. We shall come back to this problem later in Chapter 14. Here we shall simply assume that current filaments are continuous everywhere except at capacitors, where, if a current enters, it will build up charge. In fact, then, we shall

Table 13.1 **Approximate Sizes at Which Propagation Effects Become Important**

f(Hz)	$L = c/f$ (meters)	Application
60	10^6	Power
10^4	3×10^4	Telephone
10^6	300	Microwave
10^9	0.3	Radio

assume that magnetic fields can be calculated from the equations $\nabla \cdot \mathbf{B} = 0$ and $\nabla \times \mathbf{B} = \mu_0 \mathbf{J}$ in excellent approximation, but we shall not assume that $\nabla \cdot \mathbf{J} = 0$ everywhere. We shall only assume that $\nabla \cdot \mathbf{J} = 0$ wherever the charge is not "building up" or "decaying away." That is, we are assuming that charge densities change in time only at "capacitors" [and there with time constants satisfying Eq. (13.1)]. As stated previously, this will be a good approximation if Eq. (13.1) is fulfilled.

13.2 Kirchhoff's Circuit Law

In Chapter 7 we stated Kirchhoff's circuit laws for the case of steady-current circuits. Here we consider these laws in the case of quasi-static circuits. We will first take a specific circuit and analyze it, and then laws for general circuits will be stated. We consider the current loop shown in Fig. 13.1a containing a source of applied emf

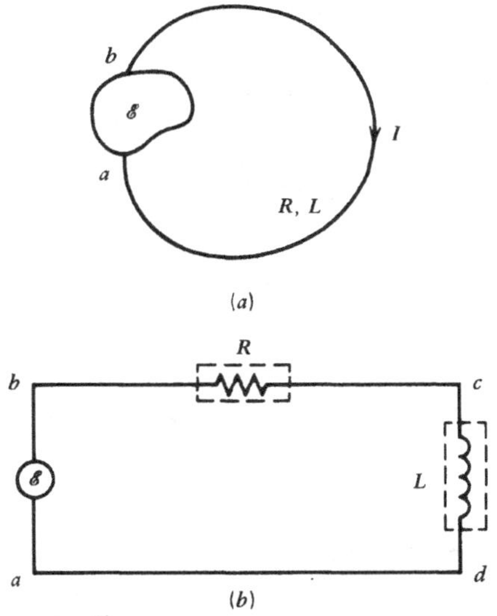

Figure 13.1 (a) A current loop containing a source of applied emf $\mathscr{E}$, which has a self-accumulate anywhere in this circuit. The current I is continuous around the loop. (b) A convential sketch of the circuit.

$\mathscr{E}$, which has a "self-accumulate" anywhere in this circuit, and the current I is continuous around the loop.

The force per unit charge, which is capable of doing work, at any point of the circuit, can be generally written as

$$\mathbf{E}_T = \mathbf{E}^s + \mathbf{E} \tag{13.3}$$

where $\mathbf{E}^s$ is the force per unit charge associated with the source of applied emf, which enables work to be done as a charge is transported completely around the circuit. It is the source (thus superscript s) of energy that enables the field $\mathbf{E} = \mathbf{E}^i + \mathbf{E}^q$ to exist. It may be attributable to batteries, generators, or fields associated with "other" circuits specifically employed to activate this circuit. By definition, $\oint \mathbf{E}^s \cdot d\mathbf{l} = \mathscr{E}$, where the line integral is taken around the circuit; (thus we use $d\mathbf{l}$ rather than $d\mathbf{r}$). Usually $\mathbf{E}^s$ is localized in space, as in a battery, so the contribution to the total closed line integral comes from only a small part of the total path length:

$$\oint \mathbf{E}^s \cdot d\mathbf{l} = \int_a^b \mathbf{E}^s \cdot d\mathbf{l} = \mathscr{E} \tag{13.4}$$

Now in the media of the circuit we assume Ohm's law to be true; that is, $\mathbf{J} = \sigma_c \mathbf{E}_T$. Therefore

$$\oint \mathbf{E}_T \cdot d\mathbf{l} = \oint \frac{\mathbf{J}}{\sigma_c} \cdot d\mathbf{l} = \oint \mathbf{E}^s \cdot d\mathbf{l} + \oint \mathbf{E}^i \cdot d\mathbf{l} + \oint \mathbf{E}^q \cdot d\mathbf{l} \tag{13.5}$$

the line integral being taken completely around the circuit (in the media). The left-hand side of Eq. (13.5) can be written in terms of the resistance of the circuit using Eq. (7.14), or

$$\oint \frac{\mathbf{J}}{\sigma_c} \cdot d\mathbf{l} = RI \tag{13.6}$$

where R is the total series resistance (including wires and emf's) of the loop and I is the continuous constant current around the loop. Note that I would not be continuous and constant if capacitance were present. The line integral of $\mathbf{E}^s$ is just the electromotive force of the source as was defined in Eq. (13.4). Moreover, $\oint \mathbf{E}^q \cdot d\mathbf{l} = 0$ and $\oint \mathbf{E}^i \cdot d\mathbf{l}$ is given by Faraday's law:

$$\oint \mathbf{E}^i \cdot d\mathbf{l} = -\frac{dF}{dt} = -L\frac{dI}{dt}$$

Therefore Eq. (13.5) becomes

$$RI = \mathscr{E} - L\frac{dI}{dt} \qquad \text{or} \qquad \mathscr{E} = L\frac{dI}{dt} + RI \tag{13.7}$$

This is the basic equation of a single, isolated, ohmic current loop containing only inductance and resistance.

Conventionally, we lump the inductance L in the symbol,  for an inductor, and the resistance R into a resistor  and sketch the circuit as in Fig. 13.1b. The assumption is that the ordinary connecting wires have negligible inductance and resistance: all the inductance of the circuit is associated with an "inductor" (like a toroid or solenoid of many turns where the B fields are relatively large) and all the resistance is in a "resistor" (of presumably low conductivity). Of course these are only approximations; the inductance and resistance of the closed loop in fact have contributions from all parts of the loop. Thus, we may formulate the "law of loop" having resistance, inductance, and emf by the simple obvious assertion that "The sum of the potential differences taken serially around loop C is zero," coupled with the statement that these potential differences are given by RI, $L(dI/dt)$, $-\mathscr{E}$ for resistors, inductors, and sources of emf, respectively. In general, we may be obliged to analyze circuits containing many loops coupled together. By "coupled" we simply mean that certain circuit elements are common to two or more loops (see Section 13.4). The method for analyzing such circuits is to write down the following Kirchhoff's circuit equations:

1. At any instant the sum of the potential drops around any closed circuit loop equals zero.
2. At any node there is no charge accumulation, so the algebraic sum of the entering currents is zero at any instant.

We can specify as many linearly independent equations as there are independent currents, and can then "solve" for all the desired currents. Of course, our equations will be linear differential equations, and their solutions may be complicated, but in principle their solution can be found uniquely if sufficient initial conditions are also specified.

13.3 Time Domain Solutions

We now treat circuits with step or impulse sources. These have transient solutions where the time dependence eventually decays to zero. We first treat RL loops, then RC loops, and finally RLC loops.

13.3.1 Series *RL* Loop

Let us now find an expression for the current I of the RL loop of Fig. 13.1*b*. Multiplying Eq. (13.7) through by the (integrating) factor $e^{(R/L)t}$, we obtain

$$\mathscr{E}e^{(R/L)t} = L\frac{dI}{dt}e^{(R/L)t} + RIe^{(R/L)t}$$

that is,

$$\frac{\mathscr{E}}{L}e^{(R/L)t} = \frac{d}{dt}(Ie^{(R/L)t}) \tag{13.8}$$

Integrating with respect to time, between times t_0 and t, we have

$$\frac{1}{L}\int_{t_0}^{t}\mathscr{E}e^{(R/L)t'}\,dt' = \int_{t_0}^{t}d(Ie^{(R/L)t'}) \equiv \left[Ie^{(R/L)t'}\right]_{t_0}^{t}$$

Calling the initial time the zero time ($t_0 \equiv 0$), we obtain the result

$$I(t)e^{(R/L)t} - I(0) = \frac{1}{L}\int_0^t \mathscr{E}e^{(R/L)t'}\,dt'$$

or

$$I(t) = I(0)e^{-(R/L)t} + \frac{e^{-(R/L)t}}{L}\int_0^t \mathscr{E}(t')e^{Rt'/L}\,dt' \tag{13.9}$$

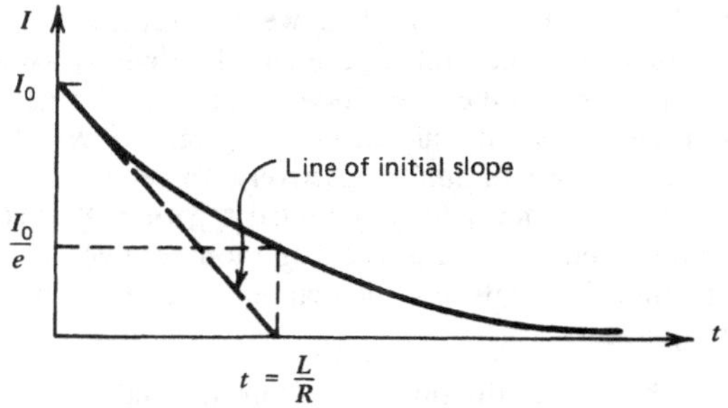

Figure 13.2 The decay of current in an *RL* circuit if the voltage source is suddenly turned to zero. The time constant of the circuit is shown as $t = L/R$.

It will be observed that this solution for $I(t)$ has two terms: one independent of $\mathscr{E}$, called the transient solution; and the other dependent on $\mathscr{E}$, called the particular solution. If $t \gg L/R$, only the particular solution persists. The transient solution is a solution for the homogeneous differential equation ($\mathscr{E} \equiv 0$) subject to the initial "condition" prevalent for the circuit at the fiducial time $t = 0$. The particular solution is one that simply satisfies the nonhomogeneous differential equation.

If $\mathscr{E} = 0$ in Eq. (13.9), then current will exist in the circuit only if $I(0) = I_0 \neq 0$. Then our solution is:

$$I(t) = I_0 e^{-(R/L)t} \tag{13.10}$$

An exponentially decaying current is observed (Fig. 13.2) that decreases to $1/e \approx 37$ percent of its initial value in a time $t = L/R$, which is called the *time constant* of this circuit. (In three time constants, the current will have decayed to $I_0/e^3 \approx I_0/(2.72)^3 \approx I_0/20 \approx 0.05I_0$). This is the behavior that would be observed if the voltage across an electromagnet was suddenly changed to zero.

We now consider the source effects in an *RL* circuit. Let $\mathscr{E} = V_0$, a constant voltage, in Eq. (13.9). We then obtain

$$I(t) = I(0)e^{-(R/L)t} + \frac{V_0}{R}[1 - e^{-(R/L)t}] \tag{13.11}$$

The grouping of first plus third terms on the right will be recognized as a solution to the homogeneous equation $L(dI/dt) + RI = 0$, subject to the initial condition of $I(0)$. It has the form $I = Ae^{-(R/L)t}$, clearly of a transient character, where $A = I(0) - V_0/R$. The second term in $I(t)$ is therefore the "particular" solution to the actual differential equation. If $I(0) = 0$, so no current is flowing initially, then $I(t)$ approaches V_0/R exponentially, again with a time constant L/R. Of course, for "long times" ($t \gg L/R$), $I \approx V_0/R$ (see Fig. 13.3*a*).

Example 13.1 Source Effects in *RL* Circuits ($\mathscr{E} = V_0$)

It will now be observed that if $\mathscr{E}$ is given by the rectangular shape (shown in Fig. 13.3*b*), perhaps obtained by opening and closing a switch in the circuit, then two cases are easily distinguished: $L/R \ll T$ and $L/R \gg T$, where T is the period of switch action. These two cases are illustrated in Fig. 13.3*b*. In the latter case the voltage across L in the circuit will eventually oscillate about $V_0/2$, the slope being approximately linear. In the former case the potential difference across the inductor follows closely the input waveform. Thus, if one desires to reproduce a fast-rising pulse, one must take care to have as small an inductance as possible.

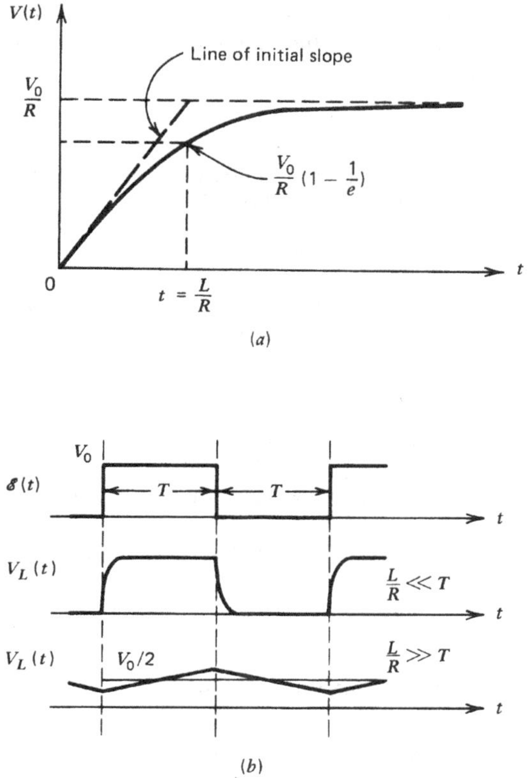

Figure 13.3 (*a*) The buildup of current in a *RL* circuit if a voltage source is suddenly turned on. (*b*) The top sketch is the emf in the *RL* circuit when a switch in the circuit is periodically opened and closed. The middle and bottom sketches are the voltages across the inductance when $L/R \ll T$ and $L/R \gg T$.

13.3.2 Series *RC* Loop

The analysis for a series *RC* loop (see Fig. 13.4) is similar to that for an *RL* loop. The capacitance of the loop is now assumed to be concentrated in the capacitor.

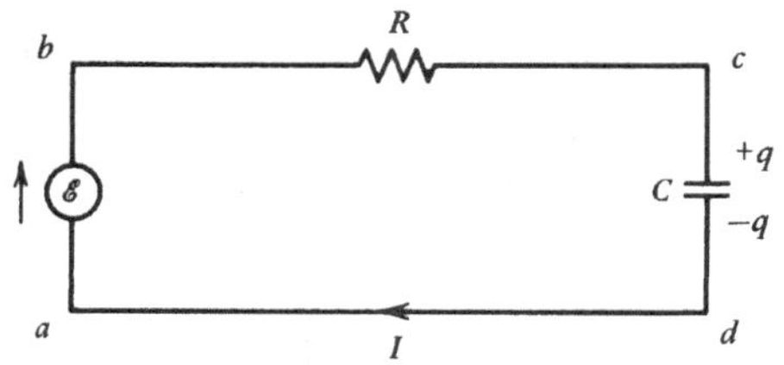

Figure 13.4 A series *RC* loop.

The inductance is deemed negligible, and the charge density is therefore assumed to change only at the capacitor. Since the sum of the potential drops around complete loop of the circuit must be zero, we have

$$\mathscr{E} = RI + \frac{q}{C} \tag{13.12}$$

where q is related to I, by $I = dq/dt$, meaning that when current flows as shown, q must increase. Thus we may also write Eq. (13.12) as

$$\mathscr{E} = R\frac{dq}{dt} + \frac{q}{C} \tag{13.13}$$

which is identical in form to Eq. (13.7). We need merely make the substitutions $L \to R$, $R \to 1/C$, and $I \to q$ in order to obtain the solution [see Eq. (13.9)]:

$$q(t) = q(0)e^{-t/RC} + \frac{e^{-t/RC}}{R}\int_0^t \mathscr{E}e^{t'/RC}\,dt' \tag{13.14}$$

where $q(0)$ is the charge on the capacitor at the initial, or fiducial, time $t = t_0 \equiv 0$. Again, we note the characteristic time, RC, and the fact that $q(0)e^{-t/RC}$ is the solution of the homogeneous equation, whereas the other term contains the particular solution of the differential equation.

If $\mathscr{E} = 0$, $q(t) = q(0)e^{-t/RC}$. The charge on the capacitor decays exponentially, with the time constant RC. The current in the circuit is

$$I = \frac{dq}{dt} = -\frac{q(0)}{RC}e^{-t/RC} = \frac{V_C(0)}{R}e^{-t/RC} \tag{13.15}$$

where $V_C(0) = q(0)/C$ is the potential drop across the capacitor at $t = 0$.

If $\mathscr{E} = V_0$, we obtain

$$q(t) = q(0)e^{-t/RC} + CV_0(1 - e^{-t/RC}) \tag{13.16}$$

Example 13.2 Integrating and Differentiating Circuits—Analog Computers

Circuits containing resistance and capacitance are often used in "integrating" or "differentiating" signals. An integrating circuit is shown in Fig. 13.5*a*. We suppose that the input signal, $\mathscr{E}$, has fluctuations in times no longer than T, a characteristic time associated with $\mathscr{E}$. If $\mathscr{E}$ were sinusoidal, T, would be its period. In the integrating circuit, $RC \equiv \tau$ must be chosen so that $\tau \gg T$. Then Eq. (13.14) may be written

$$q(t) = \frac{e^{-t/RC}}{R}\int_0^t \mathscr{E}(t')e^{t'/RC}\,dt'$$

In the interval $(0, t)$ such that $t' \le t \le T \ll RC$, we have $e^{t'/RC} \approx 1$. Then the potential across the capacitor $V_C = q/C$ is

$$V_C(t) \equiv \frac{q}{C} = \frac{1}{RC}\int_0^t \mathscr{E}(t')\,dt' \tag{13.17}$$

The output potential difference is the integral of the applied input potential difference divided by the RC time constant in this approximation. It is valid for single pulses of time duration $\ll RC$.

For a differentiating circuit, Fig. 13.5*b*, we observe the potential difference across the resistor rather than across the capacitor. Moreover, here we require the $RC \ll T$. Differentiating Eq. (13.12) with respect to t gives:

$$R\frac{dI}{dt} + \frac{I}{C} = \frac{d\mathscr{E}}{dt} \tag{13.18}$$

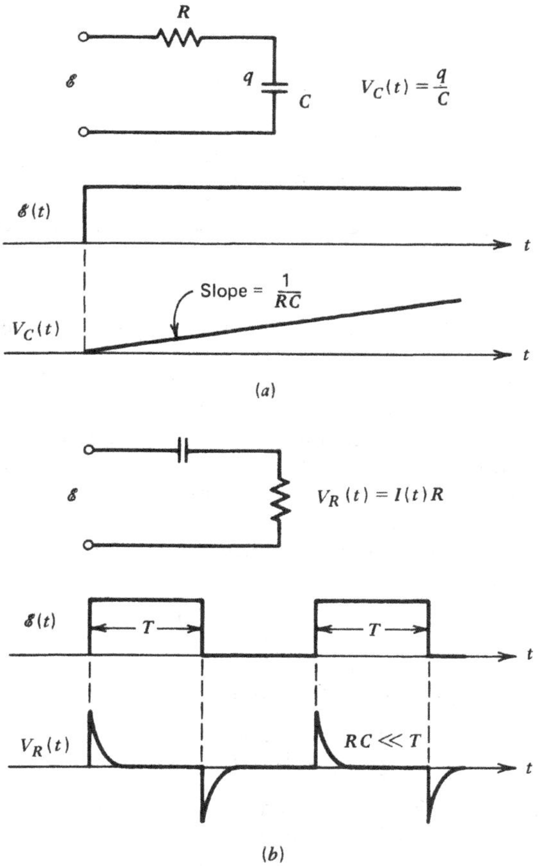

Figure 13.5 (*a*) An *RC* integrating circuit showing an input step voltage and the integrated voltage across the capacitor. (*b*) An *RC* differentiating circuit showing a periodic rectangular input voltage and the differentiated voltage across the resistor.

Multiplying by RC and taking $V_R = RI$, we get

$$RC\frac{dV_R}{dt} + V_R = RC\frac{d\mathscr{E}}{dt}$$

Since

$$\frac{dV_R}{dt} \approx \frac{\Delta V_R}{\Delta t} \approx \frac{V_R}{T}$$

then

$$V_R\left(1 + \frac{RC}{T}\right) \approx RC\frac{d\mathscr{E}}{dt}$$

Moreover, using $RC/T \ll 1$, we get

$$V_R(t) \approx RC\frac{d\mathscr{E}}{dt} \tag{13.19}$$

the result desired for the "differentiating" operation.

The results given in Eqs. (13.17) and (13.19) show that this RC network has the interesting feature that if a periodic voltage is applied to it, the resulting circuit behaves as an analog computer for appropriate RC and steady-state conditions.

13.3.3 The *RLC* Loop

We now suppose, as in Section 13.3.2, that we have a loop containing a capacitor C, but here the loop also has an inductance L, that cannot be neglected and a resistance R. The resistance R is to represent the total series resistance of the loop, including the wires and the sources of emf. The inductance of the loop L is a property of the whole circuit, but practically may be concentrated in special inductors made of wire coils. The resistance of these coils is included in R. The capacitor C is assumed to be loss-free—i.e., to have infinite resistance to the flow of current. The schematic representation of the circuit is shown in Fig. 13.6.

At any instant of time, the sum of the potential drops around the circuit is zero. Thus we have the circuit equation

$$\mathscr{E} = L\frac{dI}{dt} + RI + \frac{q}{C} \tag{13.20}$$

With the relation $I = dq/dt$, we rewrite this as

$$\frac{\mathscr{E}}{L} = \frac{d^2q}{dt^2} + \frac{R}{L}\frac{dq}{dt} + \frac{q}{LC} \tag{13.21}$$

which is a second-order, linear differential equation having constant coefficients. This is the equation of state of the single-loop circuit, and governs its behavior. A knowledge of $q(t)$ [or $I(t) = dq/dt$] completely determines the electrical characteristics of the circuit. Since R, L, and C are constants, we can write Eq. (13.21) in operator form as

$$\frac{\mathscr{E}}{L} = \left[\frac{d^2}{dt^2} + \frac{R}{L}\frac{d}{dt} + \frac{1}{LC}\right]q = \left(\frac{d}{dt} - m_1\right)\left(\frac{d}{dt} - m_2\right)q \tag{13.22}$$

where m_1 and m_2 are constants given by

$$m_1 = -\alpha + i\omega \qquad m_2 = -\alpha - i\omega \tag{13.23}$$

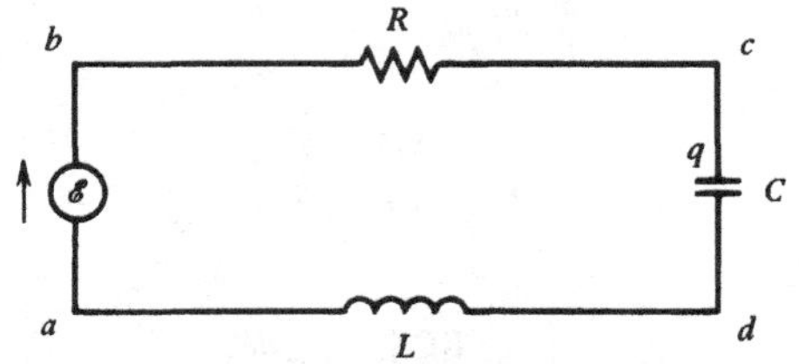

Figure 13.6 A series *RLC* circuit.

and

$$\omega = \sqrt{\frac{1}{LC} - \left(\frac{R}{2L}\right)^2} \qquad \alpha = \frac{R}{2L} \qquad i = \sqrt{-1} \tag{13.24}$$

In terms of m_1 and m_2, it is well known, and may be verified by direct substitution, that the solution of Eq. (13.21) always has the form

$$\begin{aligned} q(t) &= k_1 e^{m_1 t} + k_2 e^{m_2 t} + q_p(t) \qquad m_1 \neq m_2 \\ q(t) &= k_1 + k_2 t + q_p \qquad m_1 = m_2 \end{aligned} \tag{13.25}$$

where k_1 and k_2 are constants independent of t and $q_p(t)$ is *any* solution to all the parts of Eq. (13.21). It is the particular solution, whereas the first two terms represent the solution to the homogeneous equation obtained from Eq. (13.21) by setting $\mathscr{E} = 0$. The current in the circuit $I = dq/dt$ is

$$I = m_1 k_1 e^{m_1 t} + m_2 k_2 e^{m_2 t} + \frac{dq_p}{dt} \tag{13.26}$$

The mathematical dependence of q on time will be unique only if the charge on the capacitor and the rate of increase of charge (or current) on it are specified at some instant of time; that is, $t = 0$. These conditions are called the initial conditions. We thus must specify $q(0) \equiv q_0$ and $I(0) \equiv I_0$ to determine k_1 and k_2 uniquely.

We note that even though m_1 and m_2 are complex, $q(t)$ and $I(t)$ are real. This results because k_1 and k_2 are also complex, but in such a way as to render the solutions real when the "initial conditions" [$I(0)$ and $q(0)$] are real. Also we note that as long as $R \neq 0$ (so that $\alpha \neq 0$), the solutions have a part that is exponentially decaying away. Moreover, in the instance, when ω is real (i.e, $R^2/4L^2 < 1/LC$), the solution will have an oscillatory character associated with the decaying exponential. In the following examples, we discuss some special cases.

Examples 13.3 Discharging of a Capacitor in an *RLC* Loop—Quality Factor

We consider in this example a capacitor with an initial charge $q(0) = q_0$. It is to be short-circuited with a resistor and an inductance, as shown in Fig. 13.7*a*. The general solution of an *RLC* circuit is given by Eqs. (13.25) and (13.26). Since $\mathscr{E}(t) = 0$, then the particular solution $q_p(t)$ is zero. Moreover, since the current is zero right after the switch is closed and the current through inductors cannot change instantaneously, then $I(0) = 0$. Substituting q_0 for $q(0)$ and 0 for $I(0)$ in Eqs. (13.25) and (13.26) gives

$$\begin{aligned} k_1 &= k = \frac{m^* q_0}{m^* - m} \qquad k_2 = k^* = \frac{m q_0}{m - m^*} \\ m_1 &= m = -\alpha + i\omega \qquad m_2 = m^* \end{aligned} \tag{13.27}$$

where the star indicates complex conjugate. There are three special cases to consider corresponding to ω being real, zero, and pure imaginary, that is, corresponding to $1/LC$ >, =, $<(R/2L)^2$.

(*a*) *Oscillatory Case* (*Underdamped*). This case arises when ω is real. Thus, $\omega^* = \omega$, and

$$k = \frac{1}{2}\left(1 - i\frac{\alpha}{\omega}\right)q_0 \tag{13.28}$$

$$q(t) = \left[\frac{\sqrt{\alpha^2 + \omega^2}}{\omega}\right] \cdot [q_0 e^{-\alpha t} \cdot \cos(\omega t + \phi)]$$

where

$$\tan\phi = \frac{\text{Im}(k)}{\text{Re}(k)} = -\frac{\alpha}{\omega}$$

The scripts Im( ) and Re( ) denote the imaginary and real parts, respectively, of the (complex) quantities in parentheses. Expanding the cosine fuction, we may alternatively write

$$q(t) = q_0 e^{-\alpha t}\left[\cos\omega t + \frac{\alpha}{\omega}\sin\omega t\right] \tag{13.29}$$

The current is given by $I(t) = dq/dt$, or

$$I(t) = -q_0 e^{-\alpha t}\left(\frac{\alpha^2 + \omega^2}{\omega}\right)\sin\omega t \tag{13.30}$$

If the resistance is small, so that $\alpha/\omega \ll 1$, then

$$q(t) \approx q_0 e^{-\alpha t}\cos\omega t \qquad \text{and} \qquad I(t) \approx -q_0\omega e^{-\alpha t}\sin\omega t$$

This indicates that the current and charge are out of phase by 90°, with the current "leading" the charge by 90°. Otherwise the current "leads" the charge by the phase angle $\pi/2 - \phi$, as shown in Fig. 13.7*b*.

A measurement of the rate of decay of the oscillations is also given by parameters denoted as δ and Q, called the *logarithmic decrement* and the *quality factor*, respectively. The logarithmic decrement is defined as

$$\delta = \ln\left[\frac{I(t)}{I(t+T)}\right] \tag{13.31}$$

where $T = 2\pi/\omega$ is the period of the oscillation. The quality factor is defined by the relation

$$Q = \frac{\omega L}{R} = \frac{\omega}{2\alpha} = \frac{\omega T}{2\delta} = \frac{\pi}{\delta} \tag{13.32}$$

A small δ is associated with a large Q, and it means that the oscillations die out slowly.

We now present an alternative definition of Q, and hence δ, using energy considerations. We may note that in terms of the maximum magnetic energy, $U = \frac{1}{2}LI^2$, where I is the maximum current in any circle,

$$\frac{U(t) - U(t+T)}{U(t)} = \frac{\Delta U}{U} = 1 - \frac{I^2(t)}{I^2(t+T)}$$

or, from Eq. (13.31), $\Delta U/U = 1 - e^{-2\delta}$. For small δ, $1 - e^{-2\delta} \approx 2\delta$; thus

$$\frac{\Delta U}{U} \approx 2\delta = \frac{2\pi}{Q} \tag{13.33}$$

Thus, $Q/2\pi$ equals the maximum energy stored divided by the energy loss in any given oscillation cycle. This definition is the most frequently used for the definition of Q. A "high-Q" circuit will perform many charge or current oscillations before dying out. The maximum energy stored in the circuit will be seen to vary as $I^2 \sim \exp[-2\omega t/Q]$. Since the resistance of a circuit in which it is desired to have oscillations is often most directly associated with coils (of inductance L) placed in the circuit, one often refers to the "Q" of the coil, as this will give an upper limit to the Q of the circuit in which it is placed. A Q of approximately 2 or less will yield no oscillations, which brings us to case (b).

(b) Critically Damped Case. As $\omega \to 0$, meaning $(R/2L)^2 \to 1/LC$, we may find the correct expressions for $q(t)$ and $I(t)$ from case (a), if we note that, in this limit, $\sin\omega t/\omega \to t$ and $\cos\omega t \to 1$. Then

$$q(t) = q_0 e^{-\alpha t}[1 + \alpha t] \qquad \text{and} \qquad I(t) = -q_0(\alpha^2 t)e^{-\alpha t}. \tag{13.34}$$

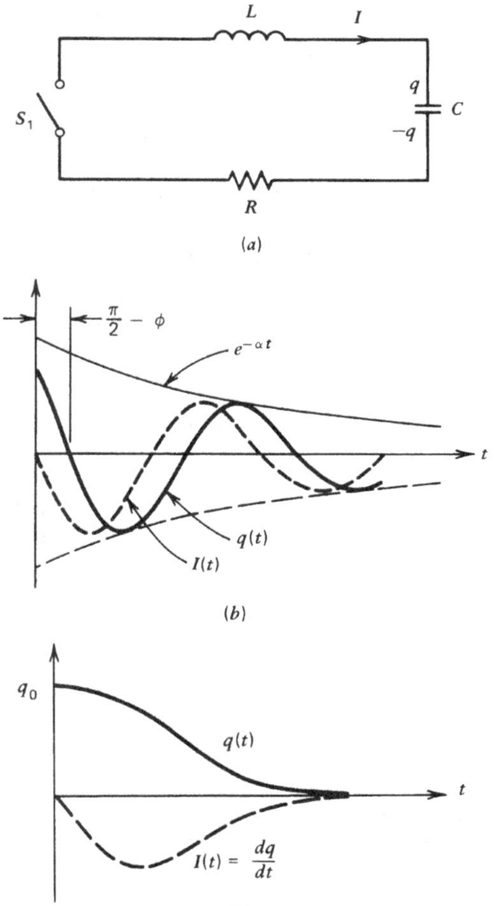

Figure 13.7 Discharging of a capacitor in an *RLC* loop. (*a*) Sketch of the *RLC* circuit. (*b*) Time behavior of the charge on the capacitor and the current in the *RLC* circuit in the oscillatory case. (*c*) Time behavior in the critically damped case.

This result is called a critically damped solution. Fig. 13.7*c* shows both I and q as a function of time. Note that this case corresponds to the case $m_1 = m_2 = -\alpha$—that is, the second form of Eq. (13.25).

(*c*) *Overdamped Case.* In this case, which arises when ω is purely imaginary,

$$m_1 = -\alpha + \beta, \qquad m_2 = -\alpha - \beta \qquad k_1 = \frac{\alpha + \beta}{2\beta} q_0 \qquad k_2 = \frac{-(\alpha - \beta)}{2\beta} q_0 \qquad \beta = +i\omega$$

are all real quantities. Equation (13.25) now becomes

$$q(t) = \frac{q_0}{2\beta} [(\alpha + \beta)e^{-(\alpha-\beta)t} - (\alpha - \beta)e^{-(\alpha+\beta)t}] \tag{13.35}$$

Differentiating q with respect to time gives the current

$$I(t) = \frac{-q_0}{2\beta}(\alpha^2 - \beta^2)(e^{+\beta t} - e^{-\beta t})e^{-\alpha t} \tag{13.36}$$

Equations (13.35) and (13.36) indicate that the current and charge decay as the sum of two exponentials, one of which falls off more rapidly than $e^{-\alpha t}$ and one less rapidly. The result is a falloff that is less rapid than in the critically damped case. In that case the falloff is at the maximum rate. By definition, $\beta \leq \alpha$, so the current always approaches zero at $t \to \infty$.

Examples 13.4 Charging a Capacitor in an *RLC* Circuit

In this example we suppose that a battery of constant output potential V_0 is connected in series with an *RLC* loop, as shown in Fig. 13.8*a*. At time zero, when switch S_1 is closed, the

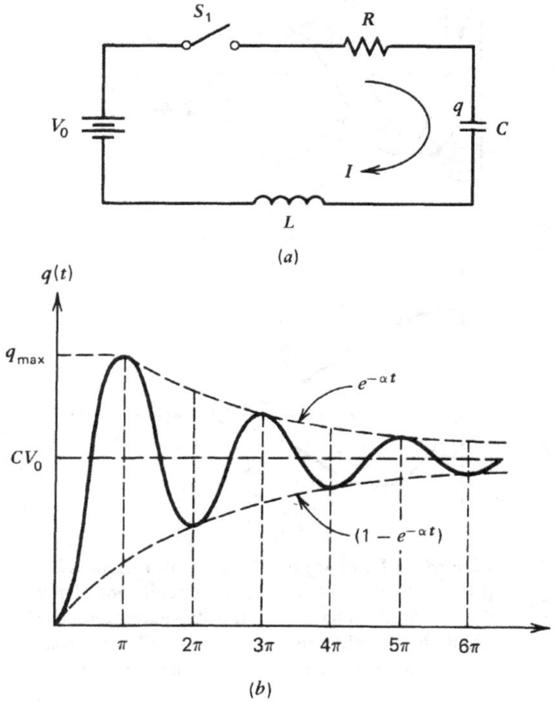

Figure 13.8 Charging of a capacitor in an *RLC* circuit. (*a*) Sketch of the circuit. (*b*) Time behavior of the charge on the capacitor.

current will be zero, and we shall assume that the capacitor had no charge. Thus, the initial conditions are that $I(0) = 0$, and $q(0) = 0$. The differential equation [see Eq. (13.21)] for the circuit is, for $t > 0$,

$$V_0 = L\frac{d^2q}{dt^2} + R\frac{dq}{dt} + \frac{q}{C}$$

whose solution is given by Eq. (13.25).

It is readily seen that we can set the particular solution equal to CV_0:

$$q_p(t) = CV_0 \tag{13.37}$$

The condition $q = 0$ and $I = 0$ at $t = 0$ can be easily applied to Eqs. (13.25) and (13.26); hence

$$q(0) = k_1 + k_2 + CV_0 = 0 \qquad \text{and} \qquad I(0) = [-\alpha + i\omega]k_1 - [\alpha + i\omega]k_2 = 0$$

Solving for k_1 and k_2, we find

$$k_1 = -\frac{CV_0}{2}\left(1 - i\frac{\alpha}{\omega}\right) \qquad \text{and} \qquad k_2 = k_1^* = -\frac{CV_0}{2}\left(1 + i\frac{\alpha}{\omega}\right) \tag{13.38}$$

We can then write $k_1 \equiv k = -|k|e^{-i\phi}$, where $\tan\phi = -\alpha/\omega$, and $|k| = (CV_0/2)\sqrt{1 + (\alpha/\omega)^2}$. Then

$$q(t) = -|k|e^{-\alpha t}[e^{i(\omega t + \phi)} + e^{-i(\omega t + \phi)}] + CV_0$$

or

$$q(t) = CV_0[1 - e^{-\alpha t}\sqrt{1 + (\alpha/\omega)^2}\cos(\omega t + \phi)] \tag{13.39}$$

The time dependence of $q(t)$ is shown in Fig. 13.8*b*. Differentiating, we also find, after trigometric simplification,

$$I(t) = CV_0 e^{-\alpha t}\left(\frac{\alpha^2 + \omega^2}{\omega}\right)\sin\omega t \tag{13.40}$$

The first maximum in $q(t)$ appears where $I(t) = 0$—that is, at $\omega t = \pi$. Since $\cos\phi = (1 + (\alpha/\omega)^2)^{-1/2}$, then

$$q_{1\max} = CV_0[1 + e^{-\alpha\pi/\omega}]$$

If $\alpha/\omega \ll 1$, $q_{1\max} \approx 2CV_0$. The potential difference across the capacitor can thus be as large as twice the applied emf. Therefore, capacitors in such circuits should have a voltage breakdown rating at least twice the applied voltage, V_0.

13.4 Coupled Circuits

So far we have only considered single loops. In this section we consider circuits containing many loops coupled together. By "coupled" we mean that certain circuit elements are common to two or more loops.

Circuits are called *capacitively* coupled when a capacitor is common between them. The circuit illustrated in Fig. 13.9 is capacitively coupled. Three currents—I_1, I_2, and I_3—are shown. It is clear from the schematic drawing that two independent

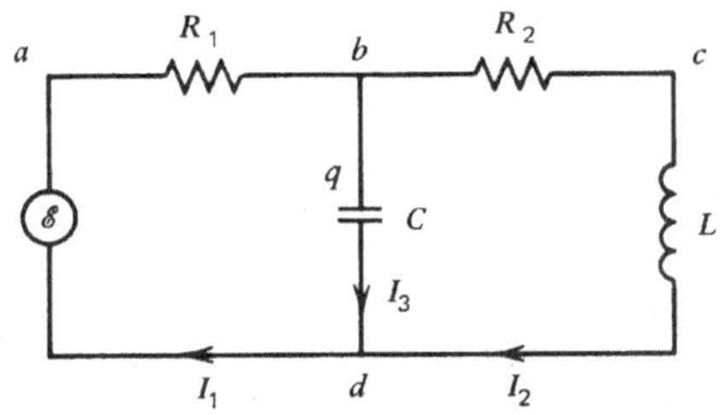

Figure 13.9 Two circuits capacitively coupled.

currents are needed to characterize this circuit. In terms of these currents, which we choose to be I_1 and I_2, we note, according to Kirchhoff's laws, that $I_3 = I_1 - I_2$. We also note that the charge on the capacitor is related to I_3 by $dq/dt = I_3 = I_1 - I_2$. There are two independent loops in this circuit. Because only two currents need be found, only two independent loop equations need be constructed. For the loop containing $\mathscr{E}$, R_1, and C (labeled *abda*), we have from Kirchhoff's laws

$$\mathscr{E} = R_1 I_1 + \frac{q}{C} \tag{13.41}$$

For the loop containing R_2, L, and C we have

$$0 = R_2 I_2 + L\frac{dI_2}{dt} - \frac{q}{C} \tag{13.42}$$

We have in both these "loop equations" taken our line integral (of potential difference) in the direction of the currents as we expect them to flow. Thus one can see from Eqs. (13.41) and (13.42) that the two loops (currents I_1 and I_2) are coupled via q/C.

Two loops of a circuit may be coupled *magnetically* by means of a mutual inductance. The situation is schematized as in Fig. 13.10*a*, where M is the coefficient of mutual inductance between two loops having self-inductances L_1 and L_2, which are shown as lumped (ideal) circuit elements. Resistors R_1 and R_2 are shown to indicate that, in practice, resistance is usually present in the coupling elements.

We seek the potential differences $\Phi(a) - \Phi(b)$ and $\Phi(c) - \Phi(d)$. (Note that these are in the same sense as the chosen currents.) Using Eq. (11.32) we write

$$\Phi(a') - \Phi(b) = L_1\frac{dI_1}{dt} + M\frac{dI_2}{dt} \tag{13.43}$$

Similarly,

$$\Phi(c) - \Phi(d') = L_2\frac{dI_2}{dt} + M\frac{dI_1}{dt} \tag{13.44}$$

where $M \equiv M_{12} = M_{21}$ is the mutual inductance of the coils. The inductance M as written may be a positive or negative geometrical coefficient, depending on how the coils are coupled. Two cases are shown in Fig. 13.10*b*. In case A, when I_1 is increasing, the self-induced potential drop is from a to b. The mutually induced potential drop in the other loop is from d to c. This correlation is denoted by the solid dots placed on the circuit terminals. By study it should be clear that when both I_1 and I_2 are increasing, counteracting induced emfs are generated (the fluxes oppose each other), so M here is negative. In case B the fluxes reinforce each other and hence M is positive. To summarize, then, for the circuit in Fig. 13.10*c*, we would write

$$\Phi(a) - \Phi(b) = R_1 I_1 + L_1\frac{dI_1}{dt} - |M|\frac{dI_2}{dt} \tag{13.45}$$

$$\Phi(c) - \Phi(d) = R_2 I_2 + L_2\frac{dI_2}{dt} - |M|\frac{dI_1}{dt} \tag{13.46}$$

If either one of the current directions were altered or the sense of one of the windings were reversed, $|M|$ would have appeared with a positive sign. Note that we express the potential drops "on each side" of the mutual inductance in the directional sense of the currents as chosen—i.e., in the sense in which we choose to

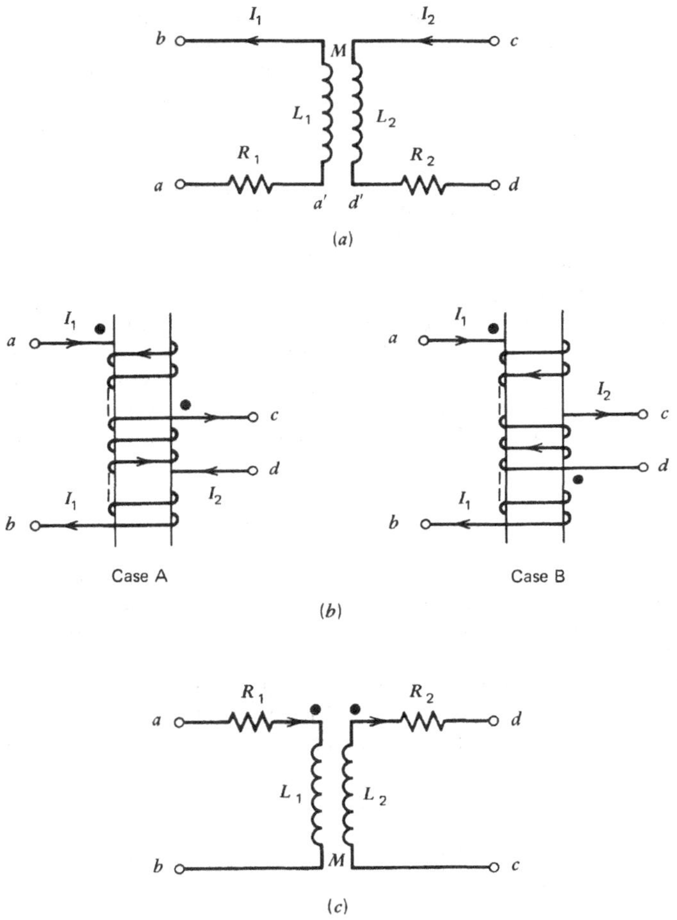

Figure 13.10 Two circuits magnetically coupled via a mutual inductance. (*a*) Sketch of the loops. (*b*) Shows two different ways of coupling the loops, with cases A and B giving negative and positive mutual inductance, respectively. (*c*) Convention for the sign of the mutual inductance using two dots.

evaluate our line integrals. The self-inductance terms in the line integrals then appear with positive signs. If we had written $\Phi(d) - \Phi(c)$ rather than $\Phi(c) - \Phi(d)$, the terms on the right-hand side of the equation would have had opposite signs.

13.5 AC Circuits—Frequency Domain

13.5.1 Phasors—Kirchhoff's Laws for Phasors

We have seen that the solution for the currents (or charge) flowing in lumped-constant circuits are obtained as solutions of *linear* differential equations with constant coefficients. In simple cases, where the applied emf's are constant, the solutions may be obtained without difficulty. A very important type of excitation, for which the solutions are also easily obtained, is a sinusoidal excitation. Thus, for any loop of a circuit, $\mathscr{E}$ may have the forms $\mathscr{E}_0 \cos \omega t$, $\mathscr{E}_0 \sin \omega t$, $\mathscr{E}_0 \cos(\omega t + \phi)$, or $\mathscr{E}_{01} \cos \omega t + \mathscr{E}_{02} \sin \omega t$, all of which are described as sinusoidal since they differ only by a constant phase factor. The $\mathscr{E}_0$'s are constant amplitudes. By "AC" we shall mean that the circuit excitations have this form: sinusoidal of angular frequency ω (or $f = \omega/2\pi$).

The reasons why ac circuit analysis is given such an important place in the study of electricity and magnetism lies in the fact that our electrical technology largely employs such excitations: It is easy to construct generators giving sinusoidal emfs. However, the importance of ac analysis transcends these reasons. Since, by Fourier analysis, any repetitive signal (excitation) may be decomposed in terms of sinusoidal components, if we understand the behavior of circuits for individual frequency components, we can understand their behavior for sums of frequency components, and thus for rather arbitrary repetitive excitations. The linearity of our equations renders this feasible. Consider Fig. 13.6, which shows a single *RLC* loop driven by a single frequency source $\mathscr{E} = \mathscr{E}_0 \cos \omega t$. The differential equation describing the circuit is

$$L\frac{dI}{dt} + RI + \frac{q}{C} = \mathscr{E}_0 \cos \omega t \tag{13.47}$$

This equation is real, and basically one is interested in solving it for I or q ignoring the initial transient solution caused by the closing of the circuit, which was discussed in detail in Section 13.3. However, the solution to this equation is most readily obtained by considering a differential equation identical to the original one except that the applied emf is put in complex form; that is, we associate with the complex emf $\hat{\mathscr{E}}e^{i\omega t}$ such that $\text{Re}(\hat{\mathscr{E}}e^{i\omega t}) = \mathscr{E}_0 \cos \omega t$. The steady-state solution of the complex linear differential equation will have a complex current and a complex charge, of the forms

$$I(t) = \hat{I}e^{i\omega t} \qquad q(t) = \hat{q}e^{i\omega t} \tag{13.48}$$

The true real current and charge are then obtained by considering the real parts only of the complex quantities via the relations

$$I(t) = \text{Re}[\hat{I}(t)e^{i\omega t}] \qquad \text{and} \qquad q(t) = \text{Re}[\hat{q}(t)e^{i\omega t}] \tag{13.49}$$

If the emf source driving the circuit had the time dependence $\mathscr{E} = \mathscr{E}_0 \sin \omega t$, instead of $\mathscr{E}_0 \cos \omega t$, then we would still represent it by $\hat{\mathscr{E}}e^{i\omega t}$. However, in this case, the real current and charge in the various elements are calculated from the imaginary part of the respective complex quantities

$$I = \text{Im}(\hat{I}e^{i\omega t}) \qquad \text{and} \qquad q = \text{Im}(\hat{q}e^{i\omega t}) \tag{13.50}$$

If $\mathscr{E} = \mathscr{E}_0 \cos(\omega t + \phi_0)$, then emf will be represented by $\hat{\mathscr{E}}e^{i(\omega t + \phi_0)}$, and the real current and charge are given by

$$I = \text{Re}(\hat{I}e^{i(\omega t + \phi_0)}) \qquad \text{and} \qquad q = \text{Re}(\hat{q}e^{i(\omega t + \phi_0)}) \tag{13.51}$$

The reason for going to the complex plane is that the algebra becomes much simpler using a complex exponential than using the real trigonometric function. We would like to show now that this procedure is rigorous. Consider the complex function $f = f_r + if_i$, where f_r and f_i are real functions of the scalar variable t. If $\mathfrak{L}$ is a real operator, and

$$\mathfrak{L}y = f = f_r + if_i$$

then y must be a complex function, and may be written as $y = y_r + iy_i$. This follows because, by definition, if f were real, y would be real. Thus we can

$$\mathcal{L}(y_r + iy_i) = f_r + if_i.$$

Since $\mathcal{L}$ is linear, $\mathcal{L}(y_r + iy_i) = \mathcal{L}(y_r) + i\mathcal{L}(y_i)$. Equating real and imaginary parts of the two sides of the latter equation yields

$$\mathcal{L}(y_r) = f_r \qquad \text{and} \qquad \mathcal{L}(y_i) = f_i \tag{13.52}$$

which proves that this procedure is rigorous.

The complex amplitude $\hat{\mathscr{E}}$, $\hat{I}$, and $\hat{q}$ may be interpreted as the complex entities $\mathscr{E}$, I, and q at $t = 0$. They are often called *phasors*. In order to have the correct solutions to the complex differential equations, certain relations must be obtained between these phasors. These relations are specified by the differential equations that I and q must satisfy. The essential nature of the relations between these phasors is seen by noting that the derivative operation d/dt is equivalent (for complex sinusoidal excitation) to multiplying by $i\omega$. The equation for complex ac excitation is

$$L\frac{dI}{dt} + RI + \frac{q}{C} = \hat{\mathscr{E}}e^{i\omega t} \tag{13.53}$$

writing $I = \hat{I}e^{i\omega t}$ and hence $q = I/i\omega$, we find that

$$i\omega L\hat{I} + \frac{\hat{I}}{i\omega C} + R\hat{I} = \hat{\mathscr{E}} \tag{13.54}$$

Thus

$$\hat{I} = \frac{\hat{\mathscr{E}}}{R + i\left(\omega L - \dfrac{1}{\omega C}\right)} \qquad \text{and} \qquad I = \frac{\hat{\mathscr{E}}e^{i\omega t}}{R + i\left(\omega L - \dfrac{1}{\omega C}\right)} \tag{13.55}$$

This result is similar to the dc relation $I = \mathscr{E}/R$, except that the denominator, $z = \hat{\mathscr{E}}/\hat{I}$, is complex and called the *complex impedance*. The associated complex impedance of the present *RLC* circuit is

$$z = R + i\left(\omega L - \frac{1}{\omega C}\right) = Ze^{i\phi} \tag{13.56}$$

where

$$Z = |z| = \sqrt{R^2 + \left(\omega L - \frac{1}{\omega C}\right)^2} \qquad \text{and} \qquad \tan\phi = \frac{\mathrm{Im}(z)}{\mathrm{Re}(z)} = \frac{\omega L - 1/\omega C}{R} \tag{13.57}$$

In terms of the impedance the real current is

$$I(t) = I_0\cos(\omega t - \phi) = \frac{\mathscr{E}_0}{Z}\cos(\omega t - \phi) \tag{13.58}$$

Using this current, we can also find that

$$q(t) = q_0\cos\left(\omega t - \phi - \frac{\pi}{2}\right) = \frac{\mathscr{E}_0}{\omega Z}\sin(\omega t - \phi) \tag{13.59}$$

For the *RLC* circuit, ϕ may lie anywhere between $-\pi/2$ and $+\pi/2$. If $-\pi/2 \le \phi < 0$, then $(\omega L - 1/\omega C) \le 0$, and the circuit is "capacitive." If $0 \le \phi \le \pi/2$, then $(\omega L - 1/\omega C) > 0$, and the circuit is "inductive." If $\phi = 0$ the circuit is resistive. The terms capacitive, inductive and resistive imply that the circuit behaves like an *RC*, *RL* or *R* circuits, respectively.

With the voltage phasors for the various elements defined the differential equations are transformed into complex *algebraic* equations relating the phasors. From the phasors, one can construct all the quantities of interest. This procedure may be described as *Kirchhoff's laws for phasors*, and is analogous to the laws of dc circuits. Thus:

1. *Mesh law* (*voltage law*): Around any closed loop in an ac circuit, the sum of the potential drop phasors must add to zero.
2. *Nodal law* (*current law*): The sum of the current phasors into all points of an ac circuit must be zero. This is especially relevant at nodes of a circuit. (At a capacitor we can consider I as the current entering a terminal and dq/dt as the current leaving that terminal, so the equation $dq/dt = I$ relates the charge on the capacitor to the incoming current.)

13.5.2 The Mesh Law

In order to facilitate the application of the mesh law, the potential-drop phasors for the different elements that may occur in any loop are now summarized. A resistor carrying a current I has the potential drop RI, and thus the "voltage" phasor, $\hat{V} = R\hat{I}$, is associated with it. A capacitor into which a current I flows such that $I = dq/dt$, has q/C for its potential drop and thus the voltage phasor $\hat{V} = \hat{q}/C$, which, using the relation $\hat{I} = i\omega\hat{q}$, becomes $\hat{V} = \hat{I}/i\omega C$. An isolated inductor of self-inductance L, that carries a current I has the potential drop $L(dI/dt)$, and so the voltage phasor $\hat{V} = i\omega L\hat{I}$ is associated with it. Associated with any mutual inductance are two circuits. The potential drop in circuit k due to a current in circuit l is given by $V_{kl} = M(dI_l/dt)$ and similarly $V_{lk} = M(dI_k/dt)$. Thus, the associated voltage phasors are $\hat{V}_{kl} = i\omega\hat{M}I_l$ and $\hat{V}_{lk} = i\omega\hat{M}I_k$.

The impedance was first introduced in Eqs. (13.56) and (13.57) with regard to a single RLC loop. Here we introduce and discuss the impedance between any two points in a passive linear circuit* through which a current flows. Because of the assumed linearity of the system, the voltage and current phasor will be linearly related, as follows:

$$\hat{V} = z\hat{I} \tag{13.60}$$

The quantity z is called the *complex impedance*. As a complex quantity, z may always be written

$$z = R + iX \tag{13.61}$$

where R, the real part of z, is called the *resistance* and X, the imaginary part of z, is called the *reactance*. We now summarize the results of the previous discussion in Table 13.2.

Table 13.2

Element	Complex Impedance	Impedance	Reactance	Voltage Phasor
R	R	R	0	$R\hat{I}$
L	$i\omega L$	ωL	ωL	$i\omega L\hat{I}$
C	$\dfrac{1}{i\omega C}$	$\dfrac{1}{\omega C}$	$-\dfrac{1}{\omega C}$	$\dfrac{\hat{I}}{i\omega C}$
M	$i\omega M$	ωM	ωM	$i\omega M\hat{I}$

* That is, a circuit containing no sources of emf.

Impedances are combined just like resistances, which follows from Kirchhoff's laws for ac circuits. Thus, if one has complex impedances z_1 and z_2, their series connection yields the impedance z, where $z = z_1 + z_2$, whereas their parallel connection yields $1/z = 1/z_1 + 1/z_2$.

With the concept of impedance, which has meaning only in the context of ac sinusoidal excitation at a particular frequency, one may state Kirchhoff's mesh law as follows:

The sum of the applied emfs (phasors) around any closed loop C in a circuit must equal the sum of all the potential drops of the form, $z\hat{I}$, once around the loop. Schematically,

$$\sum \hat{\mathscr{E}} = \sum z\hat{I} \tag{13.62}$$

13.5.3 The Nodal Method

Before we discuss the nodal method, we will introduce the current sources and the concept of admittance. Consider Fig. 13.11*a*. The circuitry to the left of A and B can be represented by a simple ideal voltage generator and a series impedance. Inasmuch as we are solely concerned with what happens electrically to the right of A and B, this is always a valid procedure for linear, sinusoidal circuits.

The fact of linearity means that the potential amplitude, $\hat{V}$, must be related to $\hat{I}$ as

$$\hat{V} = k_1 + k_2\hat{I} \tag{13.63}$$

where k_1 and k_2 are in general complex numbers that depend on what is inside the "box" whose exit terminals are A and B. Assuming that the devices inside the box

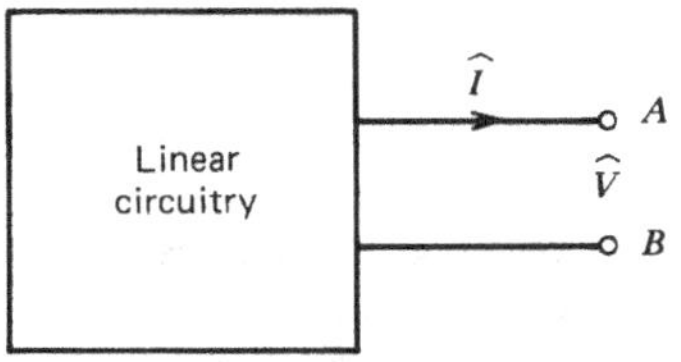

(*a*)

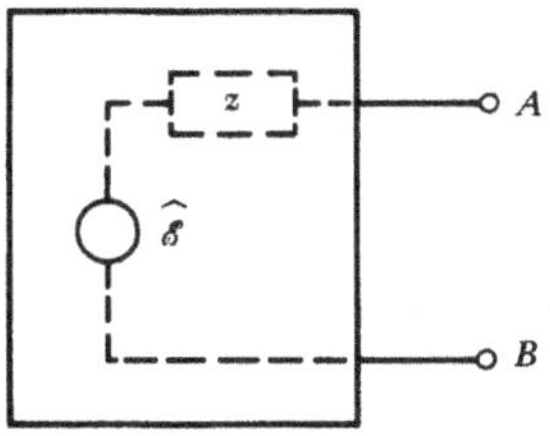

(*b*)

Figure 13.11 (*a*) Linear circuitry of single frequency ω with an output current amplitude $\hat{I}$ and voltage amplitude $\hat{V}$. (*b*) The representation of the circuit in terms of an ideal voltage generator and series impedance (Thévenin emf and output impedance).

are not themselves disturbed in the measuring process, k_1 and k_2 may be determined by two measurements:

1. Measure the potential drop across A and B when the terminals are open-circuited. Taking $\hat{I} = 0$ in Eq. (13.63) gives

$$k_1 = \hat{V} \equiv \hat{\mathscr{E}} \tag{13.64}$$

2. Short-circuit the terminals A and B, and measure the current $\hat{I}$ that flows. Taking $\hat{V} = 0$ in Eq. (13.63) gives

$$k_2 = -\frac{k_1}{\hat{I}} = \frac{-\hat{\mathscr{E}}}{\hat{I}} \equiv -z \tag{13.65}$$

Thus, substituting Eqs. (13.64) and (13.65) into Eq. (13.63), we obtain

$$\hat{V} = \hat{\mathscr{E}} - z\hat{I} \tag{13.66}$$

This relation is easily symbolized as a zero-impedance generator of emf $\hat{\mathscr{E}}$ in series with the impedance z, as shown in Fig. 13.11*b*. The quantities $\hat{\mathscr{E}}$ and z are called the *Thévenin emf* and *output impedance*, respectively.

In order to determine z and $\hat{\mathscr{E}}$ from the schematic representations of an actual circuit, we need only calculate (1) the open-circuit voltage $\hat{V}$, and (2) the impedance seen looking into the terminals A and B of the box. In the latter case, we may consider all pure emfs inside the box to be replaced by wires of zero impedance.

We note that we may write Eq. (13.66) as

$$\hat{I} = \frac{\hat{\mathscr{E}}}{z} - \frac{\hat{V}}{z} \equiv \hat{\mathscr{I}} - y\hat{V} \tag{13.67}$$

where

$$\hat{\mathscr{I}} \equiv \frac{\hat{\mathscr{E}}}{z} \qquad \text{and} \qquad y \equiv \frac{1}{z} \tag{13.68}$$

This may be schematically represented by the circuit in Fig. 13.12*a*, which is the "dual" of the Thévenin circuit shown in Fig. 13.12*b*. $\hat{\mathscr{I}}$ represents an emf that produces a constant current amplitude $\hat{\mathscr{I}}$. This is paralleled by $y = 1/z$. This is called the *Norton equivalent circuit*. The emphasis here is on the current rather than the voltage. When $\hat{\mathscr{E}}$ and z are very large, such that $\hat{\mathscr{I}}$ remains constant, then $y \to 0$, and the current $\hat{I} = \hat{\mathscr{I}}$.

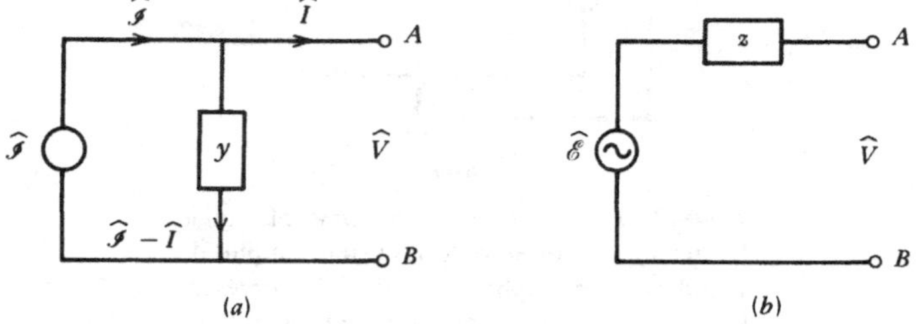

Figure 13.12 (*a*) Norton equivalent circuit, which is the dual of (*b*) the Thévenin circuit.

The reciprocal of an impedance is called *an admittance.* If, between two points of a circuit, the complex impedance is z, then the complex admittance between these points is given the notation y, and

$$y = \frac{1}{z} \tag{13.69}$$

Since z is in general complex, so is y, and we may write

$$y = G - iB \equiv \frac{1}{z} = \frac{1}{R + iX} \tag{13.70}$$

The quantity G is called the *conductance*, and B is called the *susceptance.*

The admittance is specially useful when the nodal equations (Kirchhoff's nodal law) are used to analyze a circuit, since the current phasor $\hat{I}$ is related to the voltage phasor across an element $\hat{V}$ by $\hat{I} = y\hat{V}$. Thus, for elements in parallel, having admittances y_1 and y_2, one has the nodal relation $\hat{I} = \hat{I}_1 + \hat{I}_2 = (y_1 + y_2)\hat{V}$, which may be simpler to manipulate algebraically than the equivalent relation employing the z's. Admittances combine like capacitors. Thus, for two elements in series, with admittances y_1 and y_2, one has for the total admittance $y_s = y_1 y_2/(y_1 + y_2)$, whereas for parallel elements $y_p = y_1 + y_2$.

The definition of a current source is now reiterated using the admittance terminology. In Eq. (13.67) we see that $\hat{I}$ equals the short-circuit current across A and B, and y the admittance looking into A and B when all current sources are replaced by infinite impedances.

Example 13.5 The Nodal Method

We apply Kirchhoff's nodal law to the circuits of Fig. 13.13*a* and 13.13*b*. Consider node 1 in Fig. 13.13*a*. The current $\hat{I}$ in the circuit shown is

$$\hat{I} = \hat{I}_1 + \hat{I}_2$$

Taking the voltages at nodes 1 and 2 to be $\hat{V}_1$ and $\hat{V}_2$, respectively and using $\hat{I} = y\hat{V}$, this relation between the currents becomes

$$\hat{I} = y_1(\hat{V}_1 - \hat{V}_2) + y_2(\hat{V}_1 - \hat{V}_2)$$

where the admittance, associated with R is $y_1 = 1/R$ and that associated with C is $y_2 = i\omega C$. Since $\hat{V}_1 = \mathscr{E}_0$ and taking $\hat{V}_2 = 0$ gives $\hat{I} = \mathscr{E}_0(1/R + i\omega C)$. Using Eqs. (13.48) and (13.49) to calculate the real current gives

$$I = Y\mathscr{E}_0 \cos(\omega t - \phi)$$

where

$$Y = \left(\frac{1}{R^2} + \omega^2 C^2\right)^{1/2} \qquad \text{and} \qquad \tan\phi = -\omega CR$$

Figure 13.13*b* shows a circuit with more than one node. To analyze it using the nodal method we replace the voltage source by a current source, calculate the corresponding admittances of the different elements, and assign voltages to the various independent nodes, as shown in Fig. 13.13*c*. Then we write

$$\hat{I} = y_1(\hat{V}_1 - \hat{V}_3) + y_2(\hat{V}_1 - \hat{V}_2)$$
$$0 = y_2(\hat{V}_2 - \hat{V}_1) + (y_3 + y_4)(\hat{V}_2 - \hat{V}_3)$$

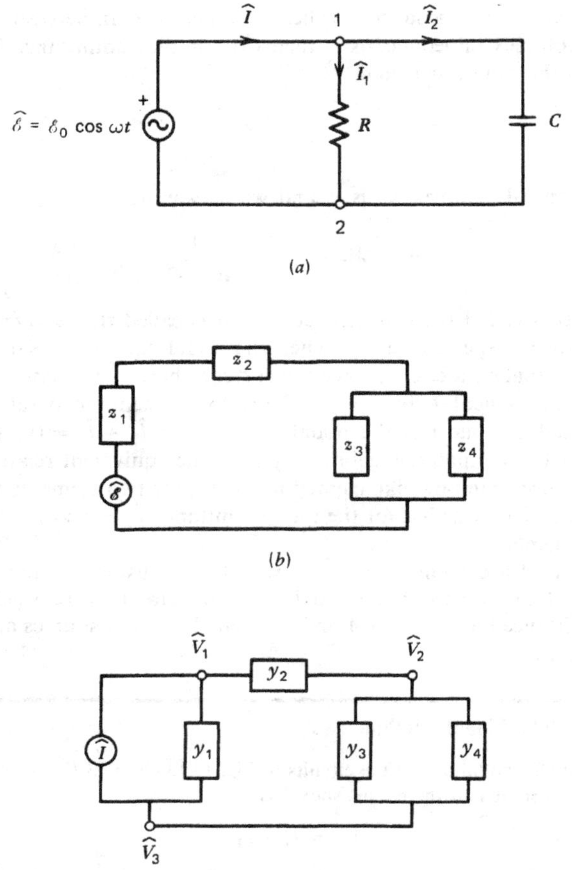

Figure 13.13 (*a*) and (*b*) Circuits for demonstrating Nodal method analysis. (*c*) The Norton equivalent circuit of (*b*).

Taking $\hat{V}_3 = 0$ (reference voltage) we obtain

$$\hat{I} = y_1\hat{V}_1 + y_2(\hat{V}_1 - \hat{V}_2)$$

$$0 = y_2(\hat{V}_2 - \hat{V}_1) + (y_3 + y_4)\hat{V}_2$$

These equations are algebraic and hence can be easily solved for the various voltages.

13.6 Power in AC circuits—Impedance Matching

Of primary importance in practical circuit considerations is a knowledge of the electric work that can be done by a source of emf or the energy absorbed by (the work done upon) elements of a circuit. The general situation is described rather well by assuming we have a source of emf, $\mathscr{E}$, applied between the terminals of some circuit, called the *load* (as shown in Fig. 13.14). We assume that $\mathscr{E}$ is given, and that

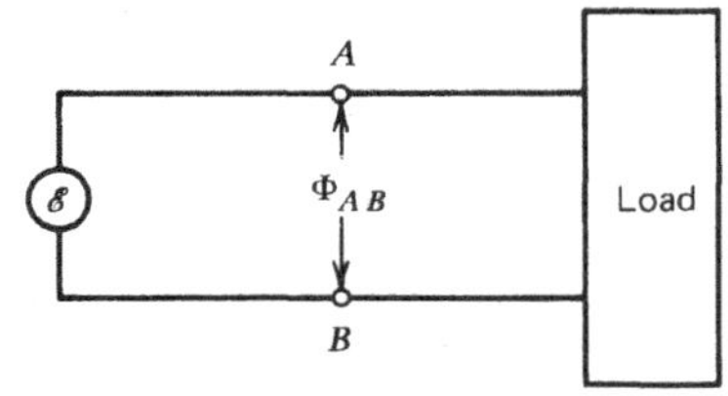

Figure 13.14 A schematic diagram of an AC source of emf showing its internal impedance and a load connected to it.

a current I flows into the load. We may take $\mathscr{E}$ as simply equal to the potential drop across the terminals to the load, $\Phi_{AB} \equiv V(t)$.

The *instantaneous* power that flows from the source to the load is given by

$$P(t) = \mathscr{E}(t)I(t) \tag{13.71}$$

This is the rate at which work is done by $\mathscr{E}$. Stating it otherwise, the rate at which energy is absorbed by the load is given by

$$P(t) = V(t)I(t) \tag{13.72}$$

In general this quantity may be positive or negative; that is, energy flow may be into or out of the load. In particular, for sinusoidal ac excitations, we may write $V(t) = V_0 \cos \omega t$ and $I(t) = I_0 \cos(\omega t - \phi)$. If the load is characterized completely by a resistance R and reactance X (that is, by an impedance $z = R + iX$), then ϕ is given simply as $\tan^{-1}(X/R)$, and I_0 is equal to V_0/Z, where $Z = |z|$. In any case,

$$P(t) = V_0 I_0 \cos \omega t \cos(\omega t - \phi) \tag{13.73}$$

Clearly if $\phi = 0$ (that is, if I and V are in phase), then $P \geq 0$. On the other hand, if I and V are out of phase by π that is $\phi = \pi$, then $P \leq 0$.

Time-averaged quantities are often of greater practical significance than the values the quantities have at any given instant. Over a period of time of length T (defined from $t = 0$ to $t = T$), the time average of a function $f(t)$ is defined by

$$\langle f(t) \rangle \equiv \frac{1}{T}\int_0^T f(t)dt \tag{13.74}$$

For sinusoidal functions, if T represents one period, $T = 2\pi/\omega$, the average over an interval T will be independent of the starting point of the interval. Moreover, the average for time intervals large compared to T will be given to good approximation by Eq. (13.74). Thus, the time average power for sinusoidal ac excitation is given by

$$\langle P \rangle = \langle V(t)I(t) \rangle = \frac{1}{T}\int_0^T V(t)I(t)dt \tag{13.75}$$

Setting $V(t) = V_0 \cos \omega t$ and $I(t) = I_0 \cos(\omega t - \phi)$ gives

$$\langle P \rangle = \frac{V_0 I_0}{2\pi}\int_0^{2\pi} \cos \omega t \cdot \cos(\omega t - \phi) d(\omega t)$$

Expanding $\cos(\omega t - \phi)$ and noting that

$$\int_0^{2\pi} \cos(\omega t)\sin(\omega t)d(\omega t) = 0 \qquad \text{and} \qquad \int_0^{2\pi} \cos^2(\omega t)d(\omega t) = \pi$$

we obtain

$$\langle P \rangle = \frac{V_0 I_0}{2} \cos \phi \tag{13.76}$$

The quantity $\cos \phi$ is called the *power factor* of the impedance or load. The average power absorbed by the load will thus be positive or negative depending on the phase angle ϕ between $V(t)$ and $I(t)$. We can see from Eq. (13.76) that if loads have no resistance, that is, if they are purely *inductive* or purely *capacitive* (or purely *reactive*), then the power factor is zero. The effective impedance of such a load is given by $z = iX$, and $\tan \phi \equiv \text{Im}(z)/\text{Re}(z) \to \pm\infty$. Therefore, $\phi \to \pm\pi/2$, and $\cos \phi = 0$. A load will be purely reactive only if it is composed of inductors and capacitors alone, having zero resistance. If the load has nonzero resistance, as is usually true (for nonsuperconductors), then $\cos \phi > 0$, and power will be dissipated in this resistance.

It should be noted that in Eqs. (13.71) to (13.76) we have used *real* quantities to calculate power or average power. It is not valid, for example, to say that $P(t) = V(t)I(t)$ if V and I are complex, and obtain the real power by taking the real part of $P(t)$. This follows because, for complex numbers z_1 and z_2, $\text{Re}(z_1 z_2) \neq \text{Re}(z_1)\cdot\text{Re}(z_2)$. However, we can find formulas, generally valid, that permit us to calculate the *average* ac power from the complex amplitudes V and I. Consider the product VI^*, where I^* is the complex conjugate of I:

$$VI^* = (Iz)I^* = |I|^2 z = |I|^2[R + iX]$$

Since $|I| \equiv I_0$, then $\text{Re}[VI^*/2] = \langle P \rangle = \frac{1}{2}I_0^2 R$. Thus

$$\langle P \rangle = \frac{1}{2}\text{Re}(VI^*) = \frac{1}{2}\text{Re}(V^*I) \qquad \text{or} \qquad \langle P \rangle = I_{\text{rms}}^2 R \tag{13.77}$$

where $I_{\text{rms}} = I_0/\sqrt{2}$ and rms stands for *root mean square.* It is useful to observe that in solving for the currents in ac circuit problems, we may indeed interpret the complex amplitudes, $\hat{V}$, $\hat{\mathscr{E}}$, and $\hat{I}$ so that their absolute values represent root mean square (rms) values. We then call these quantities *rms phasors.* We need only note that the relation $\hat{V} = z\hat{I}$ implies $|\hat{V}| = Z|\hat{I}|$. Also the fact that $|\hat{V}| = V_0 = Z|\hat{I}| = ZI_0$ implies that $V_{\text{rms}} = ZI_{\text{rms}}$. Since either of these relations implies the other, we may use either interpretation in working a problem. Of course, *all* the complex amplitudes in a problem must have the same interpretation.

As mentioned above, an ac source of emf always has some impedance associated with it, and it may be schematically represented as in Fig. 13.14. It is of interest to consider the conditions under which maximum power may be expended in the load, which we represent by the impedance $z_L = R_L + X_L(z_L$ might represent a factory, a motor, a loudspeaker, a toaster, etc.). Power enters the load through terminals A and B, and z may represent the generating station and the electric lines leading to the terminals. The average power delivered to z_L is given by

$$\langle P_L \rangle = R_L I_{\text{rms}}^2 = R_L \frac{\mathscr{E}_{\text{rms}}^2}{|z + z_L|^2}$$

Substituting for z and z_L in terms of their real and imaginary parts gives

$$\langle P_L \rangle = \frac{R_L \mathscr{E}_{\text{rms}}^2}{(R + R_L)^2 + (X + X_L)^2} \tag{13.78}$$

Each of the terms in the denominator is positive definite. It is obvious that if $X = -X_L$, the average power will be maximized with respect to variations of the circuit reactances. Thus if X_L is inductive, we should make X capacitive. If we can make $X + X_L = 0$, then

$$\langle P_L \rangle = \frac{R_L \mathscr{E}_{\text{rms}}^2}{(R + R_L)^2} \tag{13.79}$$

If we can only adjust R while keeping R_L fixed, $\langle P_L \rangle$ will be maximized by letting R be zero, since this minimizes the denominator. If, however, R is fixed and we can vary R_L, it may be seen that $\langle P_L \rangle$ will be maximized if $R_L = R$. If R_L is too large, not enough current flows; if R_L is too small, the power gets dissipated in R. When $R_L = R$, the power dissipation is equalized in the source and load impedances. Under optimum conditions the load is said to be "matched" to the source. The condition for impedance matching is

$$X = -X_L \qquad \text{and} \qquad R = R_L \qquad \text{or} \qquad z = z_L^* \tag{13.80}$$

13.7 Resonance in AC Circuits

13.7.1 Series Resonance

Series or parallel combinations of inductance, capacitance, and resistance appear often in ac circuits, and knowledge of their behavior is basic to the analysis of ac circuit properties. We discuss first the series combinations of these elements and what is called *series resonance*. As in Fig. 13.6, we assume a voltage amplitude $\hat{V}$ to appear across the series combination. If the frequency of excitation is ω, the current amplitude will be given by Eq. (13.55); that is,

$$\hat{I} = \frac{\hat{V}}{R + iX} = \frac{\hat{V}}{R + i\left(\omega L - \dfrac{1}{\omega C}\right)} \tag{13.81}$$

This equation can also be written in a polar complex form

$$\hat{I} = \frac{\hat{V}}{|Z|} e^{-i\phi}$$

where

$$Z = \left[R^2 + \left(\omega L - \frac{1}{\omega C}\right)^2\right]^{1/2} \qquad \text{and} \qquad \tan \phi = \frac{\omega^2 LC - 1}{\omega CR} \tag{13.82}$$

A discussion of any circuit is based upon what happens to the currents (or voltages) in its components as the circuit parameters are varied. In the present case, these parameters are the impedance elements R, L, and C and the excitation frequency ω. We note that when the reactance X is zero, then

$$\omega^2 = \omega_0^2 = \frac{1}{LC} \tag{13.83}$$

The circuit at this frequency is said to be *in resonance*. The frequency $\omega = \omega_0$ is called the *resonance frequency* of the circuit.

We now discuss a few features of a resonating circuit.

Phase of Current. At resonance, z becomes equal to R and hence $\tan\phi = 0$. Thus I_{rms} attains its maximum value with respect to variation of X, and the current and applied potential are instantaneously in phase.

The Voltage Across the Capacitor and the Inductance. Using Table 13.2 we write $\hat{V}_L = i\omega L\hat{I}$ and $\hat{V}_C = \hat{I}/i\omega C$. Writing $\hat{I} = \hat{V}/R$ and using the quality factor at resonance, $Q_0 = \omega_0 L/R = (\omega_0 RC)^{-1}$ [see Eqs. (13.32) and (13.83)], we obtain

$$\hat{V}_L = iQ_0\hat{V} \qquad \text{and} \qquad \hat{V}_C = -iQ_0\hat{V} \tag{13.84}$$

At resonance the complex voltages across L and C are 180° out of phase, and so cancel each other, but their real individual amplitudes are Q_0 times as large as the applied voltage amplitude.

The Energy Stored in the Circuit. The instantaneous energy stored in the inductor is given by Eq. (12.5) and in the capacitor is given by Eq. (6.44). Thus the total energy stored is

$$U = \frac{1}{2}\frac{q^2(t)}{C} + \frac{1}{2}LI^2(t) \tag{13.85}$$

The current in the circuit is easily calculated by multiplying Eq. (13.81) by $e^{i\omega t}$ and then taking the real part: $I(t) = I_0\cos(\omega t - \phi)$, where we have taken the phase of $\hat{V}$ to be zero at $t = 0(|\hat{V}| = V_0)$. The charge is similarly calculated as $q(t) = q_0\sin(\omega t - \phi)$, where $I_0 = V_0/Z$, and $q_0 = I_0/\omega = V_0/Z\omega$. Therefore

$$U(t) = \frac{1}{2C}q_0^2\sin^2(\omega t - \phi) + \frac{1}{2}LI_0^2\cos^2(\omega t - \phi) \tag{13.86}$$

At resonance, we have $Z = R$, $I = V_0/R$, and $q_0 = I_0/\omega_0$; hence the amplitudes of the sine and cosine become equal: $q_0^2/2C = I_0^2/(2C\omega_0^2) = \frac{1}{2}LI_0^2$. Thus the total energy stored at resonance becomes

$$U(t) = \frac{1}{2}\frac{q_0^2}{C} = \frac{1}{2}LI_0^2 = \frac{1}{2}\frac{LV_0^2}{R^2} \tag{13.87}$$

which is constant, and oscillates between being stored in the magnetic field (in the inductor) and in the electric field (in the capacitor).

Because of the presence of a resistor in the circuit, energy is lost at a rate of $\langle P\rangle = RI_{\text{rms}}^2$. Thus the energy lost per cycle is

$$\Delta U = RI_{\text{rms}}^2 T \tag{13.88}$$

At resonance, the ratio of the maximum instantaneous energy stored to the energy dissipated in one cycle is obtained by dividing U by ΔU [see also Eq. (13.33)]:

$$\frac{U}{\Delta U} = \frac{LI_{\text{rms}}^2}{RI_{\text{rms}}^2 T} = \frac{\omega_0 L}{R}\cdot\frac{1}{2\pi} = \frac{Q_0}{2\pi} \qquad \text{(resonance)} \tag{13.89}$$

where Q_0 is the quality factor at resonance. Thus, as was stated before, a circuit with a large Q_0 dissipates a small amount of energy per cycle compared to the stored energy.

Frequency Dependence of the Circuit. The frequency dependence of the circuit can be studied by studying the frequency dependence of the current. We first write it in terms of Q_0 and $x = \omega/\omega_0$; that is,

$$\hat{I} = \frac{\hat{V}}{Z} = \hat{I}_m\frac{1}{\sqrt{1 + Q_0^2\left(x - \frac{1}{x}\right)^2}} \tag{13.90}$$

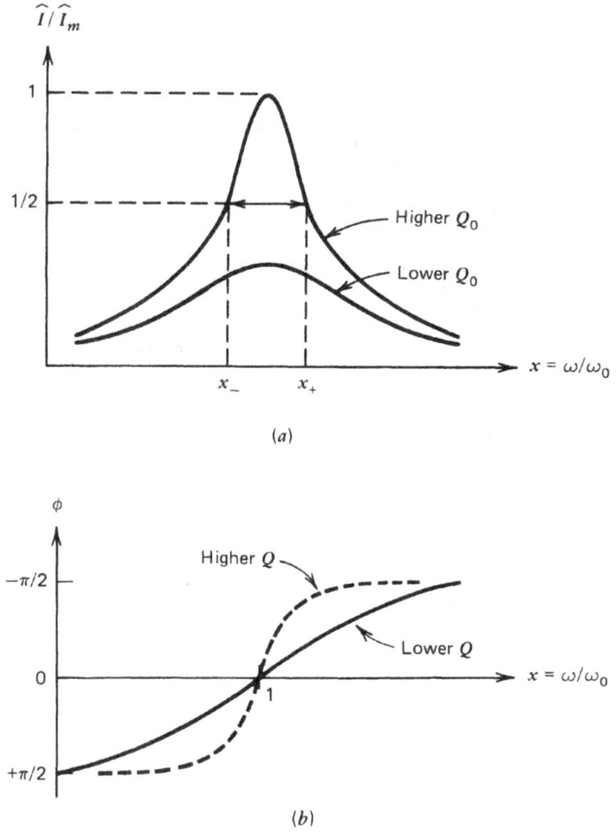

Figure 13.15 Frequency behavior of a series *RLC* resonance near the resonance frequency. (*a*) Current showing the bandwidth of the resonance. (*b*) Phase angle.

where $\hat{I}_m = \hat{V}/R$. The shape of the function is primarily dependent on the parameter Q_0. If Q_0 is small ($\ll 1$), $\hat{I}$ changes only slowly with x. If $Q_0 \gg 1$, $\hat{I}$ becomes appreciable only when $x \approx 1$—i.e., near resonance.

The average dissipation of power in the circuit is equal to RI^2_{rms}. Using $I_{\text{rms}} = (1/Z)V_{\text{rms}}$ gives

$$\langle P \rangle = \frac{RV^2_{\text{rms}}}{Z^2} = \frac{V^2_{\text{rms}}/R}{1 + \left[Q_0\left(x - \dfrac{1}{x}\right)\right]^2} \tag{13.91}$$

Figure 13.15 shows $\hat{I}/\hat{I}_m$ as a function of x. As a function of frequency, $\langle P \rangle$ is greatest when $x = 1$. It falls to one-half its maximum value when $Q_0(x - 1/x) = \pm 1$. This defines two frequencies ω_+ and ω_-, whose corresponding values of x are given by x_+ and x_-:

$$x_\pm = \pm\frac{1}{2Q_0} + \sqrt{1 + \left(\frac{1}{2Q_0}\right)^2}$$

The *bandwidth* of the resonance curve is defined as

$$\frac{\omega_+ - \omega_-}{\omega_0} = \frac{x_+ - x_-}{1} = \frac{1}{Q_0} \tag{13.92}$$

Q_0 is thus seen to be equal to the reciprocal of the relative bandwidth. A circuit with large Q_0 is *selective*; that is, it has a sharp resonance

Quality Factor Away from Resonance: Inductive and Capacitive Circuits. The energy in the circuit at any time is given by Eq. (13.86):

$$U = \frac{1}{2} LI_0^2 \left[\frac{1}{x^2} \sin^2(\omega t - \phi) + \cos^2(\omega t - \phi) \right] \tag{13.93}$$

The value of U fluctuates in time, except at resonance, when $x = 1$ (as was shown above) and the maximum value of the energy U_{max} is given by the maximum electric or magnetic energy depending on whether $x < 1$ or $x > 1$, respectively. If $x < 1$—that is, at frequencies lower than the natural frequency—the first term on the right-hand side of Eq. (13.93) has a larger maximum than the second term, and the circuit is called capacitive. In this case, $U_{max} = LI_{rms}^2/x^2 = q_0^2/2C$. Since $\Delta U = RI_{rms}^2 T$, then

$$\frac{U_{max}}{\Delta U} = \frac{\omega L}{2\pi R x^2}$$

Taking $U_{max}/\Delta U$ to be equal to $Q/2\pi$, where Q is the quality factor, gives an expression for Q:

$$Q = \frac{1}{\omega RC} \quad \text{(capacitive)} \tag{13.94}$$

If $x > 1$ (that is, at $\omega > \omega_0$) the circuit is inductive and $U_{max} = LI_{rms}^2$. Thus

$$\frac{U_{max}}{\Delta U} = \frac{\omega L}{R} \cdot \frac{1}{2\pi}$$

Equating $U_{max}/\Delta U$ to $Q/2\pi$ gives

$$Q = \frac{\omega L}{R} \quad \text{(inductive)} \tag{13.95}$$

The Q of any circuit is thus defined by $2\pi(U_{max}/\Delta U)$. The Q of an inductance is thus defined as $\omega L/R$, where R is its resistance, and the Q of a capacitance is defined by $1/\omega RC$. The inductance is presumed to be inductive and the capacitance to be capacitive. (One must be warned, however, that at very high frequencies inductors may become capacitive and capacitors inductive).

The selective frequency property of the resonant circuits makes them useful as a kind of filter (see Fig. 13.16). In this use one often "tunes" the circuit by adjusting the capacitance C. Resonance is achieved when $C = 1/\omega^2 L$, where ω is the frequency that is to be emphasized. In Fig. 13.16, if Q is large at frequency ω, only signals lying in the range $\Delta\omega \approx \omega/Q_0$ (around ω) [see Eq. (13.92)] will be passed strongly to the (high-impedance) amplifier.

It is interesting to remark that for "high-Q" circuits, the resonance frequency is approximately the natural oscillation frequency of the circuit [see Eq. (13.24)]. In the latter case,

$$\omega = \left(\omega_0^2 - \frac{R^2}{4L^2} \right)^{1/2} = \omega_0 \left[1 - \left(\frac{1}{2Q_0} \right)^2 \right]^{1/2} \approx \omega_0$$

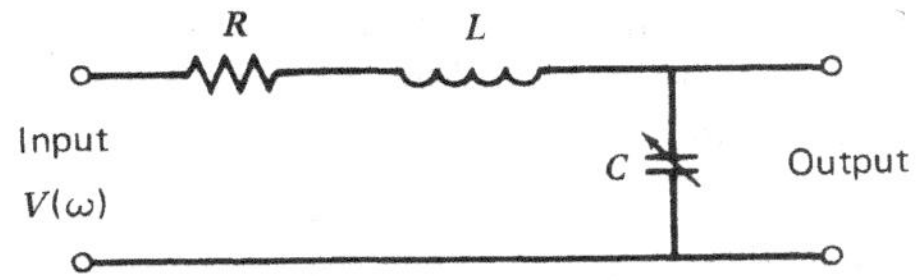

Figure 13.16 A filtering circuit.

A high-Q circuit resonates when the exciting frequency is approximately equal to its natural frequency. One then has a "reinforcing" condition on the circuit that allows it to build up large oscillations, the energy being injected into the system in phase with its free oscillation frequency. This is perhaps the most characteristic aspect of resonance. In the limit as $R \to 0$, and if the applied emf continually supplies energy to the circuit, the oscillations will grow indefinitely. In a superconducting LC circuit, oscillations may indeed persist very long times without any energy being fed into the circuit.

13.7.2 Parallel Resonance

Rather than elements R, L, and C in series, one often encounters a configuration similar to that in Fig. 13.17, where a voltage $V_0 \cos \omega t$ is applied to an inductor and capacitor in parallel. The resistance R will commonly be associated with the inductance, as shown. A resistance in the branch of the circuit having the capacitor is omitted here for simplicity and because it is often negligible in practice.

Some aspects of the behavior of this circuit can be understood qualitatively. If R is small, we effectively have an inductance and capacitance in parallel. The current $\hat{I}_L$ in the inductance, lags behind the impressed voltage by 90°, whereas the current $\hat{I}_C$ through the capacitance leads the impressed voltage by 90°. If the admittances $1/\omega L$ and ωC for the two branches differ, the total current $\hat{I} = \hat{I}_C + \hat{I}_L$ will pass mostly through the larger admittance. Since from Table 13.2 we find that $\hat{I}_C = i\omega C\hat{V}$ and $\hat{I}_L = \hat{V}/i\omega L$, then

$$\hat{I} = i\hat{V}\left(\omega C - \frac{1}{\omega L}\right) \tag{13.96}$$

If the admittances were equal, $\omega C = 1/\omega L$, then $\hat{I}_C \approx -\hat{I}_L$ and $\hat{I} = 0$ and the circuit exhibits what is called *parallel resonance*. The impedance presented to terminals AB under these conditions is infinite. This is to be contrasted with the series RLC circuit, whose impedance would approach zero for the same conditions.

To calculate the current in the circuit when R is not negligible, we calculate its total admittance y. The admittances of the resistor and the inductor are $1/R$ and

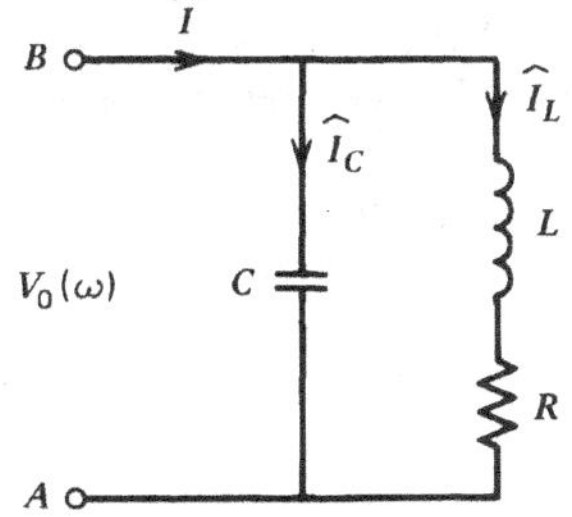

Figure 13.17 A parallel resonant circuit.

$1/i\omega L$. Since they are in series, then their total admittance is $y_L = 1/(R + i\omega L)$. The admittance of the capacitor is $y_C = i\omega C$. The complex total admittance y is calculated by adding these admittances in parallel:

$$y = \frac{1}{R + i\omega L} + i\omega C = \frac{R}{R^2 + \omega^2 L^2} + i\left(\omega C - \frac{\omega L}{R^2 + \omega^2 L^2}\right). \qquad (13.97)$$

and the current in the circuit is

$$\hat{I} = \hat{V} y = \hat{V} Y e^{-i\phi} \qquad Y = |y|$$

We will now discuss some properties of the circuit using the details of y. Resonance for a parallel RLC circuit as here discussed is not so unambiguously defined as for the series case. We have three situations to consider.

Phase Resonance. What is called *phase resonance* occurs when y becomes real, so that $\phi = 0$ and the current and voltage are in phase. From Eq. (13.97) this occurs where $x = 0$ (trivial case), and when

$$x^2 = 1 - \frac{1}{Q_0^2} \equiv x_0^2 \qquad (13.98)$$

where $x = \omega/\omega_0$. By the resonance condition, we shall usually refer to the phase resonance.

Antiresonance. What is called *antiresonance* occurs when $|y|$ is minimum, and thus $|\hat{I}|$ is also minimum (if V_0 is constant). In this case we find

$$x^2 = \sqrt{1 + \frac{2}{Q_0^2}} - \frac{1}{Q_0^2} \equiv x_0'^2 \qquad (13.99)$$

when we minimize with respect to x (or frequency) alone.

Simultaneous resonance and antiresonance at high Q_0. If $Q_0 > 10$, then x_0 and x_0', which are given in Eqs. (13.98) and (13.99) and define the resonance and antiresonance cases respectively, become nearly unity to within 0.5 percent. Hence both cases coincide within 0.5 percent.

The filtering property of series and parallel circuits means, for example, that if $Q_0 \gg 1$, frequencies differing "significantly" from the resonance frequencies will not be "passed" by the series circuit or "blocked" by the parallel circuit. The range of frequencies passed or blocked will be within a range of approximately ω_0/Q_0. If $Q_0 \gg 1$, the "bandpass" or "band stop" will be relatively narrow ($\Delta\omega/\omega_0 \ll 1$). See Examples 13.6 and 13.7 for some applications of resonant circuits.

We must emphasize, finally, that in discussing resonant circuits, we have assumed that a constant voltage amplitude V_0 was applied, and therefore the current was determined by the admittance of the circuits. In practice, we may not have such a constant voltage supply, in which case the response of the circuit may be quite different.

If we had a constant current amplitude rather than a constant voltage amplitude, the series circuit would not show resonance curves at all for the voltage amplitudes across the capacitor or inductor (see Problem 13.22). The parallel circuit in contrast, would produce very large voltages across it at resonance, with a resonance curve like that of the series circuit for current (see Problem 13.22). In any case, our impedance and admittance expressions are valid, and there is no difficulty in applying our knowledge to specific problems, remembering that parallel resonance implies a high impedance, and series resonance, a low impedance (between the input terminals).

Example 13.6 Resonant Circuits

Consider the RLC circuit shown in Fig. 13.18. The inductors and capacitors are ideal and lossless. The voltage source is $\mathscr{E} = V_0 \sin \omega t$, where $\omega = 1/\sqrt{LC}$. In order to analyze the

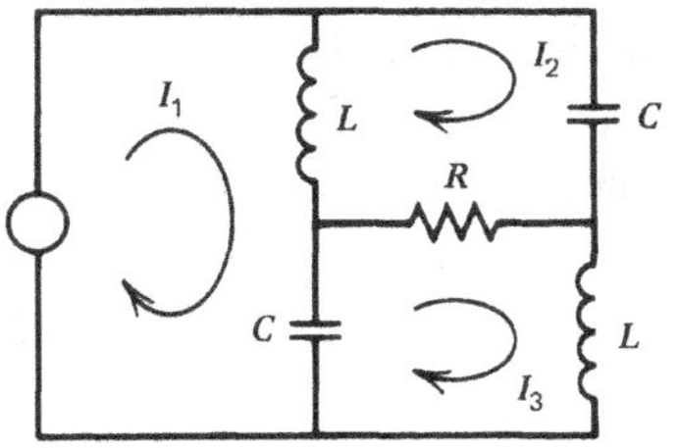

Figure 13.18 Multiloop resonant circuit.

response of this circuit we use Kirchhoff's phasor laws for the three loops using the assigned currents; that is,

$$\hat{V}_0 = i\omega L(\hat{I}_1 - \hat{I}_2) - \frac{i}{\omega C}(\hat{I}_1 - \hat{I}_3)$$

$$0 = i\omega L(\hat{I}_2 - \hat{I}_1) - \frac{i}{\omega C}\hat{I}_2 + R(\hat{I}_2 - \hat{I}_3)$$

$$0 = -\frac{i}{\omega C}(\hat{I}_3 - \hat{I}_1) + R(\hat{I}_3 - \hat{I}_2) + i\omega L\hat{I}_3$$

After dividing the first and second equations by $i\omega L$ and utilizing the relation $\omega^2 LC = 1$, we get

$$\hat{I}_3 - \hat{I}_2 = -\frac{i\hat{V}_0}{\omega L} \quad \text{and} \quad \hat{I}_1 = -\frac{iR}{\omega L}(\hat{I}_2 - \hat{I}_3) = -\frac{R\hat{V}_0}{\omega^2 L^2}$$

Multiplying $\hat{I}_3 - \hat{I}_2$ by $e^{i\omega t}$ gives the complex current in the resistor. Because the voltage source has $\sin \omega t$ time dependence, then the imaginary part of this complex current gives the real current in the resistor; that is,

$$I_3 - I_2 = \frac{V_0}{\omega L}\sin\left(\omega t - \frac{\pi}{2}\right) = \frac{-V_0}{\omega L}\cos \omega t$$

We observe that the current in the resistor is independent of the resistor itself and 90° out of phase with respect to the generator.

It is interesting to calculate the power supplied by the source. One way to calculate it is by use of the complex amplitudes of the current that passes through it and the voltage across it, as follows:

$$\langle P \rangle = \frac{1}{2}\hat{I}_1^*\hat{V} = \frac{1}{2}\frac{V_0^2 R}{\omega^2 L^2}$$

Example 13.7 Filters—Resonant Circuits

The general class of filters is too complicated to study systematically in an example, but there is a special class that can be treated here. In this class the elements are purely reactive (ideal capacitors and inductances). Consider Fig. 13.19*a*, which shows n stages of series/parallel elements in the form of a T filter. Figure 13.19*b* shows the single stage from which the filter is

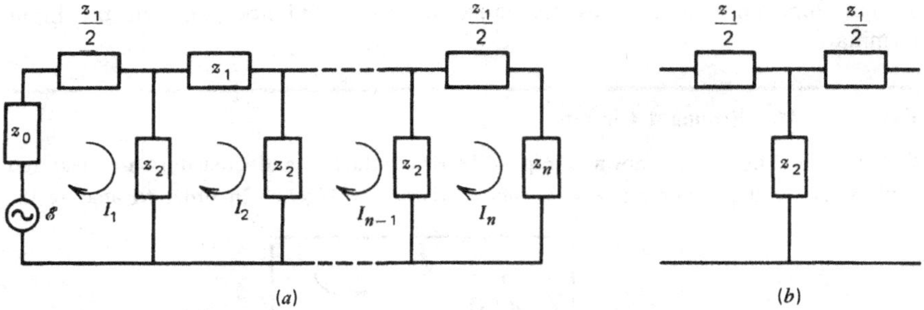

Figure 13.19 Schematic diagram of multistage filters. (*a*) The filter. (*b*) The building T block of the filter (one stage).

built. The complex impedances z_1 and z_2 represent those of a capacitor and an inductor or of an inductor and a capacitor respectively.

For the Kirchhoff's set of loop equations we obtain

$$\begin{aligned}
\mathscr{E} &= I_1 z_0 + \frac{I_1 z_1}{2} - (I_2 - I_1)z_2 \\
0 &= (I_2 - I_1)z_2 + I_2 z_1 + (I_2 - I_3)z_2 \\
&\vdots \\
0 &= (I_m - I_{m-1})z_2 + I_m z_1 + (I_m - I_{m+1})z_2 \\
&\vdots \\
0 &= (I_n - I_{n-1})z_2 + I_n\left(\frac{z_1}{2} + z_n\right)
\end{aligned} \tag{13.100}$$

One can show by substitution that the currents in the various loops are given by

$$I_m = Ae^{m\gamma + i\omega t} + Be^{-m\gamma + i\omega t} \qquad m = 1, 2, \ldots \tag{13.101}$$

where

$$\cosh(\pm\gamma) = 1 + \frac{z_1}{2z_2} = 1 - \frac{Z_1}{2Z_2} \tag{13.102}$$

Capital Z is used to indicate the magnitude of the complex impedance. In general γ can be complex even though $\cosh \gamma$ is a real quantity. That is, in general we have $\gamma = \alpha + i\beta$, where α and β are real. Using the expansion

$$\cosh \gamma = \cosh(\alpha + i\beta) = \cosh \alpha \cos \beta + i \sinh \alpha \sin \beta = \text{real quantity}$$

one can show that Eq. (13.102) can be satisfied if γ is pure imaginary; that is, $\alpha = 0$:

$$\gamma = i\beta \qquad \alpha = 0 \tag{13.103}$$

or is a complex quantity whose imaginary part is multiples of π:

$$\gamma = \alpha + in\pi \qquad \text{where } n \text{ is an integer} \tag{13.104}$$

Upon the substitution of these expressions for γ in Eq. (13.101) one can see that the current corresponding to the two expressions are drastically different. Whereas for $\gamma = i\beta$ the solution is oscillatory in nature and suffering no decay, we observe that for $\gamma = \alpha + in\pi$ the solution is a decaying one. Which solution is applicable to a given problem depends on the magnitude of Z_1/Z_2. If $Z_1/Z_2 \leq 2$, the oscillatory solution is applicable. On the other hand if $Z_1/Z_2 \geq 2$, the decaying solution becomes applicable. Since Z_1/Z_2 depends on the frequency of the source, the response of a given circuit is expected to have a frequency region in which an oscillatory solution is applicable and another region in which a decaying one is applicable.

Let us first study the nature of the two regions. When γ is pure imaginary, then

$$I_m = Ae^{i(m\beta + \omega t)} + Be^{-i(m\beta - \omega t)}$$

The second term represents a sinusoid current of amplitude B that flows from the source to the termination unattenuated and has phase shift β for each stage of the filter. The first term, on the other hand, flows from the termination to the source (reflected current). If now A and B are real and equal we have

$$I_m = 2Ae^{i\omega t} \cos m\beta$$

which is just a standing wave. If, on the other hand, $A = 0$, the filter has the property of a delay line with a frequency-dependent delay $m\beta$.

In the second type of solution, we have $\gamma = \alpha + in\pi$, with $\alpha > 0$, and

$$I_m = Ae^{m\alpha}e^{i\omega t} + Be^{-m\alpha}e^{i\omega t}$$

This case corresponds to two sinusoidal current waves, in which the second is attenuated by a factor $e^{-\alpha}$ for each stage to the right and the first is attenuated by the same factor for each stage to the left. If now $A = 0$ (no reflected wave), the filter has the characteristic of an attenuater with the total attenuation $e^{-n\alpha}$.

Let us now analyze two specific filters. In the first we take $z_1 = i\omega L$ and $z_2 = -i/\omega C$, as shown in Fig. 13.20. In this we have $\cosh \gamma = 1 - \omega^2 LC/2$. For $\omega^2 \leq 4/LC$, γ can only be

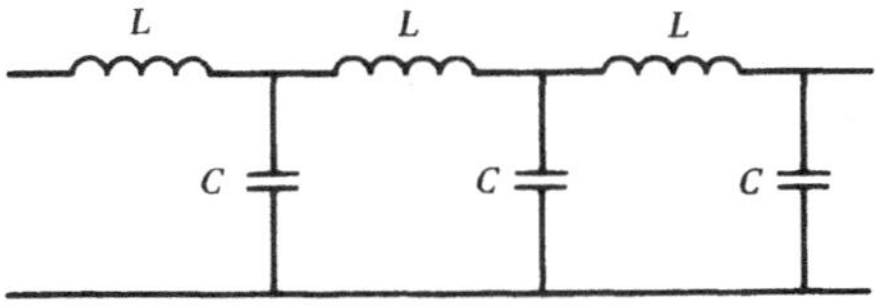

Figure 13.20 Schematic representation of a portion of a low-pass filter.

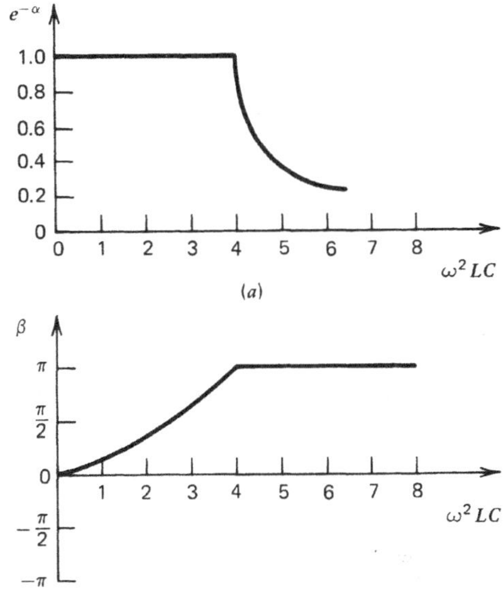

(b)

Figure 13.21 Frequency behavior per stage of a low-pass filter. (*a*) Attenuation. (*b*) Phase Shift.

pure imaginary: $\gamma = i\beta$ with $\cos \beta = 1 - \omega^2 LC/2$. This solution is called the *passband* solution, since it propagates without attenuation. For $\omega^2 \geq 4/LC$, γ can only be of the form $\gamma = \alpha + in\pi$ with $\cos \alpha = \omega^2 LC/2 - 1$, and the corresponding solution is called the *stop-band* solution since the current gets attenuated. Since this filter passes only frequencies less than $\sqrt{4/LC}$, then it is called *low-pass* filter. Figure 13.21*a* shows the attenuation per stage of this filter as a function of frequency, while Fig. 13.21*b* shows the phase shift per stage as a function of frequency.

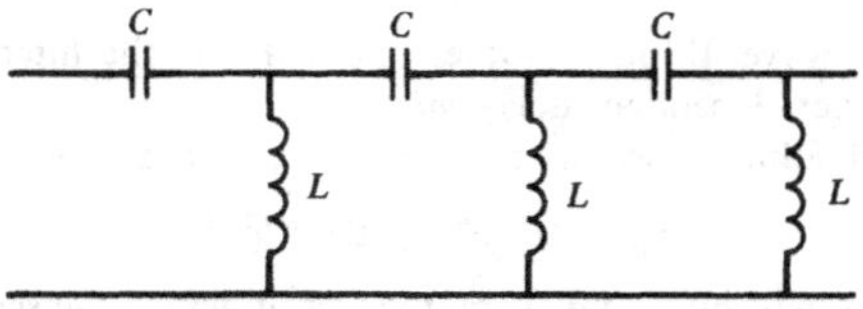

Figure 13.22 Schematic diagram of a portion of a high-pass filter.

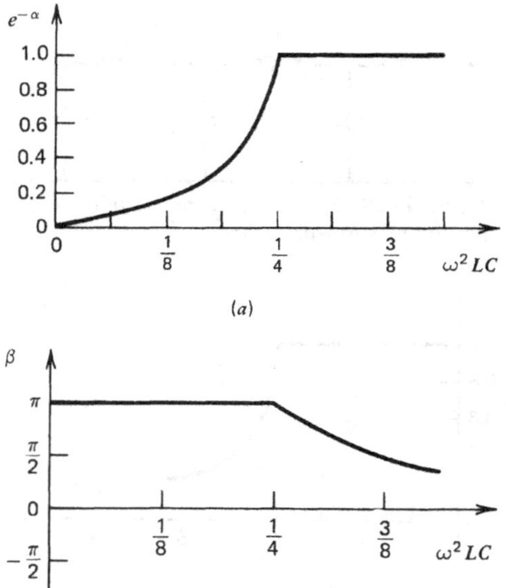

Figure 13.23 Frequency behavior per stage of a high-pass filter. (*a*) Attenuation. (*b*) Phase shift.

In the second example, which is shown in Fig. 13.22, we take $z_1 = -i/\omega C$ and $z_2 = i\omega L$, and consequently we have $\cosh \gamma = 1 - 1/2\omega^2 LC$. For $\omega^2 > 1/4LC$, we have $\alpha = 0$, $\cos \beta = 1 - 1/2\omega^2 LC$, and the corresponding solution is a *high-frequency* passband solution. On the other hand when $\omega^2 < 1/4LC$, we have $\beta = n\pi$ and $\cosh \alpha = 1 - 1/(2\omega^2 LC)$, and the solution is a low-frequency stop band. Figure 13.23*a* shows the attenuation and Fig. 13.23*b* shows the phase shift per stage of the filter as a function of frequency.

If one is interested in a filter that passes only a band of frequencies, it is then obvious that a combination of low-pass and high-pass filters should be used. Such a *bandpass* filter is shown in Fig. 13.40, in conjunction with Problem 13.25. We leave the analysis of this filter as an exercise.

Example 13.8 Resonance in Other Branches of Physics

"Resonance" equations, having the character of Eq. (13.92) occur in many branches of physics, as you will soon discover. This equation can be written as

$$\Delta\omega = \frac{\omega_0}{Q_0} = \frac{R}{L}.$$

Now, we remember that L/R has the dimensions of time, and in fact is the time $\Delta\tau$, during which the energy in a *freely* oscillating RLC circuit decays to $1/e$ of its initial value. [See Eq. (13.10).] Thus

$$\Delta\omega \cdot \Delta\tau = 1$$

The "lifetime" of the circuit times its "frequency width" is equal to unity. In quantum physics, we learn that a frequency times the constant $\hbar$ (that is, $\hbar\omega$), is interpreted as a quantum of energy. The above relation may then assume the form:

$$\hbar\Delta\omega \cdot \Delta\tau = \hbar \qquad \text{or} \qquad \Delta E \cdot \Delta\tau = \hbar$$

This is a form of the famous Heisenberg uncertainty principle. As an example in nuclear physics, ΔE might represent the width of a "resonance" or "energy" level, and $\Delta\tau$ might represent its "lifetime." In fact, a famous formula used in nuclear physics is the Breit-Wigner formula:

$$\psi(E) = \frac{\text{constant}}{1 + i\left[\dfrac{2(E - E_0)}{\Gamma}\right]}$$

which gives the behavior or "wave function" ψ of an unstable energy level as a function of energy E where E_0 is the energy of the center of the level, and Γ is a constant measured in energy units. The analog to this expression in our study of resonance is

$$\hat{I} = \frac{\hat{V}/R}{1 + iQ_0(x - 1/x)}$$

Near the resonance,

$$x - \frac{1}{x} \rightarrow 2\delta \equiv \frac{\omega - \omega_0}{\omega_0}$$

Then

$$\hat{I} \approx \frac{\hat{V}/R}{1 + i\left[\dfrac{2(\omega - \omega_0)}{\Delta\omega}\right]} \qquad \text{for } |\omega - \omega_0| \ll \omega_0$$

Both expressions have the same form. Interpreting frequencies as energies, $\hat{I}$ is thus like the wave function of a level. It determines how the "system" acts!

13.8 Summary

In this chapter we have considered electric circuits that contain elements that store energy, such as inductors and capacitors. The analysis is based on the quasi-static equations of electromagnetism. Two types of excitations will be treated: sources of constant step voltages (dc) and sources of sinusoidal voltages (ac). In the former the response is calculated in the time domain, and in the latter it is calculated in the frequency domain.

For the dc case, the current in the circuit is a slow function of time, and can be determined by applying Kirchhoff's circuit laws. The voltage, current or charge are instantaneously related in the case of R, L, and C and mutual inductance M as follows:

$$V_R = IR \qquad V_C = \frac{q}{C} \qquad V_L = L\frac{dI}{dt} \qquad V_M = M\frac{dI}{dt}$$

For an RLC circuit driven by a voltage source $\mathscr{E}(t)$ we have

$$L\frac{d^2q}{dt^2} + R\frac{dq}{dt} + \frac{q}{C} = \mathscr{E} \tag{13.21}$$

If $C \to \infty$, the circuit reduces to an RL circuit, where as if $L = 0$, the circuit reduces to an RC circuit. For a multiloop circuit one gets a set of coupled differential equations. The solution of this equation or those of other circuits has in general steady-state solution (particular solution) plus a transient solution corresponding to the solution of the equation with $\mathscr{E}(t)$ taken zero (homogeneous solution). The initial conditions are then used to determine the arbitrary constants of the homogeneous equation. The solution is oscillatory if $\omega^2 = 1/LC - R^2/4L^2 > 0$ and also decays in time with $R/2L$ decay constant. If $\omega^2 = 0$ or < 0 the solution is nonoscillatory, corresponding to the critically damped and the overdamped cases, respectively. A measure of the rate of decay of the oscillations is the Q of the circuit, given as follows:

$$Q = \frac{\omega L}{R} \tag{13.32}$$

In an ac circuit the voltage source is sinusoidal, of the form $\mathscr{E}(t) = \mathscr{E}_0 \cos(\omega t + \phi)$. If one represents $\mathscr{E}$, I, and q in complex form

$$\hat{\mathscr{E}}e^{i\omega t + \phi} \qquad \hat{I}e^{i\omega t} \qquad \hat{q}e^{i\omega t}$$

where $\hat{\mathscr{E}}$, $\hat{I}$, and $\hat{q}$ are phasors independent of time, then the voltage, current, and charge relationships for the various elements become

$$V_R = R\hat{I} \qquad V_C = \frac{\hat{I}}{i\omega C} \qquad V_L = i\omega L\hat{I} \qquad V_M = i\omega M\hat{I} \quad \text{(Table 13.2)}$$

The quantities R, $1/i\omega C$, $i\omega L$, and $i\omega M$ are referred to as the impedances of these elements: z_R, z_C, z_L and z_M. Impedances add just like resistors. The overall impedance z of a circuit is defined via the relation

$$\hat{\mathscr{E}} = z\hat{I} \tag{13.60}$$

In terms of the phasors, the differential equations of the circuits transform into algebraic equations, which can be solved simultaneously if the circuit contains more than one loop. In fact, one can write these algebraic equations directly by applying Kirchhoff's mesh and nodal laws to the phasors themselves; that is,

$$\Sigma\hat{\mathscr{E}} = \Sigma z\hat{I} \qquad \Sigma\hat{I} = 0$$

The reciprocal of an impedance, called an admittance, along with the concept of current sources replacing the voltage sources are at the heart of the implementation of the nodal method (Norton equivalent circuit).

Of primary importance in circuit considerations is a knowledge of the electric work that can be done by a source of emf or the energy absorbed (work done upon) elements of a circuit. The instantaneous power that flows from the source to the load is given by

$$P(t) = \mathscr{E}(t)I(t) \tag{13.71}$$

Also the power absorbed by the load is given by

$$P(t) = V(t)I(t) \tag{13.72}$$

where it is understood that all quantities used in these expressions are real. For ac circuits we may write $V(t) = V_0 \cos \omega t$ and $I = I_0 \cos(\omega t - \phi)$. In this case the time average value of P is

$$\langle P \rangle = \frac{1}{2} I_0 V_0 \cos \phi$$

where $\cos \phi$ is called the power factor. For loads that have no resistance $\cos \phi = 0$, and for loads that have a resistance $\cos \phi > 0$. The quantities $I_0/\sqrt{2}$ and $V_0\sqrt{2}$ are called the root mean square or the effective value of the respective quantities. In terms of complex quantities,

$$\langle P \rangle = \frac{1}{2} \operatorname{Re}(I^*V) = \frac{1}{2} \operatorname{Re}(V^*I) \tag{13.77}$$

Series or parallel combinations of inductance, capacitance, and resistance appear often in ac circuits. As a function of frequency, a series combination exhibits a resonance near $\omega = \omega_0 = \sqrt{1/LC}$, where the impedance of the circuit goes through a minimum and the current goes through a maximum. The sharpness of the resonance is given by the quality factor Q, defined above as

$$Q_0 = \frac{\omega_0 L}{R} = \frac{\omega_0}{\Delta\omega} \tag{13.89}$$

where $|\Delta\omega|$ is the full width at half maximum of the response (bandwidth). A parallel circuit also exhibits resonance at $\omega_0 = 1/\sqrt{LC}$, but with z being a maximum and I_0 being a minimum. This, however, is often called antiresonance. If the phase ϕ in parallel circuits goes through zero at some frequency, then the circuit exhibits what is called phase resonance. At high quality factors both phase and antiresonance occur simultaneously.

Resonant circuits are of practical importance because various filters are based on them.

Problems

13.1 Consider the RL circuit shown in Fig. 13.24. A voltage pulse V is applied to the circuit: $V = 0$ for $t < 0$, $V = V_0$ for $0 < t < T$ and $V = 0$ for $t > T$. (a) Use Kirchhoff's voltage law to write an equation for the current in the circuit. (b) Determine the current in the circuit for $0 < t < T$. (c) Determine the voltage across the inductance for $0 < t < T$. (d) Determine the voltage across the inductance for $t > T$.

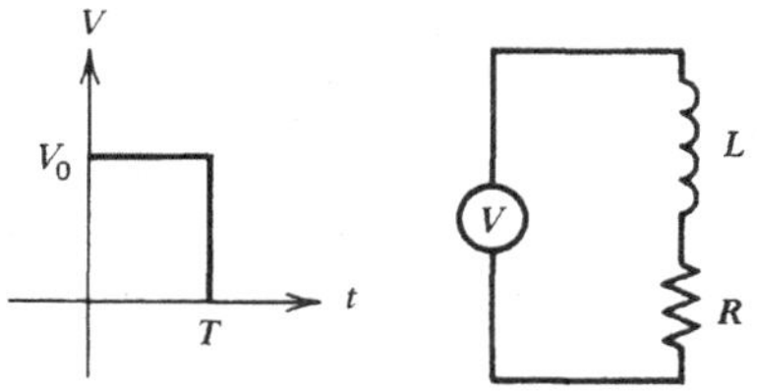

Figure 13.24 An RL circuit driven by a voltage pulse.

13.2 Determine the response of an RL circuit driven by a voltage source $V = V_0 e^{-Rt/L}$ for the case where the initial current in the circuit is zero, and when it is I_0. (When the driving voltage has the same form as the natural behavior of the circuit, the phenomenon called *resonance* appears.)

13.3 The circuit shown in Fig. 13.25 is composed of a capacitor C, a resistor R, and an inductance $L = 2R^2C$. At time $t = 0$, the charge of the capacitor is Q_0 and there is no current passing through the inductance. (a) Using the notations in the figure write down Kirchhoff's loop equations for the LC and RL loops. (b) Write down

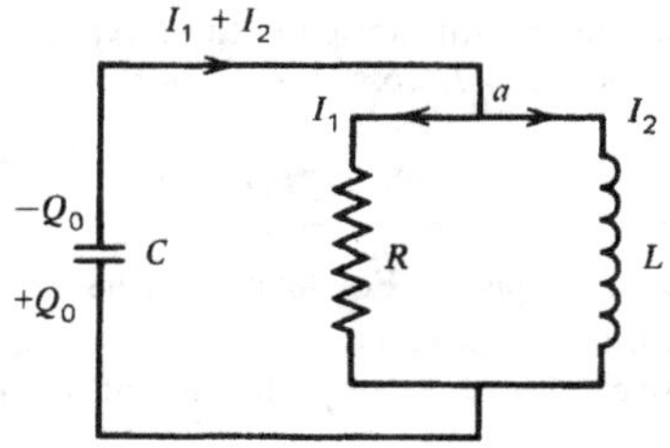

Figure 13.25

Kirchhoff's nodal equation at node a. (c) Show that the charge of the capacitor at a later time $Q = Q_0 e^{-kt}(\cos kt - \sin kt)$ where $k = 1/2RC$.

13.4 The RC network of Fig. 13.26 has the interesting feature that if any periodic source [that is, $f(t + T) = f(t)$, where T is the period] is applied, the resulting circuit behaves as an analog computer for appropriate RC and steady-state conditions. Take $R = 2\ \text{k}\Omega$ and $C = 0.05\ \mu\text{F}$. (a) Determine V_C for one period for a source of square wave of angular frequency 2×10^4 Hz. (b) Determine V_R for one period for a source of a triangle wave of frequency 10^2 Hz.

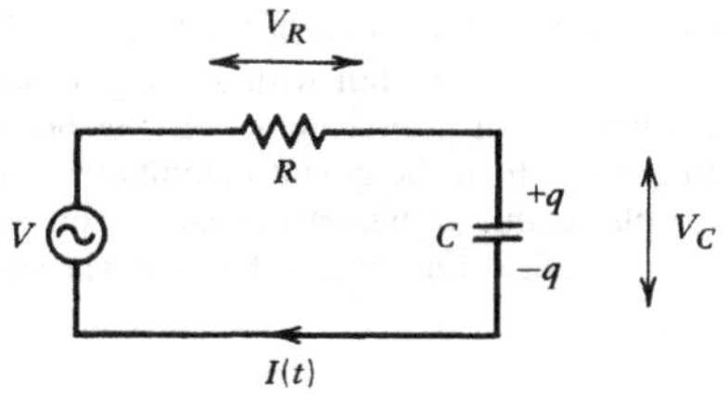

Figure 13.26

13.5 A capacitor with an initial charge q_0 is discharged through an inductance. If the current in the circuit is $I = 86.6 q_0 e^{-5t} \sin(86.6t)$. (a) Determine the quality factor of the inductance. (b) If the resistance of the inductance is $0.1\ \Omega$, determine its inductance. (c) Determine the fractional energy loss in one cycle.

13.6 A coil of inductance L and resistance R_1 is connected to a battery of voltage V_0. A resistor R_2 is connected parallel with the coil. If $L = 10$ H, $R_1 = 100\ \Omega$, and $V_0 = 20$ V, determine what R_2 should be in order to prevent the voltage across the coil from rising above 100 V when the battery circuit is suddenly opened. What is the initial rate of decrease of current in the inductance?

13.7 Consider an inductance L and a resistor R connected in series across a voltage given by $V = V_0 \sin \omega t$. (a) Find the current in the circuit and the phase angle between it and the voltage source. (b) What is the phase angle between the current in the resistance and in the inductance? (c) What is the phase angle between the voltage across the resistor and the voltage across the inductance?

13.8 Consider the RLC circuit shown in Fig. 13.27. A voltage $V = V_0 \sin \omega t$ is connected across the circuit. (a) For what frequencies will the current be a maximum and a

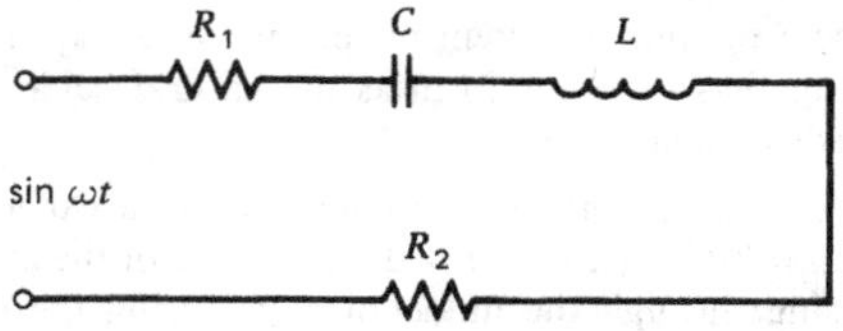

Figure 13.27

minimum? (b) What will be the value of the maximum current? (c) For what frequencies will the current be one-half its maximum value?

13.9 Figure 13.28 shows a network containing resistor R, inductance L, and capacitors C_1 and C_2. If $L < R^2C_2$, show that the network acts as a pure resistance for a current of frequency $\omega = (C_2 - L/R^2)/LC_2(C_1 + C_2)$.

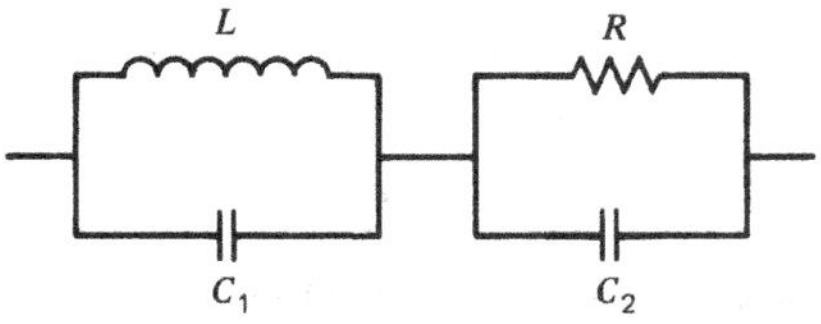

Figure 13.28

13.10 (a) Find the complex impedance z of the two-terminal network shown in Fig. 13.29. (b) A capacitor when empty has a capacitance C_0, is filled with a dielectric of a complex permittivity $\varepsilon/\varepsilon_0 = 1 - \omega_p^2/[\omega(\omega + i\gamma)]$ where ω_p and γ are constants and ω is the frequency of an external voltage source $V = V_0 \cos \omega t$. Show that the complex impedance of such a capacitor is equal to the impedance of the network in part (a) when the parameters L, C, and R are suitably chosen. Find them.

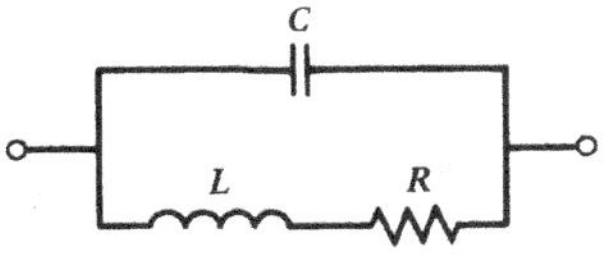

Figure 13.29

13.11 Consider the circuit shown in Fig. 13.30. (a) Write down the Kirchhoff nodal equations using the notation of the figure. (b) If $\mathscr{I}_1 = 2$ A, $\mathscr{I}_2 = 1$ A, $R_1 = 1\ \Omega$, $R_2 = 2\ \Omega$, and $R_3 = 3\ \Omega$, determine V_1 and V_2. (c) Determine the currents in the three resistors.

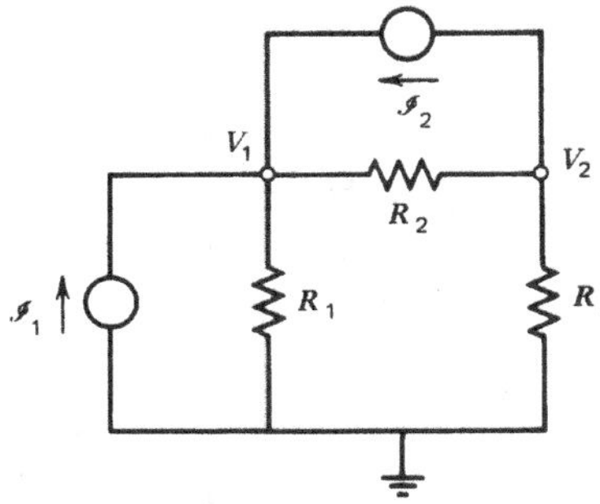

Figure 13.30

13.12 A real capacitor is approximated by a capacitance C that has a parallel leakage resistance R; it is connected in series with an ideal inductance L. (a) Determine the impedance z of this approximation. (b) Assuming that R is large, sketch $|z|$ as a function of ω.

13.13 Use the nodal method to solve for the voltage $V_2(t)$ in Fig. 13.31. Take $R = 1\ \Omega$, $C = 1$ F, $L = 1$ H, $V_0 = 10$ V, and $\omega = 1$ rad/s. What is the phase of V_2 relative to the source?

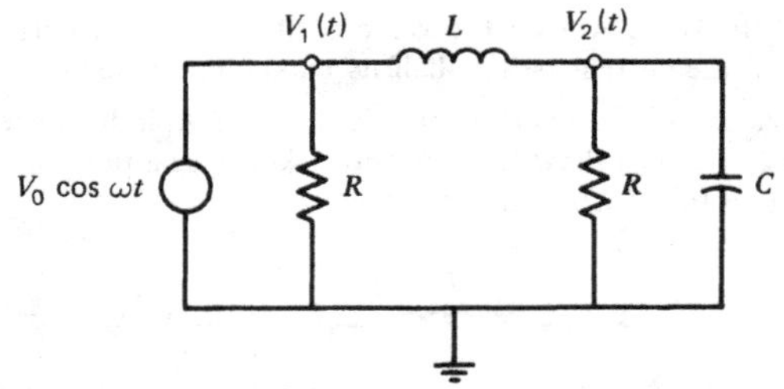

Figure 13.31

13.14 Determine the average power $\langle P \rangle$ stored per unit time in the capacitor of Problem 13.10. Find also the average heat loss $\langle Q \rangle$ per unit time and express both $\langle P \rangle$ and $\langle Q \rangle$ in terms of the potential difference between the plates.

13.15 Two circuits coupled by a mutual inductance M are shown in Fig. 13.32. Show that the ratio of power dissipated in R_2 and R_1 is given by $P_2/P_1 = \omega^2 M^2 R_2/(R_2^2 + \omega^2 L_2^2)R_1$. Draw a rough sketch of P_2/P_1 as a function of R_2 and find the value of R_2 that maximizes the ratio.

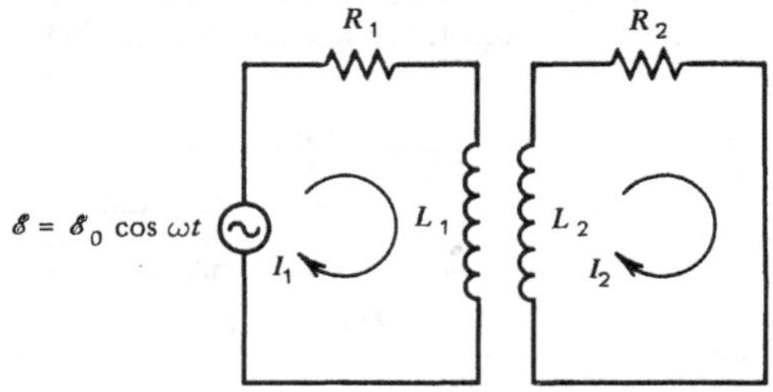

Figure 13.32

13.16 A bridge such as the one shown in Fig. 13.33 is frequently used to measure capacitance in terms of a standard capacitor C_s and two known adjustable resistors. We assume that the series resistance of the capacitors is negligible. Show that when the detector reads zero current (balance condition), then $C_x = R_a C_s / R_b$.

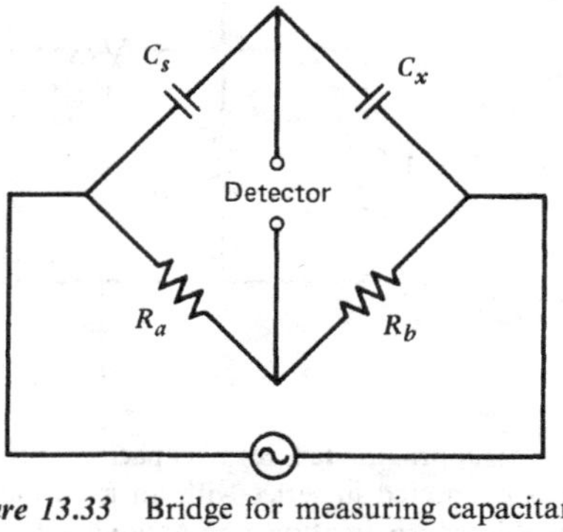

Figure 13.33 Bridge for measuring capacitance.

13.17 Figure 13.34 shows what is called a *frequency bridge*, whose balance condition depends on frequency. Show that for the detector to read zero current the following conditions must be satisfied $R_L R_4 = R_2 R_3$, and $\omega L = 1/\omega C$. If L and C are known, this bridge can serve as a frequency-measuring device.

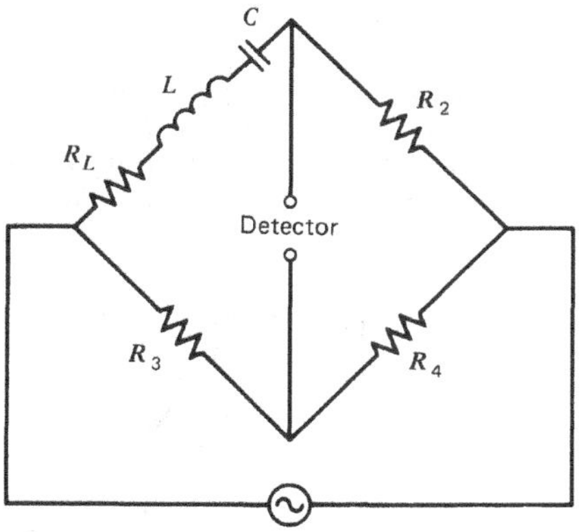

Figure 13.34 Frequency bridge.

13.18 The Maxwell bridge shown in Fig. 13.35 can be used to measure an inductance of low or moderate Q without the accurate knowledge of the generator frequency. Show that when the bridge is balanced, $R_L = R_2 R_3 / R_4$ and $L = R_2 R_3 C$. Also show that the Q of the inductance is $\omega C R_4$.

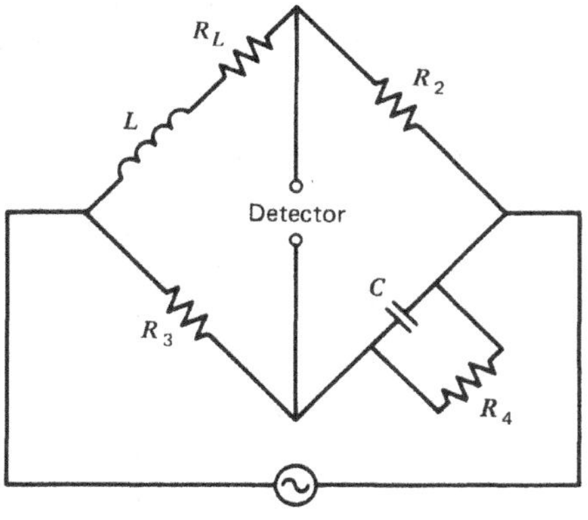

Figure 13.35 The Maxwell bridge for measuring inductance.

13.19 Consider the two circuits shown in Fig. 13.36, which are coupled capacitively ($Z = 1/\omega C$). (a) Set up a system of algebraic equations for the currents in the loops, and calculate the natural frequencies of the electrical oscillations. (b) Discuss the case when there is no coupling between the circuits (i.e., when $C = 0$) and when there is very tight coupling ($C \gg C_1$).

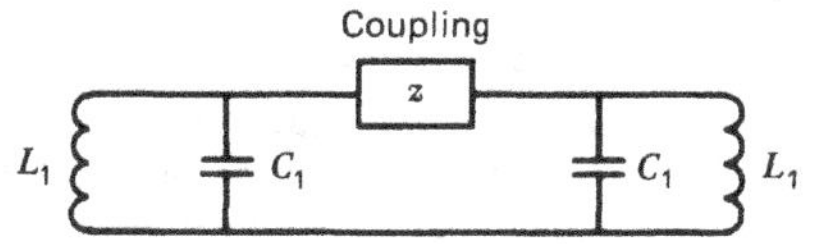

Figure 13.36

13.20 Solve the preceding problem when the coupling between the two circuits is inductive, $Z = \omega L$.

13.21 Consider the circuit shown in Fig. 13.37, which is driven by $\mathscr{E} = \mathscr{E}_0 \cos \omega t$. The frequency of the applied source $\omega^2 = 1/LC$. (a) Show that the amplitude of the current in the diagonal resistor is $2\mathscr{E}_0/[\omega(3L + CR^2)]$. (b) Determine the phase difference between this current and the applied electromotive force.

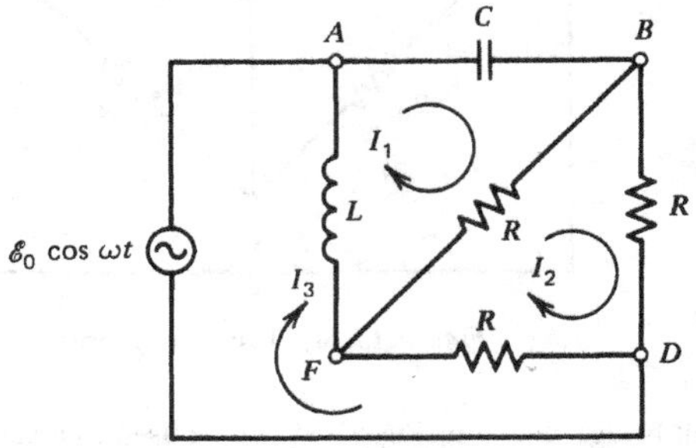

Figure 13.37

13.22 Consider a series *RLC* circuit (Fig. 13.6) and a parallel *RLC* circuit (Fig. 13.17). Each is driven by a constant-current source I of frequency ω. (a) Determine the impedances and admittances of the circuits. (b) Sketch $\hat{V}_L$ and $\hat{V}_C$ as a function of ω for the series circuit. (c) Sketch $\hat{V}_C$ as a function of ω for the parallel circuit. (d) Which circuit exhibits a resonance phenomenon?

13.23 Suppose we are given an inductor of $L = 10^{-2}$ H and a resistance of $R = 1\ \Omega$, and we wish to construct the circuit in Fig. 13.38 so that $|\hat{V}_C/V_0|$ is minimized at ω_0 of

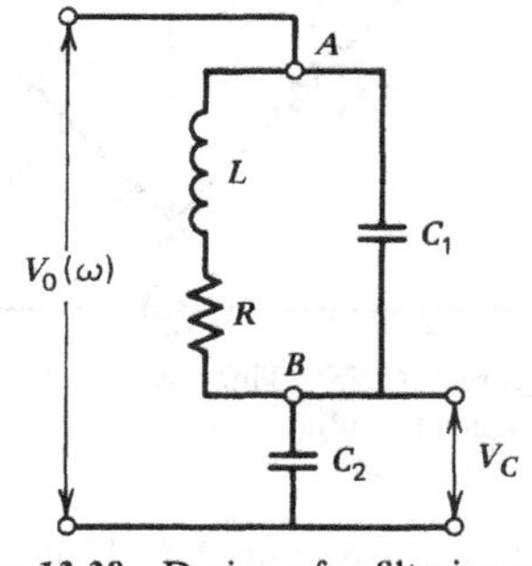

Figure 13.38 Design of a filtering circuit.

10^4 rad/s, and maximized at $\omega = 0.9\omega_0$. (a) Determine C_1 if the parallel circuit between A and B is to resonate at $\omega_0 = 10^4$ rad/s. (b) Determine the quality factor Q_0, and the impedance at the resonance in (a). Is it a phase resonance or antiresonance? (c) Determine the impedance between A and B at $\omega/\omega_0 = 0.9$. Is it inductive or capacitive? (d) Determine C_2 required to obtain series resonance at $\omega/\omega_0 = 0.9$. (e) Calculate $|V_C/V_0|$ at ω_0 and at $\omega = 0.9\omega_0$.

13.24 Consider the circuit shown in Fig. 13.39. (a) Determine the equivalent impedance z of the circuit measured at the source. (b) Find the phase angle ϕ. (c) Find the resonance frequency ω_0 of the circuit. (d) Find the average power $\langle P \rangle$ delivered by the source. (e) Determine the power loss in the resistance R for $\omega = \sqrt{1/LC}$.

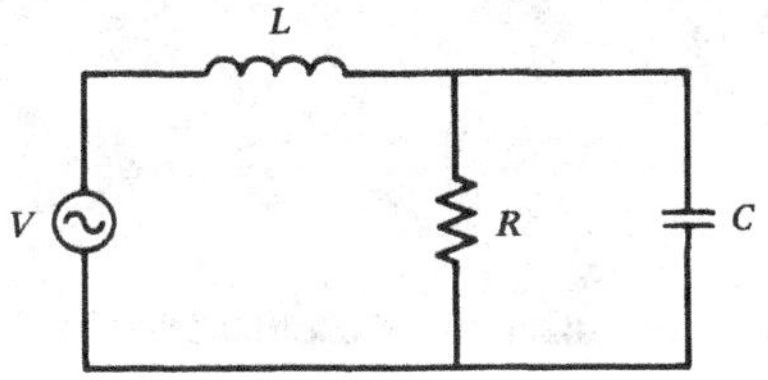

Figure 13.39

13.25 Consider the bandpass filter shown in Fig. 13.40. (a) Using the notations of Fig. 13.19, determine the effective z_1 and z_2. (b) Determine the frequency range of the band that passes without attenuation. (Find a relation between ω, L, and C.)

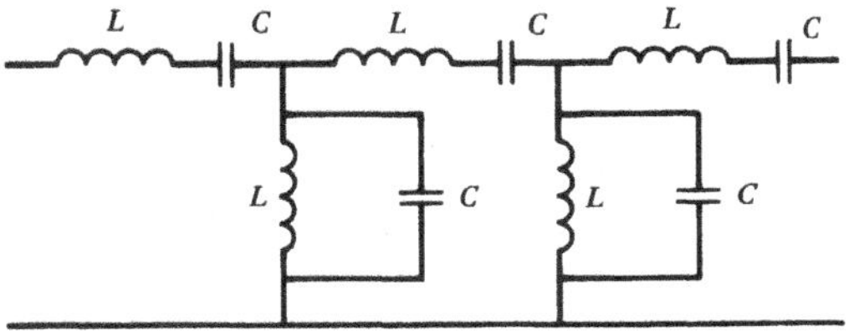

Figure 13.40 Schematic diagram of a bandpass filter.

FOURTEEN

MAXWELL'S EQUATIONS

In the first six chapters we discussed electrostatics and found that the electric field in media satisfies the equations

$$\nabla \cdot \mathbf{D} = \rho_f \qquad \text{and} \qquad \nabla \times \mathbf{E} = 0$$

Magnetostatics were treated in Chapters 8 to 10, where the magnetic field in media was found to satisfy the equations

$$\nabla \cdot \mathbf{B} = 0 \qquad \text{and} \qquad \nabla \times \mathbf{H} = \mathbf{J}_f$$

In addition to these equations relating the fields and the sources (charges and currents), we also have the mathematical expression of conservation of charge—i.e., the continuity equation (discussed in Chapter 7)—which states that at any point the current density and the charge density are related by

$$\nabla \cdot \mathbf{J}_f + \frac{\partial \rho_f}{\partial t} = 0$$

The first effect due to the time variation of the fields on these equations was introduced in Chapter 11 via Faraday's experimental law. It introduced a source in the equation for $\nabla \times \mathbf{E}$:

$$\nabla \times \mathbf{E} = -\frac{\partial \mathbf{B}}{\partial t}$$

which provided coupling between the fields. Thus the basic equations of electromagnetism as we have used them so far have the point forms:

$$\begin{aligned} &\nabla \cdot \mathbf{D} = \rho_f && \nabla \cdot \mathbf{B} = 0 \\ &\nabla \times \mathbf{E} = -\frac{\partial \mathbf{B}}{\partial t} && \nabla \times \mathbf{H} = \mathbf{J}_f \end{aligned} \tag{14.1}$$

These equations were supposed to be satisfied at every point in space at all times. They are what we called the quasi-static equations in Chapter 13, and they were used in the treatment of nonsteady currents in electric circuits.

In this chapter we show that the interdependence of the field equations is still incomplete. Modification of these equations further in the next section will yield what is called Maxwell's equations, which govern the behavior of the *classical electromagnetic field* as we believe it today. The electromagnetic wave phenomena predicted by these equations will also be discussed.

14.1 Displacement Current—Maxwell's Equations

The question now to be posed is: Are the relations given in Eq. (14.1) true? Are they consistent? The answer is that they are not consistent, and so they cannot represent the physical truth. Something is wrong. This is made obvious by noting that $\nabla\cdot(\nabla\times\mathbf{H})$ is identically zero, but the last relation in Eq. (14.1) then asserts that

$$\nabla\cdot(\nabla\times\mathbf{H}) = 0 = \nabla\cdot\mathbf{J}_f$$

In steady-state cases, as was discussed in Chapter 7, $\nabla\cdot\mathbf{J}_f = 0$. However, in cases where time variations are important, the continuity equation says that $\nabla\cdot\mathbf{J}_f = -\partial\rho_f/\partial t \neq 0$ in general. It was seen by Maxwell that the way to eliminate this contradiction was by the addition of another term to $\mathbf{J}_f$ in the last relation in Eq. (14.1). For example, since $\partial\rho_f/\partial t + \nabla\cdot\mathbf{J}_f = 0$, one has from $\nabla\cdot\mathbf{D} = \rho_f$ that

$$\frac{\partial}{\partial t}[\nabla\cdot\mathbf{D}] + \nabla\cdot\mathbf{J}_f = 0$$

that is,

$$\nabla\cdot\left[\mathbf{J}_f + \frac{\partial\mathbf{D}}{\partial t}\right] = 0$$

Therefore, if instead of the term $\mathbf{J}_f$ in Eq. (14.1), we have $\mathbf{J}_f + \partial\mathbf{D}/\partial t$, the original inconsistency is removed. With this addition, the equations of electromagnetism in point form in media become known as Maxwell's equations, and ought to be committed to memory:

$$\nabla\cdot\mathbf{D} = \rho_f \tag{14.2}$$

$$\nabla\cdot\mathbf{B} = 0 \tag{14.3}$$

$$\nabla\times\mathbf{E} = -\frac{\partial\mathbf{B}}{\partial t} \tag{14.4}$$

$$\nabla\times\mathbf{H} = \mathbf{J}_f + \frac{\partial\mathbf{D}}{\partial t} \tag{14.5}$$

These equations must be supplemented by the relations

$$\mathbf{D} = \varepsilon_0\mathbf{E} + \mathbf{P} \qquad \mathbf{H} = \frac{1}{\mu_0}\mathbf{B} - \mathbf{M} \tag{14.6}$$

which were previously introduced in Eqs. (4.35) and (9.29).

Maxwell's equations [Eqs. (14.2) to (14.5)] govern the behavior of the classical electromagnetic field as we believe it today. The added term, $\partial\mathbf{D}/\partial t$, is called the *displacement current density*. We shall see that not only does it render the equations consistent, but that it is central in the theory of radiation of electromagnetic waves; without it, such radiation would not occur in the theory. Moreover the presence of the displacement current implies that changing electric fields in space create changing magnetic fields, just as (from Faraday's law) changing **B** fields create changing **E** fields. Thus **E** and **B** are intimately related and are together called the electromagnetic (EM) field.

It will also be noted that the quasi-static equations will be in error due to neglect of the term $\partial\mathbf{D}/\partial t$. However, at time rates of change of **E** encountered for many applications (especially in circuit applications where the **E** fields are often very small, or contained in small, localized volumes), this term is small compared to the $\mathbf{J}_f$ term. That is, in these cases the actual current sources are much more important than the displacement current sources. Inside conductors (ohmic) with sinusoidally time-varying sources of frequency ω one finds that

$$\frac{\partial \mathbf{E}}{\partial t} = i\omega\mathbf{E} \qquad \text{and} \qquad \mathbf{J} = \sigma_c \mathbf{E}$$

Thus

$$\left|\frac{\varepsilon_0(\partial E/\partial t)}{J}\right| = \frac{\varepsilon_0\omega}{\sigma_c} = \frac{2\pi\varepsilon_0 f}{\sigma_c} \approx 2 \times 10^{-17} f \qquad \text{(Hz)}$$

for $\sigma_c \simeq 3 \times 10^6 (\Omega\cdot\text{m})^{-1}$ which is less than 1 percent for frequencies up to about 10^{14} Hz.

We should now note that the four Maxwell equations are not all independent. Thus, from Eq. (14.4), we get

$$\nabla\cdot\nabla\times\mathbf{E} \equiv 0 = -\nabla\cdot\frac{\partial\mathbf{B}}{\partial t} = -\frac{\partial}{\partial t}(\nabla\cdot\mathbf{B})$$

implying that $\nabla\cdot\mathbf{B} = 0$.* Similarly, from Eq. (14.5),

$$\nabla\cdot\nabla\times\mathbf{H} \equiv 0 = \nabla\cdot\mathbf{J}_f + \frac{\partial}{\partial t}(\nabla\cdot\mathbf{D})$$

Using the continuity equation, we take $\nabla\cdot\mathbf{J}_f = -\partial\rho_f/\partial t$. Thus

$$-\frac{\partial\rho_f}{\partial t} + \frac{\partial}{\partial t}(\nabla\cdot\mathbf{D}) = 0$$

or $\nabla\cdot\mathbf{D} = \rho_f$, which is Eq. (14.2).

Example 14.1 Displacement Current—Parallel-Plate Capacitor

The inconsistency of the equation $\nabla\times\mathbf{H} = \mathbf{J}_f$ is clearly seen if we try to calculate the line integral $\oint_C \mathbf{H}\cdot d\mathbf{r}$ for a situation such as that shown in the Fig. 14.1, where the current flowing, I, terminates in a capacitor or capacitive element. From Ampere's law, $\oint_C \mathbf{H}\cdot d\mathbf{r} = I$, where I is the total current through the surface S whose periphery is the closed curve C. Since S may or may not be chosen to intersect the wire carrying the current I, the value of the line integral is ambiguous. Two such surfaces S_1 and S_2 are shown in Fig. 14.1. Through S_1 a current I flows (in the wire), but through S_2 there is no current flow.

* We assume that **B** was zero in a neighborhood of any point at some time in the past.

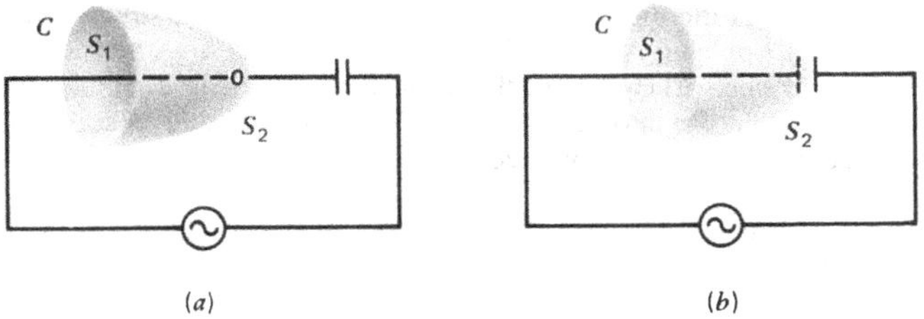

Figure 14.1 Testing Ampere's law in the presence of a capacitor using a contour C as a perimeter of two open surfaces S_1 and S_2 that are taken (a) to penetrate and (b) not to penetrate the capacitor.

The situation is clarified if one uses the integral form of Eq. (14.5), which includes the displacement current; that is,

$$\int_S (\nabla \times \mathbf{H}) \cdot d\mathbf{a} = \oint_C \mathbf{H} \cdot d\mathbf{r} = \int_S \mathbf{J}_f \cdot d\mathbf{a} + \int_S \frac{\partial \mathbf{D}}{\partial t} \cdot d\mathbf{a}.$$

Thus, if S is taken as S_1, $\partial \mathbf{D}/\partial t \approx 0$ everywhere on S_1 (assuming the fields are localized inside the capacitor), so we have $\oint \mathbf{H} \cdot d\mathbf{r} = I$. If $S = S_2$, $\mathbf{J}_f = 0$ everywhere on S_2, but

$$\int_S \frac{\partial \mathbf{D}}{\partial t} \cdot d\mathbf{a} = \frac{d}{dt}\left[\varepsilon_0 \int_{S_2} \mathbf{E} \cdot d\mathbf{a}\right]$$

Again, using the fact that $\mathbf{E}$ is zero outside the capacitor (e.g., on the surface S' shown in Fig. 14.1b), we get

$$\int_S \frac{\partial \mathbf{D}}{\partial t} \cdot d\mathbf{a} = \frac{d}{dt}\left[\varepsilon_0 \oint_{S'+S_2} \mathbf{E} \cdot d\mathbf{a}\right] = \frac{dq}{dt} = I$$

Therefore a consistent result is achieved.

Example 14.2 Charge Leakage from a Sphere—Displacement Current

As another instance where Ampere's law fails to give unambiguous results, consider the case of a sphere (conducting) which at time t has a charge $q(t)$ on its surface, embedded in a material (ohmic, homogeneous, isotropic) of conductivity σ_c, as shown in Fig. 14.2.

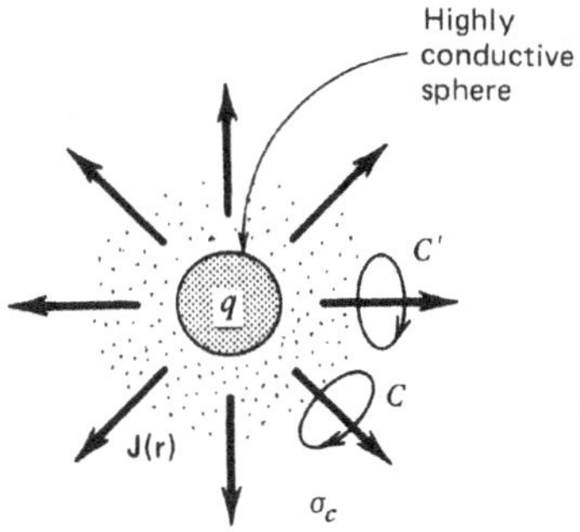

Figure 14.2 Testing of Ampere's law using an initially charged, highly conducting sphere embedded in a conducting material.

From the symmetry of the problem, **J** will be a function of r only (the distance from the center of the sphere). Applying Ampere's law to the curve C of area S as shown, we would find that $\oint \mathbf{H} \cdot d\mathbf{r} \neq 0$, since a net current flows through C. This in turn implies that $H \neq 0$. But also we observe that for another curve, C', $\oint_{C'} \mathbf{H} \cdot d\mathbf{r} \neq 0$, and in fact that, due to C' symmetry, **H** appears to be oppositely directed where the curves C and C' would touch each other. Therefore, what is H?

The answer, of course, is that $H = 0$.* With the displacement current included, we in fact have, from Eq. (14.5);

$$\oint_C \mathbf{H} \cdot d\mathbf{r} = I_C + \varepsilon_0 \frac{d}{dt} \int_S \mathbf{E} \cdot d\mathbf{a}$$

We now use the symmetry inherent in the problem. Since $I_C = \int_S \mathbf{J} \cdot d\mathbf{a}$, while the total current I flowing through all the surface of the sphere is $\oint \mathbf{J} \cdot d\mathbf{a}$, then $I_C = I\Omega/4\pi$ where Ω is the solid angle subtended by C with respect to the origin of the sphere. Also we can write

$$\int_S \mathbf{E} \cdot d\mathbf{a} = \frac{\Omega}{4\pi} \oint \mathbf{E} \cdot d\mathbf{a}$$

Thus

$$\oint_C \mathbf{H} \cdot d\mathbf{r} = \left[I \frac{\Omega}{4\pi} + \varepsilon_0 \frac{d}{dt} \left(\oint \mathbf{E} \cdot d\mathbf{a} \right) \cdot \frac{\Omega}{4\pi} \right]$$

$$\oint_C \mathbf{H} \cdot d\mathbf{r} = I \frac{\Omega}{4\pi} + \frac{dq}{dt} \frac{\Omega}{4\pi}$$

Since $I = -dq/dt$, then $\oint_C \mathbf{H} \cdot d\mathbf{r} = 0$. Thus, insertion of the displacement current gives reasonable and consistent results.

14.2 Maxwell's Equations in Simple Media—The Wave Equation

We consider now Maxwell's equations in environments where the macroscopic relations between **B** and **H** and between **D** and **E** are "simple"; that is, $\mathbf{B} = \mu \mathbf{H}$ and $\mathbf{D} = \varepsilon \mathbf{E}$. Then μ and ε are assumed to be constants independent of position and time variables. In fact, for completely arbitrary fields in material media, this condition is rarely satisfied. However, in many applications it is satisfied to sufficient accuracy to be worth detailed consideration.

Under these conditions, Maxwell's equations [Eqs. (14.2) to (14.5)] may be written as follows:

$$\nabla \cdot \mathbf{E} = \frac{\rho_f}{\varepsilon} \tag{14.7}$$

$$\nabla \cdot \mathbf{B} = 0 \tag{14.8}$$

$$\nabla \times \mathbf{E} = -\frac{\partial \mathbf{B}}{\partial t} \tag{14.9}$$

$$\nabla \times \mathbf{B} = \mu \mathbf{J}_f + \mu\varepsilon \frac{\partial \mathbf{E}}{\partial t} \tag{14.10}$$

* $\int \mathbf{H} \cdot d\mathbf{r} = 0 \rightarrow H_r = 0$ [By symmetry and uniqueness $\Rightarrow (H_\theta, H_\phi) = 0$]

We take the case where there are no *external* charge densities and currents. The currents present are *only* produced by the electric field itself via Ohm's law; that is,

$$\mathbf{J}_f = \sigma_c \mathbf{E}$$

Substituting this form of $\mathbf{J}_f$ in Eq. (14.10) gives

$$\nabla \times \mathbf{B} = \mu\sigma_c \mathbf{E} + \mu\varepsilon \frac{\partial \mathbf{E}}{\partial t} \tag{14.11}$$

In order to solve Maxwell's equations for $\mathbf{E}$ we first eliminate $\mathbf{B}$. Taking the curl of Eq. (14.9) gives:

$$\nabla \times (\nabla \times \mathbf{E}) = \nabla \times \left(-\frac{\partial \mathbf{B}}{\partial t} \right) = -\frac{\partial}{\partial t}(\nabla \times \mathbf{B}) \tag{14.12}$$

Substituting for $\nabla \times \mathbf{B}$ from Eq. (14.11), we get

$$\nabla \times (\nabla \times \mathbf{E}) = -\frac{\partial}{\partial t}\left(\mu\sigma_c \mathbf{E} + \mu\varepsilon \frac{\partial \mathbf{E}}{\partial t} \right) \tag{14.13}$$

Using the vector relation $\nabla \times (\nabla \times \mathbf{E}) = \nabla(\nabla \cdot \mathbf{E}) - (\nabla^2)\mathbf{E}$ and substituting zero for $\nabla \cdot \mathbf{E}$, we get

$$\nabla^2 \mathbf{E} = \mu\sigma_c \frac{\partial \mathbf{E}}{\partial t} + \mu\varepsilon \frac{\partial^2 \mathbf{E}}{\partial t^2} \tag{14.14}$$

Similarly, with the same procedure, we find

$$\nabla^2 \mathbf{B} = \mu\sigma_c \frac{\partial \mathbf{B}}{\partial t} + \mu\varepsilon \frac{\partial^2 \mathbf{B}}{\partial t^2} \tag{14.15}$$

or combined in

$$\nabla^2 \begin{pmatrix} \mathbf{E} \\ \mathbf{B} \end{pmatrix} - \mu\sigma_c \frac{\partial}{\partial t} \begin{pmatrix} \mathbf{E} \\ \mathbf{B} \end{pmatrix} - \mu\varepsilon \frac{\partial^2}{\partial t^2} \begin{pmatrix} \mathbf{E} \\ \mathbf{B} \end{pmatrix} = 0 \tag{14.16}$$

We thus obtain partial three-dimensional, second-order (in space and time) differential equations in $\mathbf{E}$ alone and $\mathbf{B}$ alone. These forms are standard vector *wave* equations that arise in other physical phenomena. Thus Maxwell's equations imply that $\mathbf{E}$ and $\mathbf{B}$ behave as wave amplitudes. Moreover, each equation has a term that is proportional to the time derivative of the respective field and to the conductivity of the medium; these are damping terms that cause losses in materials with nonzero conductivity. The determination of the effects caused by external sources—charges and currents $(\rho_f, \mathbf{J}_f)$—occupy a fairly large part of EM theory. We will defer such effects to Chapter 15, and take the opportunity here to note that these effects can be best obtained via equations that will be derived for the vector and scalar potentials. If also the medium is nonconducting ($\sigma_c = 0$), then these equations reduce to:

$$\nabla^2 \begin{pmatrix} \mathbf{E} \\ \mathbf{B} \end{pmatrix} - \mu\varepsilon \frac{\partial^2}{\partial t^2} \begin{pmatrix} \mathbf{E} \\ \mathbf{B} \end{pmatrix} = 0 \quad \text{(nonconducting media)} \tag{14.17}$$

which is called the undamped, three-dimensional, homogeneous wave equation. A similar derivation using the microscopic Maxwell's equations [Eq. (14.6), with $\varepsilon = \varepsilon_0$ and $\mu = \mu_0$] gives Eq. (14.17) with μ and ε replaced by μ_0 and ε_0. The vector notation used in the wave equations is highly compact. In cartesian coordinates Eq. (14.17) for $\mathbf{E}$, for example, becomes

$$\left\{ \frac{\partial^2}{\partial x^2} + \frac{\partial^2}{\partial y^2} + \frac{\partial^2}{\partial z^2} \right\} [\hat{\mathbf{x}} E_x + \hat{\mathbf{y}} E_y + \hat{\mathbf{z}} E_z] = \mu\varepsilon \frac{\partial^2}{\partial t^2} [\hat{\mathbf{x}} E_x + \hat{\mathbf{y}} E_y + \hat{\mathbf{z}} E_z]$$

which is equivalent to three *scalar-wave equations*:

$$\left(\frac{\partial^2}{\partial x^2} + \frac{\partial^2}{\partial y^2} + \frac{\partial^2}{\partial z^2}\right)E_i = \mu\varepsilon \frac{\partial^2 E_i}{\partial t^2}$$

where i stands for x, y, and z. It ought to be noted, however, that such separation of variables is not so simply achieved in other than cartesian coordinate systems.

14.3 Plane Waves in Nonconducting Media

14.3.1 The Wave Phenomenon

In this section we consider the simplest case: the homogeneous, undamped wave equation ($\mathbf{J}_f = 0$, $\rho_f = 0$, and $\sigma_c = 0$) given in Eq. (14.17). First we show that this equation has as solutions any vector function that depends on the variable $\hat{\mathbf{k}} \cdot \mathbf{r} - vt$ (*d'Alembert's solution*—see Example 14.3); that is,

$$\mathbf{E} = \mathbf{E}(\hat{\mathbf{k}} \cdot \mathbf{r} - vt) \qquad \mathbf{B} = \mathbf{B}(\hat{\mathbf{k}} \cdot \mathbf{r} - vt) \tag{14.18}$$

where

$$v = \frac{1}{\sqrt{\mu\varepsilon}} \tag{14.19}$$

which represent waves moving in the direction of $\hat{\mathbf{k}}$, and $\hat{\mathbf{k}}$ is any constant unit vector. Consider the variable $\xi \equiv \hat{\mathbf{k}} \cdot \mathbf{r} - vt$, and take $\mathbf{A}$ to stand for $\mathbf{E}$ or $\mathbf{B}$. We now transfer the differentiation with respect to x, y, z, and t in Eq. (14.17) to differentiation with respect to ξ using the chain rule:

$$\frac{\partial \mathbf{A}}{\partial x} = \frac{\partial \mathbf{A}}{\partial \xi}\frac{\partial \xi}{\partial x} = \frac{\partial \mathbf{A}}{\partial \xi}(\hat{\mathbf{k}} \cdot \hat{\mathbf{x}})$$

Hence

$$\frac{\partial^2 \mathbf{A}}{\partial x^2} = \frac{\partial^2 \mathbf{A}}{\partial \xi^2}[(\hat{\mathbf{k}} \cdot \hat{\mathbf{x}})^2] \tag{14.20}$$

Similar expressions for y and z can be written by replacing x with y and z. Thus

$$\nabla^2 \mathbf{A} = \frac{\partial^2 \mathbf{A}}{\partial \xi^2}[(\hat{\mathbf{k}} \cdot \hat{\mathbf{x}})^2 + (\hat{\mathbf{k}} \cdot \hat{\mathbf{y}})^2 + (\hat{\mathbf{k}} \cdot \hat{\mathbf{z}})^2]$$

or

$$\nabla^2 \mathbf{A} = \frac{\partial^2 \mathbf{A}}{\partial \xi^2}(\hat{\mathbf{k}} \cdot \hat{\mathbf{k}})^2 = \frac{\partial^2 \mathbf{A}}{\partial \xi^2} \tag{14.21}$$

Similarly,

$$\frac{\partial \mathbf{A}}{\partial t} = \frac{\partial \mathbf{A}}{\partial \xi}\left(\frac{\partial \xi}{\partial t}\right) = -v\frac{\partial \mathbf{A}}{\partial \xi}$$

Hence

$$\frac{\partial^2 \mathbf{A}}{\partial t^2} = v^2 \frac{\partial^2 \mathbf{A}}{\partial \xi^2} \tag{14.22}$$

Substituting Eqs. (14.21) and (14.22) into Eq. (14.17) shows that the form $\mathbf{E} = \mathbf{E}(\hat{\mathbf{k}} \cdot \mathbf{r} - vt)$ is a solution of this equation, and hence it is indeed a wave equation. The speed of the wave given by Eq. (14.19) can be written alternatively as:

$$v = \sqrt{\frac{1}{\varepsilon_0 \mu_0}} \frac{1}{\sqrt{KK_m}} = \frac{c}{\sqrt{KK_m}} \tag{14.23}$$

where $c = 1/\sqrt{(\varepsilon_0 \mu_0)} = 2.9979 \times 10^8$ m/s. Thus the speed of propagation of the wave of the electric (magnetic) field in space is given once c and the relative effective dielectric K and permeability K_m are given.

That the constant c turned out to be the speed of light in vacuum, and that v was the speed of light in material media of constants K and K_m was an essential and marvelous discovery, showing that light and electromagnetic fields are intimately related. Today, we know that light is in fact completely electromagnetic in character. Note that

$$n \equiv \frac{c}{v} = \sqrt{KK_m} \tag{14.24}$$

is called the *index of refraction*. In most materials $K_m \approx 1$, and thus $n = \sqrt{K}$. Remember, however, that although one may find the relation $\mathbf{D} = \varepsilon_0 K\mathbf{E}$ for materials in static fields, this will not guarantee a similar relation for nonstatic fields. And if such a relation is valid for such fields, there is no guarantee that the K will be the same constant as found for the static case.

A field of the form $\mathbf{E}(z - vt)$ or $\mathbf{E}(\hat{\mathbf{k}} \cdot \mathbf{r} - vt)$ is called a plane wave because at any instant the field assumes the same value at all points in any plane perpendicular to the direction of propagation—i.e., in all planes perpendicular to the z axis in the former case and in all planes perpendicular to $\hat{\mathbf{k}}$ in the latter. Note that we have not specified the functional form of $\mathbf{E}$ aside from the fact that $\mathbf{r}$ and t always must appear (aside from constant factors) in the combinations displayed.

In general, superpositions of such fields will also satisfy the wave equation. For example, a solution might be of the form $\mathbf{E} = \mathbf{E}_+(z - vt) + \mathbf{E}_-(z + vt)$, representing plane waves moving in both directions along the z axis. Whether such waves exist in a particular situation will depend upon how the waves originate—that is, upon the source of the fields and the boundary conditions imposed by the problem. The general solution of the wave equation may be regarded as a superposition of waves moving to and fro in the directions x, y, and z.

14.3.2 Interrelationships between E, B, and $\hat{\mathbf{k}}$

Thus far we have found solutions to the wave equations in simple media (Eq. 14.17) of the general form

$$\mathbf{E}(\hat{\mathbf{k}} \cdot \mathbf{r} - vt) \qquad \mathbf{B}(\hat{\mathbf{k}} \cdot \mathbf{r} - vt)$$

but have not determined the directions of $\mathbf{E}$ or $\mathbf{B}$ relative to $\hat{\mathbf{k}}$, or their interrelationships. To derive such information we shall insert these wave forms into the homogeneous Maxwell's equations. To utilize $\nabla \cdot \mathbf{E} = 0$ we write the divergence of $\mathbf{E}$ in terms of $\hat{\mathbf{k}}$ and $\xi = \hat{\mathbf{k}} \cdot \mathbf{r} - vt$. In cartesian coordinates the chain rule gives

$$\nabla \cdot \mathbf{E} = \frac{\partial E_x}{\partial \xi}\frac{\partial \xi}{\partial x} + \frac{\partial E_y}{\partial \xi}\frac{\partial \xi}{\partial y} + \frac{\partial E_z}{\partial \xi}\frac{\partial \xi}{\partial z} = 0$$

or

$$\nabla \cdot \mathbf{E} = \frac{\partial E_x}{\partial \xi}(\hat{\mathbf{k}} \cdot \hat{\mathbf{x}}) + \frac{\partial E_y}{\partial \xi}(\hat{\mathbf{k}} \cdot \hat{\mathbf{y}}) + \frac{\partial E_z}{\partial \xi}(\hat{\mathbf{k}} \cdot \hat{\mathbf{z}})$$

Thus

$$\nabla \cdot \mathbf{E} = \hat{\mathbf{k}} \cdot \frac{\partial \mathbf{E}}{\partial \xi} = \frac{\partial}{\partial \xi}(\hat{\mathbf{k}} \cdot \mathbf{E}) = 0 \tag{14.25}$$

which implies that $\hat{\mathbf{k}}$ must be perpendicular to $\partial \mathbf{E}/\partial \xi$, and consequently implies also that $\hat{\mathbf{k}}$ is perpendicular to $\mathbf{E}$; that is,*

$$\hat{\mathbf{k}} \cdot \mathbf{E} = 0 \tag{14.26}$$

Using similar procedures one can also show that $\nabla \cdot \mathbf{B} = 0$ implies, just as above,

$$\hat{\mathbf{k}} \cdot \mathbf{B} = 0 \tag{14.27}$$

Thus $\mathbf{E}$ and $\mathbf{B}$, when plane waves, are always transverse to the motion of the wave.

We now determine what $\nabla \times \mathbf{E} = -\partial \mathbf{B}/\partial t$ implies on the interrelationship. We first show that $\nabla \times \mathbf{E} = \hat{\mathbf{k}} \times (\partial \mathbf{E}/\partial \xi)$. In cartesian coordinates we write $\nabla \times \mathbf{E} = \nabla \times [\hat{\mathbf{x}}E_x + \hat{\mathbf{y}}E_y + \hat{\mathbf{z}}E_z]$. Since $\nabla \times f\mathbf{A} = \nabla f \times \mathbf{A} + f(\nabla \times \mathbf{A})$, then $\nabla \times \mathbf{E}$ becomes

$$\nabla \times \mathbf{E} = (\nabla E_x) \times (\hat{\mathbf{x}}) + (\nabla E_y) \times (\hat{\mathbf{y}}) + (\nabla E_z) \times (\hat{\mathbf{z}})$$

The gradients of the components of the field can be written in terms of $\hat{\mathbf{k}}$ and ξ using the chain rule: $\nabla E_x = \hat{\mathbf{k}}(\partial E_x/\partial \xi)$, with similar expressions for ∇E_y and ∇E_z. Therefore $\nabla \times \mathbf{E}$ becomes

$$\nabla \times \mathbf{E} = \hat{\mathbf{k}}\left(\frac{\partial E_x}{\partial \xi}\right) \times \hat{\mathbf{x}} + \hat{\mathbf{k}}\left(\frac{\partial E_y}{\partial \xi}\right) \times \hat{\mathbf{y}} + \hat{\mathbf{k}}\left(\frac{\partial E_z}{\partial \xi}\right) \times \hat{\mathbf{z}} = \hat{\mathbf{k}}\left(\frac{\partial}{\partial \xi}\right) \times \mathbf{E}$$

or, as stated above,

$$\nabla \times \mathbf{E} = \hat{\mathbf{k}} \times \frac{\partial \mathbf{E}}{\partial \xi} \tag{14.28}$$

We now write $\partial \mathbf{B}/\partial t$ in terms of ξ using the chain rule:

$$\frac{\partial \mathbf{B}}{\partial t} = \frac{\partial \mathbf{B}}{\partial \xi}\frac{\partial \xi}{\partial t} = -v\frac{\partial \mathbf{B}}{\partial \xi} \tag{14.29}$$

Substituting these expressions for $\nabla \times \mathbf{E}$ and $\partial \mathbf{B}/\partial t$ in $\nabla \times \mathbf{E} = -\partial \mathbf{B}/\partial t$ gives

$$\hat{\mathbf{k}} \times \frac{\partial \mathbf{E}}{\partial \xi} = v\frac{\partial \mathbf{B}}{\partial \xi}$$

or

$$\frac{\partial}{\partial \xi}(\hat{\mathbf{k}} \times \mathbf{E} - v\mathbf{B}) = 0 \tag{14.30}$$

which implies again, for wavelike solutions, that $\mathbf{E}$ is perpendicular to $\mathbf{B}$:

$$\hat{\mathbf{k}} \times \mathbf{E} = v\mathbf{B} \tag{14.31}$$

Thus, since $\hat{\mathbf{k}}$ and $\mathbf{E}$ are perpendicular to $\mathbf{B}$ and to each other, the vectors $\{\mathbf{E}, \mathbf{B}, \hat{\mathbf{k}}\}$ form a right-handed triad.

Finally from, $\nabla \times \mathbf{H} = \partial \mathbf{D}/\partial t$ we find, by the procedure used above, that

$$\hat{\mathbf{k}} \times \mathbf{B} = -\frac{\mathbf{E}}{v} \tag{14.32}$$

which is essentially the same as Eq. (14.31).

* If $\mathbf{E} = \hat{\mathbf{k}}f(\xi)$, then $(\partial \mathbf{E}/\partial \xi) = \hat{\mathbf{k}}(\partial f/\partial \xi)$ and $\hat{\mathbf{k}} \cdot \hat{\mathbf{k}}(\partial f/\partial \xi) \neq 0$ unless $f(\xi) =$ constant. But for waves, $f(\xi)$ is not constant. Therefore $\mathbf{E}$ can have no component along $\hat{\mathbf{k}}$.

Equations (14.31) and (14.32) also indicate that the magnitudes of **E** and **B** are not independent of each other. Taking the magnitude of either one of these equations gives

$$\frac{|\mathbf{E}|}{|\mathbf{B}|} = v \tag{14.33}$$

In conclusion, we see that **E** and **B** are (must be) *transverse waves* having the same functional dependence in ξ, and are mutually perpendicular to each other at any instant of time. (In fact, the expression $\mathbf{E} \times \mathbf{B}$ gives the direction of propagation of the wave. This is a helpful mnemonic.) But remember that these results are generally valid only in media in which K and K_m are well-defined constants.

In summary, Maxwell's differential equations [Eqs. (14.2) to (14.5)] become a set of algebraic equations in the case where there are no sources and the fields are plane waves. These relations [Eqs. (14.26), (14.27), (14.31), and (14.32)] are summarized as follows:

$$\hat{\mathbf{k}} \cdot \mathbf{E} = 0 \tag{14.34}$$

$$\hat{\mathbf{k}} \cdot \mathbf{B} = 0 \tag{14.35}$$

$$\hat{\mathbf{k}} \times \mathbf{E} = v\mathbf{B} \tag{14.36}$$

$$\hat{\mathbf{k}} \times \mathbf{B} = -\frac{\mathbf{E}}{v} \tag{14.37}$$

14.4 Sinusoidal (Monochromatic) Solutions to Maxwell's Equations

As a particularly important type of solution to Maxwell's equations, we shall consider (complex) solutions of the form:

$$\mathbf{E} = \hat{\mathbf{E}}(\mathbf{k}, \omega)e^{-i(\omega t - \mathbf{k}\cdot\mathbf{r})} \qquad \mathbf{B} = \hat{\mathbf{B}}(\mathbf{k}, \omega)e^{-i(\omega t - \mathbf{k}\cdot\mathbf{r})} \tag{14.38}$$

where $\hat{\mathbf{E}}$ and $\hat{\mathbf{B}}$ are complex vector amplitudes independent of **r** and t. The hats on $\hat{\mathbf{E}}$ and $\hat{\mathbf{B}}$ is a notation meant to signify complex vector amplitudes associated with sinusoids of angular frequency ω. If, for example, **k** is parallel to the direction $\hat{\mathbf{z}}$, then **E** and **B** lie in the x-y plane, and we can write

$$\hat{\mathbf{E}} = \hat{\mathbf{x}}E_{0x}e^{i\phi_x} + \hat{\mathbf{y}}E_{0y}e^{i\phi_y} \tag{14.39}$$

where E_{0x} and E_{0y} are the real amplitude constants, and ϕ_x and ϕ_y are phase constants. Thus, in general, for such waves Eq. (14.38) gives the real solutions:

$$\mathbf{E} = \hat{\mathbf{x}}E_{0x}\cos[\omega t - kz + \phi_x] + \hat{\mathbf{y}}E_{0y}\cos[\omega t - kz + \phi_y] \tag{14.40}$$

Similarly, for the **B** wave,

$$\mathbf{B} = \hat{\mathbf{x}}B_{0x}\cos[\omega t - kz + \phi'_x] + \hat{\mathbf{y}}B_{0y}\cos[\omega t - kz + \phi'_y] \tag{14.41}$$

where B_{0x}, B_{0y}, ϕ'_x, and ϕ'_y are analogous to E_{0x}, E_{0y}, ϕ_x and ϕ_y defined above, respectively.

The importance of solutions to Maxwell's equations such as Eqs. (14.40) and (14.41) transcends their mathematical simplicity. It derives from the fact that any wave train can be regarded as a linear superposition (for different values of ω and k) of waves of the above forms.

When dealing with monochromatic waves, concerns mentioned earlier in connection with the validity of the constitutive relations $\mathbf{D} = \varepsilon\mathbf{E}$, and $\mathbf{B} = \mu\mathbf{H}$ become somewhat less restrictive. Media are now categorized as *nondispersive* or *dispersive*. If nondispersive, ε and μ, and hence $\hat{\mathbf{E}}$ and $\hat{\mathbf{B}}$, do not depend on ω. In dispersive media ε and/or μ and hence $\hat{\mathbf{E}}$ and $\hat{\mathbf{B}}$ will be functions of ω. In this case, since ε and μ are not uniquely defined, Eqs. (14.14) to (14.17) become invalid. However, the basic Maxwell equations [Eqs. (14.2) to (14.5)] remain valid, and one can still consider solutions of these equations sinusoidal in time, involving a particular angular frequency ω. One can then define values of ε and μ for each particular frequency ω (these values are measurable). Then, instead of Eq. (14.17), for example, one finds

$$\nabla^2\mathbf{E} = \mu(\omega)\varepsilon(\omega)\frac{\partial^2\mathbf{E}}{\partial t^2} \tag{14.42}$$

from which it is easy to verify that expression of the form of Eq. (14.38) are solutions as long as $k = 2\pi/\lambda = \omega\sqrt{\varepsilon(\omega)\mu(\omega)}$.

For nondispersive media the wave velocity $v = \omega/k = c/\sqrt{KK_m}$ is independent of ω and is unique. Since real EM waves can be represented as superpositions of monochromatic waves, the real waves move through such media undistorted in space or time. On the other hand, in a dispersive medium the wave velocity will be a function of ω and will not be unique, so real waves will change form as they propagate through the medium (their components become "dispersed").

Thus, our previous results hold for monochromatic waves as long as ε and μ are scalar constants (independent of position or field strength) defined at a frequency ω. Linear, homogeneous, isotropic media generally fulfill these requirements, and will be called *chromatically simple*. Noncrystalline media (e.g., liquids and gases) are of this type.

Table 14.1 **Dispersion effect**

Medium	K (Static)	n (Sodium yellow light)
Air	1.000294	1.000293
Benzene	1.489	1.482
Water	8.94	1.333
Ethyl alcohol	5.1	1.36

We emphasize that in such media, the values of K (or K_m) to be used in the relation $v/c = 1/\sqrt{KK_m}$ are to be determined at the excitation frequency ω, and are not in general the values obtained under static ($\omega = 0$) conditions. (Table 14.1, where $K_m = 1$, illustrates this fact.) Water and ethyl alcohol have widely different values of $n(= c/v)$ under static conditions and at $\omega = 2\pi \times 5.027 \times 10^{14}$ rad/s (sodium yellow light). The overall trend of $n = \sqrt{K}$ is that as $\omega \to \infty$, $n \to 1$, but the behavior is not monotonic. The reason for this is that the value of K reflects the tendency of the molecular electric dipoles to be aligned along the oscillating EM field. Since it becomes increasingly more difficult for the dipoles to "follow the field" as the frequency of oscillation of the field increases, the polarization tends to zero and K tends to unity as $\omega \to \infty$.

The elegantly simple mathematical properties of EM monochromatic plane waves are seen in differentiation operations. Thus, if $\mathbf{E} = \hat{\mathbf{E}}e^{-i(\omega t - \mathbf{k}\cdot\mathbf{r})}$ then $\partial\mathbf{E}/\partial x = i(\mathbf{k}\cdot\hat{\mathbf{x}})\mathbf{E}$, with similar expressions for y and z coordinates. Thus

$$\nabla \cdot \mathbf{E} = i\mathbf{k}\cdot\mathbf{E} \tag{14.43}$$

$$\nabla \times \mathbf{E} = i\mathbf{k} \times \mathbf{E} \tag{14.44}$$

Similarly, for the differentiation with respect to time,

$$\frac{\partial \mathbf{E}}{\partial t} = -i\omega\mathbf{E} \tag{14.45}$$

In general, we can summarize these properties by the equivalence operations

$$\nabla * \{\ \ \} \to i\mathbf{k} * \{\ \ \} \qquad \frac{\partial}{\partial t}\{\ \ \} \to -i\omega\{\ \ \} \tag{14.46}$$

The symbol $*$ means any of the operations of ∇ upon either a vector or a scalar quantity $\{\ \ \}$. Thus, differentiation becomes equivalent to multiplication for monochromatic plane waves of the type given by Eq. (14.38).

With this equivalence, we may rewrite Maxwell's equations for monochromatic plane waves. Thus, for simple media ($\rho_f = 0$ and $\mathbf{J}_f = 0$), Eqs. (14.34) to (14.37) become

$$\begin{gathered} \mathbf{k}\cdot\hat{\mathbf{D}} = 0 \qquad \mathbf{k}\cdot\hat{\mathbf{B}} = 0 \\ \mathbf{k}\times\hat{\mathbf{E}} = \omega\hat{\mathbf{B}} \qquad \mathbf{k}\times\hat{\mathbf{B}} = -\frac{1}{v^2}\omega\hat{\mathbf{E}} \end{gathered} \tag{14.47}$$

It is now well recognized that radio waves, microwaves, blackbody radiation, light waves, X rays, and γ rays are all electromagnetic radiation. Table 14.2 gives the different kinds of EM radiation as characterized by wavelength (and frequency). What distinguishes them is their frequency or wavelength. Note that the wavelength is given by

$$\lambda = \frac{2\pi}{k} \tag{14.48}$$

because the term $e^{i(\omega t - ks)}$ has equal values at increments of distance s, given by $2\pi/k$: Thus, in free space,

$$k = \frac{2\pi}{\lambda} \equiv \frac{1}{\lambda\!\!\!^{-}} = \frac{\omega}{c} = \frac{2\pi f}{c}$$

whereas in simple media ($n \equiv$ index of refraction),

$$k = \frac{2\pi}{\lambda} = \frac{\omega}{v} \equiv \frac{n\omega}{c} = \frac{2\pi f n}{c}$$

The quantity k is sometimes called the *wave number*, as there are $k/2\pi$ wavelengths per unit distance, and $\mathbf{k}$ is called the *wave vector*.

Table 14.2 **List of Wavelengths Typical of Different Regions of the Electromagnetic Spectrum**

Type	λ (meters)	Frequency (hertz)	Description
γ rays	1.240×10^{-12}	2.418×10^{20}	1-MeV gamma photon
X rays	1.7×10^{-11}	1.8×10^{19}	75-kV tube for medical diagnosis
Ultraviolet	1.216×10^{-7}	2.465×10^{15}	First line of Lyman series (hydrogen)
Visible limit	3.8×10^{-7}	7.9×10^{14}	Approximate shortest visible wavelength
Violet	4.358×10^{-7}	6.879×10^{14}	Hg violet line
Greenish-blue	4.861×10^{-7}	6.167×10^{14}	Hg, second line of Balmer series (hydrogen)
Green	5.461×10^{-7}	5.490×10^{14}	Hg green line
Yellow	5.876×10^{-7}	5.102×10^{14}	He yellow line
Orange-yellow	5.893×10^{-7}	5.087×10^{14}	Na doublet (5890 Å and 5896 Å)
Red	6.328×10^{-7}	4.657×10^{14}	He-Ne laser light (Ne transition)
Red	6.563×10^{-7}	4.568×10^{14}	Hα first line of Balmer series (hydrogen)
Visible limit	7.8×10^{-7}	3.8×10^{14}	Approximate longest visible wavelength
Infrared	1.5×10^{-6}	2.1×10^{14}	Peak emission of blackbody at 2000 K
Infrared	3.431×10^{-5}	8.738×10^{13}	Vibration at C—H bond in CH_4
Microwave	2.600×10^{-3}	1.153×10^{11}	Lowest frequency in rotational spectrum of CO
Microwave	1.256×10^{-2}	2.387×10^{10}	Inversion line of ammonia
Microwave	3.0×10^{-2}	1.0×10^{10}	Radar
Microwave	2.111×10^{-1}	1.420×10^{9}	Discrete spectral line in general galactic radiation (hyperfine transition in ground state of hydrogen)
Radio	3.34	8.97×10^{7}	FM broadcast
Radio	5.3	5.7×10^{7}	TV
Radio	9.3	3.2×10^{7}	One of Hertz's original experiments
Radio	4.2×10^{2}	7.1×10^{5}	Standard AM broadcast

Example 14.3 D'Alembert's Solution of the Wave Equation

The procedure we used above to arrive at a solution for the Eq. (14.18) is called d'Alembert's solution. In fact this procedure has two solutions: $\mathbf{f}(\hat{\mathbf{k}} \cdot \mathbf{r} - vt)$ and $\mathbf{g}(\hat{\mathbf{k}} \cdot \mathbf{r} + vt)$, where **f** and **g** are arbitrary functions. To show the presence of two solutions we consider a special case of a scalar wave equation that depends on the variables x and y:

$$\frac{\partial^2 u}{\partial x^2} - \frac{\partial^2 u}{\partial y^2} = 0$$

Taking $\bar{x} = x + y$ and $\bar{y} = x - y$ transforms the equation to

$$4 \frac{\partial^2 u}{\partial \bar{x}\, \partial \bar{y}} = 0$$

which in general implies

$$\frac{\partial u}{\partial \bar{x}} = \Phi(\bar{x})$$

where $\Phi(\bar{x})$ is a function of $\bar{x}$ only, and which in turn implies that

$$u = \int_0^{\bar{x}} \Phi(\bar{x})d\bar{x} + f_2(\bar{y})$$

or

$$u = f_1(\bar{x}) + f_2(\bar{y}) = f_1(x + y) + f_2(x - y)$$

indicating the existence of two solutions traveling in opposite directions.

Example 14.4 Interrelationships Among E, B, and k

This example demonstrates the interrelationships between the field amplitudes and the propagation vector. Consider an electromagnetic wave in free space whose electric field is given by

$$\mathbf{E} = 60\hat{\mathbf{x}}e^{-i(10^8 t + \beta z)} \text{ V/m}$$

We now utilize the properties of **E**, **B**, and **k** to determine all the details of the wave. The direction of propagation of the wave can be determined by examining the exponent of the exponential. Comparing $-10^8 t - \beta z$ with $-\omega t + \mathbf{k}\cdot\mathbf{r}$, we conclude that the direction of propagation is $\hat{\mathbf{k}} = -\hat{\mathbf{z}}$. This direction checks with the $\hat{\mathbf{k}}\cdot\mathbf{E} = 0$ property. Also, this comparison gives $\omega/|\mathbf{k}| = c = 10^8/\beta$, thus yielding $\beta = |\mathbf{k}| = \frac{1}{3}\,\text{m}^{-1}$.

The magnetic field of this wave has a form similar to that of the electric field

$$\mathbf{B} = \mathbf{B}_0 e^{-i(10^8 t + \beta z)}$$

with $\mathbf{B}_0$ to be evaluated. Using the properties $\hat{\mathbf{k}}\cdot\mathbf{B} = 0$ and $\hat{\mathbf{k}} \times \mathbf{E} = v\mathbf{B}$, we find $-\hat{\mathbf{y}}$ to be the direction of the magnetic field. The magnitude of $\mathbf{B}_0$ can be determined from the general relation $|\mathbf{E}|/|\mathbf{B}| = v = c$, which yields $B_0 = 60/c$. Thus $\mathbf{B} = (-60/c)\hat{\mathbf{y}} \exp[i(10^8 t + \frac{1}{3}z)]$ tesla.

14.5 Polarization of Plane Waves

If the electric field of a plane wave is aligned along a fixed direction in space, it is said to be *linearly polarized*. The form of a linearly polarized plane wave is

$$\mathbf{E} \equiv \hat{\boldsymbol{\varepsilon}} E_0 e^{-i(\omega t - \mathbf{k}\cdot\mathbf{r} + \phi)} \tag{14.49}$$

where $\hat{\boldsymbol{\varepsilon}}$ is a unit vector defining the direction of the electric field (or, in other words, defining the polarization of the electric field) and ϕ is some phase angle. The amplitude of **E** is given by the *real* number E_0, and may be obtained from the relation $\mathbf{E}\cdot\mathbf{E}^* = E_0^2$. The real field is calculated by taking the real part of Eq. (14.49), or

$$\mathbf{E} = \hat{\boldsymbol{\varepsilon}} E_0 \cos[\omega t - \mathbf{k}\cdot\mathbf{r} + \phi]$$

The associated **B** vector assumes an analogous form to Eq. (14.49); that is,

$$\mathbf{B} = \hat{\boldsymbol{\beta}} B_0 e^{-i(\omega t - \mathbf{k}\cdot\mathbf{r} + \phi')} \tag{14.50}$$

where ϕ' as ϕ, is some phase angle, and $\hat{\boldsymbol{\beta}}$ is a unit vector defining the direction of the magnetic field (polarization of the magnetic field). From Eqs. (14.47), which summarize the properties of plane waves, we find that for every linearly polarized plane wave the unit vectors $\hat{\boldsymbol{\varepsilon}}$ and $\hat{\boldsymbol{\beta}}$ satisfy

$$\hat{\boldsymbol{\varepsilon}} \times \hat{\boldsymbol{\beta}} = \hat{\mathbf{k}} \tag{14.51}$$

and the magnitudes of **E** and **B** satisfy

$$|\mathbf{E}| = v|\mathbf{B}| \tag{14.52}$$

We should note that the polarization of electromagnetic waves is customarily given by the polarization of the electric field vector $\hat{\boldsymbol{\varepsilon}}$.

Other types of polarization can be constructed by the superposition of linearly polarized plane waves. Examples of these are the elliptic and circular polarizations. To see these effects consider the electric vectors of two linearly polarized, plane waves, $\mathbf{E}_1$ and $\mathbf{E}_2$, both moving in the $\hat{\mathbf{k}} \equiv \hat{\mathbf{z}}$ direction. Let $\mathbf{E}_1$ be polarized in the $\hat{\mathbf{x}}$ direction, and $\mathbf{E}_2$ in the $\hat{\mathbf{y}}$ direction. The sum of these vectors is also a plane wave $\mathbf{E} = \mathbf{E}_1 + \mathbf{E}_2$. Using the notations of Eq. (14.49), we write

$$\mathbf{E} = [\hat{\mathbf{x}}E_{0x}e^{-i\phi_x} + \hat{\mathbf{y}}E_{0y}e^{-i\phi_y}]e^{-i(\omega t - kz)} \equiv \hat{\mathbf{E}}e^{-i(\omega t - kz)} \tag{14.53}$$

The corresponding real wave is

$$\mathbf{E} = \hat{\mathbf{x}}E_{0x}\cos[\omega t - kz + \phi_x] + \hat{\mathbf{y}}E_{0y}\cos[\omega t - kz + \phi_y] \tag{14.54}$$

These resultant waves, [Eq. (14.53) and (14.54)] are said to be *elliptically polarized* if $\phi_x \neq \phi_y$. It is to be noted that if $\phi_x = \phi_y \equiv \phi$, they will be linearly polarized since **E** becomes

$$\mathbf{E} = (\hat{\mathbf{x}}E_{0x} + \hat{\mathbf{y}}E_{0y})\cos[\omega t - kz + \phi] \equiv \hat{\boldsymbol{\varepsilon}}E_0\cos[\omega t - kz + \phi]$$

where $\hat{\boldsymbol{\varepsilon}}$ is a constant unit vector in the x-y plane. The reason why the wave is said to be so polarized is that at any point of a given plane in which $z =$ constant $\equiv z_0$, the tip of the **E** vector sweeps out an ellipse in time. Consider the following examples, where the direction of **E** and its magnitude are not constant.

Example 14.5 Elliptic Polarization

Consider the plane wave given by Eq. (14.54). A special case of elliptic polarization is when $\phi_x = \phi$, and $\phi_y = \phi_x \pm \pi/2$. Then at the plane $z = z_0$,

$$\mathbf{E} = \hat{\mathbf{x}}E_{0x}\cos[\omega t - kz_0 + \phi] + \hat{\mathbf{y}}E_{0y}\sin[\omega t - kz_0 + \phi] \equiv \hat{\mathbf{x}}E_x + \hat{\mathbf{y}}E_y$$

hence we see that

$$\left(\frac{E_x}{E_{0x}}\right)^2 + \left(\frac{E_y}{E_{0y}}\right)^2 = \cos^2\Omega + \sin^2\Omega = 1 \tag{14.55}$$

where $\Omega = \omega t - kz_0 + \phi$. This equation for the components of E is the equation of an ellipse with major and minor axes E_{0x} and E_{0y} respectively (see Fig. 14.3). The sense in which E

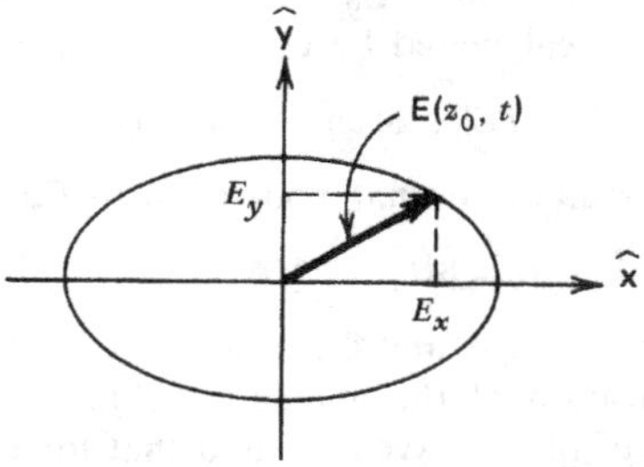

Figure 14.3 Trace of the tip of the electric field at a given point of space as a function of time for the case of elliptic polarization.

moves, clockwise (cw) or counterclockwise (ccw), can be seen by taking successive increasing values of ωt. Thus, for $\phi_y = \phi \pm \pi/2$, **E** moves clockwise for the (+), counterclockwise for the (−) sign. The ccw case is said to have *positive helicity*, the cw case *negative helicity*. Positive helicity also means that **E** rotates in space as a right-handed screw would when advancing along the direction of motion, $\hat{\mathbf{k}}$. This special elliptic polarization can also be written in complex notations using Eq. (14.53):

$$\mathbf{E} = [\hat{\mathbf{x}} E_{0x} e^{i\phi} + \hat{\mathbf{y}} E_{0y} e^{i(\phi \pm \pi/2)}] e^{-i(\omega t - kz)}$$

or

$$\mathbf{E} = [\hat{\mathbf{x}} E_{0x} \pm i\hat{\mathbf{y}} E_{0y}] e^{-i(\omega t - kz + \phi)} \tag{14.56}$$

Example 14.6 Circular Polarization

In the previous example, if we let $E_{0x} = E_{0y} \equiv E_0$ while keeping the same phase condition—that is, $\phi_x = \phi$ and $\phi_y = \phi_x \pm \pi/2$—then in complex notation we get

$$\mathbf{E} = [\hat{\mathbf{x}} \pm i\hat{\mathbf{y}}] E_0 e^{-i(\omega t - kz + \phi)} \tag{14.57}$$

and in real notation we get:

$$\mathbf{E} = E_0\{\hat{\mathbf{x}} \cos[\omega t - kz + \phi] \pm \hat{\mathbf{y}} \sin[\omega t - kz + \phi]\} \tag{14.58}$$

Hence Eq. (14.55) becomes

$$E_x^2 + E_y^2 = E_0^2 \tag{14.59}$$

which is the equation of a circle of radius E_0. One can also directly show that in real notation $\mathbf{E} \cdot \mathbf{E} = \mathbf{E}^2 = \mathbf{E}_0^2$. The magnitude of **E** remains constant, although its direction rotates. This is called *circular polarization*. Figure 14.4 shows the electric field in the $z = 0$ plane (with $\phi = 0$) for $\omega t = 0, \pi/4, \pi/2, 3\pi/4$, and π, showing that it is circularly polarized.

It is interesting that just as a circularly polarized wave may be regarded as a vector sum of linearly polarized waves, so can a linearly polarized wave be regarded as a sum of circularly polarized waves of opposite helicity. To see this we consider the following linearly polarized plane wave, which is written in the complex notation of Eq. (14.49).

$$\mathbf{E} = \hat{\boldsymbol{\varepsilon}} E = \hat{\mathbf{x}} E'_x + \hat{\mathbf{y}} E'_y = (\hat{\mathbf{x}} E_{0x} + \hat{\mathbf{y}} E_{0y}) e^{-i(\omega t - kz + \phi)} \tag{14.60}$$

so long as E'_x and E'_y are in phase. The sum of circularly polarized waves of opposite helicity is written, using Eq. (14.57), as

$$\mathbf{E} = \alpha[\hat{\mathbf{x}} - i\hat{\mathbf{y}}] + \beta[\hat{\mathbf{x}} + i\hat{\mathbf{y}}] \tag{14.61}$$

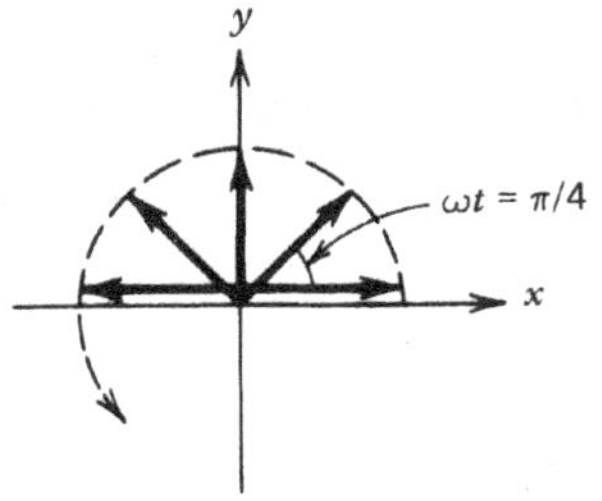

Figure 14.4 Trace of the tip of the electric field in the $z = 0$ plane as a function of time for a circularly polarized wave traveling along z.

Equations (14.60) and (14.61) are identical if we choose

$$\alpha = \frac{1}{2}[E'_x + iE'_y] \qquad \beta = \frac{1}{2}[E'_x - iE'_y] \tag{14.62}$$

14.6 Conservation of Electromagnetic Energy—Poynting's Theorem

We shall at this point derive a very important and useful result that will help us associate energy concepts with the electromagnetic field. It will provide a way for us to interpret the flow of energy with the motion of plane waves in space. It will also, more generally than heretofore, allow us to associate each point of space where the EM field exists with an energy density.

To obtain this result we shall apply the principle of conservation of energy to the EM field. The latter is formulated thus:

1. Let there be an EM field in space and particularly within the volume V whose surface is s.

2. Assume that there exists in association with this field an energy density u. Then the total EM energy in V is given by $U = \int_V u\, dv$.

3. Assume that this energy can flow such that through the surface of V there is an outward flow of energy. Thus assume that there exists a vector field, defined at all points, that gives the direction of energy flow at any point and its magnitude. Then the flux F of energy through s is

$$F \equiv \oint \mathbf{S} \cdot d\mathbf{a} \tag{14.63}$$

where $\mathbf{S}$ is the aforementioned energy flux vector.

4. Let dW/dt be the rate of work done by the EM field inside V on the charge that may exist there. The work done on this charge may be manifested as an increase in its kinetic energy, or in the kinetic energy of the matter in which it is embedded, or in the "chemical energy" of this matter, or the like. In any case, this power expended by the EM field assumes the form [see Eq. (7.74)]

$$\frac{dW}{dt} = \int_V \mathbf{J} \cdot \mathbf{E}\, dv \tag{14.64}$$

where $\mathbf{J}$ is the current density of the charges of concern.

5. Then the conservation of energy principle: [V is stationary] is expressed as:

$$-\frac{dU}{dt} = \oint \mathbf{S} \cdot d\mathbf{a} + \frac{dW}{dt}$$

or

$$-\int \frac{\partial u}{\partial t}\, dv = \oint_s \mathbf{S} \cdot d\mathbf{a} + \int_V \mathbf{J} \cdot \mathbf{E}\, dv \tag{14.65}$$

In words, if the EM field energy decreases in V (note the minus sign), the sum of the flow of EM energy out of V and the work done by $\mathbf{E}$ on the charges within V

must increase correspondingly. In particular, if $\mathbf{J} = 0$, the decrease in EM energy in V is accompanied by a flow of EM energy out of V. Note that since $\int \mathbf{S} \cdot d\mathbf{a} = \int \nabla \cdot \mathbf{S}\, dv$, and Eq. (14.65) is assumed valid for arbitrary volumes V, we have

$$\int_V \left(\frac{\partial u}{\partial t} + \nabla \cdot \mathbf{S} + \mathbf{J} \cdot \mathbf{E} \right) dv = 0$$

which yields the *differential* form of the principle of conservation of energy.

$$\mathbf{J} \cdot \mathbf{E} = -\frac{\partial u}{\partial t} - \nabla \cdot \mathbf{S} \tag{14.66}$$

The differential form is also called the *Poynting theorem*.

Our task is to find $\mathbf{S}$ and $\partial u/\partial t$. In the following, we shall assume that $\mathbf{J}$ is the free current density $\mathbf{J}_f$. If this is so, then the scalar product of $\mathbf{E}$ with Eq. (14.5) gives

$$\mathbf{J}_f \cdot \mathbf{E} = \left[\nabla \times \mathbf{H} - \frac{\partial \mathbf{D}}{\partial t} \right] \cdot \mathbf{E} = \mathbf{E} \cdot \nabla \times \mathbf{H} - \mathbf{E} \cdot \frac{\partial \mathbf{D}}{\partial t}$$

Using the vector identity $\nabla \cdot (\mathbf{E} \times \mathbf{H}) = (\nabla \times \mathbf{E}) \cdot \mathbf{H} - (\nabla \times \mathbf{H}) \cdot \mathbf{E}$ gives

$$\mathbf{J}_f \cdot \mathbf{E} = -\nabla \cdot (\mathbf{E} \times \mathbf{H}) + (\nabla \times \mathbf{E}) \cdot \mathbf{H} - \mathbf{E} \cdot \frac{\partial \mathbf{D}}{\partial t}$$

Substituting $\nabla \times \mathbf{E} = -\partial \mathbf{B}/\partial t$ gives

$$\mathbf{J}_f \cdot \mathbf{E} = -\nabla \cdot (\mathbf{E} \times \mathbf{H}) - \mathbf{H} \cdot \frac{\partial \mathbf{B}}{\partial t} - \mathbf{E} \cdot \frac{\partial \mathbf{D}}{\partial t} \tag{14.67}$$

From comparison of this result with Eq. (14.66) we identify

$$\mathbf{S} = \mathbf{E} \times \mathbf{H} \tag{14.68}$$

$$\frac{\partial u}{\partial t} = \mathbf{H} \cdot \frac{\partial \mathbf{B}}{\partial t} + \mathbf{E} \cdot \frac{\partial \mathbf{D}}{\partial t}$$

The vector $\mathbf{S}$ as identified in Eq. (14.68) is called *Poynting's vector*. It is our energy flow density of the EM field. Note, however, that this choice of $\mathbf{S}$ is not unique; e.g., we could add the curl of any vector function to $\mathbf{S}$ without changing the basic relation given in Eq. (14.66) The identification of $\partial u/\partial t$ in Eq. (14.66) is made reasonably by considering that in simple media ($\mathbf{B} = \mu \mathbf{H}$, $\mathbf{D} = \varepsilon \mathbf{E}$),

$$\mathbf{H} \cdot \frac{\partial \mathbf{B}}{\partial t} = \frac{\mathbf{B}}{\mu} \cdot \frac{\partial \mathbf{B}}{\partial t} = \frac{1}{2\mu} \frac{\partial}{\partial t} (B^2)$$

and hence $u_M \equiv (1/2\mu)B^2$, and that

$$\mathbf{E} \cdot \frac{\partial \mathbf{D}}{\partial t} = \varepsilon \mathbf{E} \cdot \frac{\partial \mathbf{E}}{\partial t} = \frac{1}{2} \varepsilon \frac{\partial}{\partial t} (E^2)$$

and hence $u_E = \frac{1}{2}\varepsilon E^2$. Therefore $u = u_M + u_E$. Also one can write u as

$$u = \tfrac{1}{2}\mathbf{E} \cdot \mathbf{D} + \tfrac{1}{2}\mathbf{H} \cdot \mathbf{B} \tag{14.69}$$

These are just expressions previously encountered for assembling static charge and current distributions [see Eqs. (6.59) and (12.23)]. [In free space, $\mathbf{D} = \varepsilon_0 \mathbf{E}$ and

$\mathbf{B} = \mu_0 \mathbf{H}$, and thus $u_E = \frac{1}{2}\varepsilon_0 E^2$ and $u_M = (1/2\mu_0)B^2$.] More generally in nonsimple media, however, we simply assume the identification $(\partial u_E/\partial t) = \mathbf{E}\cdot(\partial \mathbf{D}/\partial t)$, so that*

$$u_E = \int_{0,0}^{\mathbf{E},t} \frac{\partial u_E}{\partial t}\,dt = \int_{0,0}^{\mathbf{E},t} \mathbf{E}\cdot\frac{\partial \mathbf{D}}{\partial t}\,dt = \int_0^{\mathbf{E}} \mathbf{E}\cdot d\mathbf{D} \tag{14.70}$$

and $(\partial u_M/\partial t) = \mathbf{H}\cdot(\partial \mathbf{B}/\partial t)$, so that

$$u_M = \int_0^{\mathbf{B}} \mathbf{H}\cdot\frac{\partial \mathbf{B}}{\partial t}\,dt = \int_0^{\mathbf{B}} \mathbf{H}\cdot d\mathbf{B} \tag{14.71}$$

The vector $\mathbf{D}$ is regarded here as a function of $\mathbf{E}$ and the time. Similarly, $\mathbf{H}$ is a function of $\mathbf{B}$ and time.

We shall at this point interpret $\mathbf{S}$ as the energy flow for EM waves. Consider a linearly polarized, monochromatic, plane wave moving along $\hat{\mathbf{k}}$ in free space or in simple dielectric media ($\sigma_c = 0$). Since $\hat{\mathbf{k}} \times \mathbf{E} = v\mathbf{B}$, then $E = vB$. Thus

$$\mathbf{S} = \hat{\mathbf{k}}\,\frac{E^2}{v\mu} \tag{14.72}$$

or, in terms of μ and ε of the medium,

$$\mathbf{S} = \hat{\mathbf{k}}\,\frac{E^2}{\sqrt{\mu/\varepsilon}} = \hat{\mathbf{k}}\,\frac{E^2}{\eta} \tag{14.73}$$

The quantity $\eta = \sqrt{\mu/\varepsilon}$ has the units of ohms and is called the *intrinsic impedance* of the material. In free space $\sqrt{\mu/\varepsilon} \to \sqrt{\mu_0/\varepsilon_0} = 377\ \Omega$, which is the impedance of free space. Now we calculate the energy density $u = u_E + u_M$; as follows:

$$u = \frac{1}{2}\varepsilon E^2 + \frac{1}{2\mu}B^2 = \varepsilon E^2 = \frac{E^2}{\mu v^2} \tag{14.74}$$

Thus, the EM energy is equally shared with the electric and magnetic fields $u_E = u_M = \frac{1}{2}u$. Moreover, if we take v as a vector in the propagation direction, then comparing Eqs. (14.72) and (14.74) gives

$$u\mathbf{v} = \mathbf{S} \tag{14.75}$$

The EM energy density times the velocity of the waves is seen to be equal to the Poynting vector. It is as if the energy per unit volume is moving with a velocity $\mathbf{v}$, so through any element of area $\mathbf{da}$ the energy flow is $u\mathbf{v}\cdot d\mathbf{a} = \mathbf{S}\cdot d\mathbf{a}$. This is analogous to the relationship between the charge density and the current density [see Eq. (7.6)], where

$$\rho\mathbf{v} = \mathbf{J}$$

with the assignment $u \to \rho$ and $\mathbf{S} \to \mathbf{J}$.

We now introduce a quantity that is useful in dealing with electromagnetic radiation. The quantity is the magnitude of the time average of the Poynting vector; it is called the *intensity of radiation* $|\langle \mathbf{S}\rangle|$, or simply $\langle S\rangle$. We note that, for a plane wave of $\mathbf{E} = E_0\hat{\mathbf{\varepsilon}}\cos(\omega t - \mathbf{k}\cdot\mathbf{r})$,

$$\langle \mathbf{S}\rangle = \hat{\mathbf{k}}\,\frac{E_0^2}{\eta}\langle\cos^2[\omega t - \mathbf{k}\cdot\mathbf{r}]\rangle = \frac{\hat{\mathbf{k}}E_{\mathrm{rms}}^2}{\eta} \tag{14.76}$$

* We assume that $u_E = 0$ when $\mathbf{E} = 0$ and $u_M = 0$ when $\mathbf{B} = 0$.

where η is the impedance of the medium of permittivity and permeability ε and μ. The intensity is then $\langle S\rangle \equiv \sqrt{\langle \mathbf{S}\rangle \cdot \langle \mathbf{S}\rangle} = E_{\mathrm{rms}}^2/\eta$. The intensity can also be related to the EM energy density. Using Eq. (14.75) we get

$$\langle S\rangle = \langle u\rangle v \tag{14.77}$$

We note that all the terms in the Poynting theorem: energy density u, Poynting vector $\mathbf{S}$, and $\mathbf{J}\cdot\mathbf{E}$ utilize, as defined above, real electric and magnetic fields. In many cases, however, the fields are more readily calculable in complex notations, and thus it is useful to calculate $\langle \mathbf{S}\rangle$ using the complex notation of $\mathbf{E}$ and $\mathbf{B}$ directly. We start from the definition of $\mathbf{S}$ using the real fields:

$$\langle \mathbf{S}\rangle = \langle \mathrm{Re}\{\mathbf{E}\} \times \mathrm{Re}\{\mathbf{H}\}\rangle$$

Using the relation $\mathrm{Re}\,\mathbf{A} = \frac{1}{2}(\mathbf{A} + \mathbf{A}^*)$, we get

$$\langle \mathbf{S}\rangle = \frac{1}{4}[\langle \mathbf{E} \times \mathbf{H}\rangle + \langle \mathbf{E}^* \times \mathbf{H}^*\rangle + \langle \mathbf{E} \times \mathbf{H}^* + \mathbf{E}^* \times \mathbf{H}\rangle]$$

Writing $\mathbf{E} = \mathbf{E}(r)e^{-i\omega t}$ and $\mathbf{H} = \mathbf{H}(r)e^{-i\omega t}$ and averaging over a period $T = 2\pi/\omega$, the first two terms become zero, leaving the result

$$\langle \mathbf{S}\rangle = \frac{1}{4}(\mathbf{E} \times \mathbf{H}^* + \mathbf{E}^* \times \mathbf{H}) = \frac{1}{2}\mathrm{Re}(\mathbf{E} \times \mathbf{H}^*) = \frac{1}{2}\mathrm{Re}(\mathbf{E}^* \times \mathbf{H}) \tag{14.78}$$

Similarly, one can show that

$$\langle u\rangle = \frac{1}{4}\mathrm{Re}[\mathbf{E}\cdot\mathbf{D}^* + \mathbf{H}\cdot\mathbf{B}^*] \tag{14.79}$$

Example 14.7 Fields of a Laser Beam

Laser beams (He–Ne laser) of 100 W/mm^2 and whose EM fields are continuous waves (CW) are easily available. We shall calculate the rms EM fields for these beams, and the average energy density in such a beam. From (14.77), we get: $\langle u\rangle = \langle S\rangle/c = 100/(10^{-6} \times 3 \times 10^8) = 0.33$ J/m^3. From Eqs. (14.76) and (14.77) we get $E_{\mathrm{rms}}^2 = c\eta\langle u\rangle = \eta\langle S\rangle = 377 \times 10^8$ V^2/m^2, or $E_{\mathrm{rms}} = 19.4 \times 10^4$ V/m, and $B_{\mathrm{rms}} = E_{\mathrm{rms}}/c = 6.5 \times 10^{-4}$ T ($=6.5$ G).

Example 14.8 Steady Current Flow in a Wire

Consider a segment of a wire or radius a and conductivity σ_c with a potential difference V across its length l and a constant current I flowing through it, as shown in Fig. 14.5. We would like to discuss the flow from the point of view of the Poynting theorem. The electric field is directed along the wire and is equal to V/l. The magnetic field at the surface is tangential to the curved surface and from Ampere's law it is $B = \mu_0 I/2\pi a$.

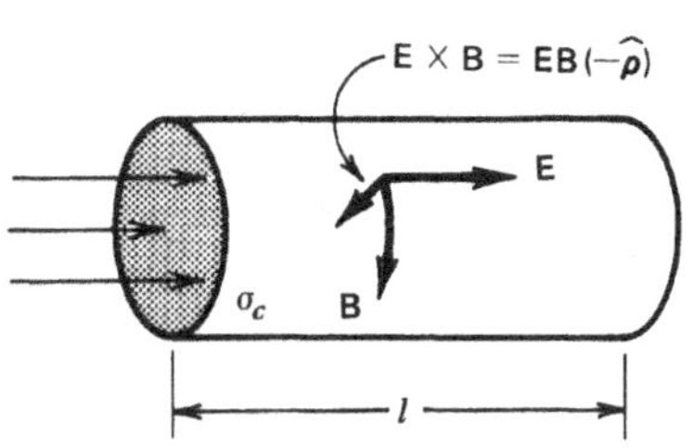

Figure 14.5 Application of the Poynting theorem to current flow in a wire.

Since the fields are steady, then $\partial u/\partial t$ is zero; hence the conservation of energy expressed by the Poynting's theorem [Eq. (14.65)] asserts that $-\oint \mathbf{S}\cdot d\mathbf{a} = \int_V \mathbf{E}\cdot\mathbf{J}\,dv$. The energy flow (into) through the closed surface of the segment rises from the round surface (rs)

$$-\oint \mathbf{S}\cdot d\mathbf{a} = \int_{\text{rs}} \frac{EB\,da}{\mu_0} = \frac{EB}{\mu_0}2\pi a l$$

Since from Ampere's law $2\pi aB = \mu_0 I = \mu_0 \pi a^2 \sigma_c E$, the energy flow becomes

$$-\oint \mathbf{S}\cdot d\mathbf{a} = \frac{EB}{\mu_0}2\pi a l = \sigma_c E^2 \pi a^2 l$$

We now calculate the volume integral over the segment $\int \mathbf{E}\cdot\mathbf{J}\,dv$. Using Ohm's law, $\mathbf{J} = \sigma_c\mathbf{E}$, we get

$$\int \mathbf{E}\cdot\mathbf{J}\,dv = \int \sigma_c E^2\,dv = \sigma_c E^2 \pi a^2 l$$

which agrees with the Poynting theorem. By using the relations $E = V/l$ and $I = \pi a^2 J = \pi a^2 \sigma_c E$, the latter is seen to be equal to $RI^2 = VI$, where $R = l/\sigma_c \pi a^2$ is the resistance of the segment of wire.

The interpretation of this from the view of **S** is that the EM energy flows from outside into the segment through the sides as seen in Fig. 14.5. Since the energy source (the emf) creates the fields that establish the charge and the currents, the EM energy thus seems to flow from the energy source into space. Inside the wire the energy is dissipated as heat (i.e., RI^2). But energy also appear to be transported along the wire with the charges. Does energy enter from the sides or the ends of the wire? Which interpretation is true for the flow of energy cannot be resolved, and in fact it matters little for such static problems. Conservation of energy must be maintained in either case.

Example 14.9 Steady Current Flow in a Coaxial Cable

As another illustration of Poynting's theorem, we consider a coaxial cable of negligible resistance (Fig. 14.6). If this cable is inserted between a source of constant emf and some load, a steady current I will flow down the cable. If the emf provides a constant potential difference V, it will supply power to the cable of magnitude VI.

Let us calculate the rate at which energy passes down the cable using Poynting's theorem. If we assume the space between the inner and outer conductors (or radii a and b, respectively) to be vacuum, we know that the electric and magnetic fields and hence the Poynting vector (with end effects neglected) are given by

$$\mathbf{E} = \hat{\boldsymbol{\rho}}\frac{V}{\ln(b/a)\,\rho} \qquad \mathbf{B} = \hat{\boldsymbol{\phi}}\frac{\mu_0 I}{2\pi\rho} \qquad a \le \rho \le b$$

$$\mathbf{S} = \frac{\mathbf{E}\times\mathbf{B}}{\mu_0} = \hat{\boldsymbol{\rho}}\times\hat{\boldsymbol{\phi}}\frac{VI}{2\pi\ln(b/a)}\frac{1}{\rho^2} = \frac{IV\hat{\mathbf{z}}}{2\pi\rho^2\ln(b/a)}$$

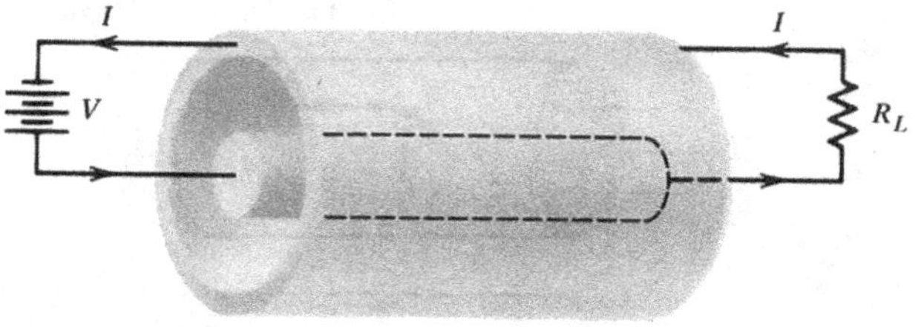

Figure 14.6 Application of the Poynting theorem to current flow in a coaxial cable.

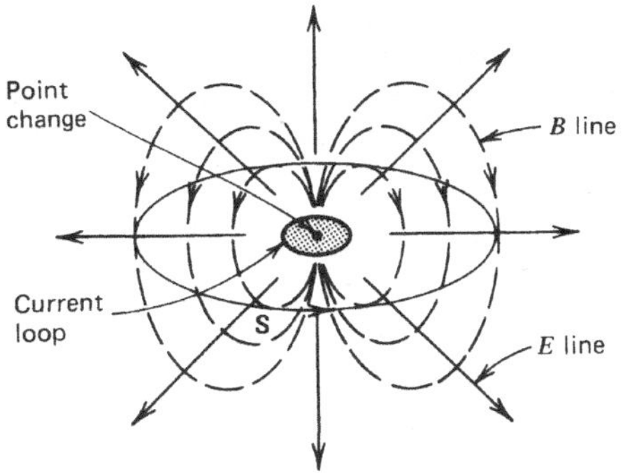

Figure 14.7 Application of Poynting theorem to a superposition of a static point charge and a magnetic dipole.

Note that **S** points along the axis of the cable; thue EM energy is flowing down the cable. Since $\mathbf{E} = 0$ everywhere except for $a \leq \rho \leq b$, we can get the total EM power flow by integrating **S** over the cross-sectional area of the cable between the inner and outer conductors

$$\int \mathbf{S} \cdot d\mathbf{a} = \int_a^b \frac{VI}{2\pi \ln(b/a)} \frac{2\pi\rho \, d\rho}{\rho^2} = VI$$

As expected, the energy provided by the emf flows down the cable without attenuation. This energy finally will be absorbed in the load resistance R_L.

In practice, the conductors of the cable will have a finite resistance, so that energy will also be dissipated as heat in them. This means that there will be a flow of energy into the conductors of the cable. This in turn implies that there will be a component of **E** parallel to the axis of the cable, so that **S** can point into the conductors and energy can flow into the conductors to replace the energy dissipated there.

The interpretation of the Poynting vector as giving the flow of energy density has peculiar effects, especially in static problems, that cannot be resolved. See the two previous examples. The fact that Ponyting's theorem is true for such problems does not guarantee that **S** really is an energy flow. In the example of a magnetic dipole and a point charge superimposed statistically in space (see Fig. 14.7), which is a static problem with constant **E** and **B** fields, it seems as if **S** is flowing around the symmetry axis of the dipole*. It is hard to believe this is happening, and we cannot verify it experimentally. What is clear is that through any sphere containing the dipole, the integrated energy flow is zero.

We have illustrated Poynting's theorem for static problems. Its main utility however is in nonstatic problems, especially where one wishes to calculate the EM radiation flowing from some energy source (see Chapter 15).

14.7 Plane Monochromatic Waves in a Conducting Medium

In this section we consider Maxwell's equations in the presence of conducting media. In order to simplify the discussion we take the space to be charge-free and

* See R. Feynman, R. Leighton, and M. Sands, *The Feynman Lectures on Physics*, (Reading, Mass.: Addison Wesley, 1966) p. 27–8.

external-current-free such that the currents existing in the material are induced only by the electromagnetic wave itself; that is, we take $\mathbf{J} = \sigma_c \mathbf{E}$. Moreover, we take the medium to be simple; that is, $\mathbf{D} = \varepsilon \mathbf{E}$ and $\mathbf{B} = \mu \mathbf{H}$, where ε and μ are constants. The electromagnetic fields in such media satisfy the damped-wave equation [Eq. (14.16)]:

$$\nabla^2 \mathbf{E} - \mu \sigma_c \frac{\partial \mathbf{E}}{\partial t} - \mu \varepsilon \frac{\partial^2 \mathbf{E}}{\partial t^2} = 0 \tag{14.16}$$

Since we are interested in plane monochromatic waves, we take $\mathbf{E}(\mathbf{r}, t) \to \mathbf{E}(\mathbf{r}) e^{-i\omega t}$; hence

$$\nabla^2 \mathbf{E} + (\mu \varepsilon \omega^2 + i \mu \sigma_c \omega) \mathbf{E} = 0$$

or, in terms of $n_0 = \sqrt{K K_m}$,

$$\nabla^2 \mathbf{E} + n_0^2 \frac{\omega^2}{c^2} \left(1 + i \frac{\sigma_c}{\varepsilon \omega}\right) \mathbf{E} = 0 \tag{14.80}$$

This equation can alternatively be written in terms of a *complex refractive index* $\hat{n}$ (the hat indicating complex), as follows:

$$\nabla^2 \mathbf{E} + \hat{n}^2 \frac{\omega^2}{c^2} \mathbf{E} = 0 \tag{14.81}$$

where, using the real quantities n and k,

$$\hat{n} = n_0 \left(1 + i \frac{\sigma_c}{\varepsilon \omega}\right)^{1/2} = n + ik \tag{14.82}$$

We now consider plane waves traveling along the z axis; thus we take $|\mathbf{E}|$ independent of the coordinates x and y, and ∇^2 to be d^2/dz^2. Thus Eq. (14.81) becomes

$$\frac{d^2}{dz^2} \mathbf{E} + \frac{\hat{n}^2 \omega^2}{c^2} \mathbf{E} = 0 \tag{14.83}$$

In the absence of the conducting properties of the material, this equation has the simple solution $\mathbf{E} = \mathbf{E}_0 e^{in_0(\omega/c)z} = \mathbf{E}_0 e^{ik_0 z}$, where $k_0 = n_0 \omega / c = \omega / v$ is the wave number, a real quantity. Because of the presence of the complex factor $1 + i\sigma_c/\omega\varepsilon$, the wave number will be complex. We take in this case

$$\mathbf{E}(z) = \mathbf{E}_0 e^{i\hat{K}z} \tag{14.84}$$

where $\hat{K}$ is a constant complex wave number vector; it is independent of z and $\mathbf{E}_0$. Substituting this form of $\mathbf{E}(z)$ in Eq. (14.83) gives

$$\hat{K}^2 = n_0^2 \frac{\omega^2}{c^2} \left(1 + \frac{i\sigma_c}{\varepsilon \omega}\right) \tag{14.85}$$

Writing $\hat{K}^2$ in complex polar form gives

$$\hat{K}^2 = n_0^2 \frac{\omega^2}{c^2} \left(1 + \frac{\sigma_c^2}{\varepsilon^2 \omega^2}\right)^{1/2} e^{i\phi} \tag{14.86}$$

where

$$\tan \phi = \frac{\sigma_c}{\varepsilon \omega} \tag{14.87}$$

Taking the square root of Eq. (14.86) gives

$$\hat{K} = \frac{n_0\omega}{c}\left(1 + \frac{\sigma_c^2}{\varepsilon^2\omega^2}\right)^{1/4} e^{i\phi/2}$$

or

$$\hat{K} = \frac{n_0\omega}{c}\left(1 + \frac{\sigma_c^2}{\varepsilon^2\omega^2}\right)^{1/4} (\cos\phi/2 + i\sin\phi/2) \tag{14.88}$$

To find $\cos(\phi/2)$ and $\sin(\phi/2)$ from $\tan\phi$, and hence explicitly in terms of $\sigma_c/\omega\varepsilon$, we use the relations:

$$\cos\frac{\phi}{2} = \frac{(1 + \cos\phi)^{1/2}}{\sqrt{2}} \quad \text{and} \quad \sin\frac{\phi}{2} = \frac{(1 - \cos\phi)^{1/2}}{\sqrt{2}}$$

Now

$$\cos\phi = (1 + \tan^2\phi)^{-1/2}$$

Thus

$$\cos\frac{\phi}{2} = \frac{1}{\sqrt{2}}\left[1 + \left(1 + \frac{\sigma_c^2}{\varepsilon^2\omega^2}\right)^{-1/2}\right]^{1/2}$$

$$\sin\frac{\phi}{2} = \frac{1}{\sqrt{2}}\left[1 - \left(1 + \frac{\sigma_c^2}{\varepsilon^2\omega^2}\right)^{-1/2}\right]^{1/2} \tag{14.89}$$

Substituting these explicit forms of $\sin(\phi/2)$ and $\cos(\phi/2)$ in Eq. (14.88) gives

$$\hat{K} = \frac{\omega}{c}n + i\frac{\omega}{c}k \tag{14.90}$$

where

$$\binom{n}{k} = \frac{n_0}{\sqrt{2}}\left[\left(1 + \frac{\sigma_c^2}{\varepsilon^2\omega^2}\right)^{1/2} \pm 1\right]^{1/2} \tag{14.91}$$

Thus the electric field takes the following form:

$$\mathbf{E} = \mathbf{E}_0 e^{-(\omega/c)kz} e^{-i\omega(t - nz/c)} \tag{14.92}$$

The quantity n is interpreted as the *refractive index* of the medium. It is interesting to note that it depends on the conductivity of the medium and the frequency of the wave. The quantity k is associated with the decay of the wave as it travels in the medium. The *attenuation constant* $\omega k/c$ also depends on σ_c and ω. It is customary to refer to the distance δ at which the field goes down to $1/e$ of its initial value as the *skin depth*, where $\omega k\delta/c = 1$ or

$$\delta = \frac{c}{\omega k} \tag{14.93}$$

The magnetic field corresponding to the electric field can now be calculated from Maxwell's equation. One interesting effect that arises in conducting media is the fact that the E and B fields are out of phase. This can be seen from Maxwell's curl equation (Eq. (14.4)). We consider a plane wave polarized along the x axis: $E_0\hat{\mathbf{x}}e^{-i\omega t + i\hat{K}z}$. Taking $\mathbf{B} = \mu\mathbf{H}$ and $\mathbf{D} = \varepsilon\mathbf{E}$ gives:

$$i\hat{K}\hat{\mathbf{z}} \times \mathbf{E} = i\omega\mu\mathbf{H} \tag{14.94}$$

or

$$\mathbf{H} = \hat{\mathbf{y}}\frac{\hat{K}}{\mu\omega}E_0 e^{-i\omega t + i\hat{K}z} = H_0 \hat{\mathbf{y}} e^{-i\omega t + i\hat{K}z} \tag{14.95}$$

Substituting for $\hat{K}$ from Eq. (14.85) gives

$$H_0 = \sqrt{\frac{\varepsilon}{\mu}}\left(1 + \frac{i\sigma_c}{\varepsilon\omega}\right)^{1/2} E_0 \tag{14.96}$$

This equation indicates that the magnetic field and the electric field are out of phase. Moreover, the ratio E_0/H_0 is characteristic of the medium; it is the *intrinsic complex impedance* of the medium $\hat{\eta}$ and appears to be frequency dependent:

$$\hat{\eta} = \sqrt{\frac{\mu}{\varepsilon}}\frac{1}{(1 + i\sigma_c/\varepsilon\omega)^{1/2}} \tag{14.97}$$

The phase of η can be calculated by expressing it in complex polar notations:

$$\hat{\eta} = |\eta| e^{-i\phi/2} \tag{14.98}$$

where

$$|\hat{\eta}| = \frac{\sqrt{\mu/\varepsilon}}{[1 + (\sigma_c/\varepsilon\omega)^2]^{1/4}} \qquad \text{and} \qquad \tan\phi = \frac{\sigma_c}{\varepsilon\omega} \tag{14.99}$$

The above treatment shows that the general dependence of the propagation on the frequency and the properties of the medium is quite complicated. There are, however, some special limiting cases that are interesting to consider.

Good Conductors. The conductivity of metals and even semiconductors is very large such that the ratio $\sigma_c/\varepsilon\omega$ is very much larger than unity even at frequencies as high as the optical frequencies. In this limit, one neglects 1 relative to $\sigma_c/\varepsilon\omega$ (and takes $K_m = 1$); thus

$$n = k = \sqrt{\frac{KK_m}{2}}\sqrt{\frac{\sigma_c}{\varepsilon\omega}} = \sqrt{\frac{\sigma_c}{2\varepsilon_0\omega}} \tag{14.100}$$

Substituting this result in Eq. (14.93) gives the following expression for the skin depth.

$$\delta = \frac{c}{\omega}\sqrt{\frac{\varepsilon\omega}{\sigma_c}}\sqrt{\frac{2}{KK_m}} = \sqrt{\frac{2}{\mu_0\sigma_c\omega}} \tag{14.101}$$

which shows that in this limit δ does not depend on the dielectric properties of the material. The fact that the skin depth is inversely proportional to the square root of the product, the frequency and the conductivity implies that waves of higher frequencies do not penetrate as much as those of lower frequencies. The refractive index of metals is enhanced by the factor $\sqrt{(\sigma_c/2\varepsilon\omega)}$ over what is expected from the dielectric properties of the material: $\sqrt{KK_m}$. This increase causes a reduction in the wavelength of the radiation in the material by the same factor.

Another feature of the propagation in conductors is the phase relationship between the magnetic and electric fields. Equation (14.99) shows that $\tan\phi$ becomes very large when $\sigma_c/\varepsilon\omega \gg 1$; hence $\phi \to \pi/2$. Therefore the phase difference between the **E** and the **B** fields in a perfect conductor is $\pi/4$.

Good Insulators. Good insulators have small conductivities such that the ratio $\sigma_c/\omega\varepsilon$ is much less than unity even at very low frequencies. For strictly dc fields ($\omega = 0$), however, the conductivity has to be strictly zero. Taking this limit in Eq. (14.91) gives

$$n = \sqrt{KK_m} \qquad k = \frac{1}{2}\sqrt{KK_m}\,\frac{\sigma_c}{\varepsilon\omega} \tag{14.102}$$

The phase between the E and B field in good insulators is derivable from Eq. (14.98). Keeping the lowest order in $\sigma_c/\varepsilon\omega$ gives

$$\tan\phi \approx \phi = \frac{\sigma_c}{\varepsilon\omega} \tag{14.103}$$

Thus the phase angle between the E and B fields is:

$$\frac{\phi}{2} = \frac{1}{2}\frac{\sigma_c}{\varepsilon\omega} \tag{14.104}$$

Example 14.10 Frequency Dependence of the Optical Properties of Conducting Material

In this example we examine the optical properties of conducting materials at an electromagnetic frequency ω. Consider a material of conductivity σ_c, permittivity ε, and permeability μ. The refractive index n and the absorption coefficient k of this material are functions of σ_c, ω, μ, and ε, and are given in Eq. (14.91). Figures 14.8 and 14.9 give plots of n and k as a function of ω. It shows that as ω becomes large, and hence $\sigma_c/\omega\varepsilon$ becomes small, the refractive index of the material becomes governed by the dielectric and magnetic properties (ε, μ): $n \to \sqrt{KK_m}$. The absorption coefficient also becomes governed by the dielectric properties in the same limit; that is, it drops to zero. It is clear from the figures that the conducting nature of the material dominates at low frequencies, where $\sigma_c/\varepsilon\omega \gg 1$.

The skin depth of the material $\delta = c/\omega k$ becomes independent of frequency and approaches $n_0 c\sigma_c/2\varepsilon$ at high frequencies. At low frequencies it takes the expression $\sqrt{2/\mu_0\sigma_c\omega}$.

The characteristic impedance of the material

$$|\hat{\eta}| = \sqrt{\frac{\mu}{\varepsilon}}\Big/\left[1 + \left(\frac{\sigma_c}{\varepsilon\omega}\right)^2\right]^{1/4}$$

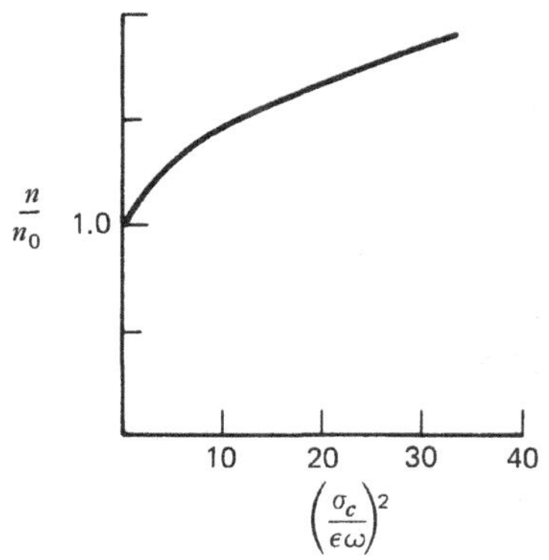

Figure 14.8 Frequency behavior of the optical constant n of a conducting material.

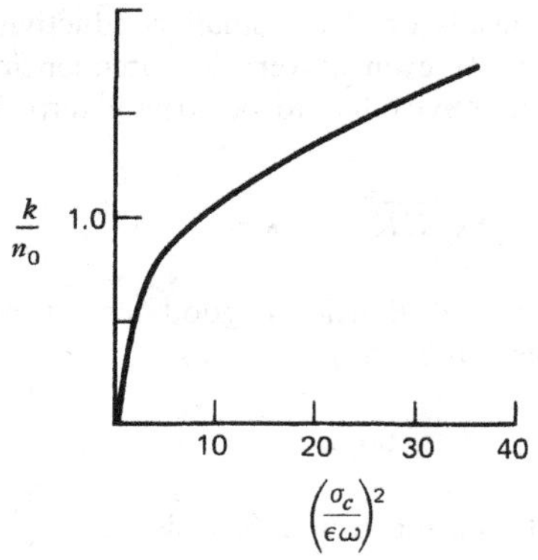

Figure 14.9 Frequency behavior of the optical constant k of a conducting material.

approaches

$$|\hat{\eta}| \to \sqrt{\frac{\mu}{\varepsilon}} \qquad \text{and} \qquad |\hat{\eta}| \to \sqrt{\frac{\mu}{\varepsilon}}\left(\frac{\sigma_c}{\varepsilon\omega}\right)^{-1/2}$$

at high and low frequencies, respectively. This shows that $|\hat{\eta}|$ decreases in conductors.

Example 14.11 Power Dissipation in a Conductor

A monochromatic plane wave of frequency ω, polarized in the $\hat{\mathbf{x}}$ direction, travels along the z axis in a highly conducting medium occupying $z \geq 0$ space (see Fig. 14.10). The medium is of permittivity ε, permeability μ, and conductivity σ_c. From Eqs. (14.92) and (14.100), the electric field at a distance z in the material is given by

$$\mathbf{E} = E_0 \hat{\mathbf{x}}\, e^{-z(1-i)/\delta - i\omega t} \tag{14.105}$$

where $\delta = \sqrt{(2/\mu_0\omega\sigma_c)}$ is the skin depth. The current density in the medium is $\mathbf{J} = \sigma_c \mathbf{E}$; thus

$$\mathbf{J} = E_0 \sigma_c \hat{\mathbf{x}}\, e^{-z(1-i)/\delta - i\omega t} \tag{14.106}$$

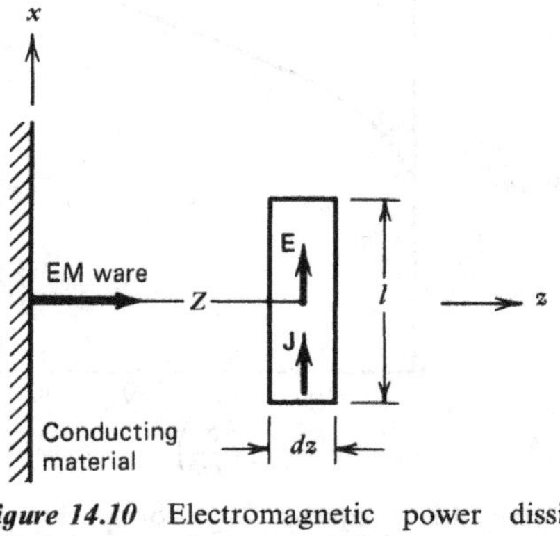

Figure 14.10 Electromagnetic power dissipation in a conducting material.

The current dI in a slab of height h, length l along the current, and depth along propagation dz is along the x axis and is given by

$$dI = E_0\sigma_c h\, dz\, e^{-z(1-i)/\delta - \omega t} \tag{14.107}$$

Thus the total current in the medium is

$$I = \int_0^\infty dI = E_0\sigma_c h e^{-i\omega t}\int e^{-z(1-i)/\delta}\, dz$$

which gives

$$I = \frac{E_0\sigma_c h\delta e^{-i\omega t}}{1-i} = \frac{E_0\sigma_c h\delta e^{-i(\omega t - \pi/4)}}{\sqrt{2}} \tag{14.108}$$

The root mean square of I can be calculated from Eq. (14.108) as follows:

$$I_{\text{rms}} = \frac{E_0\sigma_c h\delta}{2} \tag{14.109}$$

Let us now calculate the power dissipated in the medium. To do this we need to calculate the resistance of the slab considered above. Since the resistance of a rectangular conductor of length l, area A, and conductivity σ_c is $R = l/\sigma_c A$, then the resistance of the slab is

$$R = \frac{l}{\sigma_c h\, dz} \tag{14.110}$$

The average power dissipated in the slab is given in terms of dI and R by

$$dP = \frac{1}{2}(dI)dI^* R \tag{14.111}$$

Substituting for dI and R from Eqs. (14.107) and (14.110) we get

$$dP = \frac{1}{2}\frac{E_0^2\sigma_c^2 h^2 l}{\sigma_c h} e^{-2z/\delta}\, dz. \tag{14.112}$$

Integrating dP gives

$$P = \frac{E_0^2\sigma_c hl}{2}\int_0^\infty e^{-2z/\delta}\, dz = \frac{1}{4}E_0^2\sigma_c hl\delta \tag{14.113}$$

An effective resistance can be associated with this power loss by writing the average power in the form

$$P = R_{\text{eff}} I_{\text{rms}}^2$$

Substituting for I_{rms} from Eq. (14.109) gives

$$R_{\text{eff}} = \frac{l}{(h\delta)\sigma_c} = \frac{l}{A\sigma_c} \tag{14.114}$$

where A is the area of a slab of height h and width equal to the skin depth.

14.8 Summary

The quasi-static equations of electromagnetism $\nabla\cdot\mathbf{D} = \rho_f$, $\nabla\cdot\mathbf{B} = 0$, $\nabla\times\mathbf{E} = -\partial\mathbf{B}/\partial t$, and $\nabla\times\mathbf{H} = \mathbf{J}_f$ are modified for the last time by modifying $\nabla\times\mathbf{H} = \mathbf{J}_f$ to include the effect of the variation of $\mathbf{D}$ with time. The addition of the displacement current $\partial\mathbf{D}/\partial t$ to $\mathbf{J}_f$ by Maxwell yielded what are called Maxwell's equations, which govern the behavior of the classical electromagnetic field as we believe it today; that is,

$$\nabla\cdot\mathbf{D} = \rho_f \qquad \nabla\cdot\mathbf{B} = 0 \qquad \nabla\times\mathbf{E} = -\frac{\partial\mathbf{B}}{\partial t} \qquad \nabla\times\mathbf{H} = \frac{\partial\mathbf{D}}{\partial t} + \mathbf{J}_f \quad \text{(14.2)–(14.5)}$$

In a linear medium of μ, ε, and σ_c, Maxwell's equations give for $\rho_f = 0$, and $\mathbf{J}_f = \sigma_c \mathbf{E}$ (currents are produced only by E itself),

$$\left[\nabla^2 - \mu\sigma_c \frac{\partial}{\partial t} - \mu\varepsilon \frac{\partial^2}{\partial t^2}\right]\begin{pmatrix}\mathbf{E}\\ \mathbf{B}\end{pmatrix} = 0 \tag{14.16}$$

which are three-dimensional wave equations with some damping (term proportional to σ_c). Even in vacuum, Maxwell's equations predict wave phenomena for $\mathbf{E}$ and $\mathbf{B}$. The speed v of the wave is

$$v = \frac{1}{\sqrt{\mu\varepsilon}} = \frac{1}{n\sqrt{\mu_0 \varepsilon_0}} = \frac{c}{n} \qquad (14.19),(14.24)$$

where $c = 1/\sqrt{\mu_0 \varepsilon_0} = 3 \times 10^8$ m/s, which is the speed of light in vacuum, and $n = \sqrt{KK_m}$ is the refractive index of the medium.

One solution of the wave equation in media of $\sigma_c = 0$, is a plane wave where $\mathbf{E}$ and $\mathbf{B}$ depend only on one cartesian coordinate, say in the $\hat{\mathbf{k}}$ direction, $\mathbf{E}$ and $\mathbf{B}$ will be functions of $\hat{\mathbf{k}} \cdot \mathbf{r} - vt$. Such waves are also called transverse waves. Maxwell's equations require the following interrelationships between $\hat{\mathbf{k}}$, $\mathbf{E}$, and $\mathbf{B}$:

$$\hat{\mathbf{k}} \cdot \mathbf{B} = 0 \qquad \hat{\mathbf{k}} \cdot \mathbf{E} = 0 \qquad \hat{\mathbf{k}} \times \mathbf{E} = v\mathbf{B} \qquad |\mathbf{E}| = v|\mathbf{B}| \qquad (14.34),(14.37)$$

That is, these three vectors are normal to each other. One form of solution is the sinusoidal or monochromatic solution, where $\mathbf{E}$ and $\mathbf{B}$ are given by a sine function or a cosine function of $\hat{\mathbf{k}} \cdot \mathbf{r} - vt$. Using complex notations we write

$$\mathbf{E} = E_0 \hat{\mathbf{e}} e^{-i(\omega t - \mathbf{k}\cdot\mathbf{r} + \phi)} \qquad \mathbf{B} = B_0 \hat{\boldsymbol{\beta}} e^{-i(\omega t - \mathbf{k}\cdot\mathbf{r} + \phi')} \qquad (14.49),(14.50)$$

where $\mathbf{k}$, a wave vector, specifies the direction of propagation; $\hat{\mathbf{e}}$ and $\hat{\boldsymbol{\beta}}$ are unit vectors; ω is the frequency of oscillation of the wave; $k = \omega/v$ is the wave number, which can also be expressed in terms of the wavelength of the wave $k = 2\pi/\lambda$; and (ϕ, ϕ') and (E_0, B_0) are constant phases and amplitudes. It is customary to refer to the direction of the $\mathbf{E}$ of the wave as the direction of polarization of the wave. If $\hat{\mathbf{e}}$ stays along one direction, the wave is called a plane or linearly polarized wave. Combinations of linear polarization can result in elliptic or circular polarization.

The rate at which energy is carried by the wave is given by the Poynting vector $\mathbf{S}$ as follows:

$$\mathbf{S} = \mathbf{E} \times \mathbf{H} \tag{14.68}$$

If the wave strikes a region that has electric currents $\mathbf{J}$, then the wave is expected to do work on these currents at a rate

$$\frac{dP}{dv} = \mathbf{J} \cdot \mathbf{E} \tag{7.75}$$

Just as in electrostatics and magnetostatics, the wave at any instant of time has stored electromagnetic energy with a density of

$$u = \frac{1}{2}\mathbf{E} \cdot \mathbf{D} + \frac{1}{2}\mathbf{H} \cdot \mathbf{B} \tag{14.69}$$

In fact, one can show that energy is conserved, and express it in the form of the Poynting theorem at any point in space; that is,

$$\nabla \cdot \mathbf{S} + \frac{\partial u}{\partial t} + \mathbf{J} \cdot \mathbf{E} = 0 \tag{14.66}$$

One can show that

$$\mathbf{S} = u\mathbf{v} = \frac{\hat{\mathbf{k}} E^2}{\eta} \qquad (14.73),(14.75)$$

where $\eta = \sqrt{\mu/\varepsilon}$ is called the impedance of the space, and $\hat{\mathbf{k}}$ is a unit vector along the direction of propagation. For plane monochromatic waves, the time average of the intensity of the radiation

$$\langle \mathbf{S} \rangle = \frac{1}{2}\,\mathrm{Re}(\mathbf{E}^* \times \mathbf{H}) \tag{14.78}$$

and

$$\langle u \rangle = \frac{1}{4}\,\mathrm{Re}(\mathbf{E}^* \cdot \mathbf{D} + \mathbf{H}^* \cdot \mathbf{B}) \tag{14.79}$$

In a conducting medium a plane monochromatic wave suffers damping (attenuation). The effect of a nonzero σ_c can be derived by the modification of the above solutions through the introduction of an effective complex permittivity and hence wave vector $\hat{\mathbf{K}}$ and refractive index $\hat{n}$; that is,

$$\hat{K}^2 = \frac{n_0^2 \omega^2}{c^2}\left(1 + \frac{i\sigma_c}{\varepsilon\omega}\right) = \hat{n}^2 \frac{\omega^2}{c^2}$$

where the hat is used to indicate the complex property. Taking the square root of $\hat{K}^2$ and writing in terms of a real and imaginary parts we get

$$\hat{K} = \frac{\omega}{c} n + i\frac{\omega}{c} k \tag{14.90}$$

where

$$\begin{pmatrix} n \\ k \end{pmatrix} = \frac{n_0}{\sqrt{2}}\left[\left(1 + \frac{\sigma_c^2}{\varepsilon^2\omega^2}\right)^{1/2} \pm 1\right]^{1/2} \tag{14.91}$$

The corresponding **E** field propagating along z takes the form

$$\mathbf{E} = \mathbf{E}_0 e^{-\omega k z/c} e^{-i\omega(t - nz/c)} \tag{14.92}$$

Thus n is interpreted as an effective real refractive index of the medium with additional dependence on σ_c and ω (dispersive effect). The inverse of the quantity $\omega k/c$ is interpreted as an absorption length or skin depth δ, which is frequency-dependent:

$$\delta = \frac{c}{\omega k} \tag{14.93}$$

For high conductivity and/or low frequencies such that $\sigma_c/\varepsilon\omega \gg 1$, we find (for nonmagnetic materials) that

$$n \approx k \approx \sqrt{\frac{\sigma_c}{2\varepsilon_0 \omega}} \quad \text{and} \quad \delta = \sqrt{\frac{2}{u_0 \sigma_c \omega}} \tag{14.100}$$

Inside conducting materials, **E** and **H** are out of phase, with

$$\frac{E_0}{H_0} = \hat{\eta} = \sqrt{\frac{\mu}{\varepsilon}}\left(1 + \frac{i\sigma_c}{\varepsilon\omega}\right)^{-1/2} \tag{14.96),(14.97}$$

is the complex impedance of the medium.

Problems

14.1 An ac generator is connected to a parallel-plate capacitor. The plates are circular disks of area A. The charge on the plates is $q = q_0 \sin \omega t$. Neglect edge effects. (a) Calculate the conduction and displacement currents. How do they compare? (b) What is the direction of the magnetic field inside the capacitor? (c) Calculate the magnitude of the magnetic field inside the capacitor.

14.2 A coaxial, cylindrical capacitor has inner and outer radii of 0.5 and 0.6 cm, respectively, and a length of 50 cm. The material between the cylinders has a dielectric constant of 6.7. The cylinders are kept at a potential difference $V = 250 \sin 377t$ volts. Determine the displacement current I_D and the conduction current I_c. Compare the two currents.

14.3 The space between two concentric spherical conducting shells of inner and outer radii a and b is filled with a dielectric for which $K = 8.5$. Given an applied voltage $V = 150 \sin 500t$ volts. Obtain the conduction and displacement currents and compare.

14.4 Show explicitly that the fields (a) $E = Ace^{x-ct}$ and $B = Ae^{x-ct}$, and (b) $E = Ac \ln(x + ct)$ and $B = -A \ln(x + ct)$ are solutions of the one-dimensional wave equation given by Eq. (14.17).

14.5 (a) In free space we have $\mathbf{E} = E_0 \hat{\mathbf{y}} \sin(\omega t - kz)$. Find **D**, **B**, and **H**. (b) In free space we have $\mathbf{H} = H_0 \hat{\mathbf{x}} e^{-i(\omega t + kz)}$. Determine **E**.

14.6 An electromagnetic wave propagates in a ferrite material whose dielectric and magnetic constants are $K = 10$ and $K_M = 1000$. Find the speed of propagation, and the wavelength of a wave of frequency 100 MHz.

14.7 Consider a linearly polarized, plane electromagnetic wave $\mathbf{E} = E_0 \hat{\mathbf{x}} e^{i(kz - \omega t)}$, where $k = \omega/c$ and E_0 is real. (a) Calculate the energy density u and Poynting vector **S**, and show that u moves along with the wave. (b) Determine the time average of **S** when the averaging is done over an infinite time, and again over one period.

14.8 If a wave is added to that of Problem 14.7 such that one now has

$$\mathbf{E} = E_0 \hat{\mathbf{x}} e^{i(kz - \omega t)} + E_0' \hat{\mathbf{x}} e^{i(kz - \omega t + \phi)}$$

where E_0' is real, find u, **S**, and $\langle \mathbf{S} \rangle$ for the resultant wave and show that in general none are equal to the sum of the corresponding quantities for the separate waves.

14.9 Assume that a wave is added to that of Problem 14.7 such that one now has

$$\mathbf{E} = E_0 \hat{\mathbf{x}} e^{i(kz - \omega t)} + E_0' \hat{\mathbf{y}} e^{i(kz - \omega t + \phi)}$$

where E_0' is real. (a) Show that in this case u, **S**, and $\langle \mathbf{S} \rangle$ for the resultant wave are the sums of the corresponding quantities of the individual waves. (b) Determine the polarization of the resultant electric wave. (c) How are the resultant electric and magnetic waves related? (d) If $E_0' = 2E_0$ and $\phi = \pi/4$, determine and plot the locus of the tip of **E** in the $z = 0$ plane.

14.10 Assume that a wave is added to that of Problem 14.7 such that one now has

$$\mathbf{E} = E_0 \hat{\mathbf{x}} e^{i(kz - \omega t)} + E_0 \hat{\mathbf{x}} e^{i(-kz - \omega t)}$$

(a) Determine u and **S**. (b) Discuss how the energy is sometimes all in the electric field, and sometimes all in the magnetic field. (c) Does **S** account for this transfer? (d) Find the fixed planes perpendicular to the z axis such that no energy flows across any of them.

14.11 When a linearly polarized EM wave is incident at 45° on a reflecting mirror, the electric field of the waves in front of the mirror may be written as

$$\mathbf{E} = \hat{\mathbf{x}} e^{i(ky - \omega t)} + \hat{\mathbf{y}} e^{i(kx - \omega t)}$$

(a) Write **E** in the form $\mathbf{E} = \mathbf{E}_0 e^{-i\omega t}$. (b) Draw an array of ellipses (showing the electric field) at points whose coordinates are $kx = m\pi/4$ and $ky = n\pi/4$, where m and n take on the values 0, 1, 2, 3, and 4 independently. (This array was used to explain with enough detail the pattern in front of the mirror in the experiment done by Wiener.)

14.12 The energy flow associated with sunlight, striking the surface of the earth in a normal direction is 1.4 kW/m². (a) If the corresponding electromagnetic wave is taken to be a plane polarized monochromatic wave, determine the maximum values of E, H, and B. (b) Taking the distance from the earth to the sun as 1.5×10^{11} m, find the total power radiated by the sun.

14.13 (a) Given $\mathbf{E} = 50\hat{\mathbf{x}} \cos(\omega t - kz)$ volts per meter in free space. Find the average power passing through a circular area of diameter 5 m in the plane where $z = 5$ m. (b) Given that in free space

$$\mathbf{E} = 2 \times 10^2 \hat{\boldsymbol{\theta}} \sin\theta \cos(\omega t - kr)/r \text{ (volts per meter)}$$

and

$$\mathbf{H} = 0.53\hat{\boldsymbol{\phi}} \sin\theta \cos(\omega t - kr)/r \text{ (amperes per meter)}$$

Determine the average power passing through a hemispherical shell of radius $r = 10^2$ m and $0 \le \theta \le \pi/2$.

14.14 Moist soil has a conductivity σ_c of $10^{-3}\ (\Omega \cdot \text{m})^{-1}$ and a dielectric constant K of 2.5. If the electric field in the soil is $E_0 \sin\omega t$ where $E_0 = 6 \times 10^{-6}$ V/m and $\omega = 0.9 \times 10^9$ rad/s, determine the displacement and conduction current density in the soil.

14.15 An electromagnetic wave with a frequency of 10^6 Hz "travels" along the z axis in an aluminum medium located at $z \ge 0$. The conductivity of aluminum is $38.2 \times 10^6\ (\Omega \cdot \text{m})^{-1}$ and its relative permeability $K_m = 1$. Just inside the conductor at $z = +0$, the electric field amplitude is $E_0\hat{\mathbf{x}}$. (a) Write down an expression for the electric field inside the conductor. (b) Find the skin depth, wave velocity, and wavelength of the wave in aluminum. (c) Determine the corresponding magnetic field. (d) Find the phase difference between the electric and magnetic fields at each fixed location in aluminum.

14.16 A plane monochromatic wave of frequency 10^9 Hz polarized along the $\hat{\mathbf{x}}$ axis travels along the z axis in a partially conducting material of permittivity $\varepsilon = 18\varepsilon_0$, permeability $\mu = 800\mu_0$, and conductivity σ_c such that $\sigma_c/\varepsilon\omega = 1$. Just inside the material at $z = +0$, the intensity of the wave is 1 W/m^2. (a) Write down an expression for the electric field inside the conductor. (b) Determine the skin depth and the wave velocity (c) What is the intensity at $z = 1$ mm. (d) Determine the ratio of E/H. (e) Calculate the amplitude of the E field at $z = +0$. (f) Determine the phase between E and H.

14.17 Calculate the intrinsic impedance and the wave velocity for a conducting medium in which $\sigma_c = 5 \times 10^7\ (\Omega \cdot \text{m})^{-1}$ and $\mu = \mu_0$ at a frequency $\omega = 2\pi \times 10^8$ rad/s.

14.18 An electromagnetic wave of frequency 3×10^8 Hz travels in a partially conducting medium of dielectric constant $K = 20$, magnetic constant $K_m = 1000$, and conductivity $\sigma_c = 2(\Omega \cdot \text{m})^{-1}$. (a) Determine the complex refractive index of the medium at the frequency of the wave. (b) Determine the effective refractive index and absorption (attenuation) constant. (c) What is the characteristic impedance of the medium?

14.19 In a rectangular duct having a cross section of 1 m^2, the portion of space corresponding to $y > 0$ is filled with a gas of conductivity $\sigma_c = 10^4\ (\Omega \cdot \text{m})^{-1}$. A magnetic field of uniform strength is established in the vacuum space $y < 0$. Neglect the displacement current and assume that B and E inside the gas depend on y only and are along z and x, respectively. (a) Show that the B field for $t > 0$ and $y > 0$ satisfies the equation

$$\mu_0 \sigma_c \frac{\partial B}{\partial t} = \frac{\partial^2 B}{\partial y^2}$$

(b) Using a substitution $\xi = \sqrt{\mu_0 \sigma_c}\, y/\sqrt{4t}$, find an expression for B, and then determine it, using the boundary conditions. (c) Find the *magnetic* pressure at $y = 0$, and then find the magnetic force on the portion of gas contained between $y = 0$ and $y = 1$ at $t = 0$. (d) What will be the magnitude of the *magnetic* force at $t = 1$ s? *Hint*: You may use the error functions to express your answers.

$$[\operatorname{erfc}(z) = 1 - \frac{2}{\sqrt{\pi}} \int_0^z e^{-z^2}\, dz,]$$

FIFTEEN
RADIATION

In this chapter we discuss the generation of electromagnetic radiation by means of moving charges. We will find that it can be produced only if the charges undergo acceleration. A charge that is not moving produces a static electric field and a zero magnetic field in its inertial frame. A uniformly moving one, on the other hand, produces an electric and a magnetic field, each of which has $1/r^2$ radial dependence. Thus the Poynting vector has $1/r^4$ radial dependence, and consequently its integral over a closed surface vanishes at large distances, indicating that radiation is not generated. An accelerating charge produces additional electric and magnetic fields, each of which is proportional to the acceleration and has $1/r$ radial dependence and hence the corresponding Poynting vector has $1/r^2$ radial dependence, and consequently the integrated Poynting vector is finite at large distances. This indicates that accelerated charges produce radiation.

There are a number of procedures for the calculation of radiation. The general problem will be treated by obtaining, for given time-dependent charge and current distributions, the scalar and vector potentials from which the fields are then obtained. A multipole expansion of the potentials will be derived; this expansion in turn is useful in deriving the fields associated with slowly moving accelerated charges and with antennas.

In regions where the charge and the current distributions vanish, the problem will be treated by solving the wave equations of the electric and magnetic fields directly. Some of the formalism used in solving Laplace's equation in Chapter 3 will be used to generate a multipole expansion. Emission from time-dependent electric and magnetic dipoles, as well as some scattering problems, can be treated with this method.

15.1 Wave Equation of the Potentials with Sources—Gauge Transformations

In Chapter 14 we solved Maxwell's equation in the absence of external charge distributions, $\rho_f = 0$, and external current distributions $\mathbf{J}_f = 0$. In Section 14.8 we considered a case involving electric currents; however, these currents were not external but were produced by the impinging wave itself as a result of the conducting property of the material: $\mathbf{J} = \sigma_c \mathbf{E}$. In this section the case involving external charge and current sources will be studied. Consider an infinite medium of permittivity ε, permeability μ, and conductivity $\sigma_c = 0$, in which there exists the charge and current distributions $\rho(\mathbf{r}, t)$ and $\mathbf{J}(\mathbf{r}, t)$ which are functions of space and time. There are two approaches for obtaining the fields produced by the distributions: the field or the potential approach. We will follow the potential approach because it involves less computation. In fact, we have previously seen, that the potential approach is also more convenient in the cases of electrostatics and magnetostatics.

We start by utilizing two of Maxwell's equations [Eqs. (14.3) and (14.4)] in order to find relations between the fields $\mathbf{E}(\mathbf{r}, t)$ and $\mathbf{B}(\mathbf{r}, t)$ and the vector and scalar potentials $\mathbf{A}(\mathbf{r}, t)$ and $\Phi(\mathbf{r}, t)$. These relations are not expected to be identical to what we have known in static situations. Since $\nabla \cdot \mathbf{B} = 0$, then $\mathbf{B}$ can be written as a curl of the vector potential $\mathbf{A}$:

$$\mathbf{B} = \nabla \times \mathbf{A}. \tag{15.1}$$

Substituting this expression in Eq. (14.4) gives

$$\nabla \times \mathbf{E} = -\frac{\partial}{\partial t}(\nabla \times \mathbf{A}) \tag{15.2}$$

Interchanging differentiation with respect to time and space gives

$$\nabla \times \left(\mathbf{E} + \frac{\partial \mathbf{A}}{\partial t}\right) = 0 \tag{15.3}$$

Since the curl of the vector $(\mathbf{E} + \partial \mathbf{A}/\partial t)$ is zero, then it should be equal to the gradient of a scalar potential Φ:

$$\mathbf{E} + \frac{\partial \mathbf{A}}{\partial t} = -\nabla \Phi(\mathbf{r}, t)$$

or

$$\mathbf{E} = -\nabla \Phi(\mathbf{r}, t) - \frac{\partial \mathbf{A}}{\partial t} \tag{15.4}$$

Equations (15.1) and (15.4) give the sought relations between the fields and the potentials. The relation between $\mathbf{B}$ and $\mathbf{A}$ is identical to the magnetostatic case, except that $\mathbf{A}$ is now allowed to be a function of time. Equation (15.4), on the other hand, shows a departure from electrostatics; the electric field is not anymore a gradient of a scalar. It is the sum of the gradient of $\Phi(\mathbf{r}, t)$ and the derivative of $\mathbf{A}$ with respect to time. The latter contribution removes the conservative nature of the electric field and it is a direct outcome of Faraday's experimental law of induction (see Chapter 11).

Equations (15.1) and (15.4) can now be substituted in Maxwell's equations [Eqs. (14.2) and (14.5)] in order to derive the differential equations satisfied by $\mathbf{A}$ and Φ. Substituting $\mathbf{H} = \mathbf{B}/\mu$, $\mathbf{D} = \varepsilon \mathbf{E}$ and Eqs. (15.1) and (15.4) into Eq. (14.5) gives

$$\frac{1}{\mu} \nabla \times (\nabla \times \mathbf{A}) = -\varepsilon \frac{\partial}{\partial t}\left(\nabla \Phi + \frac{\partial \mathbf{A}}{\partial t}\right) + \mathbf{J}_f \tag{15.5}$$

Using $\nabla \times (\nabla \times \mathbf{A}) = \nabla(\nabla \cdot \mathbf{A}) - \nabla^2 \mathbf{A}$ and interchanging differentiation with respect to time and space in Eq. (15.5) gives

$$\nabla^2 \mathbf{A} - \varepsilon\mu \frac{\partial^2 \mathbf{A}}{\partial t^2} - \nabla(\nabla \cdot \mathbf{A}) - \varepsilon\mu \nabla \frac{\partial \Phi}{\partial t} = -\mu \mathbf{J}_f. \tag{15.6}$$

Now we turn to Eq. (14.2). Substituting $\mathbf{D} = \varepsilon \mathbf{E}$ and Eq. (15.4) into it gives

$$-\varepsilon \nabla \cdot \left(\nabla \Phi + \frac{\partial \mathbf{A}}{\partial t} \right) = \rho_f$$

which upon interchanging differentiation with respect to time and space gives

$$\nabla^2 \Phi + \frac{\partial}{\partial t} \nabla \cdot \mathbf{A} = -\frac{\rho_f}{\varepsilon} \tag{15.7}$$

It is apparent from Eqs. (15.6) and (15.7) that so far we have not obtained equations for **A** and Φ that are independent of each other as was the case in the field approach. Since **E** and **B** determine the forces acting on the charges and hence are more directly related to the physical world, then for given **E** and **B**, **A** cannot be uniquely defined by just one relation (namely, $\nabla \times \mathbf{A} = \mathbf{B}$); thus it has some degree of arbitrariness. In fact, for **A** to be uniquely defined, its divergence must be specified. (In fact, not all physicists agree on the statement that **E** and **B** are more directly related to the physical world than the potentials **A** and Φ.*) The act of specifying $\nabla \cdot \mathbf{A}$ is called a *gauge condition.* The *Lorentz gauge*, for example, requires

$$\nabla \cdot \mathbf{A} + \varepsilon\mu \frac{\partial \Phi}{\partial t} = 0 \tag{15.8}$$

which relates $\nabla \cdot \mathbf{A}$ to the rate of change of the scalar potential with respect to time. The substitution of Lorentz gauge in Eqs. (15.6) and (15.7) removes the coupling terms in these equations and hence results in a *wave* equations for each of **A** and Φ:

$$\nabla^2 \mathbf{A} - \varepsilon\mu \frac{\partial^2 \mathbf{A}}{\partial t^2} = -\mu \mathbf{J}_f \tag{15.9}$$

$$\nabla^2 \Phi - \varepsilon\mu \frac{\partial^2 \Phi}{\partial t^2} = -\frac{\rho_f}{\varepsilon} \tag{15.10}$$

Another useful gauge is the so-called *Coulomb or transverse gauge* (see the previous discussion of this gauge in Section 8.5); in this gauge one takes

$$\nabla \cdot \mathbf{A} = 0 \tag{15.11}$$

Substituting this into Eq. (15.7) gives a Poisson equation for Φ (not a wave equation):

$$\nabla^2 \Phi = \frac{-\rho_f}{\varepsilon} \tag{15.12}$$

with a solution

$$\Phi(\mathbf{r}, t) = \frac{1}{4\pi\varepsilon} \int \frac{\rho_f(\mathbf{r}', t)}{|\mathbf{r} - \mathbf{r}'|} dv' \tag{15.13}$$

The scalar potential in this gauge is just the *instantaneous* Coulomb potential due to the charge density $\rho(\mathbf{r}, t)$, from which the name Coulomb gauge is derived. It has the same form as the static potential.

* Not all physicists agree that the **E** and **B** fields are more real than the potentials **A** and Φ. The Aharonov–Bohm effect, which was first introduced in 1958, states that contrary to the conclusions of classical mechanics, there exist effects of potentials on charged particles even in the region where all the fields (and therefore all the forces on the particles) vanish. See Y. Aharonov and D. Bohm, *Physical Review*, vol. 115, p. 485, 1959 and A. Tonomura *et al.*, *Physical Review Letters*, vol. 51, p. 331, 1983.

The vector potential in this gauge satisfies the equation

$$\nabla^2 \mathbf{A} - \varepsilon\mu \frac{\partial^2 \mathbf{A}}{\partial t^2} = -\varepsilon\mu \nabla \frac{\partial \Phi}{\partial t} - \mu \mathbf{J}_f$$

Now we show that the term $\nabla \frac{\partial \Phi}{\partial t}$ is related to $\mathbf{J}_f$. Using $\partial \rho_f(\mathbf{r}, t)/\partial t = -\nabla \cdot \mathbf{J}_f$, then

$$\mathbf{J}_l = -\varepsilon \nabla \frac{\partial \Phi}{\partial t} = \frac{\nabla}{4\pi} \int \frac{\nabla \cdot \mathbf{J}_f(\mathbf{r}', t)}{|\mathbf{r} - \mathbf{r}'|} dv'$$

and one can show that

$$\mathbf{J}_f = \mathbf{J}_l + \mathbf{J}_t \tag{15.14}$$

where

$$\mathbf{J}_t = \frac{1}{4\pi} \nabla \times \nabla \times \int \frac{\mathbf{J}_f}{|\mathbf{r} - \mathbf{r}'|} dv' \tag{15.15}$$

Thus

$$\nabla^2 \mathbf{A} - \varepsilon\mu \frac{\partial^2 \mathbf{A}}{\partial t^2} = -\mu \mathbf{J}_t \tag{15.16}$$

Clearly $\nabla \times \mathbf{J}_t = 0$, and $\nabla \cdot \mathbf{J}_l = 0$, and therefore $\mathbf{J}_t$ and $\mathbf{J}_l$ are called the *transverse* and *longitudinal* parts of the current, which is the basis of the name transverse gauge.

The Coulomb or transverse gauge is often utilized in cases where there are no charge or current distributions: $\rho_f = 0$ and $\mathbf{J}_f = 0$. Thus in this case one takes $\Phi = 0$, and $\mathbf{A}$ satisfies the equation

$$\nabla^2 \mathbf{A} - \varepsilon\mu \frac{\partial^2 \mathbf{A}}{\partial t^2} = 0 \tag{15.17}$$

Thus the fields in this gauge are derivable from a single potential:

$$\mathbf{B} = \nabla \times \mathbf{A} \qquad \text{and} \qquad \mathbf{E} = -\frac{\partial \mathbf{A}}{\partial t} \tag{15.18}$$

It is instructive to consider the gauge condition in the presence of conducting materials. We take $\rho_f = 0$ and $\mathbf{J} = \sigma_c \mathbf{E}$. Using the same procedure we used above to arrive at Eqs. (15.6) and (15.7), we get

$$\nabla^2 \mathbf{A} - \varepsilon\mu \frac{\partial^2 \mathbf{A}}{\partial t^2} - \sigma_c \varepsilon \frac{\partial \mathbf{A}}{\partial t} - \nabla \left[\nabla \cdot \mathbf{A} + \varepsilon\mu \frac{\partial \Phi}{\partial t} + \sigma_c \mu \Phi \right] = 0$$

$$\nabla^2 \Phi + \frac{\partial}{\partial t} \nabla \cdot \mathbf{A} = 0$$

In order to decouple these equations we choose the following gauge:

$$\nabla \cdot \mathbf{A} = -\varepsilon\mu \frac{\partial \Phi}{\partial t} - \sigma_c \mu \Phi$$

which reduces them to the damped-wave equations:

$$\nabla^2 \mathbf{A} - \sigma_c \varepsilon \frac{\partial \mathbf{A}}{\partial t} - \varepsilon\mu \frac{\partial^2 \mathbf{A}}{\partial t^2} = 0 \tag{15.19}$$

$$\nabla^2 \Phi - \sigma_c \mu \frac{\partial \Phi}{\partial t} - \varepsilon\mu \frac{\partial^2 \Phi}{\partial t^2} = 0 \tag{15.20}$$

with the damping terms $\sigma_c \varepsilon(\partial \mathbf{A}/\partial t)$ and $\sigma_c \mu(\partial \Phi/\partial t)$ caused by the losses in the conducting medium.

As we have seen above, the Lorentz condition is a relation between the vector and the scalar potentials. It is commonly used, first, because it results in the reduction of the coupled equations satisfied by the potentials $\mathbf{A}$ and Φ to independent wave equations and puts them on equivalent footing and, second, because it fits naturally into the considerations of the theory of special relativity (see Chapter 17). In view of its importance and the great simplification achieved by imposing such a gauge, we should examine whether it is always possible to impose it without introducing or causing any nonphysical effects. That is, we would like to see whether it is possible at all to find potentials associated with given electric and magnetic fields and, at the same time, satisfy the Lorentz condition. To answer this question we start with a given set of $\mathbf{E}$, $\mathbf{B}$, $\mathbf{A}$, and Φ such that Eqs. (15.1) and (15.4) are satisfied: $\mathbf{B} = \nabla \times \mathbf{A}$, and $\mathbf{E} = -\partial \mathbf{A}/\partial t - \nabla \Phi$. Since the curl of a gradient of a scalar is zero, then the potential $\mathbf{A}$ given by

$$\mathbf{A}' = \mathbf{A} + \nabla \psi \tag{15.21}$$

gives the same physical field $\mathbf{B}$ where ψ is any scalar function of space and time. However, because the electric field depends on the derivative of $\mathbf{A}'$ with respect to time, the scalar potential will have to be changed in order to give the same physical electric field. Thus we write

$$\mathbf{E} = -\frac{\partial \mathbf{A}'}{\partial t} - \nabla \Phi'$$

Substituting Eq. (15.21) and using Eq. (15.4) we get

$$-\frac{\partial \mathbf{A}}{\partial t} - \nabla \Phi = -\frac{\partial \mathbf{A}}{\partial t} - \frac{\partial}{\partial t} \nabla \psi - \nabla \Phi'$$

which gives

$$\Phi' = \Phi - \frac{\partial}{\partial t} \psi \tag{15.22}$$

The transformation of Eq. (15.21) and (15.22) is called a *gauge transformation*, and the invariance of the fields under such transformations is called *gauge invariance*.

Now that the new potentials $\mathbf{A}'$ and Φ' give the same physical fields $\mathbf{E}$ and $\mathbf{B}$ as those given by the original potentials $\mathbf{A}$ and Φ we check if $\mathbf{A}'$ and Φ' satisfy the Lorentz condition:

$$\nabla \cdot \mathbf{A}' + \varepsilon\mu \frac{\partial \Phi'}{\partial t} = 0$$

Substituting Eqs. (15.21) and (15.22) in the above Lorentz condition gives

$$\nabla^2 \psi - \varepsilon\mu \frac{\partial^2 \psi}{\partial t^2} = -\left(\nabla \cdot \mathbf{A} + \varepsilon\mu \frac{\partial \Phi}{\partial t}\right) \tag{15.23}$$

There are two cases to consider. If the original potentials $\mathbf{A}$ and Φ satisfy the Lorentz condition, that is, $\nabla \cdot \mathbf{A} + \varepsilon\mu(\partial \Phi/\partial t) = 0$, then the new potentials $\mathbf{A}'$ and Φ' will satisfy the Lorentz condition provided that ψ satisfies the equation

$$\nabla^2 \psi - \varepsilon\mu \frac{\partial^2 \psi}{\partial t^2} = 0 \tag{15.24}$$

which is the homogeneous scalar wave equation. On the other hand, if the original potentials did not satisfy the Lorentz condition—that is, $\nabla \cdot \mathbf{A} + \varepsilon\mu(\partial\Phi/\partial t) \neq 0$—the new potentials, however, can be chosen to satisfy it provided we choose ψ such that it satisfies the scalar wave equation with $-[\nabla \cdot \mathbf{A} + \varepsilon\mu(\partial\Phi/\partial t)]$ taken as an inhomogeneous source. In other words, ψ must satisfy Eq. (15.23). As a result we conclude that it is always possible to impose the Lorentz condition while still maintaining the physical effects (namely, **E**, **B**, and hence the forces on charged particles).

Example 15.1 The Lorentz Gauge

This example discusses the Lorentz gauge and the gauge transformations. The given **E** and **B** fields are derivable from two pairs of scalar and vector potentials: $(\Phi, \mathbf{A})$ and $(\Phi_0, \mathbf{A}_0)$. The potentials are related as follows:

$$\Phi(\mathbf{r}, t) = \Phi_0(\mathbf{r}, t) + \omega\frac{\cos(kr - \omega t)}{r}$$

$$\mathbf{A}(\mathbf{r}, t) = \mathbf{A}_0(\mathbf{r}, t) + \left[k\frac{\cos(kr - \omega t)}{r} - \frac{\sin(kr - \omega t)}{r^2}\right]\hat{\mathbf{r}}$$

First we note that the potentials can be written as

$$\Phi(\mathbf{r}, t) = \Phi_0(\mathbf{r}, t) - \frac{\partial}{\partial t}\frac{\sin(kr - \omega t)}{r}$$

$$\mathbf{A}(r, t) = \mathbf{A}_0(\mathbf{r}, t) + \hat{\mathbf{r}}\frac{\partial}{\partial r}\left[\frac{\sin(kr - \omega t)}{r}\right]$$

Thus these potentials indicate, using the notation of Eqs. (15.21) and (15.22), that

$$\Phi(\mathbf{r}, t) = \Phi_0 - \frac{\partial}{\partial t}\psi \qquad \mathbf{A}(\mathbf{r}, t) = \mathbf{A}_0 + \nabla\psi \qquad \text{with } \psi = \frac{\sin(kr - \omega t)}{r} \tag{15.25}$$

Let us substitute **A** and Φ in the Lorentz condition, as follows:

$$\nabla \cdot \mathbf{A} + \varepsilon\mu\frac{\partial\Phi}{\partial t} = \nabla \cdot \mathbf{A}_0 + \varepsilon\mu\frac{\partial\Phi_0}{\partial t} + \nabla^2\psi - \varepsilon\mu\frac{\partial^2\psi}{\partial t^2}$$

Using

$$\nabla^2 = \frac{1}{r^2}\frac{\partial}{\partial r}r^2\frac{\partial}{\partial r} \qquad \text{and} \qquad \varepsilon\mu = v^{-2} = \frac{k^2}{\omega^2}$$

one can easily show that

$$\nabla^2\psi - \varepsilon\mu\frac{\partial^2\psi}{\partial t^2} = 0$$

Hence if $\mathbf{A}_0$ and Φ_0 satisfy the Lorentz condition, then **A** and Φ will also satisfy the condition.

15.2 Retarded Potentials

Our aim in this section is to solve the wave equations given by Eqs. (15.9) and (15.10), which were derived using the Lorentz gauge. Because of the presence of the sources in these equations, the solution consists of two contributions: a solution of the homogeneous equation and a particular solution of the inhomogeneous equation. The overall solution must then be made to satisfy the prescribed boundary conditions. Since the homogeneous (sourceless) problem was treated in detail in Chapter 14, we will concentrate here on the effect of the sources. The method to be used has some resemblance to the method we used for the solution of Poisson's equation for static situations [see Section 3.6 and Eqs. (15.12) and (15.13)]. However, because of the second time derivatives of the potentials in the present problem one should not get carried away. For example, it is wrong to say that the solution to the time-dependent problem is given by the static solution with the time dependence being accounted for by its insertion in the sources (*instantaneous solution*). This reasoning is faulty because of the fact that what one observes at a distance r and corresponding time t is not caused by the parameters of the sources at that same time. The electromagnetic disturbance (in vacuum) has a finite speed, c; hence, it takes a period of time equal to r/c to reach the observer. Therefore, what the observer measures at time t is correlated to the sources at what is called the *retarded time*, or $t - r/c$.

Let us now solve the wave equation of the scalar potential for a given $\rho(\mathbf{r}, t)$. We will solve it first for a point charge q that is located at the *origin* at time t. It is to be cautioned now that the time dependence we are considering for the point charge does not arise from the *motion* of the charge. The charge is a fictitious mathematical entity; it is not in motion, but its magnitude is a function of time. The fields produced by the motion of charges are more complicated than what we are considering here. Once the potential produced by the fictitious point charge is obtained it will be generalized by summing over all the actual charge distribution. It is important to note, however, that the potential of the *actual charge* distribution can be specialized to the case of a moving point charge. This will be done later in the chapter.

At points away from the origin and in vacuum Eq. (15.10) becomes

$$\nabla^2\Phi - \frac{1}{c^2}\frac{\partial^2\Phi}{\partial t^2} = 0 \qquad r \neq 0 \tag{15.26}$$

Because of the spherical symmetry of a point charge, the potential is expected to have only a radial spatial dependence. Using only the radial part of the Laplacian, Eq. (15.26) becomes

$$\left(\frac{1}{r^2}\frac{\partial}{\partial r}r^2\frac{\partial}{\partial r} - \frac{1}{c^2}\frac{\partial^2}{\partial t^2}\right)\Phi(r, t) = 0. \tag{15.27}$$

The solution of this equation must have some resemblance to the solution of the static equation

$$\frac{1}{r^2}\frac{\partial}{\partial r}r^2\frac{\partial}{\partial r} = 0$$

namely, $1/r$ dependence. Hence we take

$$\Phi(r, t) = \frac{\psi(r, t)}{r} \tag{15.28}$$

where $\psi(r, t)$ is a function of r and t that needs to be evaluated. Substituting Eq. (15.28) into Eq. (15.27) gives the following equation for $\psi(r, t)$:

$$\frac{\partial^2\psi}{\partial r^2} - \frac{1}{c^2}\frac{\partial^2\psi}{\partial t^2} = 0 \tag{15.29}$$

Note that the transformation shown in Eq. (15.28) got rid of the first derivative in the original equation, thus transforming the latter to the one-dimensional radial wave equation. The one-dimensional wave equation can be solved using d'Alembert's procedure (see Section 14.4 and Example 14.3). We define two variables u and v in terms of r and t:

$$u = r + ct \qquad \text{and} \qquad v = r - ct \tag{15.30}$$

Hence

$$\frac{\partial \psi}{\partial r} = \left(\frac{\partial \psi}{\partial u} + \frac{\partial \psi}{\partial v}\right) \qquad \text{and} \qquad \frac{\partial \psi}{\partial t} = c\left(\frac{\partial \psi}{\partial u} - \frac{\partial \psi}{\partial v}\right) \tag{15.31}$$

Substituting the partial derivatives in Eq. (15.29) gives

$$-4c^2 \frac{\partial^2 \psi}{\partial u\, \partial v} = 0 \tag{15.32}$$

which upon integration with respect to u gives $\partial\psi/\partial v = h(v)$, where h is an arbitrary function that depends on v only. The function h can now be integrated with respect to v:

$$\psi = \int_0^v h(v)dv + g(u) \tag{15.33}$$

where $g(u)$ is an arbitrary function that depends on u only. The integral is taken to be $f(v)$. Thus

$$\psi = f(r - ct) + g(r + ct) \tag{15.34}$$

The second contribution to ψ is not physical since it implies an incoming wave; it is propagating inward towards the point charge (origin) from infinity. The first contribution, however, is physical since it describes an outgoing wave; it is propagating outward from the sources (origin) to infinity. Hence we drop g and keep f; that is,

$$\psi = f(r - ct) \tag{15.35}$$

$$\Phi(\mathbf{r}, t) = \frac{f(r - ct)}{r} \tag{15.36}$$

We now evaluate the function f using the information (boundary condition) given at $r = 0$; namely, at the origin the total charge is q (equivalent to the application of Gauss' law in electrostatics). To utilize this information a limiting process should be used because the potential blows up at $r = 0$. Thus integrating Eq. (15.10) over a small volume Δv that contains the origin gives

$$\int_{\Delta v} \left(\nabla^2 \Phi - \frac{1}{c^2}\frac{\partial^2 \Phi}{\partial t^2}\right) dv = -\frac{1}{\varepsilon_0}\int_{\Delta v} \rho_f \, dv = -\frac{q}{\varepsilon_0} \tag{15.37}$$

which, upon the substitution for Φ from (Eq. 15.36), gives

$$\int_{\Delta v} f \nabla^2 \frac{1}{r}\, dv = -\frac{q}{\varepsilon_0} \tag{15.38}$$

In the limit r becomes very small, $f(r - ct)$ becomes $f(-ct)$, which can then be taken outside the integral:

$$f \int_{\Delta v} \nabla^2 \frac{1}{r}\, dv = -\frac{q}{\varepsilon_0} \tag{15.39}$$

Since $\nabla^2(1/r)$ is $-4\pi\delta(\mathbf{r})$, where $\delta(\mathbf{r})$ is the Dirac delta function, then the integral gives $-4\pi f = -q/\varepsilon_0$, or f near $r = 0$ is $q(t)/4\pi\varepsilon_0$. Therefore

$$f(r - ct) = \frac{q(t - r/c)}{4\pi\varepsilon_0} \tag{15.40}$$

$$\Phi(\mathbf{r}, t) = \frac{1}{4\pi\varepsilon_0}\frac{q(t - r/c)}{r} \tag{15.41}$$

It is to be noted that, as we remarked at the beginning of this section, the potential observed at some time t, $\Phi(r, t)$, correlates with the magnitude or the charge at the earlier time $t - r/c$. This time

$$t' = t - \frac{r}{c} \tag{15.42}$$

is called the *retarded time*. The corresponding potential is called the *retarded potential.*

Equation (15.41) can be easily generalized to the case of a given charge distribution $\rho(\mathbf{r}, t)$; that is, it can be generalized to

$$\Phi(\mathbf{r}, t) = \frac{1}{4\pi\varepsilon_0}\int \frac{\rho(\mathbf{r}', t')}{|\mathbf{r} - \mathbf{r}'|}\, dv' \tag{15.43}$$

where

$$t' = t - \frac{|\mathbf{r} - \mathbf{r}'|}{c} \tag{15.44}$$

is the retarded time for a given element of the distribution seated at $\mathbf{r}'$.

Using a similar procedure, one can solve Eq. (15.9) for the vector potential. This becomes obvious when it is realized that each cartesian component of Eq. (15.9) is a scalar equation identical to the wave equation [see Eq. (15.10)] satisfied by the scalar potential. For example the z component of Eq. (15.9) in vacuum is

$$\nabla^2 A_z - \varepsilon_0\mu_0 \frac{\partial^2 A_z}{\partial t^2} = -\mu_0 J_{fz} \tag{15.45}$$

The x and y components of $\mathbf{A}$ satisfy analogous equations. Thus, following Eqs. (15.10) and (15.43), we write

$$A_z(\mathbf{r}, t) = \frac{\mu_0}{4\pi}\int \frac{J_z(\mathbf{r}', t')}{|\mathbf{r} - \mathbf{r}'|}\, dv' \tag{15.46}$$

Hence $\mathbf{A} = A_x\hat{\mathbf{x}} + A_y\hat{\mathbf{y}} + A_z\hat{\mathbf{z}}$ is easily obtained, as follows:

$$\mathbf{A}(\mathbf{r}, t) = \frac{\mu_0}{4\pi}\int \frac{\mathbf{J}(\mathbf{r}', t')}{|\mathbf{r} - \mathbf{r}'|}\, dv' \tag{15.47}$$

where $t' = t - |\mathbf{r} - \mathbf{r}'|/c$ is the retarded time, and $\mathbf{A}(\mathbf{r}, t)$ is the retarded vector potential. Let us note that in the Coulomb gauge, the scalar potential is an instantaneous one [given in Eq. (15.13)], whereas the vector potential is a retarded one.

Having determined the retarded scalar and vector potentials for a given charge distribution and a given current distribution, one can determine the magnetic and electric fields using Eqs. (15.1) and (15.4), respectively. For example, the electric field corresponding to the scalar potential given in Eq. (15.41) is

$$-\nabla\Phi = -\frac{\hat{\mathbf{r}}}{4\pi\varepsilon_0}\frac{\partial}{\partial r}\frac{q(t - r/c)}{r} \tag{15.48}$$

or

$$-\nabla\Phi = \frac{\hat{\mathbf{r}}}{4\pi\varepsilon_0}\left[\frac{q(t-r/c)}{r^2} + \frac{\dot{q}}{cr}\right] \tag{15.49}$$

where $\dot{q}$ is the derivative of q with respect to its argument $(t - r/c)$.

Example 15.2 Retarded Potential of an Infinite, Straight, Filamentary Current

A constant current I_0 is started at $t = 0$ in an infinite, straight, filamentary conducting wire, as shown in Fig. 15.1. The scalar potential produced by the current at the observation point

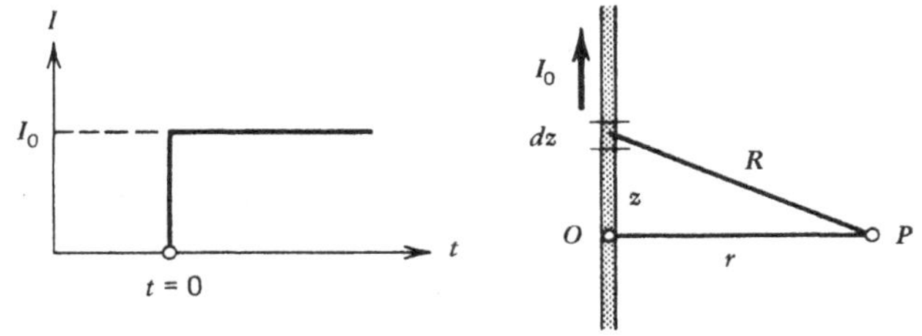

Figure 15.1 Calculation of the retarded vector potential of filamentary current that is suddenly turned on at $t = 0$.

is zero since the charge density $\rho_f = 0$ everywhere in the wire. The vector potential can be calculated using the retarded vector potential given in Eq. (15.47). Using the filamentary approximation $\mathbf{J}\,dv = I\,d\mathbf{l}$ and taking $I(t) = I_0$ for $t \geq 0$ and $I(t) = 0$ for $t < 0$, we write

$$\mathbf{A} = \frac{\mu_0}{4\pi}\hat{\mathbf{z}}\int_{-\infty}^{\infty}\frac{I(t - R/c)}{R}\,dz \tag{15.50}$$

Because of the finite speed of the disturbance, the retarded current is restricted. A given element at distance z from the origin will contribute only if its distance from the point of observation is less than ct. Thus $I(t - R/c) = 0$ when $|z|^2 \geq c^2t^2 - r^2$ and I_0 when $|z|^2 < c^2t^2 - r^2$. Thus Eq. (15.50) becomes $\mathbf{A} = 0$ when $r \geq ct$ and

$$\mathbf{A}(\mathbf{r}, t) = \frac{\mu_0 I_0 \hat{\mathbf{z}}}{4\pi}\int\frac{dz}{(z^2 + r^2)^{1/2}} \qquad r < ct \tag{15.51}$$

Upon integration we get $2\ln[z + (z^2 + r^2)^{1/2}]$ for the integral; hence

$$\begin{aligned} \mathbf{A}(\mathbf{r}, t) &= \frac{\mu_0 I\hat{\mathbf{z}}}{2\pi}\ln\left(\frac{\sqrt{c^2t^2 - r^2} + ct}{r}\right) && r < ct \\ \mathbf{A} &= 0 && r \geq ct \end{aligned} \tag{15.52}$$

It is interesting to note that, as $t \to \infty$, this potential reduces to the steady-state time-independent potential that was calculated in Example 8.9. In this limit, $\sqrt{c^2t^2 - r^2} \approx ct \to \infty$; thus

$$\mathbf{A}(\mathbf{r}, t) = -\frac{\mu_0 I}{2\pi}\ln r + \text{a very large constant} \tag{15.53}$$

Example 15.3 Retarded Scalar Potential of an Electric Dipole

An electric dipole of moment $\hat{\mathbf{K}}p(t)$, located at the origin and with an arbitrary direction $\hat{\mathbf{K}}$, varies with time. The electric field produced is calculable using the result of Problem 15.4. That is,

$$\mathbf{E} = \nabla \times \nabla \times (\hat{\mathbf{K}}\Phi) = \nabla\nabla \cdot (\hat{\mathbf{K}}\Phi) - \nabla^2 \hat{\mathbf{K}}\Phi \tag{15.54}$$

where Φ is a scalar function that satisfies the homogeneous wave equation. The standard solution of this wave equation is given by Eq. (15.36), which upon substitution into Eq. (15.54) gives

$$\mathbf{E} = \frac{\hat{\mathbf{K}} \cdot \mathbf{r}}{r}\left(\frac{3f}{r^4} + \frac{3f'}{cr^3} + \frac{f''}{c^2 r^2}\right)\mathbf{r} - \left(\frac{f}{r^3} + \frac{f'}{cr^2} + \frac{f''}{c^2 r}\right)\hat{\mathbf{K}} \tag{15.55}$$

where a single prime means differentiation with respect to the arguments. At distances close to the dipole, the retarded effects become negligible and the dominant part of this field should be identical to the instantaneous field of the same form as the static field of a dipole of moment $\hat{\mathbf{K}}p$. As $r \to 0$, Eq. (15.55) reduces to

$$\mathbf{E} = 3f\frac{\hat{\mathbf{K}} \cdot \mathbf{r}}{r^5}\mathbf{r} - \frac{f}{r^3}\hat{\mathbf{K}} \qquad r \to 0 \tag{15.56}$$

Comparing this field with the field of a static dipole along the $\hat{\mathbf{K}}$ direction:

$$\mathbf{E} = \frac{1}{4\pi\varepsilon_0}\left[3p\frac{\hat{\mathbf{K}} \cdot \mathbf{r}}{r^5}\mathbf{r} - \frac{p}{r^3}\hat{\mathbf{K}}\right]$$

yields $f = p/4\pi\varepsilon_0$, and hence $f(t - r/c) = p(t - r/c)/4\pi\varepsilon_0$. Hence the retarded potential of an electric dipole associated with the electric field given by Eq. (15.54) is

$$\Phi = \frac{p(t - r/c)}{4\pi\varepsilon_0 r} \tag{15.57}$$

15.3 Spherical Waves and Field Wave Equations—Multipole Expansion for Slowly Moving Distributions

In Chapter 14 we discussed the propagation of plane waves in regions of charge- and current-free space in which we assumed the waves to depend only on a *rectangular* coordinate (plane waves). This restriction is not unphysical; for example, the radiation emitted by charge and current distributions when observed at large distances compared to their dimensions is quite adequately described by plane waves. In this section we again study the propagation in regions of charge- and current-free space but with no restriction on the distances from the sources, and hence we allow the radiation fields to depend on more than one coordinate. The developed theory will also be applicable to another physical situation where the radiation depends on more than coordinates: the scattering of plane waves by conducting or dielectric objects (e.g., conducting or refractive spheres). See Example 15.6 for this application.

These physical situations can be best described via the wave equations of the electric and magnetic fields in *spherical polar* coordinates. Later on, in Section 15.4, we will use the potential equations to arrive at similar results. Let us consider the wave equation of the electric field away from external sources and in the absence of conducting materials (for vacuum):

$$\nabla^2 \mathbf{E} - \frac{1}{c^2}\frac{\partial^2 \mathbf{E}}{\partial t^2} = 0 \tag{14.17}$$

We are interested in monochromatic radiation; that is, we take

$$\mathbf{E} \to \mathbf{E}e^{-i\omega t} \tag{15.58}$$

which upon substitution transforms the wave equation to what is known as the *vector Helmholtz equation*:

$$\nabla^2 \mathbf{E} + \frac{\omega^2}{c^2} \mathbf{E} = 0 \tag{15.59}$$

Because of the vector nature of **E**, this equation is not easily solvable since it does not separate in spherical coordinates. However we show below that it can be transformed to a scalar equation and hence becomes separable using the transformation

$$\mathbf{E} = \mathbf{r} \times \nabla\psi \tag{15.60}$$

Since $\nabla \times \mathbf{r} = 0$, and using the relation $\nabla \times (f\mathbf{A}) = f\nabla \times \mathbf{A} - \mathbf{A} \times \nabla f$, this transformation can also be written as

$$\mathbf{E} = -\nabla \times (\mathbf{r}\psi) \tag{15.61}$$

Substituting this form into the Helmholtz equation gives

$$-\nabla^2[\nabla \times (\mathbf{r}\psi)] - \frac{\omega^2}{c^2} \nabla \times (\mathbf{r}\psi) = 0 \tag{15.62}$$

The first term in this equation can now be simplified. The Laplacian of a vector is given in Eq. (1.67): $\nabla^2\mathbf{E} = -\nabla \times \nabla \times \mathbf{E} + \nabla\nabla \cdot \mathbf{E}$. Because **E** is written as a curl of the vector $\mathbf{r}\psi$, then $\nabla \cdot \mathbf{E}$ is zero (the divergence of a curl of a vector is identically zero). Thus Eq. (15.62) becomes

$$\nabla \times \left[\nabla \times \nabla \times (\mathbf{r}\psi) - \frac{\omega^2}{c^2} \mathbf{r}\psi\right] = 0 \tag{15.63}$$

Using $\nabla \times (\mathbf{r}\psi) = -\mathbf{r} \times \nabla\psi$, then Eq. (15.63) becomes

$$\nabla \times \left[\nabla \times (\mathbf{r} \times \nabla\psi) + \frac{\omega^2}{c^2} \mathbf{r}\psi\right] = 0 \tag{15.64}$$

The first term in the bracket of this equation can be simplified using the identity $\nabla \times (\mathbf{A} \times \mathbf{B}) = (\mathbf{B} \cdot \nabla)\mathbf{A} + \mathbf{A}(\nabla \cdot \mathbf{B}) - (\mathbf{A} \cdot \nabla)\mathbf{B} - \mathbf{B}(\nabla \cdot \mathbf{A})$. Thus

$$\nabla \times (\mathbf{r} \times \nabla\psi) = \mathbf{r}\nabla^2\psi - \nabla\psi(\nabla \cdot \mathbf{r}) + (\nabla\psi \cdot \nabla)\mathbf{r} - (\mathbf{r} \cdot \nabla)\nabla\psi$$

Using the fact that $\nabla \cdot \mathbf{r} = 3$ and $(\nabla\psi \cdot \nabla)\mathbf{r} = \nabla\psi$, we get

$$\nabla \times (\mathbf{r} \times \nabla\psi) = \mathbf{r}\nabla^2\psi - 2\nabla\psi - (\mathbf{r} \cdot \nabla)\nabla\psi \tag{15.65}$$

The term $(\mathbf{r} \cdot \nabla)\nabla\psi$ can be transformed into the gradient of a scalar function using the vector identity $\nabla(\mathbf{A} \cdot \mathbf{B}) = (\mathbf{B} \cdot \nabla)\mathbf{A} + \mathbf{B} \times (\nabla \times \mathbf{A}) + (\mathbf{A} \cdot \nabla)\mathbf{B} + \mathbf{A} \times (\nabla \times \mathbf{B})$, and using the fact that $\nabla \times \mathbf{r} = 0$. Thus

$$(\mathbf{r} \cdot \nabla)\nabla\psi = \nabla(\mathbf{r} \cdot \nabla\psi) - \nabla\psi \tag{15.66}$$

which upon substitution into Eq. (15.65) gives

$$\nabla \times (\mathbf{r} \times \nabla\psi) = \mathbf{r}\nabla^2\psi - \nabla(\psi - \mathbf{r} \cdot \nabla\psi) \tag{15.67}$$

Substituting this result into Eq. (15.64) and using the fact that the curl of a gradient of a scalar is identically zero, we get

$$\nabla \times \mathbf{r}\left(\nabla^2\psi + \frac{\omega^2}{c^2}\psi\right) = 0 \tag{15.68}$$

This equation can be identically satisfied if $\nabla^2\psi + (\omega^2/c^2)\psi$ is a function of r only or if it is identically zero. Thus we take

$$\nabla^2\psi + \frac{\omega^2}{c^2}\psi = 0 \tag{15.69}$$

which shows that ψ satisfies a homogeneous scalar equation. It is called the *scalar Helmholtz equation* as opposed to the vector Helmholtz equation satisfied by the electric field itself. Note that, in the limit $\omega \to 0$, it reduces to Laplace's equation, $\nabla^2\psi = 0$, whose solution was discussed in Chapter 3 in various dimensions using series expansions (multipole solution). We will use some of the formalism followed in that chapter to solve the scalar Helmholtz equation (multipole solution). But before we solve for ψ, we will discuss the nature of the fields. Two cases arise when we try to construct the electromagnetic fields.

1. Transverse Electric (TR). In this case, one starts with the Helmholtz equation for the **E** field. The dependence of the **E** field on the spatial coordinates in this case is

$$\mathbf{E} = \mathbf{r} \times \nabla\psi \tag{15.60}$$

which shows that **E** is perpendicular to the radial direction or tangent to the spherical surface through the point of interest and with center at the origin. The **B** field corresponding to this case is calculated from Maxwell's equation: $\nabla \times \mathbf{E} = -\partial\mathbf{B}/\partial t$. Taking the time dependence of **E** and **B** of the form $e^{-i\omega t}$ (monochromatic radiation), then

$$\nabla \times \mathbf{E} = i\omega\mathbf{B} \tag{15.70}$$

Hence

$$\mathbf{B} = -\frac{i}{\omega}\nabla \times \mathbf{E} = -\frac{i}{\omega}\nabla \times (\mathbf{r} \times \nabla\psi) \tag{15.71}$$

2. Transverse Magnetic (TM). In this case, we start with the vector Helmholtz equation for the **B** field

$$\nabla^2\mathbf{B}' + \frac{\omega^2}{c^2}\mathbf{B}' = 0 \tag{15.72}$$

The use of the prime on **B** is to emphasize that it is not the same one given in Eq. 15.71. The dependence of the **B′** field on the spatial coordinates is taken as

$$\mathbf{B}' = \frac{1}{c}\mathbf{r} \times \nabla\psi \tag{15.73}$$

which implies that the magnetic field is perpendicular to the radial direction or tangent to the spherical surface through the point of interest and with center at the origin. The **E′** field corresponding to this case is calculated from Maxwell's $\nabla \times \mathbf{H} = \partial\mathbf{D}/\partial t$. Taking the time dependence of **E′** and **B′** of the form $e^{-i\omega t}$ gives

$$\nabla \times \mathbf{B}' = -\frac{i\omega}{c^2}\mathbf{E}' \tag{15.74}$$

Hence

$$\mathbf{E}' = \frac{ic}{\omega}\nabla \times (\mathbf{r} \times \nabla\psi) \tag{15.75}$$

With the fields defined in terms of ψ, we turn our attention to obtaining solutions for it using some of the formalism we used in solving Laplace's equation in spherical coordinates. Using $k = \omega/c$ and writing ∇^2 in spherical coordinates, we obtain

$$\frac{1}{r^2}\frac{\partial}{\partial r}\left(r^2\frac{\partial\psi}{\partial r}\right) + \frac{1}{r^2 \sin\theta}\frac{\partial}{\partial\theta}\left(\sin\theta\frac{\partial\psi}{\partial\theta}\right) + \frac{1}{r^2\sin^2\theta}\frac{\partial^2\psi}{\partial\phi^2} + k^2\psi = 0 \quad (15.76)$$

This equation can be solved using the method of separation of variables, which was discussed in Chapter 3 for the solution of electrostatic boundary value problems. Now although the procedural steps that will be used here are identical to those followed in the electrostatic problems, the presence of the term $k^2\psi$ in the present equation, which is a direct result of propagation, will make the solution more complicated. Although we will solve the three-dimensional equation (general case), the applications presented in this book will have azimuthal symmetry. Moreover, all the applications we will consider in this book utilize the lowest two terms in this expansion: $(l, m) = (0, 0)$ and $(1, 0)$, where l and m are separation constants. These terms are as follows:

$$\psi_{00} = C_0^0 \frac{e^{ikr}}{ikr} \qquad \psi_{10} = -C_1^0 \frac{e^{ikr}}{kr}\left(1 + \frac{i}{kr}\right)\cos\theta \quad (15.77)$$

where C_0^0 and C_1^0 are constants. See Example 15.4 for the solution of the general case

Example 15.4 General Solution of the Scalar Helmholtz Equation

In the separation-of-variables method we take ψ as a product of three functions, each of which depends on a single variable only:

$$\psi(r, \theta, \phi) = R(r)P(\theta)F(\phi) \quad (15.78)$$

Substituting this product into Eq. (15.76) and dividing by $\psi/r^2 \sin^2\theta$, where $r \neq 0$ and $\theta \neq 0$, we get

$$\frac{1}{R}\sin^2\theta\frac{d}{dr}\left(r^2\frac{dR}{dt}\right) + \frac{1}{P}\sin\theta\frac{d}{d\theta}\left(\sin\theta\frac{dP}{d\theta}\right) + \frac{1}{F}\frac{d^2F}{d\phi^2} + k^2r^2\sin^2\theta = 0 \quad (15.79)$$

The third term $[(d^2F/d\theta^2)/F]$ depends only on ϕ, whereas the rest of the terms in the equation do not depend on ϕ; therefore, it should be set equal to a separation constant: $-m^2$. Thus,

$$\frac{1}{F}\frac{d^2F}{d\phi^2} = -m^2 \quad (15.80)$$

or

$$\frac{d^2F_m}{d\phi^2} + m^2F_m = 0 \quad (15.81)$$

where we used the subscript m to indicate the dependence of F on m. Inserting the separation constant $-m^2$ into Eq. (15.79) and dividing by $\sin^2\theta$, where $\theta \neq 0$, we get

$$\frac{1}{R}\frac{d}{dr}r^2\frac{dR}{dr} + k^2r^2 + \frac{1}{P\sin\theta}\frac{d}{d\theta}\sin\theta\frac{dP}{d\theta} - \frac{m^2}{\sin^2\theta} = 0 \quad (15.82)$$

Since the first two terms of this equation depend only on r, whereas the last two terms depend only on θ, we set the first pair equal to a constant: $l(l+1)$. Thus

$$\frac{1}{\sin\theta}\frac{d}{d\theta}\sin\theta\frac{dP_l^m}{d\theta}+\left[l(l+1)-\frac{m^2}{\sin^2\theta}\right]P_l^m=0 \tag{15.83}$$

$$\frac{d}{dr}r^2\frac{dR_l}{dr}-[l(l+1)-k^2r]R_l=0 \tag{15.84}$$

where we used the subscript l to indicate the dependence of R and P on l and the superscript m to indicate the dependence of P on m.

We now proceed to solve Eqs. (15.81), (15.83), and (15.84). The solution of Eq. (15.81) is

$$F_m(\phi)=e^{\pm im\phi} \tag{15.85}$$

The magnitude of m has to be restricted in order to make these solutions single-valued functions of ϕ. For the solution to make sense physically, it should be the same after a rotation of 2π, or

$$F_m(\phi+2\pi)=F_m(\phi) \tag{15.86}$$

which requires m to be a positive integer or zero. An important property of these solutions that we need to draw our attention to is the fact that they are orthogonal

$$\int_0^{2\pi}F_{m_1}F_{m_2}^*\,d\phi=2\pi\delta_{m_1m_2} \tag{15.87}$$

where $\delta_{m_1m_2}$ is the Kronecker delta function.

Equation (15.83) depends on both of the separation constants $l(l+1)$ and m^2. For the special case of $m=0$, it reduces to the Legendre's equation we encountered in the solution of electrostatic boundary value problems of azimuthal symmetry [see Eq. (3.22)]. It was found that this special equation has solutions that behave well for all values of θ, including 0 and π, only if l is a positive integer. This requirement is also necessary in the present situation where m is nonzero. However, for a given l, the m values are restricted to $m\le l$. With these conditions, Eq. (15.83) is called the *associated Legendre's equation.* Its solutions are called the associated Legendre's polynomials; for a given l and m they can be calculated from the Legendre's polynomial of the same l, P_l, using the generating relation

$$P_l^m(\eta)=(1-\eta^2)^{m/2}\frac{d^m}{d\eta^m}P_l(\eta) \tag{15.88}$$

where $\eta=\cos\theta$. Table 15.1 gives the explicit dependence on θ of a few of these polynomials. An important property of these polynomials is the fact that they are orthogonal to each other; that is,

Table 15.1

$P_0^0=1$	$P_2^0=\frac{1}{2}[3\cos^2\theta-1]$
$P_1^0=\cos\theta$	$P_2^1=3\cos\theta[1-\cos^2\theta]^{1/2}$
$P_1^1=\sin\theta$	$P_2^2=3(1-\cos^2\theta)$

$$\int_{-1}^{1}P_l^m(\eta)P_{l'}^{m'}(\eta)d\eta=\frac{2}{2l+1}\frac{(l+m)!}{(l-m)!}\delta_{ll'}\delta_{mm'} \tag{15.89}$$

where $(l+m)!$ means the factorial of $(l+m)[=(l+m)(l+m-1)\ldots1]$.

With the separation constants determined, the radial eqation can now be solved for different values of l. Taking $\xi = kr$, Eq. (15.84) becomes

$$\frac{d}{d\xi}\xi^2\frac{d}{d\xi}R_l - [l(l+1) - \xi^2]R_l = 0 \tag{15.90}$$

We use the substitution $R_l = \chi_l/\sqrt{\xi}$ to write this equation in the well-known form called the *Bessel equation*:

$$\xi^2\frac{d^2\chi_l}{d\xi^2} + \xi\frac{d\chi_l}{d\xi} - \left[\left(l+\frac{1}{2}\right)^2 - \xi^2\right]\chi_l = 0 \tag{15.91}$$

The Bessel equation has two linearly independent solutions for each l. These solutions are known as the *Bessel* and the *Neumann* functions and are designated as $J_{l+1/2}(\xi)$ and $N_{l+1/2}(\xi)$ respectively. Both of these solutions are well known and have been numerically tabulated. They are of half-integral order and often called the *cylindrical* Bessel functions. The corresponding solutions for R_l, however, are called the *spherical* Bessel and Neumann functions:

$$j_l(kr) = \frac{\pi}{\sqrt{2kr}}J_{l+1/2}(kr) \qquad \eta_l(kr) = \frac{\pi}{\sqrt{2kr}}N_{l+1/2}(kr) \tag{15.92}$$

Table 15.2 shows the explicit dependence on r of a few of these functions. It also shows that j_l and η_l involve combinations of sine and cosine functions. It is easy to show that the following combinations of these functions involve complex exponentials:

Table 15.2

l	$j_l(kr)$	$\eta_l(kr)$
0	$\dfrac{\sin kr}{kr}$	$-\dfrac{1}{kr}\cos kr$
1	$\dfrac{\sin kr}{(kr)^2} - \dfrac{\cos kr}{kr}$	$-\dfrac{\cos kr}{(kr)^2} - \dfrac{\sin kr}{kr}$
2	$\left[\dfrac{3}{(kr)^3} - \dfrac{1}{kr}\right]\sin kr - \dfrac{3\cos kr}{(kr)^2}$	$-\left[\dfrac{3}{(kr)^3} - \dfrac{1}{kr}\right]\cos kr - \dfrac{3\sin kr}{(kr)^2}$

$$h_l^{(1)}(kr) = j_l(kr) + i\eta_l(kr) \qquad \text{and} \qquad h_l^{(2)}(kr) = j_l(kr) - i\eta_l(kr) \tag{15.93}$$

The complex functions $h_l^{(1)}$ and $h_l^{(2)}$ are called *Hankel functions.* For real kr, such as the case here, $h_l^{(1)} = h_l^{(2)*}$. Table 15.3 shows the explicit r dependence of few of the Hankel functions:

Table 15.3

l	$h_l^{(1)}(kr)$	$h_l^{(2)}(kr)$
0	$\dfrac{e^{ikr}}{ikr}$	$-\dfrac{e^{-ikr}}{ikr}$
1	$-\dfrac{e^{ikr}}{kr}\left(1+\dfrac{i}{kr}\right)$	$-\dfrac{e^{-ikr}}{kr}\left(1-\dfrac{i}{kr}\right)$
2	$i\dfrac{e^{ikr}}{kr}\left[1+\dfrac{3i}{kr}-\dfrac{3}{(kr)^2}\right]$	$-\dfrac{ie^{-ikr}}{kr}\left[1-\dfrac{3i}{kr}-\dfrac{3}{(kr)^2}\right]$

Because the asymptotic property of the Hankel functions is useful in radiation problems, we state here that as $kr \to \infty$, these functions become

$$h_l^{(1)}(kr) = h_l^{(2)*}(kr) \simeq (-i)^{l+1}\frac{e^{ikr}}{kr} \tag{15.94}$$

The most general solution of the scalar wave equation can now be written:

$$\psi(\mathbf{r}, t) = \sum_{l,m} e^{-i\omega t}[C_l^m h_l^{(1)}(kr) + D_l^m h_l^{(2)}(kr)]P_l^m(\cos\theta)e^{\pm im\phi} \tag{15.95}$$

We are interested in disturbances produced by sources located near the origin, and hence in outgoing waves (propagating from small r towards infinity). Taking the limit $kr \to \infty$ in Eq. (15.95) and using Eq. (15.94) give:

$$\psi(\mathbf{r}, t) = \sum_{l,m}\left[C_l^m(-i)^{l+1}\frac{e^{i(kr-\omega t)}}{kr} + D_l^m(i)^{l+1}\frac{e^{-i(kr+\omega t)}}{kr}\right]P_l^m(\theta)e^{\pm im\phi} \tag{15.96}$$

The first term represents an *outgoing wave*, whereas the second term represents an *incoming wave* (propagating from infinity towards the origin) produced by sources at infinity. Therefore the second term does not represent a physical solution. We consequently take $D_l^m = 0$ and retain C_l^m. Thus

$$\psi(\mathbf{r}, t) = \sum_{l,m} C_l^m h_l^{(1)}(kr)P_l^m(\cos\theta)e^{\pm im\phi} \tag{15.97}$$

See the corresponding considerations following Eq. (15.34).

In Eq. (15.77) the very lowest solution, ψ_{00}, depends only on the radial distance and therefore does not contribute to the electromagnetic fields. This is due to the fact that $\nabla\psi_{00}$ is radial; hence $\mathbf{r} \times \nabla\psi_{00} = 0$. The next solution ψ_{10}, however, contributes to the EM fields. For the TE case, Eqs. (15.60), (15.71), and (15.77) give

$$\nabla\psi_{10} = \left[\hat{\mathbf{r}}\left(\frac{i}{kr} - \frac{2}{k^2r^2} - \frac{2i}{k^3r^3}\right)ke^{ikr}\cos\theta - \hat{\boldsymbol{\theta}}\left(\frac{1}{k^2r^2} + \frac{i}{k^3r^3}\right)ke^{ikr}\sin\theta\right]E_0 \tag{15.98}$$

$$\mathbf{E}(\mathbf{r}) = \mathbf{r} \times \nabla\psi_{10} = -\hat{\boldsymbol{\phi}}E_0\left(\frac{1}{kr} + \frac{i}{k^2r^2}\right)e^{ikr}\sin\theta \tag{15.99}$$

$$\mathbf{B}(\mathbf{r}) = -\frac{i}{\omega}\nabla \times \mathbf{E} = \frac{ikE_0}{\omega}e^{ikr}\left[2\left(\frac{1}{k^2r^2} + \frac{i}{k^3r^3}\right)\cos\theta\,\hat{\mathbf{r}} - \left(\frac{i}{kr} - \frac{1}{k^2r^2} - \frac{i}{k^3r^3}\right)\sin\theta\,\hat{\boldsymbol{\theta}}\right] \tag{15.100}$$

where E_0 is a constant. The electric field turns out to be in the $\hat{\boldsymbol{\phi}}$ direction, which is perpendicular to the radial direction as we supposed in the TE case. The magnetic field, on the other hand, has one component along $\hat{\boldsymbol{\theta}}$ and one component along $\hat{\mathbf{r}}$.

For the TM case, Eqs. (15.73), (15.75), and (15.77), give

$$\mathbf{B}'(\mathbf{r}) = -\hat{\boldsymbol{\phi}}\,B_0\left(\frac{1}{kr} + \frac{i}{k^2r^2}\right)e^{ikr}\sin\theta \tag{15.101}$$

$$\mathbf{E}'(\mathbf{r}) = icB_0e^{ikr}\left[2\left(\frac{1}{k^2r^2} + \frac{i}{k^3r^3}\right)\cos\theta\,\hat{\mathbf{r}} - \left(\frac{i}{kr} - \frac{1}{k^2r^2} - \frac{i}{k^3r^3}\right)\sin\theta\,\hat{\boldsymbol{\theta}}\right] \tag{15.102}$$

where B_0 is a constant. Equations (15.99) to (15.102) show that the roles of E and B are interchanged in the TE and TM cases. The fields $\mathbf{E}(\mathbf{r}, t)$ and $\mathbf{B}(\mathbf{r}, t)$ can now be calculated by multiplying the spatial parts of the fields with $e^{-i\omega t}$ and taking the real parts. Thus, for the TM case,

$$\mathbf{B}'(\mathbf{r}, t) = -\hat{\boldsymbol{\phi}} B_0\left[\frac{\cos(\omega t - kr)}{kr} + \frac{\sin(\omega t - kr)}{k^2r^2}\right]\sin\theta$$

$$\mathbf{E}'(\mathbf{r}, t) = 2cB_0\left[\frac{\sin(\omega t - kr)}{k^2r^2} - \frac{\cos(\omega t - kr)}{k^3r^3}\right]\cos\theta\,\hat{\mathbf{r}} + B_0\left[\frac{\cos(\omega t - kr)}{kr} + \frac{\sin(\omega t - kr)}{k^2r^2} - \frac{\cos(\omega t - kr)}{k^3r^3}\right]\sin\theta\,\hat{\boldsymbol{\theta}} \tag{15.103}$$

A similar procedure can be used to calculate the **E** and **B** field for the TE case.

The nature of the fields will now be discussed. We consider an electric dipole, which is placed at the origin along the z axis and whose magnitude varies sinusoidally with time: $\mathbf{p}(t) = p_0\hat{\mathbf{z}}\cos\omega t$. In order to determine the EM fields produced by it, we examine the fields in Eqs. (15.99) to (15.102). For example, let us examine the fields close to the origin. Taking $kr \to 0$, the retardation effects become unimportant. In the TE case the following terms dominate.

$$\mathbf{E}(\mathbf{r}) = -\hat{\boldsymbol{\phi}}\frac{iE_0}{k^2r^2}\sin\theta \tag{15.104}$$

$$\mathbf{B}(\mathbf{r}) = -\frac{kE_0}{\omega k^3r^3}(2\cos\theta\,\hat{\mathbf{r}} + \sin\theta\,\hat{\boldsymbol{\theta}}) \tag{15.105}$$

In the TM case, on the other hand, the following terms dominate as $kr \to 0$.

$$\mathbf{E}'(\mathbf{r}) = -\frac{cB_0}{k^3r^3}(2\cos\theta\,\hat{\mathbf{r}} + \sin\theta\,\hat{\boldsymbol{\theta}}) \tag{15.106}$$

$$\mathbf{B}'(\mathbf{r}) = -\hat{\boldsymbol{\phi}} i\frac{B_0}{k^2r^2}\sin\theta \tag{15.107}$$

The electric field $\mathbf{E}'$ of the TM case has the spatial dependence of an electric dipole [see Eq. (2.46)]. On the other hand, the magnetic field **B** of the TE case has the spatial dependence of a magnetic dipole [see (Eq. 8.98)]. Therefore we conclude that the fields of the TE case are produced by a magnetic dipole, and the fields of the TM case are produced by an electric dipole.

Comparing Eq. (15.106) with the field of a static electric dipole [Eq. (2.46)] gives

$$B_0 = -\frac{p_0k^3}{4\pi\varepsilon_0 c} \tag{15.108}$$

Thus

$$\mathbf{E}'(\mathbf{r}) = \frac{ip_0k^3}{4\pi\varepsilon_0}e^{ikr}\left[2\left(\frac{1}{k^2r^2} + \frac{i}{k^3r^3}\right)\cos\theta\,\hat{\mathbf{r}} - \left(\frac{i}{kr} - \frac{1}{k^2r^2} - \frac{i}{k^3r^3}\right)\sin\theta\,\hat{\boldsymbol{\theta}}\right] \tag{15.109}$$

$$\mathbf{B}'(\mathbf{r}) = \hat{\boldsymbol{\phi}}\frac{p_0k^3}{4\pi\varepsilon_0 c}e^{ikr}\left(\frac{1}{kr} + \frac{i}{k^2r^2}\right)\sin\theta \tag{15.110}$$

Similarly, one can follow the same procedure in the case of TE and show that the fields described by Eqs. (15.99) and (15.100) are due to a magnetic dipole, directed along the z axis, $\mathbf{m} = m_0\hat{\mathbf{z}}\cos\omega t$; that is,

$$E_0 = \frac{m_0k^3}{4\pi\varepsilon_0 c} \tag{15.111}$$

The above treatment shows that the fields of an electric (magnetic) dipole are quite complicated. The electric field of an electric dipole for example has terms varying as $1/r$, $1/r^2$, and $1/r^3$, whereas the corresponding magnetic field has terms varying as $1/r$ and $1/r^2$. Full discussions of these fields will be given in Sections 15.4.1, 15.5 and 15.6. At this point we will investigate the electromagnetic fields of the electric dipole at large distances compared to the wavelength of the radiation; that is, $kr \gg 1$. Taking $kr \gg 1$ in Eqs. (15.109) and (15.110), multiplying by $e^{-i\omega t}$, and taking the real parts, we find that

$$\begin{aligned} \mathbf{E}'(\mathbf{r}, t) &= \frac{p_0 k^3}{4\pi\varepsilon_0} \frac{\cos(\omega t - kr)}{kr} \sin\theta\, \hat{\boldsymbol{\theta}} \\ \mathbf{B}'(\mathbf{r}, t) &= \frac{p_0 k^3}{4\pi\varepsilon_0 c} \frac{\cos(\omega t - kr)}{kr} \sin\theta\, \hat{\boldsymbol{\phi}} \end{aligned} \tag{15.112}$$

These fields are called the *radiation fields*; they vary as $1/r$. The electric field is in the $\hat{\boldsymbol{\theta}}$ direction, whereas the magnetic field is in the $\hat{\boldsymbol{\phi}}$ direction; that is, each of them is normal to $\hat{\mathbf{r}}$ and to each other. Moreover, they are related by the simple expression

$$\mathbf{E}' = c\hat{\mathbf{r}} \times \mathbf{B}' \tag{15.113}$$

The Poynting vector, $\mathbf{S} = \mathbf{E} \times \mathbf{H}$, in the region is

$$\mathbf{S} = \frac{p_0^2 k^4}{(4\pi\varepsilon_0)^2 \mu_0 c r^2} \cos^2(\omega t - kr) \sin^2\theta\, \hat{\mathbf{r}} \tag{15.114}$$

The total power radiated can be calculated by integrating $\mathbf{S}$ over a sphere of radius r:

$$P = \oint \mathbf{S} \cdot \hat{\mathbf{n}}\, da = \frac{p_0^2 k^4}{8\pi\mu_0 \varepsilon_0^2 c} \cos^2(\omega t - kr) \int_0^\pi \sin^3\theta\, d\theta$$

or

$$P = \frac{p_0^2 k^4}{6\pi\varepsilon_0^2 \mu_0 c} \cos^2(\omega t - kr) \tag{15.115}$$

The average power radiated is obtained by averaging over time. Since averaging $\cos^2(\omega t - kr)$ gives $\frac{1}{2}$, then

$$\langle P \rangle = \frac{p_0^2 k^4}{12\pi\varepsilon_0^2 \mu_0 c} = \frac{p_0^2 k^4 c}{12\pi\varepsilon_0} \tag{15.116}$$

Example 15.5 Dipoles Near Conducting Planes—Method of Images

The radiation from electric or magnetic dipoles is affected when a conducting plane is brought near these dipoles. This is due to the induced charges and currents in the conductor. In Chapters 3 and 8, these effects were studied in static and steady-state situations, using the method of images. In this example we consider these effects in the presence of time dependence.

Consider a system of charges of density $\rho(\mathbf{r}, t)$ and current density $\mathbf{J}(\mathbf{r}, t)$ near a conducting plane that lies in the x-y plane. The reflection of the system in the plane is governed by the fact that each point at $\mathbf{r}$ with coordinates (x, y, z) transforms to $\mathbf{r}'$ with coordinates $(x, y, -z)$. Moreover, from the method of images, the reflection results in a change of sign of the charge density: $\rho(\mathbf{r}, t)$ becomes $-\rho(\mathbf{r}', t)$.

The reflection law of the current density can now be deduced from the above transformations. The current is related to the density ρ and velocity $\mathbf{v}$ of the charges by Eq. (7.6): $\mathbf{J}(\mathbf{r}, t) = \rho(\mathbf{r}, t)\mathbf{v}$. Therefore the reflection of $\mathbf{J}$ in the plane results in the current density $\mathbf{J}'$ such that

$$\mathbf{J}'(x, y, -z) = -\rho(x, y, -z)\frac{d\mathbf{r}}{dt}(x, y, -z) \tag{15.117}$$

or

$$J'_x = -J_x \qquad J'_y = -J_y \qquad J'_z = J_z \tag{15.118}$$

The electric dipole moment of the distribution transforms in the same fashion as the current since it is defined in terms of the product of the charge density and $\mathbf{r}$; that is, $\mathbf{p}(\mathbf{r}, t) = \int \rho\mathbf{r}\, dv$. Therefore

$$p'_x = -p_x \qquad p'_y = -p_y \qquad p'_z = p_z \tag{15.119}$$

The magnetic dipole moment, however, transforms differently because it is defined in terms of the cross product of the current density and $\mathbf{r}$; that is, $\mathbf{m} = \frac{1}{2}\int \mathbf{J} \times \mathbf{r}\, dv$. Thus

$$m'_x = m_x \qquad m'_y = m_y \qquad m'_z = -m_z \tag{15.120}$$

If a time-dependent electric dipole is placed at a distance b from a large conducting plate such that $b \ll \lambda$, where λ is the wavelength of the emitted radiation, then the dipole will have an image dipole of magnitude given by Eq. (15.119). Because the distance between the dipole and its image is *much smaller* than λ, then one can add their dipole moments vectorially first. Thus

$$\mathbf{p}_T = \mathbf{p}'(t) + \mathbf{p}(t) = 2\hat{\mathbf{z}}p_z = 2\hat{\mathbf{z}}p\cos\alpha \tag{15.121}$$

where α is the angle between the dipole and the normal to the plate (z axis). Since the power radiated by a dipole is proportional to the square of the dipole moment, then the power emitted is

$$\text{Power} = 4P_0\cos^2\alpha \tag{15.122}$$

where P_0 is the power radiated by the dipole in the absence of the plate. Thus when the dipole is normal to the plate, the power emitted is $4P_0$; when the dipole is parallel to the plate, the power is zero. In the latter, radiation from higher-order charge distributions such as the quadrupole radiation will dominate.

Example 15.6 Superposition of Dipole Radiation Fields

We treat in this example the superposition of the radiation fields of two dipoles. Consider two dipoles, one of which is located along the x axis and has the time dependence $\mathbf{p}_1(t) = p_0\hat{\mathbf{x}}\cos\omega t$. The other dipole is located in the x-y plane at an angle $\phi = \phi_0$ from the x-axis and has the time dependence $p_2(t) = p_0\sin\omega t$, as shown in Fig. 15.2*a*. We take the distance between them to be much *smaller than the wavelength* of the radiation, and hence we can take both of them to be located at the origin.

The electromagnetic radiation fields produced at large distances by an electric dipole lying along the z axis are given by Eqs. (15.112). The fields produced by a dipole lying along the x axis can be obtained from these results by making a coordinate transformation that takes the z axis into the x axis. The result is

$$\mathbf{E}_1(\mathbf{r}, t) = \frac{p_0k^3}{4\pi\varepsilon_0kr}[\hat{\boldsymbol{\phi}}\sin\phi - \hat{\boldsymbol{\theta}}\cos\phi\cos\theta]e^{-i(\omega t - kr)} \tag{15.123}$$

$$\mathbf{B}_1(\mathbf{r}, t) = -\frac{p_0k^3}{4\pi\varepsilon_0ckr}[\hat{\boldsymbol{\phi}}\cos\phi\cos\theta + \hat{\boldsymbol{\theta}}\sin\phi]e^{-i(\omega t - kr)} \tag{15.124}$$

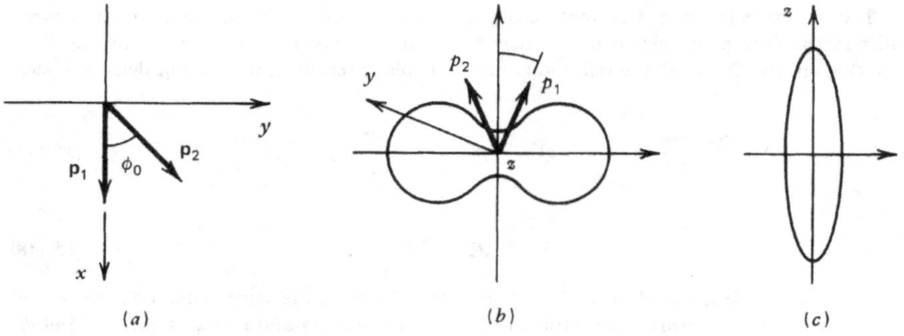

Figure 15.2 Superposition of dipole radiation fields. (*a*) Schemetic representation of two dipoles in the $x - y$ planes. (*b*) Angular distribution of the power radiated in the $x - y$ plane. (*c*) The same distribution in the plane normal to $x - y$ plane and symmetric with the dipoles.

For the second dipole, we have

$$\mathbf{E}_2(\mathbf{r}, t) = \frac{p_0 k^3}{4\pi\varepsilon_0 kr} e^{i\pi/2}[\hat{\boldsymbol{\phi}} \sin(\phi - \phi_0) - \hat{\boldsymbol{\theta}} \cos(\phi - \phi_0)\cos\theta]e^{-i(\omega t - kr)} \tag{15.125}$$

$$\mathbf{B}_2(\mathbf{r}, t) = -\frac{p_0 k^3}{4\pi\varepsilon_0 ckr} e^{i\pi/2}[\hat{\boldsymbol{\phi}} \cos(\phi - \phi_0)\cos\theta + \hat{\boldsymbol{\theta}} \sin(\phi - \phi_0)]e^{-i(\omega t - kr)} \tag{15.126}$$

The total field produced by the dipoles is the vector sums $\mathbf{B} = \mathbf{B}_1 + \mathbf{B}_2$ and $\mathbf{E} = \mathbf{E}_1 + \mathbf{E}_2$. Using $i = e^{i\pi/2}$ gives

$$\mathbf{E} = \frac{p_0 k^3}{4\pi\varepsilon_0 kr}(\hat{\boldsymbol{\phi}}[\sin\phi + i\sin(\phi - \phi_0)] - \hat{\boldsymbol{\theta}}[\cos\phi + i\cos(\phi - \phi_0)]\cos\theta)e^{-i(\omega t - kr)} \tag{15.127}$$

$$\mathbf{B} = -\frac{p_0 k^3}{4\pi\varepsilon_0 ckr}(\hat{\boldsymbol{\phi}}[\cos\phi + i\cos(\phi - \phi_0)]\cos\theta + \hat{\boldsymbol{\theta}}[\sin\phi + i\sin(\phi - \phi_0)])e^{-i(\omega t - kr)} \tag{15.128}$$

The time average of the Poynting vector per solid angle can be calculated by taking the cross product $\frac{1}{2}\,\mathrm{Re}(\mathbf{E} \times \mathbf{H}^*)$. This gives

$$\frac{d\langle P\rangle}{d\Omega} = \frac{p_0^2 k^4}{32\pi^2\varepsilon_0^2 \mu_0 c}[\sin^2\phi + \cos^2\phi\cos^2\theta + \sin^2(\phi - \phi_0) + \cos^2(\phi - \phi_0)\cos^2\theta]$$

or

$$\frac{d\langle P\rangle}{d\Omega} = \frac{p_0^2 k^4 c}{32\pi^2\varepsilon_0}(2 - [\cos^2\phi + \cos^2(\phi - \phi_0)]\sin^2\theta) \tag{15.129}$$

The angular distribution of the power radiated is a maximum in the directions that are perpendicular to the dipole moments: $\theta = 0$ and $\theta = \pi$. As an example we take the case where the angle between the dipoles is $\phi_0 = 45°$. Figure 15.2*b* shows the distribution in the plane of the dipoles (x-y plane), and Fig. 15.2*c* shows the distribution in a plane normal to the x-y plane and defined by $\phi = 22.5°$.

The question of the polarization of the radiated wave and its dependence on the angles is interesting. We will leave these properties as an exercise (see Problem 15.11).

***Example 15.7 Scattering of EM Wave by a Small Sphere**

In this example we sketch the application of the above theory (multipole expansion) to the problem of scattering of plane EM waves by small charge-free and current-free spheres.

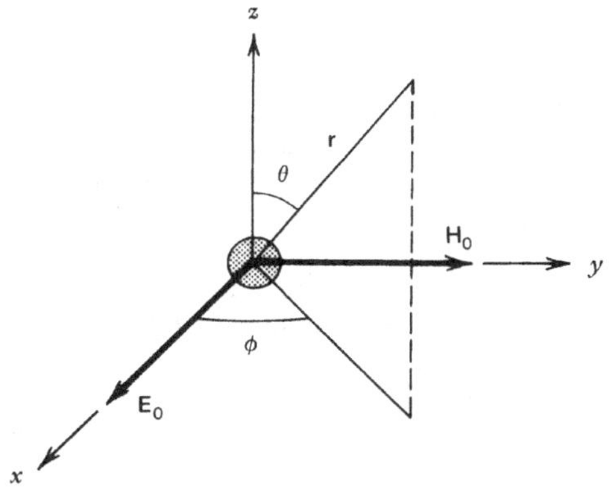

Figure 15.3 Scattering of an electromagnetic wave by a small sphere of given electric and magnetic polarizabilities.

A linearly polarized, plane, monochromatic wave is incident on a sphere, whose radius R is much smaller than the wavelength of the incident wave: $R \ll \lambda$. Because of this the external EM fields in the neighborhood of the sphere may be taken as uniform; hence we may treat the scattering of the wave in the radiation zone from the point of view of the radiation emitted by the induced electric and magnetic dipole moments of the sphere.

Let us take the electric and magnetic polarizabilities of the sphere to be β_e and β_m, respectively. The directions of propagation and polarization of the incident wave are taken along the z and x axes, as shown in Fig. 15.3. The electric dipole moment of the sphere is given by $\mathbf{p} = \beta_e \mathbf{E}$, where the electric field is evaluated at the center of the sphere. Since R is much less than λ, then $kR \ll 1$. Thus

$$\mathbf{p}(t) = \beta_e E_0 \hat{\mathbf{x}} e^{-i(\omega t - kr)}\big|_{\text{sphere}} = \beta_e E_0 \hat{\mathbf{x}} e^{-i\omega t} \tag{15.130}$$

Similarly, the magnetic dipole moment, of the sphere is

$$\mathbf{m}(t) = \beta_m H_0 \hat{\mathbf{y}} e^{-i\omega t} \tag{15.131}$$

The *radiation* fields produced by an electric dipole moment along the x axis were given in Example 15.6 [see Eqs. (15.123) and (15.124)]. Thus we write:

$$\begin{aligned} \mathbf{E}_e &= \frac{\beta_e E_0 k^2}{4\pi\varepsilon_0 r}[\hat{\boldsymbol{\phi}} \sin\phi - \hat{\boldsymbol{\theta}} \cos\phi \cos\theta]e^{-i(\omega t - kr)} \\ \mathbf{B}_e &= -\frac{\beta_e E_0 k^2}{4\pi\varepsilon_0 cr}[\hat{\boldsymbol{\phi}} \cos\phi \cos\theta + \hat{\boldsymbol{\theta}} \sin\phi]e^{-i(\omega t - kr)} \end{aligned} \tag{15.132}$$

The fields produced by the magnetic dipole can also be shown to be:

$$\begin{aligned} \mathbf{E}_m &= \frac{\beta_m H_0 k^2}{4\pi\varepsilon_0 cr}[-\hat{\boldsymbol{\phi}} \sin\phi \cos\theta + \hat{\boldsymbol{\theta}} \cos\phi]e^{-i(\omega t - kr)} \\ \mathbf{B}_m &= -\frac{\beta_m H_0 k^2}{4\pi\varepsilon_0 c^2 r}[\hat{\boldsymbol{\phi}} \cos\phi + \hat{\boldsymbol{\theta}} \sin\phi \cos\theta]e^{-i(\omega t - kr)} \end{aligned} \tag{15.133}$$

The total fields $\mathbf{E} = \mathbf{E}_e + \mathbf{E}_m$ and $\mathbf{B} = \mathbf{B}_e + \mathbf{B}_m$, can be then used to calculate the scattered power as a function of angles. They also can be used to discuss the polarization of the scattered wave. We leave these and other calculations as exercises (see Problem 15.14).

15.4 Radiation from Antennas

In this section we consider the radiation emitted by the so-called *simple antennas*, which are short, straight, conducting wires that carry time-dependent currents. We will first start with a very short antenna (*differential antenna*) such that its length is much smaller than the wavelength of the radiation emitted (*dipole approximation*). The radiation from longer ones (of length equal to one-half of the wavelength of the radiation), will then be constructed by summing the fields of the very short ones. Then we will discuss the radiation from an array of half-wave antennas in which interference effects similar to those encountered in optics arise.

Since we are interested in very small antennas observed at large distances, we can use the method of field multipole expansion (spherical waves) directly. However, we elect to use the potential method of Section 15.2 and specialize it to large distances. After doing this, we will show that both methods give identical results.

15.4.1 Differential Antennas—Electric Dipole Fields

Consider a very short length of wire carrying a harmonically varying current: $I = I_0 \cos \omega t$ (see Fig. 15.4*a*). By "short" we mean that its length is much smaller than the wavelength of radiation emitted. For frequency of oscillation ω, the wavelength of the emitted radiation is $\lambda = 2\pi c/\omega$. Thus we take

$$l \ll \lambda = \frac{2\pi c}{\omega} \tag{15.134}$$

In terms of the period of oscillations, $T = 2\pi/\omega$, this condition becomes

$$l \ll cT \tag{15.135}$$

This restriction on the length of the antenna is known as the *dipole approximation.* For example, for $\omega = 10^{10}$ rad/s, $\lambda = 19$ cm, and this condition is satisfied when l is of the order of 1 mm.

The vector potential produced by the wire will be calculated using the retarded potential [see Eq. (15.47)]. Using the filamentary approximation $\mathbf{J}\, dv = I\, d\mathbf{l}$ we write

$$\mathbf{A} = \frac{\mu_0 \hat{\mathbf{z}}}{4\pi} \int_{-l/2}^{l/2} \frac{I_0 \cos[\omega(t - R'/c)]}{R'} dz' \tag{15.136}$$

where $\mathbf{R}' = \mathbf{r} - \hat{\mathbf{z}}z'$ and hence $R' = r(1 - 2(z'/r)\cos\theta + z'^2/r^2)^{1/2}$. We are interested in determining $\mathbf{A}$ at points where $r \gg l$; hence one can take $R' \approx r$ in the denominator. In the numerator, however, one cannot neglect the variation along the wire unless one invokes the dipole approximation in addition to the condition $r \gg l$. That is, the argument of the cosine must satisfy $\omega T \gg (\omega z'/c)\cos\theta$ or

$$cT \gg z' \cos\theta \tag{15.137}$$

where T is the time for which appreciable change in the amplitude of oscillations takes place (i.e., the period). Equation (15.137) is just the dipole approximation assumed above, since $z' \cos\theta \leq l/2$. Thus

$$\mathbf{A} = \frac{\mu_0 \hat{\mathbf{z}}}{4\pi} I_0 \frac{\cos[\omega(t - r/c)]}{r} \int_{-l/2}^{l/2} dz'$$

or

$$\mathbf{A} = \frac{\mu_0 l \hat{\mathbf{z}}}{4\pi} \frac{I_0 \cos[\omega(t - r/c)]}{r} \tag{15.138}$$

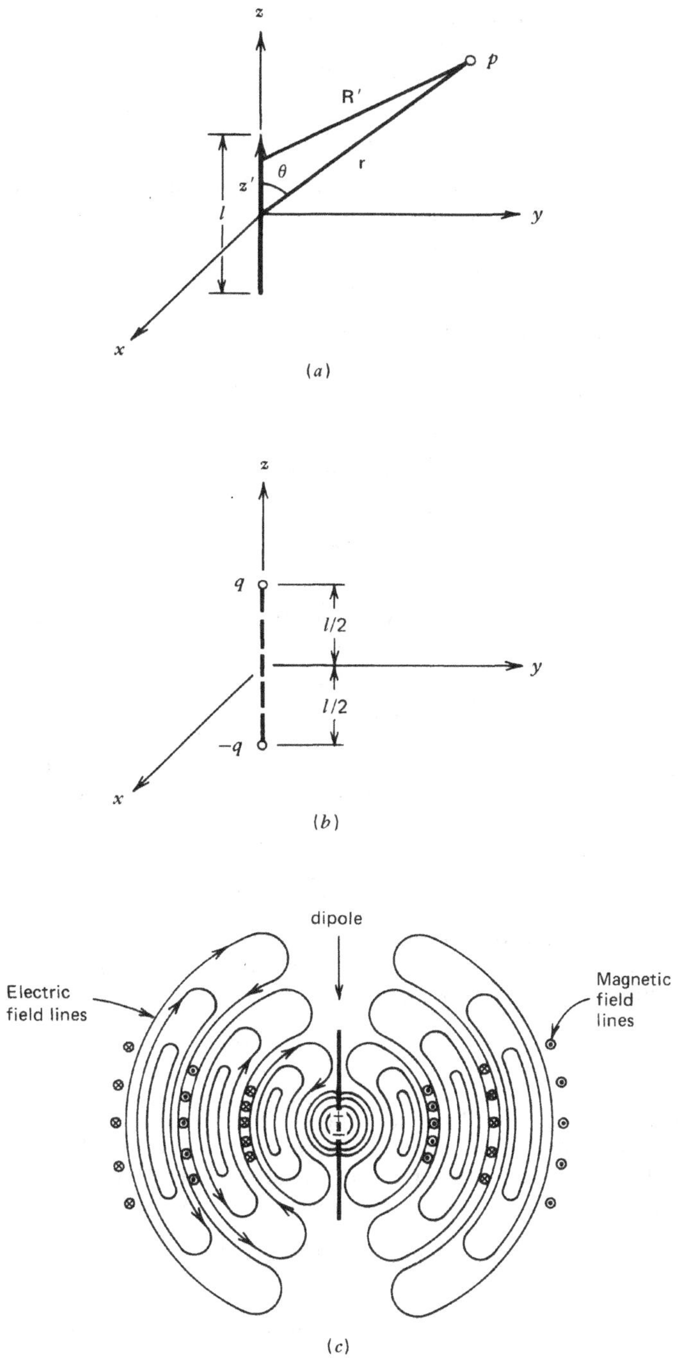

Figure 15.4 (*a*) Differential antenna. (*b*) Schematic of the equivalent oscillating dipole. (*c*) Field lines of an oscillating dipole.

It is to be noted that the above procedure is not restricted to only harmonic variations in the current and charge distributions. For any given time dependence that satisfies the dipole approximation, $I(t)$, we get

$$\mathbf{A} = \frac{\mu_0 l\hat{\mathbf{z}}}{4\pi r} I(t - r/c) \tag{15.139}$$

It is sometimes useful to use the components of **A** in spherical coordinates. These are easily calculated from Eq. (15.139) as

$$\mathbf{A} = \frac{\mu_0 l}{4\pi r} I\left(t - \frac{r}{c}\right)(\cos\theta\,\hat{\mathbf{r}} - \sin\theta\,\hat{\boldsymbol{\theta}}) \tag{15.140}$$

The scalar retarded potential can now be calculated from **A** using the Lorentz gauge:

$$\frac{1}{c^2}\frac{\partial\Phi(\mathbf{r}, t)}{\partial t} = -\nabla\cdot\mathbf{A}$$

Substituting Eq. (15.139) into this equation gives

$$\frac{\partial\Phi(\mathbf{r}, t)}{\partial t} = -\frac{\mu_0 c^2}{4\pi} l \frac{\partial}{\partial z}\frac{I(t - r/c)}{r} \tag{15.141}$$

To solve for Φ, one can carry out the differentiation with respect to z and then integrate over time. Here, however, we choose to do it in a slightly different way, by writing

$$\Phi(\mathbf{r}, t) = -\frac{l}{4\pi\varepsilon_0}\int^t \left[\frac{\partial}{\partial z}\frac{I(t - r/c)}{r}\right]dt \tag{15.142}$$

where $c^2\mu_0$ was replaced by $1/\varepsilon_0$. We interchange the order of differentiation and integration and note that the integral $\int I(t - r/c)dt$ is just the charge $q(t - r/c)$; therefore

$$\Phi(\mathbf{r}, t) = -\frac{l}{4\pi\varepsilon_0}\frac{\partial}{\partial z}\left[\frac{q(t - r/c)}{r}\right] \tag{15.143}$$

where $q(t - r/c)/4\pi\varepsilon_0 r$ is the retarded scalar potential of a point charge q placed at the origin (see Section 15.2 for the derivation of such a potential). It is interesting to note that this result can also be written in the form [see Eqs. (2.45) and (2.68)]:

$$\Phi(\mathbf{r}, t) = \frac{1}{4\pi\varepsilon_0}(\mathbf{l}\cdot\nabla)\frac{q(t - r/c)}{r} \tag{15.144}$$

This form shows that the scalar potential of a differential antenna is the differential of the retarded monopole potential of charge q. Because of the dependence of the charge on the distance of the point of observation, then the retarded scalar potential will depend on the current. This can be seen by realizing that

$$\frac{\partial}{\partial z} q\left(t - \frac{r}{c}\right) = -\frac{1}{c}\dot{q}\frac{\partial r}{\partial z} = -\frac{zI(t - r/c)}{cr}$$

where $\dot{q} = I$ is the derivative of q with respect to its argument. Thus

$$\Phi(\mathbf{r}, t) = \frac{l}{4\pi\varepsilon_0}\left[\frac{zq(t - r/c)}{r^3} + \frac{z}{r^2}\frac{I(t - r/c)}{c}\right] \tag{15.145}$$

In the case of harmonic variations, the result gives:

$$\Phi(\mathbf{r}, t) = \frac{lq_0 k}{4\pi\varepsilon_0} \frac{z}{r^2} \left[\frac{\cos(\omega t - kr)}{kr} - \sin(\omega t - kr) \right] \tag{15.146}$$

where $k = \omega/c$, $q_0 = I_0/\omega$ and $q = q_0 \cos \omega t$.

We would like to note that as the point of observation approaches the wire, then the first term in Eq. (15.145) dominates and $q(t - r/c)$ may be taken as $q(t)$. Thus $\Phi(\mathbf{r}, t)$ reduces to

$$\Phi(\mathbf{r}, t) = \frac{1}{4\pi\varepsilon_0} \frac{\mathbf{p}(t) \cdot \hat{\mathbf{r}}}{r^3} \qquad kr \ll 1 \tag{15.147}$$

which is just the instantaneous (scalar) potential of a dipole of moment $\mathbf{p}(t) = lq(t)\hat{\mathbf{z}}$, where the retardation effects are unimportant. Figure 15.4*b* shows the equivalent oscillating dipole schematically.

With the vector and scalar potentials calculated, we can now calculate the corresponding electric and magnetic fields. We consider the case where the current is varying harmonically. Substituting Eq. (15.138) in $\mathbf{B} = \nabla \times \mathbf{A}$ gives:

$$\mathbf{B} = \frac{\mu_0 I_0 l}{4\pi kr} \hat{\boldsymbol{\phi}} \frac{\omega^2}{c^2} \left[\cos(\omega t - kr) + \frac{1}{kr} \sin(\omega t - kr) \right] \sin\theta \tag{15.148}$$

which shows that the $\mathbf{B}$ field is entirely along the $\hat{\boldsymbol{\phi}}$ direction. The electric field produced by the wire can be calculated by substituting Eqs. (15.138) and (15.146) in $\mathbf{E} = -\partial \mathbf{A}/\partial t - \nabla\Phi$. Writing ∇ in spherical coordinates gives

$$\mathbf{E} = -\left(\frac{\partial A_r}{\partial t} + \frac{\partial \Phi}{\partial r} \right)\hat{\mathbf{r}} - \left(\frac{\partial A_\theta}{\partial t} + \frac{1}{r}\frac{\partial \Phi}{\partial \theta} \right)\hat{\boldsymbol{\theta}} - \left(\frac{\partial A_\phi}{\partial t} + \frac{1}{r \sin\theta}\frac{\partial \Phi}{\partial \phi} \right)\hat{\boldsymbol{\phi}}$$

Thus

$$E_\phi = 0$$

$$E_\theta = -\frac{lI_0\omega^2}{4\pi\varepsilon_0 c^3} \left[\left(\frac{1}{k^3 r^3} - \frac{1}{kr} \right) \cos(\omega t - kr) - \frac{1}{k^2 r^2} \sin(\omega t - kr) \right] \sin\theta \tag{15.149}$$

$$E_r = \frac{2lI_0\omega^2}{4\pi\varepsilon_0 c^3} \left[\frac{\sin(\omega t - kr)}{k^3 r^3} - \frac{\cos(\omega t - kr)}{k^2 r^2} \right] \cos\theta$$

These fields are exactly the fields of an electric dipole obtained previously using the spherical-wave method (field multipole expansion) for ($l = 1$, $m = 0$) if we make the identification

$$p_0 = \frac{lI_0}{\omega} = lq_0 = \frac{4\pi\varepsilon_0 c}{k^3} B_0 \tag{15.150}$$

where p_0 is taken as the amplitude of the dipole moment of the charge-current distribution as shown schematically in Fig. 15.4*b*. In addition to the previous discussions of dipole fields, we now make the following comments.

1. Equations (15.148) and (15.149) show that the fields of a differential antenna, the simplest of all, are quite complicated. The electric field has terms varying as $1/r$, $1/r^2$, and $1/r^3$, and the magnetic field has terms varying as $1/r$ and $1/r^2$. The terms varying with odd powers of r^{-1} are proportional to $\cos(\omega t - kr)$, whereas the terms varying with even powers of r^{-1} are proportional to $\sin(\omega t - kr)$, which indicates a phase difference of $\pi/2$ between these two sets.

2. The direction of the fields is also interesting. The **B** field is purely in the $\hat{\boldsymbol{\phi}}$ direction, or transverse magnetic (TM), whereas the **E** field has components in the $\hat{\mathbf{r}}$ and $\hat{\boldsymbol{\theta}}$ directions. Because the radiation emitted by EM fields is given by the cross product **E** × **H**, then the $\hat{\mathbf{r}}$ component of the electric field does not contribute to the radiation. Moreover, because E_θ depends on $\sin\theta$, then there will be no radiation in the $\theta = 0$ and $\theta = \pi$ directions (along the direction of the dipole).

3. The radial dependence of dipole fields has three distinct regions: the *static*, *induction*, and *radiation* regions. Some aspects of these regions were discussed in Section 15.3. Here we discuss them in more detail. The *static* (instantaneous) region is defined by the condition $kr \ll 1$, where the electric field is dominated by $1/r^3$ radial dependence. The radiation region, on the other hand, is defined by the condition $kr \gg 1$. In this region **E** and **B** are dominated by $1/r$ radial dependence and are called *radiation* fields. The intermediate region, where kr is of the order of unity, is called the *induction* region and the electric field is dominated by $1/r^2$ dependence. Thus we write

$$\mathbf{E}_R = \frac{lI_0\omega^2}{4\pi\varepsilon_0 c^3}\frac{1}{kr}\cos(\omega t - kr)\sin\theta\hat{\boldsymbol{\theta}} \qquad kr \gg 1$$

$$\mathbf{E}_I = \frac{lI_0\omega^2}{4\pi\varepsilon_0 c^3}\frac{\sin(\omega t - kr)}{k^2r^2}[2\cos\theta\hat{\mathbf{r}} + \sin\theta\hat{\boldsymbol{\theta}}] \qquad kr \approx 1 \tag{15.151}$$

$$\mathbf{E}_S = -\frac{lI_0\omega^2}{4\pi\varepsilon_0 c^3}\frac{\cos(\omega t - kr)}{k^3r^3}[2\cos\theta\hat{\mathbf{r}} + \sin\theta\hat{\boldsymbol{\theta}}] \qquad kr \ll 1$$

where the subscripts R, I, and S stand for radiation, induction, and static regions, respectively. Figure 15.4*c* gives the **E** and **B** field lines of an oscillating dipole.

These regions are of practical importance in communication. Radio transmission utilizes the radiation field. Some specialized military applications, for example, utilize the static region (see Example 15.8).

With the fields produced by a differential antenna (an electric dipole) calculated, the power radiated by it can be calculated. In Section 15.3 the power radiated by an electric dipole of moment $p_0\hat{\mathbf{z}}\cos\omega t$ [see Eq. (15.116)] was found to be $\langle P\rangle = p_0^2k^4c/12\pi\varepsilon_0$. Using the identification between p_0 and the current in the antenna [Eq. (15.150)], $p_0 = lI_0/\omega$, then we can immediately write for the antenna

$$\langle P\rangle = \frac{l^2I_0^2k^4c}{12\pi\varepsilon_0\omega^2} = \frac{1}{2}RI_0^2 \tag{15.152}$$

where

$$R = \frac{l^2\omega^2}{6\pi\varepsilon_0 c^3} \tag{15.153}$$

is an effective radiation resistance of the antenna. This resistance can also be written in terms of the ratio l/λ:

$$R = \frac{2\pi}{3}\sqrt{\frac{\mu_0}{\varepsilon_0}}\left(\frac{l}{\lambda}\right)^2 = 789\left(\frac{l}{\lambda}\right)^2 \text{ ohms} \tag{15.154}$$

15.4.2 Radiation from a Half-Wave Antenna

We now consider the radiation from a realistic antenna. The one we considered in Section 15.4.1 was so short that its length was much less than the wavelength of the

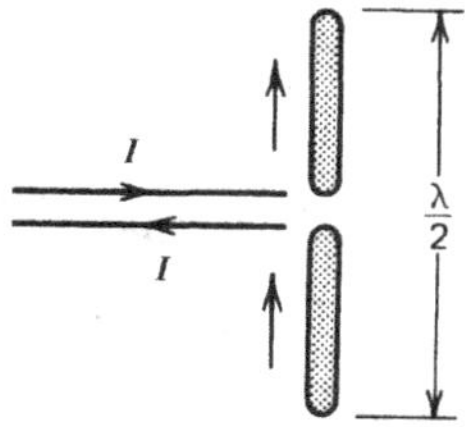

Figure 15.5 Half-wave antenna.

emitted radiation. At this point we consider one whose length is comparable to the wavelength (Fig. 15.5). In the very short case the magnitude of the current was taken not to vary along it. In the case of the long one the current must vary along its length if we hope to satisfy the required null at its ends. We will assume the current to vary harmonically with the distance along the wire. Although this is a good approximation, however, it is not exact because the losses due to the radiation will tend to cause some departure from the harmonic dependence.

We take the current in the antenna to be of the form

$$I(z', t) = I_0 \cos \omega t \cos kz' \tag{15.155}$$

The time dependence is exactly the dependence we took in the short antenna. The spatial dependence $\cos kz'$ vanishes at $z' = \pm\pi/2k = \pm\lambda/4$; thus with the antenna length taken as $\lambda/2$ and its center taken at the origin, this current distribution satisfies the null condition at its ends. Moreover, it is a standing wave since the time and spatial variations enter via the product of their harmonic dependences.

We will use the following procedure for obtaining the fields produced by the antenna. It will be subdivided into many very short sections. The fields of each are just those of a differential one. Note that one cannot calculate an effective dipole moment of this antenna by adding the dipole moments of all of its differential elements because some of them are separated by distances of order λ. A summation (integration) of the differential fields must be performed. Because the resulting fields are very complicated, we will only concern ourselves here with the contributions of the fields that give rise to the radiation—the fields that vary as $1/r$. Thus the radiation field of an element of length dz' located at z' and which is at a distance $\mathbf{R} = \mathbf{r} - \hat{\mathbf{z}}z'$ from the observation point (see Fig. 15.6) is obtained from Eqs. (15.148), (15.149), and (15.155):

$$dE_\theta = \frac{I_0 \omega^2}{4\pi\varepsilon_0 c^3} \frac{1}{kR} \cos(\omega t - kR) \sin\theta' \cos kz' \, dz' \tag{15.156}$$

$$dB_\phi = \frac{\mu_0 I_0}{4\pi kR} \frac{\omega^2}{c^2} [\cos(\omega t - kR)] \sin\theta' \sin kz' \, dz' \tag{15.157}$$

The total fields are determined by integrating these differential ones over z'. Let us consider the electric field first:

$$E_\theta = \frac{I_0 \omega^2}{4\pi\varepsilon_0 c^3} \int \frac{\cos(\omega t - kR)}{kR} \cos kz' \sin\theta' \, dz' \tag{15.158}$$

Because R and θ' depend on z', the above integral is quite complicated. However, because we are interested in the case where $r \gg \lambda \approx z'$ (that is, the distance of the point of observation is much larger than the dimension of the antenna), then the angle θ' can be approximated by θ for all the elements and the denominator kR can

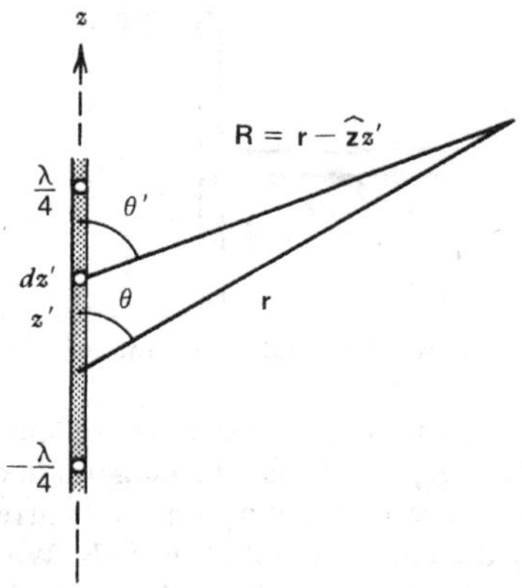

Figure 15.6 Calculation of the radiation fields of a half-wave antenna.

be taken as kr. The latter approximation, however, cannot be used in the argument of $\cos(\omega t - kR)$; it is because kR changes by amounts $\simeq \pi/2$ which although much smaller that kR, they change the phase of the cosine term drastically. Some approximation, however, can still be made by neglecting terms of the order z'^2:

$$kR \approx k(r - z' \cos \theta)$$

Thus E_θ becomes

$$E_\theta = \frac{I_0 \omega^2 \sin \theta}{4\pi\varepsilon_0 c^3 kr} \int \cos[\omega t - kr + kz' \cos \theta] \cos kz' \, dz'$$

In order to separate the dependence of r from z' we expand the cosine term; thus

$$E_\theta = \frac{I_0 \omega^2 \sin \theta}{4\pi\varepsilon_0 c^3 k^2 r} \left[\cos(\omega t - kr) \int_{-\pi/2}^{\pi/2} \cos(kz' \cos \theta) \cos kz' \, dkz' \right.$$

$$\left. - \sin(\omega t - kr) \int_{-\pi/2}^{\pi/2} \sin(kz' \cos \theta) \cos kz' \, dkz' \right] \tag{15.159}$$

where the integration is taken over kz'. The second integral vanishes because the integrand is an odd function of kz'. The first integral gives (see a table of integrals):

$$E_\theta = \frac{I_0}{2\pi\varepsilon_0 rc} \cos(\omega t - kr) \frac{\cos[(\pi/2)\cos \theta]}{\sin \theta} \tag{15.160}$$

Similarly, the integration of Eq. (15.157) gives

$$B_\phi = \frac{\mu_0 I_0}{2\pi r} \cos(\omega t - kr) \frac{\cos[(\pi/2)\cos \theta]}{\sin \theta} \tag{15.161}$$

The Poynting vector $\mathbf{S}$ is $\mathbf{E} \times \mathbf{B}/\mu_0 = E_\theta B_\phi \hat{\mathbf{r}}/\mu_0$; thus

$$\mathbf{S} = \frac{I_0^2}{4\pi^2 \varepsilon_0 c r^2} \cos^2(\omega t - kr) \frac{\cos^2[(\pi/2)\cos \theta]}{\sin^2 \theta} \hat{\mathbf{r}} \tag{15.162}$$

Integrating $\mathbf{S}$ over a sphere of radius r gives the power radiated:

$$P(t) = \frac{I_0^2}{2\pi\varepsilon_0 c} \cos^2(\omega t - kr) \int_0^\pi \frac{\cos^2[(\pi/2)\cos \theta]}{\sin^2 \theta} \sin \theta \, d\theta \tag{15.163}$$

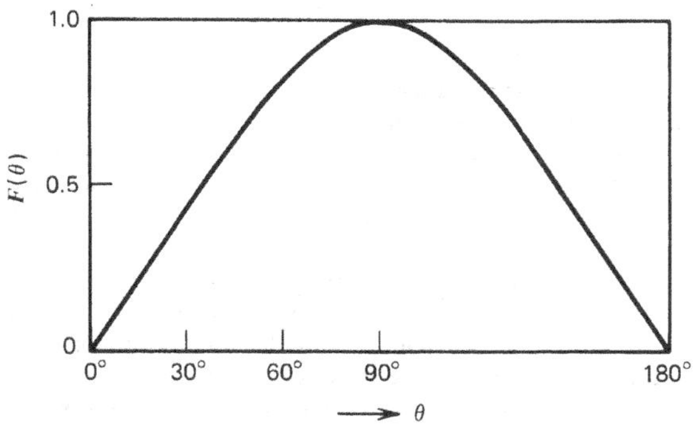

Figure 15.7 Angular dependence of the radiation field of a half-wave antenna.

The integral over θ does not have a closed form. A numerical integration yields 0.82. Thus

$$P(t) = 0.82 \frac{I_0^2}{2\pi\varepsilon_0 c} \cos^2(\omega t - kr) \tag{15.164}$$

To get the average power, we average the $\cos^2$ term over time. Since the average of a $\cos^2$ term is $\frac{1}{2}$, then

$$\langle P \rangle = 0.82 \frac{I_0^2}{4\pi\varepsilon_0 c} = \frac{1}{2} R I_0^2 \tag{15.165}$$

where $R = 73.1\ \Omega$ is an effective radiation resistance.

It is interesting to discuss the polarization and the angular dependence of the radiation. Equation (15.160) shows that the electric field is in the θ direction; thus the polarization is always in the plane containing the dipole. This property is called the *vertical* polarization. Equation (15.162) shows that the radiation is independent of ϕ, with the angular dependence shown in Fig. 15.7 and given by $F(\theta) = \cos^2[(\pi \cos \theta)/2]/\sin^2 \theta$. This angular function is zero at $\theta = 0$ and $\theta = \pi$, and unity at $\theta = \pi/2$. For the other values of θ, the distribution is a smooth function of θ.

Example 15.8 Communication from an Aircraft Carrier

In this example we discuss the importance of the various components of the dipole field by considering the communication with a carrier. The static field ($1/r^3$ dependence) is suitable for high-security communication because of its short-range nature as compared with the $1/r$ radiation field. Let us take the desired working signal of the electric field to be $1\ \mu$V/cm ($1\ \mu\text{V} = 10^{-6}$ V) and the minimum detectable signal to be $10^{-2}\ \mu$V/cm. If the communication is desired to be from a distance $r_1 = 2$ km, then the distance r_2 at which the field becomes undetectable is $E_2/E_1 = (r_2/r_1)^3 = 100$. Thus $r_2 = (800)^{1/3} \simeq 9.8$ km. For a radiation field, however, the distance at which the field becomes undetectable is 200 km, which shows the attractiveness of the static field. In order to minimize the security risk radiation field, the frequency of radiation can be chosen such that $kr_2 = 2\pi r_2/\lambda$ is much less than one. Since the ratio of the static field to the radiation field goes as $(kr)^{-2}$, this makes the field predominantly static.

Example 15.9 Antenna Array

The radiation from a half-wave antenna was found to be uniform in horizontal directions ($\hat{\boldsymbol{\phi}}$ direction) and a continuous function of θ. In many applications, it is desirable to concentrate a certain amount of power into some specified directions in order to enhance the reception, improve the resolution and minimize interference effects with other activities. These desired properties can be achieved by use of an array of half wave dipoles.

Consider Fig. 15.8, where we show a linear array of N half-wave dipoles, which are equally spaced and normal to the x-y plane, with the separation equal to half the wavelength of the radiation. The currents in all the dipoles are equal and in phase. The electric field produced by the array at a point P in the radiation zone is the vector sum of the individual fields produced by each of the dipoles. Because the various dipoles are at different distances from P, then the corresponding fields at P have different amplitudes and phases. For a given configuration there are definite relationships among the amplitudes and among the phases. We note that since the changes in the distances are quite small, the variation in the amplitudes of the various fields can be neglected. The corresponding variations in the phases however, is very important and can cause complete destruction of the radiation. From Fig. 15.8 we see that the distance of the mth dipole from the point of observation, is larger than the corresponding distance of the first dipole by $(m-1)\delta = (m-1)d \sin \phi$. The corresponding phase difference is $2\pi(m-1)\delta/\lambda$. Thus the electric field $\mathbf{E}_m$ of the mth dipole in terms of the field of the first one, $\mathbf{E}_0$, is

$$\mathbf{E}_m = \mathbf{E}_0 e^{2\pi(i(m-1)\delta/\lambda)} \tag{15.166}$$

Summing $\mathbf{E}_m$ over m gives the total field

$$\mathbf{E} = \mathbf{E}_0 \sum_{m=1}^{N} e^{2\pi i(m-1)\delta/\lambda} = \mathbf{E}_0 \frac{\sin(\frac{1}{2}N\alpha)}{\sin(\frac{1}{2}\alpha)} e^{i(N-1)\pi\delta/\lambda} \tag{15.167}$$

where $\alpha = 2\pi\delta/\lambda$. Using $\delta = d \sin \phi = \frac{1}{2}\lambda \sin \phi$, gives

$$|\mathbf{E}| = |\mathbf{E}_0| \frac{\sin\left(\frac{\pi}{2} N \sin\phi\right)}{\sin\left(\frac{\pi}{2} \sin \phi\right)} \tag{15.168}$$

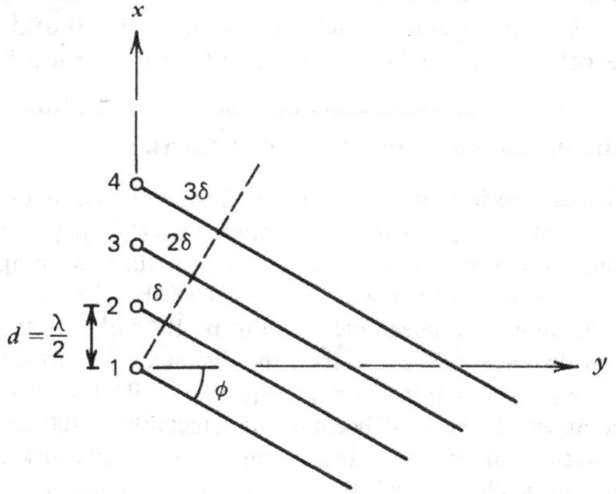

Figure 15.8 Array of half-wave antennas with separation equal to $\lambda/2$.

This result shows that the field depends on ϕ, thus resulting in concentration of the power in certain directions. We leave the details of the pattern of the radiation as an exercise (see Problem 15.18).

Example 15.10 Vertical Quarter-Wave Antenna Near a Grounded Plate

When the length of a half-wave antenna is inconveniently long, a combination of a quarter-wave antenna and the ground can be used. Such a combination is used in low-frequency applications where the wavelength of the radiation is large. Consider Fig. 15.9, in which a $\lambda/4$ antenna is placed vertically at a small distance compared to λ above the ground. The radiation pattern due to the antenna is going to be due to the direct radiation produced by the various elements of the antenna and the reflected waves. The reflected waves, appear to originate from a $\lambda/4$ image antenna just below the ground. The antenna and the image form a half-wave antenna; hence the pattern of the power radiated is given by Eqs. (15.162) to (15.165).

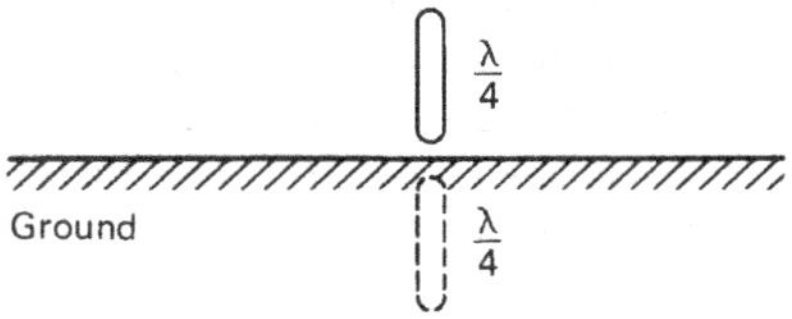

Figure 15.9 Quarter-wave antenna erected vertically and close to the ground.

15.5 Multipole Expansion of the Retarded Potentials—Radiation from Slowly Moving Charges—Electric Dipole

The potentials produced by charge and current distributions are given by the retarded potentials in Eqs. (15.43) and (15.47). For general distributions, the integrals are generally very complicated where closed forms may not exist. Some analytical progress using some approximations, however, may be achieved in certain distributions. These distributions are defined by two properties: (1) We take the largest dimension of the distribution to be much smaller than the dominant wavelength of the radiation (dipole approximation), and (2) the distributions are bounded in a volume whose largest dimension is much smaller than the distance of the distribution from the point of observation. Because we are considering charges in motion, condition 2 implies that the speeds of the charges have to be small compared to the speed of light, so that in the time it takes the radiation to reach the point of observation, the distribution continues to have a dimension much smaller than the distance to the point of observation.

We start by expanding the scalar potential:

$$\Phi(\mathbf{r}, t) = \frac{1}{4\pi\varepsilon_0} \int \frac{\rho(\mathbf{r}', t')}{|\mathbf{r} - \mathbf{r}'|} dv' \tag{15.169}$$

where $t' = t - |\mathbf{r} - \mathbf{r}'|/c$, $\mathbf{r}$ is the distance of the point of observation with respect to the origin, and $\mathbf{r}'$ is the distance of a charge element in a volume dv' with respect to the origin. Since $r'/r \ll 1$, we expand $|\mathbf{r} - \mathbf{r}'|$ and $1/|\mathbf{r} - \mathbf{r}'|$ in powers of r'/r and keep the lowest terms; that is,

$$|\mathbf{r} - \mathbf{r}'| = (r^2 + r'^2 - 2\mathbf{r}\cdot\mathbf{r}')^{1/2} \approx r - \frac{\mathbf{r}'\cdot\mathbf{r}}{r} + \cdots \tag{15.170}$$

and

$$\frac{1}{|\mathbf{r}-\mathbf{r}'|} \approx \frac{1}{r}\left[1+\frac{\mathbf{r}'\cdot\mathbf{r}}{r^2}\right]+\cdots \tag{15.171}$$

Substituting these expansions in $\Phi(\mathbf{r}, t)$ gives

$$\Phi(\mathbf{r}, t) = \frac{1}{4\pi\varepsilon_0 r}\int \rho\left(\mathbf{r}', t-\frac{r}{c}+\frac{\mathbf{r}'\cdot\mathbf{r}}{cr}\right)\left[1+\frac{\mathbf{r}'\cdot\mathbf{r}}{r^2}\right]dv' \tag{15.172}$$

We now expand $\rho(\mathbf{r}', t - r/c + \mathbf{r}'\cdot\mathbf{r}/cr)$ in the neighborhood of the argument $(\mathbf{r}', t - r/c)$ using the Taylor series expansion:

$$\rho\left(\mathbf{r}', t-\frac{r}{c}+\frac{\mathbf{r}'\cdot\mathbf{r}}{cr}\right) \approx \rho\left(\mathbf{r}', t-\frac{r}{c}\right)+\frac{\mathbf{r}'\cdot\mathbf{r}}{cr}\left.\frac{\partial\rho}{\partial t}\right|_{\mathbf{r}', t-r/c}+\cdots \tag{15.173}$$

The physical basis for this kind of approximate expansion is that we take the distribution to be located at the origin when the retardation effects are to be calculated. That is, all elements of the distribution regardless of their position $\mathbf{r}'$ will have the same retarded time $t' = t - r/c$. Substituting Eq. (15.173) into Eq. (15.172) and keeping terms up to order r'/r, we get

$$\Phi(\mathbf{r}, t) = \frac{1}{4\pi\varepsilon_0}\left[\frac{Q}{r}+\frac{\mathbf{r}\cdot\mathbf{p}(t-r/c)}{r^3}+\frac{\mathbf{r}\cdot\dot{\mathbf{p}}(t-r/c)}{cr^2}\right] \tag{15.174}$$

where

$$Q = \int \rho\left(\mathbf{r}', t-\frac{r}{c}\right)dv' \tag{15.175}$$

is the total charge of the distribution, which is a constant quantity independent of time, and

$$\mathbf{p}\left(t-\frac{r}{c}\right) = \int \mathbf{r}'\rho\left(\mathbf{r}', t-\frac{r}{c}\right)dv' \tag{15.176}$$

is the retarded electric dipole moment of the distribution, and $\dot{\mathbf{p}}(t - r/c) = d\mathbf{p}/dt$. Equation (15.174) represents the first three expansion terms. Higher-order terms such as the quadrupole terms and so on depend on high moments of the distribution (higher powers of r'/r), thus making their contribution negligible compared to the terms we already calculated. However, the next higher-order term will dominate if the calculated terms vanish due to some symmetry.

A similar procedure can now be used to expand $\mathbf{A}(\mathbf{r}, t)$ in powers of r'/r. Because the current density is defined through the relation $\mathbf{J} = \rho\mathbf{v} = \rho(d\mathbf{r}'/dt)$, then the zero order of the expansion of $\mathbf{J}$ is of first order in r'/r; hence to make the expansion consistent with the above expansion of $\Phi(\mathbf{r}, t)$, we should retain the zero order only. In the vector potential expression,

$$\mathbf{A}(\mathbf{r}, t) = \frac{\mu_0}{4\pi}\int \frac{\mathbf{J}(\mathbf{r}', t-|\mathbf{r}-\mathbf{r}'|/c)}{|\mathbf{r}-\mathbf{r}'|}dv' \tag{15.177}$$

we retain the lowest order of $1/|\mathbf{r}-\mathbf{r}'|$, or $1/r$, and the lowest order of the expansion of $\mathbf{J}(\mathbf{r}', t - r/c + \mathbf{r}\cdot\mathbf{r}'/c)$ near the argument $(\mathbf{r}', t - r/c)$, $\mathbf{J}(\mathbf{r}', t - r/c)$. Therefore $\mathbf{A}$ becomes

$$\mathbf{A}(\mathbf{r}, t) = \frac{\mu_0}{4\pi r}\int \mathbf{J}\left(\mathbf{r}', t-\frac{r}{c}\right)dv' \tag{15.178}$$

which shows that, in this approximation, **A** depends on just the volume integral of the current density. In order to make the result more convenient for the calculations of the EM fields, we rewrite **A** in terms of the dipole moment of the charge distribution. One can show that (see Example 15.11)

$$\int \mathbf{J}\, dv = \frac{d\mathbf{p}}{dt} \tag{15.179}$$

and hence

$$\mathbf{A}(\mathbf{r}, t) = \frac{\mu_0}{4\pi r}\dot{\mathbf{p}}\left(t - \frac{r}{c}\right) \tag{15.180}$$

Example 15.11

Consider the divergence of the product of the coordinate x and the current density **J** (that is, $x\mathbf{J}$). Using Eq. (1.57), we can write as $\nabla\cdot(x\mathbf{J}) = (\hat{\mathbf{x}}\cdot\mathbf{J}) + x\nabla\cdot\mathbf{J}$. Integrating $\nabla\cdot(x\mathbf{J})$ over a volume V and using the divergence theorem gives

$$\oint_S x\mathbf{J}\cdot\hat{\mathbf{n}}\, da = \int_V \hat{\mathbf{x}}\cdot\mathbf{J}\, dv + \int_V x\nabla\cdot\mathbf{J}\, dv \tag{15.181}$$

where S is the surface bounding the volume V. Multiplying Eq. (15.181) by $\hat{\mathbf{x}}$, writing similar expressions for the products $y\mathbf{J}$ and $z\mathbf{J}$, and summing, we obtain

$$\oint_S \mathbf{r}\mathbf{J}\cdot\hat{\mathbf{n}}\, da = \int_V \mathbf{J}\, dv + \int_V \mathbf{r}\nabla\cdot\mathbf{J}\, dv \tag{15.182}$$

Far away from the distribution, the surface integral vanishes; hence $\int \mathbf{J}\, dv = -\int \mathbf{r}\nabla\cdot\mathbf{J}\, dv$. Using the continuity relation, we write $\nabla\cdot\mathbf{J} = -\partial\rho/\partial t$; thus

$$\int \mathbf{J}\, dv = \int \mathbf{r}\frac{\partial\rho}{\partial t}\, dv = \frac{d}{dt}\int \mathbf{r}\rho\, dv \tag{15.183}$$

Using the fact that $\mathbf{p} = \int \mathbf{r}\rho\, dv$, we get Eqs. (15.179) and (15.180). Note that in the steady-state case we have $\int \mathbf{J}\, du = 0$ as was given in Eq. (8.99).

We should note that the vector and the scalar potentials calculated above satisfy the Lorentz condition. Therefore, one could have calculated $\Phi(\mathbf{r}, t)$ from **A** and the Lorentz condition (see Problem 15.17).

With the potentials obtained, **E** and **B** can now be calculated from $\mathbf{B} = \nabla \times \mathbf{A}$ and $\mathbf{E} = -\nabla\Phi - \partial\mathbf{A}/\partial t$. We take the case $Q = 0$. Taking the curl of Eq. (15.180) gives

$$\mathbf{B} = \frac{\mu_0}{4\pi}\nabla \times \frac{\dot{\mathbf{p}}}{r} = \frac{\mu_0}{4\pi}\left(\nabla\frac{1}{r}\right) \times \dot{\mathbf{p}} + \frac{\mu_0}{4\pi}\frac{1}{r}\nabla \times \dot{\mathbf{p}} \tag{15.184}$$

Since $\dot{\mathbf{p}}$ is not a function of θ or ϕ, then $\nabla \times \dot{\mathbf{p}} = \hat{\mathbf{r}} \times \partial\dot{\mathbf{p}}/\partial r = -(\hat{\mathbf{r}}/c) \times \ddot{\mathbf{p}}$; thus

$$\mathbf{B}(\mathbf{r}, t) = -\frac{\mu_0}{4\pi r^2}\hat{\mathbf{r}} \times \dot{\mathbf{p}} - \frac{\mu_0}{4\pi c r}\hat{\mathbf{r}} \times \ddot{\mathbf{p}} \tag{15.185}$$

To calculate **E**, we evaluate $-\partial\mathbf{A}/\partial t$ and $-\nabla\Phi$. The first one is easy to calculate; that is,

$$\frac{\partial\mathbf{A}}{\partial t} = \frac{\mu_0}{4\pi r}\ddot{\mathbf{p}}\left(t - \frac{r}{c}\right) \tag{15.186}$$

The calculation of $-\nabla\Phi$, however, is more complicated because Φ involves the scalar product of two vectors. Using the vector relations given in Eq. (1.59) and the property $\nabla \times \mathbf{r} = 0$, we get

$$-\nabla\Phi = -\frac{1}{4\pi\varepsilon_0}\left[\frac{1}{cr}\hat{\mathbf{r}}\cdot\nabla\dot{\mathbf{p}} + \frac{1}{r^2}\hat{\mathbf{r}}\cdot\nabla\mathbf{p} + \dot{\mathbf{p}}\cdot\nabla\frac{\hat{\mathbf{r}}}{cr} + \mathbf{p}\cdot\nabla\frac{\hat{\mathbf{r}}}{r^2} + \frac{1}{cr}\hat{\mathbf{r}}\times(\nabla\times\dot{\mathbf{p}}) + \frac{\hat{\mathbf{r}}}{r^2}\times(\nabla\times\mathbf{p})\right]. \tag{15.187}$$

Noting that $\mathbf{p}$ and $\dot{\mathbf{p}}$ are functions of $t - r/c$, then the ∇ operation on them can be replaced by $\hat{\mathbf{r}}(\partial/\partial r)$. Thus $\mathbf{E}$ becomes

$$\mathbf{E} = \mathbf{E}_R + \mathbf{E}_I + \mathbf{E}_S \tag{15.188}$$

where

$$\mathbf{E}_R = \frac{1}{4\pi\varepsilon_0 c^2 r}[\hat{\mathbf{r}}(\hat{\mathbf{r}}\cdot\ddot{\mathbf{p}}) - \ddot{\mathbf{p}}] \tag{15.189}$$

$$\mathbf{E}_I = \frac{1}{4\pi\varepsilon_0}\frac{1}{cr^2}[3\hat{\mathbf{r}}(\hat{\mathbf{r}}\cdot\dot{\mathbf{p}}) - \dot{\mathbf{p}}] \tag{15.190}$$

$$\mathbf{E}_S = \frac{1}{4\pi\varepsilon_0}\frac{1}{r^3}[3\hat{\mathbf{r}}(\hat{\mathbf{r}}\cdot\mathbf{p}) - \mathbf{p}] \tag{15.191}$$

The $\mathbf{E}_R$ field is called the *radiation field*. It has $1/r$ dependence; hence it contributes to the radiation. Moreover, it depends on the second derivative of the dipole moment (and thus depends on the acceleration of the charge distribution). The E_I field is called the *intermediate* (*induction*) field; it has $1/r^2$ dependence and it depends on the first derivative of the dipole moment (and thus it depends on the velocity of the distribution). Finally E_S is called the *static field*; it has $1/r^3$ dependence and it depends on the dipole moment itself hence on the position of the distribution. In fact, if we take $\mathbf{p} = p_0\hat{\mathbf{z}}\cos\omega t$, then the above fields give exactly the fields of Eq. (15.151), which are produced by a differential dipole antenna with sinusoidal current excitation or by a sinusoidally varying electric dipole (TM) [see Eq. (15.103)].

The magnetic field given in Eq. (15.185) can be written as the sum of the fields $\mathbf{B_R}$ and $\mathbf{B}_I$ (that is, $\mathbf{B} = \mathbf{B}_R + \mathbf{B}_I$), where

$$\mathbf{B}_R = -\frac{\mu_0}{4\pi cr}\hat{\mathbf{r}}\times\ddot{\mathbf{p}} \tag{15.192}$$

$$\mathbf{B}_I = -\frac{\mu_0}{4\pi r^2}\hat{\mathbf{r}}\times\dot{\mathbf{p}} \tag{15.193}$$

The contribution $\mathbf{B}_R$ is a *radiation field*; it depends on $\ddot{\mathbf{p}}$ and has $1/r$ radial dependence. The contribution $\mathbf{B}_I$ is an *induction field*; it depends on $\dot{\mathbf{p}}$ and has $1/r^2$ radial dependence. In fact, the $\mathbf{B}$ field can be shown to be related to the $\mathbf{E}_I + \mathbf{E}_R$ through a simple cross product with $\hat{\mathbf{r}}$:

$$\mathbf{B} = \frac{1}{c}\hat{\mathbf{r}}\times(\mathbf{E}_I + \mathbf{E}_R) \tag{15.194}$$

Although we calculated the power radiation by an electric dipole in the case of sinusoidal time variations, it is instructive to determine the power in the general nonsinusoidal case where $\mathbf{p} = \mathbf{p}(t)$. First we note that the radiation fields $\mathbf{E}_R$ and $\mathbf{B}_R$ can be related via a simple cross product:

$$\mathbf{E}_R = \frac{1}{4\pi\varepsilon_0 c^2 r}\hat{\mathbf{r}} \times (\hat{\mathbf{r}} \times \ddot{\mathbf{p}}) = -c\hat{\mathbf{r}} \times \mathbf{B}_R \tag{15.195}$$

The Poynting vector $\mathbf{S} = \mathbf{E} \times \mathbf{H}$ is then

$$\mathbf{S} = -\frac{c}{\mu_0}(\hat{\mathbf{r}} \times \mathbf{B}_R) \times \mathbf{B}_R = \frac{c}{\mu_0}\hat{\mathbf{r}}|\mathbf{B}_R|^2. \tag{15.196}$$

Substituting Eq. (15.192) in this equation gives

$$\mathbf{S} = \frac{\mu_0}{16\pi^2 c r^2}\hat{\mathbf{r}}|\hat{\mathbf{r}} \times \ddot{\mathbf{p}}|^2 \tag{15.197}$$

Let us consider the case where $\ddot{\mathbf{p}}$ is along the z axis. In this case $\mathbf{S}$ becomes

$$\mathbf{S} = \frac{\hat{\mathbf{r}}|\ddot{\mathbf{p}}|^2 \sin^2\theta}{16\pi^2\varepsilon_0 c^3 r^2} \tag{15.198}$$

Figure 15.10 is a schematic diagram of the radiation fields, showing the Poynting vector for this dipole for three points of observation—one along the z axis, one in the x-y plane, and one at arbitrary (θ, ϕ). The total power radiated is calculated by integrating $\mathbf{S}$ over a sphere with a radius r and center at the origin. The result is

$$P = \frac{1}{4\pi\varepsilon_0}\frac{2}{3}\frac{|\ddot{\mathbf{p}}|^2}{c^3} \tag{15.199}$$

This result shows that the power radiated is proportional to the square of the second derivative of the dipole moment of the distribution; hence the distribution can radiate only if it is accelerated.

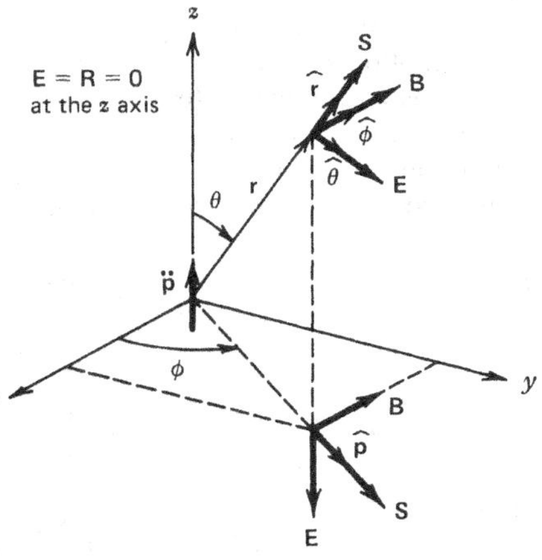

Figure 15.10 Schematic diagram of the radiation fields of a time-dependent electric dipole whose second derivative is along z, at three selected orientations.

Equation (15.199) can be specialized to the case of a single charge undergoing an accelerated motion. The speed of the charge is required to be much smaller than the speed of light since Eq. (15.199) was derived under this condition. Consider a charge q whose position is given by $\mathbf{r}'$, which is measured with respect to an arbitrary origin. The dipole moment of the charge is $\mathbf{p} = q\mathbf{r}'$; hence $\ddot{\mathbf{p}} = q\ddot{\mathbf{r}}' = q\dot{\mathbf{v}}$, which upon substitution in Eq. (15.199) gives

$$P = \frac{q^2}{4\pi\varepsilon_0 c^3}\frac{2}{3}|\ddot{\mathbf{r}}'|^2 = \frac{q^2}{4\pi\varepsilon_0 c^3}\frac{2}{3}|\dot{\mathbf{v}}|^2 \tag{15.200}$$

The power radiated from a fast-moving accelerated charge will be treated in the following section.

15.6 The Lienard-Weichert Potential—Fast-Moving Point Charges

The retarded potentials describe the potentials of arbitrary charge and current distributions $\rho(\mathbf{r}, t')$ and $\mathbf{J}(\mathbf{r}', t')$. In the case where the charge and the current distributions are bounded (the speeds of the various elements of the distribution are much smaller than the speed of light) and the dominant wavelength of radiation is much larger than the actual size of the distribution (dipole approximation), the multipole expansion is quite adequate, as was seen in the previous section. In this section we remove these restrictions and consider fast-moving charge and current distributions without invoking the dipole approximation.

For simplicity we consider the case of a point charge q in arbitrary motion, whose trajectory is completely described by the radius vector $\mathbf{r}'(t')$. We will calculate the retarded scalar potential of the charge using Eq. (15.43):

$$\Phi(\mathbf{r}, t) = \frac{1}{4\pi\varepsilon_0}\int \frac{\rho(\mathbf{r}', t')}{|\mathbf{r} - \mathbf{r}'|}\, dv'$$

Since we are dealing with a point charge, then there is no direct variation of ρ with r', and hence the charge density can be replaced by a Dirac delta function. However, because of the appreciable motion of the charge, it will essentially look like an extended charge distribution as a function of time. Two effects need to be considered: The distance between the charge and the point of observation $\mathbf{R}(t') = \mathbf{r} - \mathbf{r}'(t')$ varies with time, and the argument of the delta function describing the position of the charge depends on time because of the retardation effect. Thus we write

$$\rho(\mathbf{r}', t) = q\delta\left(t' - t + \frac{|\mathbf{r} - \mathbf{r}'(t')|}{c}\right) = q\delta(t'') \tag{15.201}$$

where

$$t'' = t' - t + \frac{|\mathbf{r} - \mathbf{r}'|}{c} = t' - t + \frac{|\mathbf{R}|}{c} \tag{15.202}$$

The evaluation of the potential $\Phi(\mathbf{r}, t)$ then requires an integration with respect to the retarded time t' over the entire volume that contains the charge; namely, from $-\infty$ to ∞. Thus

$$\Phi(\mathbf{r}, t) = \frac{q}{4\pi\varepsilon_0}\int_{-\infty}^{\infty} \frac{\delta(t' - t + |\mathbf{r} - \mathbf{r}'(t')|/c)}{|\mathbf{r} - \mathbf{r}'(t')|}\, dt' \tag{15.203}$$

The integration can be easily done if we recall an important property of the Dirac delta function:

$$\psi(x) = \int g(x)\delta(f(x))dx = \left.\frac{g(x)}{df(x)/dx}\right|_{f(x)=0} \tag{15.204}$$

Thus

$$\Phi(\mathbf{r}, t) = \left.\frac{q}{4\pi\varepsilon_0 R(t')} \cdot \frac{dt'}{dt''}\right|_{t''=0} \tag{15.205}$$

Since t is constant as far as t' is concerned, then

$$\frac{dt''}{dt'} = 1 - \frac{1}{c}\frac{d}{dt'} R(t') \tag{15.206}$$

But $d|\mathbf{r} - \mathbf{r}'(t')| = \nabla'|\mathbf{r} - \mathbf{r}'(t')| \cdot d\mathbf{r}' = -\nabla|\mathbf{r} - \mathbf{r}'(t')| \cdot d\mathbf{r}'$. Thus

$$\frac{dt''}{dt'} = 1 + \frac{1}{c}\nabla|\mathbf{r} - \mathbf{r}'(t')| \cdot \frac{d\mathbf{r}'}{dt'}$$

Taking $\nabla|\mathbf{r} - \mathbf{r}'(t')| = -\mathbf{R}(t')/R(t')$ and calling $(1/c)d\mathbf{r}'/dt' = \boldsymbol{\beta}(t')$ the velocity of the charge normalized to the speed of light, we obtain

$$\frac{dt''}{dt'} = \frac{R(t') - \boldsymbol{\beta}(t') \cdot \mathbf{R}(t')}{R(t')} \tag{15.207}$$

$$\Phi(\mathbf{r}, t) = \left.\frac{q}{4\pi\varepsilon_0} \frac{1}{R(t') - \boldsymbol{\beta}(t') \cdot \mathbf{R}(t')}\right|_{t''=0} \tag{15.208}$$

But the condition $t'' = 0$ implies that $t' = t - |\mathbf{r} - \mathbf{r}'(t')|/c$ (the retarded time). Thus

$$\Phi(\mathbf{r}, t) = \frac{q}{4\pi\varepsilon_0} \frac{1}{R(t') - \boldsymbol{\beta}(t') \cdot \mathbf{R}(t')} \tag{15.209}$$

The vector potential of a moving charge can be calculated using the retarded vector potential

$$\mathbf{A}(\mathbf{r}, t) = \frac{\mu_0}{4\pi} \int \frac{\mathbf{J}(\mathbf{r}', t')}{|\mathbf{r} - \mathbf{r}'|} dv'$$

Since the current density is equal to the product of the charge density and the velocity of the charge element, then an analogous derivation to that of the scalar potential can be used to determine **A**. The result is

$$\mathbf{A}(\mathbf{r}', t) = \frac{q}{4\pi\varepsilon_0 c} \frac{\boldsymbol{\beta}(t')}{R(t') - \boldsymbol{\beta}(t') \cdot \mathbf{R}(t')} \tag{15.210}$$

which is just the product of $\boldsymbol{\beta}/c$ and the scalar potential; that is,

$$\mathbf{A}(\mathbf{r}, t) = \frac{\boldsymbol{\beta}(t')\Phi(\mathbf{r}, t)}{c} \tag{15.211}$$

The potentials Φ and **A** given by Eqs. 15.209–15.211 are called the *Lienard-Weichert potentials* in memory of the two scientists who developed this treatment. The corresponding electric and magnetic fields can be calculated from $\mathbf{B} = \nabla \times \mathbf{A}$, and $\mathbf{E} = -\nabla\Phi - \partial\mathbf{A}/\partial t$. The actual calculation is quite complicated, and therefore we will give only the result here:

$$\mathbf{E} = \mathbf{E}_v + \mathbf{E}_a \qquad \mathbf{B} = \mathbf{B}_v + \mathbf{B}_a \tag{15.212}$$

where

$$\mathbf{E}_v = \frac{q}{4\pi\varepsilon_0 K^3}(\mathbf{R} - R\boldsymbol{\beta})(1 - \beta^2)$$

$$\mathbf{E}_a = \frac{q}{4\pi\varepsilon_0 c^2 K^2}\left[\frac{(\mathbf{R}\cdot\mathbf{a})(\mathbf{R} - R\boldsymbol{\beta})}{K} - R\mathbf{a}\right]$$

$$\mathbf{B}_v = \frac{q}{4\pi\varepsilon_0 c K^3}[(\boldsymbol{\beta}\times\mathbf{R})(1 - \beta^2)] \tag{15.213}$$

$$\mathbf{B}_a = \frac{q}{4\pi\varepsilon_0 c^3 K^2}\left[\frac{(\mathbf{R}\cdot\mathbf{a})(\boldsymbol{\beta}\times\mathbf{R})}{K} + (\mathbf{a}\times\mathbf{R})\right]$$

and

$$K = R(t') - \boldsymbol{\beta}(t')\cdot\mathbf{R}(t')$$

The acceleration $\mathbf{a}$ of the charge is given by

$$\mathbf{a} = c\frac{\partial\boldsymbol{\beta}}{\partial t} \tag{15.214}$$

and all the quantities are evaluated at the retarded time $t' = t - |\mathbf{r} - \mathbf{r}'|/c$. It is to be noted that the above fields satisfy the following relations:

$$\mathbf{B} = \frac{\hat{\mathbf{n}}(t')\times\mathbf{E}}{c} \tag{15.215}$$

$$\mathbf{B}_v = \frac{\hat{\mathbf{n}}(t')}{c}\times\mathbf{E}_v \qquad \text{and} \qquad \mathbf{B}_a = \frac{\hat{\mathbf{n}}(t')}{c}\times\mathbf{E}_a \tag{15.216}$$

where $\hat{\mathbf{n}}(t') = \mathbf{R}(t')/R(t')$.

The fields $\mathbf{E}_v$ and $\mathbf{B}_v$ are called *velocity* fields; they fall off as R^{-2} and move together with the charge. The $\mathbf{E}_a$ and $\mathbf{B}_a$ fields are called *acceleration* fields; they fall off as R^{-1}, and the total power associated with them is independent of R. This implies that the radiation fields are not rigidly attached to the charge producing them. In the wave zone (that is, at large distances from the charge) the power associated with the velocity fields is negligible compared to the radiation from the acceleration fields (radiation fields).

We would like to note that since the time relevant to the motion of the charge is t' whereas $t = t' + |\mathbf{r} - \mathbf{r}'(t')|/c$ is relevant to the observer, then we expect the *power* radiated in a given solid angle to differ from the two points of view. The difference comes about because although the total energy radiated is the same, the factor dt/dt' is not unity.

The energy radiated in a solid angle $d\Omega$ and measured by the observer in an interval dt at time t is $-dW = \mathbf{S}\cdot\hat{\mathbf{n}}R^2\,dR\,dt$. Because this energy was radiated by the charge at the retarded time $t' = t - R(t')/c$, then the power radiated per unit solid angle is

$$-\frac{dW}{dt'\,dR} = \frac{dP}{d\Omega} = \mathbf{S}\cdot\hat{\mathbf{n}}R^2\frac{dt}{dt'} \tag{15.217}$$

Differentiating $t' = t - R(t')/c$ with respect to t' gives

$$1 = \frac{dt}{dt'} - \frac{1}{c}\frac{dR(t')}{dt'} \tag{15.218}$$

Using Eqs. (15.206) and (15.207) we find that

$$\frac{1}{c}\frac{d}{dt'}R(t') = -\frac{\boldsymbol{\beta}(t')\cdot\mathbf{R}}{R}$$

Hence

$$\frac{dt}{dt'} = 1 - \frac{\boldsymbol{\beta}\cdot\mathbf{R}}{R} = 1 - \beta\cos\theta \tag{15.219}$$

where θ is the angle between $\boldsymbol{\beta}$ and $\mathbf{R}$.

For slowly moving charges or distributions (bounded distributions) such as all the cases considered prior to the present section, one takes $r' \ll r$, and hence $t = t' + r/c$, or $dt/dt' = 1$. Consequently the power distributions become identical. For fast-moving distributions (unbounded), the power distributions are expected to be different (see Example 15.13).

Example 15.12 Radiation Damping

It is now clear that the motion of a charged particle is influenced by the presence of electromagnetic fields via the Coulomb and Lorentz forces. At the same time a charged particle that undergoes acceleration generates electromagnetic radiation. In this example we consider the effect of the radiated fields on the motion of the accelerated charge producing them. This phenomenon is called *radiation damping* or *radiation reaction.*

We consider a charge q moving in an external field with a velocity $v \ll c$ and acceleration $\mathbf{a} = \dot{\mathbf{v}}$. The force due to the external field is taken to be a restoring force: $\mathbf{F}_e = -k\mathbf{r}$, where $\mathbf{r}$ is the position of the charge with respect to the origin and k is a positive quantity.

The power radiated by the charge is given by Eq. (15.200), as follows:

$$P = -\frac{dW}{dt} = \frac{1}{4\pi\varepsilon_0}\frac{2q^2}{3}\frac{|\dot{\mathbf{v}}|^2}{c^3}$$

Corresponding to the radiation losses, there exists an effective reaction force $\mathbf{F}_r$ such that

$$\frac{dW}{dt} = \mathbf{F}_r\cdot\mathbf{v} = -\frac{1}{4\pi\varepsilon_0}\frac{2}{3}q^2\frac{|\dot{\mathbf{v}}|^2}{c^3}$$

which when integrated over a cycle gives

$$\oint \mathbf{F}_r\cdot\mathbf{v}\,dt = -\frac{1}{4\pi\varepsilon_0}\frac{2}{3}\frac{q^2}{c^3}\int|\dot{\mathbf{v}}|^2\,dt$$

Integrating the right-hand side of this equation by parts gives

$$\int \mathbf{F}_r\cdot\mathbf{v}\,dt = -\frac{1}{4\pi\varepsilon_0}\frac{2}{3}\frac{q^2}{c^3}\dot{\mathbf{v}}\cdot\mathbf{v}\big|_{\text{cycle}} + \frac{1}{4\pi\varepsilon_0}\frac{2}{3}q^2\int\ddot{\mathbf{v}}\cdot\mathbf{v}\,dt$$

The quantity $\dot{\mathbf{v}}\cdot\mathbf{v}$ evaluated over a cycle gives zero, and therefore

$$\int\left(\mathbf{F}_r - \frac{1}{4\pi\varepsilon_0}\frac{2}{3}q^2\ddot{\mathbf{v}}\right)\cdot\mathbf{v}\,dt = 0 \tag{15.220}$$

Thus over a cycle one can take, for the radiation reaction force,

$$\mathbf{F}_r = \frac{1}{4\pi\varepsilon_0}\frac{2}{3}q^2\ddot{\mathbf{v}} = \frac{1}{4\pi\varepsilon_0}\frac{2}{3}q^2\frac{d^3\mathbf{r}}{dt^3} \tag{15.221}$$

The motion of the charge of mass m under the influence of the external and the reaction forces is therefore governed by the following equation of motion

$$m\frac{d^2\mathbf{r}}{dt^2} = \mathbf{F}_e + \mathbf{F}_r = -k\mathbf{r} + \frac{q^2}{6\pi\varepsilon_0}\left(\frac{d^3\mathbf{r}}{dt^3}\right) \tag{15.222}$$

Example 15.13 Stoppage of a Fast Charge

In this example we calculate the energy radiated during the stoppage of a fast-moving charge. The deceleration is taken to be uniform and along the direction of the initial motion. The electric field in the radiation zone is given by Eq. (15.213), or

$$\mathbf{E}_a = \frac{q}{4\pi\varepsilon_0 c^2 K^3}\left[(\mathbf{R}\cdot\mathbf{a})\mathbf{R} - R^2\mathbf{a} - R(\mathbf{R}\cdot\mathbf{a})\boldsymbol{\beta} + R(\mathbf{R}\cdot\boldsymbol{\beta})\mathbf{a}\right] \tag{15.223}$$

When $\boldsymbol{\beta}$ and $\mathbf{a}$ are in the same direction, the last two terms cancel each other, and $\mathbf{E}_a$ becomes

$$\mathbf{E}_a = \frac{q}{4\pi\varepsilon_0 c^2 K^3}\left[(\mathbf{R}\cdot\mathbf{a})\mathbf{R} - R^2\mathbf{a}\right] \tag{15.224}$$

Using $\mathbf{B}_a = \hat{\mathbf{n}}(t') \times \mathbf{E}_a/c$ in the expression for the Poynting vector gives

$$\mathbf{S} = \frac{1}{\mu_0 c}\mathbf{E}_a \times (\hat{\mathbf{n}}(t') \times \mathbf{E}_a) = \frac{1}{\mu_0 c}E_a^2\hat{\mathbf{n}}(t') \tag{15.225}$$

Squaring Eq. (15.224) gives

$$\mathbf{E}_a^2 = \frac{q^2}{(4\pi\varepsilon_0)^2 c^4 K^6}R^2(R^2a^2 - (\mathbf{R}\cdot\mathbf{a})^2) \tag{15.226}$$

Taking the angle between $\mathbf{a}$ and $\mathbf{R}$ to be θ and substituting Eq. (15.226) in Eq. (15.225) gives

$$\mathbf{S} = \frac{q^2a^2R^4}{(4\pi\varepsilon_0)^2\mu_0 c^5 K^6}\sin^2\theta\,\hat{\mathbf{n}} \tag{15.227}$$

Substituting Eqs. (15.227) and (15.219) in Eq. 15.217 and noting that $K = R(1 - \beta\cos\theta)$ gives

$$\frac{dP}{d\Omega} = \frac{q^2a^2\sin^2\theta}{(4\pi\varepsilon_0)^2\mu_0 c^5(1 - \beta\cos\theta)^5} \tag{15.228}$$

This equation gives the angular distribution of the radiation of the charge. We will leave further studies of the pattern as an exercise (see Problem 15.24). The evaluation of the total power and energy radiated also will be left as exercises (see Problems 15.22 and 15.23).

Example 15.14 Slowly Moving Accelerated Charge

Consider a charge q moving with a small velocity $|\mathbf{v}| \ll c$ and a small acceleration $d\mathbf{v}/dt$ such that it stays bounded in a region whose largest dimension is R_0. Let the point of observation be at a large distance from the origin $r \gg R_0$. The retarded time $t' = t - |\mathbf{r} - \mathbf{r}'(t')/c|$ can be approximated by $t' = t - r/c$ since $r' < R_0 \ll r$. The distance of the charge from the point of observation, $\mathbf{R}(t') = \mathbf{r}(t') - \mathbf{r}'(t')$, can also be taken equal to $\mathbf{r}(t')$ for all times. The quantity $1/K^n = 1/(R(t') - \mathbf{R}(t')\cdot\boldsymbol{\beta}(t'))^n$, where n is an integer, can be approximated by $1/K^n = (1 + n\boldsymbol{\beta}\cdot\hat{\mathbf{r}})/r^n$, in which we kept the lowest-order term in $\boldsymbol{\beta} = \mathbf{v}/c$. Substituting these approximate expressions in Eqs. (15.213) gives the following:

$$\begin{aligned}
\mathbf{E}_v &= \frac{q}{4\pi\varepsilon_0 r^2}\left[\hat{\mathbf{r}}\left(1 + 3\frac{\hat{\mathbf{r}}\cdot\mathbf{v}}{c}\right) - \frac{\mathbf{v}}{c}\right] \\
\mathbf{E}_a &= \frac{q}{4\pi\varepsilon_0 c^2 r}\left[\left(\hat{\mathbf{r}}\cdot\frac{d\mathbf{v}}{dt'}\right)\hat{\mathbf{r}} - \frac{d\mathbf{v}}{dt'}\right] = \frac{q}{4\pi\varepsilon_0 c^2 r}\hat{\mathbf{r}} \times \left(\hat{\mathbf{r}} \times \frac{d\mathbf{v}}{dt'}\right) \\
\mathbf{B}_v &= \frac{\mu_0}{4\pi}\frac{q\mathbf{v}\times\mathbf{r}}{r^3} \qquad \mathbf{B}_a = \frac{q}{4\pi\varepsilon_0 c^3 r}\frac{d\mathbf{v}}{dt'} \times \hat{\mathbf{r}}
\end{aligned} \tag{15.229}$$

with the expressions evaluated at the retarded time $t' = t - r/c$.

We will now discuss the nature of the various fields. The first term of $\mathbf{E}_v$, $q\hat{\mathbf{r}}/4\pi\varepsilon_0 r^2$, is just the Coulomb field of q taken at the origin. In fact the first two terms are radial and can be looked at as a Coulomb field of an effective velocity dependent charge $Q = q(1 + 3\hat{\mathbf{r}}\cdot\mathbf{v}/c)$. The last contribution to $\mathbf{E}_v$ is in the direction of the velocity itself. The field $\mathbf{B}_v$ in this approximation, on the other hand, is just the Biot-Savart field.

The acceleration fields, $\mathbf{E}_a$ and $\mathbf{B}_a$ are proportional to $1/r$, and thus they dominate at large distances. One can estimate the distance r_0 beyond which this happens. Since $v/c \ll 1$, then at $r = r_0$ we can take

$$E_v \cong E_a \cong \frac{q}{4\pi\varepsilon_0 r_0^2} = \frac{q}{4\pi\varepsilon_0 c^2 r_0}\frac{dv}{dt'}$$

or

$$r_0 = \frac{c^2}{dv/dt'}$$

Taking $v^2 \approx 2(dv/dt')R_0$, then one can write

$$r_0 \approx R_0\left(\frac{c^2}{v^2}\right)$$

We now discuss the radiation emitted by the charge. If one makes the identification $q(d\mathbf{v}/dt) = \ddot{\mathbf{p}}$, where $\mathbf{p} = q\mathbf{r}'$ is the dipole moment of the charge, then one can easily show that the Poynting vector and the total power radiated by the charge are the same as those given by Eqs. (15.198) and (15.200).

Example 15.15 Pulsed Emission

If a fast-moving charge of velocity $\mathbf{v} = v_0\hat{\mathbf{z}}$ stops ($\mathbf{v}_f = 0$) in a time interval τ under the influence of a constant deceleration $\mathbf{a} = -a_0\hat{\mathbf{z}}$, then an observer located at distance r and angle θ will detect radiation from the charge during a time interval that is not necessarily equal to τ. From Eq. (15.219) we find that

$$dt = \left[1 - \frac{v(t')}{c}\cos\theta\right]dt'$$

Taking $v(t') = v_0 - a_0 t'$ and integrating gives

$$t = t' - \frac{v_0}{c}t'\cos\theta + \frac{1}{2c}a_0 t'^2\cos\theta$$

If at $t' = \tau$, $v(t') = v_f = 0$, then $a_0\tau = v_0$; hence

$$t = \tau\left(1 - \frac{v_0}{2c}\cos\theta\right) \tag{15.230}$$

This result indicates that the measured radiation pulse at the observation point depends on the angle θ, with the two equal ($t = \tau$) at $\theta = \pi/2$.

Example 15.16 Self-Force of a Slow Accelerated Charge

In this example we consider a particle which has a spherically symmetric charge and undergoing an accelerated motion. If the motion is translational—that is, if the acceleration is along the velocity and if the motion is bounded ($v/c \ll 1$)—then we can use the results of Example 15.14 to determine the force exerted by an element dq_1 on another element dq_2 of the particle. We will consider only the effect of the part of the electric field that depends on the acceleration, and write

$$d\mathbf{F}(t) = -dq_2\,d\mathbf{E} = -\frac{dq_1\,dq_2}{4\pi\varepsilon_0 c^2 r}\left[\left(\hat{\mathbf{r}}\cdot\frac{d\mathbf{v}}{dt'}(t')\right)\hat{\mathbf{r}} - \frac{d\mathbf{v}}{dt'}(t')\right] \tag{15.231}$$

where r is the distance between the two elements and $\hat{\mathbf{r}} = \mathbf{r}/r$, and $t' = t - r/c$ is the retarded time. The total force can be calculated by integrating $d\mathbf{F}(t)$ over all the elements, but because $v(t')$ and t' depend on r as a result of the retardation effects, this operation is not straightforward in general. Let us here consider the effect of retardation to the lowest order in $t' - t$; that is, when expanding $d\mathbf{v}(t')/dt'$ we write

$$\frac{d\mathbf{v}(t')}{dt'} = \frac{d\mathbf{v}(t)}{dt} - \frac{r}{c}\frac{d^2\mathbf{v}(t)}{dt} + \cdots \tag{15.232}$$

Substituting the expansion in Eq. (15.231) and integrating, we solve for $\mathbf{F} = \mathbf{F}_1 + \mathbf{F}_2$, as follows:

$$\mathbf{F}_1 = -\frac{1}{4\pi\varepsilon_0 c^2}\int \frac{dq_1 dq_2}{r}\left[\left(\hat{\mathbf{r}}\cdot\frac{d\mathbf{v}(t)}{dt}\right)\hat{\mathbf{r}} - \frac{d\mathbf{v}(t)}{dt}\right] = -\frac{4}{3}\frac{U_0}{c^2}\frac{d\mathbf{v}}{dt}$$

with

$$U_0 = \frac{1}{8\pi\varepsilon_0}\int \frac{dq_1\, dq_2}{r}$$

the electrostatic energy of the charge, and

$$\mathbf{F}_2 = \frac{1}{4\pi\varepsilon_0 c^3}\int dq_1\, dq_2\left[\left(\hat{\mathbf{r}}\cdot\frac{d^2\mathbf{v}}{dt^2}\right)\hat{\mathbf{r}} - \frac{d^2\mathbf{v}}{dt^2}\right] = \frac{q^2}{6\pi\varepsilon_0 c^3}\frac{d^2\mathbf{v}}{dt^2}$$

In both integrals we took $d\mathbf{v}/dt$ and $d^2\mathbf{v}/dt^2$ to be along the z axis, and wrote $\hat{\mathbf{r}} = \sin\theta\cos\phi\,\hat{\mathbf{x}} + \sin\theta\sin\phi\,\hat{\mathbf{y}} + \cos\theta\,\hat{\mathbf{z}}$, and $\hat{\mathbf{r}}\cdot d\mathbf{v}/dt = \cos\theta\, d\mathbf{v}/dt$. The integration over angles then gives nonvanishing contributions (4/3) along $\hat{\mathbf{z}}$ only. Thus

$$\mathbf{F} = -\frac{4}{3}\frac{U_0}{c^2}\frac{d\mathbf{v}}{dt} + \frac{q^2}{6\pi\varepsilon_0 c^3}\frac{d^2\mathbf{v}}{dt^2} \tag{15.233}$$

The factor $4U_0/3c^2$ has the units of mass and if we call it m_0', an effective "electromagnetic mass of the particle" (see Example 6.1), then $\mathbf{F}_1$ has the nature of inertial forces. The force $\mathbf{F}_2$, on the other hand, is that of radiation damping, which was discussed in Example 15.12. It is independent of the structure of the charge and hence remains constant as the size of the charge vanishes. Higher-order correction to Eq. (15.232) can be shown to vanish, however, in the point charge limit. In the same limit, m_0' becomes infinite inasmuch as it depends roughly on $1/r$.

15.7 Summary

Radiation from a moving charge distribution can be calculated by two methods. In one method the potentials and then the fields (and hence the radiation) are calculated. In the second method, the fields are calculated directly without a need for the potentials. Using $\nabla\cdot\mathbf{B} = 0$ and $\nabla\times\mathbf{E} = -\partial\mathbf{B}/\partial t$, we arrive at

$$\mathbf{B} = \nabla\times\mathbf{A} \quad \text{and} \quad \mathbf{E} = -\nabla\Phi - \frac{\partial\mathbf{A}}{\partial t} \tag{15.1),(15.4}$$

Using $\nabla\cdot\mathbf{E} = \rho/\varepsilon$ and $\nabla\times\mathbf{H} = \partial\mathbf{D}/\partial t + \mathbf{J}_f$, as well as the above relations between the fields and the potentials, we find that one can arrive at independent differential equations for $\mathbf{A}$ and Φ if some extra conditions called gauges are used. In the Lorentz gauge, one requires

$$\nabla\cdot\mathbf{A} + \varepsilon\mu\frac{\partial\Phi}{\partial t} = 0 \tag{15.8}$$

The potentials in this gauge satisfy the following inhomogeneous wave equations

$$\left(\nabla^2 - \varepsilon\mu\frac{\partial^2}{\partial t^2}\right)\begin{pmatrix}\mathbf{A}\\ \Phi\end{pmatrix} = -\begin{pmatrix}\mu\mathbf{J}_f\\ \rho_f/\varepsilon\end{pmatrix} \tag{15.9),(15.10}$$

In the Coulomb gauge (transverse gauge) we require

$$\nabla \cdot \mathbf{A} = 0 \tag{15.11}$$

resulting in the following equations for the potentials:

$$\nabla^2 \Phi = -\frac{\rho_f}{\varepsilon} \tag{15.12}$$

$$\nabla^2 \mathbf{A} - \mu\varepsilon \frac{\partial^2 \mathbf{A}}{\partial t^2} = -\mu \mathbf{J}_t \tag{15.16}$$

where

$$\mathbf{J}_t = \frac{1}{4\pi} \nabla \times \nabla \times \int \frac{\mathbf{J}_f}{|\mathbf{r} - \mathbf{r}'|} dv' \tag{15.15}$$

is the transverse component of $\mathbf{J}$, having the property $\nabla \times \mathbf{J}_t = 0$. Note that $\mathbf{J}_l = \mathbf{J}_f - \mathbf{J}_t$ is longitudinal where $\nabla \cdot \mathbf{J}_l = 0$.

The fields $\mathbf{E}$ and $\mathbf{B}$ derived from given $\mathbf{A}$ and Φ can still be the same (gauge invariance) under gauge transformations of the form

$$\mathbf{A}' = \mathbf{A} + \nabla\psi \qquad \text{and} \qquad \Phi' = \Phi - \frac{\partial \psi}{\partial t} \tag{15.21),(15.22}$$

where ψ is a scalar function. For $\mathbf{A}'$ and Φ' to satisfy the Lorentz condition, ψ must satisfy the equation

$$\nabla^2 \psi - \varepsilon\mu \frac{\partial^2 \psi}{\partial t^2} = -\left(\nabla \cdot \mathbf{A} + \varepsilon\mu \frac{\partial \Phi}{\partial t}\right) \tag{15.23}$$

which reduces to the homogeneous wave equation if $\mathbf{A}$ and Φ already satisfy the Lorentz condition:

$$\nabla^2 \psi - \varepsilon\mu \frac{\partial^2 \psi}{\partial t^2} = 0 \tag{15.24}$$

Thus it is always possible to satisfy the Lorentz gauge.

The solution of the wave equations in the Lorentz gauge gives the retarded scalar and vector potentials (in vacuum); that is,

$$\Phi(\mathbf{r}, t) = \frac{1}{4\pi\varepsilon_0} \int \frac{\rho(\mathbf{r}', t')}{|\mathbf{r} - \mathbf{r}'|} dv' \tag{15.43}$$

$$\mathbf{A}(\mathbf{r}, t) = \frac{\mu_0}{4\pi} \int \frac{\mathbf{J}(\mathbf{r}', t')}{|\mathbf{r} - \mathbf{r}'|} dv' \tag{15.47}$$

where t' is the retarded time, given by

$$t' = t - \frac{|\mathbf{r} - \mathbf{r}'|}{c} \tag{15.44}$$

Thus it is seen that apart from the introduction of the retarded time, the potentials are identical in form to what we encountered in the absence of propagation.

In the Coulomb gauge, Φ satisfies exactly Poisson's equation of electrostatics except that ρ_f is now a function of time. The solution for Φ is therefore the instantaneous Coulomb expression

$$\Phi(\mathbf{r}, t) = \frac{1}{4\pi\varepsilon_0} \int \frac{\rho(r', t)}{|\mathbf{r} - \mathbf{r}'|} dv' \tag{15.13}$$

The vector potential in the Coulomb gauge, however, is the retarded vector potential with $\mathbf{J}_f$ replaced by the transverse component $\mathbf{J}_t$. The fields are then calculated from $\mathbf{A}$ and Φ using $\mathbf{B} = \nabla \times \mathbf{A}$, and $\mathbf{E} = -\nabla\Phi - \partial\mathbf{A}/\partial t$.

In regions where $\rho_f = 0$ and $\mathbf{J}_f = 0$, one can directly solve the wave equations of either **E** (or **B**) and use Maxwell's equations to calculate **B** (or **E**). Thus in vacuum and for monochromatic radiation ($\mathbf{E} \to \mathbf{E}e^{-i\omega t}$), we get

$$\nabla^2\mathbf{E} + k^2\mathbf{E} = 0 \qquad k = \frac{\omega}{c} \tag{15.59}$$

This vector equation is called the Helmholtz vector equation. We should note that it is hard to solve directly. The following relation, however, transforms it to the Helmholtz scalar equation:

$$\mathbf{E} = \mathbf{r} \times \nabla\psi \qquad \nabla^2\psi + k^2\psi = 0 \qquad \text{(TE)} \quad (15.60),(15.69)$$

where ψ is a scalar function. The electric field is fixed to be perpendicular to **r**, and as such it is called the transverse electric (TE) case. The **B** field in this case takes the form

$$\mathbf{B} = -\frac{i}{\omega}\nabla \times (\mathbf{r} \times \nabla\psi) \qquad \text{(TE)} \tag{15.71}$$

If the roles of **E** and **B** are interchanged in this procedure, then we get the transverse magnetic (TM) case; that is,

$$\mathbf{B}' = \frac{\mathbf{r} \times \nabla\psi}{c} \qquad \mathbf{E}' = \frac{ic}{\omega}\nabla \times (\mathbf{r} \times \nabla\psi) \qquad \text{(TM)} \quad (15.73),(15.75)$$

with ψ satisfying the above Helmholtz scalar equation.

A multipole expansion of ψ using the method of separation of variables gives in principle an infinite series solution. The lowest order terms are

$$\psi_{00} = C_0^0 \frac{e^{ikr}}{ikr} \qquad \text{and} \qquad \psi_{10} = -C_1^0 \frac{e^{ikr}}{kr}\left(1 + \frac{i}{kr}\right)\cos\theta \tag{15.77}$$

where C_0^0 and C_1^0 are constants. The function ψ_{00} is spherically symmetric; hence it produces no fields, whereas ψ_{10} has azimuthal symmetry, and the fields produced by it in the TE and TM cases correspond to those of a magnetic dipole and an electric dipole along the z axis respectively. Such fields are complicated since they involve retardation effects.

Applications of either the potential technique or the field technique include oscillatory dipoles in vacuum and in the presence of conducting planes. Scattering of electromagnetic waves by small conducting spheres, differential and quarter-wave antennas, slowly moving bounded charge distributions, and fast-moving charges.

The potentials, and hence the fields, produced by a differential antenna of length l carrying a periodic current $I(t)$ of period T are those of an electric dipole when the point of observation at distances $r \gg l$, and if $l \ll cT = \lambda$ (dipole approximation). The effective dipole moment is $p = lI_0/\omega$, where $\omega = 2\pi/T$ and I_0 is the amplitude of the current. The total average power radiated by the antenna is

$$\langle P \rangle = \frac{1}{2}RI_0^2 \tag{15.152}$$

where $R = 789(l/\lambda)^2$ ohms is an effective radiation resistance. The fields of a longer antenna can be determined by integrating the fields of a differential antenna. For a half-wave antenna the effective radiation resistance is 73.1 Ω.

For slowly moving bounded charge distributions (that is, $v \ll c$ and $r' \ll r$) the leading terms of the potentials are

$$\Phi(\mathbf{r}, t) = \frac{1}{4\pi\varepsilon_0}\left[\frac{Q}{r} + \frac{\hat{\mathbf{r}}\cdot\mathbf{p}}{r^2} + \frac{\hat{\mathbf{r}}\cdot\dot{\mathbf{p}}}{cr}\right] \tag{15.174}$$

$$\mathbf{A}(\mathbf{r}, t) = \frac{\mu_0}{4\pi r}\dot{\mathbf{p}} = \frac{\mu_0}{4\pi r}\int J\left(\mathbf{r}', t - \frac{r}{c}\right)dv' \tag{15.178}$$

where Q is the total charge, $\mathbf{p} = \mathbf{p}(t - r/c) = \int \mathbf{r}'\rho(\mathbf{r}', t - r/c)dv'$ is the retarded dipole moment, and $\dot{\mathbf{p}}$ is the derivative of $\mathbf{p}$ with respect to its argument. Thus the electric and magnetic fields of such a distribution are those of a point charge, and of an electric dipole to lowest order. The radiation fields, and hence $\mathbf{S}$, are

$$\mathbf{B}_R = -\frac{\mu_0}{4\pi cr}\hat{\mathbf{r}} \times \ddot{\mathbf{p}} \quad \text{and} \quad \mathbf{E}_R = -c\hat{\mathbf{r}} \times \mathbf{B}_R \qquad (15.192),(15.195)$$

$$\mathbf{S} = \frac{\mu_0}{16\pi^2 cr^2}\hat{\mathbf{r}}|\hat{\mathbf{r}} \times \ddot{\mathbf{p}}|^2 = \frac{\hat{\mathbf{r}}|\ddot{\mathbf{p}}|^2 \sin^2\theta}{16\pi\varepsilon_0 c^3 r^2} \qquad (15.197)$$

Integrating over an entire sphere gives the total power radiated:

$$P = \frac{1}{4\pi\varepsilon_0}\frac{2}{3}\frac{|\ddot{\mathbf{p}}|^2}{c^3} \qquad (15.199)$$

For a slowly moving point charge, $\ddot{\mathbf{p}} = q\dot{\mathbf{v}}$,

$$P = \frac{1}{4\pi\varepsilon_0}\frac{2}{3}\frac{q^2|\dot{\mathbf{v}}|^2}{c^3} \qquad (15.200)$$

For an unlocalized point charge q moving at a speed $v \approx c$, the multipole expansion cannot be used. The retarded potential, however, can be integrated for this charge exactly. The result is

$$\Phi(\mathbf{r}, t) = \frac{q}{4\pi\varepsilon_0}\frac{1}{R(t') - \boldsymbol{\beta}(t')\cdot\mathbf{R}(t')} \qquad \mathbf{A}(\mathbf{r}, t) = \frac{\boldsymbol{\beta}(t')}{c}\Phi(\mathbf{r}, t) \quad (15,209),(15.211)$$

where $\boldsymbol{\beta}(t') = \mathbf{v}(t')/c$, t' is the retarded time, and $R(t') = |\mathbf{r} - \mathbf{r}'|$. The fields of the charge are quite complicated. They are related to each other by

$$\mathbf{B} = \frac{\hat{\mathbf{n}}(t')}{c} \times \mathbf{E} \qquad (15.215)$$

where $\hat{\mathbf{n}}(t') = \mathbf{R}(t')/R(t')$. The power radiated in a solid angle $d\Omega$ as measured by the observer is not the same power that was radiated by the charge since $t \neq t'$. They are related by the factor

$$\frac{dt}{dt'} = 1 - \frac{\boldsymbol{\beta}\cdot\mathbf{R}}{R} = 1 - \beta\cos\theta \qquad (15.219)$$

where θ is the angle between $\boldsymbol{\beta}$ and $\mathbf{R}$.

Problems

15.1 Show by direct substitution that $\Phi = f(t - r/c)/r$ is a solution of the scalar wave equation.

15.2 (a) Determine the vector potential at point (0, 0) of the current loop shown in Fig. 15.11, which carries a current $I(t)$. (b) Evaluate $\mathbf{A}$ and $\mathbf{E}$ if $I(t) = \alpha t$.

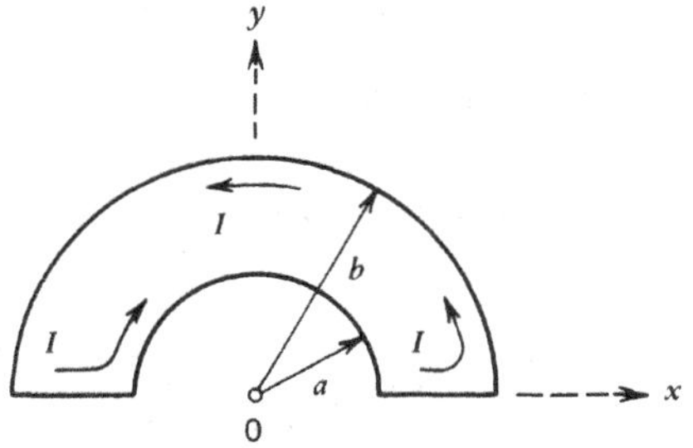

Figure 15.11

15.3 In a medium an electromagnetic field is described in terms of the vector and scalar potentials $\mathbf{A} = \hat{\mathbf{z}}(\frac{1}{2})(x^2 + y^2)\sin \alpha t + \nabla\Psi$ and $\Phi = -\partial\Psi/\partial t$, where Ψ is an arbitrary function and α is a constant. (a) Find **E** and **B**. (b) Show that if one puts $\Psi \equiv 0$, the Lorentz condition is satisfied and **A** and Φ will satisfy the wave equation [Eqs. (15.9) and (15.10)]. (c) Find the force on the medium due to the field as a function of position and time.

15.4 Show that there is a possible solution of Maxwell's equation in free space of the form (a) $\mathbf{E} = \nabla \times \nabla \times (\hat{\mathbf{K}}\Phi)$, $\mathbf{B} = (1/c^2)\nabla \times (\hat{\mathbf{K}}\, \partial\Phi/\partial t)$ or (b) $\mathbf{B} = \nabla \times \nabla(\hat{\mathbf{K}}\Phi)$, $\mathbf{E} = -\nabla \times (\hat{\mathbf{K}}\, \partial\Phi/\partial t)$, where $\hat{\mathbf{K}}$ is a unit constant vector and Φ satisfies the scalar wave equation. (See Example 15.4.)

15.5 An electric charge Q is distributed in a continuous spherically symmetric distribution in a bounded region of space. The distribution undergoes radial oscillations. Show that

$$\mathbf{E} = \frac{1}{4\pi\varepsilon_0}\frac{Q}{r^2}\hat{\mathbf{r}}$$

and $\mathbf{B} = 0$.

15.6 A current I flows in a wire described by the following parametric equations

$$x = \frac{1 + t^4}{(1 + t^2)^2} \qquad y = \frac{2t}{(1 + t^2)^{3/2}} \qquad z = \frac{2t^3}{(1 + t^2)^2} \qquad -\infty < t < \infty$$

where t is just a parameter. Show that the retarded potential at the origin is equal to that at an infinite distance from the wire.

15.7 Determine the vector potential of two infinitely long, parallel wires, 1 and 2, at distances ρ_1 and ρ_2 from an observation point P in the plane of the wires. The distance between the wires is d. The current I_0 in wire 1 is switched on at $t = 0$ and the current I_0 in wire 2 is switched on at $t = t_0$. Derive an equation for the time at which the vector potentials of the wires at P will be equal. Plot the time dependence of the vector potential of wire 1.

15.8 Determine the equation of the line of force produced by a time-dependent electric dipole $\hat{\mathbf{z}}p(t)$ (discussed in Example 15.3) in a plane containing the origin and the z axis. Compare the result with the lines of force of a static electric dipole.

15.9 An electric dipole $p(t) = p_0\hat{\mathbf{z}} \cos \omega t$ is placed at the origin. (a) Determine the fields and the radiation along the dipole. (b) Repeat part (a) for the direction normal to the dipole. (c) Give a sketch of the radiation as a function of θ. (d) Plot the total power radiated as a function of the frequency of oscillations, ω.

15.10 A time-dependent magnetic dipole $\mathbf{m} = \mathbf{m}_0 \cos \omega t$ is placed at a distance b from a large conducting plate such that $b \ll \lambda$, where λ is the wavelength of the emitted radiation. Determine the power radiated when the dipole is (a) parallel and (b) normal to the plate. How do they compare to the power emitted in the absence of the plate?

15.11 Consider two electric dipoles placed at the origin in the x-y plane. One of the dipoles is $\mathbf{p}_1 = p_0\hat{\mathbf{x}} \cos \omega t$ and the other is $\mathbf{p}_2 = p_0(\cos \phi_0\hat{\mathbf{x}} + \sin \phi_0\hat{\mathbf{y}})\sin \omega t$ (see Example 15.6). Determine the polarization of the radiation in the following cases: (a) $\theta = 90°$, (b) $\theta = 0, \pi$ and $\phi_0 = \pi/2$, (c) $\theta = 0, \pi$, and $\phi_0 \neq \pi/2$.

15.12 Derive Eqs. (15.123) to (15.126).

15.13 Derive Eq. (15.133).

15.14 An electromagnetic wave polarized along the x axis and propagating along the z axis is scattered by a very small sphere. Calculate the power radiated along the x axis (see Example 15.7).

15.15 A conducting wire placed at the origin along the z axis carries a current $I_0 \sin \omega t$, where $\omega = 5 \times 10^{10}$ rad/s, and $I_0 = 1$ A. (a) Give a length for the wire such that the dipole approximation holds. (b) Determine the dipole moment of the wire. (c) Determine the effective radiation resistance of the antenna.

15.16 A linear antenna of length l carries a standing current distribution of amplitude I_0 and frequency ω. The antenna has n current half-waves and the ends of the antenna are nodes (see Section 15.4.2 for a half-wave antenna). (a) Write an expression for the current in the antenna. (b) Determine the electric dipole moment per unit length. (c) Determine the magnetic field produced by the antenna in the radiation zone. (d) Determine the power radiated by the antenna.

15.17 Derive Eq. (15.174) from the vector potential given by Eq. (15.178) and the Lorentz condition.

15.18 Plot the radiation pattern of four half-wave dipoles arranged in the same way the antenna array of Example 15.9 is arranged.

15.19 A half-wave antenna is placed near ground at distance small compared to λ at an angle 60° with the normal to ground. Determine the pattern of the emitted radiation.

15.20 Consider an electric dipole of charges q and $-q$ placed at a distance l from each other. The magnitude of each of the charges varies sinusoidally with time at a frequency ω. (a) Determine the dipole moment of the charges. (b) Determine the current flowing between them. (c) Determine the radiation fields of the charges using Eqs. (15.189) and (15.192). What are these fields along the line joining the charges?

15.21 A charged particle of charge q and acceleration $a\hat{\mathbf{z}}$ is initially moving at low velocities $v_0\hat{\mathbf{z}}$, where $v_0 \ll c$, and a is a constant. (a) Determine the angular distribution of the radiation; how does it depend on v_0? (b) Determine the total power and energy radiated.

15.22 A charged particle of charge q and acceleration $a\hat{\mathbf{z}}$ is initially moving with velocity $v_0\hat{\mathbf{z}}$. (a) Write down the angular dependence of the power radiated. (b) Determine the total power radiated.

15.23 A charged particle of charge q and initial velocity $v_0\hat{\mathbf{z}}$ is uniformly decelerated along its velocity. Using the result of Problem 15.22, determine the total energy radiated.

15.24 Show that the maximum radiation emitted by the charge of Problem 15.23 is observed at an angle θ_0, (with respect to $\mathbf{v}$) which is given by $\cos\theta_0 = (1/4\beta)[(1 + 24\beta^2)^{1/2} - 1]$, where $\beta = v_0/c$. Discuss the limits $\beta \to 0$ and $\beta \to 1$.

15.25 An electric dipole, located at the origin, is rotating in the x-y plane with an angular frequency ω. (a) Decompose the dipole into two oscillating dipoles along the x and y axes. (b) What is the phase difference between the two components of the dipole? (c) Use the result of Example 15.6 to find the angular distribution of the power radiated by the dipole. (d) Use Eq. (15.197) to determine the quantity in part (c).

SIXTEEN

ELECTROMAGNETIC BOUNDARY VALUE PROBLEMS

We have already studied Maxwell's equation in two important cases. In one case, which was treated in Chapter 14, we studied the propagation of plane waves in an infinite space of one type of material with prescribed properties (ε, μ, and σ_c), free of external charge and current distributions. In the second case, which was treated in Chapter 15, we had again an infinite space of one type of material with prescribed properties (ε, μ, and $\sigma_c = 0$—(that is, nonconducting), and charge and current sources. The radiation in the form of nonplane waves from these distributions was derived. In this chapter we study another class of problems, an extension of the first case, wherein we deal with the propagation of plane waves in a space free of external charge and current distributions but filled with several types of materials of different dielectric, magnetic, and conducting properties. Although the treatment of the field equations in each of the materials individually is exactly the same as the procedures we followed in Chapter 14, it is necessary to have some prescribed boundary conditions at the interface to allow matching of the solutions. Applications of this class of problems involve reflection and refraction from dielectric or conducting boundaries, transmission through thin films, propagation in waveguides, and resonant cavities.

16.1 Boundary Conditions on the Fields

The behavior of the static electric and magnetic fields across boundaries were previously derived. It was found that the equation $\nabla \cdot \mathbf{B} = 0$ implies that the component of **B** normal to the boundary is continuous ($B_{1n} = B_{2n}$). Also it was found that the equation $\nabla \cdot \mathbf{D} = \rho_f$ implies that the component of **D** normal to the boundary is discontinuous, with the discontinuity given by $D_{2n} - D_{1n} = \sigma$. The curl equations $\nabla \times \mathbf{E} = 0$ and $\nabla \times \mathbf{H} = \mathbf{J}_f$ on the other hand gave us conditions on the tangential components of the **E** and **H** vectors. It was found that the component of **E** tangent

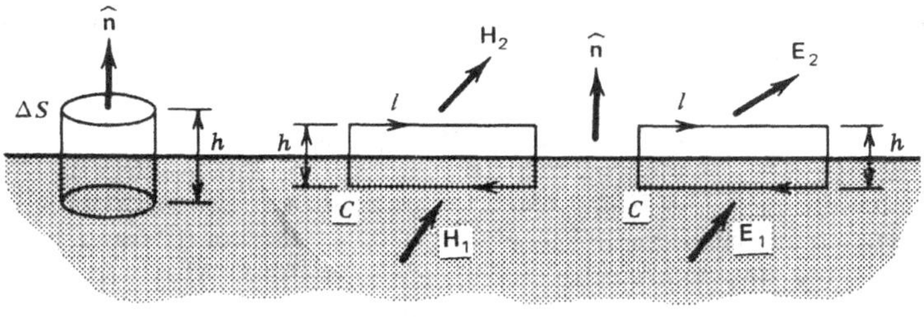

Figure 16.1 Obtaining the boundary conditions on the field vectors and current density as governed by Maxwell's equations at the interface between two media (dielectric or conducting) (*a*) A pillbox-shaped surface at the interface is used to obtain conditions on the normal components of **B**, **D**, and **J**. (*b*) and (*c*) Rectangular paths shown at the interface are used to obtain conditions on the tangential components of **H** and **E**.

to the boundary is continuous ($\mathbf{E}_{1t} = \mathbf{E}_{2t}$) and the component of **H** tangent to the boundary is discontinuous ($\mathbf{H}_{2t} - \mathbf{H}_{1t} = \mathbf{K}_f \times \hat{\mathbf{n}}$). Finally, in the presence of conductors it was found that the continuity equation under steady-state conditions, $\nabla \cdot \mathbf{J} = 0$, implies that the component of the current normal to the boundary is continuous ($J_{1n} = J_{2n}$).

The boundary conditions on the electromagnetic fields governed by Maxwell's equation can be deduced from these equations. Maxwell's divergence equations

$$\nabla \cdot \mathbf{B}(t) = 0 \qquad \text{and} \qquad \nabla \cdot \mathbf{D}(t) = \rho_f(t)$$

are exactly of the same form as two of the equations of magnetostatics and electrostatics, except that **B**, **D**, and ρ_f have time dependence. Therefore we deduce, using Fig. 16.1*a*, that

$$B_{1n}(t) = B_{2n}(t) \tag{16.1}$$

$$D_{2n}(t) - D_{1n}(t) = \sigma_f(t) \tag{16.2}$$

On the other hand, Maxwell's curl equations,

$$\nabla \times \mathbf{E} = -\frac{\partial \mathbf{B}}{\partial t} \qquad \nabla \times \mathbf{H} = \frac{\partial \mathbf{D}}{\partial t} + \mathbf{J}_f$$

are quite different from the static curl equations: $\nabla \times \mathbf{E} = 0$ and $\nabla \times \mathbf{H} = \mathbf{J}_f$; it will, however, turn out that the conditions are exactly the condition of the static case with the time dependence retained in the fields and the current:

$$\mathbf{H}_{2t}(t) - \mathbf{H}_{1t}(t) = \mathbf{K}_f(t) \times \hat{\mathbf{n}} \tag{16.3}$$

$$\mathbf{E}_{1t}(t) = \mathbf{E}_{2t}(t) \tag{16.4}$$

To prove these conditions on the tangential components of $\mathbf{E}(t)$ and $\mathbf{H}(t)$ we follow the same procedure followed in the static case utilizing Fig. 16.1*b* and 16.1*c*. For example, in the case of **H** we integrate $\nabla \times \mathbf{H} = \mathbf{J}_f + \partial \mathbf{D}/\partial t$ over a surface area, ΔS, bounded by the rectangular curve C shown in Fig. 16.1*a*. The rectangle, which has a length l and width h is partially immersed in medium 2. Thus

$$\int_{\Delta S} \nabla \times \mathbf{H} \cdot d\mathbf{a} = \int_{\Delta S} \mathbf{J}_f \cdot d\mathbf{a} + \int_{\Delta S} \frac{\partial \mathbf{D}}{\partial t} \cdot d\mathbf{a}$$

By means of Stokes' theorem, the left-hand side of this equation can be transformed to a line integral over the curve C. Thus

$$\oint_C \mathbf{H}\cdot d\mathbf{l} = \int_{\Delta S} \mathbf{J}_f \cdot d\mathbf{a} + \int_{\Delta S} \frac{\partial \mathbf{D}}{\partial t}\cdot d\mathbf{a}$$

To determine the change in H in passing through the interface we take the width of the rectangle to be very small. When this is done, the contribution to the line integral from the sides normal to the boundary vanishes. Moreover, if $\partial\mathbf{D}/\partial t$ is bounded, then its integral over ΔS vanishes in this limit. Thus

$$(\mathbf{H}_2 - \mathbf{H}_1)\cdot\mathbf{l} = \lim_{h\to 0} \mathbf{J}_f \cdot \Delta\mathbf{S} = (\mathbf{K}_f \times \hat{\mathbf{n}})\cdot\mathbf{l}$$

where $\mathbf{K}_f$ is the *surface* current density. Thus,

$$(\mathbf{H}_2 - \mathbf{H}_1)_t = \mathbf{K}_f \times \hat{\mathbf{n}} \tag{16.5}$$

or

$$\hat{\mathbf{n}} \times (\mathbf{H}_2(t) - \mathbf{H}_1(t)) = \mathbf{K}_f(t) \tag{16.6}$$

A similar integration of $\nabla \times \mathbf{E} = -\partial\mathbf{B}/\partial t$ over the rectangle shown in Fig. 16.1*b* yields the condition in Eq. (16.4) provided $\partial\mathbf{B}/\partial t$ is bounded. Finally, integrating the continuity condition $\nabla\cdot\mathbf{J} = -\partial\rho/\partial t$ over the column of the pillbox shown in Fig. 16.1*a*, using the divergence theorem and taking the limit $h\to 0$ gives [see Eq. (7.4)]

$$(\mathbf{J}_2 - \mathbf{J}_1)\cdot\hat{\mathbf{n}}\,\Delta S = -\int_{h\to 0} \frac{\partial\rho}{\partial t}\,dv = -\frac{\partial\sigma}{\partial t}\Delta S$$

or

$$J_{2n} - J_{1n} = -\frac{\partial\sigma}{\partial t} \tag{16.7}$$

Because $\mathbf{J} = \sigma_c\mathbf{E}$ for ohmic materials and $\mathbf{D} = \varepsilon\mathbf{E}$, for linear materials, Eqs. (16.2) and (16.7) give two conditions for the normal component of the electric field at the interface. If the fields in the media are to be physical, they have to satisfy both conditions simultaneously. Moreover, if only monochromatic radiation is considered, the surface charge density will vary as $e^{-i\omega t}$, then $\partial\sigma_f/\partial t = -i\omega\sigma_f$. Thus Eqs. (16.2) and (16.7) become

$$\varepsilon_2 E_{2n} - \varepsilon_1 E_{1n} = \sigma_f \tag{16.8}$$

$$\sigma_{2c} E_{2n} - \sigma_{1c} E_{1n} = i\omega\sigma_f \tag{16.9}$$

16.1.1 Special Cases: Normal Component

Partially Conducting Materials. For an arbitrary nonzero σ_f, Eqs. (16.8) and (16.9) can be combined by eliminating σ_f. Thus

$$\varepsilon_1\left(1 + i\frac{\sigma_{c1}}{\varepsilon_1\omega}\right)E_{1n} = \varepsilon_2\left(1 + i\frac{\sigma_{c2}}{\varepsilon_2\omega}\right)E_{2n}$$

or

$$\left(1 + i\frac{\sigma_{c1}}{\varepsilon_1\omega}\right)D_{1n} = \left(1 + i\frac{\sigma_{c2}}{\varepsilon_2\omega}\right)D_{2n} \tag{16.10}$$

This result indicates that not only does the magnitude of the normal components of **E** and **D** change across the boundary but so do their phases. Also we can say that the normal component of **D**, in general, will not be continuous because free charge will necessarily build up at the interface.

Dielectric-Dielectric Interface. When both materials have zero conductivity, then Eq. (16.9) indicates that $\sigma_f = 0$; hence Eq. (16.8) becomes

$$D_{1n} = D_{2n} \tag{16.11}$$

Conductor-Dielectric Interface. If medium 1 is highly conducting (that is, σ_{c1} is very large), then E_{1n} has to be zero so that $\sigma_{1c}E_{1n}$ in Eq. (16.9) may stay finite. Thus

$$E_{1n} = 0 \qquad \text{and} \qquad D_{2n} = \sigma_f \tag{16.12}$$

Partially Conducting Media with $\varepsilon_1\sigma_{c2} = \varepsilon_2\sigma_{c1}$. If the dielectric and conducting properties of the media satisfy the special condition

$$\frac{\varepsilon_1}{\sigma_{c1}} = \frac{\varepsilon_2}{\sigma_{c2}}$$

then the interface will not support any surface charges ($\sigma_f = 0$) regardless of the magnitude of the electric field or the current, and in this special case D_n will be continuous.

16.1.2 Special Cases: Tangential Component

Dielectric-Dielectric Interface. If the materials are nonconducting, then $\sigma_{c1} = \sigma_{c2} = 0$; hence there will be no induced currents, and consequently Eq. (16.5) or (16.6) gives

$$\mathbf{H}_{1t} = \mathbf{H}_{2t} \tag{16.13}$$

Partially Conducting Media. When both materials are partially conducting, then currents are induced in the materials. However, they will produce no surface current density because the conductivities are finite. Consequently Eq. (16.5) again gives, in the absence of external surface currents,

$$\mathbf{H}_{1t} = \mathbf{H}_{2t} \tag{16.13}$$

Conductor-Dielectric Interface. When one of the media is highly conducting, then surface currents can be supported (see Example 14.11), and hence the H field will be discontinuous across the interface. To determine the correct boundary condition in this case we use the boundary condition on the curresponding electric field in addition to Maxwell's equations. The fields in region 1 are related to each other by

$$\nabla \times \mathbf{H}_1 = \mathbf{J}_{f1} + \frac{\partial \mathbf{D}_1}{\partial t}$$

Taking $\mathbf{D}_1 = \varepsilon_1 \mathbf{E}_1 e^{-i\omega t}$ and $\mathbf{J}_{f1} = \sigma_{c1}\mathbf{E}_1$ in this equation gives

$$\mathbf{E}_1 = \frac{\nabla \times \mathbf{H}_1}{\sigma_{c1} - i\omega\varepsilon_1} \tag{16.14}$$

On the other hand in the same region $\nabla \times \mathbf{E}_1 = -\partial \mathbf{B}_1/\partial t$. Taking $\mathbf{B}_1 = \mu_1 \mathbf{H}_1 e^{-i\omega t}$ gives

$$\mathbf{H}_1 = \frac{\nabla \times \mathbf{E}_1}{i\omega\mu_1} \tag{16.15}$$

Table 16.1 **Summary of the Boundary Conditions***

Interface	$\mathbf{E}_t$	D_n	$\mathbf{H}_t$	B_n
Dielectric-dielectric	$\mathbf{E}_{1t} = \mathbf{E}_{2t}$	$D_{1n} = D_{2n}$	$\mathbf{H}_{1t} = \mathbf{H}_{2t}$	$B_{1n} = B_{2n}$
Dielectric-highly	$\mathbf{E}_{1t} = 0$	$D_{1n} = 0$	$\mathbf{H}_{1t} = 0$	$B_{1n} = 0$
conducting $\sigma_{c1} \to \infty$	$\mathbf{E}_{2t} = 0$	$D_{2n} = \sigma$	$\mathbf{H}_{2t} = \mathbf{K} \times \hat{\mathbf{n}}$	$B_{2n} = 0$
Conducting-conducting	$\mathbf{E}_{1t} = \mathbf{E}_{2t}$	$\left(1 + i\frac{\sigma_{c1}}{\varepsilon_1 \omega}\right)D_{1n} = \left(1 + i\frac{\sigma_{c2}}{\varepsilon_2 \omega}\right)D_{2n}$	$\mathbf{H}_{1t} = \mathbf{H}_{2t}$	$B_{1n} = B_{2n}$

* We assume that there is no external surface charges or surface currents at the interface.

Equation (16.14) implies that in a highly conducting medium (σ_{c1} is very large), and if $\mathbf{H}_1$ is differentiable and bounded, the electric field $\mathbf{E}_1$ is zero. This means that both the tangential and normal components are zero; that is,

$$\mathbf{E}_{1t} = 0 \quad \text{and} \quad D_{1n} = \varepsilon_1 E_{1n} = 0 \qquad (\sigma_{c1} \to \infty) \tag{16.16}$$

When the electric field in a highly conducting medium is zero, the $\mathbf{H}$ field in turn will be zero according to Eq. (16.15). This means that both tangential and normal components of $\mathbf{H}$ are zero; that is,

$$\mathbf{H}_{1t} = 0 \quad \text{and} \quad B_{1n} = \mu_1 H_{1n} = 0 \qquad (\sigma_{c1} \to \infty) \tag{16.17}$$

Substituting this result in Eq. (16.5) gives

$$\mathbf{H}_{2t} = \mathbf{K} \times \hat{\mathbf{n}} \qquad (\sigma_{c1} = \infty) \tag{16.18}$$

In Table 16.1 we summarize the boundary conditions discussed in this section for the case where there is no external free surface charge or surface current at the interface.

16.2 Propagation Across a Plane Interface of Nonconducting (Dielectric) Materials

The boundary conditions derived above in Section 16.1 can now be used to study the propagation of plane electromagnetic waves across a plane interface between two nonpermeable, nonconducting materials. We will first treat normal incidence, then oblique incidence.

16.2.1 Normal Incidence

Consider an electromagnetic wave of frequency ω and amplitude E_1 polarized along the x axis and propagating (along the z axis) normal to the plane interface of two simple (linear) dielectric materials of refractive indices n_1 and n_2, as shown in Fig. 16.2. At the interface, the wave will be partially reflected and the rest is transmitted if there are no losses in medium 2. The reflection phenomenon is a well-known one and indeed necessary if the boundary conditions are to be satisfied. Hence we

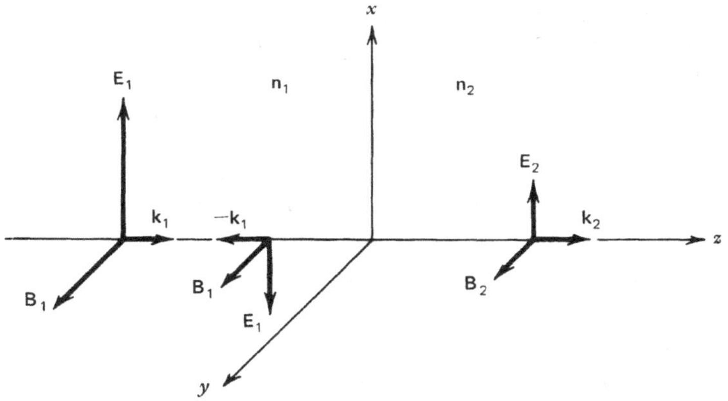

Figure 16.2 Transmission and reflection from a dielectric interface at normal incidence.

assume the presence of three waves. The frequency of the reflected and transmitted waves will be taken equal to the incident frequency. Such a case will turn out to be a must if the materials are linear (simple) and if the boundary conditions are to be satisfied at all times. Thus we write

Incident wave:	$E_1\hat{\mathbf{x}}e^{i(k_1z-\omega t)}$	$B_1\hat{\mathbf{y}}e^{i(k_1z-\omega t)}$
Reflected wave:	$-E'_1\hat{\mathbf{x}}e^{-i(k_1z+\omega t)}$	$B'_1\hat{\mathbf{y}}e^{-i(k_1z+\omega t)}$
Transmitted wave:	$E_2\hat{\mathbf{x}}e^{i(k_2z-\omega t)}$	$B_2\hat{\mathbf{y}}e^{i(k_2z-\omega t)}$

where the amplitude of the B field is related to the amplitude of the E field by $B = E/v = nE/c$ and $k = n\omega/c$. We now apply the boundary conditions.

1. The continuity of the tangential component of the electric field at $z = 0$ gives

$$E_1 - E'_1 = E_2 \tag{16.19}$$

This condition reaffirms the requirement that the frequencies have to be equal if the boundary condition is to be satisfied at all times.

2. The continuity of the tangential components of $\mathbf{H} = \mathbf{B}/\mu_0$ at $z = 0$ gives:

$$B_1 + B'_1 = B_2 \tag{16.20}$$

or, using $B = nE/c$,

$$n_1(E_1 + E'_1) = n_2E_2 \tag{16.21}$$

Solving Eqs. (16.19) and (16.21) simultaneously for E'_1 and E_2 gives

$$r_{12p} = \frac{E'_1}{E_1} = \frac{n_2 - n_1}{n_2 + n_1} \qquad t_{12p} = \frac{E_2}{E_1} = \frac{2n_1}{n_2 + n_1} \tag{16.22}$$

where r_{12p} is the *Fresnel reflection coefficient* and t_{12p} is the *Fresnel transmission coefficient*. In terms of the intensities of the reflected and transmitted waves we now define the quantities reflectance and transmittance, R and T, as follows:

$$R = \frac{\langle S'_1\rangle}{\langle S_1\rangle} \qquad T = \frac{\langle S_2\rangle}{\langle S_1\rangle} \tag{16.23}$$

where $\langle S_1 \rangle$, $\langle S_1' \rangle$ and $\langle S_2 \rangle$ are the time averages of the intensities of the incident, reflected and transmitted waves. Since, $\langle S \rangle = n\langle E^2 \rangle/\mu c$, then Eqs. (16.22) and (16.23) give

$$R = r_{12p}^2 \qquad T = \frac{n_2}{n_1} t_{12p}^2 \tag{16.24}$$

Since in the above treatment the interface was taken to be lossless, we expect the energy to be conserved. Indeed, if we use the explicit expressions for r_{12p} and t_{12p}, we will find that $T + R = 1$.

Let us now consider a more general polarization of the incident wave (elliptical polarization):

$$\mathbf{E} = [E_{1x}\hat{\mathbf{x}} + E_{1y}\hat{\mathbf{y}}]e^{i(k_1 z - \omega t)}$$

We note that the plane of incidence (plane defined by the normal to the interface, and the wave vector of the incident wave) is not unique in the normal incidence case. In the oblique incidence case, however, it is unique. We call the component of the wave polarized in the plane of incidence *p polarization* and the component polarized normal to the plane of incidence, *s polarization* and use them to label the Fresnel coefficients. Since in the case of normal incidence both the s and the p polarizations are tangent to the plane of the interface, then there is no distinction between their respective transmission coefficients. As will be shown below we have

$$r_{12s} = -r_{12p} \qquad t_{12p} = t_{12s} \tag{16.25}$$

16.2.2 Oblique Incidence—Phase Matching

The normal incidence is a special case of the phenomenon of reflection and transmission at an interface. In this subsection we consider the more general case of oblique incidence, in which we will find, contrary to the normal incidence case, that the Fresnel coefficients for s and p polarizations are different. Such a difference is very useful in practical applications, such as producing polarized waves.

***p* Polarization.** We will first analyze the propagation of p polarization. Consider Fig. 16.3, which shows this case schematically, with an incident, a reflected, and a transmitted wave. The reflected and transmitted waves are taken to propagate in the

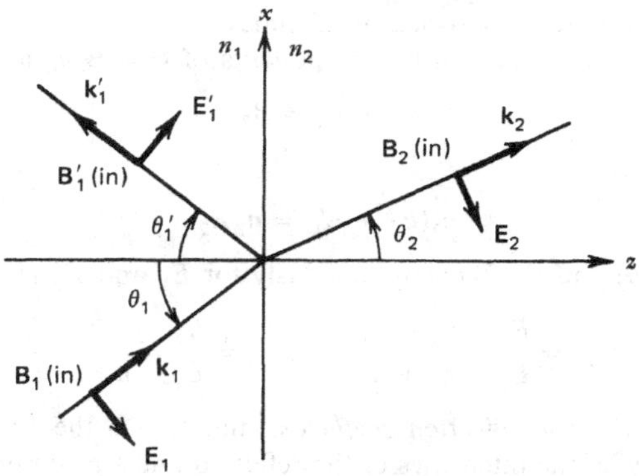

Figure 16.3 Reflection and refraction at oblique incidence at a dielectric interface with the incident electric field in the plane of incidence (p polarization), the plane of the page ($x - z$ plane).

plane of incidence which is the plane of $\mathbf{k}_1$ and the normal to the interface (the plane of the page). Moreover, as we did in the normal-incidence case, we take the frequencies of the three to be identical.* Thus for the electric and magnetic fields we write

$$\text{Incident wave:} \qquad \mathbf{E}_1 e^{i(\mathbf{k}_1\cdot\mathbf{r}-\omega t)} \qquad \frac{n_1}{c}\hat{\mathbf{k}}_1 \times \mathbf{E}_1 e^{i(\mathbf{k}_1\cdot\mathbf{r}-\omega t)} \tag{16.26}$$

$$\text{Reflected wave:} \qquad \mathbf{E}'_1 e^{i(\mathbf{k}'_1\cdot\mathbf{r}-\omega t)} \qquad \frac{n_1}{c}\hat{\mathbf{k}}'_1 \times \mathbf{E}'_1 e^{i(\mathbf{k}'_1\cdot\mathbf{r}-\omega t)} \tag{16.27}$$

$$\text{Transmitted wave:} \qquad \mathbf{E}_2 e^{i(\mathbf{k}_2\cdot\mathbf{r}-\omega t)} \qquad \frac{n_2}{c}\hat{\mathbf{k}}_2 \times \mathbf{E}_2 e^{i(\mathbf{k}_2\cdot\mathbf{r}-\omega t)} \tag{16.28}$$

where $\mathbf{k}_1$, $\mathbf{k}'_1$, and $\mathbf{k}_2$ are the propagation vectors of the waves with corresponding unit vectors $\hat{\mathbf{k}}_1$, $\hat{\mathbf{k}}'_1$, and $\hat{\mathbf{k}}_2$ and corresponding angles θ_1, θ'_1, and θ_2 with respect to the normal to the interface.

We now apply the boundary conditions. (1) At the interface ($z = 0$), the tangential component of the electric field is continuous. Thus

$$E_1 \cos\theta_1\, e^{i(\mathbf{k}_1\cdot\boldsymbol{\rho})} - E'_1 \cos\theta'_1\, e^{i(\mathbf{k}'_1\cdot\boldsymbol{\rho})} = E_2 \cos\theta_2\, e^{i(\mathbf{k}_2\cdot\boldsymbol{\rho})} \tag{16.29}$$

where $\boldsymbol{\rho}$ is the radius vector in the plane of the interface. (2) At the interface ($z = 0$), the tangential component of the $\mathbf{H}$ field is continuous. Thus

$$n_1 E_1 e^{i\mathbf{k}_1\cdot\boldsymbol{\rho}} + n_1 E'_1 e^{i\mathbf{k}'_1\cdot\boldsymbol{\rho}} = n_2 E_2 e^{i\mathbf{k}_2\cdot\boldsymbol{\rho}} \tag{16.30}$$

We note that these two conditions on the tangential components are sufficient to determine the fields everywhere; the conditions on the normal components will give redundant information (show this). Note how these equations differ from the normal incidence case by observing the phase factors $e^{i\mathbf{k}\cdot\boldsymbol{\rho}}$. For these two conditions to be satisfied at any point on the interface, the phase factors have to be first matched and then the amplitudes have to be related appropriately. Thus *phase matching* requires

$$\mathbf{k}_1\cdot\boldsymbol{\rho} = \mathbf{k}'_1\cdot\boldsymbol{\rho} = \mathbf{k}_2\cdot\boldsymbol{\rho} \tag{16.31}$$

and the amplitudes are related as follows:

$$E_1\cos\theta_1 - E'_1\cos\theta'_1 = E_2\cos\theta_2 \qquad n_1(E_1 + E'_1) = n_2 E_2 \tag{16.32}$$

where θ_1, θ'_1 and θ_2 are the angles of incidence, reflection, and refraction respectively. Let us first discuss the consequences of phase matching. Since $\boldsymbol{\rho} = -\hat{\mathbf{n}} \times (\hat{\mathbf{n}} \times \boldsymbol{\rho})$, where $\hat{\mathbf{n}}$ ($\hat{\mathbf{z}}$ in this case) is a unit vector normal to the interface and away from material 1, then

$$\mathbf{k}\cdot\boldsymbol{\rho} = -\mathbf{k}\cdot[\hat{\mathbf{n}} \times (\hat{\mathbf{n}} \times \boldsymbol{\rho})] = -(\mathbf{k} \times \hat{\mathbf{n}})\cdot(\hat{\mathbf{n}} \times \boldsymbol{\rho})$$

Hence Eq. (16.31) becomes

$$(\mathbf{k}_1 \times \hat{\mathbf{n}})\cdot(\hat{\mathbf{n}} \times \boldsymbol{\rho}) = (\mathbf{k}'_1 \times \hat{\mathbf{n}})\cdot(\hat{\mathbf{n}} \times \boldsymbol{\rho}) = (\mathbf{k}_2 \times \hat{\mathbf{n}})\cdot(\hat{\mathbf{n}} \times \boldsymbol{\rho})$$

or

$$\mathbf{k}_1 \times \hat{\mathbf{n}} = \mathbf{k}'_1 \times \hat{\mathbf{n}} = \mathbf{k}_2 \times \hat{\mathbf{n}} \tag{16.33}$$

The vector nature of this result reaffirms our assumptions that $\mathbf{k}_1$, $\mathbf{k}'_1$, and $\mathbf{k}_2$ are coplanar; its magnitude, on the other hand, gives

$$k_1 \sin\theta_1 = k'_1 \sin\theta'_1 = k_2 \sin\theta_2 \tag{16.34}$$

* This is only valid in the case of linear (simple) materials. When the materials have dielectric constants that are functions of the field (such as what are called nonlinear crystals), harmonics of the frequency of the incident wave may be generated.

Because both incident and reflected waves travel in the same medium, their propagation vectors must have the same magnitudes; that is, $k_1 = k_1'$. Hence the equality $k_1 \sin \theta_1 = k_1' \sin \theta_1'$ gives $\theta_1 = \theta_1'$. This relation is the *law of reflection*: The angle of incidence equals the angle of reflection. Since $k = \omega/v = n\omega/c$, the other equality of (Eq. 16.34), $k_1 \sin \theta_1 = k_2 \sin \theta_2$, gives

$$n_1 \sin \theta_1 = n_2 \sin \theta_2 \tag{16.35}$$

which is the *law of refraction*, or *Snell's law*.

The Fresnel reflection and transmission coefficients can now be determined by solving Eq. (16.32) simultaneously for E_1'/E_1 and E_2/E_1; that is,

$$r_{12p} = \frac{n_2 \cos \theta_1 - n_1 \cos \theta_2}{n_2 \cos \theta_1 + n_1 \cos \theta_2} \qquad t_{12p} = \frac{2n_1 \cos \theta_1}{n_2 \cos \theta_1 + n_1 \cos \theta_2} \tag{16.36}$$

Using Snell's law one can eliminate the refractive indices from the expressions

$$r_{12p} = \frac{\tan(\theta_1 - \theta_2)}{\tan(\theta_1 + \theta_2)} \qquad t_{12p} = \frac{2 \cos \theta_1 \sin \theta_2}{\sin(\theta_1 + \theta_2)\cos(\theta_1 - \theta_2)} \tag{16.37}$$

s Polarization. In the s polarization case, the electric field of the incident wave is normal to the plane of incidence, and therefore the magnetic field is in the plane of incidence, as shown in Fig. 16.4. Note that in this case the polarization of the magnetic field changes upon reflection and transmission instead of that of the electric field, as is the case in the p polarization. As in the p polarization case, however, the frequencies of the three waves are taken the same, and the propagation vectors are taken coplanar (in the plane of incidence). The continuity of the tangential components of **E** and **H** gives

$$E_1 e^{i(\mathbf{k}_1 \cdot \boldsymbol{\rho})} + E_1' e^{i(\mathbf{k}_1' \cdot \boldsymbol{\rho})} = E_2 e^{i(\mathbf{k}_2 \cdot \boldsymbol{\rho})} \tag{16.38}$$

$$n_1 E_1 \cos \theta_1 e^{i(\mathbf{k}_1 \cdot \boldsymbol{\rho})} - n_1 E_1' \cos \theta_1' e^{i(\mathbf{k}_1' \cdot \boldsymbol{\rho})} = n_2 E_2 \cos \theta_2 e^{i(\mathbf{k}_2 \cdot \boldsymbol{\rho})} \tag{16.39}$$

We first invoke phase matching: $\mathbf{k}_1 \cdot \boldsymbol{\rho} = \mathbf{k}_1' \cdot \boldsymbol{\rho} = \mathbf{k}_2 \cdot \boldsymbol{\rho}$. This is the same equality we encountered in the case of p polarization. Hence the law of reflection and Snell's law apply here too. Moreover, this equality reaffirms the assumption that $\mathbf{k}_1$, $\mathbf{k}_1'$, and $\mathbf{k}_2$ are coplanar. Phase matching reduces Eqs. (16.38) and (16.39) to

$$E_1 + E_1' = E_2 \tag{16.40}$$

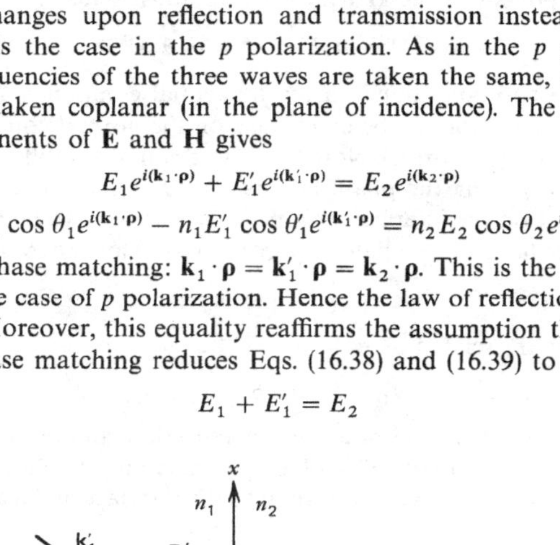

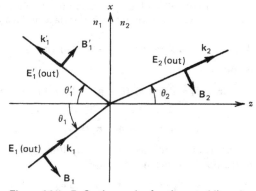

Figure 16.4 Reflection and refraction at oblique incidence at a dielectric interface with the incident electric field perpendicular to the plane of incidence (s polarization), the plane of the page ($x - z$ plane).

and

$$n_1 E_1 \cos\theta_1 - n_1 E_1' \cos\theta_1 = n_2 E_2 \cos\theta_2 \tag{16.41}$$

The Fresnel reflection and transmission coefficients are determined by solving these equations simultaneously for E_1'/E_1 and E_2/E_1, respectively; that is,

$$r_{12s} = \frac{n_1 \cos\theta_1 - n_2 \cos\theta_2}{n_1 \cos\theta_1 + n_2 \cos\theta_2} \qquad t_{12s} = \frac{2n_1 \cos\theta_1}{n_1 \cos\theta_1 + n_2 \cos\theta_2}. \tag{16.42}$$

Using Snell's law, the refractive indices can be eliminated from the expressions for r_{12s} and t_{12s}; as follows:

$$r_{12s} = \frac{\sin(\theta_2 - \theta_1)}{\sin(\theta_2 + \theta_1)} \qquad t_{12s} = \frac{2\cos\theta_1 \sin\theta_2}{\sin(\theta_2 + \theta_1)} \tag{16.43}$$

We will now examine some limiting cases of the above results.

Normal Incidence. At normal incidence (that is, $\theta_1 = \theta_2 = 0$), Eqs. (16.36) and (16.42) give:

$$r_{12p} = -r_{12s} = \frac{n_2 - n_1}{n_2 + n_1} \qquad t_{12p} = t_{12s} = \frac{2n_1}{n_2 + n_1} \tag{16.44}$$

This result is what we stated previously in Eq. (16.25). The reflection and the transmission coefficients for both s and p are the same since they depend on the square of r_{12} and t_{12}, respectively. As we noted before, the plane of incidence in the normal incidence case is not defined and thus the two reflection coefficients should be the same.

Grazing-Angle Incidence. In the *grazing-angle incidence*, the incoming wave strikes the interface at an angle $\theta_1 \approx \pi/2$; thus Eqs. (16.36) and (16.42) give

$$r_{12p} = r_{12s} = -1 \qquad t_{12p} = t_{12s} = 0 \tag{16.45}$$

Hence

$$R_p = R_s = 1 \qquad T_p = T_s = 0 \tag{16.46}$$

indicating total reflection for both polarizations.

Now if medium 1 is air ($n_1 = 1$) and medium 2 is ordinary glass ($n_2 = 1.5$), then $R_p = R_s = 0.04$ at normal incidence. Thus the reflectivity of a dielectric surface increases considerably as the incidence varies from normal to grazing.

Critical Incidence—Total Internal Reflection. What is called *critical-angle incidence* occurs when the wave is incident on the interface from the region of higher refractive index and when the refraction angle θ_2 is $\pi/2$. Consider Eqs. (16.36) and (16.42). Taking $n_1 > n_2$, and taking θ_2 to be equal to the largest possible value (that, is $\pi/2$), we obtain

$$r_{12p} = 1 \quad \text{and} \quad r_{12s} = 1 \tag{16.47}$$

or

$$R_p = R_s = 1 \tag{16.48}$$

This indicates that the intensity of the reflected wave is the same as the intensity of the incident wave or in other words the wave has suffered a *total internal reflection.*

The incidence angle at which total internal reflection takes place is called the critical angle θ_c and it can be determined from Snell's law. Taking $\theta_2 = \pi/2$ in Eq. (16.35) gives

$$n_1 \sin \theta_c = n_2 \qquad \text{or} \qquad \theta_c = \sin^{-1} \frac{n_2}{n_1} \tag{16.49}$$

Beyond the Critical-Angle Incidence. Because $|\sin \theta| \leq 1$, Snell's law puts restrictions on the possible angles of incidence where straightforward refraction takes place. For example, at incident angles greater than the critical angle, we find that $n_1 \sin \theta_1 = n_2 \sin \theta_2$ implies that

$$\sin \theta_2 = \frac{\sin \theta_1}{\sin \theta_c} = \alpha > 1 \tag{16.50}$$

This condition cannot be satisfied if θ_2 is a real angle. In fact if $\sin \theta_2 > 1$, then $\cos \theta_2$ is pure imaginary; that is,

$$\cos \theta_2 = \sqrt{1 - \sin^2 \theta_2} = i\beta \tag{16.51}$$

where β is a real quantity. Substituting Eq. (16.51) into Eqs. (16.36) and (16.42) gives

$$\hat{r}_{12p} = \frac{n_2 \cos \theta_1 - in_1\beta}{n_2 \cos \theta_1 + in_1\beta} \qquad \hat{r}_{12s} = \frac{n_1 \cos \theta_1 - in_2\beta}{n_1 \cos \theta_1 + in_2\beta} \tag{16.52}$$

where the hats on top of $\hat{r}_{12p}$ and $\hat{r}_{12s}$ are used to indicate that both of these quantities are complex. Complex Fresnel coefficients also arise in the reflection from conducting media (see the upcoming section). Since the reflectance R_p is defined by $\langle S_1' \rangle / \langle S_1 \rangle = |\mathbf{E}_1'^* \times \mathbf{H}_1'| / |\mathbf{E}_1^* \times \mathbf{H}_1|$, then it is as follows:

$$R_p = \hat{r}_{12p}\hat{r}_{12p}^* = |r_{12p}|^2 \qquad \text{and} \qquad R_s = \hat{r}_{12s}\hat{r}_{12s}^* = |r_{12s}|^2 \tag{16.53}$$

Therefore, Eqs. (16.52) and (16.53) indicate (show it) that regardless of the value of β, and hence for any $\theta_1 > \theta_c$, the reflectance for both p and s polarizations reaches unity; that is,

$$R_p = R_s = 1 \qquad \theta_1 > \theta_c \tag{16.54}$$

Thus total internal reflection takes place for all angles greater or equal to the critical angle. Figure 16.5 shows the reflectance for s and p polarization with incidence on the denser of the two transparent media. Figure 16.6, on the other hand, shows the reflectances with incidence on the lighter of the two media, which exhibits the phenomenon of critical incidence.

The concept of total internal reflection has very useful applications. The total reflection prism and the light pipe, shown in Fig. 16.7, are among these applications. When the incidence on face AB of the prism is normal, then the angle of incidence on the hypotenuse face is 45°. Since the critical angle of the crown glass and air interface is 41°, then there is total internal reflection at the hypotenuse. In the second example, the light may propagate in solid dielectric cylinders, as shown if it enters at a large angle appropriate for total internal reflections. When a well-collimated beam of light enters a cylinder made of plastic, the beam may make it through the pipe (see Problem 16.7).

16.2.3 Polarization by Reflection and Refraction—Brewster Angle

The electromagnetic waves emitted by most of the ordinary sources (such as gas discharges, arcs, and incandescent lamps) are unpolarized; that is, the direction of

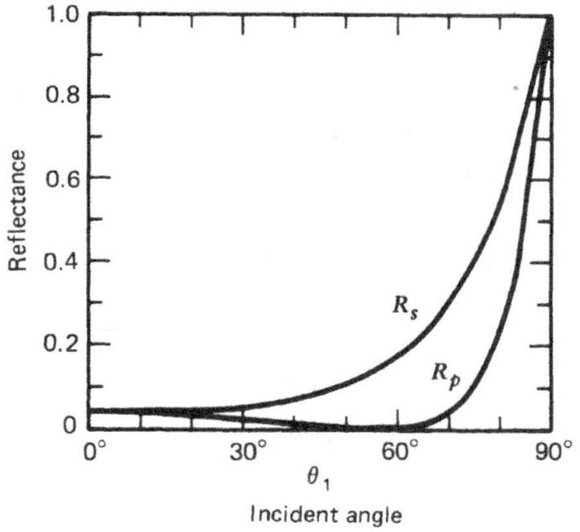

Figure 16.5 The reflectence from a higher refractive index material making a plane interface for both *s* and *p* polarizations, showing the phenomenon of Brewster angle.

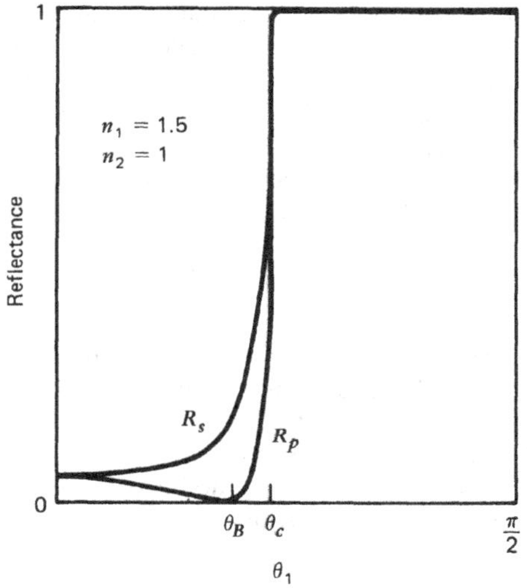

Figure 16.6 The reflectence from a lower refractive index material forming a plane interface for both *s* and *p* polarizations, showing the phenomena of Brewster angle and the critical angle.

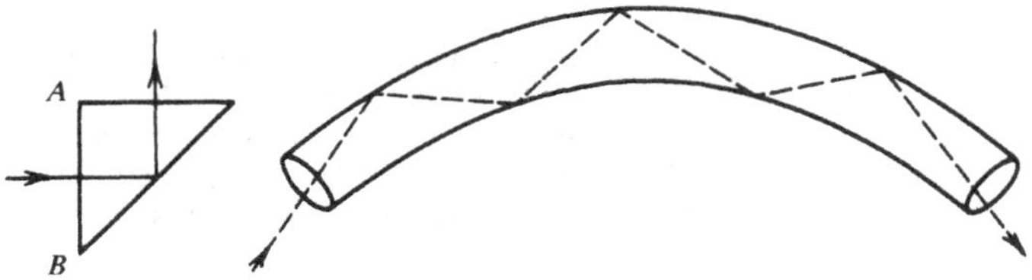

Figure 16.7 Schematic of total internal reflection prisms and light pipes.

polarization is symmetric around the direction of propagation. The light is thus said to be *natural*. In this subsection we show that unpolarized waves can be polarized by treating them in several ways: suitable reflections and/or refractions at certain interfaces between transparent materials. The polarization effect by reflection can be seen by examining the reflection coefficients for p and s waves:

$$r_{12p} = \frac{\tan(\theta_1 - \theta_2)}{\tan(\theta_1 + \theta_2)} \qquad r_{12s} = \frac{\sin(\theta_2 - \theta_1)}{\sin(\theta_2 + \theta_1)}$$

For $\theta_1 + \theta_2 = \pi/2$, $\tan(\theta_1 + \theta_2) = \infty$; hence r_{12p} vanishes. According to Snell's law, $n_1 \sin\theta_1 = n_2 \sin\theta_2$, then the incident angle θ_B at which r_{12p} vanishes is given by

$$n_1 \sin\theta_B = n_2 \sin\left(\frac{\pi}{2} - \theta_B\right) = n_2 \cos\theta_B$$

or

$$\tan\theta_B = \frac{n_2}{n_1} \tag{16.55}$$

The angle θ_B is called the *Brewster angle*, after the discoverer of this phenomenon. It is to be noted that while r_{12p} vanishes at this angle, the reflection coefficient of the s waves does not ($r_{12s} = \cos 2\theta_B \neq 0$). Hence the reflected waves will be purely polarized normal to the plane of incidence. The Brewster angle effect is widely used in the construction of lasers (see Example 16.2), and in various applications, which include the study of structure of atoms and properties of surfaces, among others.

An unpolarized wave can also be polarized by passing through transparent materials. Consider the ratio of t_{12s} and t_{12p}. From Eqs. (16.37) and (16.43) we find

$$\frac{t_{12s}}{t_{12p}} = \cos(\theta_1 - \theta_2) \qquad \text{or} \qquad \frac{T_s}{T_p} = \cos^2(\theta_1 - \theta_2) \tag{16.56}$$

The magnitude of the ratio is equal to 1 for normal incidence ($\theta_1 = \theta_2 = 0$), decreases by increasing the angle of incidence; however, it never vanishes. Repeating this transmission N times gives the ratio $\cos^{2N}(\theta_1 - \theta_2)$ which can be made very small by taking N large. Thus unpolarized waves can be polarized to a certain degree by appropriately choosing N. This method is of particular importance in the infrared part of the electromagnetic spectrum, and in cases where the incident intensity is very high, such as in intense laser beams.

Example 16.1 Boundary Conditions—Current Sheet

The boundary conditions discussed in Section 16.2 can be used to determine the magnetic field when passing across a discontinuity (see Fig. 16.8). Consider, for example, a current sheet of surface density $\mathbf{K} = -10\hat{\mathbf{y}}$ amperes per meter, located in the plane $z = 0$. The permeabilities of the materials filling the $z < 0$ and $z > 0$ spaces are $5\mu_0$ and $2\mu_0$, respectively. The magnetic field in the region $z > 0$ is $\mathbf{H}_2 = 15\hat{\mathbf{x}} + 8\hat{\mathbf{z}}$ A/m. The $\mathbf{B}$ field in region 2 is $\mathbf{B}_2 = 5\mu_0\mathbf{H}_2 = (75\hat{\mathbf{x}} + 40\hat{\mathbf{z}})\mu_0$ teslas. The fields in region 1 can now be determined using the boundary conditions given by Eqs. (16.1) and (16.5). From $B_{2n} = B_{1n}$, we find $B_{1n} = 40\mu_0\hat{z}$ T and hence $H_{1n} = 20\hat{\mathbf{z}}$ A/m. From $(\mathbf{H}_2 - \mathbf{H}_1)_t = \mathbf{K} \times \hat{\mathbf{n}}$, where $\hat{\mathbf{n}}$ is a unit vector normal to the interface and points away from material 1, we find

$$(\mathbf{H}_2 - \mathbf{H}_1)_t = -10\hat{\mathbf{y}} \times \hat{\mathbf{z}} = -10\hat{\mathbf{x}} \text{ A/m}$$

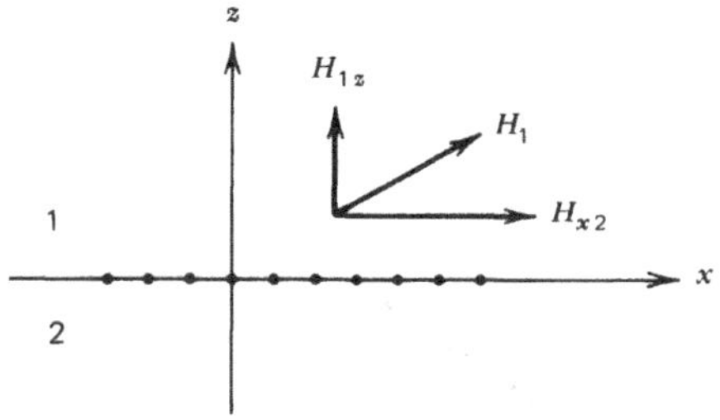

Figure 16.8 Utilization of the boundary conditions to determine the fields across a current sheet.

Thus $H_{1y} = H_{2y} = 0$ and $H_{1x} = H_{2x} + 10 = 25$. Therefore

$$\mathbf{H}_1 = H_n\hat{\mathbf{z}} + H_{1x}\hat{\mathbf{x}} = 20\hat{\mathbf{z}} + 25\hat{\mathbf{x}} \text{ A/m}$$

and

$$\mathbf{B}_1 = \mu_0(40\hat{\mathbf{z}} + 50\hat{\mathbf{x}}) \text{ T}$$

Example 16.2 Design of a He–Ne Plasma Tube—Polarization Effects

The Brewster angle concept is widely used in the construction of lasers for the purpose of producing an output of a single linear polarization. Figure 16.9 shows a schematic diagram of a He–Ne laser. A He–Ne discharge tube is placed between two highly reflecting parallel mirrors (99.9 percent reflectivity). Consider a beam of light bouncing between the mirrors. In every round trip, the beam suffers losses due to reflections at the windows of the discharge tube and due to other effects, and gains more intensity as it passes through the atomic

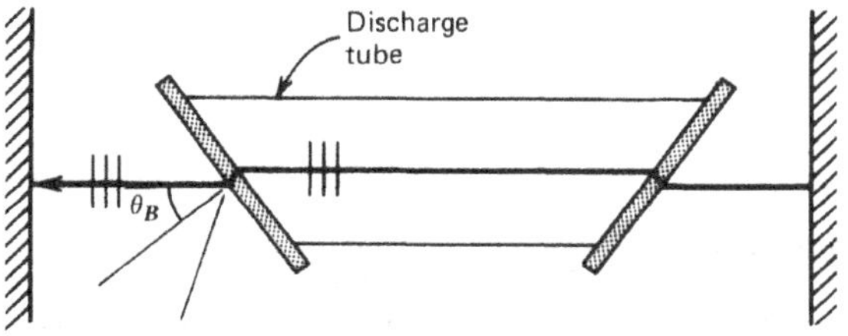

Figure 16.9 Schematic diagram of a He-Ne laser utilizing the Brewster angle phenomenon in order to produce polarized laser beams.

concentration in the discharge. For the laser to work, the gain must exceed the losses. By constructing the windows of the discharge tube at Brewster's angle, as shown, virtually all reflection from the p polarization of the laser light at the surface of the window can be eliminated. The other linear polarization (s polarization), however, suffers high reflection at the windows and therefore gets attenuated below the intensity required for laser oscillation to occur. As a result, only the p polarization bounces between the mirrors many times, with each one resulting in further gain; hence the output of the laser becomes p polarized.

Example 16.3 Rotation of Polarization by Refraction

A light wave falls obliquely at an angle θ_1 on a plane interface between two dielectric media of refractive indices n_1 and n_2 as shown in Fig. 16.10a. The wave is linearly polarized with the

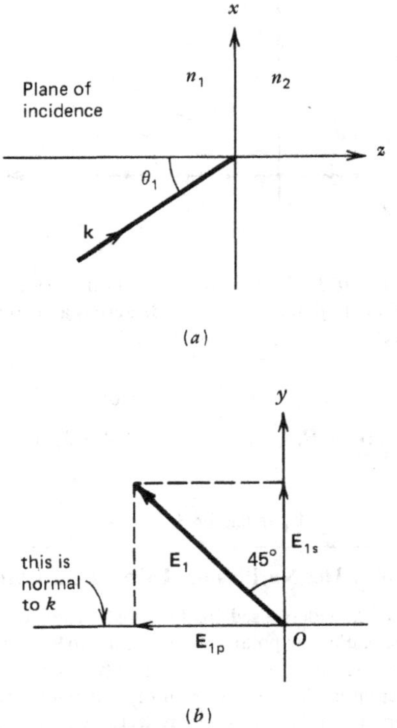

Figure 16.10 Showing the concept of rotation of polarization after refraction (*a*) The plane of incidence is the plane of the page, (*b*) the plane of the page is the plane normal to the plane of incidence and contains the **E** field.

direction of polarization being at 45° to the plane of incidence (see Fig. 16.10*b*). In Fig. 16.10*b*, the plane of the paper is taken as the interface plane; and the plane of incidence is normal to the plane of the paper, with Ox being their intersection. The amplitudes of the incident p and s polarizations are equal and given by $E_{1p} = E_1/\sqrt{2}$ and $E_{1s} = E_1/\sqrt{2}$. After passing the interface we find, from Eqs. (16.37) and (16.43),

$$E_{2p} = t_{12p}E_{1p} = \frac{2\cos\theta_1 \sin\theta_2}{\sin(\theta_1+\theta_2)\cos(\theta_1-\theta_2)}\frac{E_1}{\sqrt{2}}$$

$$E_{2s} = t_{12s}E_{1s} = \frac{2\cos\theta_1 \sin\theta_2}{\sin(\theta_1+\theta_2)}\frac{E_1}{\sqrt{2}}$$

Thus the angle θ' that the polarization of the transmitted wave makes with the plane of incidence is given by

$$\frac{E_{2s}}{E_{2p}} = \cos(\theta_1-\theta_2) = \tan\theta' \qquad \text{or} \qquad \theta' = \tan^{-1}[\cos(\theta_1-\theta_2)]$$

Therefore, the polarization has undergone a rotation $\delta\theta = \theta' - \pi/4$.

Example 16.4 Polarization After Total Reflection

In this example we discuss the state of polarization of a wave after it suffers a total internal reflection from a surface of a dielectric. Consider a linearly polarized wave incident from a region of refractive index n_1 onto a material of refractive index $n_2 < n_1$. We use the notation of Figs. 16.3 and 16.4. Multiplying the numerator and denominator of Eq. (16.52) by the complex conjugate of the denominator gives

$$\hat{r}_{12p} = \frac{E_1'}{E_1} = \frac{(n_2 \cos\theta_1 - in_1\beta)^2}{n_2^2 \cos^2\theta_1 + n_1^2\beta^2} \tag{16.57}$$

The phase of E_1' with respect to E_1, ϕ_p, is defined as follows:

$$\frac{E_1'}{E_1} = |\hat{r}_{12p}|e^{i\phi_p} \quad \text{or} \quad \left(\frac{E_1'}{E_1}\right)^{1/2} = |\hat{r}_{12p}|^{1/2}e^{i\phi_p/2} \tag{16.58}$$

Comparing the second form of Eq. (16.58) with Eq. (16.57), we find that

$$\tan\frac{\phi_p}{2} = -\frac{n_1\beta}{n_2\cos\theta_1} \tag{16.59}$$

Using Eq. (16.51) and Snell's law gives β in terms of θ_1. Hence

$$\tan\frac{\phi_p}{2} = \frac{(\sin^2\theta_1 - n_2^2/n_1^2)^{1/2}}{(n_2^2/n_1^2)\cos\theta_1} \tag{16.60}$$

A similar procedure yields

$$\tan\frac{\phi_s}{2} = \frac{(\sin^2\theta_1 - n_2^2/n_1^2)^{1/2}}{\cos\theta_1} \tag{16.61}$$

Since ϕ_p is in general different from ϕ_s, then the reflected wave is in general elliptically polarized.

We should note that the concept of total internal reflection has an important practical application in optics. When it is incorporated with *nonsimple* dielectric materials, one can construct polarizing prisms; they are called total internal reflection polarizers.

Example 16.5 Transport of Energy in Total Reflection

In this example we investigate the transmitted power in the case of total internal reflection. From the first glance we conclude that the transmittance should be zero since the reflectance is 1 [see (Eq. 16.54)]. However, when we examine $\hat{t}_{12}$ for this case we find that it is not zero; that is,

$$\hat{t}_{12p} = \frac{2n_1\cos\theta_1}{n_2\cos\theta_1 + in_1\beta} \neq 0 \quad \text{and} \quad \hat{t}_{12s} = \frac{2n_1\cos\theta_1}{n_1\cos\theta_1 + in_2\beta} \neq 0$$

This apparent contradiction can be resolved if we examine the propagation in medium 2. The electric field in this region has the form given in Eq. (16.28); that is, $\mathbf{E}_2 \exp[i(\mathbf{k}_2\cdot\mathbf{r} - \omega t)]$. According to the discussion following Eq. (16.51), at an incident angle beyond the critical angle ($\theta_1 > \theta_c$), the refraction angle $\hat{\theta}_2$ is complex, with its cosine purely imaginary. Expanding the scalar product $\mathbf{k}_2\cdot\mathbf{r}$ and using $\cos\hat{\theta}_2 = i\beta$ gives the following expression for the electric field.

$$\mathbf{E}_2 e^{i(k_2\cos\hat{\theta}_2 z + k_2\sin\hat{\theta}_2 x - \omega t)} = \mathbf{E}_2 e^{-\beta k_2 z} e^{i(\alpha k_2 x - \omega t)} \tag{16.62}$$

where $\alpha = \sin\theta_1/\sin\theta_c = (1+\beta^2)^{1/2}$. Thus, for $\theta_1 > \theta_c$, the wave is propagating along the x axis (interface between the two materials), and is *attenuated in the z direction* in the conducting medium. The corresponding magnetic field will also propagate *along the surface* and be attenuated in the medium. The degree of penetration of the fields is governed by the attenuation constant βk_2. In spite of the fact that the fields seem to penetrate the second medium, the *average* Poynting vector normal to the interface vanishes, indicating that there is no power transmitted in the medium (see Problem 16.2).

Example 16.6 Waves Incident on a Slab of a Dielectric or a Layered Interface—Antireflection Coatings and Dielectric Mirrors

A dielectric slab of refractive index n_2, thickness d, and with plane parallel faces, lies between two nonpermeable media of refractive indices n_1 and n_3 as shown in Fig. 16.11. A linearly

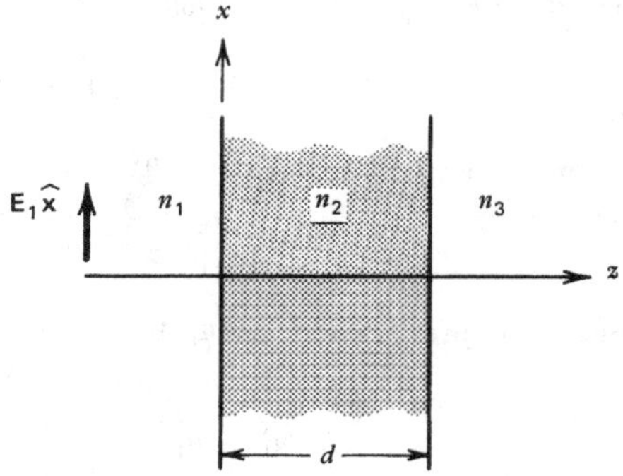

Figure 16.11 Reflection and transmission from a dielectric slab at normal incidence that may be used to explain the phenomena of antireflection coatings and dielectric mirrors.

polarized wave is normally incident on the layered interface. As the wave strikes the $z = 0$ interface, a fraction of the intensity will be reflected due to the discontinuity in the refractive index. Thus, in the $z < 0$ region, two counterpropagating waves exist. The transmitted wave in the region $0 < z < d$ suffers also some reflection at $z = d$ interface; hence two counterpropagating waves exist in this region. In the $z > d$ region, however, only a wave propagating along z exists. Thus we write the following expressions for the fields in the three regions:

$$E_1\hat{x}e^{i(k_1z-\omega t)} - E_1'\hat{x}e^{-i(k_1z+\omega t)} \qquad B_1\hat{y}e^{i(k_1z-\omega t)} + B_1'\hat{y}e^{-i(k_1z+\omega t)} \qquad z < 0$$

$$E_2\hat{x}e^{i(k_2z-\omega t)} - E_2'\hat{x}e^{-i(k_2z+\omega t)} \qquad B_2\hat{y}e^{i(k_2z-\omega t)} + B_2'\hat{y}e^{-i(k_2z+\omega t)} \qquad 0 < z < d$$

$$E_3\hat{x}e^{i(k_3z-\omega t)} \qquad B_3\hat{y}e^{i(k_3z-\omega t)} \qquad z > d$$

where the amplitude of the B field is related to the amplitude of the E field by $B_i = n_iE_i/c$, and $k_i = n_i\omega/c$.

We now apply the continuity of the tangential components of the E field and the tangential components of the H field at $z = 0$ and $z = d$. Because of the nonpermeable nature of the materials, the continuity of the tangential component of H is equivalent to the continuity of the tangential component of B. These conditions give

$$E_1 - E_1' = E_2 - E_2' \qquad n_1(E_1 + E_1') = n_2(E_2 + E_2')$$

$$E_2e^{i\theta_2} - E_2'e^{-i\theta_2} = E_3e^{i\theta_3} \qquad n_2(E_2e^{i\theta_2} + E_2'e^{-i\theta_2}) = n_3E_3e^{i\theta_3}$$

where $\theta_2 = n_2\omega d/c$ and $\theta_3 = n_3\omega d/c$. These equations can be solved simultaneously. For example, one can show that

$$\frac{E_1}{E_3} = \frac{1}{2}\left[\left(1 + \frac{n_3}{n_1}\right)\cos\theta_2 - i\left(\frac{n_2}{n_1} + \frac{n_3}{n_2}\right)\sin\theta_2\right]e^{i\theta_3}$$

The transmitted intensity, therefore, is given by

$$\frac{E_1^2}{E_3^2} = \frac{1}{4}\left(1 + \frac{n_3}{n_1}\right)^2 \cos^2 \theta_2 + \frac{1}{4}\left(\frac{n_2}{n_1} + \frac{n_3}{n_2}\right)^2 \sin^2 \theta_2$$

(See Problem 16.5 for further details of this result.)

This layered interface has an important application in optics. For example, if the medium n_3 is part of an optical system (e.g., a lens) and medium n_1 is air ($n_1 = 1$), then one can choose n_2 and d such that the reflected wave is minimal for a given wavelength λ_0. The optical coating (medium n_2) is called *antireflection coating*. The reflectivity of a glass surface ($n_3 = 1.5$) can be decreased by depositing on it a thin, transparent film of cryolite ($n_2 = 1.36$) or of magnesium fluoride ($n_2 = 1.35$) of thickness d such that $n_2 d = \lambda_0/4$.

If in general one takes $n_1 < n_2 < n_3$, one can show that in normal incidence the reflectivity is minimal for $n_2 d = (2m + 1)\lambda_0/4$, where m is an integer or zero and is given by

$$R_{\text{min}} = \frac{(n_1 n_3 - n_2^2)^2}{(n_1 n_3 + n_2^2)^2}$$

Also, one can show that for $n_2 d = m\lambda_0/2$, the reflectivity is maximal and is given by

$$R_{\text{max}} = \frac{(n_1 - n_3)^2}{(n_1 + n_3)^2}$$

Now the reflectivity of a medium of refractive index n_3 with respect to a medium of index n_1 can be enhanced by introducing between them a thin layer of index $n_2 > n_3$. For example if an appropriate layer of stibnite ($n \approx 3$) or zinc sulfide ($n \approx 2.3$) is deposited on glass, the reflectivity of glass can be increased. A series of films can produce extremely high reflectivities (*dielectric mirrors*), which far exceed the best reflectivities of metallic mirrors (to be discussed in the following section) at optical wavelengths. This property of dielectric mirrors enables them to withstand very high power densities (from high-power lasers) at which metallic mirrors fail (get damaged because of residual absorption). (See Problem 16.6.)

16.3 Propagation Across a Plane Interface of a Conductor and a Dielectric—Complex Fresnel Coefficients

So far we have considered reflection and transmission at a plane interface between two dielectric media. In this section we consider the case where one of the media is conducting. Reflection from metallic interfaces is of considerable practical use; mirrors at optical frequencies and microwave frequencies are based on this principle, with the conductivity of the material used taken very large. For simplicity we will first treat the case of normal incidence, then treat the case of oblique incidence.

16.3.1 Normal Incidence

In Chapter 14 we discussed the propagation of electromagnetic waves inside conducting media. We found that the propagation is described through use of a complex refractive index and a complex propagation vector, both of which are frequency-dependent [Eq. (14.82)].

$$\hat{n} = n_0\left(1 + i\frac{\sigma_c}{\varepsilon\omega}\right)^{1/2} = n + ik \quad \text{and} \quad \hat{k} = \frac{\omega}{c}\hat{n} \tag{14.82}$$

where the hat indicates a complex quantity. The real part of $\hat{n}$, n, accounts for the refractive nature of the medium while the imaginary part, k, accounts for the absorption in the media. We will find in this section that the reflection coefficient from this interface approaches unity as the absorption becomes very weak, and hence provides the concept of *high-reflectivity mirrors.*

Consider an electromagnetic wave of frequency ω and amplitude E_1 polarized along the x axis and propagating along the z axis normal to the plane interface of one dielectric material of permittivity ε_1 (refractive index n_1) and a conducting material of permittivity ε_2 and conductivity σ_{c2} (complex refractive index $\hat{n}_2$). The propagation can be described schematically by Fig. 16.2 (used previously to describe propagation across a dielectric interface), with $\hat{n}_2$ replacing n_2 and $\hat{k}_2 = \hat{n}_2\omega/c$ replacing k_2.

The treatment of this problem is very similar to that of nonconducting media (Section 16.2.1). This statement is based on the fact that the boundary conditions satisfied by the fields are the same in both cases: the tangential components of the E and H are continuous [see Eqs. (16.4) and (16.13)]. In fact, the previous results of Eqs. (16.22) apply to the present problem with n_2 replaced by $\hat{n}_2$; that is,

$$\hat{r}_{12} = \frac{\hat{n}_2 - n_1}{\hat{n}_2 + n_1} \qquad \hat{t}_{12} = \frac{2n_1}{\hat{n}_2 + n_1} \tag{16.63}$$

where the hats over $\hat{r}_{12}$ and $\hat{t}_{12}$ indicate complex quantities. Other convenient forms of these equations can be written by substituting for $\hat{n}_2$ in terms of its real and imaginary parts: $\hat{n}_2 = n_2 + ik_2$. Thus

$$\hat{r}_{12} = \frac{(n_2 - n_1) + ik_2}{(n_2 + n_1) + ik_2} \qquad \hat{t}_{12} = \frac{2n_1}{(n_2 + n_1) + ik_2} \tag{16.64}$$

Because of the complex nature of $\hat{r}_{12}$ and $\hat{t}_{12}$, the reflected and transmitted waves are expected to have phase shifts relative to the incoming wave. In order to determine the shifts, we write the Fresnel coefficients in spherical polar form:

$$\hat{r}_{12} = |\hat{r}_{12}|e^{i\phi_r} \qquad \hat{t}_{12} = |\hat{t}_{12}|e^{i\phi_t} \tag{16.65}$$

where

$$|\hat{r}_{12}| = \left[\frac{(n_2 - n_1)^2 + k_2^2}{(n_2 + n_1)^2 + k_2^2}\right]^{1/2} \qquad |\hat{t}_{12}| = \frac{2n_1}{[(n_2 + n_1)^2 + k_2^2]^{1/2}} \tag{16.66}$$

$$\tan\phi_r = \frac{2n_1k_2}{n_2^2 - n_1^2 + k_2^2} \qquad \tan\phi_t = -\frac{k_2}{n_2 + n_1}$$

It is clear from these results that the reflected and transmitted electric fields are phase shifted relative to the incident electric field. These shifts are just ϕ_r for the reflected field and ϕ_t for the transmitted wave.

Once the Fresnel coefficients are calculated, then the reflection and transmission coefficients can be calculated. Because of the complex nature of the coefficients we should be careful when calculating T. Using the definitions given by Eqs. (16.23) and (16.64) we get

$$R = r_{12}r_{12}^* = |\hat{r}_{12}|^2 \tag{16.67}$$

and

$$T = 1 - R \tag{16.68}$$

Since the transmitted wave will eventually be absorbed in the conducting medium, it is customary to call T the absorption, A; hence we write $A = 1 - R$. Explicitly in terms of the properties of the material, Eqs. (16.67) and (16.64) give for R, and hence for A,

$$A = \frac{4n_1 n_2}{(n_2 + n_1)^2 + k_2^2} \qquad R = 1 - \frac{4n_1 n_2}{(n_2 + n_1)^2 + k_2^2} \tag{16.69}$$

It is interesting to discuss the case of highly conducting materials such as metals ($\sigma_c/\varepsilon\omega \gg 1$). In this case we have $n_2 \approx \sqrt{\sigma_c/2\varepsilon_0\omega} \gg 1$ [see Eq. (14.100)]. Equation 16.69 shows that in this limit, A approaches zero and R approaches unity as follows:

$$A = \frac{2}{k_2} \qquad R = 1 - \frac{2}{k_2} \tag{16.70}$$

or

$$A = 2\sqrt{\frac{2\varepsilon_0\omega}{\sigma_c}} \qquad R = 1 - 2\sqrt{\frac{2\varepsilon_0\omega}{\sigma_c}} \tag{16.71}$$

This relation is called the *Hagen-Rubens* formula. For the materials classified as conductors, this relation is valid at frequencies as high as 10^{15} Hz (optical and infrared frequencies). For moderately good conductors, this relation is accurate at frequencies below the microwave frequencies. Let us use it to determine the reflectivity of aluminum. Taking $38.2 \times 10^6\ (\Omega\cdot\text{m})^{-1}$ for the conductivity and 10^9 rad/s for ω, we find that $R = 0.99976$ and $A = 2.4 \times 10^{-4}$. However at $\omega = 10^{15}$ rad/s, $A = 0.24$, thus showing that aluminum is not highly reflecting at optical frequencies.

16.3.2 Oblique Incidence

The oblique incidence case is much more complicated than the corresponding case of a dielectric-dielectric interface. This complication is due to the fact that both the propagation vector and the angle of the refraction are complex. Although presenting a drawing of the schematic of the propagation is not possible with complex angles, its mathematical analysis should be identical to the dielectric-dielectric case with n_2 and θ_2 replaced by the corresponding complex quantities $\hat{n}_2$ and $\hat{\theta}_2$. Hence Snell's law becomes

$$\hat{n}_2 \sin\hat{\theta}_2 = n_1 \sin\theta_1 \tag{16.72}$$

and the Fresnel coefficients become

$$\hat{r}_{12p} = \frac{\hat{n}_2\cos\theta_1 - n_1\cos\hat{\theta}_2}{\hat{n}_2\cos\theta_1 + n_1\cos\hat{\theta}} \qquad \hat{t}_{12p} = \frac{2n_1\cos\theta_1}{\hat{n}_2\cos\theta_1 + n_1\cos\hat{\theta}} \tag{16.73}$$

$$\hat{r}_{12s} = \frac{n_1\cos\theta_1 - \hat{n}_2\cos\hat{\theta}_2}{n_1\cos\theta_1 + \hat{n}_2\cos\hat{\theta}} \qquad \hat{t}_{12s} = \frac{2n_1\cos\theta_1}{n_1\cos\theta_1 + \hat{n}_2\cos\hat{\theta}_2} \tag{16.74}$$

The reflectance and the transmittance are calculated easily from these coefficients using Eqs. (16.67) and (16.68).

Equations (16.72) to (16.74) are compact and therefore convenient for the calculation of the transmitted or reflected powers; however, it is not convenient to use them as such for the determination of directions of propagation since they involve complex angles and products of complex quantities. It would be very desirable to describe the propagation with real rules, which involve real angles. If this is accomplished, then it will be possible to define a real angle of refraction in the conducting medium and hence make it possible to draw a real diagram of the propagation.

In order to do this, we examine phase matching at the boundary. Replacing $\mathbf{k}_2$ by $\hat{\mathbf{k}}_2$ in Eqs. (16.31) and (16.33) gives

$$\hat{\mathbf{k}}_2 \cdot \boldsymbol{\rho} = \mathbf{k}_1 \cdot \boldsymbol{\rho} \tag{16.75}$$

and

$$\hat{\mathbf{k}}_2 \times \hat{\mathbf{z}} = \mathbf{k}_1 \times \hat{\mathbf{z}} \tag{16.76}$$

where $\hat{\mathbf{z}} = \hat{\mathbf{n}}$ is a unit vector normal to the interface. Writing $\hat{\mathbf{k}}_2$ in terms of its real and imaginary parts $\mathbf{K}_r$ and $\mathbf{K}_i$, $\hat{\mathbf{k}}_2 = \mathbf{K}_r + i\mathbf{K}_i$, equating the real parts, and equating the imaginary parts of Eq. (16.76), we get

$$\mathbf{k}_1 \times \hat{\mathbf{z}} = \mathbf{K}_r \times \hat{\mathbf{z}} \qquad \mathbf{K}_i \times \hat{\mathbf{z}} = 0 \tag{16.77}$$

These equations give conditions on both the directions and magnitudes of the real and imaginary parts of $\hat{\mathbf{k}}_2$.

1. Both $\mathbf{K}_r$ and $\mathbf{K}_i$ are in the plane of incidence.
2. The second relation implies that $\mathbf{K}_i$ is along $\hat{\mathbf{n}}$ or normal to the interface:
$$\mathbf{K}_i = K_i \hat{\mathbf{z}} \tag{16.78}$$
3. The magnitude of the first relation, on the other hand, implies that
$$k_1 \sin \theta_1 = K_r \sin \phi \tag{16.79}$$
where ϕ is the real angle between $\mathbf{K}_r$ and the normal to the interface.

Figure 16.12*a* illustrates the propagation, and shows the planes of constant amplitude and the planes of constant phase. Because the planes of constant phase and constant amplitude are normal to $\mathbf{K}_r$ and $\mathbf{K}_i$, respectively, they make an angle ϕ with each other. However, at normal incidence these planes are coincident as shown in Fig. 16.12*b*.

In order to determine ϕ and the magnitudes of K_r and K_i, we need to determine more relations among them and the parameters of the propagation: k_1, θ_1, and the properties of the conducting material. Since $\hat{\mathbf{k}}_2$ makes an angle $\hat{\theta}$ with the z axis and at the same time its real component makes an angle ϕ with the z axis, whereas its imaginary part is along the z axis, then

$$(K_r \sin \phi)\hat{\mathbf{x}} + (K_r \cos \phi + iK_i)\hat{\mathbf{z}} = (\hat{k}_2 \sin \hat{\theta}_2)\hat{\mathbf{x}} + (\hat{k}_2 \cos \hat{\theta}_2)\hat{\mathbf{z}}$$

Equating the x and z components on both sides and using Eq. (16.72), we get

$$K_r \sin \phi = \hat{k}_2 \sin \hat{\theta}_2 = k_1 \sin \theta_1 \tag{16.80}$$

$$K_r \cos \phi + iK_i = \hat{k}_2 \cos \hat{\theta}_2 \tag{16.81}$$

These equations (which include Snell's law), can now be solved for K_r, K_i and ϕ. The real and imaginary parts of Eq. (16.81) give

$$K_r \cos \phi = \mathrm{Re}(\hat{k}_2 \cos \hat{\theta}_2) \tag{16.82}$$

$$K_i = \mathrm{Im}(\hat{k}_2 \cos \hat{\theta}_2) \tag{16.83}$$

We now evaluate the real and imaginary parts of $\hat{k}_2 \cos \hat{\theta}_2$. Writing $\cos \hat{\theta}_2$ in terms of $\sin \hat{\theta}_2$:

$$\hat{k}_2 \cos \hat{\theta}_2 = (\hat{k}_2^2 - \hat{k}_2^2 \sin^2 \hat{\theta}_2)^{1/2}$$

using

$$\hat{k}_2^2 = \mu_0 \varepsilon_2 \omega^2 + i\mu_0 \sigma_c \omega \qquad \text{and} \qquad \hat{k}_2 \sin \hat{\theta}_2 = k_1 \sin \theta_1,$$

then

$$\hat{k}_2 \cos \hat{\theta}_2 = [(\mu_0 \varepsilon_2 - \mu_0 \varepsilon_1 \sin^2 \theta_1)\omega^2 + i\mu_0 \sigma_c \omega]^{1/2}$$

or, in terms of a modified dielectric property of the conducting material, $\bar{\varepsilon}_2$,

$$\hat{k}_2 \cos \hat{\theta}_2 = [\mu_0 \bar{\varepsilon}_2 \omega^2 + i\mu_0 \sigma_c \omega]^{1/2} \tag{16.84}$$

where

$$\bar{\varepsilon}_2 = \varepsilon_2 - \varepsilon_1 \sin^2 \theta_1 \tag{16.85}$$

Thus we can write

$$\hat{k}_2 \cos \hat{\theta}_2 = \frac{\omega}{c} [\bar{n} + i\bar{k}] \tag{16.86}$$

where $\bar{n}$ and $\bar{k}$ are calculated from the optical constants of the material given by Eq. (14.82) with ε_2 replaced by $\bar{\varepsilon}_2$. Thus

$$K_r \cos \phi + iK_i = \frac{\omega}{c} (\bar{n} + i\bar{k}) \tag{16.87}$$

which upon equating its real parts on both sides and its imaginary parts on both sides gives

$$K_r \cos \phi = \frac{\omega}{c} \bar{n} \qquad \text{and} \qquad K_i = \frac{\omega}{c} \bar{k} \tag{16.88}$$

The first of these equations can now be solved with Eq. (16.79) for K_r. Squaring these equations and adding them gives

$$K_r = \frac{\omega}{c} \sqrt{\bar{n}^2 + n_1^2 \sin^2 \theta_1} \tag{16.89}$$

Thus one can define a real effective refractive index of the conducting medium

$$N(\theta_1) = \sqrt{\bar{n}^2 + n_1^2 \sin^2 \theta_1} \tag{16.90}$$

In terms of $N(\theta_1)$, Eqs. (16.79) and (16.88) become

$$N(\theta_1) \sin \phi = n_1 \sin \theta_1 \tag{16.91}$$

and

$$N(\theta_1) \cos \phi = \bar{n} \tag{16.92}$$

Equation (16.91) is the analog of Snell's law of dielectric materials and is drawn schematically in Fig. 16.12*a*; it can be used to determine the angle of refraction ϕ once the effective refractive index $N(\theta_1)$ is calculated. However, it is a complicated relationship because $N(\theta_1)$ has a complicated dependence on the properties of the conducting material and the incidence angle. In addition, Eq. (16.92) can also be used to determine the angle of refraction. Because of the complicated nature of these equations, it is not useful to discuss them any further; however, some special cases (highly conducting materials) that are of great practical importance will be discussed in a number of examples.

Now that we determined real rules for the propagation of EM waves across a conducting interface, it is interesting to express the fields of the transmitted wave (in the conducting material) in terms of real quantities. Thus we write

$$\mathbf{E}_2 e^{i(\hat{\mathbf{k}}_2 \cdot \mathbf{r} - \omega t)} = \mathbf{E}_2 e^{i(\mathbf{K}_r \cdot \mathbf{r} - \omega t) - K_i z} \tag{16.93}$$

where $\mathbf{K}_r$ makes the angle ϕ with the z axis. The skin depth is $\delta = 1/K_i = c\omega/\bar{k}$.

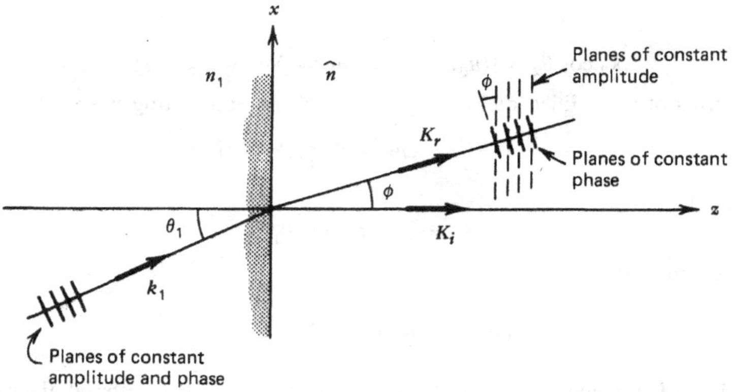

(a)

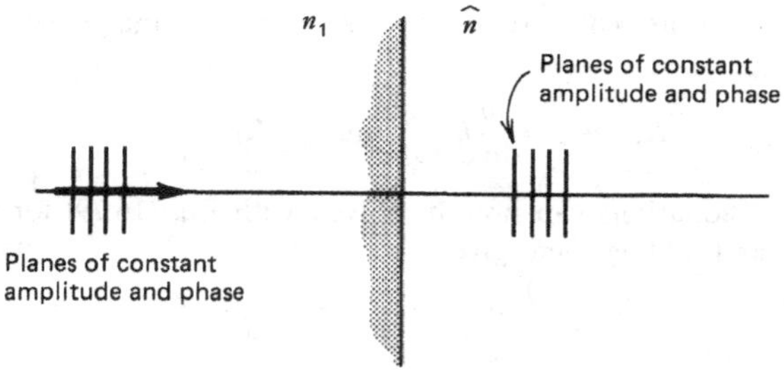

(b)

Figure 16.12 Schematic representation of reflection and refraction from a conducting interface showing the plane of constant amplitude and constant phase. (*a*) Contrary to the dielectric interface, at oblique incidence these planes are at an angle ϕ with respect to each other (angle of refraction) (*b*) At normal incidence these planes are coincident.

Example 16.7 Optical Constants of a Metal at Oblique Incidence

Consider an electromagnetic wave of frequency $\omega = 2\pi$ GHz. It is incident from a dielectric medium of $\varepsilon_1 = 10\varepsilon_0$ on a conducting interface of conductivity $\sigma_c = 0.5\ (\Omega \cdot \text{m})^{-1}$, dielectric constant $\varepsilon_2 = 11.5\varepsilon_0$, with the angle of incidence 30°. From Eq. (16.85), the effective permittivity of the conducting material is $\bar{\varepsilon}_2 = \varepsilon_2 - \varepsilon_1 \sin^2\theta_1 = 9\varepsilon_0$. The optical constants of the material $\bar{n}$ and $\bar{k}$ are calculated from Eq. (14.82) with ε taken to be equal to $\bar{\varepsilon}_2$. This gives $\bar{n} = 3.3$ and $\bar{k} = 1.38$. The effective refractive index of the material $N(\theta_1)$ is calculated from Eq. (16.90), and the result is $N(\theta_1) = 3.66$. Substituting for $N(\theta_1)$, θ_1 in $N(\theta_1)\sin\phi = n_1 \sin\theta_1$ gives 25.6° for the angle of refraction. Note that $N(\theta_1)\cos\phi = \bar{n}$ would give the same result.

Let us calculate the angle of refraction in the limit $\sigma_c = 0$. In this limit, the material is nonconducting; hence Snell's law: $n_2 \sin\theta_2 = n_1 \sin\theta_1$ gives $\theta_2 = 27.8°$. This shows that $\phi < \theta_2$, indicating that the conducting property of the material causes a reduction in the degree of refraction.

Example 16.8 Fresnel Coefficients for a Highly Conducting Interface—Metallic Mirrors

Simplified expressions for the propagation across a plane interface between a dielectric and a highly conducting material can be determined from the results of Section 16.3.2. In order to deduce the results, we need to note that in this case $\sigma_c/\varepsilon\omega \gg 1$, and consequently the complex refractive index $|\hat{n}|$ is much larger than unity. If $|\hat{n}| \gg 1$, then the complex form of Snell's law (Eq. 16.72)—that is, $n_1 \sin\theta_1 = \hat{n}_2 \sin\hat{\theta}_2$—requires that $\hat{\theta}_2 \approx 0$ in order for $\hat{n}_2 \sin\hat{\theta}_2$ to stay finite. Moreover, there are two limiting regimes that need to be treated separately: near-normal incidence, $\pi/2 - \theta_1 \gg 1/|\hat{n}_2|$, and near-grazing-angle incidence, $\pi/2 - \theta_1 \ll 1/|\hat{n}_2|$.

(*i*) *Near-Normal Incidence.* Taking $\hat{\theta}_2 \approx 0$ we find that Eqs. (16.73) and (16.74) reduce to

$$\hat{r}_{12p} = \frac{\hat{n}_2 \cos\theta_1 - n_1}{\hat{n}_2 \cos\theta_1 + n_1} \qquad \hat{r}_{12s} = \frac{n_1 \cos\theta_1 - \hat{n}_2}{n_1 \cos\theta_1 + \hat{n}_2}$$
$$\hat{t}_{12p} = \frac{2n_1 \cos\theta_1}{\hat{n}_2 \cos\theta_1 + n_1} \qquad \hat{t}_{12s} = \frac{2n_1 \cos\theta_1}{n_1 \cos\theta_1 + \hat{n}_2} \tag{16.94}$$

It is useful to write these results in terms of the real and imaginary parts of $\hat{n}_2$. Using Eq. (14.82) we write $\hat{n}_2 = n_2 + ik_2$; thus

$$\hat{r}_{12p} = \frac{(n_2 - n_1/\cos\theta_1) + ik_2}{(n_2 + n_1/\cos\theta_1) + ik_2} \qquad \hat{r}_{12s} = \frac{(n_1 \cos\theta_1 - n_2) - ik_2}{(n_1 \cos\theta_1 + n_2) + ik_2}$$

For a highly conducting material, the relations given in Eq. (16.93) can be reduced further by taking $\hat{n}_2 \gg n_1 \cos\theta_1$ and $\hat{n}_2 \cos\theta_1 \gg n_1$; then

$$\hat{r}_{12p} = 1 - \frac{2n_1}{\hat{n}_2 \cos\theta_1} \qquad \hat{r}_{12s} = \frac{2n_1 \cos\theta_1}{\hat{n}_2} - 1$$
$$\hat{t}_{12p} = \frac{2n_1}{\hat{n}_2} \qquad \hat{t}_{12s} = \frac{2n_1 \cos\theta_1}{\hat{n}_2} \tag{16.95}$$

(*ii*) *Near-Grazing-Angle Incidence.* For near-grazing-angle incidence, where $\phi_0 = \pi/2 - \theta_1 \ll 1/|\hat{n}_2|$ (that is, $\cos\theta_1 \approx \phi_0$), one can show that (see Problem 16.9) the Fresnel coefficients for p polarization take on the expressions

$$\hat{r}_{12p} = \frac{\phi_0 - n_1/\hat{n}_2}{\phi_0 + n_1/\hat{n}_2} \qquad \hat{t}_{12p} = \frac{2n_1\phi_0/\hat{n}_2}{\phi_0 + 1/\hat{n}_2} \tag{16.96}$$

For s polarization we have

$$\hat{r}_{12s} = \frac{2n_1 \cos\theta_1}{\hat{n}_2} - 1 \qquad \hat{t}_{12s} = \frac{2n_1 \cos\theta_1}{\hat{n}_2} \tag{16.97}$$

which are identical to the expressions for near-normal incidence given above in Eqs. (16.95). Finally, in the limit of very high conductivity (*perfect* or *ideal* conductor, $\hat{n}_2 \to \infty$), then $\hat{r}_{12s} \to -1$, and $\hat{r}_{12p} \to 1$ regardless of the variation in the angle of incidence.

Example 16.9 Reflection Coefficient of a Highly Conducting Surface—Metallic Mirrors

In the previous example the Fresnel coefficients for propagation across a highly conducting plane interface were calculated. Using these coefficients, we can now examine the reflectivity from the interface as a function of the angle of incidence. Substituting from Eqs. (16.95) or (16.77) in $R_s = \hat{r}_{12s}\hat{r}^*_{12s}$ and keeping the lowest order in $\hat{n}_2^{-1}$ gives:

$$R_s = 1 - 4n_1\eta_2 \cos\theta_1 \tag{16.98}$$

where $\hat{\eta}_2 = \eta_2 + i\eta'_2 = \hat{n}_2^{-1}$. This result indicates that the reflectivity of electromagnetic waves of s polarization is nearly equal to unity for all angles of incidence, and it reaches its minimum at normal incidence ($\theta_1 = 0$).

The reflectivity of the p polarization for near-normal and grazing angle incidence take on different expressions and can be calculated using $R_p = \hat{r}_{12p}\hat{r}^*_{12p}$ and Eqs. (16.95) and (16.96), respectively. The results (see Problem 16.10) are as follows:

$$R_p = 1 - \frac{4n_1\eta_2}{\cos\theta_1} \qquad \text{(near-normal incidence)}$$

$$R_p = \frac{(\phi_0 - n_1\eta_2)^2 + n_1^2\eta_2'^2}{(\phi_0 + n_1\eta_2)^2 + n_1^2\eta_2'^2} \qquad \text{(grazing incidence)} \tag{16.99}$$

This shows that the reflectivity of metal changes very little as the incidence is varied from normal to grazing and is always very high. Figure 16.13 shows the reflectances at air-

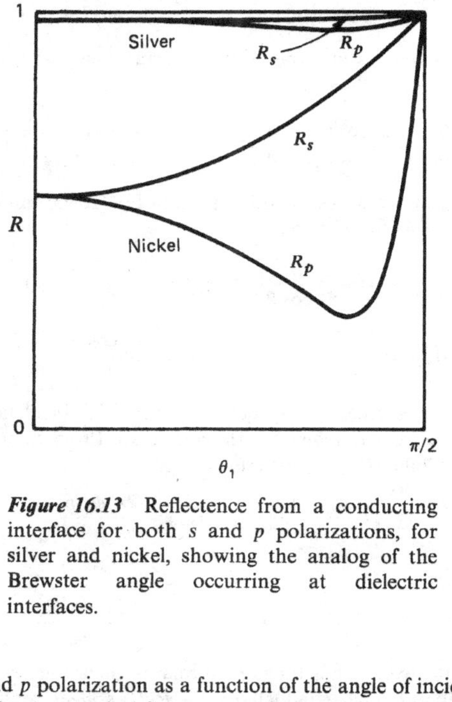

Figure 16.13 Reflectence from a conducting interface for both s and p polarizations, for silver and nickel, showing the analog of the Brewster angle occurring at dielectric interfaces.

metal interface for s and p polarization as a function of the angle of incidence, as governed by Eqs. (16.73) and (16.74), at an optical frequency such that $\hat{n}_2 = 0.05 + 3i$ for silver and $\hat{n}_2 = 2 + 3i$ for nickel. In both cases, R_s continues to rise monotonically as the angle of incidence increases from zero to $\pi/2$. On the other hand, R_p goes through a minimum in both cases. This effect is analogous to the Brewster angle at a dielectric interface, and it is examined analytically in the following example.

Example 16.10 Reflection from Metals—The Analog of the Brewster Angle

When an electromagnetic wave is incident on the boundary of two dielectrics at the Brewster angle, the reflectivity of the p waves vanishes. In this example, we consider the analog of this effect in the case of the reflection from a metal (see Fig. 16.13). The angle ϕ_0 at which R_p goes through a minimum can be determined by invoking the condition $\partial R/\partial\phi_0 = 0$. This effect in fact occurs at large incidence angles that is at grazing incidence. The determination of ϕ_0 can be achieved by using Eq. (16.99) for R_p, however, it is easier to use the form $R_p = \hat{r}_{12p}\hat{r}^*_{12p}$ along with Eq. (16.96):

$$R_p = \frac{\phi_0 - n_1\hat{\eta}_2}{\phi_0 + n_1\hat{\eta}_2} \cdot \frac{\phi_0 - n_1\hat{\eta}^*_2}{\phi_0 + n_1\hat{\eta}^*_2}$$

where again $\hat{\eta}_2 = 1/\hat{n}_2$. The condition $\partial R/\partial\phi_0 = 0$ then gives

$$\phi_0^2(\hat{\eta}^*_2 + \hat{\eta}_2) - n_1^2|\hat{\eta}_2|^2(\hat{\eta}^*_2 + \hat{\eta}_2) = 0$$

Thus

$$\phi_0 = \phi_m = \frac{\pi}{2} - \theta_m = n_1|\hat{\eta}_2| = \frac{n_1}{|\hat{n}_2|} \tag{16.100}$$

The reflectivity of the interface at this incidence angle can be easily shown to take on the following minimum value (see Problem 16.11):

$$R_p = \frac{|\hat{\eta}_2| - \eta_2}{|\hat{\eta}_2| + \eta_2} \tag{16.101}$$

Let us compare these results with the data given in Fig. 16.13 for nickel. Since $\hat{n}_2 = 2 + 3i$, then $|\hat{n}_2| = 3.6$, $|\hat{\eta}_2| = 0.2778$, $\eta_2 = \text{Re}\,\hat{\eta}_2 = 0.1538$, which upon substitution in Eqs. (16.100) and (16.101) give $\phi_m = 15.8°$ or $\theta_m = 74.2°$ and $R_p = 0.278$. From the figure we read $\theta_m = 74.1°$ and $R_p = 0.287$.

Example 16.11 Polarization by Reflection at a Metallic Interface

We consider in this example some polarization effects caused by reflection at a metallic interface. These effects will be examined at the angle of incidence that minimizes the reflectivities—that is, the angle analogous to the Brewster angle, given by Eq. (16.100). The reflected s wave has a phase shift δ_s with respect to the incoming s wave given by $\hat{r}_{12s} = |\hat{r}_{12s}|e^{i\delta_s}$. Writing $1/\hat{n}_2 = \hat{\eta}_2 = \eta_2 + i\eta'_2$, then, from (Eq. 16.95),

$$\tan\delta_s = \frac{2\eta'_2\cos\theta_1}{-1 + 2\eta_2\cos\theta_1} \tag{16.102}$$

The reflected p wave has a shift δ_p with respect to the incoming p wave given by $\hat{r}_{12p} = |\hat{r}_{12p}|e^{i\delta_p}$. Taking $|\hat{r}_{12p}| = R_p^{1/2}$, where R_p is given by Eq. (16.101), and using Eq. (16.96), we get

$$\tan\delta_p = -\frac{2\phi_m n_1\eta'_2}{\phi_m^2 - n_1^2|\hat{\eta}_2|^2} \to \infty \tag{16.103}$$

Thus $\delta_p = \pi/2$. The polarization of the reflected wave can now be determined; however, we will leave further discussions of the polarization as an exercise (see Problem 16.12).

Example 16.12 Transmission of a Metal Foil

In this example we examine the transmission of electromagnetic waves through the metallic foil shown in Fig. 16.14. An electromagnetic wave of amplitude $E_0\hat{\mathbf{x}}$ and frequency ω is incident normally on a silver foil of conductivity σ_c and thickness d, which is large enough that multiple reflections can be neglected. We take the amplitudes of the electric fields inside the foil just at the front surface to be $E_1\hat{\mathbf{x}}$ and just at the back surface to be $E_2\hat{\mathbf{x}}$. The transmitted field amplitude is taken to be $E_3\hat{\mathbf{x}}$. These amplitudes can be determined in terms of the incident amplitude by using the boundary conditions. To do so we need to determine the complex refractive index of silver at normal incidence. Using Eq. (14.82) we find, for $\sigma_c/\varepsilon_0\omega \gg 1$ and $K_m = 1$, that

$$\hat{n}_2 = \sqrt{i}\sqrt{\frac{\sigma_c}{\varepsilon_0\omega}} = \sqrt{\frac{\sigma_c}{\varepsilon_0\omega}}\,e^{i\pi/4} = (1 + i)\sqrt{\frac{\sigma_c}{2\varepsilon_0\omega}}$$

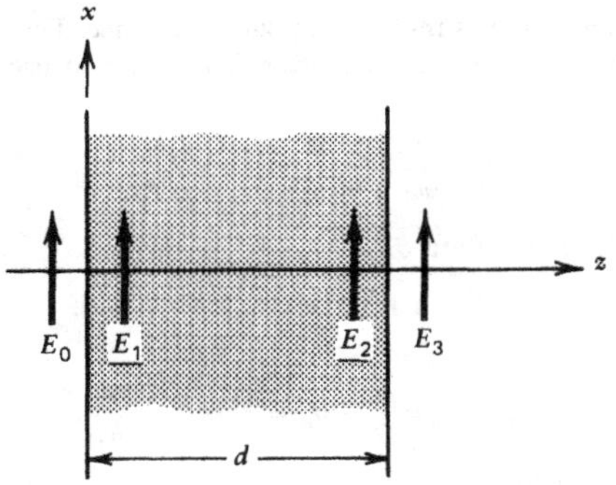

Figure 16.14 Reflection and transmission from a metallic foil, thick enough such that multiple reflections can be neglected.

Thus Eq. (16.22) gives

$$\frac{E_1}{E_0} = \frac{2n_1}{\hat{n}_2 + n_1} = \frac{2}{1 + \sqrt{\dfrac{\sigma_c}{\varepsilon_0 \omega}}\, e^{i\pi/4}}$$

As the wave travels in silver it suffers exponential attenuation governed by its skin depth $\delta = \sqrt{2/\mu_0 \sigma_c \omega}$:

$$\frac{E_2}{E_1} = e^{-d/\delta}$$

The transmitted field can now be calculated using Eq. (16.22) as follows:

$$\frac{E_3}{E_2} = \frac{2\hat{n}_2}{\hat{n}_2 + n_1} = \frac{2}{1 + \sqrt{\dfrac{\varepsilon_0 \omega}{\sigma_c}}\, e^{-i\pi/4}}$$

For $\omega = 400\pi \times 10^6$ rad/s, $\sigma_c = 61.7 \times 10^6\ (\Omega \cdot \text{m})^{-1}$ and $d = 10^{-5}$ m, we find $\sqrt{\sigma_c/\varepsilon_0 \omega} = 7.45 \times 10^3$, and $\delta = 4.5 \times 10^{-6}$ m. Hence

$$\frac{E_1}{E_0} = 2.68 \times 10^{-5} e^{-i\pi/4} \qquad \frac{E_2}{E_0} = 2.9 \times 10^{-6} e^{-i\pi/4}$$

$$\frac{E_3}{E_0} = 0.57 \times 10^{-5} e^{-i\pi/4}$$

Thus we find that the transmitted field and the field in the foil are 45° out of phase relative to the incident field. The neglect of multiple reflections for a given d can now be justified since the skin depth and the refractive index of the foil are calculated. This will, however, be left as an exercise.

16.4 Waveguides and Cavity Resonators

In the previous sections we consider the propagation of electromagnetic waves across single-plane interfaces of dielectric and conducting materials. (Examples 16.6 and 16.12, however, dealt with multiple boundary problems.) In this section we consider the propagation in the presence of more than one boundary (multiple boundaries). A section of space bounded on all sides by metal walls is one example; it is called a cavity resonator. A section of space bounded by a conducting material in the form of a long open pipe is another example; it is called a waveguide.

Here we will consider only waveguides made of hollow metal cylinders and cavities made of hollow metal boxes, with fields only inside the hollow. Other guiding structures, however, are possible, including parallel-wire transmission lines and dielectric cylinders of high dielectric constants. Moreover, we will assume that the material is highly conducting (*perfect* or *ideal* conductors), such that the reflectivity of s and p waves is unity for all incident angles. In this limit the skin depth is taken zero, with no fields existing inside the material and hence no power loss occurring. For a discussion of power loss in *real* conducting guides and cavities, see J. D. Jackson, *Classical Electrodynamics*, 2nd Ed. (New York: Wiley, 1975), Chapter 8.

Waveguides and cavity resonators confine all electromagnetic radiation within their walls; hence they eliminate the effect of inductive and capacitive couplings, and also prevent the power loss due to radiation. As a result they are used to transport electric energy by injecting electromagnetic waves at one end and picking them up at the other end (using antennas).

We have seen that the propagation across a single boundary is governed by Snell's law and the Fresnel coefficients. In situations involving more than one boundary, it is solved for by superimposing the individual solutions at single boundaries. Because of the superposition of various solutions, interference effects play an important role in waveguides.

16.4.1 Propagation Between Two Conducting Plates (Metallic Mirrors)

For the purpose of simplifying the discussion of waveguides, we will first treat a simpler boundary value problem—namely, the propagation of electromagnetic waves in a region in space that is bounded only at two sides by two parallel, highly conducting plates (perfect conductors). Take the plates at $y = 0$ and $y = a$ as shown in Fig. 16.15, with the material in between of permittivity and permeability equal to

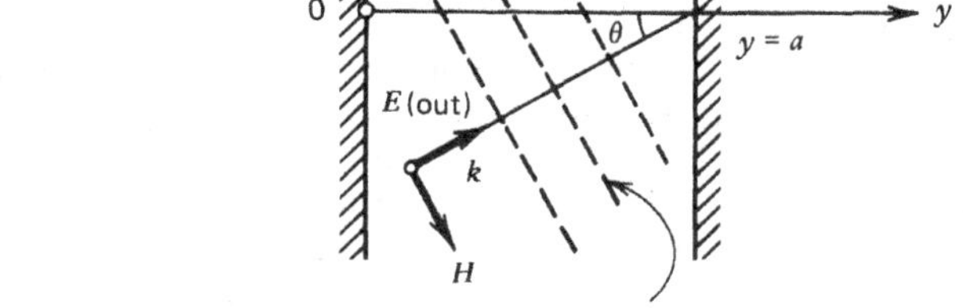

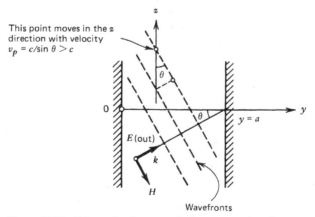

Figure 16.15 Schematic diagram of the propagation between two large, parallel, highly conducting plates, showing the wavefronts and the phase velocity along z.

those of vacuum, ε_0 and μ_0, respectively, and free of charges and currents. Let a wave of frequency ω (free-space wavelength $\lambda_0 = c/2\pi\omega = 2\pi/k$), wave vector $\mathbf{k}$ (in the y-z plane), and polarized along x be incident on the $y = a$ plate with an angle of incidence θ. Since the wave is polarized along the x axis (s polarization), the electric field is tangent to the plate. This case is called *transverse electric* (TE). If the polarization of the electric field were in the y-z plane and that of the H field along the x (p polarization), the case would have been called *transverse magnetic* (TM); see Problems 16.16 and 16.17. Although we have chosen at this stage an *arbitrary* direction for the wave vector, we will find out below that for the waves to exist between the plates, the angle θ will have to be restricted to a set of discrete values. We should note that in the absence of one of the plates, however, the angle of incidence is not restricted to discrete values.

The electric field of the wave between the plates is the sum of the fields of the incident and reflected waves. Because the plates are highly conducting (perfect conductors), the Fresnel coefficient $\hat{r}_{12s}$ is -1 (see Example 16.8), meaning that the relected wave has the same amplitude as the incident wave but with a phase difference π:

$$\mathbf{E} = E'\hat{\mathbf{x}}e^{i(\mathbf{k}_1\cdot\mathbf{r}-\omega t)} - E'\hat{\mathbf{x}}e^{i(\mathbf{k}_1\cdot\mathbf{r}-\omega t)} \tag{16.104}$$

Using $\mathbf{r} = \hat{\mathbf{z}}z + \hat{\mathbf{y}}y$, $\mathbf{k}_1 = k\cos\theta\,\hat{\mathbf{y}} + k\sin\theta\,\hat{\mathbf{z}}$, and $\mathbf{k}_1' = -k\cos\theta\,\hat{\mathbf{y}} + k\sin\theta\,\hat{\mathbf{z}}$, the $\mathbf{E}$ field becomes

$$\mathbf{E} = E'\hat{\mathbf{x}}[e^{ik\cos\theta\,y} - e^{-ik\cos\theta\,y}]e^{i(k\sin\theta\,z-\omega t)}$$

or

$$\mathbf{E} = E_0\hat{\mathbf{x}}\sin(k\cos\theta\,y)e^{i(k\sin\theta\,z-\omega t)} \tag{16.105}$$

where $E_0 = 2iE'$. The electric field as given above is not yet completely determined because the correct boundary conditions are not yet satisfied (imposed) despite the fact that the amplitude E_0, frequency ω, and the magnitude of the wave vector k are externally controlled. The direction of the wave vector and hence the angle θ are left undetermined at this stage, so we may choose them appropriately to satisfy the boundary conditions. At a highly conducting surface, the tangential electric field vanishes [Eq. (16.16)]. This condition is already satisfied at the $y = 0$ plate; it is equivalent to the use of the Fresnel coefficient $\hat{r}_{12s} = -1$ in writing down Eq. (16.104). The same boundary condition at $y = a$ however gives us the condition $\sin(ka\cos\theta) = 0$, which implies

$$ka\cos\theta = n\pi \qquad \text{or} \qquad \cos\theta = \frac{n\pi}{ka} \tag{16.106}$$

where n is a positive integer. Thus, for given k and a, there are a number of discrete directions (*modes*) possible for the propagation between the plates and the corresponding wave is labeled TE_n. Note that only the boundary conditions on $\mathbf{E}$ were utilized to determine a unique solution since the boundary conditions on $\mathbf{H}$ give redundant information (see Example 16.17 for the conditions on $\mathbf{H}$).

With the solution uniquely determined between the plates, we can now discuss some features of the propagation. We will determine effective wavelengths for the y and z directions, and discuss the phase and group velocities for each mode of the wave. We start by writing Eq. (16.105) in the form

$$\mathbf{E} = E_0\hat{\mathbf{x}}\sin(k_c\,y)e^{i(k_g z-\omega t)} \tag{16.107}$$

where

$$k_g = k \sin\theta \qquad \text{and} \qquad k_c = k \cos\theta \tag{16.108}$$

Noting that one can write k_g and k_c in terms of effective wavelengths as follows

$$k_g = \frac{2\pi}{\lambda_g} \qquad \text{and} \qquad k_c = \frac{2\pi}{\lambda_c} \tag{16.109}$$

and using $\lambda_0 = 2\pi/k = c/2\pi\omega$, the free-space wavelength, give the following expressions for the wavelengths.

$$\lambda_c = \frac{\lambda_0}{\cos\theta} \qquad \lambda_g = \frac{\lambda_0}{\sin\theta} \tag{16.110}$$

The wavelengths and the wave numbers can also be written in terms of the integer n by using Eq. (16.106); that is,

$$\begin{aligned} \lambda_c &= \frac{2a}{n} \qquad \lambda_g = \left(\frac{1}{\lambda_0^2} - \frac{n^2}{4a^2}\right)^{-1/2} \\ k_c &= \frac{\pi n}{a} \qquad k_g = 2\pi\left(\frac{1}{\lambda_0^2} - \frac{n^2}{4a^2}\right)^{1/2} \end{aligned} \tag{16.111}$$

Also one can write the following two relations.

$$\begin{aligned} \left(\frac{2\pi}{\lambda_g}\right)^2 &= \left(\frac{2\pi}{\lambda_0}\right)^2 - \left(\frac{2\pi}{\lambda_c}\right)^2 = \left(\frac{2\pi}{\lambda_0}\right)^2 - \left(\frac{n\pi}{a}\right)^2 \\ k_g^2 &= k^2 - k_c^2 = k^2 - \left(\frac{n\pi}{a}\right)^2 \end{aligned} \tag{16.112}$$

The physical importance of λ_c and λ_g is derived from the fact that for a given n and a, there is an upper-limit cutoff condition on λ_0, the wavelength of the waves in free space, that can propagate between the plates. This can be seen using Eq. (16.111) or (16.112). When $(2\pi/\lambda_0)^2 - (2\pi/\lambda_c)^2 < 0$, λ_g becomes imaginary. Consequently the exponent $ik_g z$ becomes negative for $z > 0$, thus causing the wave to decay exponentially. Thus if the wave is to propagate between the plates without attenuation, we should have $\lambda_0 < \lambda_c = 2a/n$. The names *cutoff wavelength* and *waveguide wavelength* are therefore coined for λ_c and λ_g respectively. Moreover, since $\sin\theta \le 1$, then λ_g is larger than the vacuum wavelength λ_0.

We now discuss the phase and group velocities of the wave. We start out by noting that, in free space, the phase velocity with which planes of constant phase move and the group velocity with which energy in the wave move are both equal to c. When the waves are confined as in the present case, these velocities become no longer equal to c. The phase velocity can be calculated from the definition $v_p = dz/dt$ using the planes of constant phase $2\pi z/\lambda_g - \omega t = \text{constant}$. Thus (see Fig. 16.15)

$$v_p = \frac{\omega\lambda_g}{2\pi} = \frac{c}{\sin\theta} \tag{16.113}$$

Since $\sin\theta \le 1$, $v_p \ge c$. In fact at $\theta = 0$ (that is, exactly at cutoff), $\lambda_c \to \lambda_0$, and v_p becomes infinite. This result may seem to contradict the theory of relativity, which states that energy cannot be propagated with velocities larger than the speed of light. This apparent contradiction can be cleared when we realize that the electromagnetic energy between the plates does not propagate with the phase velocity but rather with the group velocity, which we will show below to be smaller than c, thus remaining in complete harmony with the theory of relativity.

The group velocity can be calculated from the definition $u\mathbf{v}_g = \mathbf{S}$ [Eq. (14.75)], or from its average form, $\mathbf{v}_g\langle u\rangle = \langle\mathbf{S}\rangle$, where u is the energy density stored in the electromagnetic field and $\mathbf{S}$ is the Poynting vector. Substituting Eq. (16.107) in $\nabla \times \mathbf{E} = -\partial\mathbf{B}/\partial t$ and taking the time dependence of $\mathbf{B}$ to be $e^{-i\omega t}$, we find that

$$\mathbf{B} = \frac{E_0}{\omega}\,[k_g\hat{\mathbf{y}}\sin(k_c y) + ik_c\hat{\mathbf{z}}\cos(k_c y)]e^{i[k_g z - \omega t]} \tag{16.114}$$

From Eqs. (14.68) to (14.71) we write $u = \frac{1}{2}(\mathbf{E}\cdot\mathbf{D} + \mathbf{B}\cdot\mathbf{H})$, and $\mathbf{S} = \mathbf{E} \times \mathbf{H}$. The time average of $\mathbf{S}$ and u can be calculated using Eqs. (14.78) and (14.79). Thus

$$\langle u\rangle = \frac{1}{4}\left[\varepsilon_0|E|^2 + \frac{|B|^2}{\mu_0}\right] \qquad \text{and} \qquad \langle\mathbf{S}\rangle = \frac{1}{2\mu_0}\,\text{Re}[\mathbf{E}^* \times \mathbf{B}]$$

Substituting for $\mathbf{E}$ and $\mathbf{B}$ from Eqs. (16.107) and (16.114) gives

$$\langle u\rangle = \frac{\varepsilon_0}{4}|E_0|^2\left\{\left[1 + \left(\frac{k_c}{k}\right)^2\right]\sin^2(k_c y) + \left(\frac{k_g}{k}\right)^2\cos^2(k_c y)\right\}$$

$$\langle S\rangle = \frac{\varepsilon_0 c}{2}|E_0|^2\hat{\mathbf{z}}\,\frac{k_g}{k}\sin^2(k_c y)$$

The average energy density and Poynting vector per unit length across the space between the plates can be calculated by integrating $\langle u\rangle$ and $\langle\mathbf{S}\rangle$ over y and dividing by a. Noting that

$$\frac{1}{a}\int_0^a \sin^2(k_c y)dy = \frac{1}{a}\int_0^a \cos^2(k_c y)dy = \frac{1}{2}$$

then

$$\frac{\langle u\rangle}{a} = \frac{1}{4}\varepsilon_0|E_0|^2 \qquad \frac{\langle\mathbf{S}\rangle}{a} = \frac{1}{4}\varepsilon_0|E_0|^2 c\,\frac{k_g}{k}\,\hat{\mathbf{z}}$$

Thus the group velocity $\mathbf{v}_g$ is

$$\mathbf{v}_g = \frac{\langle\mathbf{S}\rangle/a}{\langle u\rangle/a} = c\,\frac{k_g}{k}\,\hat{\mathbf{z}} = c\,\frac{\lambda_0}{\lambda_g}\,\hat{\mathbf{z}}$$

or

$$\mathbf{v}_g = c\sin\theta\,\hat{\mathbf{z}} \tag{16.115}$$

Note that the group velocity is smaller than the speed of light in vacuum, as we stated above. Moreover, it is interesting to note that $v_p v_g = c^2$. The detailed dependence of the energy, intensity, and group and phase velocities of the modes is the subject matter of Problem 16.18.

Finally, we should note that in the so-called *transverse electromagnetic* (TEM) case waves with both E_z and $H_z = 0$ can propagate between the plates. In this case there is no restriction on ω nor on θ (see Problem 16.19).

16.4.2 Waveguides

In the previous subsection we considered the propagation of electromagnetic waves between two large, parallel, metallic plates. Here we consider the propagation inside

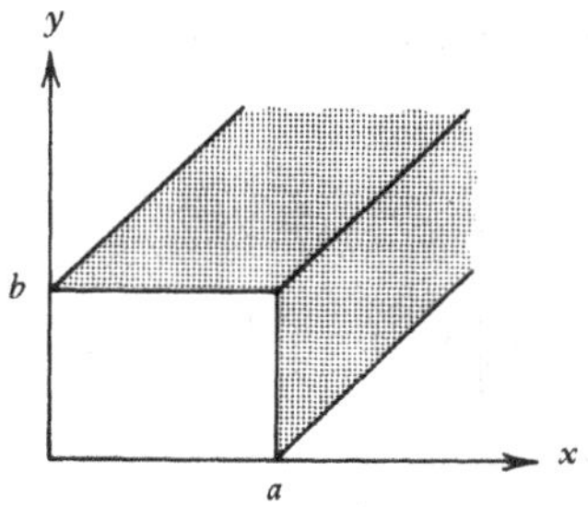

Figure 16.16 A rectangular waveguide with propagation along the z axis.

a metallic pipe. For the sake of simplicity, we will analyze rectangular waveguides only. Consider a rectangular waveguide of a cross section of dimensions a and b and axis along the z, axis as shown in Fig. 16.16. In order to facilitate the analysis, we assume a wavelength λ_g associated with the propagation along z in analogy with the wavelength λ_g we encountered in the propagation between two parallel plates. Thus we take the z dependence of the fields of the form $e^{(ik_g z)}$.

We will consider the TE case; that is, we take $\mathbf{E} = \hat{\mathbf{x}}E_x + \hat{\mathbf{y}}E_y$. The three components of the $\mathbf{H}$ field will, however, be nonzero; that is, $\mathbf{H} = \hat{\mathbf{x}}H_x + \hat{\mathbf{y}}H_y + \hat{\mathbf{z}}H_z$. (See Example 16.14 for the TM case.) In the absence of charge and current distributions inside the waveguide, the electric and magnetic field satisfy the homogeneous undamped vector wave equation

$$\nabla^2\begin{pmatrix}\mathbf{E}\\ \mathbf{H}\end{pmatrix} - \frac{1}{c^2}\frac{\partial^2}{\partial t^2}\begin{pmatrix}\mathbf{E}\\ \mathbf{H}\end{pmatrix} = 0$$

In cartesian coordinates this equation separates into five scalar wave equations. It will be sufficient to solve the equation for one of the components, followed by the use of Maxwell's equations directly for the determination of the rest of them. Thus we consider

$$\nabla^2 H_z - \frac{1}{c^2}\frac{\partial^2 H_z}{\partial t^2} = 0 \tag{16.116}$$

Taking the time dependence and the z dependence of H_z of the form $e^{i(k_g z - \omega t)}$, and expanding ∇^2 in cartesian coordinates, this equation can be transformed to

$$\frac{\partial^2 H_z}{\partial x^2} + \frac{\partial^2 H_z}{\partial y^2} + (k^2 - k_g^2)H_z = 0 \tag{16.117}$$

Before we solve this equation we turn to Maxwell's curl equations to write H_x, H_y, E_x, and E_y in terms of H_z. Taking the z dependence and the time dependence of all the field components of the form $e^{i(k_g z - \omega t)}$ we find that $\nabla \times \mathbf{E} + \partial\mathbf{B}/\partial t = 0$ gives, among other relations,

$$E_x = \frac{\mu_0 \omega}{k_g} H_y \qquad \text{and} \qquad E_y = -\frac{\mu_0 \omega}{k_g} H_x \tag{16.118}$$

and that $\nabla \times \mathbf{H} - \partial\mathbf{D}/\partial t = 0$ gives, among other relations,

$$i\varepsilon_0 \omega E_x + \frac{\partial H_z}{\partial y} - ik_g H_y = 0 \qquad \text{and} \qquad i\varepsilon_0 \omega E_y + ik_g H_x - \frac{\partial H_z}{\partial x} = 0 \tag{16.119}$$

Combining these equations gives, among others, the following relations between H_x and H_z and between H_y and H_z.

$$ik_g\left[1-\left(\frac{k}{k_g}\right)^2\right]H_x=\frac{\partial H_z}{\partial x}$$
$$ik_g\left[1-\left(\frac{k}{k_g}\right)^2\right]H_y=\frac{\partial H_z}{\partial y} \tag{16.120}$$

Thus Eqs. (16.118) and (16.120) give E_x, E_y, H_x, and H_y in terms of H_z.

We now solve Eq. (16.117) for H_z. We use the method of separation of variables. Substituting $H_z(x, y) = h_1(x)h_2(y)$, and dividing by H_z, we get

$$-\frac{1}{h_1}\frac{d^2h_1}{dx^2}=\frac{1}{h_2}\frac{d^2h_2}{dy^2}+k^2-k_g^2$$

Equating both sides to the separation constant k_x^2 gives

$$\frac{d^2h_1}{dx^2}+k_x^2h_1=0 \qquad \text{and} \qquad \frac{d^2h_2}{dy^2}+k_y^2h_2=0 \tag{16.121}$$

where

$$k_x^2+k_y^2-k^2-k_g^2=0 \tag{16.122}$$

Equation (16.121), together with the boundary conditions, specifies what is called an *eigenvalue problem*, and Eq. (16.122) is called the *eigenvalue condition.* The equation for h_1 has the solutions $\sin k_x x$ and $\cos k_x x$, and the equation for h_2 has the solution $\sin k_y y$ and $\cos k_y y$. Thus the most general solution for $H_z(x, y, z)$ is

$$\begin{aligned}H_z(x, y, z) = [&A_1 \sin k_x x \sin k_y y + A_2 \sin k_x x \cos k_y y\\ &+ A_3 \cos k_x x \sin k_y y + A_4 \cos k_x x \cos k_y y]e^{ik_g z}\end{aligned}$$

where A_i are constants to be evaluated from the boundary condition. Using Eqs. (16.118) to (16.120), we obtain

$$\begin{aligned}E_x &= \frac{-i\mu_0\omega k_y}{k_g^2}\left[1-\left(\frac{k}{k_g}\right)^2\right]^{-1}[A_1 \sin k_x x \cos k_y y - A_2 \sin k_x x \sin k_y y\\ &\quad + A_3 \cos k_x x \cos k_y y - A_4 \cos k_x x \sin k_y y]e^{ik_g z}\\ E_y &= \frac{i\mu_0\omega k_x}{k_g^2}\left[1-\left(\frac{k}{k_g}\right)^2\right]^{-1}[A_1 \cos k_x x \sin k_y y + A_2 \cos k_x x \cos k_y y\\ &\quad - A_3 \sin k_x x \sin k_y y - A_4 \sin k_x x \cos k_y y]e^{ik_g z}\end{aligned} \tag{16.123}$$

We now apply the boundary conditions and note that one can use those of the electric field or of the magnetic field since either set is sufficient to determine a unique solution. We elect to use those of **E**, but for a discussion of those of **H** see Example 16.17. At the surface of the conductors, the tangential electric field vanishes. Taking $E_y = 0$ at $x = 0$ and $x = a$ gives $A_1 = A_2 = 0$, and $k_x = m\pi/a$, where m is a positive integer. On the other hand, taking $E_x = 0$ at $y = 0$ and $y = b$ gives

$A_1 = A_3 = 0$ and $k_y = n\pi/b$, where n is a positive integer. That is, only A_4 is nonzero. Substituting these values in the above equations gives all the field components inside the guide. For example, the most general expression of H_z is

$$H_z(x, y, z) = \sum_{n,m} A_{mn} \cos\left(\frac{m\pi}{a} x\right)\cos\left(\frac{n\pi}{b} y\right)e^{(2\pi i/\lambda_g)z} \tag{16.124}$$

where A_{mn} are constants. Each term of the infinite sum is called a mode, labeled by two integers m and n. Each of these terms is called a TE_{mn} mode, where TE stands for transverse electric, as was mentioned above.

A relation between λ_g for each mode of the guide and the free-space wavelength of the radiation, analogous to the one determined for the two parallel plates (Eq. 16.112), can now be determined by substituting for k_x and k_y in Eq. (16.122). Thus

$$\left(\frac{2\pi}{\lambda_g}\right)^2 = \left(\frac{2\pi}{\lambda_0}\right)^2 - \left(\frac{m\pi}{a}\right)^2 - \left(\frac{n\pi}{b}\right)^2 = \left(\frac{2\pi}{\lambda_0}\right)^2 - \left(\frac{2\pi}{\lambda_c}\right)^2 \tag{16.125}$$

where

$$\frac{1}{\lambda_c^2} = \left(\frac{m}{2a}\right)^2 + \left(\frac{n}{2b}\right)^2 \tag{16.126}$$

defines the cutoff wavelength λ_c.

In the above discussion we determined only the fields of TE or TM (see Example 16.14) propagation in rectangular waveguides. However, one can show in general that these, and linear combinations, are all the cases that can propagate in such guides and in pipes. For example, one can show that this is true for TEM modes wherein both E_z and $H_z = 0$ for all x, y, and z. (See Problem 16.22.)

16.4.3 Cavity Resonators

In this subsection we consider electromagnetic oscillations in a section of space bounded on all sides by metal walls—i.e., a *cavity resonator*. The cavity will be taken in the form of a parallelepiped of dimensions a, b, and c. Because the space is bounded on all sides, the cavity will not support any propagating waves; the solutions will be *standing waves* in the three orthogonal directions.

For monochromatic waves—that is, waves that have $e^{-i\omega t}$ time dependence, the electric field satisfies the equations $\nabla^2\mathbf{E} + (\omega^2/c^2)\mathbf{E} = 0$ and $\nabla \cdot \mathbf{E} = 0$. In cartesian coordinates, each component of $\mathbf{E}$ satisfies the scalar wave equation. We consider the equation for E_x first:

$$\left(\frac{d^2}{dx^2} + \frac{d^2}{dy^2} + \frac{d^2}{dz^2} + \frac{\omega^2}{c^2}\right)E_x = 0 \tag{16.127}$$

Using the method of separation of variables, one can show that in general E_x has the solution

$$E_x = (A_1 \sin k_x x + A_2 \cos k_x x)(B_1 \sin k_y y + B_2 \cos k_y y)(C_1 \sin k_z z + C_2 \cos k_z z) \tag{16.128}$$

where

$$\frac{\omega^2}{c^2} = k_x^2 + k_y^2 + k_z^2 \tag{16.129}$$

and A_1, A_2, B_1, B_2, C_1, and C_2 are constants to be evaluated from the boundary conditions, and k_x, k_y, and k_z are the separation constants that also are to be found from the boundary conditions (e.g., those of **E**) and from $\nabla \cdot \mathbf{E} = 0$. Since at the surface of a highly conducting material the tangential electric field vanishes, we subject the solution of Eq. (16.128) to the condition $E_x = 0$ at $z = 0$, $z = c$, $y = 0$, and $y = b$, which requires $C_2 = B_2 = 0$, $k_z = n\pi/c$ and $k_y = m\pi/b$, where m and n are positive integers or zeros. Thus

$$E_x = E_{0x} F_1(x) \sin\left(\frac{m\pi}{b} y\right) \sin\left(\frac{n\pi}{c} z\right) \tag{16.130}$$

where

$$F_1(x) = A_1 \sin k_x x + A_2 \cos k_x x \tag{16.131}$$

A similar procedure for the solution of the scalar wave equation for E_y subject to the appropriate boundary conditions ($E_y = 0$ at $z = 0$, c, and at $x = 0$, a) gives

$$E_y = E_{0y} F_2(y) \sin\left(\frac{l\pi}{a} x\right) \sin\left(\frac{n\pi}{c} z\right) \tag{16.132}$$

where l is a positive integer or zero and we used the same n as in E_x, and

$$F_2(y) = A_1' \sin k_y' y + A_2' \cos k_y' y \tag{16.133}$$

where k_y' is a constant to be evaluated. Finally, the same procedure is used to solve for E_z subject to the boundary conditions $E_z = 0$ at $x = 0$, $x = a$, $y = 0$, and $y = b$, with the result

$$E_z = E_{0z} F_3(z) \sin\left(\frac{l\pi}{a} x\right) \sin\left(\frac{m\pi}{b} y\right) \tag{16.134}$$

where we used the same l and m as in E_x and E_y, and

$$F_3(z) = A_1'' \sin k_z'' z + A_2'' \cos k_z'' z \tag{16.135}$$

where k_z'' is a constant to be evaluated.

The rest of the unknowns—that is, F_1, F_2, and F_3—and a relation between the amplitudes of the fields in the three directions are now to be determined from Maxwell's divergence equation: $\nabla \cdot \mathbf{E} = 0$. Substituting Eqs. (16.130), (16.132), and (16.134) in $\nabla \cdot \mathbf{E} = 0$ gives

$$E_{0x} \frac{dF_1}{dx} \sin\left(\frac{m\pi}{b} y\right) \sin\left(\frac{n\pi}{c} z\right) + E_{0y} \frac{dF_2}{dy} \sin\left(\frac{l\pi}{a} x\right) \sin\left(\frac{n\pi}{c} z\right)$$
$$+ E_{0z} \frac{dF_3}{dz} \sin\left(\frac{l\pi}{a} x\right) \sin\left(\frac{m\pi}{b} y\right) = 0 \tag{16.136}$$

For this equation to be identically zero for all values of x, y, and z, requires that $A_1 = A_1' = A_1'' = 0$, $A_2 = A_2' = A_2'' = 1$, $k_x = l\pi/a$, $k_y' = m\pi/b$, and $k_z'' = n\pi/c$; hence

$$F_1 = \cos\left(\frac{l\pi}{a} x\right) \qquad F_2 = \cos\left(\frac{m\pi}{b} y\right) \qquad F_3 = \cos\left(\frac{n\pi}{c} z\right) \tag{16.137}$$

Moreover, for Eq. (16.136) to be satisfied, the following relation should be satisfied.

$$\frac{l\pi}{a} E_{0x} + \frac{m\pi}{b} E_{0y} + \frac{n\pi}{c} E_{0z} = 0 \tag{16.138}$$

The electric fields allowed in the cavity are therefore given by Eqs. (16.130), (16.132), (16.134), (16.137), and (16.138).

The frequency of the oscillations that can be supported in the cavity [see Eq. (16.129)] becomes

$$\frac{\omega^2}{c^2} = \frac{l^2\pi^2}{a^2} + \frac{m^2\pi^2}{b^2} + \frac{n^2\pi^2}{c^2} \tag{16.139}$$

Equation (16.138) shows that two of the constants E_{0x}, E_{0y}, and E_{0z} are independent, and hence the allowed oscillations for l, m, $n \neq 0$ are in general *doubly degenerate*. This degeneracy means that for each allowed frequency there corresponds two allowed oscillations that have different configurations of the electromagnetic fields. However, if one of the integers is zero, then the degeneracy is removed (see Problem 16.24).

The corresponding magnetic field in the cavity can be calculated from Maxwell's equation $\nabla \times \mathbf{E} = -\mu_0\, \partial\mathbf{H}/\partial t$. Taking the time dependence of $\mathbf{H}$ of the form $e^{-i\omega t}$ gives $\nabla \times \mathbf{E} = i\mu_0\omega\mathbf{H}$. Thus

$$H_x = \frac{-i}{\mu_0\omega}\left(\frac{m\pi}{b}E_{0z} - \frac{n\pi}{c}E_{0y}\right)\sin\frac{l\pi x}{a}\cos\frac{m\pi y}{b}\cos\frac{n\pi z}{c}$$

$$H_y = \frac{-i}{\mu_0\omega}\left(\frac{n\pi}{c}E_{0x} - \frac{l\pi}{a}E_{0z}\right)\cos\frac{l\pi x}{a}\sin\frac{m\pi y}{b}\cos\frac{n\pi z}{c}$$

$$H_z = \frac{-i}{\mu_0\omega}\left(\frac{l\pi}{a}E_{0y} - \frac{m\pi}{b}E_{0x}\right)\cos\frac{l\pi x}{a}\cos\frac{m\pi y}{b}\sin\frac{n\pi z}{c} \tag{16.140}$$

Note that the equations $\nabla\cdot\mathbf{B} = 0$ and $\nabla \times \mathbf{H} = \partial\mathbf{D}/\partial t$ do not give any new information on the fields.

We should note that in the above discussion we calculated the most general allowed transverse electric fields in a rectangular cavity TE_{lmn}. If one takes $m = 0$, then E_x, E_z, and H_y vanish, leaving E_y, H_x, and H_z nonzero. However if any two of l, m, and n vanish, then all the components of $\mathbf{E}$ and $\mathbf{H}$ vanish. Hence the lowest mode has only one of them zero. For example, we find that if $n = 0$, all components of the electromagnetic field vanish identically; thus $n = 1$ is the lowest mode. Now if $l = m = 0$, then again all components vanish.

One can now easily make this general solution specific to various types of propagation. Thinking of z as the old waveguide axis of propagation, we then consider two types: TM modes where $H_z \equiv 0$ and TE modes where $E_z = 0$ for all possible values of m, n, and l. Note that these special modes are decided relative to the rather arbitrary (in this case) z axis. Thus for the TE mode we take $F_{0z} = 0$ by definition. Consequently Eq. (16.138) becomes

$$\frac{l\pi}{a}E_{0x} + \frac{m\pi}{b}E_{0y} = 0 \qquad \text{(TE)} \tag{16.141}$$

On the other hand, we take $H_{0z} = 0$ by definition for TM modes. In this case we require that (from Eq. (16.140))

$$\frac{l\pi}{a}E_{0y} - \frac{m\pi}{b}E_{0x} = 0 \qquad \text{(TM)} \tag{16.142}$$

The above results show that the $\mathbf{E}$ and $\mathbf{H}$ fields in the cavity are always $\pi/2$ out of phase at any point. Therefore, the standing waves in the cavity transmit no energy (see Problem 16.26). Finally we reiterate that in the above derivative the walls of the cavity were taken to be infinitely conducting; therefore, the tangential E field and the currents on the walls are zero. As a result, there are no power losses. In any real cavity, however, the walls have a finite conductivity which results in establishing currents in the wall, hence causing power dissipation as a result of the I^2R term and consequently causing a net flow of energy out of the cavity as the currents are attenuated. This loss will dampen and broaden each sharp resonant response into a broad curve similar to the response of a series RLC resonant circuit. Moreover, the concept of the Q factor used to describe the quality (sharpness) of resonances in circuit theory can be applied to cavities. [See J. D. Jackson, *Classical Electrodynamics*, 2nd ed. (New York: Wiley, 1975) p. 356.]

Example 16.13 Phase and Group Velocity of Radiation Between Parallel Plates

We have previously shown that the phase and group velocities of radiation between two parallel metallic plates are $c/\sin\theta$ and $c\sin\theta$, respectively [Eqs. (16.113) and (16.115)]. In general these velocities are given by the relations:

$$v_p = \frac{\omega}{k_g} \quad \text{and} \quad v_g = \frac{d\omega}{dk_g} \tag{16.143}$$

The first relation can be easily checked by substituting Eq. (16.108) in $v_p = \omega/k_g$, giving

$$v_p = \frac{\omega}{k\sin\theta} = \frac{c}{\sin\theta}$$

The second relation can be checked using Eq. (16.112): $k_g^2 = k^2 - k_c^2 = \omega^2/c^2 - k_c^2$, or $\omega = c(k_g^2 + k_c^2)^{1/2}$. Thus

$$v_g = \frac{d\omega}{dk_g} = ck_g(k_g^2 + k_c^2)^{-1/2}$$

Replacing $(k_g^2 + k_c^2)^{-1/2}$ by c/ω, and k_g by $k\sin\theta$ gives $v_g = c\sin\theta$.

Example 16.14 TM Waves in a Rectangular Waveguide

We have previously determined the TE fields inside a rectangular guide by taking $E_z \equiv 0$. The magnetic field is then allowed to have a longitudinal component, H_z. In this example we discuss the transverse magnetic case. We take $H_z = 0$ and allow the electric field to have a longitudinal component. Thus, in analogy with the TE case, we solve the equation

$$\frac{d^2E_z}{dx^2} + \frac{d^2E_z}{dy^2} + \left[\frac{\omega^2}{c^2} - \left(\frac{2\pi}{\lambda_g}\right)^2\right]E_z = 0 \tag{16.144}$$

subject to the boundary condition that on the surface of a highly conducting material the tangential electric field vanishes. Thus

$$E_z(x, y, z) = A\sin\frac{m\pi x}{a}\sin\frac{n\pi y}{b}\,e^{2\pi iz/\lambda_g} \tag{16.145}$$

This result shows that the lowest nonvanishing mode of TM waves in a wave guide is TM_{11}, since TM_{10} vanishes identically, whereas the lowest nonvanishing TE mode is the TE_{10}. The rest of the components of **E** and **H** and their properties can now be determined from Maxwell's equations. We leave this as an exercise.

Example 16.15 Field Patterns Between Two Conducting Plates

We discuss in this example the possible field patterns between two parallel, highly conducting plates. Equation (16.107) gives such a field for the case of TE waves. Consider the lowest possible mode of Eqs. (16.107) and (16.114), where $n = 1$; that is,

$$\mathbf{E} = E_0\hat{\mathbf{x}} \sin\frac{\pi y}{a} e^{i(k_g z - \omega t)} \qquad k_g = 2\pi\left(\frac{1}{\lambda_0^2} - \frac{1}{4a^2}\right)^{1/2}$$

$$\mathbf{B} = \frac{E_0}{\omega}\left[k_g\hat{\mathbf{y}} \sin\frac{\pi y}{a} + i\frac{\pi}{a}\hat{\mathbf{z}} \cos\frac{\pi y}{a}\right]e^{i(k_g z - \omega t)}$$

The electric field is entirely in the x direction; Figs. 16.17a and 16.17b show its dependence on y and z. Figure 16.17a also gives the end view of the plots showing higher density of E field lines in the middle compared to the sides.

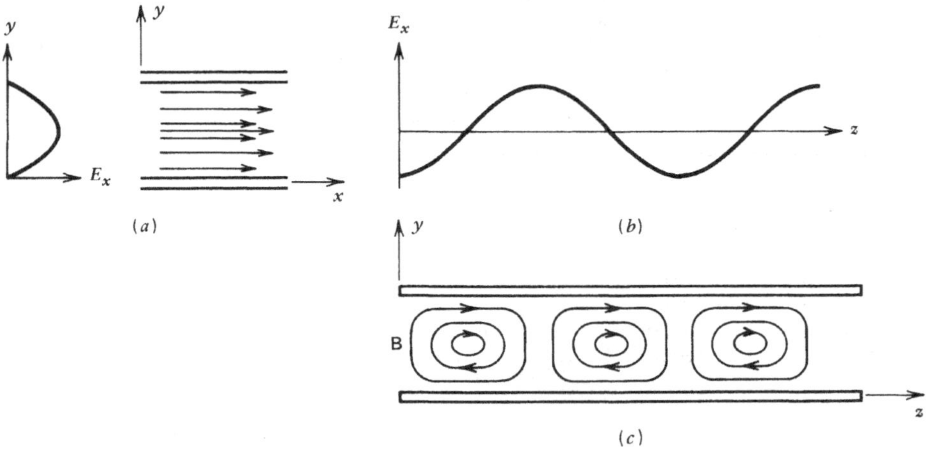

Figure 16.17 Field patterns between two large, highly conducting plates. (a) An end view showing the dependence of $|\mathbf{E}|$ on the direction normal to the plates (y direction). (b) The dependence of $|\mathbf{E}|$ on the direction of propagation. (c) The dependence of $|\mathbf{B}|$ on the direction of propagation and on the direction normal to the plates (z and y).

The B field has components along y and z. Moreover, the z dependence of both components is $\pi/2$ out of phase, thus making the determination of the pattern more difficult than that of the **E** field. The **B** field is expected to have closed lines. Near the surfaces of the plates ($y = 0$ and a), $\sin \pi y/a$ vanishes, and the field becomes entirely tangential. Near $y = a/2$, $\cos \pi y/a$ vanishes and the field becomes entirely normal to the plates (y direction). Figure 16.14c gives a few **B** field lines.

Example 16.16 Density of Allowed Modes in a Cavity Resonator

The density of allowed electromagnetic oscillations per frequency interval $d\omega$ in a cavity resonator discussed in Section 16.4.3 can be calculated from the wave numbers in the three orthogonal directions: $k_x = l\pi/a$, $k_y = m\pi/b$, and $k_z = n\pi/c$. In a frequency interval $d\omega$, hence in wave number intervals dk_x, dk_y, and dk_z, the number of oscillations is $dN = dl\,dm\,dn = (abc/\pi^3)dk_x\,dk_y\,dk_z$. In spherical coordinates we can write $dk_x\,dk_y\,dk_z = k^2\,dk\,d\cos\theta\,d\phi$, where θ and ϕ specify the direction of $\mathbf{k}$; thus $dN = (V/\pi^3)k^2\,dk\,d\cos\theta\,d\phi$ where

V is the volume of the cavity. Integrating over angles gives $dN = (4V/\pi^2)k^2\,dk$, writing $k = \omega/c$, then $k^2\,dk = (\omega^2/c^3)d\omega$; hence

$$\frac{dN}{d\omega} = \frac{4V}{\pi^2 c^3}\omega^2$$

which indicates that for a given $d\omega$, the density is larger at higher frequencies. For example the density grows by a factor of approximately 10^{10} when the frequency is changed from a microwave frequency to an optical frequency. This effect played a key role in design considerations in the extension of masers (operating on microwave frequencies) to lasers (operating on optical frequencies).*

Example 16.17 Boundary Conditions on *H* in a Waveguide

In the above discussions we determined the fields inside waveguides and cavity resonators by imposing boundary conditions on the electric field. In this example, we derive an equivalent set of boundary conditions on the magnetic field that can alternatively be used to derive the fields. We consider the wave guide discussed in Section 16.4.2. From Eq. 16.17, the normal component of the magnetic field vanishes on the surface of a highly conducting material, $H_n = 0$. This condition means that $H_x = 0$ at $x = 0$ and $x = a$, and $H_y = 0$ at $y = 0$ and $y = b$. When these conditions are combined with Eqs. 16.120 we get: $\partial H_z/\partial x = 0$ at $x = 0$ and $x = a$ and $\partial H_z/\partial y = 0$ at $y = 0$ and $y = b$. Note that the last two conditions require that the x and y dependence of H_z consist entirely of cosine functions (see Eq. 16.140).

16.5 Summary

Consider a linearly polarized electromagnetic wave of frequency ω and amplitude E_1 incident from a region of refractive index n_1 onto a plane interface with a material of refractive index n_2, and with θ_1 the angle of incidence. If the polarization of the wave is in the plane of incidence (p polarization), then the angles of reflection and refraction θ_1' and θ_2 are related to θ_1 by the laws of reflection and refraction (Snell's law) as follows (see Figs. 16.2–16.3):

$$\theta_1' = \theta_1 \qquad n_1 \sin\theta_1 = n_2 \sin\theta_2 \tag{16.35}$$

The amplitude of the reflected and refracted waves normalized to E_1, $E_1'/E_1 = r_{12p}$ and $E_2/E_1 = t_{12p}$ are

$$r_{12p} = \frac{\tan(\theta_1 - \theta_2)}{\tan(\theta_1 + \theta_2)} \qquad t_{12p} = \frac{2\cos\theta_1 \sin\theta_2}{\sin(\theta_1 + \theta_2)\cos(\theta_1 - \theta_2)} \tag{16.37}$$

The coefficients r_{12p} and r_{12s} are called Fresnel coefficients. If the polarization of the wave is normal to the plane of incidence (s polarization), we have the same laws of reflection and refraction, but the Fresnel coefficients become

$$r_{12s} = \frac{\sin(\theta_2 - \theta_1)}{\sin(\theta_2 + \theta_1)} \qquad t_{12s} = \frac{2\cos\theta_1 \sin\theta_2}{\sin(\theta_2 + \theta_1)} \tag{16.43}$$

If $\theta_1 = 0$, the incidence is called normal incidence. In this case $\theta_1 = \theta_1' = \theta_2 = 0$, and

$$r_{12p} = -r_{12s} = \frac{n_2 - n_1}{n_2 + n_1} \qquad t_{12p} = t_{12s} = \frac{2n_1}{n_2 + n_1} \tag{16.22}$$

The reflectance and transmittance, R and T, are defined in terms of the Poynting vectors and in terms of Fresnel coefficients as follows:

$$R_p = \frac{\langle S_1' \rangle}{\langle S_1 \rangle} = r_{12p}^2 \qquad T_p = \frac{\langle S_2 \rangle}{\langle S_1 \rangle} = \frac{n_2}{n_1} t_{12p}^2 \qquad (16.23),(16.24)$$

* A. L. Schawlow and C. H. Townes, *Physical Review*, vol. 112, p. 1940, 1958.

with the conservation of power expressed as $R + T = 1$. Similar expressions hold for s polarization.

If $\theta_1 \approx \pi/2$, the incidence is called grazing angle incidence. In this case $r_{12p} = r_{12s} = -1$ and $t_{12p} = t_{12s} = 0$; hence $R_p = R_s = 1$ and $T_p = T_s = 0$, indicating total reflection in both cases.

If $n_1 > n_2$, and

$$\theta_1 \geq \theta_c = \sin^{-1}(n_2/n_1) \tag{16.49}$$

the incidence is called critical angle incidence. At $\theta_1 = \theta_c$, θ_2 reaches its maximum value of $\pi/2$, and $R_p = R_s = 1$(total internal reflection). For $\theta_1 > \theta_c$, the angle θ_2 becomes complex, where $\cos\theta_2 = i\beta = i(\sin^2\theta_1/\sin^2\theta_c - 1)^{1/2}$, a pure imaginary quantity. Hence the Fresnel coefficients also become complex; that is,

$$\hat{r}_{12p} = \frac{n_2\cos\theta_1 - in_1\beta}{n_2\cos\theta_1 + in_1\beta} \qquad \hat{r}_{12s} = \frac{n_1\cos\theta_1 - in_2\beta}{n_1\cos\theta_1 + in_2\beta} \tag{16.52}$$

with the reflectance defined as

$$R_p = \hat{r}_{12p}\hat{r}^*_{12p} = |\hat{r}_{12p}|^2 = 1 \qquad R_s = |\hat{r}_{12s}|^2 = 1$$

If

$$\theta_1 = \theta_B = \tan^{-1}\frac{n_2}{n_1} \tag{16.55}$$

then $\theta_1 + \theta_2 = \pi/2$, and the incidence is called Brewster angle incidence. In this case r_{12p} vanishes, and

$$R_p = 0 \qquad R_s = \frac{n_2}{n_1}\cos^2 2\theta_B$$

This effect can be used to polarize electromagnetic waves.

In the case of a conducting interface, n_2 and θ_2 become complex; hence we use a hat on both of them to indicate so: $\hat{n}$, $\hat{\theta}_2$. The mathematical formalism and all results are identical to the case of dielectric-dielectric interface when $\hat{n}_2$ and $\hat{\theta}_2$ are used. However the physical implications in terms of real quantities differ drastically from the nonconducting case. Snell's law becomes

$$N(\theta_1)\sin\phi = n_1\sin\theta_1 \qquad \text{or} \qquad N(\theta_1)\cos\phi = \bar{n} \tag{16.91),(16.92}$$

where ϕ is the angle of refraction,

$$N(\theta_1) = (\bar{n}^2 + n_1^2\sin^2\theta_1)^{1/2} \tag{16.90}$$

and $\bar{n}$ is the optical constant of the conducting material with the effective permittivity being

$$\bar{\varepsilon}_2 = \varepsilon_2 - \varepsilon_1\sin^2\theta_1 \tag{16.85}$$

The amount of absorption depends on the distance of penetration along the normal to the interface. The absorption length (skin depth) of the material *along the normal* to the interface is the inverse of $\omega\bar{k}/c$ where again $\bar{k}$ is the optical constant of the medium with the above effective permittivity of the medium. Thus the planes of constant phase make an angle ϕ with the planes of constant amplitudes.

As in the dielectric case, there is an analog of the Brewster angle, except that it is not as drastic. Also, at high conductivities and/or low frequencies, the differences between s and p coefficients become negligible. In the ideal case $\sigma_c/\varepsilon\omega \to \infty$, we find $\hat{r}_{12p} = 1$ and $\hat{r}_{12s} = -1$ (ideal conductor), which indicates that $R_p = R_s = 1$. In general we have at a conducting interface

$$R = \hat{r}_{12}\hat{r}^*_{12} = |\hat{r}_{12}|^2 \qquad \text{and} \qquad A = T = 1 - R \tag{16.67),(16.68}$$

where A is the absorption.

When electromagnetic waves are restricted to regions of space bounded by highly conducting materials—two parallel plates or a pipe—there are certain conditions to be satisfied if propagation of these waves is to occur. For two parallel plates separated by a distance a, and where n is an integer, there exists a cutoff wavelength $\lambda_c = 2a/n$ at which waves of vacuum wavelength $\lambda_0 > \lambda_c$ cannot propagate. The effective wavelength of the propagating wave, called λ_g, becomes longer than the vacuum wavelength and is given by

$$\frac{1}{\lambda_g^2} \equiv \frac{1}{\lambda_0^2} - \frac{1}{\lambda_c^2} \qquad \lambda_c = \frac{2a}{n} \tag{16.111}$$

Because λ_c has a set of discrete values, then λ_g will also have a corresponding set of discrete values. The phase and group velocities of the propagating wave are

$$v_p = \frac{c}{\sin\theta} \qquad v_g = c\sin\theta \qquad \sin\theta = \frac{\lambda_0}{\lambda_g} \qquad \text{(16.113),(16.115)}$$

Therefore

$$v_p v_g = c^2$$

For a rectangular pipe of dimensions a and b, we have for the cutoff wavelength for TE_{mn} modes

$$\frac{1}{\lambda_c^2} = \left(\frac{m}{2a}\right)^2 + \left(\frac{n}{2b}\right)^2 \tag{16.126}$$

where n and m are integers.

The frequency of oscillations that can be supported in a rectangular cavity of dimensions a, b, and c is

$$\frac{\omega^2}{c^2} = \frac{l^2\pi^2}{a^2} + \frac{m^2\pi^2}{b^2} + \frac{n^2\pi^2}{c^2} \tag{16.139}$$

Problems

16.1 The regions $z < 0$ and $z > 0$ are filled with materials of permeabilities $1.5\mu_0$ and $5\mu_0$ respectively. The magnetic fields in the regions, in teslas, are $\mathbf{B}_1 = 2.4\hat{\mathbf{x}} + 10\hat{\mathbf{z}}$ and $\mathbf{B}_2 = 25\hat{\mathbf{x}} - 17.5\hat{\mathbf{y}} + 10\hat{\mathbf{z}}$. Determine the current distribution at $z = 0$.

16.2 Consider a linearly polarized electromagnetic wave of frequency ω incident from a region of refractive index n_1 onto a plane interface with a material of refractive index $n_2 < n_1$. The angle of incidence θ_1 is larger than the critical angle θ_c. (a) Determine the direction of propagation and the phase velocity of the E field in the n_2 medium. (b) Determine the direction and coefficient of attenuation of this wave. (c) Show that the average transmitted Poynting vector vanishes.

16.3 Consider the propagation examined in Problem 16.2. Take the direction of the polarization of the wave to be at 45° to the plane of incidence. The phases of the p and s polarization, ϕ_p and ϕ_s, were calculated in Example 16.4. (a) Determine the angles of incidence for which $\phi_p = \phi_s$. (b) Determine the maximum phase difference. (c) Can $\phi_p - \phi_s = \pi/2$ and hence the reflected wave become circularly polarized?

16.4 The region $0 \le z \le z_0$ is filled with a dielectric material of permittivity $\varepsilon = 4\varepsilon_0$ and permeability $\mu = \mu_0$. A linearly polarized wave of amplitude $E_0\hat{\mathbf{x}}$ and angular frequency ω is incident normally on the interface at $z = 0$, from the region $z < 0$. Show that the ratio of the reflected intensity to the incident intensity in the $z < 0$ region is

$$\left[1 + \frac{16}{9}\csc^2\left(\frac{2\omega z_0}{c}\right)\right]^{1/2}$$

16.5 Consider the layered interface analyzed in Example 16.6. Calculate the transmitted Poynting's vector and sketch its frequency dependence for the cases $n_1 = 1$, $n_2 = 2$, $n_3 = 3$; $n_1 = 2$, $n_2 = 4$, $n_3 = 1$; and $n_1 = 3$, $n_2 = 2$, $n_3 = 1$.

16.6 Consider the layered interface analyzed in Example 16.6 where $n_1 < n_3 < n_2$. Determine the reflected Poynting's vector and determine dn_2 such that it is (a) minimal and (b) maximal for $\omega = \omega_0$.

16.7 Consider an optical fiber such as the one shown in Fig. 16.5. The refractive indices of the fiber and the surrounding material are 1.55 and 1. Calculate the largest angle between the fiber axis and the direction of a light ray that would allow the propagation along the fiber. Cladding materials are also used to surround the fiber; calculate the change in the angle when the cladding material has a refractive index equal to 1.52 and 1.50.

16.8 A material of conductivity σ_c and permittivity ε fills the region $z \geq 0$. A plane monochromatic wave of frequency ω is normally incident on the conducting interface. Show that the ratio of the reflected amplitude to the incident amplitude is $(\hat{n}_2 - 1)/(\hat{n}_2 + 1)$, where $\hat{n} = n_2 + ik_2 = \sqrt{\varepsilon/\varepsilon_0}(1 + i\sigma_c/\varepsilon\omega)^{1/2}$.

16.9 A p polarization wave is incident from a material of refractive index n_1 onto a highly conducting, plane interface. The angle of incidence θ_1 is such that the incidence is near the grazing-angle incidence ($\pi/2 - \theta_1 = \phi_0 \ll 1$). Show that the Fresnel coefficients are given by Eqs. (16.96).

16.10 Derive both parts of Eq. (16.99).

16.11 Derive Eq. (16.101).

16.12 An electromagnetic wave is incident on a highly conducting, plane interface with an angle of incidence $\pi/2 - \phi_m$, which minimizes the reflectivity of the p waves (see Example 16.10). (a) Show that the p wave suffers a phase change of $\pi/2$ upon reflection. (b) Show that the s wave suffers a phase change of π upon reflection. (c) What should be the ratio of the incident amplitudes of p and s polarization in order for the reflected wave to be (1) circularly polarized and (2) linearly polarized?

16.13 An electromagnetic wave of amplitude $H_0\hat{\mathbf{x}} = 3\hat{\mathbf{x}}$ (A/m) and frequency $\omega = 400\pi \times 10^6$ rad per sec is incident normally on a foil of silver of thickness 15 μm and conductivity $\sigma_c = 62 \times 10^6\ (\Omega\cdot\text{m})^{-1}$. Determine the field amplitudes just after it enters the foil, just after it exists the foil, and just before it exists the foil.

16.14 Two infinite copper parallel plates are 8 cm apart. A TE wave of frequency 3×10^9 Hz propagates in the y-z plane, where y is normal to the plates. (a) Determine the cutoff wavelength for TE_2 mode. (b) Determine the waveguide wavelength for the TE_1 mode.

16.15 An electromagnetic wave of angular frequency ω and wave number k is propagating unattenuated in the z direction between two perfectly conducting plates located at $y = 0$ and $y = a$. (a) Using $\nabla \cdot \mathbf{E} = 0$, show that the amplitude of the electric field is independent of x. (b) Derive a differential equation for the amplitude of the electric field as a function of y, and determine the electric field as a function of space and time. (c) Find the smallest value for ω such that the wave can propagate without attenuation.

16.16 Two perfectly conducting plates are located at $y = 0$ and $y = a$. Consider a TM wave propagating between the plates with $H_x = H_0 \cos(ky \cos\beta)\cos(kz \sin\beta - \omega t)$, and $H_y = H_z = 0$ where β, H_0, and k are constants. (a) Determine an expression for $\mathbf{E}$ between the plates. (b) Determine the relation between β, a, and k. (c) Show that the time average of the Poynting vector is entirely in the z direction. (d) Calculate the average power crossing the rectangle bounded by $x = 0$, $x = a$, $y = 0$, and $y = b$.

16.17 Consider the propagation of TM waves between two perfectly parallel conducting plates at $y = 0$ and $y = a$. (a) Determine the $\mathbf{E}$ and $\mathbf{B}$ fields propagating in the y-z plane between the plates. (b) What is the lowest mode? (c) Sketch the field patterns of the lowest mode.

16.18 Consider the propagation of a TE wave between the plates discussed in Section 16.4.1. (a) Calculate the average energy density and the average Poynting vector per unit width for the lowest two modes. (b) Calculate also v_p and v_g for the same modes.

16.19 (a) Show that TEM waves wherein $E_z = H_z = 0$ can exist between two large, parallel conducting planes placed at $x = 0$ and $x = a$. As an example, choose $\mathbf{H} = H_0(z)\hat{\mathbf{y}}$ and $\mathbf{E} = E_0(z)\hat{\mathbf{x}}$. (b) Show that this solution has no restriction on ω and can even work for direct current flow ($\omega = 0$).

16.20 Determine the phase and group velocities of electromagnetic waves in rectangular waveguides with perfectly conducting walls. Determine their dependence on the free-space wavelength.

16.21 Consider a rectangular wave guide of dimensions a and b along x and y coordinates and of axis along z. (a) If $a > b$, show that the largest cutoff wavelength of the TE_{mn} mode occurs for $m = 1$ and $n = 0$. (b) If $a = 2b$, make a table of the cutoff frequency for $m, n \leq 4$. (c) What modes will propagate in the wavelength ranges $1.1a$–$2a$ and $0.41a$–$2a$.

16.22 Consider the possibility of having a TEM wave in a waveguide wherein both E_z and $H_z = 0$ for all x, y, and z. (a) Show that ω is not restricted to a discrete set of frequencies. (b) Determine H_x and H_y for a rectangular guide of dimensions a and b along x and y, respectively. (c) Use Maxwell's equations to show that such solutions of the wave equation cannot be satisfied for real mode numbers m and n. (Note that there are solutions for TEM waves in coaxial cylindrical guides and in parallel-plane geometry. See Problem 16.19.)

16.23 Show that the $m = n = 0$ solution in a rectangular waveguide does not satisfy all of Maxwell's equations and the boundary conditions even if it does satisfy the wave equation. [Note that H_x and H_y are not necessarily zero for this special case (TE mode)].

16.24 A cavity resonator with perfectly conducting walls has a rectangular shape of sides a, b, and c. The allowed modes in the cavity are characterized by the integers l, m, and n, respectively. (a) Determine the degeneracy of the allowed modes when l, m, and $n \neq 0$. (b) Determine the degeneracy when either l, m, or n is zero. (c) Discuss the order of the degeneracy when a, b, and c are in the ratio of integers.

16.25 Consider TE oscillations in a rectangular cavity resonator of sides a, b, and c where $E_z \equiv 0$. (a) Determine the fields of the allowed modes. (b) Determine the lowest cavity mode. (c) Calculate the spectrum of resonance frequencies when $a = b = c$. (d) Show that the fields of part (a) can be obtained by specializing the fields derived in Section 16.4.3 for the general case.

16.26 Show that the fields in a cavity resonator such as the one treated in Section 16.4.3 transmit no energy.

SEVENTEEN

SPECIAL THEORY OF RELATIVITY—ELECTRODYNAMICS

The electromagnetic interaction between two fast-moving, charged particles occupies an important position in modern physics. It is usually analyzed in two steps. In the first step, the fields produced by one of the charges are calculated, and in the second step, the interaction between the fields and the other charge is determined. Since we calculated the fields produced by time-dependent distributions of charges and currents in Chapter 15, we analyze here some aspects of the motion of charged particles in external electromagnetic fields. The interaction of each charge with its own field (self-interaction), however, will not be considered here.

In the course of this treatment, however, we face some difficulty, because the field measurement is not unique since it depends on the state of the motion of the observer. It is the aim of this chapter to answer the following questions. Do the equations governing electromagnetism (Maxwell's equations) depend on the motion of the observer? How do the potentials and the fields measured by an observer relate to those measured by another observer? In this book, however, we will not go beyond the answers to these questions. (We will not formulate relativistic mechanics.)

The answers to these questions lie in the heart of the special theory of relativity, whose history of development is quite interwoven with that of electromagnetism. In fact, the research for the unification of electricity, magnetism, and optics made the adoption of special relativity a necessity. Lorentz and Poincaré established the ground work of the theory in their studies of electromagnetism, and Einstein's work placed the theory on a more general basis.

17.1 Galilean Transformation and the Wave Equation

The principle of relativity (Galilean relativity) was first introduced by Newton in his discussion of the laws of motions: The motion of objects in a given region of space are the same among themselves, whether that region is at rest or moves uniformly in a linear motion (straight line).

In order to see the validity of this principle, we take two frames S and S' with parallel axes and origins coinciding originally at $t = 0$, and attached respectively to the region when it is at rest, and when it is in uniform motion along z with velocity v. The spatial coordinates and time of an event occurring in this space as measured in S and S' are related by what are called the *Galilean* transformations:

$$x' = x \qquad y' = y$$
$$z' = z - vt \tag{17.1}$$
$$t' = t$$

It is easy to show that if we substitute this transformation into Newton's laws as stated in S, we find that these laws transform to the same laws in S'. That is, Newton's laws are said to be *invariant* under a Galilean transformation, and hence it is impossible to determine the absolute velocity of any reference frame by mechanical experiments.

On the other hand, it has been well known that the propagation of *mechanical* waves in stationary media has a fixed velocity with respect to the medium and appears more complicated when observed in a reference frame moving with respect to the medium. This can be easily seen by examining the one-dimensional wave equation of mechanical waves, which have a fixed propagation velocity c relative to the stationary medium (under Galilean transformation).

$$\frac{\partial^2 \Phi}{dz^2} - \frac{1}{c^2}\frac{\partial^2 \Phi}{\partial t^2} = 0$$

Taking $\partial/\partial z = \partial/\partial z'$, $\partial/\partial t = \partial/\partial t' - v(\partial/\partial z')$, we find that

$$\frac{\partial^2 \Phi}{\partial z'^2} - \frac{1}{c^2}\frac{\partial^2 \Phi}{\partial t'^2} + \frac{v^2}{c^2}\frac{\partial^2 \Phi}{\partial z'^2} - \frac{2v}{c^2}\frac{\partial^2 \Phi}{\partial t'\,\partial z'} = 0$$

which shows that the transformed equation is no longer of the same form as the original one (it is not invariant).

Because the investigations into the phenomena of electricity and magnetism, and of light, led to the formulation of the laws of motion of charged particles—Maxwell's equations—it was logical to examine these laws under Galilean transformation. Since Maxwell's equations lead *directly* to a scalar wave equation governing each cartesian component of **E** and **B**, it was realized that the Galilean transformation will not keep the electromagnetic wave equation invariant. Hence the transformation is not the correct one for the laws of motion of charged particles. This failure and other experimental findings (see Section 17.3) led to the finding of the correct transformation, which is introduced in the following section.

17.2 Lorentz Transformation

By studying the transformation properties of Maxwell's equations, Lorentz discovered what is now known as the *Lorentz transformation*, which leaves the equations in the same form. To describe the transformation, we consider two coordinate systems S and S', where S' is moving with uniform velocity v along the z axis relative to S. The motion is such that the coordinate axes stay parallel and that at

$t = 0$ they coincide. The relation between the space and time coordinates of the two systems (x', y', z', t') and (x, y, z, t) are related by the Lorentz transformation:

$$x' = x \qquad y' = y$$

$$z' = \frac{z - vt}{(1 - v^2/c^2)^{1/2}} \qquad t' = \frac{t - (v/c^2)z}{(1 - v^2/c^2)^{1/2}} \tag{17.2}$$

It is customary to use the abbreviations

$$\beta = \frac{v}{c} \qquad \text{and} \qquad \gamma = \left(1 - \frac{v^2}{c^2}\right)^{-1/2} = (1 - \beta^2)^{-1/2} \tag{17.3}$$

To get the *inverse transformation* equations, giving x, y, z, and t, in terms of x', y', z', and t', we merely interchange the prime signs and replace v with $-v$ in Eq. (17.2) or (17.3). The relation between the velocities in S and S' can be derived from the above transformations. Consider Fig. 17.1. If there is a velocity vector $\mathbf{u}'$ in the S' system that makes polar angles θ' and ϕ' with z' of the S' system with components $u'_x = \partial x'/\partial t'$, $u'_y = \partial y'/\partial t'$, and $u'_z = \partial z'\partial t'$, then the corresponding components in the S system $u_x = \partial x/\partial t$, $u_y = \partial y/\partial t$, and $u_z = \partial z/\partial t$ can be determined from Eq. (17.2) or rather the inverse transformation. Using the chain rule,

$$dx = dx' \qquad dy = dy' \tag{17.4}$$

$$dz = \left(1 - \frac{v^2}{c^2}\right)^{-1/2} [dz' + v\,dt']$$

$$dt = \left(1 - \frac{v^2}{c^2}\right)^{-1/2} \left[dt' + \left(\frac{v}{c}\right)^2 dz'\right] \tag{17.5}$$

Thus substituting Eqs. (17.4) and (17.5) in $u_i = dx_i/dt$ gives

$$u_x = \frac{dx}{dt} = \frac{dx'}{dt} = \left(1 - \frac{v^2}{c^2}\right)^{1/2} \frac{dx'}{dt' + (v/c)^2 dz'} = \left(1 - \frac{v^2}{c^2}\right)^{1/2} \frac{u'_x}{1 + vu'_z/c^2} \tag{17.6}$$

$$u_y = \frac{(1 - v^2/c^2)^{1/2} u'_y}{1 + vu'_z/c^2} \qquad u_z = \frac{u'_z + v}{1 + vu'_z/c^2} \tag{17.7}$$

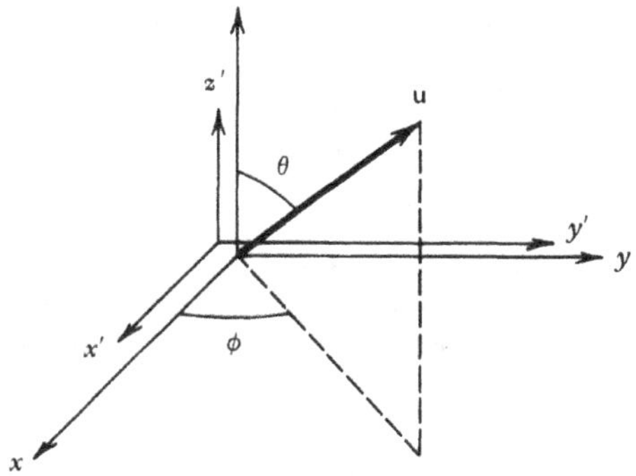

Figure 17.1 Two coordinate systems moving relative to each other with uniform velocity v along the z axis. The systems coincide at $t = 0$ and stay parallel.

The magnitude of **u** and the polar angles θ and ϕ that it makes with the z axis in terms of θ' and ϕ' can be determined from Eqs. (17.6) and (17.7). The results are (see Problem 17.3) $\phi = \phi'$ and

$$\tan\theta = \left(1 - \frac{v^2}{c^2}\right)^{1/2} \frac{u'\sin\theta'}{u'\cos\theta' + v} \tag{17.8}$$

$$u = \frac{[u'^2 + v^2 + 2u'v\cos\theta' - (u'^2v^2/c^2)\sin^2\theta']^{1/2}}{1 + u'v/c^2\cos\theta'} \tag{17.9}$$

The inverse transformation giving (θ', u') in terms of (θ, u) can be obtained by interchanging the prime signs and changing the sign of v.

We will now make the following comments about Lorentz coordinate and velocity transformations.

Limit of Small v/c (Galilean Transformation). The Lorentz transformation given by Eq. (17.2) reduces to the Galilean transformation in the range of low speeds such that $v/c \ll 1$. Neglecting terms of order v^2/c^2 or higher in these equations gives

$$x = x' \qquad y = y' \qquad z = z' + vt \qquad t = t' \tag{17.10}$$

These are the Galiean coordinate transformations in Eq. (17.1). It is to be noted that the Galilean transformation can also be looked at as the Lorentz transformation in the limit where the speed of light is infinite ($c \to \infty$).

The Lorentz velocity transformations reduce to the Galilean velocity transformation in the limit when both the velocity of the coordinate system v and the velocity u are small compared to the speed of light ($v/c \ll 1$, $u/c \ll 1$, and $uv/c^2 \ll 1$). In this limit Eqs. (17.6) and (17.7) give $u_x = u'_x$, $u_y = u'_y$ and $u_z = u'_z + v$.

As we mentioned above the Galilean transformations between two frames of reference in relative motion had been long known in classical physics before the introduction of Lorentz transformation. It was known that Newton's laws of motion were unaffected by a Galilean transformation, with the result that it is impossible to determine the absolute velocity of any reference frame by mechanical means.

Maximum Possible Speed for Any Observer. It is clear from (Eqs. 17.2) that no speed can exceed c, the speed of light. The case $v > c$ makes the factor $(1 - v^2/c^2)^{1/2}$ imaginary, hence making the time or space coordinates imaginary.

The Lorentz velocity transformation also shows that it is impossible to obtain a velocity greater than that of light by adding two velocities even if each is very close to it. This can be seen using Eqs. (17.8) and (17.9). For the parallel velocity case ($\theta' = \theta = 0$), Eq. (17.9) gives

$$u = \frac{u' + v}{1 + u'v/c^2} \tag{17.11}$$

which shows that even as u' and v approach c, u approaches c without exceeding it.

Constancy of the Speed of Light. As the Lorentz transformations stand, they imply that for all reference frames moving with uniform speeds there is only one value of the speed of light. That is, the speed of light is the same for all inertial frames [see (Eq. 17.11)].

Invariance of the wave equation. We analyze what happens to the electromagnetic, scalar, homogeneous wave equation under a Lorentz transformation. Consider the equation

$$\nabla^2\Phi - \frac{1}{c^2}\frac{\partial^2\Phi}{\partial t^2} = \frac{\partial^2\Phi}{\partial x^2} + \frac{\partial^2\Phi}{\partial y^2} + \frac{\partial^2\Phi}{\partial z^2} - \frac{1}{c^2}\frac{\partial^2\Phi}{\partial t^2} = 0 \tag{17.12}$$

where Φ is a scalar function that stands for the scalar potential, the cartesian components of the vector potential, or the cartesian components of the electric field or magnetic field. The derivatives with respect to x, y, z, and t can be written in terms of the primed quantities using the Lorentz transformation and the chain rule. This gives

$$\frac{\partial^2}{\partial x^2} = \frac{\partial^2}{\partial x'^2} \qquad \frac{\partial^2}{\partial y^2} = \frac{\partial^2}{\partial y'^2} \qquad \frac{\partial^2}{\partial z^2} = \left(\frac{\partial}{\partial z'} - \frac{v}{c^2}\frac{\partial}{\partial t'}\right)^2\left(1 - \frac{v^2}{c^2}\right)^{-1} \tag{17.13}$$

$$\frac{\partial^2}{\partial t^2} = \left(\frac{\partial}{\partial t'} - v\frac{\partial}{\partial z'}\right)^2\left(1 - \frac{v^2}{c^2}\right)^{-1} \tag{17.14}$$

Substituting for the second derivatives in Eq. 17.12 gives

$$\nabla^2\Phi - \frac{1}{c^2}\frac{\partial^2\Phi}{\partial t^2} = \nabla'^2\Phi - \frac{1}{c^2}\frac{\partial^2\Phi}{\partial t'^2} = 0$$

which shows that the form of the wave equation is retained in the primed system; and since it is homogeneous we can interchange Φ and Φ' and write

$$\nabla^2\Phi' - \frac{1}{c^2}\frac{\partial^2\Phi'}{\partial t^2} = 0 \qquad \Phi \to \Phi' \tag{17.15}$$

This result indicates that the homogeneous scalar wave equation is *invariant* under a Lorentz transformation, whereas it is not invariant under a Galilean transformation, as we showed in Section 17.1.

Invariance of the "Length" Element and Proper Time. First we define what we mean by the "length" element and the *proper time*. In classical physics where electromagnetic interactions are not involved (Galilean relativity), the space and time coordinates are not interrelated; that is, the infinitismal element of distance, $ds = (dx^2 + dy^2 + dz^2)^{1/2}$, and the infinitesimal element of time, dt, are separately invariant under Galilean transformations [see Eq. (17.1)]. Thus

$$ds^2 = ds'^2 \qquad \text{and} \qquad dt^2 = dt'^2 \tag{17.16}$$

In physics where electromagnetic interactions are involved (Lorentzian relativity), time and space coordinates are interrelated. We define a differential "length" element by

$$ds^2 = dx^2 + dy^2 + dz^2 - c^2\,dt^2 \tag{17.17}$$

It is easy to show that this length is invariant under a Lorentz transformation (see Problem 17.5). The concept of an invariant "length" element, that is,

$$dx^2 + dy^2 + dz^2 - c^2\,dt^2 = dx'^2 + dy'^2 + dz'^2 - c^2\,dt'^2 \tag{17.18}$$

leads to the concept of the invariant proper time. We define the proper time as the time measured in the moving reference frame—that is, in S'. Thus for an event occurring in S', $dx' = dy' = dz' = 0$ and dt' is called the proper time. Using Eq. (17.18) we find, for this case,

$$-c^2\,dt'^2 = ds^2 = dx^2 + dy^2 + dz^2 - c^2\,dt^2 \tag{17.19}$$

Because ds is an invariant quantity, then the proper time $dt' = (1/c)ds$ is an invariant quantity.

Time Dilation. Equation (17.19) can now be written in terms of the velocity of the S' frame: Writing $dx/dt = v_x$, and so forth, it becomes

$$dt = dt'\left(1 - \frac{v^2}{c^2}\right)^{-1/2} \tag{17.20}$$

where $v^2 = v_x^2 + v_y^2 + v_z^2$. We should note that, in some books on relativity, the proper time t' is often labeled as τ_0. Equation (17.20) shows what is called the *time-dilation* effect. It shows that, for example, the time measured by a clock in S' is less than the time measured by a clock in the nonmoving frame S. Thus a clock moving relative to an observer runs more slowly than one that is at rest relative to the observer.

Length Contraction. Consider an object at rest in the reference frame S', which is moving with speed v parallel to the z axis relative to a frame S. The dimension along the z' axis measured in S' is $l' = z_2' - z_1'$. The Lorentz transformation can be used to determine the corresponding length as measured in S. Taking the time at which z_2 and z_1 are observed to be the same gives

$$z_2 - vt = \left(1 - \frac{v^2}{c^2}\right)^{1/2} z_2' \qquad \text{and} \qquad z_1 - vt = \left(1 - \frac{v^2}{c^2}\right)^{1/2} z_1' \tag{17.21}$$

Thus $l = z_2 - z_1$ is as follows:

$$l = \left(1 - \frac{v^2}{c^2}\right)^{1/2} l' \tag{17.22}$$

We should note that in some books on relativity the symbol l_0 is used for l'.

This result shows that the length of the object along the z axis gets contracted as seen in reference S. This effect is known as the *FitzGerald-Lorentz contraction rule.* Although the points z_2 and z_1 were measured simultaneously at time t, the points z_2' and z_1' in S' are not measured simultaneously.

Phase, Wave Vector, and Frequency of Electromagnetic Waves. The phase ϕ of a plane electromagnetic wave (Fig. 17.2) at time t and position $\mathbf{r}$ relative to a reference frame S is

$$\phi = \mathbf{k}\cdot\mathbf{r} - \omega t \tag{17.23}$$

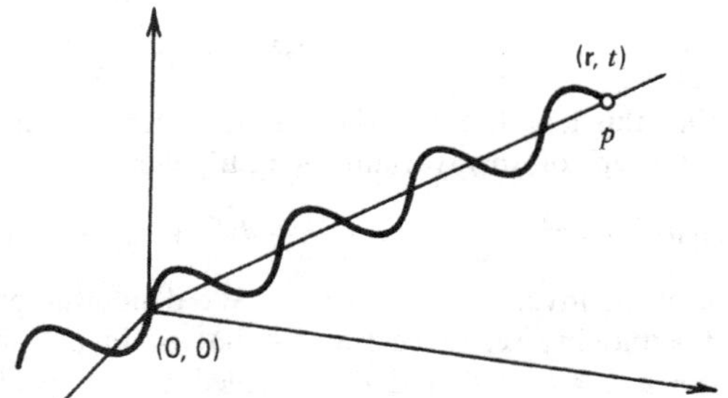

Figure 17.2 The waveform of a plane electromagnetic wave at time t, showing the number of crest reaching the observation point P from the origin in the time period from zero to t.

where $\mathbf{k}$ is the wave vector of the wave. An observer P located at a distance $\mathbf{r}$ from the origin counts the number of wave crests passing him or her. The crests are labeled as follows: The crest that passes the origin at $t = 0$ is 1, followed at a later time by 2, and so on. When the crest labeled 1 reaches the observer, he starts counting by counting it as 1. At time t, he will have counted $N = \phi/2\pi$ wave crests or $N = (1/2\pi)(\mathbf{k}\cdot\mathbf{r} - \omega t)$. Because N is just a number, then it should be independent of the coordinate system being used, and hence invariant under a Lorentz transformation (see Problem 17.6). The invariance means that

$$\mathbf{k}\cdot\mathbf{r} - \omega t = \mathbf{k}'\cdot\mathbf{r}' - \omega' t' \tag{17.24}$$

where the prime designates that the quantities are measured in a reference frame S' moving with velocity v parallel to the z axis relative to S, and coinciding at $t = 0$. Using the Lorentz transformation [Eq. (17.2)] one can show that

$$k'_x = k_x \qquad k'_y = k_y \qquad k'_z = \left(1 - \frac{v^2}{c^2}\right)^{-1/2}\left(k_z - \frac{v}{c^2}\,\omega\right) \tag{17.25}$$

$$\omega' = \left(1 - \frac{v^2}{c^2}\right)^{-1/2}(\omega - vk_z) \tag{17.26}$$

In spherical polar coordinates, we take $\mathbf{k}$ and $\mathbf{k}'$ to make angles θ and θ' relative to the z and z' axes (direction of v), respectively. Since $k = \omega/c$, $k' = \omega'/c$, $k_z = k\cos\theta$, and $k'_z = k'\cos\theta'$, then one can show that

$$\omega' = \frac{\omega}{(1 - v^2/c^2)}\left(1 - \frac{v}{c}\cos\theta\right) \tag{17.27}$$

The numerator gives the customary *Doppler shift* valid at low speeds ($v/c \ll 1$). The factor $(1 - v^2/c^2)^{-1/2}$, on the other hand, modifies the customary shift at higher speeds. Similarily one can show that

$$\tan\theta' = \frac{(1 - v^2/c^2)^{1/2}\sin\theta}{\cos\theta - v/c} \tag{17.28}$$

which gives the relations between the directions of the wave vectors in the two frames (see Problem 17.7). Also, one can show that

$$\omega' = \frac{(1 - v^2/c^2)^{1/2}}{1 + (v/c\cos\theta')}\,\omega \tag{17.29}$$

Example 17.1 Time Dilation

(*i*) *Rocket Ship.* A rocket of mass 10 tons is designed to travel to a distant star and return to earth with a constant velocity equal to $v = (1 - \frac{1}{2}\times 10^{-4})c$. The distance between the earth and the star is $d = 5$ light-years. The time taken by the rocket to make the round trip as measured by a clock at rest on the earth's surface is $t = 2d/v \approx 10$ years.

To estimate the amount of food, drinks, and other supplies needed for the trip, one needs the time duration of the trip from the point of view of a clock on the rocket: $t' = \sqrt{1 - v^2/c^2}\,t = 0.01 \times 10 = 0.1$ year $= 1.2$ months. Thus the mission control needs to stock supplies on the rocket for only 1.2 months rather than for 10 years. This is relatively inexpensive; however, it takes a lot of energy to get the rocket to the above speed.

(*ii*) *Decay of Muons.* When a muon, an elementary particle of mass $m = 207m_e$, where m_e is the mass of the electron, is produced *at rest* in the laboratory, an observer at rest in the laboratory measures its lifetime, which is a measure of how long the particle lives before it decays to something else to be $t' = 1.52 \times 10^{-6}$ s. We used t' since this is the proper time measured in the rest frame of the event.

To test the time-dilation concept, one has to find the change in the lifetime when the particle is produced in motion with respect to the laboratory. The production of this particle with $v \approx 0.98c$, in fact, takes place above the earth in cosmic-ray collisions with nuclei in the atmosphere. The observer in the laboratory on earth who applies the time-dilation concept, would predict a longer lifetime t for the moving particle. That is,

$$t = \frac{t'}{(1 - v^2/c^2)^{1/2}} = 5t' = 7.6 \times 10^{-6} \text{ s}$$

Thus if the particle indeed lives longer, the laboratory observer will find that the particle travels longer distances than what one would expect without the application of the concepts of relativity. Experimental observations were performed and it was found that indeed the particles travel longer distances, thus confirming the concept of time dilation. How do these decay events appear to an observer moving with the decaying muons? This question is left as an exercise (see Problem 17.1).

Example 17.2 Transverse Doppler Effect—Red Shift

A laser source is moving with velocity v_0 along the z axis with respect to a detector. The laser beam is directed along the x axis according to an observer at rest with respect to the detector, as shown in Fig. 17.3. The frequency of the laser beam measured in the rest frame of the laser is ω_0. Let us take the rest frame of the laser to be frame S' and the rest frame of the detector to be S. When the beam hits the detector, the detector measures a frequency that is different from ω_0, and we will take it to be ω. We can use Eq. (17.26) to determine ω, that is,

$$\omega_0 = \left(1 - \frac{v_0^2}{c^2}\right)^{-1/2} (\omega - v_0 k_z) \tag{17.30}$$

Since the wave vector of the laser radiation is along x, then $k_z = 0$. Thus

$$\omega = \left(1 - \frac{v_0^2}{c^2}\right)^{1/2} \omega_0 \tag{17.31}$$

The same result can be also determined from Eq. (17.27). Since the angle θ between the propagation and the direction of motion of S' relative to S is $\pi/2$, then $\cos\theta = 0$, and hence Eq. (17.27) reduces to the above result.

The result of Eq. (17.31) is referred to as the *transverse Doppler effect*, or the so-called *red shift*. Note that it is the same whether the detector is approaching or departing the source. Moreover, the leading term in the expansion is quadratic (second order in v/c); that is, it depends on v_0^2/c^2, which makes it generally smaller than the first-order effect. This *relativistic* transverse Doppler shift has been observed spectroscopically with atoms in motion (in 1938). The direction of the propagation as measured by the detector can be calculated from Eq. (17.25) or from Eq. (17.28); see Problem 17.8.

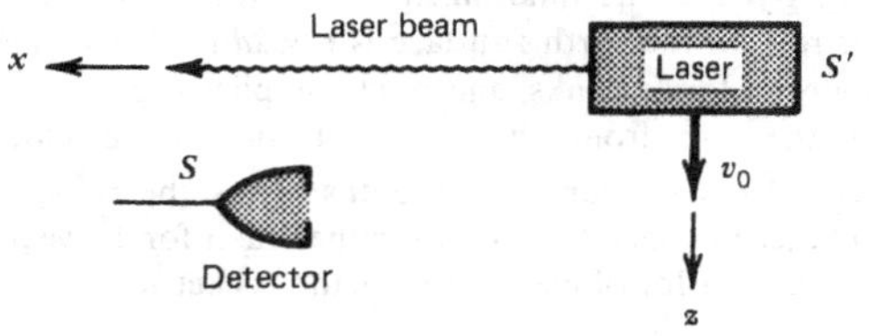

Figure 17.3 Transverse (relativistic) Doppler effect. A laser is moving normal to its own output beam relative to a stationary detector.

Example 17.3 Solid-Angle Transformation

A beam of light subtends a solid angle $d\Omega$ as measured by an observer in frame S. We are interested in calculating the corresponding solid angle of the beam observed by an observer S' moving along the z axis with velocity v relative to the first observer.

Since $d\Omega = \sin\theta\, d\theta\, d\phi$, then $d\Omega'$ in S' is $d\Omega' = \sin\theta'\, d\theta'\, d\phi'$. The angle ϕ' seen by observer S' is the same as seen by observer S. The angle θ' can be calculated from Eq. (17.28). Using trigonometric relations, one can show that

$$\cos\theta' = \frac{\cos\theta - v/c}{1 - (v/c)\cos\theta} \qquad \sin\theta' = \frac{(1 - v^2/c^2)^{1/2}\sin\theta}{1 - v/c} \tag{17.32}$$

Thus

$$d\Omega' = \frac{1 - v^2/c^2}{[1 - (v/c)\cos\theta]^2}\, d\Omega \tag{17.33}$$

For $v/c = 0.95$, this result shows that $d\Omega'/d\Omega$ drops from its maximum value at $\theta = 0$ (namely, 39) to its minimum value at $\theta = \pi$ (namely, 0.024). Moreover it can be shown that $\int d\Omega' = 4\pi$ for any velocity v.

Example 17.4 Reflection Laws from a Moving Mirror

A mirror S' is moving relative to the laboratory frame S with a velocity v along the normal to its surface. According to an observer at rest with respect to the mirror, a linearly polarized wave of frequency ω'_1, and wave vector $\mathbf{k}'_1$ is obliquely incident on its surface at an angle θ'_1 with respect to its normal, as shown in Fig. 17.4. He also sees the wave get reflected at an angle θ'_2 with respect to the normal and with a frequency ω'_2 and wave vector $\mathbf{k}'_2$. According to the laws of reflection in S', the wave gets reflected at an angle $\theta'_2 = \pi - \theta'_1$, with a frequency $\omega'_2 = \omega'_1$.

The laws of reflection from the moving mirror, however, are not the same according to an observer at rest in the laboratory. These can be calculated using Eqs. (17.27) and (17.28). First, let us call the frequencies of the incident and reflected waves seen by an observer S at rest with respect to the laboratory ω_1 and ω_2. Moreover, let us call the angles of incidence and reflection relative to the normal in the S frame θ_1 and θ_2.

From Eq. (17.28) or (17.32) we write

$$\cos\theta'_2 = \frac{\cos\theta_2 - \beta}{1 - \beta\cos\theta_2} \qquad \cos\theta'_1 = \frac{\cos\theta_1 - \beta}{1 - \beta\cos\theta_1} \tag{17.34}$$

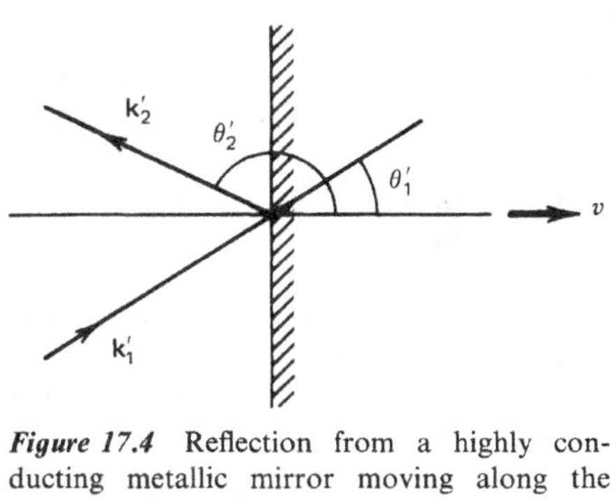

Figure 17.4 Reflection from a highly conducting metallic mirror moving along the normal to its surface.

Taking $\cos\theta_1' = -\cos\theta_2'$ in these equations, we get

$$\frac{\cos\theta_2 - \beta}{1 - \beta\cos\theta_2} = -\frac{\cos\theta_1 - \beta}{1 - \beta\cos\theta_1}$$

Rearranging the terms gives

$$\cos\theta_2 = -\frac{(1 + \beta^2)\cos\theta_1 - 2\beta}{1 - 2\beta\cos\theta_1 + \beta^2} \tag{17.35}$$

We now determine the relation between ω_1 and ω_2. Using (Eq. 17.27) we write

$$\omega_1' = \frac{\omega_1}{\sqrt{1 - \beta^2}}(1 - \beta\cos\theta_1) \qquad \omega_2' = \frac{\omega_2}{\sqrt{1 - \beta^2}}(1 - \beta\cos\theta_2)$$

Taking $\omega_1' = \omega_2'$ gives

$$\omega_2 = \frac{1 - \beta\cos\theta_1}{1 - \beta\cos\theta_2}\omega_1 \qquad \text{or} \qquad \omega_2 = \frac{1 - 2\beta\cos\theta_1 + \beta^2}{1 - \beta^2}\omega_1 \tag{17.36}$$

17.3 Postulates of Special Relativity

Although the research carried out by Lorentz in electrodynamics, which resulted in his discovery of the Lorentz transformation, carries the basis for the theory of special relativity, the far-reaching consequences of the theory were not actually realized by him. It was Einstein and Poincaré who saw through all the bits and pieces of the new experimental evidence and theoretical advances resulting in the emergence of the theory of special relativity.

In this section we discuss this theory, but before presenting its postulates we give a brief description of the chronological unfolding of research preceding its introduction. This will give us more appreciation of the importance of the theory and will show us how it solved many of the problems and cleared up much of the confusion in physics.

In 1865 Maxwell modified the existing equations of electromagnetism, in order to eliminate the contradiction present in them, by introducing the displacement current. A triumphant result of his modification is the explanation of the nature of light; the new set, Maxwell's equations, correlated light and electromagnetism and showed that light is just an electromagnetic wave.

With the wave nature of light established, researchers directed their attention to the study of its propagation. Influenced by the previously established facts that mechanical waves needed a medium to propagate in, researchers assumed that light needed a medium through which to propagate. In view of the fact that light propagates over all space, it was necessary to assume that the medium, which was called "ether," permeated all space. Moreover, the medium was assumed to be of negligible density and to have negligible interaction with matter.

This special medium, however, caused some problems. What about the state of motion of ether? How do Maxwell's equations transform under the existing Galilean transformation? The existence of an ether implied that the laws of electromagnetism were not invariant under Galilean coordinate transformation; there was a preferred coordinate system in which the ether was at rest. This contradicted what had been known that laws of mechanics were the same in inertial coordinate systems—systems moving uniformly relative to one another.

In order to remove this contradiction and to avoid setting electromagnetism apart from the rest of physics, a number of solutions were introduced. These solutions included the following:

1. Assume that the velocity of light is c with respect to a reference frame in which the source is at rest.
2. Assume that the preferred reference frame for light is the one in which the medium through which the light is propagating (ether) is at rest.
3. Assume that astronomical bodies drag (carry along) the ether despite the fact that it has a very small interaction with matter.

These solutions were abandoned soon after their introduction as a result of the findings of three major experiments:

1. Observation of the aberration of starlight—that is, the shift in the apparent position of distant stars. This observation contradicted the hypothesis that the velocity of light is determined by the transmitting medium or that the ether is dragged along by the earth.

2. Measurements of the velocity of light in moving fluids in 1859 (Fizeau experiment). This experiment showed that if ether existed, it would have to be dragged to some extent by small objects with the degree of dragging depending on the refractive indices of the objects. This result makes the either-drag hypothesis very superficial and not physical.

3. Measurement of the motion of the earth relative to a preferred reference frame—the ether—in 1887 (Michelson-Morely experiment). This experiment showed that there is no evidence for relative motion through the ether. This implies two alternatives: either there is no preferred frame or the earth drags the ether with it. Although this negative result can be explained by the ether-drag hypothesis, this explanation however is inconsistent with the first two experiments.

Influenced by the negative result of the Michelson-Morely experiments, Poincaré reintroduced the principle of relativity, which states that the laws of physics are the same in all frames moving with uniform velocities relative to each other. Simultaneously and independently Einstein formulated special relativity in a general and complete way, obtaining the results of Lorentz and its implications. These postulates are as follows:

1. The laws of nature are the same in all Galilean reference frames—frames moving with uniform velocities relative to one another.
2. The velocity of light is constant, independent of the motion of the source and the observer. This postulate allows the derivation of the Lorentz transformations, which are the correct ones for the electromagnetic phenomena. (See Problem 17.12).

17.4 Geometry of Space-Time (Four-Dimensional Space)—Four-Vectors and Four Tensors

In applying special relativity to physical applications, it is more convenient to treat time, t, on equal footing to the three spatial coordinates x, y, z and introduce what is called a *space-time* geometry (*four-dimensional space*). Moreover, the three-dimensional vectors as we know them in three-dimensional space will be generalized to what is called *four-vectors.*

17.4.1 Three-Dimensional Space—Euclidean Space

Before we introduce the four-dimensional space we will say a few words about the three-dimensional space. Consider a coordinate system x, y, and z, with origin O. The length of the position vector $\mathbf{R}$ of a point whose coordinates are x_1, x_2, and x_3, which stand for x, y, and z, respectively, is given by

$$R^2 = x_1^2 + x_2^2 + x_3^2 \tag{17.37}$$

After any transformation of the coordinate system, the coordinates of the point change to (x'_1, x'_2, x'_3). The transformation is called *linear* if the new coordinates are given by a linear combination of the old coordinates; that is,

$$x'_i = \sum_{j=1}^{3} a_{ij} x_j \tag{17.38}$$

where the a_{ij} coefficients are constants independent of the coordinates, and i and j run from 1 to 3. Furthermore, the transformation is called *orthogonal* if it is a real one (the a coefficients are real) and the length of the vector remains unchanged; that is,

$$R'^2 = x_1'^2 + x_2'^2 + x_3'^2 = R^2 \tag{17.39}$$

Since a rotation of the coordinate system about any of the coordinates x_1, x_2, and x_3 or about any combination of them, is a linear transformation and retains the length of the vector, then such rotations are *linear orthogonal* transformations (see Example 17.5).

Finally we define what is called a *Euclidean space.* A space is called Euclidean if two conditions are met. First, the length of a position vector is the sum of the squares of the coordinates of the endpoint of the vector as given in Eq. (17.37). Second, the rotations in the space are real.

17.4.2 Four-Dimensional Space—Minkowski Space

We now consider a space where the coordinates of a point are given by the three spatial coordinates x, y, and z, and a fourth coordinate based on time, ct, where c is the speed of light. The length of the position vector $\mathbf{R}$ of a point whose coordinates are (x, y, z, ct) is *not* given by $R^2 = x^2 + y^2 + z^2 + c^2t^2$, as one would simply conclude using Eq. (17.37). In fact, it is given [see (Eq. 17.17)] as follows:

$$R^2 = x^2 + y^2 + z^2 - c^2t^2 \tag{17.40}$$

Let us determine the effect of a Lorentz transformation on R. We consider a frame moving along the z axis with velocity v. Using Eq. (17.2), one can easily show that

$$R'^2 = R^2 \tag{17.41}$$

or

$$x'^2 + y'^2 + z'^2 - c^2t'^2 = x^2 + y^2 + z^2 - c^2t^2 \tag{17.42}$$

Equations (17.41) and (17.42) indicate that the length of the vector in four-dimensional space remains unchanged under a Lorentz transformation. Thus the Lorentz transformation is a linear orthogonal transformation. The fact that the length of a vector in this space is not the sum of the squares of the coordinates of the endpoint is, however, troublesome, and indicates that the four-dimensional space is not Euclidean. In order to make it mathematically look like a Euclidean space, Minkowski introduced a mathematical trick by replacing the minus sign in front of c^2t^2 of Eq. (17.40) by i^2, where $i = \sqrt{-1}$ is the unit imaginary number, and wrote

$$R^2 = x^2 + y^2 + z^2 + (ict)^2 \tag{17.43}$$

or

$$R^2 = x_1^2 + x_2^2 + x_3^2 + x_4^2 \tag{17.44}$$

where x_1, x_2, x_3 stand for x, y, z, and

$$x_4 = ict \tag{17.45}$$

Using the Minkowski notation and the abbreviations of Eq. (17.3), $\beta = v/c$ and $\gamma = (1 - \beta^2)^{-1/2}$, the Lorentz transformations of Eq. 17.2 become

$$x'_1 = x_1 \qquad x'_2 = x_2 \tag{17.46}$$

$$x'_3 = \gamma x_3 + i\beta\gamma x_4 \tag{17.47}$$

$$x'_4 = -i\beta\gamma x_3 + \gamma x_4 \tag{17.48}$$

Moreover, they can be written in the form of Eq. 17.38; that is,

$$x'_\mu = \sum_{\nu=1}^{4} a_{\mu\nu} x_\nu \tag{17.49}$$

where μ and ν run from 1 through 4, and $a_{\mu\nu}$ are constants independent of the coordinates. Note that it is customary to use Greek indices (that is, μ and ν) to label four-dimensional vectors and Latin indices (i and j) to label three-dimensional vectors. Comparing Eqs. (17.46) to (17.48) with Eq. (17.49) one finds that the $a_{\mu\nu}$ coefficients are $a_{11} = 1$, $a_{22} = 1$, $a_{33} = \gamma$, $a_{34} = i\beta\gamma$, $a_{43} = -i\beta\gamma$, $a_{44} = \gamma$, and the rest of the coefficients are zero. It is convenient to write $a_{\mu\nu}$ in a matrix form; that is,

$$a_{\mu\nu} = \begin{array}{c} \overset{\nu}{\rightarrow} \\ \mu\downarrow \begin{bmatrix} 1 & 0 & 0 & 0 \\ 0 & 1 & 0 & 0 \\ 0 & 0 & \gamma & i\beta\gamma \\ 0 & 0 & -i\beta\gamma & \gamma \end{bmatrix} \end{array} \tag{17.50}$$

The inverse transformation where x_μ is given in terms of the four coordinates of the moving frame can be determined by inverting Eq. (17.49).

$$x_\mu = \sum_{\nu=1}^{4} x'_\nu a_{\nu\mu} \tag{17.51}$$

Using the matrix of Eq. (17.50) gives

$$\begin{aligned} x_1 &= x'_1 \qquad x_2 = x'_2 \\ x_3 &= \gamma x'_3 - i\beta\gamma x'_4 \\ x_4 &= i\beta\gamma x'_3 + \gamma x'_4 \end{aligned} \tag{17.52}$$

These results show that the inverse transformation can be derived by the interchange of the primed with the unprimed and taking $\beta \rightarrow -\beta$.

It is clear that the Minkowski space is complex since one of the coordinates, $x_4 = ict$, is pure imaginary, and hence any transformation in it is complex.

Moreover, the transformation in this space is orthogonal [see Eq. (17.44)] and may be looked at as a rotation through a complex angle in the $x_3 - x_4$ plane (see Example 17.8). The orthogonality property [see Eq. (17.41)] requires the coefficients of the transformation $a_{\mu\nu}$ given in Eq. (17.50) to satisfy the relation

$$\sum_{\mu=1}^{4} a_{\mu\nu} a_{\mu\lambda} = \delta_{\nu\lambda}$$

where $\delta_{\nu\lambda}$ is equal to 1 for $\nu = \lambda$ and equal to 0 for $\nu \neq \lambda$ (see the proof in Example 17.6).

Although the mathematical form of this space looks exactly like a Euclidean space, it is not physically so because of its complex nature as compared to the real nature of the Euclidean space. In fact, this space is known as Minkowski space after Minkowski, who introduced the mathematical trick given in Eq. (17.45). We should, however, note that many of the properties of the Minkowski space may be derived mathematically by treating it as a Euclidean space.

17.4.3 Vector Properties of Four-Dimensional Space

In the previous subsection we introduced the four-dimensional space in which the space and time coordinates are put on the same footing. In here we present some aspects of the vector algebra of this space in order to allow us to manipulate and recast the equations of electromagnetism from the point of view of the new space.

Four-Dimensional Vectors—Four-Vectors. In three-dimensional (spatial) space, quantities are classified as scalars, vectors, and tensors of various ranks. In the four-dimensional space, on the other hand, quantities are classified as Lorentz scalars (or scalars), Lorentz four-vectors (or four-vectors), and Lorentz tensors (or four-tensors). In four-space, the coordinates of a point are represented by (x_1, x_2, x_3, x_4), and for brevity they are represented by x_μ where μ runs from 1 to 4. The four coordinates form a *four-dimensional vector*. In general a four-dimensional vector A_μ is a set of four quantities, A_1, A_2, A_3, and A_4, that transform as the components x_1, x_2, x_3, and x_4 under a Lorentz transformation. Four-dimensional vectors are also called *four-vectors* or *world vectors*. Using Eqs. (17.46) to (17.48) and Eq. (17.51) we have

$$A'_\mu = \sum_{\nu=1}^{4} a_{\mu\nu} A_\nu \qquad A_\mu = \sum_{\nu=1}^{4} A'_\nu a_{\nu\mu} \tag{17.53}$$

or

$$\begin{aligned} &A'_1 = A_1 \qquad A'_2 = A_2 \\ &A'_3 = \gamma A_3 + i\beta\gamma A_4 \qquad A'_4 = -i\beta\gamma A_3 + \gamma A_4 \end{aligned} \tag{17.54}$$

A *world scalar*, on the other hand, is a quantity that remains unchanged under a Lorentz transformation. Finally a point in space-time (in Minkowski space) is called a *world point*, and a trajectory of a point particle in this space is called a *world line*.

Four-Dimensional Gradient. When a Lorentz scalar Φ is differentiated with respect to x'_μ, the resulting components, $\partial\Phi/\partial x'_\mu$, where $\mu = 1, 2, 3$, and 4 transform as a four-vector. This can be shown as follows. Using the chain rule one writes

$$\frac{\partial \Phi}{\partial x'_\mu} = \sum_{\nu=1}^{4} \frac{\partial \Phi}{\partial x_\nu} \frac{\partial x_\nu}{\partial x'_\mu} \tag{17.55}$$

But

$$x_\nu = \sum_{\mu=1}^{4} a_{\mu\nu} x'_\mu$$

and thus $\partial x_\nu / \partial x'_\mu = a_{\mu\nu}$, where $a_{\mu\nu}$ are the coefficients of Lorentz transformation. Hence

$$\frac{\partial \Phi}{\partial x'_\mu} = \sum_{\nu=1}^{4} a_{\mu\nu} \frac{\partial \Phi}{\partial x_\nu} \tag{17.56}$$

or

$$A'_\mu = \sum_{\nu=1}^{4} a_{\mu\nu} A_\nu$$

where

$$A_\mu = \frac{\partial \Phi}{\partial x_\mu} \tag{17.57}$$

This result indicates that $\partial \Phi / \partial x_\mu$ is a four-vector. This operation is analogous to the gradient operation ∇ of the three-dimensional space. The symbol $\Box$ is sometimes used for the four-dimensional analog of ∇. Thus we write

$$\text{grad } \Phi = \Box \Phi = \frac{\partial \Phi}{\partial x_\mu} \tag{17.58}$$

Four-Dimensional Divergence. When a four-vector, A_μ, is differentiated with respect to x_μ, the resulting components, $\partial A_\mu / \partial x_\mu$ or their sum, where $\mu = 1, 2, 3$, and 4, transform as a scalar under a Lorentz transformation. This can be shown using a similar procedure to the one followed in the gradient case. Using the chain rule, one writes

$$\frac{\partial A'_\nu}{\partial x'_\nu} = \sum_{\mu=1}^{4} \frac{\partial A'_\nu}{\partial x_\mu} \frac{\partial x_\mu}{\partial x'_\nu} \tag{17.59}$$

But

$$x_\mu = \sum_{\nu=1}^{4} x'_\nu a_{\nu\mu}$$

and hence $\partial x_\mu / \partial x'_\nu = a_{\nu\mu}$. Thus Eq. (17.59) becomes

$$\frac{\partial A'_\nu}{\partial x'_\nu} = \sum_{\mu=1}^{4} \frac{\partial A'_\nu}{\partial x_\mu} a_{\nu\mu} \tag{17.60}$$

Summing Eq. (17.60) over ν, we get

$$\sum_{\nu=1}^{4} \frac{\partial A'_\nu}{\partial x'_\nu} = \sum_{\nu=1}^{4} \sum_{\mu=1}^{4} \frac{\partial A'_\nu}{\partial x_\mu} a_{\nu\mu} \tag{17.61}$$

Interchanging the order of summation over μ and ν and using the fact that the coefficients $a_{\mu\nu}$ are independent of x_μ, we get

$$\sum_{\nu=1}^{4} \frac{\partial A'_\nu}{\partial x'_\nu} = \sum_{\mu=1}^{4} \frac{\partial}{\partial x_\mu} \sum_{\nu=1}^{4} A'_\nu a_{\nu\mu} \tag{17.62}$$

From Eq. (17.53),

$$A_\mu = \sum_{\nu=1}^{4} A'_\nu a_{\nu\mu},$$

and thus Eq. (17.62) becomes

$$\sum_{\nu=1}^{4} \frac{\partial A'_\nu}{\partial x'_\nu} = \sum_{\mu=1}^{4} \frac{\partial A_\mu}{\partial x_\mu} \tag{17.63}$$

The sum on either side is called the *four-dimensional divergence* of a four-vector. It is analogous to the three-dimensional divergence defined in Chapter 1:

$$\nabla \cdot \mathbf{F} = \sum_{i=1}^{3} \frac{\partial F_i}{\partial x_i}.$$

The symbol $\square \cdot$ is sometimes used to indicate the four-dimensional divergence. Thus

$$\text{div } A_\mu = \square \cdot A_\mu = \sum_{\mu=1}^{4} \frac{\partial A_\mu}{\partial x_\mu} \tag{17.64}$$

We also note that Eq. (17.63) indicates that the four-dimensional divergence of a four-vector is a Lorentz scalar.

Four-Dimensional Laplacian or d'Alembertian Operator. In three dimensions, the Laplacian ∇^2 was defined by taking the divergence of a gradient of a scalar, or $\nabla \cdot (\nabla \Phi) = \nabla^2 \Phi$. The four-dimensional *Laplacian or d'Alembertian* operator can be determined also using the same procedure. Taking the four-dimensional divergence of the four-dimensional gradient of a scalar Φ, we get $\square \cdot \square\Phi = \square^2\Phi$. Explicitly,

$$\square^2\Phi = \sum_{\mu=1}^{4} \frac{\partial^2 \Phi}{\partial x_\mu^2} = \left(\nabla^2 - \frac{1}{c^2}\frac{\partial}{\partial t^2}\right)\Phi \tag{17.65}$$

Using procedures similar to these we used to derive the transformation properties of the four-dimensional gradient and divergence, one can show that the four-dimensional Laplacian of a Lorentz scalar is a Lorentz scalar too. That is, one can show that

$$\square'^2\Phi = \sum_{\nu=1}^{4} \frac{\partial^2 \Phi}{\partial x_\nu'^2} = \square^2\Phi = \sum_{\mu=1}^{4} \frac{\partial^2 \Phi}{\partial x_\mu^2} \tag{17.66}$$

Moreover, if $\square^2$ operates on some other function, such as a four-vector A_μ, the resulting quantity retains the transformation properties of the function operated on. That is, $\square^2 A_\mu$ is a four-vector.

Four-Tensors. Lorentz four-vectors are also called tensors of the *first rank* in a four-dimensional space. Higher-rank tensors are defined in an analogous way. A second rank tensor $T_{\mu\nu}$, for example, is a set of 16 quantities; it is labeled by two indices and transforms according to the law

$$T'_{\mu\nu} = \sum_{\lambda,\sigma=1}^{4} a_{\mu\lambda} a_{\nu\sigma} T_{\lambda\sigma} \tag{17.67}$$

where $a_{\mu\lambda}$ and $a_{\nu\sigma}$ are the 4×4 Lorentz transformation coefficient matrix [Eq. (17.50)].

A second-rank tensor, which we will encounter in the next section, results from the differentiation of a four-vector A_μ with respect to x_ν, where both μ and ν run from 1 through 4. We can show, using the same procedures used above, that the resulting 16 components $\partial A_\mu/\partial x_\nu$ transform according to the law

$$\frac{\partial A_\mu}{\partial x_\nu} = \sum_{\lambda,\sigma=1}^{4} a_{\mu\lambda} a_{\nu\sigma} \frac{\partial A'_\lambda}{\partial x'_\sigma} \tag{17.68}$$

One can construct what is called *symmetric* or *antisymmetric* tensor from $\partial A_\mu/\partial x_\nu$. For example,

$$T_{\mu\nu} = \frac{\partial A_\mu}{\partial x_\nu} + \frac{\partial A_\nu}{\partial x_\mu} \tag{17.69}$$

is a symmetric tensor; it satisfies the relation $T_{\mu\nu} = T_{\nu\mu}$. On the other hand, the tensor

$$T_{\mu\nu} = \frac{\partial A_\mu}{\partial x_\nu} - \frac{\partial A_\nu}{\partial x_\mu} \tag{17.70}$$

is an antisymmetric tensor since $T_{\mu\nu} = -T_{\nu\mu}$. Note that the four diagonal components of an antisymmetric tensor $T_{\mu\mu}$ are zeros, and therefore it has only six independent components. The symmetric tensor, on the other hand, has nine independent components.

Example 17.5 Rotation in Three Dimensions

We discuss in this example the properties of rotation in the three-dimensional space. Consider a cartesian system with origin O. If the system is rotated about the z axis through an angle θ as shown in Fig. 17.5, then the x, y, and z axes will be transformed into the x', y', and z' axes, where

$$x' = x\cos\theta + y\sin\theta \qquad y' = -x\sin\theta + y\cos\theta \qquad z' = z \tag{17.71}$$

Writing this transformation in the form of Eq. (17.38), one can easily show that the a coefficients are: $a_{11} = \cos\theta$, $a_{12} = \sin\theta$, $a_{21} = -\sin\theta$, $a_{22} = \cos\theta$, $a_{33} = 1$, and $a_{13} = a_{23} = a_{31} = a_{32} = 0$.

Now we show that this rotation is an orthogonal transformation. The length of the position vector in the original coordinate system $\mathbf{R}$ is given by $R = (x^2 + y^2 + z^2)^{1/2}$. In the new

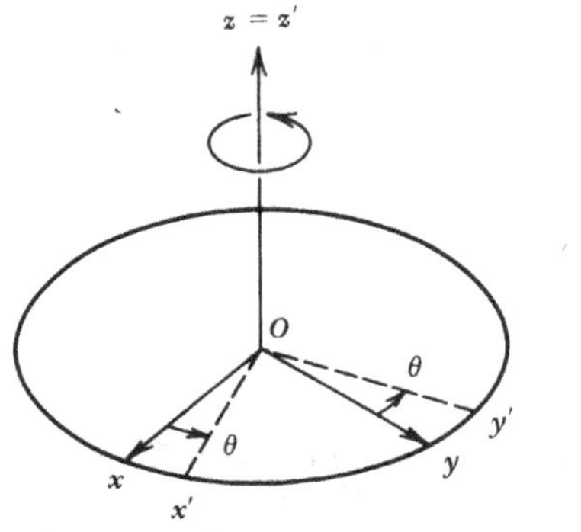

Figure 17.5 Rotation in three-dimensional space about the z axis.

system (that is, the primed system) it becomes $\mathbf{R}'$, where $R' = (x'^2 + y'^2 + z'^2)^{1/2}$. Substituting for x', y', and z' in terms of x, y, and z gives

$$R' = [(x \cos \theta + y \sin \theta)^2 + (-x \sin \theta + y \cos \theta)^2 + z^2]^{1/2}$$

Expanding the squares and noting that $\cos^2 \theta + \sin^2 \theta = 1$, we get $R' = (x^2 + y^2 + z^2)^{1/2} = R$. Thus the rotation is orthogonal.

Example 17.6 Lorentz Transformation as an Orthogonal Transformation

Equation (17.41) means that the Lorentz transformation is an orthogonal transformation. In this example we will determine the implication of this condition on the coefficients $a_{\mu\nu}$. Taking

$$x'_\mu = \sum_{\nu=1}^{4} a_{\mu\nu} x_\nu$$

then Eq. (17.41) is equivalent to

$$\sum_{\mu=1}^{4} x'^2_\mu = \sum_{\nu=1}^{4} x^2_\nu \tag{17.72}$$

The left-hand side is

$$\sum_{\mu=1}^{4} x'^2_\mu = \sum_{\mu=1}^{4} \left(\sum_{\nu=1}^{4} a_{\mu\nu} x_\nu \right) \left(\sum_{\lambda=1}^{4} a_{\mu\lambda} x_\lambda \right)$$

or

$$\sum_{\mu=1}^{4} x'^2_\mu = \sum_{\mu,\nu,\lambda=1}^{4} a_{\mu\nu} a_{\mu\lambda} x_\nu x_\lambda \tag{17.73}$$

Equations (17.72) and (17.73) agree only if

$$\sum_{\mu=1}^{4} a_{\mu\nu} a_{\mu\lambda} = \delta_{\nu\lambda} \tag{17.74}$$

where $\delta_{\nu\lambda}$ is the Kronecker delta function; it is equal to zero if $\nu \neq \lambda$ and equal to 1 if $\nu = \lambda$. Equation (17.74) is the condition that defines the orthogonality of the transformation.

Example 17.7 Four-Dimensional Differential Volume

The four-dimensional differential volume is defined as $d4x = -i\, dx_1\, dx_2\, dx_3\, dx_4$. The transformation law of this volume element can be derived from the transformation of x_μ. One can show the well-known relation $d4x' = J\, d4x$, where J is the *Jacobian determinant*.

$$J = \frac{\partial(x'_1, x'_2, x'_3, x'_4)}{\partial(x_1, x_2, x_3, x_4)}$$

The Jacobian determinant $\partial(q, p)/\partial(u, v)$ is defined as the matrix:

$$\frac{\partial(q, p)}{\partial(u, v)} = \begin{bmatrix} \partial q/\partial u & \partial p/\partial u \\ \partial q/\partial v & \partial p/\partial v \end{bmatrix}$$

The Jacobian determinant J can be shown to be just the determinant of $a_{\mu\nu}$, and since $|a_{\mu\nu}| = 1$, then

$$d4x' = d4x \tag{17.75}$$

indicating that the "four-volume" is a Lorentz scalar.

Example 17.8 The Lorentz Transformation as a Complex Rotation In Four-Dimensional Space

We now show that the Lorentz transformation can be interpreted as a rotation in the x_3-x_4 plane through an imaginary angle. Assume the angle of rotation is θ; then one can write, along the lines of Eqs. (17.71),

$$x'_3 = x_3 \cos\theta + x_4 \sin\theta \tag{17.76}$$

Comparing Eq. (17.76) with Eq. (17.47), one finds that

$$\cos\theta = \gamma \qquad \text{and} \qquad \sin\theta = i\beta\gamma$$

or

$$\tan\theta = i\beta = \frac{iv}{c} \tag{17.77}$$

which shows that the angle of rotation is not a real angle.

17.5 Relativistic Electrodynamics—Covariance of Electrodynamics

We are now in a position to cast the equations of electromagnetism using the four-dimensional formalism. A formulation of these equations that treats the space and time coordinates on the same footing is called a *covariant* formulation.

Continuity Equation. The continuity equation relating the charge and current density is

$$\nabla\cdot\mathbf{J} + \frac{\partial\rho}{\partial t} = 0$$

or using $x_4 = ict$

$$\nabla\cdot\mathbf{J} + ic\frac{\partial\rho}{\partial x_4} = 0 \tag{17.78}$$

Since the four-dimensional divergence [Eq. (17.64)] is given by $\Box\cdot A_\mu = \nabla\cdot\mathbf{A} + \partial A_4/\partial x_4$, then Eq. (17.78) can be cast in covariant form by introducing the four-vector current density (see Problem 17.14); that is,

$$J_\mu = (\mathbf{J}, ic\rho) = (J_1, J_2, J_3, ic\rho) \tag{17.79}$$

With this four-vector, Eq. (17.78) takes the following covariant form:

$$\Box\cdot J_\mu = \sum_{\mu=1}^{4}\frac{\partial J_\mu}{\partial x_\mu} = 0 \tag{17.80}$$

Potential Wave Equations and Lorentz Condition. The wave equations for the vector potential **A** and the scalar potential **Φ** are given by Eqs. (15.9) and (15.10). Note that

$$\Box^2 = \nabla^2 - \frac{1}{c^2}\frac{\partial^2}{\partial t^2}$$

is the differential operator of the left-hand side of both of the wave equations. Moreover, since **J** and ρ are the space and time components of the four-vector J_μ, then **A** and **Φ** will have to combine to form a four-vector of the form

$$A_\mu = \left(\mathbf{A}, \frac{i\Phi}{c}\right) = \left(A_1, A_2, A_3, \frac{i\Phi}{c}\right) \tag{17.81}$$

Using Eqs. (17.79) and (17.81) for the four-vectors of the current density and potential, the wave equations take the following single covariant form

$$\Box^2 A_\mu = \sum_{\nu=1}^{4} \frac{\partial^2 A_\mu}{\partial x_\nu^2} = -\mu_0 J_\mu \tag{17.82}$$

The Lorentz condition that relates the vector potential to the scalar potential

$$\nabla \cdot \mathbf{A} + \varepsilon_0 \mu_0 \frac{\partial \Phi}{\partial t} = 0$$

hence takes a similar covariant form to the covariant form of the continuity equation

$$\Box \cdot A_\mu = \sum_{\mu=1}^{4} \frac{\partial A_\mu}{\partial x_\mu} = 0 \tag{17.83}$$

Maxwell's Equations. Maxwell's equations in vacuum are $\nabla \cdot \mathbf{E} = \rho/\varepsilon_0$, $\nabla \times \mathbf{B} = \mu_0 \mathbf{J} + (1/c^2)\partial \mathbf{E}/\partial t$, $\nabla \cdot \mathbf{B} = 0$, and $\nabla \times \mathbf{E} = -\partial \mathbf{B}/\partial t$. Examining these equations shows that there is no obvious way of writing them in covariant form as was the case in the continuity equation, wave equations, and the Lorentz condition. This may indicate that the **E** and **B** fields do not combine to form four-vectors such as the above cases, but perhaps higher-rank four-tensors.

Maxwell's equations are not a good starting point for showing this statement, but rather the following equations, which are derived from them: $\mathbf{B} = \nabla \times \mathbf{A}$, and $\mathbf{E} = -\nabla\Phi - \partial \mathbf{A}/\partial t$. These were derived from Maxwell's equations $\nabla \cdot \mathbf{B} = 0$ and $\nabla \times \mathbf{E} = -\partial \mathbf{B}/\partial t$; they relate the fields individually to the potentials **A** and Φ, which have already been combined in a four-vector form. Using Eq. (17.81) one can show that the components E_1 and B_1 take the following explicit expressions.

$$B_1 = \frac{\partial A_3}{\partial x_2} - \frac{\partial A_2}{\partial x_3} \tag{17.84}$$

$$iE_1 = \frac{\partial A_1}{\partial x_4} - \frac{\partial A_4}{\partial x_1} \tag{17.85}$$

Similarly, one can write explicit expressions for E_2, E_3, B_2, and B_3. It is clear that the electric and magnetic fields are elements of the second-rank, antisymmetric, *field-strength tensor* $F_{\mu\nu}$ similar to the one defined in Eq. (17.70); that is,

$$F_{\mu\nu} = \frac{\partial A_\mu}{\partial x_\nu} - \frac{\partial A_\nu}{\partial x_\mu} \tag{17.86}$$

where A_μ is the four-vector potential of Eq. (17.81). Under a Lorentz transformation $F_{\mu\nu}$ transforms according to Eq. (17.67), as follows:

$$F'_{\mu\nu} = \sum_{\lambda,\sigma}^{4} a_{\mu\lambda} a_{\nu\sigma} F_{\lambda\sigma} \tag{17.87}$$

Explicitly, the field strength tensor is

$$F_{\mu\nu} = \begin{matrix} & \overset{\nu}{\rightarrow} \\ \mu\downarrow & \begin{bmatrix} 0 & B_3 & -B_2 & -iE_1/c \\ -B_3 & 0 & B_1 & -iE_2/c \\ B_2 & -B_1 & 0 & -iE_3/c \\ iE_1/c & iE_2/c & iE_3/c & 0 \end{bmatrix} \end{matrix} \tag{17.88}$$

With the electric and magnetic fields combined in a four-tensor we can now proceed to write Maxwell's equations in a covariant form. Let us consider the equations:

$$\nabla \cdot \mathbf{E} = \frac{\rho}{\varepsilon_0} \tag{17.89}$$

$$\nabla \times \mathbf{B} - \frac{1}{c^2}\frac{\partial \mathbf{E}}{\partial t} = \mu_0 \mathbf{J} \tag{17.90}$$

Expanding the three-dimensional divergence in Eq. (17.89) gives

$$\frac{\partial E_1}{\partial x_1} + \frac{\partial E_2}{\partial x_2} + \frac{\partial E_3}{\partial x_3} = \frac{J_4}{ic\varepsilon_0} \tag{17.91}$$

Substituting for E_1, E_2, and E_3 in terms of $F_{\mu\nu}$ of Eq. (17.88) replaces Eq. (17.91) by

$$-ic\frac{\partial F_{41}}{\partial x_1} - ic\frac{\partial F_{42}}{\partial x_2} - ic\frac{\partial F_{43}}{\partial x_3} = \frac{J_4}{ic\varepsilon_0} \tag{17.92}$$

Adding $-ic\,\partial F_{44}/\partial x_4 \equiv 0$ to this equation gives

$$\sum_{\nu=1}^{4} \frac{\partial F_{\mu\nu}}{\partial x_\nu} = \mu_0 J_\mu \qquad \mu = 4 \tag{17.93}$$

Equation (17.93) is the covariant form of Eq. (17.89). A similar procedure gives the covariant form of Eq. (17.90) to be Eq. (17.93) with $\mu = 1, 2,$ and 3. Thus

$$\sum_{\nu=1}^{4} \frac{\partial F_{\mu\nu}}{\partial x_\nu} = \mu_0 J_\mu \qquad \mu = 1, 2, 3, 4 \tag{17.94}$$

represent Eqs. (17.89) and (17.90).

Similarly, the covariant form of the homogeneous Maxwell's equations

$$\nabla \cdot \mathbf{B} = 0 \qquad \text{and} \qquad \nabla \times \mathbf{E} = -\frac{\partial \mathbf{B}}{\partial t} \tag{17.95}$$

can be shown to be given by

$$\frac{\partial F_{\mu\nu}}{\partial x_\lambda} + \frac{\partial F_{\nu\lambda}}{\partial x_\mu} + \frac{\partial F_{\lambda\mu}}{\partial x_\nu} = 0 \qquad \mu \neq \nu \neq \lambda = 1, 2, 3, 4 \tag{17.96}$$

Each term in Eq. (17.96) is a tensor of *third rank* because $F_{\mu\nu}$ is a tensor of second rank, and its differentiation with respect to x_λ, where $\lambda \neq \mu \neq \nu$, increases the rank by 1.

As an application of the covariant formulation of the E and B fields in terms of four-tensors, we use the transformation property of the field strength tensor [Eq. (17.88)] to derive the transformation laws of these fields. Consider an observer at rest in reference frame S, where the electric and magnetic fields of an electromagnetic wave are $\mathbf{E}$ and $\mathbf{B}$. We now calculate the field $\mathbf{E}'$ and $\mathbf{B}'$ as observed by an observer moving with velocity v along the z axis. According to Eq. (17.87) we have

$$F'_{\mu\nu} = \sum_{\lambda,\sigma}^{4} a_{\mu\lambda} a_{\nu\sigma} F_{\lambda\sigma}$$

Since [from Eq. (17.88)], $B_1 = F_{23}$, then

$$B'_1 = F'_{23} = \sum_{\lambda,\sigma}^{4} a_{2\lambda} a_{3\sigma} F_{\lambda\sigma} \tag{17.97}$$

Expanding the sums over λ and σ and using the matrix given in Eq. (17.50), one finds that many of the terms in Eq. (17.97) vanish, and $B_1' = a_{22}(a_{33}F_{23} + a_{34}F_{24})$ or

$$B_1' = \gamma B_1 + \frac{\beta\gamma}{c} E_2$$

On the other hand, the B_3' component, which is the component along the direction of the motion, is invariant under the Lorentz transformation. This can be shown as follows:

$$B_3' = F_{12}' = \sum_{\lambda,\sigma}^{4} a_{1\lambda} a_{2\sigma} F_{\lambda\sigma} = a_{11} a_{22} F_{12} = B_3$$

Using similar procedures, one can show that the transformation of all of the field components are

$$\begin{aligned} E_1' &= \gamma(E_1 - c\beta B_2) \qquad & B_1' &= \gamma\left(B_1 + \frac{\beta}{c} E_2\right) \\ E_2' &= \gamma(E_2 + c\beta B_1) \qquad & B_2' &= \gamma\left(B_2 - \frac{\beta}{c} E_1\right) \\ E_3' &= E_3 & B_3' &= B_3 \end{aligned} \tag{17.98}$$

The relations in Eq. (17.98) indicate that the components of the fields parallel to the direction of the motion stay unchanged under a Lorentz transformation. These equations can be generalized to the case where the frame S' is moving with an arbitrary velocity $\mathbf{v}$ relative to S:

$$\begin{aligned} \mathbf{E}_\parallel' &= \mathbf{E}_\parallel \qquad & \mathbf{B}_\parallel' &= \mathbf{B}_\parallel \\ \mathbf{E}_\perp' &= \gamma[\mathbf{E}_\perp + \mathbf{v} \times \mathbf{B}_\perp] \qquad & \mathbf{B}_\perp' &= \gamma\left[\mathbf{B}_\perp - \frac{\mathbf{v} \times \mathbf{E}_\perp}{c^2}\right] \end{aligned} \tag{17.99}$$

where $\parallel$ stands for the component parallel to $\mathbf{v}$ and $\perp$ stands for the component normal to $\mathbf{v}$.

The transformation of the fields can also be expressed in vector form, as follows:

$$\mathbf{E}' = \gamma[\mathbf{E} + \mathbf{v} \times \mathbf{B}] - (\gamma - 1)\mathbf{v}\frac{(\mathbf{v} \cdot \mathbf{E})}{v^2}$$

$$\mathbf{B}' = \gamma\left[\mathbf{B} - \frac{\mathbf{v} \times \mathbf{E}}{c^2}\right] - (\gamma - 1)\mathbf{v}\frac{(\mathbf{v} \cdot \mathbf{B})}{v^2} \tag{17.100}$$

It is often required to determine the fields in the unprimed system in terms of the fields in the primed system. The resulting relations are the *inverse transformation.* One can easily show that the inverse transformation can be obtained from Eq. (17.98) to Eq. (17.100) by the interchange of primed and unprimed quantities and $\mathbf{v} \to -\mathbf{v}$. Finally, we note that the transformations given by Eqs. (17.98) to (17.100) indicate that the **E** and **B** fields have no independent existence; they are completely interrelated.

Example 17.9 $\mathbf{E}\cdot\mathbf{B}$ and $E^2 - c^2B^2$ Are Lorentz Scalars

In this example we use Eq. (17.98) to show that $\mathbf{E}\cdot\mathbf{B}$ is a Lorentz scalar; that is, we show that $\mathbf{E}'\cdot\mathbf{B}' = \mathbf{E}\cdot\mathbf{B}$. Substituting for the primed components from Eq. (17.98), we get

$$\mathbf{E}'\cdot\mathbf{B}' = \gamma^2(E_1 - c\beta B_2)\left(B_1 + \frac{\beta}{c}E_2\right) + \gamma^2(E_2 + c\beta B_1)\left(B_2 - \frac{\beta}{c}E_1\right) + E_3B_3$$

Expanding the terms, we get

$$\mathbf{E}' \odot \mathbf{B}' = (\gamma^2 - \gamma^2\beta^2)[E_1B_1 + E_2B_2] + E_3B_3 = \mathbf{E}\cdot\mathbf{B}$$

Using similar procedures, we can show that $E^2 - c^2B^2$ is also a Lorentz scalar (see Problem 17.15).

Example 17.10 A Charged, Current-Carrying Cylinder

An infinitely long cylinder is uniformly charged with a density λ per unit length. Also, the cylinder carries a uniformly distributed current I as shown in Fig. 17.6. An observer S' is traveling in a direction parallel to the axis of the cylinder with a velocity $\mathbf{v}$.

According to an observer S at rest with respect to the cylinder (Fig. 17.6*a*), the electric and magnetic fields are

$$\mathbf{E} = E_\rho\hat{\boldsymbol{\rho}} = \frac{\lambda}{2\pi\rho\varepsilon_0}\hat{\boldsymbol{\rho}} \qquad \mathbf{B} = B_\phi\hat{\boldsymbol{\phi}} = \frac{\mu_0 I}{2\pi\rho}\hat{\boldsymbol{\phi}}$$

The fields observed by S' can be calculated using the Lorentz transformation. With ρ, ϕ, and z taken to be the coordinates x_1, x_2, and x_3, respectively, Eq. (17.99) gives

$$E'_\rho = \gamma(E_\rho - vB_\phi) = \frac{\gamma}{2\pi\rho}\left(\frac{\lambda}{\varepsilon_0} - v\mu_0 I\right) \tag{17.101}$$

$$B'_\phi = \gamma\left(B_\phi - \frac{v}{c^2}E_\rho\right) = \frac{\gamma}{2\pi\rho}\left(\mu_0 I - \frac{v}{c^2}\frac{\lambda}{\varepsilon_0}\right) \tag{17.102}$$

Now we show that at certain velocities the observer S' can observe only either a pure electric or a pure magnetic field. We first consider the case $\lambda c/I < 1$ (current-like cylinder). When the observer S' is moving with velocity $v = \lambda/\varepsilon_0\mu_0 I$, then the electric and magnetic fields become (see Fig. 17.6*b*)

$$B'_\phi = \frac{\mu_0 I}{2\pi\rho}\left(1 - \frac{\lambda^2c^2}{I^2}\right)^{1/2} \qquad E'_\rho = 0 \tag{17.103}$$

where we used $\gamma = (1 - \lambda^2c^2/I^2)^{-1/2}$. In this case the cylinder behaves as a current-carrying conductor. We now consider the case where $\lambda c/I > 1$ (charge-like cylinder). In this case and when S' is moving with velocity $v = I/\lambda$, the electric and magnetic fields become (see Fig. 17.6*c*)

$$E'_\rho = \frac{\lambda}{2\pi\rho\varepsilon_0}\left(1 - \frac{I^2}{\lambda^2c^2}\right)^{1/2} \qquad B'_\phi = 0 \tag{17.104}$$

where we used $\gamma = (1 - I^2/\lambda^2c^2)^{-1/2}$. In this case, the cylinder behaves as a charged conductor.

The last case we discuss is when $\lambda c/I = 1$. In this case there is no reference frame in which the electromagnetic field is pure electric or pure magnetic. As one can see from Eqs. (17.103) and (17.104), the velocity of the frame in this case approaches the velocity of light and both fields approach zero.

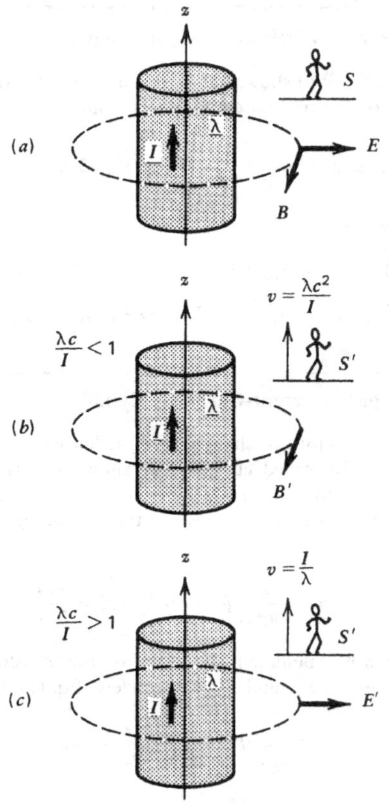

Figure 17.6 The fields of a long, stationary cylinder which simultaneously has a uniform charge density, and a uniform current measured in intertial frames moving along the cylinder. Observer is moving relative to the cylinder at velocity (*a*) 0, (*b*) $\lambda c^2/I$, (*c*) I/λ.

Example 17.11 Parallel E and B Fields

Assume that a uniform electromagnetic wave of **E** and **B** fields exists in a reference frame S. We are interested in finding the velocity **v** of the reference frame S' relative to the frame S in which the fields are parallel, as shown in Fig. 17.7.

There are an infinite number of frames in which the parallelism can be achieved. To see this, let us consider a frame S' moving with velocity **v** in which **E**′ and **B**′ are parallel; then the fields will stay parallel when observed in any other frame moving relative to S' in a direction along the common direction of **E**′ and **B**′ [see Eq. (17.99)]. Thus we need only to determine the velocity of the frame S' that moves normal to the plane of **E** and **B**. For example, if **E** and **B** are in the x-y plane, then a frame moving along the z axis with the appropriate velocity **v** as shown in Fig. 17.7 can make **E**′ and **B**′ parallel. Now, if **E**′ and **B**′ turn out to be along the x axis, then S' can also acquire any velocity along the x axis without changing the parallelism.

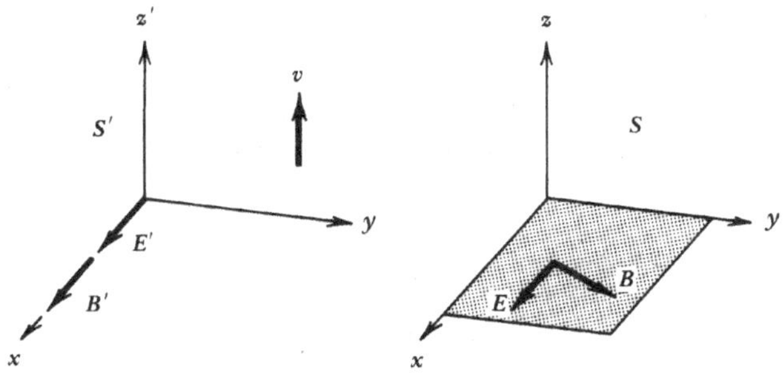

Figure 17.7 Observation of an electromagnetic field (**E**, **B**) in two reference frames moving relative to each other, showing that **E** and **B** appear parallel in one of them.

Thus taking $\mathbf{E}' \times \mathbf{B}' = 0$, and $\mathbf{v}$ to be along $\mathbf{E} \times \mathbf{B}$—that is, normal to the plane of **E** and **B**—one can show by crossing the inverse of Eq. (17.100) that (see Problem 17.16)

$$\frac{\mathbf{v}}{c} = \hat{\mathbf{n}} \frac{E^2 + c^2B^2 - \sqrt{(E^2 - c^2B^2)^2 + 4c^2(\mathbf{E} \cdot \mathbf{B})^2}}{2c|\mathbf{E} \times \mathbf{B}|} \tag{17.105}$$

where $\hat{\mathbf{n}}$ is a unit vector normal to the plane of **E** and **B**.

The magnitude of the fields **E**′ and **B**′ can also be determined from the Lorentz transformation. We leave this as an exercise (see Problem 17.16).

Example 17.12 Field-Strength Tensor in the Presence of Dielectric and Magnetic Materials

In the presence of materials that have dielectric and magnetic properties, the fields **E**, **D**, **H**, and **B** are related to the polarization and magnetization vectors of the material, **P** and **M**, by

$$\mathbf{E} = \frac{(\mathbf{D} - \mathbf{P})}{\varepsilon_0} \qquad \mathbf{B} = \mu_0(\mathbf{H} + \mathbf{M}) \tag{17.106}$$

Replacing the components E_i and B_i of the field strength tensor [Eq. (17.88)] by their values, $(D_i - P_i)/\varepsilon_0$ and $\mu_0(H_i + M_i)$, then one can easily show that $F_{\mu\nu}$ can be expressed in terms of two tensors $H_{\mu\nu}$ and $M_{\mu\nu}$, where

$$H_{\mu\nu} = \begin{bmatrix} 0 & H_3 & -H_2 & -icD_1 \\ -H_3 & 0 & H_1 & -icD_2 \\ H_2 & -H_1 & 0 & -icD_3 \\ icD_1 & icD_2 & icD_3 & 0 \end{bmatrix} \tag{17.107}$$

$$M_{\mu\nu} = \begin{bmatrix} 0 & M_3 & -M_2 & icP_1 \\ -M_3 & 0 & M_1 & icP_2 \\ M_2 & -M_1 & 0 & icP_3 \\ -icP_1 & -icP_2 & -icP_3 & 0 \end{bmatrix} \tag{17.108}$$

and

$$F_{\mu\nu} = \mu_0(H_{\mu\nu} + M_{\mu\nu}) \tag{17.109}$$

Note that Eq. (17.109) combines the relations $\mathbf{D} = \varepsilon_0 \mathbf{E} + \mathbf{P}$ and $\mathbf{B} = \mu_0(\mathbf{H} + \mathbf{M})$ into one single relation.

Both $H_{\mu\nu}$ and $M_{\mu\nu}$ are second-rank antisymmetric four-tensors (Lorentz tensors). They transform under a Lorentz transformation according to Eq. (17.67) or Eq. (17.87). Using the same procedure we followed in deriving the transformation laws of **E** and **B**, we get

$$\mathbf{P}'_{\parallel} = \mathbf{P}_{\parallel} \qquad \mathbf{M}'_{\parallel} = \mathbf{M}_{\parallel}$$

$$\mathbf{P}'_{\perp} = \gamma\left(\mathbf{P}_{\perp} - \frac{\mathbf{v}}{c^2} \times \mathbf{M}_{\perp}\right) \qquad \mathbf{M}'_{\perp} = \gamma(\mathbf{M}_{\perp} + \mathbf{v} \times \mathbf{P}_{\perp}) \tag{17.110}$$

Further applications of these transformations are found in Problems 17.20, 17.21, and 17.23.

Example 17.13 The Potentials of a Uniformly Moving Charge

A point charge q is moving with a uniform velocity **v** along the z axis relative to an observer S at rest with respect to the laboratory (see Fig. 17.8*a*). According to an observer S' at rest with respect to q, the potentials produced by the charge are

$$\Phi' = \frac{q}{4\pi\varepsilon_0 r'} \qquad \mathbf{A}' = \mathbf{0} \tag{17.111}$$

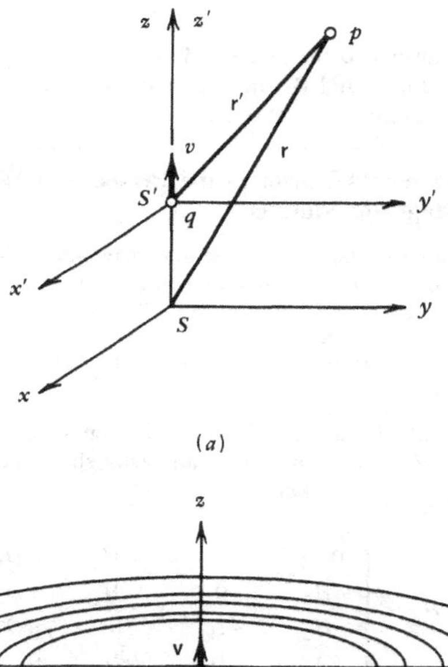

(*a*)

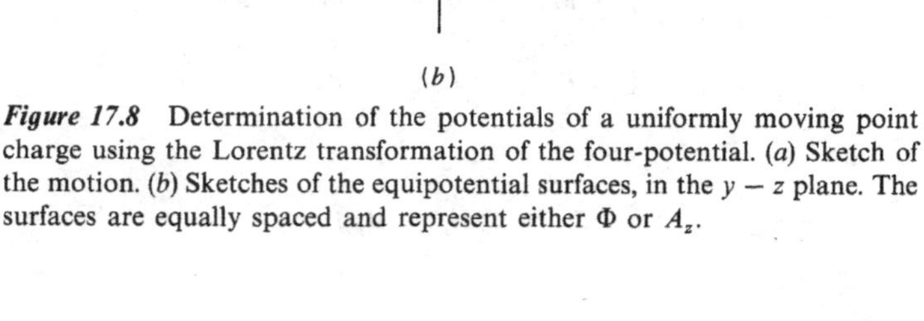

(*b*)

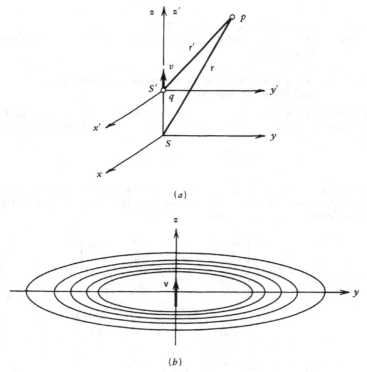

Figure 17.8 Determination of the potentials of a uniformly moving point charge using the Lorentz transformation of the four-potential. (*a*) Sketch of the motion. (*b*) Sketches of the equipotential surfaces, in the $y - z$ plane. The surfaces are equally spaced and represent either Φ or A_z.

where $\mathbf{r}' = \gamma(z - vt)\hat{z} + x\hat{\mathbf{x}} + y\hat{\mathbf{y}}$. According to the observer S, the potentials are not the same; they can be calculated from Eq. (17.54) by using the transformation properties of the four-vector $A_\mu = (\mathbf{A}, i\Phi/c)$. Using the inverse transformation of Eq. (17.54) we get $A_4 = \gamma A_4'$, $A_3 = -i\beta\gamma A_4'$ and $A_1 = A_2 = 0$. Thus

$$\Phi = \gamma\Phi' \quad \text{and} \quad A_3 = \frac{\beta\gamma}{c}\Phi' = \frac{\beta}{c}\Phi \tag{17.112}$$

Substituting for Φ' from Eq. (17.111) gives

$$\Phi = \frac{\gamma q}{4\pi\varepsilon_0 r'} \qquad A_z = \frac{\mu_0 v\gamma}{4\pi}\frac{q}{r'} \tag{17.113}$$

It is convenient to introduce the quantity $\mathbf{R}^*$ as follows:

$$\mathbf{R}^* = \frac{\mathbf{r}'}{\gamma} \tag{17.114}$$

In terms of $\mathbf{R}^*$, Eqs. (17.113) can be represented as

$$\Phi = \frac{q}{4\pi\varepsilon_0 R^*} \qquad A_z = \frac{\mu_0 v}{4\pi}\frac{q}{R^*} \tag{17.115}$$

Figure 17.8*b* illustrates a few equally spaced equipotentional surfaces in the y-z planes, they represent either Φ or A_z. We observe that the potentials have larger gradients in the direction perpendicular to the motion than in the direction along it.

Example 17.14 The Fields Produced by a Uniformly Moving Charge

Consider the uniformly moving charge discussed in Example 17.13. The fields produced by this charge can be calculated from the scalar and the vector potentials of the charge Φ and $\mathbf{A}$ using the relations $\mathbf{E} = -\nabla\Phi - \partial\mathbf{A}/\partial t$, and $\mathbf{B} = \nabla \times \mathbf{A}$. These potentials were calculated in Example 17.13 using the Lorentz transformation. Here, however, we will not calculate the fields from the potentials but from the Lorentz transformation directly.

According to the observer S' at rest with respect to the charge, the fields are given by (see the previous example)

$$\mathbf{E}' = \frac{q\mathbf{r}'}{4\pi\varepsilon_0 r'^3} \qquad \mathbf{B}' = 0 \tag{17.116}$$

where $\mathbf{r}' = x\hat{\mathbf{x}} + y\hat{\mathbf{y}} + \gamma(z - vt)\hat{\mathbf{z}} = \rho\hat{\boldsymbol{\rho}} + \gamma(z - vt)\hat{\mathbf{z}}$. In the frame S, the fields can be calculated using Eq. (17.99). Since v is along the z axis, then the parallel and normal directions are in the z and ρ directions respectively. Thus the inverse transformation of Eq. (17.99) gives, for the electric fields,

$$E_z = E_z' \qquad E_\rho = \gamma E_\rho' \tag{17.117}$$

In terms of the quantity $\mathbf{R}^*$ defined in Eq. (17.114), the fields become

$$E_z = \frac{q}{4\pi\varepsilon_0}\frac{(z - vt)}{\gamma^2(R^*)^3} \qquad E_\rho = \frac{q}{4\pi\varepsilon_0}\frac{\rho}{\gamma^2(R^*)^3} \tag{17.118}$$

In terms of vector notations, the total field $\mathbf{E} = E_z\hat{\mathbf{z}} + E_\rho\hat{\boldsymbol{\rho}}$ is

$$\mathbf{E} = \frac{q}{4\pi\varepsilon_0}\frac{\mathbf{R}}{(R^*)^3}(1 - \beta^2) \tag{17.119}$$

where $\mathbf{R} = (z - vt)\hat{\mathbf{z}} + \rho\hat{\boldsymbol{\rho}}$ is the vector position of the charge in the S frame. Moreover, one can easily show that Eq. (17.119) can be written as

$$\mathbf{E} = \frac{q}{4\pi\varepsilon_0}\frac{\mathbf{R}}{\gamma^2 R^3[1 - (v^2/c^2)\sin^2\theta]^{3/2}} \tag{17.120}$$

where θ is the angle between $\mathbf{R}$ and $\mathbf{v}$.

The transformation properties of the magnetic field give

$$B_z = B'_z = 0 \qquad \mathbf{B}_\perp = \frac{\gamma}{c^2}\mathbf{v} \times \mathbf{E}'_\perp = \frac{\mathbf{v}}{c^2} \times \mathbf{E}_\perp \tag{17.121}$$

Since $\mathbf{v} = v\hat{\mathbf{z}}$, $\mathbf{E}_\perp = E_\rho\hat{\boldsymbol{\rho}}$, and $\hat{\mathbf{z}} \times \hat{\boldsymbol{\rho}} = \hat{\boldsymbol{\phi}}$ is a unit vector in the ϕ direction, then

$$\mathbf{B}_\perp = \frac{v}{c^2} E_\rho \hat{\boldsymbol{\phi}} = \frac{\mu_0 v q}{4\pi} \frac{\rho}{\gamma^2 (R^*)^3} \hat{\boldsymbol{\phi}} \tag{17.122}$$

Figure 17.9*a* gives the **E** field lines of the charge. They are radial straight lines originating from the present position of the charge, which is a remarkable property since the fields came from retarded (earlier) times. Moreover the figure shows that the lines are concentrated in a direction normal to **v**. (see Fig. 17.8*b*). Figures 17.9*b* and 17.9*c* give the **B** field lines as a

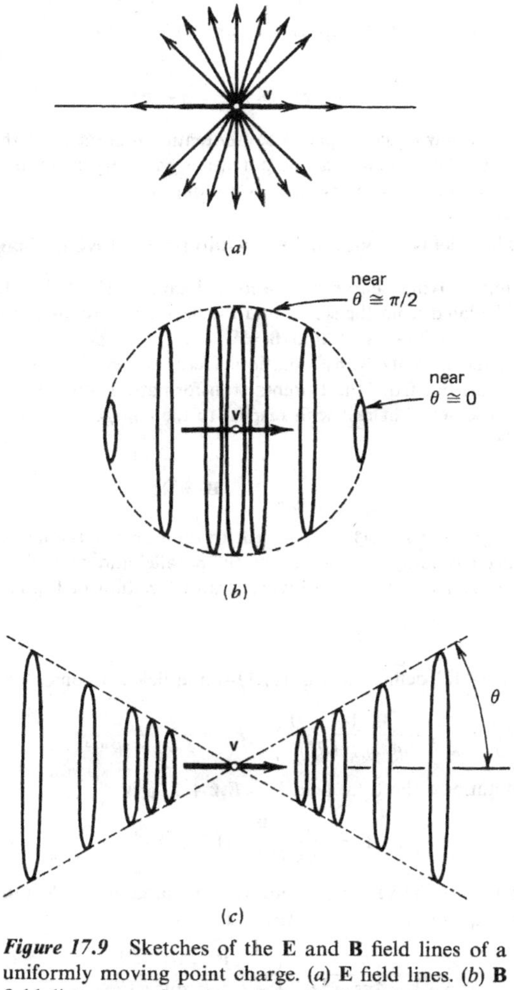

Figure 17.9 Sketches of the **E** and **B** field lines of a uniformly moving point charge. (*a*) **E** field lines. (*b*) **B** field lines as a function of θ. (*c*) **B** field lines as a function of r.

function of θ and r, respectively. Both show that the magnetic field lines are circles with centers along the z axis—the charge path. Moreover, they show that the lines are concentrated near the charge and in directions perpendicular to $\mathbf{v}$.

Example 17.15 Electric Dipole in Uniform Motion

An electric dipole $\mathbf{p} = p_0\hat{\mathbf{z}}$ is moving with a uniform velocity $\mathbf{v}$ along the z axis relative to an observer S at rest with respect to the laboratory, as shown in Fig. 17.10.

According to an observer S' at rest with respect to $\mathbf{p}$, the potentials produced by the dipole are

$$\Phi' = \frac{1}{4\pi\varepsilon_0}\frac{\mathbf{p}\cdot\mathbf{r}'}{r'^3} \qquad \mathbf{A}' = 0 \tag{17.123}$$

where $\mathbf{r}' = \gamma(z - vt)\hat{\mathbf{z}} + x\hat{\mathbf{x}} + y\hat{\mathbf{y}}$. The potentials in S can be calculated from 17.123 using the inverse transformation [see Eq. (17.54)]. Thus $\Phi = \gamma\Phi'$, and $A_3 = \beta\gamma\Phi'/c$. Substituting for Φ' from 17.123 and using the convenient quantity R^* of Eq. (17.114), we get

$$\Phi = \frac{1}{4\pi\varepsilon_0}\frac{\mathbf{p}\cdot\mathbf{R}^*}{\gamma(R^*)^3} \qquad A_z = \frac{v}{c^2}\Phi \tag{17.124}$$

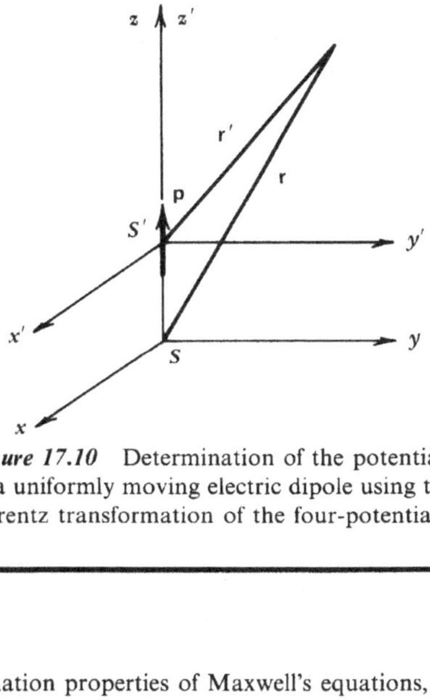

Figure 17.10 Determination of the potentials of a uniformly moving electric dipole using the Lorentz transformation of the four-potential.

17.6 Summary

By Studying the transformation properties of Maxwell's equations, Lorentz discovered what is now known as the Lorentz transformation. The transformation mixes the space and time variables, and thus puts them on the same footing in the form (x, y, z, ict) where $i = \sqrt{-1}$. The transformation, of course, yields at low velocities (with respect to the speed of light) the well-known Galilean transformation according to which space and time stay independent from each other. The new four-dimensional (space-time) space is complex (physically non-Euclidean) and is called Minkowski space, but mathematically has the same form as a Euclidean space. The implications of the transformation extend not only to geometry, but also to mechanics, electromagnetism, and electrodynamics, among other disciplines.

The transformation applies to any four-vector in this space: These vectors include the coordinate four-vectors x_μ (**r**, ict), potential four-vector A_μ (**A**, $i\Phi/c$), and current four-vector J_μ (**J**, $ic\rho$), which are relevant to electromagnetism. Moreover, four-tensors can be defined in this space, an example of which is the second-rank tensor defined by two consecutive Lorentz transformations. The electric and magnetic field components constitute the elements of such a tensor $F_{\mu\nu}$:

$$F_{\mu\nu} = \frac{\partial A_\mu}{\partial x_\nu} - \frac{\partial A_\nu}{\partial x_\mu} \tag{17.86}$$

The Lorentz transformation to a system moving with a velocity **v** in the z direction is

$$a_{\mu\nu} = \begin{bmatrix} 1 & 0 & 0 & 0 \\ 0 & 1 & 0 & 0 \\ 0 & 0 & \gamma & i\beta\gamma \\ 0 & 0 & -i\beta\gamma & \gamma \end{bmatrix} \tag{17.50}$$

where $\beta = v/c$, and $\gamma = (1 - \beta^2)^{-1/2}$, and

$$B'_\mu = \sum_{\nu=1}^{4} a_{\mu\nu} B_\nu \qquad \text{and} \qquad T'_{\mu\nu} = \sum_{\lambda,\sigma=1}^{4} a_{\mu\lambda} a_{\nu\sigma} T_{\lambda\sigma} \tag{17.53),(17.67}$$

where B_μ and $T_{\mu\nu}$ stand for any vector and any second-rank tensor.

Some of the implications or facts that are built into the transformations are as follows: (1) The speed of propagation of electromagnetic radiation in vacuum c is constant in every uniformly moving coordinate system (inertial systems). (2) The transformation leaves $x^2 + y^2 + z^2 - c^2t^2$ invariant. (3) The wave equation and Maxwell's equations are unchanged (covariant) under Lorentz transformation (requiring fields transforming as tensor elements, and potentials transforming as vector components). (4) The various relations in electromagnetism take on the following forms in four-space notations.

Continuity relation: $\square \cdot J_\mu = 0$ (17.80)

Lorentz condition: $\square \cdot A_\mu = 0$ (17.83)

Potential wave equation: $\square^2 A_\mu = -\mu_0 J_\mu$ (17.82)

Maxwell's equations:

$$\sum_{\nu=1}^{4} \frac{\partial F_{\mu\nu}}{\partial x_\nu} = \mu_0 J_\mu \tag{17.94}$$

$$\frac{\partial F_{\mu\nu}}{\partial x_\lambda} + \frac{\partial F_{\nu\lambda}}{\partial x_\mu} + \frac{\partial F_{\lambda\mu}}{\partial x_\nu} = 0 \quad \text{for } \mu \neq \nu \neq \lambda = 1, 2, 3, 4 \tag{17.96}$$

where the symbols $\square$, $\square\cdot$, and $\square^2$ are the analog quantities of gradient, divergence, and Laplacian of three-dimensional space

$$\square\Phi = \frac{\partial \Phi}{\partial x_\mu} \qquad \square \cdot A_\mu = \sum_{\mu=1}^{4} \frac{\partial A_\mu}{\partial x_\mu} \qquad \square^2\Phi = \left(\nabla^2 - \frac{1}{c^2}\frac{\partial^2}{\partial t^2}\right)\Phi \qquad \begin{matrix}(17.58)\\(17.64)\\(17.66)\end{matrix}$$

Explicitly the field components transform as follows:

$$\begin{aligned} \mathbf{E}'_\parallel &= \mathbf{E}_\parallel & \mathbf{E}'_\perp &= \gamma(\mathbf{E}_\perp + \mathbf{v} \times \mathbf{B}_\perp) \\ \mathbf{B}'_\parallel &= \mathbf{B}_\parallel & \mathbf{B}'_\perp &= \gamma\left(\mathbf{B}_\perp - \frac{\mathbf{v} \times \mathbf{E}_\perp}{c^2}\right) \end{aligned} \tag{17.99}$$

where $\parallel$ stands for the component parallel to **v** and $\perp$ stands for the component normal to **v**. Using these transformations, one can determine the fields produced by a uniformly moving charge from the fields produced by it in a frame in which the charge is at rest.

Problems

17.1 Consider a high-speed muon produced in collisions between cosmic-ray particles from outer space and nuclei in the atmosphere (see Example 17.1). Analyze the decay from the point of view of an observer moving with the decaying muon. How do these decay events appear to an observer fixed on earth?

17.2 Two rulers of the same length l_0 when they are at rest move in opposite directions with uniform velocities at relative velocity v. Observers 1 and 2 are at rest with respect to rulers 1 and 2, respectively, as shown in Fig. 17.11. (a) Determine the time interval Δt between the instants at which the left and right ends of the rulers pass each other as observed by observers 1 and 2. (b) What is the order in which the ends pass each other according to observers 1 and 2, and to a third observer with respect to whom both rulers move with equal velocities in opposite direction?

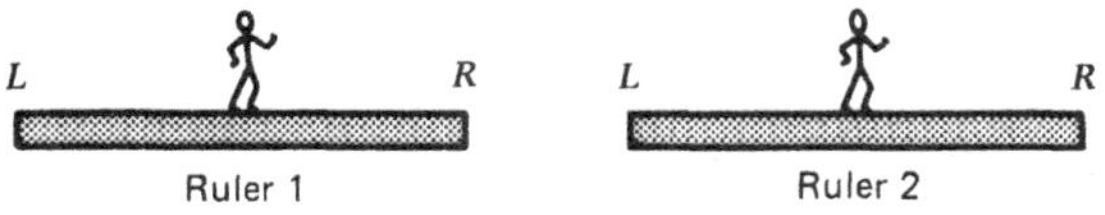

Figure 17.11

17.3 Verify the velocity transformation in spherical coordinates given by Eqs. (17.8) and (17.9).

17.4 Three reference frames S, S', and S'' are moving relative to each other uniformly while keeping their corresponding axes parallel. The frame S'' is moving relative to S' with a velocity v_1 along the z' axis, and S' is moving relative to S with a velocity v_2 along the z axis. (a) Show that the frames S'' and S are connected by a single Lorentz transformation with a velocity v given by

$$v = \frac{v_1 + v_2}{1 + v_1 v_2/c^2}$$

17.5 Show that the four-dimensional "length" element given in Eq. (17.17) is a Lorentz invariant.

17.6 Show that the number of wave crests in an electromagnetic wave is a Lorentz invariant quantity.

17.7 Verify Eq. (17.28), which gives the relation between the directions of the wave vectors of an electromagnetic wave observed in two frames. Show also that it can be derived from Eq. (17.8).

17.8 Consider the transverse Doppler effect treated in Example 17.2. Determine the direction of propagation of the wave in the system in which the source is at rest.

17.9 Assume that the stars are uniformly distributed. Determine their distribution as measured by an observer moving with a velocity $v \approx c$.

17.10 A mirror moves along the normal to its plane with a velocity $v \approx c$ with respect to the laboratory. A light wave of frequency ω_1 is normally incident on the mirror according to an observer at rest in the laboratory. Determine the frequency of the reflected wave when the mirror is approaching and when it is receding, as measured by the same observer.

17.11 Determine the law of reflection of a plane monochromatic wave from a mirror moving in a direction parallel to its plane.

17.12 Determine the Lorentz transformation from the second postulate of special relativity—constancy of the speed of light.

17.13 The wavelength of a light wave measured in the frame in which the source is at rest is λ_0. Determine the wavelength of the wave measured by an observer moving with velocity v relative to the source for both cases: an approaching and a receding observer.

17.14 (a) It is established experimentally that $\rho\, dx_1\, dx_2\, dx_3$, where ρ is the charge density, is a Lorentz invariant. This is known as the invariance of electric charge. Use this invariance to derive the transformation property of ρ. Does it make J_μ a legitimate four-vector? (b) The number of electrons per unit length in a cylindrical conductor is λ_0. When a current is established in the wire, the electrons drift along the wire at the drift speed v. Find the linear density λ of the drifting electrons in the wire in terms of λ_0, v, and c, as measured by an observer at rest with respect to the wire. Assume that the electrons were at rest before the current was established.

17.15 Show that $E^2 - c^2B^2$ is a Lorentz scalar.

17.16 A uniform electromagnetic wave of **E** and **B** fields is observed by an observer in a frame S. In a reference frame S' the fields $\mathbf{E}'$ and $\mathbf{B}'$ are parallel. (a) Show that S' is moving relative to S at a velocity **v** given by Eq. (17.105). (b) Determine the fields E' and B' in terms of the field invariants $E^2 - c^2B^2$ and $\mathbf{E}\cdot\mathbf{B}$.

17.17 In a reference frame S, the electric and magnetic fields of an electromagnetic wave **E** and **B** are perpendicular to each other. Determine the velocity of a reference frame S' in which (a) $\mathbf{E}' = 0$ and (b) $\mathbf{B}' = 0$.

17.18 Consider a uniformly moving point charge. (a) Show that the motion reduces **E** along the line of motion as compared to the Coulomb field of a static charge. (b) Show that **E** is "compressed" in the direction of motion. (c) Interpret these properties in relation to the transformation $\mathbf{E}_\parallel = \mathbf{E}'_\parallel$.

17.19 Consider a uniformly moving point charge (see Example 17.13). (a) Determine $-\nabla\Phi$ and sketch the field lines corresponding to this field on Fig. 17.8*b*. (b) Determine $-\partial\mathbf{A}/\partial t$ and sketch the field lines corresponding to this field also on Fig. 17.8*b*. (c) Show that the strength of $-\partial\mathbf{A}/\partial t$ is enough to straighten out (make radial) the total electric field $\mathbf{E} = -\nabla\Phi - \partial\mathbf{A}/\partial t$ and to reduce it along the motion as shown in Fig. 17.9*a*.

17.20 Determine the **E** and **B** fields of an electric dipole moving along the z axis with a velocity **v** relative to the laboratory using the Lorentz transformation of the fields. The dipole moment in its rest frame is $\mathbf{p} = p_0\hat{\mathbf{z}}$.

17.21 Determine the Lorentz transformation for the electric and magnetic dipole moments **p** and **m**. [Hint: use the transformation formula for **P** and **M**].

17.22 A conducting bar of length l moves with constant velocity **v** normal to itself in a magnetic field **B** (see Example 11.5). Determine the induced emf in the bar using (a) the transformation of the fields and (b) the transformation of the potentials.

17.23 A plane, rectangular, wire loop carries a current I' as measured by an observer S' at rest with respect to the loop. The loop is moving relative to the laboratory S with a velocity **v**, as shown in Fig. 17.12. The lengths of sides 1 and 2 are a and b. (*a*) Calculate the charge on all sides as observed in the S frame. (*b*) Determine the electric dipole moment of the loop in S. (*c*) Determine the magnetic dipole moment in S.

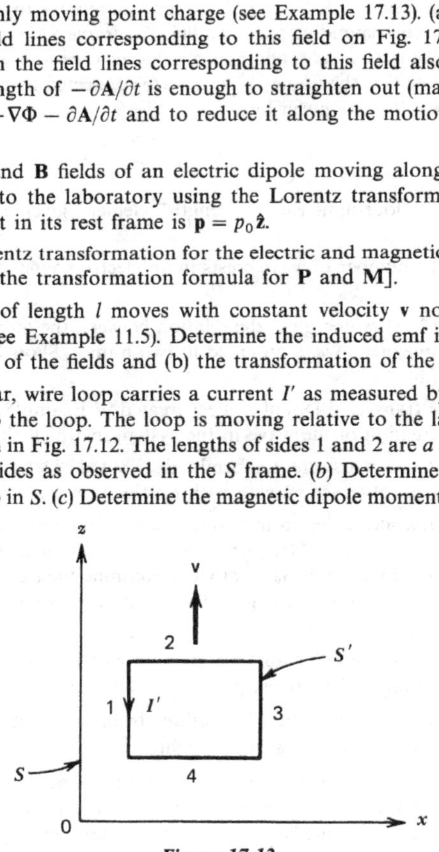

Figure 17.12

APPENDIX I
SYSTEMS OF UNITS

If one examines the history of the development of science and the system of units that served science, one would observe that the development of the systems reflects the various stages of the development of science itself. The electrostatic units (esu) system was, for example, introduced to suit the formulation of electrostatics and the electric properties of matter. The electromagnetic units (emu) system, on the other hand, was introduced to suit the field of magnetism and the magnetic properties of matter, a field that was studied separately from electricity.

When the union between the fields of electricity and magnetism was established, and the field of electromagnetic theory emerged, new systems of units that suited the new theory emerged. These include the well-known Système International (SI), which is also known as MKSA, and the Gaussian system.

The subsequent introduction of quantum mechanics and the emergence of the fields of atomic, plasma, nuclear, and particle physics added a great diversity to science and to the kind of considerations in choosing systems of units.

Because the SI system utilizes the practical units of amperes, volts, and ohms, it has become the universal system in electromagnetic theory and its engineering applications. On the other hand, the Gaussian system has become the widely used system in the other fields of physics. Because electromagnetic interactions play some roles in the other fields of physics, a physicist must be familiar with the formulation of electromagnetic theory in both the SI and Gaussian system and should be able to convert from one to the other.

The purpose of this appendix is not to give thorough discussions of the various systems, but to give a brief account that is enough to allow the student to recognize the system of units that a given relation is written in, and hence to choose the numbers that should be plugged in if a numerical answer is sought. Moreover, it will give guidance to how to convert a given equation from one system to the other.

I.1 Force Laws—Origin of Systems

All through this book the SI system has been used. This system is based on the four quantities—the meter, kilogram, second, and ampere, whereas the Gaussian system is based on the centimeter, gram, and second. In order to see the origin of these systems, and others, one needs to examine the two force laws: Coulomb's law between two charges and the magnetic force between two current-carrying wires per unit length

$$F = \frac{1}{4\pi\varepsilon_0}\frac{qq'}{r^2} \qquad \frac{dF}{dl} = \frac{\mu_0}{2\pi}\frac{II'}{r} \qquad \text{(SI, MKSA)} \tag{I.1}$$

which can be rewritten in more general forms as

$$F = C_e\frac{qq'}{r^2} \qquad \frac{dF}{dl} = 2C_m\frac{II'}{r} \tag{I.2}$$

where q and I stand for charge and current, F for force, l and r for length and distance, and the numerical values $1/(4\pi\varepsilon_0)$ and $\mu_0/4\pi$ are what we used in the SI system, and whose ratio respectively is $1/\varepsilon_0\mu_0 = c^2$ the square of the speed of light.

If one always uses the same set of mechanical units (units of force, energy, power,...) i.e., mass × distance/time^2, mass × distance^2/time^2, mass × distance^2/time^3 and realizing that $I = q/t$, when t represents time, then other systems of units can be introduced with freedom of choice of the numerical values and the dimensions of the constants C_e and C_m as long as their ratio always satisfies the relation:

$$\frac{C_e}{C_m} = c^2 \tag{I.3}$$

The use of 4π explicitly in C_e and C_m of a system (such as the SI system) is a convenience that makes no numerical factors of 4π appear in Maxwell's equations, and for this reason such a system is called a rationalized system of units.

We will now use the above outline to discuss the esu and emu systems very briefly because they are not presently in use, and less so the Gaussian system because it is widely used.

I.2 Electrostatic and Electromagnetic Systems

In the esu system, the units of length, mass and time are chosen as the centimeter, gram, and second (cgs), and hence the unit of force is the dyne = 10^{-7} newton. In the system, we take $C_e = 1$ and consequently $C_m = 1/c^2$. The force laws then become:

$$F = \frac{qq'}{r^2} \qquad \frac{dF}{dl} = \frac{2II'}{c^2 r} \qquad \text{(esu)} \tag{I.4}$$

The unit of charge in this system is the statcoulomb. Its definition comes from the statement that two point charges are one statcoulomb each if the force of interaction at 1 centimeter of separation is 1 dyne. The statampere per square centimeter, the unit of current density, is defined via $\mathbf{J} = \rho\mathbf{v}$ where ρ is in statcoulombs per cubic centimeter and $\mathbf{v}$ is in centimeters per second. The unit of the electric field, statvolts per centimeter is defined by the relation $\mathbf{F} = q\mathbf{E}$, where $\mathbf{F}$ is in dynes and q is in statcoulombs. Finally the esu unit of the magnetic field $\mathbf{B}$ is defined by the relation $\mathbf{F} = \mathbf{J} \times \mathbf{B}$.

In the emu system, the units of length, mass, time, and hence force are taken as centimeter, gram, second, and dyne, respectively just as in the esu system. In the system we take $C_m = 1$ and consequently $C_e = c^2$. The force laws then become

$$F = c^2\frac{qq'}{r^2} \qquad \frac{dF}{dl} = \frac{2II'}{r} \qquad \text{(emu)} \tag{I.5}$$

The units of current and current density I and J are abampere and abampere per square centimeter, respectively. The definition comes from the statement that two very long filamentary currents are one abampere each if the force of interaction per unit length at 1 centimeter separation is 2 dynes per centimeter. The unit of charge is the abampere-second, whereas that of charge density $\rho = \mathbf{J}/\mathbf{v}$ is the abcoulomb per cubic centimeter. One can go on and on with this procedure to define quantities like abvolt, abfarad, and so on. Here, however, we note that the unit of $\mathbf{B}$ comes from $\mathbf{F} \times \mathbf{J} \times \mathbf{B}$ and is called the gauss $= 10^{-4}$ tesla. Consequently the unit of magnetic flux is gauss × centimeters^2, which is given the name 1 maxwell.

I.3 Gaussian System

In the Gaussian system the units of length, mass, time, and hence force are centimeter, gram, second, and dyne, respectively. That is, it is a cgs system just like the esu and emu systems. Each of the constants C_e and C_m have more than one value; in fact the system is a mixed system in which ρ, $\mathbf{J}$, and $\mathbf{E}$ are measured in esu units, and $\mathbf{B}$ is measured in emu units.

In this system, Maxwell's equations take on the form given in the accompanying table.

Maxwell's Equations

Microscopic	Macroscopic
$\nabla \cdot \mathbf{E} = 4\pi\rho$	$\nabla \cdot \mathbf{D} = 4\pi\rho_f$
$\nabla \cdot \mathbf{B} = 0$	$\nabla \cdot \mathbf{B} = 0$
$\nabla \times \mathbf{E} = -\dfrac{1}{c}\dfrac{\partial \mathbf{B}}{\partial t}$	$\nabla \times \mathbf{E} = -\dfrac{1}{c}\dfrac{\partial \mathbf{B}}{\partial t}$
$\nabla \times \mathbf{B} = \dfrac{4\pi}{c}\mathbf{J} + \dfrac{1}{c}\dfrac{\partial \mathbf{E}}{\partial t}$	$\nabla \times \mathbf{H} = \dfrac{4\pi}{c}\mathbf{J}_f + \dfrac{1}{c}\dfrac{\partial \mathbf{D}}{\partial t}$

The field vectors are related as follows:

$$\mathbf{D} = \mathbf{E} + 4\pi\mathbf{P} \qquad \text{and} \qquad \mathbf{H} = \mathbf{B} - 4\pi\mathbf{M} \tag{I.6}$$

The constitutive relations in simple materials are written

$$\mathbf{D} = \varepsilon\mathbf{E} \qquad \mathbf{B} = \mu\mathbf{H} \qquad \mathbf{J} = \sigma_c\mathbf{E} \tag{I.7}$$

$$\mathbf{P} = \chi\mathbf{E} \qquad \text{and} \qquad \mathbf{M} = \chi_m\mathbf{H} \tag{I.8}$$

Equations (I.6) to (I.8) imply that

$$\varepsilon = 1 + 4\pi\chi \qquad \text{and} \qquad \mu = 1 + 4\pi\chi_m \tag{I.9}$$

From the relations

$$\nabla \cdot \mathbf{B} = \mathbf{0} \qquad \text{and} \qquad \nabla \times \mathbf{E} = -\frac{1}{c}\frac{\partial \mathbf{B}}{\partial t},$$

we find that the fields relate to the potentials in this system according to:

$$\mathbf{B} = \nabla \times \mathbf{A} \qquad \mathbf{E} = -\nabla\Phi - \frac{1}{c}\frac{\partial \mathbf{A}}{\partial t} \tag{I.10}$$

The Lorentz force acting on a charge q moving in a region of **E** and **B** fields is

$$\mathbf{F} = q\left(\mathbf{E} + \frac{\mathbf{v}}{c} \times \mathbf{B}\right) \tag{I.11}$$

The electromagnetic energy density associated with the fields, and the Poynting vector in this system are

$$u = \frac{1}{8\pi}(\mathbf{E}\cdot\mathbf{D} + \mathbf{B}\cdot\mathbf{H}) \qquad \mathbf{S} = \frac{c}{4\pi}(\mathbf{E} \times \mathbf{H}) \tag{I.12}$$

As we mentioned above, the MKSA (SI) system is widely used in electrical engineering applications. This is because it uses practical units including the ampere, volt, ohm, henry, and farad. We can now see this point if we examine Table I.1, which gives the numerical relationships of SI units to Gaussian units. For practical reasons the statampere = 0.33×10^{-9} ampere is too small and the abampere = 10 amperes is slightly too large. The statvolt = 300 volts, staohm = 9×10^{11} ohms, stathenry = 9×10^{11} henrys, are all too large. On the other hand the statfarad = 1.1×10^{-12} farad is much too small. We should, however, note that the unit tesla of SI is too large for practical purposes whereas the gauss = 10^{-4} tesla is more of a practical unit.

Table I.2 gives the symbolic relationships of SI variables to Gaussian variables. The table is useful for quick conversion from SI to Gaussian or Gaussian to SI.

Table I.1

Quantity	SI (MKSA)	Gaussian
Length	1 meter (m)	10^2 centimeters (cm)
Mass	1 kilogram (kg)	10^3 grams (g)
Time	1 second (s)	1 second (s)
Force	1 newton (N)	10^5 dynes (dyn)
Work, energy	1 joule (J)	10^7 ergs (erg)
Power	1 watt (W)	10^7 ergs per second
Capacitance (C)	1 farad (F)	9×10^{11} statfarads
Charge (q)	1 coulomb (C)	3×10^9 statcoulombs
Charge density (ρ)	1 coulomb per cubic meter (C/m^3)	3×10^3 statcoulomb/cm^3
Conductivity (σ_c)	1 (ohm-meter)$^{-1}$ ($\Omega\cdot$m)$^{-1}$	9×10^9 (statohm-cm)$^{-1}$
Current (I)	1 ampere (A)	3×10^9 statamperes = 10^{-1} abampere
Current density (J)	1 ampere per square meter (A/m^2)	3×10^5 statampere/cm^2
Displacement (D)	1 coulomb per square meter (C/m^2)	$12\pi \times 10^5$ statvolt/cm
Electric field (E)	1 volt per meter (V/m)	$\frac{1}{3} \times 10^{-4}$ statvolt/cm
Inductance (L)	1 henry (H)	$\frac{1}{9} \times 10^{-11}$ stathenry
Magnetic intensity (H)	1 ampere per meter (A/m)	$4\pi \times 10^{-3}$ oersted (Oe)
Magnetic flux (F)	1 weber (Wb)	10^8 maxwells (Mx)
Magnetic field (B)	1 Wb/m^2 = 1 tesla (T)	10^4 gauss (G)
Magnetization (M)	1 ampere per meter (A/m)	10^{-3} oersted (Oe)
Polarization (P)	1 coulomb per square meter (C/m^2)	3×10^5 statvolt/cm
Potential (Φ)	1 volt (V)	$\frac{1}{300}$ statvolt
Resistance (R)	1 ohm (Ω)	$\frac{1}{9} \times 10^{-11}$ statohm

Table I.2

Quantity	SI (MKSA)	Gaussian
Capacitance	C	$4\pi\varepsilon_0 C$
Charge	q	$(4\pi\varepsilon_0)^{1/2}q$
Charge density	$\rho\ (\sigma, \lambda)$	$(4\pi\varepsilon_0)^{1/2}\rho\ (\sigma, \lambda)$
Conductivity	σ_c	$4\pi\varepsilon_0\sigma_c$
Current	I	$(4\pi\varepsilon_0)^{1/2}I$
Current density	$\mathbf{J}\ (\mathbf{K})$	$(4\pi\varepsilon_0)^{1/2}\mathbf{J}, (\mathbf{K})$
Dielectric constant	K	ε
Dipole moment (electric)	$\mathbf{p}$	$(4\pi\varepsilon_0)^{1/2}\mathbf{p}$
Dipole moment (magnetic)	$\mathbf{m}$	$(4\pi/\mu_0)^{1/2}\mathbf{m}$
Displacement vector	$\mathbf{D}$	$(\varepsilon_0/4\pi)^{1/2}\mathbf{D}$
Electric field	$\mathbf{E}$	$(4\pi\varepsilon_0)^{-1/2}\mathbf{E}$
Inductance	L	$(4\pi\varepsilon_0)^{-1}L$
Magnetic intensity (**H**)	$\mathbf{H}$	$(4\pi\mu_0)^{-1/2}\mathbf{H}$
Magnetic flux	F	$(\mu_0/4\pi)^{1/2}F$
Magnetic field (**B**)	$\mathbf{B}$	$(\mu_0/4\pi)^{1/2}\mathbf{B}$
Magnetization	$\mathbf{M}$	$(4\pi/\mu_0)^{1/2}\mathbf{M}$
Permeability	μ	(1) $K_m\mu_0$, then (2) $K_m \rightarrow \mu$
Permeability (relative)	K_m	μ
Permittivity	ε	(1) $K\varepsilon_0$, then (2) $K \rightarrow \varepsilon$
Polarization	$\mathbf{P}$	$(4\pi\varepsilon_0)^{1/2}\mathbf{P}$
Resistance	R	$(4\pi\varepsilon_0)^{-1}R$
Resistivity	ρ	$(4\pi\varepsilon_0)^{-1}\rho$
Scalar potential	Φ	$(4\pi\varepsilon_0)^{-1/2}\Phi$
Speed of light	$(\mu_0\varepsilon_0)^{-1/2}$	c
Susceptibility	$\chi\ (\chi_m)$	$4\pi\chi(\chi_m)$
Vector potential	$\mathbf{A}$	$(\mu_0/4\pi)^{1/2}\mathbf{A}$

APPENDIX II

DIVERGENCE, CURL, GRADIENTS, AND LAPLACIAN

Cartesian Coordinates

$$\nabla \cdot \mathbf{A} = \frac{\partial A_x}{\partial x} + \frac{\partial A_y}{\partial y} + \frac{\partial A_z}{\partial z}$$

$$\nabla \times \mathbf{A} = \left(\frac{\partial A_z}{\partial y} - \frac{\partial A_y}{\partial z}\right)\hat{\mathbf{x}} + \left(\frac{\partial A_x}{\partial z} - \frac{\partial A_z}{\partial x}\right)\hat{\mathbf{y}} + \left(\frac{\partial A_y}{\partial x} - \frac{\partial A_x}{\partial y}\right)\hat{\mathbf{z}}$$

$$\nabla \Phi = \frac{\partial \Phi}{\partial x}\,\hat{\mathbf{x}} + \frac{\partial \Phi}{\partial y}\,\hat{\mathbf{y}} + \frac{\partial \Phi}{\partial z}\,\hat{\mathbf{z}}$$

$$\nabla^2 \Phi = \frac{\partial^2 \Phi}{\partial x^2} + \frac{\partial^2 \Phi}{\partial y^2} + \frac{\partial^2 \Phi}{\partial z^2}$$

Cylindrical Coordinates

$$\nabla \cdot \mathbf{A} = \frac{1}{\rho}\frac{\partial}{\partial \rho}(\rho A_\rho) + \frac{1}{\rho}\frac{\partial A_\phi}{\partial \phi} + \frac{\partial A_z}{\partial z}$$

$$\nabla \times \mathbf{A} = \left(\frac{1}{\rho}\frac{\partial A_z}{\partial \phi} - \frac{\partial A_\phi}{\partial z}\right)\hat{\boldsymbol{\rho}} + \left(\frac{\partial A_\rho}{\partial z} - \frac{\partial A_z}{\partial \rho}\right)\hat{\boldsymbol{\phi}} + \frac{1}{\rho}\left[\frac{\partial}{\partial \rho}(\rho A_\phi) - \frac{\partial A_\rho}{\partial \phi}\right]\hat{\mathbf{z}}$$

$$\nabla \Phi = \frac{\partial \Phi}{\partial \rho}\,\hat{\boldsymbol{\rho}} + \frac{1}{\rho}\frac{\partial \Phi}{\partial \phi}\,\hat{\boldsymbol{\phi}} + \frac{\partial \Phi}{\partial z}\,\hat{\mathbf{z}}$$

$$\nabla^2 \Phi = \frac{1}{\rho}\frac{\partial}{\partial \rho}\left(\rho\frac{\partial \Phi}{\partial \rho}\right) + \frac{1}{\rho^2}\frac{\partial^2 \Phi}{\partial \phi^2} + \frac{\partial^2 \Phi}{\partial z^2}$$

Spherical Coordinates

$$\nabla \cdot \mathbf{A} = \frac{1}{r^2}\frac{\partial}{\partial r}(r^2 A_r) + \frac{1}{r \sin\theta}\frac{\partial}{\partial\theta}(A_\theta \sin\theta) + \frac{1}{r\sin\theta}\frac{\partial A_\phi}{\partial\phi}$$

$$\nabla \times \mathbf{A} = \frac{1}{r\sin\theta}\left[\frac{\partial}{\partial\theta}(A_\phi \sin\theta) - \frac{\partial A_\theta}{\partial\phi}\right]\hat{\mathbf{r}} + \frac{1}{r}\left[\frac{1}{\sin\theta}\frac{\partial A_r}{\partial\phi} - \frac{\partial}{\partial r}(rA_\phi)\right]\hat{\boldsymbol{\theta}}$$

$$+ \frac{1}{r}\left[\frac{\partial}{\partial r}(rA_\theta) - \frac{\partial A_r}{\partial\theta}\right]\hat{\boldsymbol{\phi}}$$

$$\nabla\Phi = \frac{\partial\Phi}{\partial r}\hat{\mathbf{r}} + \frac{1}{r}\frac{\partial\Phi}{\partial\theta}\hat{\theta} + \frac{1}{r\sin\theta}\frac{\partial\Phi}{\partial\phi}\hat{\phi}$$

$$\nabla^2\Phi = \frac{1}{r^2}\frac{\partial}{\partial r}\left(r^2\frac{\partial\Phi}{\partial r}\right) + \frac{1}{r^2\sin\theta}\left(\sin\theta\frac{\partial\Phi}{\partial\theta}\right) + \frac{1}{r^2\sin^2\theta}\frac{\partial^2\Phi}{\partial\phi^2}$$

APPENDIX III

SOME FUNDAMENTAL CONSTANTS OF PHYSICS

Constant	Symbol	Computational Value	Best (1973) Value	
			Value[a]	Uncertainty[b]
Speed of light in a vacuum	c	3.00×10^{8} m/s	2.99792458	
Elementary charge	e	1.60×10^{-19} C	1.6021892	2.9
Electron rest mass	m_e	9.11×10^{-31} kg	9.109534	5.1
Permitivity constant	ε_0	8.85×10^{-12} F/m	8.854187818	0.008
Permeability constant	μ_0	12.6×10^{-7} H/m	4π (exactly)	—
Proton rest mass	m_p	1.67×10^{-27} kg	1.6726485	5.1
Avogadro constant	N_A	6.02×10^{23}/mol	6.022045	5.1
Boltzmann constant	k	1.38×10^{-23} J/K	1.380662	32.0
Molar volume of ideal gas at STP[c]	V_m	2.24×10^{-2} m^3/mol	2.241383	31.0
Bohr radius	a_0	5.29×10^{-11} m	5.2917706	0.82
Bohr magneton	μ_B	9.27×10^{-24} J/T	9.274078	3.9

[a] Same unit and power of ten as the computational value.
[b] Parts per million.
[c] Standard temperature and pressure (STP) = 0°C and 1.0 atm.
Source: The values in this table were selected from a longer list developed by E. Richard Cohen and B. N. Taylor, *Journal of Physical and Chemical Reference Data*, Vol. 2, no. 4, 1973.

APPENDIX IV

SOME SI DERIVED UNITS WITH SPECIAL NAMES

Quantity	SI Unit Name	SI Unit Symbol	Expression in terms of other units	Expression in terms of SI base units
Frequency	hertz	Hz		s^{-1}
Force	newton	N		$m \cdot kg/s^2$
Energy, work, quantity of heat	joule	J	$N \cdot m$	$kg \cdot m^2/s^2$
Power, radiant flux	watt	W	J/s	$kg \cdot m^2/s^3$
Quantity of electricity, electric charge	coulomb	C		$A \cdot s$
Electric potential, potential difference, electromotive force	volt	V	W/A	$kg \cdot m^2/A \cdot s^3$
Capacitance	farad	F	C/V	$A^2s^4/kg \cdot m^2$
Electric resistance	ohm	Ω	V/A	$kg \cdot m^2/A^2 \cdot s^3$
Magnetic flux	weber	Wb	$V \cdot s$	$kg \cdot m^2/A \cdot s^2$
Magnetic field	tesla	T	Wb/m^2	$kg/A \cdot s^2$
Inductance	henry	H	Wb/A	$kg \cdot m^2/A^2 \cdot s^2$

ANSWERS TO ODD-NUMBERED PROBLEMS

Chapter 1

1.1 $\pm(3\hat{\mathbf{x}} - 2\hat{\mathbf{y}} + 6\hat{\mathbf{z}})/7$ **1.3** $2x + 3y + 6z = -28$ **1.5** $7\hat{\mathbf{x}} - 3\hat{\mathbf{y}} + 8\hat{\mathbf{z}}$
1.9 (a) $a = 4$, $b = 2$, and $c = -1$;
(b) $\Phi = \frac{1}{2}(x^2 - 3y^2 + 2z^2 + 4xy + 8xz - 2yz)$
1.11 $\Phi = \ln(a/r)$ **1.13** 3, $3V$ where V is the volume enclosed by S
1.15 (a) $\hat{\mathbf{n}} = (x\hat{\mathbf{x}} + y\hat{\mathbf{y}})/(x^2 + y^2)^{1/2} = \hat{\boldsymbol{\rho}}$; (b) 90
1.19 $1/r^2$, $n(n+1)r^{n-2}$, 0 **1.21** $\hat{\mathbf{x}}\hat{\mathbf{x}} + \hat{\mathbf{y}}\hat{\mathbf{y}} + \hat{\mathbf{z}}\hat{\mathbf{z}}$

Chapter 2

2.1 172 N **2.3** $6.39 \times 10^8 \lambda\ V/m\hat{\boldsymbol{\rho}}$

2.5 $\mathbf{E} = \dfrac{\sigma z}{2\varepsilon_0}\left(\dfrac{1}{z} - \dfrac{1}{(a^2 + z^2)^{1/2}}\right)\hat{\mathbf{z}}$ **2.7** $\mathbf{E} = (32\hat{\mathbf{x}}/3 + 8\hat{\mathbf{y}} + 16\hat{\mathbf{z}})/4\pi\varepsilon_0$

2.9 (a) $\mathbf{E} = \sigma\hat{\mathbf{x}}/\varepsilon_0$, 0, $-\sigma\hat{\mathbf{x}}/\varepsilon_0$ for $x > 1$, $-1 < x < 1$, and $x < -1$, respectively; (b) $\mathbf{E} = -\sigma\hat{\mathbf{x}}/\varepsilon_0$, 0 for $|x| < 1$ and $|x| > 1$, respectively

2.11 $\mathbf{E} = \dfrac{5}{4\varepsilon_0 r}[1 - 2e^{-2r}(r^2 + r + \frac{1}{2})]\hat{\mathbf{r}}$

2.13 (a) $1.1 \times 10^{-7} C/\varepsilon_0 \sin(\mathrm{V} \cdot \mathrm{m})$; (b) $2\pi/\varepsilon_0 \sin(\mathrm{V} \cdot \mathrm{m})$; (c) 0; (d) 0
2.15 (a) $\rho = \varepsilon_0(18\sin^2\theta + 2\cos^2\theta)/r\sin\theta$; (b) $\rho_0 = \alpha\varepsilon_0$ for $a \le \rho \le b$

2.17 (a) $\Phi = \dfrac{\lambda}{2\pi\varepsilon_0}\ln\left|\dfrac{\sqrt{l^2 + r_1^2} + l}{r_1}\right|$; (b) $\Phi = -\dfrac{\lambda}{2\pi\varepsilon_0}\ln r_1 + c$;
(c) $\Phi_2 - \Phi_1 = -(\lambda/2\pi\varepsilon_0)\ln(r_2/r_1)$ same as that of an infinite wire

2.19 (a) $\mathbf{E}_1 = \rho_0\mathbf{r}/3\varepsilon_0$; (b) $\mathbf{E}_2 = \rho_0'\mathbf{r}'/3\varepsilon_0$; (c) $\mathbf{E} = \rho_0 z_0\hat{\mathbf{z}}/3\varepsilon_0$; (d) $\Phi = -\rho_0 z_0 r'\cos\theta'/3\varepsilon_0$ **2.23** $\mathbf{p} = \pi\sigma R^3\hat{\mathbf{z}}$, $\hat{\mathbf{z}}$ normal to the face
2.25 $\mathbf{F} = (q/4\pi\varepsilon_0 r^3)[3(\mathbf{p}\cdot\hat{\mathbf{r}})\hat{\mathbf{r}} - \mathbf{p}]$, $\tau = -qp\hat{\boldsymbol{\phi}}\sin\theta/(4\pi\varepsilon_0 r^3)$
2.27 (a) $\mathbf{F} = 6qdl(d^2+l^2)^{-5/2}(p_x\hat{\mathbf{z}} + p_z\hat{\mathbf{x}})/4\pi\varepsilon_0$; (b) $\mathbf{F} = (p_1/4\pi\varepsilon_0 l^4)[3\hat{\mathbf{x}}(\mathbf{p}\cdot\hat{\mathbf{z}}) + 3\hat{\mathbf{z}}(\mathbf{p}\cdot\hat{\mathbf{x}})]$ with $\mathbf{p}_1 = 2q\mathbf{d}\mathbf{l}^5(\mathbf{d}^2+\mathbf{l}^2)^{-5/2}\hat{\mathbf{x}}$
2.29 (a) Monopole moment = 0, $p = 0$, $Q_{zz} = 2qz_0^2/9$; (b) monopole moment = 0, $p = 0$, $Q_{33} = qR^2/6$

Chapter 3

3.1 (a) $\Phi = V\dfrac{\ln\tan\theta/2}{\ln\tan\theta_1/2}$, $\mathbf{E} = -\dfrac{V\hat{\boldsymbol{\theta}}}{r\sin\theta\ln(\tan\theta_1/2)}$; (b) $\sigma = \dfrac{\varepsilon_0 V}{r\ln(\tan\theta_1/2)}$

3.3 $\mathbf{E} = -V\hat{\boldsymbol{\theta}}/r\sin\theta\ln\left(\dfrac{\tan\theta_1/2}{\tan\theta_2/2}\right)$,

$$\sigma = -\varepsilon_0 V/r\sin\theta_1\ln\left(\frac{\tan\theta_1/2}{\tan\theta_2/2}\right) \text{ at } \theta_1$$

$$\sigma = \varepsilon_0 V/r\sin\theta_2\ln\left(\frac{\tan\theta_1/2}{\tan\theta_2/2}\right) \text{ at } \theta_2$$

3.5 $$\mathbf{E} = \hat{\mathbf{r}}\left[\frac{q}{4\pi\varepsilon_0 r^2} - \frac{\delta q}{4\pi\varepsilon_0(R_2^3 - R_1^3)}\left(1 + 2\frac{R_1^3}{r^3}\right)\cos\theta\right]$$
$$+\hat{\boldsymbol{\theta}}\left[\frac{\delta q}{4\pi\varepsilon_0(R_2^3 - R_1^3)}\left(1 - \frac{R_1^3}{r^3}\right)\sin\theta\right];\quad \sigma(R_1,\theta) = \frac{q}{4\pi}\left(\frac{1}{R_2^2} - \frac{3\delta\cos\theta}{(R_2^3 - R_1^3)}\right),$$
$$\sigma(R_2,\theta) = \frac{q}{4\pi}\left(-\frac{1}{R_2^2} + \frac{\delta(1 + 2R_1^3/R_2^3)}{R_2^3 - R_1^3}\cos\theta\right);\quad -q$$

3.7 $$\Phi_1 = \frac{2\sigma_0 R}{3\varepsilon_0}\left[2 - \left(\frac{r}{R}\right)\cos\theta + \frac{1}{5}\left(\frac{r}{R}\right)^2\frac{1}{2}(3\cos^2\theta - 1)\right] \quad \text{for } r \le R$$
$$\Phi_2 = \frac{2\sigma_0 R}{3\varepsilon_0}\left[2\left(\frac{R}{r}\right) - \left(\frac{R}{r}\right)^2\cos\theta + \frac{1}{5}\left(\frac{R}{r}\right)^3\frac{1}{2}(3\cos^2\theta - 1)\right] \quad \text{for } r \ge R$$

3.9 $$V(x,y) = \frac{4V_0}{\pi}\sum_{m\text{ odd}}\frac{1}{m\sinh\frac{m\pi z_0}{y_0}}\sinh\frac{m\pi z}{y_0}\sin\frac{m\pi y}{y_0}$$

3.11 $$\sigma = -\frac{q}{4\pi R^2}\left(1 + \frac{3}{2}\frac{l}{R}\cos\theta\right),\quad F = -\frac{q^2}{4\pi\varepsilon_0}\frac{l}{2R^3}$$

3.13 (a) $$\mathbf{F} = -\frac{p^2\left(\frac{R}{z_0}\right)^3\hat{\mathbf{r}}}{4\pi\varepsilon_0(z_0 - R^2/z_0)^4};\quad \text{(b) } \mathbf{F} = \frac{q\mathbf{p}}{4\pi\varepsilon_0 z_0^3}$$

3.15 $$F = -\frac{p^2 R z_0}{2\pi\varepsilon_0(z_0^2 - R^2)^4}(z_0^2 + 2R^2),\quad \tau = 0$$

3.17 (a) $\sigma(\phi) = -\lambda(\gamma_- - \gamma_+)\dfrac{R(\gamma_- + \gamma_+) + \gamma_-\gamma_+}{(R^2 + 2\gamma_+ R\cos\phi + \gamma_+^2)(R^2 + 2\gamma_- R\cos\phi + \gamma_-^2)}$ at $-\lambda$ cylinder where ϕ is measured with respect to the x axis and origin at its center, $\gamma_- = -((\Delta/2)^2 + R^2)^{1/2} - \Delta/2$; $\gamma_+ = ((\Delta/2)^2 + R^2)^{1/2} - \Delta/2$

(b) $\dfrac{dF}{dl} = -(\lambda^2/4\pi\varepsilon_0)\left[\left(\dfrac{\Delta}{2}\right)^2 + R^2\right]^{-1/2}$

3.19 $\Phi = \dfrac{\beta}{\varepsilon_0}(R_2 - R_1)$ for $r \le R_1$,

$-\dfrac{\beta}{2\varepsilon_0}(r + R_1^2/r - 2R_2)$ for $R_1 \le r \le R_2$, $\dfrac{\beta}{2\varepsilon_0}(R_2^2 - R_1^2)/r$ for $r \ge R_2$

3.21 $\Phi(z < -z_0) = \dfrac{\rho_0 z_0}{\varepsilon_0}z + \dfrac{\pi\rho_0 z_0^2}{2\varepsilon_0} + B_0$,

$\Phi(z \ge z_0) = -\dfrac{\rho_0 z_0}{\varepsilon_0}z + \dfrac{\pi\rho_0 z_0^2}{2\varepsilon_0} + B_0$,

$\Phi(-z_0 \le z \le z_0) = \dfrac{\rho z_0^2}{\varepsilon_0}\cos(\pi z/z_0) + B_0$ where B_0 is a constant

3.23 $\phi(x, y, z) = \dfrac{\sigma_0}{2\varepsilon_0\alpha}e^{\mp\alpha z}\cos a_1 x\cos a_2 y$ with $\alpha = \sqrt{a_1^2 + a_2^2}$; the minus sign for $z > 0$ and the plus sign for $z < 0$ **3.25** $-\dfrac{q}{\pi a^3}e^{-2r/a} + q\delta(\mathbf{r})$

Chapter 4

4.1 (a) $\rho_p = 0$; $\sigma_P = -P$ and $P\cos\theta$ on flat and curved surfaces; (b) $Q_p = 0$; (c) electrical neutrality; (d) $2\pi R^3 P\hat{\mathbf{z}}/3$, same using both methods
4.3 $\rho_P = -(3ax^2 + b + a + cy)$;
$\sigma_P = [ax^4 + (b + a)x^2 + (cx^2 + px)\sqrt{R^2 - x^2}]/R$
4.5 (a) $\mathbf{E} = q\hat{\mathbf{r}}/2\pi(\varepsilon_0 + \varepsilon)r^2$; (b) $\sigma_f = q\varepsilon_0/[2\pi a^2(\varepsilon_0 + \varepsilon)]$ in the vacuum side, $\sigma_f = q\varepsilon/[2\pi a^2(\varepsilon_0 + \varepsilon)]$ in the dielectric side; (c) $\sigma_P = -q(\varepsilon - \varepsilon_0)/[2\pi a^2(\varepsilon + \varepsilon_0)]$
4.7 (a) $\mathbf{D} = q\hat{\mathbf{r}}/4\pi r^2$; (b) $\rho_P = -q\alpha/4\pi r^2$ **4.9** (a) $\mathbf{E} = \lambda\hat{\boldsymbol{\rho}}/2\pi\varepsilon_0\rho$ where $\lambda = q/L$; (b) with $\varepsilon = \alpha/\rho$ we have $\mathbf{E} = \lambda\hat{\boldsymbol{\rho}}/2\pi\alpha$ **4.11** $K_1\tan\theta_2 = K_2\tan\theta_1$
4.13 $\sigma_P = -A(\varepsilon_1 - \varepsilon_0)(2\theta + \cot\theta)/r$ and $-Aa^2(\varepsilon_2 - \varepsilon_0)\cot\theta/r^3$ in the sphere and in the shell, respectively. $\sigma_P = A(2\varepsilon_0 - \varepsilon_1 - \varepsilon_2)\theta$ at $r = a$, and $\sigma_P = Aa^2(\varepsilon_2 - \varepsilon_0)\theta/b^2$ at $r = b$. $\sigma_f = A\theta(\varepsilon_1 + \varepsilon_2)$
4.15 If $\varepsilon_1 > \varepsilon_2$, then the axes of the cylinder and the disk become parallel to the field (stable equilibrium). If $\varepsilon_1 < \varepsilon_2$, then the axes become perpendicular.
4.17 (a) $F = -q^2/16\pi\varepsilon_0 d^2$; (b) same as (a)
4.19 $\Phi_1 = P_0\rho\cos\phi/2\varepsilon_0$, $\mathbf{E}_1 = -P_0\hat{\mathbf{x}}/2\varepsilon_0$ in the cylinder;
$\Phi_2 = P_0\rho_0^2\cos\phi/2\varepsilon_0\rho$, $\mathbf{E}_2 = P_0\rho_0^2(\cos\phi\hat{\boldsymbol{\rho}} + \sin\phi\hat{\boldsymbol{\phi}})/2\varepsilon_0\rho^2$
4.21 (a) $[(x + \Delta x)^{n+1} - x^{n+1}]\hat{\mathbf{x}}/\Delta x(n + 1)$; (b) $[(x + \Delta x)^n - x^n]\hat{\mathbf{x}}/\Delta x$; (c) same as (b)

4.23 (a) $\Phi = -\dfrac{\rho x^2}{2\varepsilon} + \dfrac{(2\varepsilon + \varepsilon_0)}{2\varepsilon(\varepsilon + \varepsilon_0)}d\rho x$ for $0 \le x \le d$

$\Phi = \rho d(2d - x)/2(\varepsilon + \varepsilon_0)$ for $d \le x \le 2d$

(b) $\frac{d\mathbf{F}}{da} = \frac{\rho^2 d^2 (2\varepsilon + \varepsilon_0)^2}{8\varepsilon(\varepsilon + \varepsilon_0)^2}\hat{\mathbf{x}}$ at $x = 0$, $\frac{d\mathbf{F}}{da} = -\frac{\varepsilon_0 \rho^2 d^2}{8(\varepsilon + \varepsilon_0)^2}\hat{\mathbf{x}}$

Chapter 5

5.1 $\alpha = 1.3 \times 10^{-39}\ C^2 m^2 V^{-1}$

5.3 (a) $\mathbf{p}_i = \alpha \mathbf{E} = \frac{\alpha}{4\pi\varepsilon_0 R^3}(3(\mathbf{p}\cdot\hat{\mathbf{r}})\hat{\mathbf{r}} - \mathbf{p})$;

(b) $U = -p_i \cdot \mathbf{E} = -\frac{\alpha}{(4\pi\varepsilon_0 R^3)^2}(3(\mathbf{p}\cdot\hat{\mathbf{r}})^2 + p^2)$

5.5 (a) $\alpha = \frac{1}{2(E_0/p_0 + 1/4\pi\varepsilon_0 R^3)}$; (b) $\alpha' = \frac{1}{E_0/p_0 + \frac{1}{4\pi\varepsilon_0}\left(\frac{1}{R_0^3} - \frac{1}{R^3}\right)}$

5.7 $\alpha = 4\pi\varepsilon_0 R_0^3 + \frac{p_0}{E_0}\left(\cot h\eta - \frac{1}{\eta}\right)$ where $\eta = p_0 E_0/kT$,

$\langle P\rangle = 4\pi\varepsilon_0 R_0^3 E_0 + \frac{p_0^2 E_0}{3kT}$

5.9 $\frac{N\alpha}{3\varepsilon_0} = \frac{K-1}{K+2} = 0.00096 \neq 1$; thus the gas is not ferroelectric; $\frac{N\alpha}{3\varepsilon_0} = 0.367$ $\neq 1$; thus the liquid is not ferroelectric

Chapter 6

6.1 $3Q^2/20\pi\varepsilon_0 R$

6.3 $q_1 = -\frac{2qa}{l}$, $q_2 = -\frac{qa}{l}\left(1 - \frac{2a}{l}\right)$, $q_3 = \frac{qa^2}{l^3}\left(3 - \frac{2a}{l}\right)$

6.5 (a) $P_{ii} = \frac{1}{4\pi\varepsilon_0 a}$; (b) $P_{ij} = \frac{1}{4\pi\varepsilon_0 l}$;

(c) $q_1 = q/8$, $q_2 = q/2$, $q_3 = q/4$, $q_4 = q/8$

6.7 (a) $\mathbf{E}_2 = \frac{Q}{4\pi\varepsilon_0 r^2}\hat{\mathbf{r}}$ $a \le r \le b$ and $r \ge c$, otherwise it is zero;

(b) $V = \frac{Q}{4\pi\varepsilon_0}\left(\frac{1}{c} + \frac{1}{a} - \frac{1}{b}\right)$; (c) $4\pi\varepsilon_0\left(\frac{1}{c} + \frac{1}{a} - \frac{1}{b}\right)^{-1}$;

(d) $4\pi\varepsilon_0 c, 4\pi\varepsilon_0\left(\frac{1}{a} - \frac{1}{b}\right)^{-1}$

6.9 $\Delta C = 0$ in first order in δ; $\Delta C \neq 0$ for higher-order corrections

6.11 (a) $Q_1 = -qR_1/d$; (b) $\Phi_1 = \frac{q}{4\pi\varepsilon_0 d}$; (c) same results

6.15 (a) $u = 18\varepsilon_0 V_0^2/\rho^2$; (b) $U = 1.5 \times 10^{-9}$ J **6.17** $U = \frac{50\varepsilon_0}{3}J$

6.19 (a) $\Phi = \dfrac{V\phi}{\beta}$; (b) $\sigma = \dfrac{\varepsilon_0 V}{\beta\rho}$ at $\phi = \beta$ and $-\sigma$ at $\phi = 0$,
$Q = \dfrac{\varepsilon_0 Vh}{\beta}\ln(\rho_2/\rho_1)$ at $\phi = \beta$ and $-Q$ at $\phi = 0$; (c) $C = \dfrac{\varepsilon_0 h}{\beta}\ln(\rho_2/\rho_1)$;
(d) $\tau = -\dfrac{\varepsilon_0 h V^2}{2\beta}$

6.21 (a) Parallel to the field: stable, perpendicular to the field: unstable;
(b) $W = -12\pi\varepsilon_0 a^6 E_0^2/l^3$

6.23 (a) $W = -\dfrac{bq^2}{8\pi\varepsilon_0(r^2 - b^2)}$; (b) yes;
(c) $-\dfrac{bq^2}{8\pi\varepsilon_0(r^2 - b^2)} + \dfrac{q}{4\pi\varepsilon_0}\left(\dfrac{Q}{r} + \dfrac{bq}{2r^2}\right)$

Chapter 7

7.1 3.1 cm/s **7.3** $\dfrac{1}{2\pi\sigma_{c_1} l}\ln\left(\dfrac{c}{a}\right) + \dfrac{1}{2\pi\sigma_{c_2}}\ln\left(\dfrac{b}{c}\right)$

7.5 (a) $Q = Q_0 e^{-(\sigma_c/\varepsilon_0 K)t}$; (b) $\dfrac{Q_0^2 d}{2A\varepsilon_0 K} = \dfrac{1}{2}\dfrac{Q_0^2}{C}$; (c) $\tau = 3.8 \times 10^{-24}$ s

7.7 (c) $\mathbf{I} = \alpha E_0 a^2 \hat{\mathbf{x}}$ **7.11** 1.4×10^5 A **7.13** $Q = \sum_{i,k} I_i I_k R_{ik}$

Chapter 8

8.1 (a) $md^2\mathbf{r}/dt^2 = q(\mathbf{E} + \mathbf{v} \times \mathbf{B})$; (b) smallest $B = 5.69 \times 10^3$ T
8.3 $\mathbf{F}/l = -(I\mu_0 K_0 \hat{\mathbf{y}}/\pi)\tan^{-1}(W/2h)$, $\mathbf{F}/l = -I\mu_0 K_0 \hat{\mathbf{y}}/2$
8.5 $\nabla \times \mathbf{B} = 0$ for $\rho \neq 0$, $\nabla \times \mathbf{B} \to \infty$ as $\rho \to 0$ (filamentary current)
8.9 (a) $\mathbf{B} = (\mu_0 I_1/2\pi\rho)\hat{\boldsymbol{\phi}}$; (b) $F = \mu_0 I_1 z_0 \ln(\rho_2/\rho_1)/2\pi$; (c) $\Delta\mathbf{A} = \mu_0 I_1 \ln(\rho_2/\rho_1)/2\pi$; (d) $\mathbf{F}$ on ab is $\mu_0 I_1 I_2 \hat{\mathbf{z}} \ln(\rho_2/\rho_1)/2\pi$, $\mathbf{F}$ on bc is $-\mu_0 I_1 I_2 z_0 \hat{\mathbf{y}}/2\pi\rho_1$
8.11 $\mathbf{A} = -(\mu_0 I \ln\rho)\hat{\mathbf{z}}/2\pi$ **8.13** $\mathbf{B} = 2\sqrt{2}\,\mu_0 I\hat{\mathbf{z}}/\pi a$
8.15 $\mathbf{B} = \mu_0 IN\hat{\mathbf{z}}/4R$
8.17 (a) $\mathbf{B} = \mu_0 IR^2\hat{\mathbf{z}}/2(z^2 + R^2)^{3/2} + 3\mu_0 IR^2 z\rho\hat{\boldsymbol{\rho}}/4(z^2 + R^2)^{5/2}$
8.19 (a) $\mathbf{B} = -B_0[\hat{\mathbf{x}}z/b + \hat{\mathbf{z}}(1 + x/b)]$;
(b) $\mathbf{F} = eB_0[-\dot{y}(1 + x/b)\hat{\mathbf{x}} + \dot{x}(1 + x/b)\hat{\mathbf{y}} + \dot{y}\hat{\mathbf{z}}z/b]$
8.21 (a) $\mathbf{m} = \pi R^2 NI\hat{\mathbf{z}}$; (b) zero; (c) $\boldsymbol{\tau} = \pi R^2 NIB_0(\hat{\mathbf{y}} - \hat{\mathbf{x}})/\sqrt{2}$
8.23 (a) $\mathbf{m} = \pi a^2 I(\hat{\mathbf{x}} + \hat{\mathbf{z}}) = 10^{-4}\pi(\hat{\mathbf{x}} + \hat{\mathbf{z}})$ A · m^2;
(b) $\mathbf{B} = \pi \times 10^{-11}[(27/25 - 1)\hat{\mathbf{x}} + 36\hat{\mathbf{y}}/25 - \hat{\mathbf{z}}]/5^3$ T
8.25 (a) $\mathbf{m} = 4\pi\rho\omega R^5\hat{\mathbf{z}}/15$; (b) $\mathbf{m} = 4\pi\sigma\omega R^4 z/3$

Chapter 9

9.1 (a) $\rho_m = -2(a_2 x + a_1 y)$, $\sigma_m = [a_2 x^3 + (a_1 y^2 + b_1)y]/r$; (b) $\mathbf{J}_m = 0$,
$\mathbf{K}_m = \hat{\mathbf{x}}z(a_1 y^2 + b_1)/r - \hat{\mathbf{y}}za_2 x^2/r + \hat{\mathbf{z}}[a_2 x^2 y - x(a_1 y^2 + b_1)]/r$

9.3 (a) $\rho_m = 0$, $\sigma_m = 0$ on flat surfaces, $\sigma_m = M_0 \cos\phi$ on the ribbon; (b) $\mathbf{J}_m = 0$, $\mathbf{K}_m = -M_0\hat{\mathbf{y}}$ on top, $= M_0\hat{\mathbf{y}}$ on bottom and $= M_0\hat{\mathbf{z}}\sin\phi$ on the ribbon; (c) $\mathbf{H} = -\mathbf{M}T/4R$, $\mathbf{B} = \mu_0\mathbf{M}(1 - T/4R)$
9.5 (a) $\nabla \times \mathbf{M} = \mathbf{J}_m = \nabla \times \mathbf{M}_0$, which does not depend on M_1
9.7 (a) $\mathbf{B} = \mu_0 I\hat{\boldsymbol{\phi}}(\rho^2 - \rho_1^2)/(\rho_2^2 - \rho_1^2)2\pi\rho$ for $\rho_1 < \rho < \rho_2$, $\mathbf{B} = \mu_0 I\hat{\boldsymbol{\phi}}/2\pi\rho$ for $\rho > \rho_2$, $\mathbf{B} = 0$ for $\rho < \rho_1$; (b) $\mathbf{A} = \mu_0 I\hat{\mathbf{z}}(-\rho^2 + 2\rho_1^2 \ln\rho)/4\pi(\rho_2^2 - \rho_1^2)$ for $\rho_1 < \rho < \rho_2$; $\mathbf{A} = -\mu_0 I\hat{\mathbf{z}}\ln\rho/2\pi$ for $\rho > \rho_2$; (c) $\mathbf{B}$ is the same as in (a)
9.9 $\mathbf{B} = 9kB_0\hat{\mathbf{z}}/[(2k + 1)(k + 2) - 2(R_1/R_2)^3(k - 1)^2]$ where $k = \mu_2/\mu_1$; when $\mu_2 \ll \mu_1$, $k \to 0$, and $\mathbf{B} = 0$: a shielding effect
9.11 $\mathbf{B}_1 = \mu_2\mu_0 M_0\hat{\mathbf{x}}/(\mu_0 + \mu_2)$ for $\rho < a$,
$\mathbf{B}_2 = \mu_0\mu_2 M_0\rho_0^2(\cos\phi\hat{\boldsymbol{\rho}} + \sin\phi\hat{\boldsymbol{\phi}})/(\mu_0 + \mu_2)\rho^2$ for $\rho > a$
9.13 (a) $\mathbf{B} = \mu_0(\mathbf{H} + \mathbf{M}/3)$; (b) $\mathbf{B} = \mu_0\mathbf{H}$; (c) $\mathbf{B} = \mu_0(\mathbf{H} + \mathbf{M}/2)$
9.15 $\mathbf{m} = q\omega R^2\hat{\mathbf{z}}/3$
9.19 $dF/dl = \mu_0 I^2 b/4\pi(a^2 - b^2)$
9.21 $\mathbf{B} = \mu_0\mathbf{M}(1 - \rho_0^2/8L^2 - l_g/\rho_0) \simeq \mu_0\mathbf{M}$ **9.23** 1.24 T
9.25 (a) $\mathbf{A} = \pi t^2 BR^2\hat{\mathbf{z}}(z^2 + R^2)^{-3/2}/8$;
(b) $\pi t^2 BR^2\hat{\mathbf{z}}\{[(z - d)^2 + R^2]^{-3/2} + [(z + d)^2 + R^2]^{-3/2}\}/8$

Chapter 10

10.1 (a) $\mathbf{m} = -\dfrac{e\omega\rho^2}{2}\hat{\mathbf{z}}$; (b) $= -\dfrac{e\rho^2}{2}\left(\omega + \dfrac{eB}{2m_e}\right)\hat{\mathbf{z}}$;
(c) $\langle r\rangle = \langle r^2\rangle^{1/2} = 7.8 \times 10^{-11}$ m **10.3** 3.67 K
10.5 (a) 3.3×10^{-3}; (b) 2.63×10^{-3} ($\mathrm{A \cdot m^2}$);
(c) $M = 1.8 \times 10^6$ A/m (T), $m = 18\ \mathrm{A \cdot m^2}$
10.7 $M_s = 8.44 \times 10^5$ A/m, $\gamma = 2654$

Chapter 11

11.1 $\mathbf{E} = 0.05\rho\hat{\boldsymbol{\phi}}$ for $\rho \le \rho_0$, $\mathbf{E} = 0.05\rho_0^2\hat{\boldsymbol{\phi}}/\rho$ for $\rho \ge \rho_0$
11.3 (a) $A_\phi = \dfrac{R^2 B_0}{2\rho}$ for $\rho \ge R$, $A_\phi = \frac{1}{2}\rho B_0$ for $\rho \le R$; (b) $E_\phi = -\dfrac{R^2}{2\rho}\dfrac{dB_0}{dt}$ for $\rho \ge R$, $E_\phi = -\dfrac{\rho}{2}\dfrac{dB_0}{dt}$ for $\rho \le R$; (c) $\mathbf{J} = \sigma_c E_\phi\hat{\boldsymbol{\phi}}$
11.5 (a) $\varepsilon = \dfrac{\mu_0 IA}{2\pi}\dfrac{v}{l(l + a)}$; (b) $M = \dfrac{\mu_0 b}{2\pi}\ln\left(1 + \dfrac{a}{l}\right)$
11.7 (a) $-7.5\cos\omega t$; (b) $-7.5\cos\omega t - 2.25\sin\omega t$
11.9 $I_2 = \dfrac{\mu_0 I_1 A_1 A_2}{4\pi r^3}\omega\sin\omega t$ **11.13** $\dfrac{\mu_0}{8\pi} + \dfrac{\mu}{2\pi}\ln(b/a)$
11.15 2.37×10^{-6} H/m **11.17** (a) $\dfrac{\mu_0\pi R_1^2 R_2^2}{2h^3}$; (b) $\mu_0 R\left[\ln\left(\dfrac{8R}{h}\right) - 2\right]$

Chapter 12

12.1 (a) 62.5 J; (b) 25 Webers

12.3 (a) $\mathbf{B} = \dfrac{\mu_0 I\rho}{2\pi a^2}\hat{\boldsymbol{\phi}}$, $U/l = \dfrac{\mu_0 I^2}{16\pi}$; (b) $L/l = \mu_0/8\pi$

12.5 (a) 2.3×10^{-4} W; (b) $I_{max} = 4 \times 10^{-2}$ A; (c) 2.5×10^{-3} H

12.7 (a) $U = \dfrac{\pi R^2 \mu_0 N^2 I^2}{2l}$, $F = \mu_0 N I^2/2l$ radial; (b) same as (a)

12.9 (a) $\dfrac{\mu_0}{2\pi} I_1 \ln \rho \hat{\mathbf{z}}$; (b) $\dfrac{U}{l} = \dfrac{\mu_0}{2\pi} I_1 I_2 \ln R$; (c) $F/l = -\dfrac{\mu_0}{2\pi}\dfrac{I_1 I_2}{R}$

12.11 $F = -\dfrac{3\mu_0 a^2 b^2 I_1 I_2}{2l^4}$ along axis

Chapter 13

13.1 (a) $V = RI + L\dfrac{dI}{dt}$; (b) $I = \dfrac{V_0}{R}(1 - e^{-(R/L)t})$; (c) $V_0 e^{-Rt/L}$;
(d) $V_0(e^{-Rt/L} - e^{-R(t-T)/L})$ **13.5** (a) $Q = 8.66$; (b) 0.01 H; (c) 0.52

13.7 (a) $I = \dfrac{V_0}{(R^2 + \omega^2 L^2)^{1/2}} \sin(\omega t + \phi)$, $\tan\phi = \omega L/R$; (b) 0;
(c) $-\pi/2$, V_R lags V_L

13.11 (b) $V_1 = 2$, $V_2 = 0$ V; (c) 2, 1, and 0, A in R_1, R_2, and R_3

13.13 $V_2 = 4.48\cos(t - 63.5°) = 4.48\cos(t - 1.1)$, $\phi = -1.1$ radians

13.15 $R_2 = \omega\sqrt{L_2}$

13.19 (a) $\omega^2 = \dfrac{C_1 + 2C}{L_1 C_1 (C_1 + 2C)}$, $\dfrac{1}{L_1(C_1 + 2C)}$;
(b) $\omega^2 = \dfrac{1}{L_1 C_1}$, $\dfrac{1}{L_1 C_1}$ for $C = 0$ $\omega^2 = \dfrac{1}{L_1 C_1}$ and 0 for $C \gg C_1$

13.21 (b) $\pi/2$

13.23 (a) $C_1 = 10^{-6}$ F; (b) $Q_0 = 100$, $Z = 10^4$; it is both since $Q_0 \gg 1$;
(c) $z \simeq 25(1 + 18i)$, inductive; (d) $C_2 = 0.25 \times 10^{-6}$ F;
(e) 4×10^{-2}, 450 at ω_0 and 0.9 ω_0, respectively

13.25 (a) $z_1 = i\left(\omega L - \dfrac{1}{\omega C}\right)$, $z_2 = -\dfrac{i\omega L}{\omega^2 LC - 1}$; (b) $0 \le \dfrac{(\omega^2 LC - 1)^2}{\omega^2 LC} \le 4$

Chapter 14

14.1 (a) $I_D = I_C = q_0\omega\cos\omega t$; (b) $\mathbf{B} = B\hat{\boldsymbol{\phi}}$, z axis is normal to plates;
(c) $\mu_0 q_0 \omega\rho\cos(\omega t)/2A$

14.3 $I_D = I_C = 7.1 \times 10^{-5}\left(\dfrac{ba}{b-a}\right)\cos 500t$

14.5 (a) $\mathbf{D} = \varepsilon_0\mathbf{E}$, $\mathbf{B} = -\mathbf{E}/c$, $\mathbf{H} = -\mathbf{E}/c\mu_0$; (b) $\mathbf{E} = \mathbf{H}/c\varepsilon_0 = c\mu_0\mathbf{H}$

14.7 (a) $u = \varepsilon_0 E_0^2 \cos^2(kz - \omega t)$, $\mathbf{S} = \varepsilon_0 c E_0^2 \cos^2(kz - \omega t)\hat{\mathbf{z}}$, $\mathbf{S} = u\mathbf{v}$, where $|\mathbf{v}| = c$; (b) $\langle \mathbf{S} \rangle = \frac{1}{2}\varepsilon_0 c E_0^2 \hat{\mathbf{z}}$ for both
14.9 (b) Elliptical polarization ($E_0 \neq E_0'$); (c) $\mathbf{E}$ is normal to $\mathbf{B}$; (d) $\mathbf{E} = E_0\hat{\mathbf{x}} \cos \omega t + 2E_0\hat{\mathbf{y}} \cos(\omega t - \pi/4)$
14.11 (a) $\mathbf{E} = \mathbf{E}_0 e^{-i\omega t}$, $\mathbf{E}_0 = \hat{\mathbf{x}} e^{iky} + \hat{\mathbf{y}} e^{ikx}$; (b) $m = n$ linearly polarized at 45°, $m = n \pm 4$ linearly polarized at 135°, $m = n \pm 2$ circularly polarized, $m = n \pm 1$ elliptically polarized along $y = \pm x$ axis.
14.13 (a) 65.2 W; (b) 222 W
14.15 (a) $\mathbf{E} = E_0\hat{\mathbf{x}} \exp[-z/\delta - i\omega(t - nz/c)]$; (b) $\delta = 8.15\ \ 10^{-5}$ m, $v = c/n = 511.5$ m/s, $\lambda = 5.12 \times 10^{-4}$ m; (c) $\mathbf{B} = (1 + i)\hat{\mathbf{y}}E/\delta\omega$; (d) $\pi/4$
14.17 $\hat{\eta} = 3.98 \times 10^{-3} e^{-i\pi/4}\ \Omega$, $v = 4.47 \times 10^3$ m/s
14.19 (b) $B = 2\left(1 - \frac{2}{\sqrt{\pi}} \int_0^{\xi} e^{-\xi^2} d\xi\right)$; (c) $P_m = 8 \times 10^5$ N/m², $F = 8 \times 10^5$ N; (d) $F = 2 \times 10^5$ N

Chapter 15

15.3 (a) $\mathbf{E} = -\frac{1}{2}\hat{\mathbf{z}}\alpha\rho^2 \cos \alpha t$, $\mathbf{B} = -\rho\hat{\boldsymbol{\phi}} \sin \alpha t$;
(c) $\mathbf{F}(x, t) = \frac{\hat{\rho}}{\mu}\left(2\rho - \frac{1}{2}\varepsilon\mu\alpha^2\rho^3\right)\sin^2 \alpha t$ N/m³
15.7 $\mathbf{A}_1 = \mu_0 I_0 \hat{\mathbf{z}} \ln\left(\sqrt{c^2t^2 - \rho_1^2} + ct/\rho_1\right)/4\pi$,
$\mathbf{A}_2 = \mu_0 I_0 \hat{\mathbf{z}} \ln\left(\sqrt{c^2t'^2 - \rho_2^2} + ct'/\rho_2\right)/4\pi$
$\mathbf{A} = \mathbf{A}_1 + \mathbf{A}_2$ if $\rho_1 < ct$ and $\rho_2 < ct'$;
$\mathbf{A} = \mathbf{A}_1$ if $\rho_1 < ct$, and $\rho_2 \geq ct'$; $\mathbf{A} = \mathbf{A}_2$ if $\rho_1 > ct$ and $\rho_2 < ct'$; and $\mathbf{A} = 0$ if $\rho_1 > ct$ and $\rho_2 > ct'$ where $t' = t - t_0$; the vector potential will be equal when $\left(\sqrt{c^2t^2 - \rho_1^2} + ct\right)\rho_2 = \left(\sqrt{c^2t'^2 - \rho_2^2} + ct'\right)\rho_1$
15.9 (a) $\mathbf{E} = p_0\hat{\mathbf{z}}[\cos(\omega t - kz)/z^3 - \omega \sin(\omega t - kz)/cz^2]/2\pi\varepsilon_0$, $\mathbf{B} = 0$, $\mathbf{S} = 0$; (b) $\mathbf{E} = -p_0\hat{\mathbf{z}}[\cos(\omega t - kx)/x^3 - \omega \sin(\omega t - kx)/cx^2 - \omega^2 \cos(\omega t - kx)/c^2x]/4\pi\varepsilon_0$, $\mathbf{B} = p_0\omega\hat{\mathbf{y}}[k \cos(\omega t - kx) + \sin(\omega t - kx)/x]/4\pi\varepsilon_0 c^2 x$, $\mathbf{S} = (p_0/4\pi\varepsilon_0)^2(k\omega^3/c^4\mu_0)\hat{\mathbf{x}} \cos^2(\omega t - kx)/x^2$; (d) $p_0\omega^4 \cos^2(\omega t - kx)/6\pi\varepsilon_0 c^3$
15.11 (a) Linear in the $\hat{\boldsymbol{\phi}}$ direction; (b) circular in the direction $-\hat{\mathbf{x}} - i\hat{\mathbf{y}}$; (c) elliptical
15.15 (a) $l \ll 3.8$ cm; take $l = 0.1$ mm; (b) $p \simeq 3 \times 10^{-34}$ C.m; (c) $R = 6 \times 10^{-4}\ \Omega$
15.19 The pattern of a quarter wave antenna placed vertically close to ground, i.e., a vertical half-wave antenna
15.21 (a) $dP/d\Omega = q^2a^2\sin^2\theta/16\pi^2\varepsilon_0 c^3 r^2$, independent of v_0; (b) $P = q^2a^2/6\pi\varepsilon_0 c^3$, $E = q^2 a v_0/6\pi\varepsilon_0 c^3$
15.23 $E = q^2a/(6\pi\varepsilon_0 c^2)\int_0^{\beta_0}(1 - \beta^2)^{-3} d\beta$ where $\beta_0 = v_0/c$
15.25 (a) $\mathbf{p}(t) = p_0 \sin \omega t\hat{\mathbf{x}} + p_0 \cos \omega t\hat{\mathbf{y}}$; (b) $\pi/2$; (c) $dP/d\Omega = p_0^2k^4c(1 + \cos^2\theta)/8\pi^2\varepsilon_0$

Chapter 16

16.1 $(3.5\hat{\mathbf{x}} + 3.4\hat{\mathbf{y}})/\mu_0$ A/m
16.3 (a) $\theta_1 = \sin^{-1} n_2/n_1$ or $\pi/2$; (b) $(\phi_p - \phi_s)_{\max} = (1 - n_2^2/n_1^2)/(n_2/n_1)$; (c) yes provided $1 - n_2^2/n_1^2 \geq 2n_2/n_1$ and $n_2/n_1 \leq 0.414$
16.5 $T = 3/[4 - 15\sin^2(2\omega d/c)]$, $8/[9 + 72\sin^2(4\omega d/c)]$, $1/[4/27 - 45\sin^2(2\omega d/c)/1296]$
16.7 49.82, −0.96, −1.63 **16.13** 6, 3.14×10^{-4}, 0.21 A/m
16.15 (a) $\nabla \cdot \mathbf{E} = 0 = \partial E_x/\partial x = 0$; (b) $d^2E_x/dy^2 + (\omega^2/c^2 - k^2)E_x = 0$, $\mathbf{E} = E_0\hat{x}\sin(n\pi y/a)e^{i(kz-\omega t)}$; (c) $\omega > \dfrac{\pi c}{a}$
16.17 (a) $\mathbf{B} = B_0\hat{\mathbf{x}}\cos(ky\cos\theta)e^{i(kz\sin\theta-\omega t)}$ $\mathbf{E} = -cB_0[\hat{\mathbf{y}}\sin\theta\cos(ky\cos\theta) - \hat{\mathbf{z}}i\cos\theta\sin(ky\cos\theta)]e^{i(kz\sin\theta-\omega t)}$ where $\cos\theta = \dfrac{n\pi}{ka}$ and n is an integer; (b) $n = 0$
16.19 (a) $\mathbf{E} = \hat{\mathbf{x}}E_0e^{i(kz-\omega t)}$, $\mathbf{H} = \hat{\mathbf{y}}(E_0/\mu_0 c)e^{i(kz-\omega t)}$ E_y, H_x and $\partial H_y/\partial x = 0$ at $x = 0$ and $x = a$; (b) the solution satisfies the wave equation and the boundary conditions without restriction on ω
16.21 (a) $\lambda_c = \left[\left(\dfrac{m}{2a}\right)^2 + \left(\dfrac{n}{2b}\right)^2\right]^{-1/2} \leq 2a$ which is given by $m = 1$ and $n = 0$; (b) ω_{mn} (normalized to ω_{10}): $\omega_{01} = 2$, $\omega_{02} = 4$, $\omega_{03} = 6$, $\omega_{10} = 1$, $\omega_{12} = 2.24$, $\omega_{13} = 4.13$, $\omega_{20} = 2$, $\omega_{21} = 2.84$, $\omega_{22} = 4.48$, $\omega_{30} = 3$, $\omega_{31} = 3.61$, $\omega_{32} = 5$, $\omega_{40} = 4$, $\omega_{41} = 4.48$, $\omega_{42} = 5.66$
16.25
(a) $\mathbf{E} = \hat{\mathbf{x}}E_{ox}\cos\left(\dfrac{l\pi}{a}x\right)\sin\left(\dfrac{m\pi}{b}y\right)\sin\left(\dfrac{n\pi}{c}z\right) + \hat{\mathbf{y}}E_{oy}\sin\left(\dfrac{l\pi}{a}x\right)\cos\left(\dfrac{m\pi}{b}y\right)\sin\left(\dfrac{n\pi}{c}z\right)$
where $(E_{oy}/E_{ox}) = -(lb/ma)$;

$$\mathbf{H} = -\frac{iE_{ox}}{\mu_0\omega}\left[\frac{nl\pi b}{mac}\hat{\mathbf{x}}\sin\left(\frac{l\pi}{a}x\right)\cos\left(\frac{m\pi}{b}y\right)\cos\left(\frac{n\pi}{c}z\right)\right.$$
$$\left. + \frac{n\pi}{c}\hat{\mathbf{y}}\cos\left(\frac{l\pi}{a}x\right)\sin\left(\frac{m\pi}{b}y\right)\cos\left(\frac{n\pi}{c}z\right) - 2\frac{m\pi}{b}\hat{\mathbf{z}}\cos\left(\frac{l\pi}{a}x\right)\cos\left(\frac{m\pi}{b}y\right)\sin\left(\frac{n\pi}{c}z\right)\right];$$

(b) TE_{111}; (c) $\omega^2/c^2 = (\pi^2/a^2)(l^2 + m^2 + n^2)$

Chapter 17

17.1 The life time is 1.52×10^{-6} s. The distance to earth contracts by a factor of 5, thus giving the same result.
17.9 $dN/d\Omega' = \dfrac{N_0}{4\pi}\dfrac{1-\beta^2}{(1+\beta\cos\theta')^2}$ **17.11** $\theta_1 = \theta_2$, and $\omega_1 = \omega_2$
17.13 $\lambda = \lambda_0\sqrt{\dfrac{1-\beta}{1+\beta}}$, $\lambda = \lambda_0\sqrt{\dfrac{1+\beta}{1-\beta}}$
17.17 (a) when $E < cB$, $\mathbf{v} = \mathbf{B} \times \mathbf{E}/B^2$ and $E' = 0$, $c\mathbf{B}' = \mathbf{B}\sqrt{c^2B^2 - E^2}/B$. When $E > cB$, $\mathbf{v} = c^2\mathbf{E} \times \mathbf{B}/E^2$ and $B' = 0$,

$\mathbf{E}' = \mathbf{E}\sqrt{E^2 - c^2B^2}/E$. In the latter case $\mathbf{B}'$ will also be zero in any frame moving in the direction of $\mathbf{E}'$.

17.19 (a) $-\nabla\Phi = \dfrac{q}{4\pi\varepsilon_0 R^{*3}}[(z - vt)\hat{\mathbf{z}} + (x\hat{\mathbf{x}} + y\hat{\mathbf{y}})(1 - \beta^2)]$; (b) $-\dfrac{\partial \mathbf{A}}{\partial t} = -\dfrac{\mu_0 q v \mathbf{v}}{4\pi R^{*3}}(z - vt)$; (c) $\mathbf{E} = -\nabla\Phi - \dfrac{\partial \mathbf{A}}{\partial t} = \dfrac{q\mathbf{R}}{4\pi\varepsilon_0\gamma^2 R^{*3}}$ which is radial from the present position of the charge.

17.21 $\mathbf{p}'_{\parallel} = \gamma p_{\parallel}$, $\mathbf{m}'_{\parallel} = \gamma m_{\parallel}$ $\quad p'_{\perp} = \mathbf{p}_{\perp} - \dfrac{\mathbf{v}}{c^2} \times \mathbf{m}_{\perp}$, $\mathbf{m}'_{\perp} = \mathbf{m}_{\perp} + \mathbf{v} \times \mathbf{p}_{\perp}$

17.23 (a) $q_1 = -q_3 = -vaI'/c^2$, sides two and four are unchanged; (b) $q_3 b$; (c) $m = m'$.

INDEX

A CATALOG OF SELECTED

DOVER BOOKS

IN SCIENCE AND MATHEMATICS

Physics

THEORETICAL NUCLEAR PHYSICS, John M. Blatt and Victor F. Weisskopf. An uncommonly clear and cogent investigation and correlation of key aspects of theoretical nuclear physics by leading experts: the nucleus, nuclear forces, nuclear spectroscopy, two-, three- and four-body problems, nuclear reactions, beta-decay and nuclear shell structure. 896pp. 5 3/8 x 8 1/2. 0-486-66827-4

QUANTUM THEORY, David Bohm. This advanced undergraduate-level text presents the quantum theory in terms of qualitative and imaginative concepts, followed by specific applications worked out in mathematical detail. 655pp. 5 3/8 x 8 1/2. 0-486-65969-0

ATOMIC PHYSICS AND HUMAN KNOWLEDGE, Niels Bohr. Articles and speeches by the Nobel Prize–winning physicist, dating from 1934 to 1958, offer philosophical explorations of the relevance of atomic physics to many areas of human endeavor. 1961 edition. 112pp. 5 3/8 x 8 1/2. 0-486-47928-5

COSMOLOGY, Hermann Bondi. A co-developer of the steady-state theory explores his conception of the expanding universe. This historic book was among the first to present cosmology as a separate branch of physics. 1961 edition. 192pp. 5 3/8 x 8 1/2. 0-486-47483-6

LECTURES ON QUANTUM MECHANICS, Paul A. M. Dirac. Four concise, brilliant lectures on mathematical methods in quantum mechanics from Nobel Prize-winning quantum pioneer build on idea of visualizing quantum theory through the use of classical mechanics. 96pp. 5 3/8 x 8 1/2. 0-486-41713-1

THE PRINCIPLE OF RELATIVITY, Albert Einstein and Frances A. Davis. Eleven papers that forged the general and special theories of relativity include seven papers by Einstein, two by Lorentz, and one each by Minkowski and Weyl. 1923 edition. 240pp. 5 3/8 x 8 1/2. 0-486-60081-5

PHYSICS OF WAVES, William C. Elmore and Mark A. Heald. Ideal as a classroom text or for individual study, this unique one-volume overview of classical wave theory covers wave phenomena of acoustics, optics, electromagnetic radiations, and more. 477pp. 5 3/8 x 8 1/2. 0-486-64926-1

THERMODYNAMICS, Enrico Fermi. In this classic of modern science, the Nobel Laureate presents a clear treatment of systems, the First and Second Laws of Thermodynamics, entropy, thermodynamic potentials, and much more. Calculus required. 160pp. 5 3/8 x 8 1/2. 0-486-60361-X

QUANTUM THEORY OF MANY-PARTICLE SYSTEMS, Alexander L. Fetter and John Dirk Walecka. Self-contained treatment of nonrelativistic many-particle systems discusses both formalism and applications in terms of ground-state (zero-temperature) formalism, finite-temperature formalism, canonical transformations, and applications to physical systems. 1971 edition. 640pp. 5 3/8 x 8 1/2. 0-486-42827-3

QUANTUM MECHANICS AND PATH INTEGRALS: Emended Edition, Richard P. Feynman and Albert R. Hibbs. Emended by Daniel F. Styer. The Nobel Prize–winning physicist presents unique insights into his theory and its applications. Feynman starts with fundamentals and advances to the perturbation method, quantum electrodynamics, and statistical mechanics. 1965 edition, emended in 2005. 384pp. 6 1/8 x 9 1/4. 0-486-47722-3

Physics

INTRODUCTION TO MODERN OPTICS, Grant R. Fowles. A complete basic undergraduate course in modern optics for students in physics, technology, and engineering. The first half deals with classical physical optics; the second, quantum nature of light. Solutions. 336pp. 5 3/8 x 8 1/2. 0-486-65957-7

THE QUANTUM THEORY OF RADIATION: Third Edition, W. Heitler. The first comprehensive treatment of quantum physics in any language, this classic introduction to basic theory remains highly recommended and widely used, both as a text and as a reference. 1954 edition. 464pp. 5 3/8 x 8 1/2. 0-486-64558-4

QUANTUM FIELD THEORY, Claude Itzykson and Jean-Bernard Zuber. This comprehensive text begins with the standard quantization of electrodynamics and perturbative renormalization, advancing to functional methods, relativistic bound states, broken symmetries, nonabelian gauge fields, and asymptotic behavior. 1980 edition. 752pp. 6 1/2 x 9 1/4. 0-486-44568-2

FOUNDATIONS OF POTENTIAL THERY, Oliver D. Kellogg. Introduction to fundamentals of potential functions covers the force of gravity, fields of force, potentials, harmonic functions, electric images and Green's function, sequences of harmonic functions, fundamental existence theorems, and much more. 400pp. 5 3/8 x 8 1/2. 0-486-60144-7

FUNDAMENTALS OF MATHEMATICAL PHYSICS, Edgar A. Kraut. Indispensable for students of modern physics, this text provides the necessary background in mathematics to study the concepts of electromagnetic theory and quantum mechanics. 1967 edition. 480pp. 6 1/2 x 9 1/4. 0-486-45809-1

GEOMETRY AND LIGHT: The Science of Invisibility, Ulf Leonhardt and Thomas Philbin. Suitable for advanced undergraduate and graduate students of engineering, physics, and mathematics and scientific researchers of all types, this is the first authoritative text on invisibility and the science behind it. More than 100 full-color illustrations, plus exercises with solutions. 2010 edition. 288pp. 7 x 9 1/4. 0-486-47693-6

QUANTUM MECHANICS: New Approaches to Selected Topics, Harry J. Lipkin. Acclaimed as "excellent" (*Nature*) and "very original and refreshing" (*Physics Today*), these studies examine the Mössbauer effect, many-body quantum mechanics, scattering theory, Feynman diagrams, and relativistic quantum mechanics. 1973 edition. 480pp. 5 3/8 x 8 1/2. 0-486-45893-8

THEORY OF HEAT, James Clerk Maxwell. This classic sets forth the fundamentals of thermodynamics and kinetic theory simply enough to be understood by beginners, yet with enough subtlety to appeal to more advanced readers, too. 352pp. 5 3/8 x 8 1/2. 0-486-41735-2

QUANTUM MECHANICS, Albert Messiah. Subjects include formalism and its interpretation, analysis of simple systems, symmetries and invariance, methods of approximation, elements of relativistic quantum mechanics, much more. "Strongly recommended." – *American Journal of Physics.* 1152pp. 5 3/8 x 8 1/2. 0-486-40924-4

RELATIVISTIC QUANTUM FIELDS, Charles Nash. This graduate-level text contains techniques for performing calculations in quantum field theory. It focuses chiefly on the dimensional method and the renormalization group methods. Additional topics include functional integration and differentiation. 1978 edition. 240pp. 5 3/8 x 8 1/2. 0-486-47752-5